Optical Inspection of Microsystems

OPTICAL SCIENCE AND ENGINEERING

Founding Editor
Brian J. Thompson
University of Rochester
Rochester, New York

1. Electron and Ion Microscopy and Microanalysis: Principles and Applications, *Lawrence E. Murr*
2. Acousto-Optic Signal Processing: Theory and Implementation, *edited by Norman J. Berg and John N. Lee*
3. Electro-Optic and Acousto-Optic Scanning and Deflection, *Milton Gottlieb, Clive L. M. Ireland, and John Martin Ley*
4. Single-Mode Fiber Optics: Principles and Applications, *Luc B. Jeunhomme*
5. Pulse Code Formats for Fiber Optical Data Communication: Basic Principles and Applications, *David J. Morris*
6. Optical Materials: An Introduction to Selection and Application, *Solomon Musikant*
7. Infrared Methods for Gaseous Measurements: Theory and Practice, *edited by Joda Wormhoudt*
8. Laser Beam Scanning: Opto-Mechanical Devices, Systems, and Data Storage Optics, *edited by Gerald F. Marshall*
9. Opto-Mechanical Systems Design, *Paul R. Yoder, Jr.*
10. Optical Fiber Splices and Connectors: Theory and Methods, *Calvin M. Miller with Stephen C. Mettler and Ian A. White*
11. Laser Spectroscopy and Its Applications, *edited by Leon J. Radziemski, Richard W. Solarz, and Jeffrey A. Paisner*
12. Infrared Optoelectronics: Devices and Applications, *William Nunley and J. Scott Bechtel*
13. Integrated Optical Circuits and Components: Design and Applications, *edited by Lynn D. Hutcheson*
14. Handbook of Molecular Lasers, *edited by Peter K. Cheo*
15. Handbook of Optical Fibers and Cables, *Hiroshi Murata*
16. Acousto-Optics, *Adrian Korpel*
17. Procedures in Applied Optics, *John Strong*
18. Handbook of Solid-State Lasers, *edited by Peter K. Cheo*
19. Optical Computing: Digital and Symbolic, *edited by Raymond Arrathoon*
20. Laser Applications in Physical Chemistry, *edited by D. K. Evans*
21. Laser-Induced Plasmas and Applications, *edited by Leon J. Radziemski and David A. Cremers*
22. Infrared Technology Fundamentals, *Irving J. Spiro and Monroe Schlessinger*
23. Single-Mode Fiber Optics: Principles and Applications, Second Edition, Revised and Expanded, *Luc B. Jeunhomme*
24. Image Analysis Applications, *edited by Rangachar Kasturi and Mohan M. Trivedi*
25. Photoconductivity: Art, Science, and Technology, *N. V. Joshi*
26. Principles of Optical Circuit Engineering, *Mark A. Mentzer*
27. Lens Design, *Milton Laikin*
28. Optical Components, Systems, and Measurement Techniques, *Rajpal S. Sirohi and M. P. Kothiyal*

29. Electron and Ion Microscopy and Microanalysis: Principles and Applications, Second Edition, Revised and Expanded, *Lawrence E. Murr*
30. Handbook of Infrared Optical Materials, *edited by Paul Klocek*
31. Optical Scanning, *edited by Gerald F. Marshall*
32. Polymers for Lightwave and Integrated Optics: Technology and Applications, *edited by Lawrence A. Hornak*
33. Electro-Optical Displays, *edited by Mohammad A. Karim*
34. Mathematical Morphology in Image Processing, *edited by Edward R. Dougherty*
35. Opto-Mechanical Systems Design: Second Edition, Revised and Expanded, *Paul R. Yoder, Jr.*
36. Polarized Light: Fundamentals and Applications, *Edward Collett*
37. Rare Earth Doped Fiber Lasers and Amplifiers, *edited by Michel J. F. Digonnet*
38. Speckle Metrology, *edited by Rajpal S. Sirohi*
39. Organic Photoreceptors for Imaging Systems, *Paul M. Borsenberger and David S. Weiss*
40. Photonic Switching and Interconnects, *edited by Abdellatif Marrakchi*
41. Design and Fabrication of Acousto-Optic Devices, *edited by Akis P. Goutzoulis and Dennis R. Pape*
42. Digital Image Processing Methods, *edited by Edward R. Dougherty*
43. Visual Science and Engineering: Models and Applications, *edited by D. H. Kelly*
44. Handbook of Lens Design, *Daniel Malacara and Zacarias Malacara*
45. Photonic Devices and Systems, *edited by Robert G. Hunsberger*
46. Infrared Technology Fundamentals: Second Edition, Revised and Expanded, *edited by Monroe Schlessinger*
47. Spatial Light Modulator Technology: Materials, Devices, and Applications, *edited by Uzi Efron*
48. Lens Design: Second Edition, Revised and Expanded, *Milton Laikin*
49. Thin Films for Optical Systems, *edited by Francoise R. Flory*
50. Tunable Laser Applications, *edited by F. J. Duarte*
51. Acousto-Optic Signal Processing: Theory and Implementation, Second Edition, *edited by Norman J. Berg and John M. Pellegrino*
52. Handbook of Nonlinear Optics, *Richard L. Sutherland*
53. Handbook of Optical Fibers and Cables: Second Edition, *Hiroshi Murata*
54. Optical Storage and Retrieval: Memory, Neural Networks, and Fractals, *edited by Francis T. S. Yu and Suganda Jutamulia*
55. Devices for Optoelectronics, *Wallace B. Leigh*
56. Practical Design and Production of Optical Thin Films, *Ronald R. Willey*
57. Acousto-Optics: Second Edition, *Adrian Korpel*
58. Diffraction Gratings and Applications, *Erwin G. Loewen and Evgeny Popov*
59. Organic Photoreceptors for Xerography, *Paul M. Borsenberger and David S. Weiss*
60. Characterization Techniques and Tabulations for Organic Nonlinear Optical Materials, *edited by Mark G. Kuzyk and Carl W. Dirk*
61. Interferogram Analysis for Optical Testing, *Daniel Malacara, Manuel Servin, and Zacarias Malacara*
62. Computational Modeling of Vision: The Role of Combination, *William R. Uttal, Ramakrishna Kakarala, Spiram Dayanand, Thomas Shepherd, Jagadeesh Kalki, Charles F. Lunskis, Jr., and Ning Liu*
63. Microoptics Technology: Fabrication and Applications of Lens Arrays and Devices, *Nicholas Borrelli*
64. Visual Information Representation, Communication, and Image Processing, *edited by Chang Wen Chen and Ya-Qin Zhang*
65. Optical Methods of Measurement, *Rajpal S. Sirohi and F. S. Chau*
66. Integrated Optical Circuits and Components: Design and Applications, *edited by Edmond J. Murphy*

67. Adaptive Optics Engineering Handbook, *edited by Robert K. Tyson*
68. Entropy and Information Optics, *Francis T. S. Yu*
69. Computational Methods for Electromagnetic and Optical Systems, *John M. Jarem and Partha P. Banerjee*
70. Laser Beam Shaping, *Fred M. Dickey and Scott C. Holswade*
71. Rare-Earth-Doped Fiber Lasers and Amplifiers: Second Edition, Revised and Expanded, *edited by Michel J. F. Digonnet*
72. Lens Design: Third Edition, Revised and Expanded, *Milton Laikin*
73. Handbook of Optical Engineering, *edited by Daniel Malacara and Brian J. Thompson*
74. Handbook of Imaging Materials: Second Edition, Revised and Expanded, *edited by Arthur S. Diamond and David S. Weiss*
75. Handbook of Image Quality: Characterization and Prediction, *Brian W. Keelan*
76. Fiber Optic Sensors, *edited by Francis T. S. Yu and Shizhuo Yin*
77. Optical Switching/Networking and Computing for Multimedia Systems, *edited by Mohsen Guizani and Abdella Battou*
78. Image Recognition and Classification: Algorithms, Systems, and Applications, *edited by Bahram Javidi*
79. Practical Design and Production of Optical Thin Films: Second Edition, Revised and Expanded, *Ronald R. Willey*
80. Ultrafast Lasers: Technology and Applications, *edited by Martin E. Fermann, Almantas Galvanauskas, and Gregg Sucha*
81. Light Propagation in Periodic Media: Differential Theory and Design, *Michel Nevière and Evgeny Popov*
82. Handbook of Nonlinear Optics, Second Edition, Revised and Expanded, *Richard L. Sutherland*
83. Polarized Light: Second Edition, Revised and Expanded, *Dennis Goldstein*
84. Optical Remote Sensing: Science and Technology, *Walter Egan*
85. Handbook of Optical Design: Second Edition, *Daniel Malacara and Zacarias Malacara*
86. Nonlinear Optics: Theory, Numerical Modeling, and Applications, *Partha P. Banerjee*
87. Semiconductor and Metal Nanocrystals: Synthesis and Electronic and Optical Properties, edited by *Victor I. Klimov*
88. High-Performance Backbone Network Technology, *edited by Naoaki Yamanaka*
89. Semiconductor Laser Fundamentals, *Toshiaki Suhara*
90. Handbook of Optical and Laser Scanning, *edited by Gerald F. Marshall*
91. Organic Light-Emitting Diodes: Principles, Characteristics, and Processes, *Jan Kalinowski*
92. Micro-Optomechatronics, *Hiroshi Hosaka, Yoshitada Katagiri, Terunao Hirota, and Kiyoshi Itao*
93. Microoptics Technology: Second Edition, *Nicholas F. Borrelli*
94. Organic Electroluminescence, *edited by Zakya Kafafi*
95. Engineering Thin Films and Nanostructures with Ion Beams, *Emile Knystautas*
96. Interferogram Analysis for Optical Testing, Second Edition, *Daniel Malacara, Manuel Sercin, and Zacarias Malacara*
97. Laser Remote Sensing, *edited by Takashi Fujii and Tetsuo Fukuchi*
98. Passive Micro-Optical Alignment Methods, *edited by Robert A. Boudreau and Sharon M. Boudreau*
99. Organic Photovoltaics: Mechanism, Materials, and Devices, *edited by Sam-Shajing Sun and Niyazi Serdar Saracftci*
100. Handbook of Optical Interconnects, *edited by Shigeru Kawai*
101. GMPLS Technologies: Broadband Backbone Networks and Systems, *Naoaki Yamanaka, Kohei Shiomoto, and Eiji Oki*

102. Laser Beam Shaping Applications, *edited by Fred M. Dickey, Scott C. Holswade and David L. Shealy*
103. Electromagnetic Theory and Applications for Photonic Crystals, *Kiyotoshi Yasumoto*
104. Physics of Optoelectronics, *Michael A. Parker*
105. Opto-Mechanical Systems Design: Third Edition, *Paul R. Yoder, Jr.*
106. Color Desktop Printer Technology, *edited by Mitchell Rosen and Noboru Ohta*
107. Laser Safety Management, *Ken Barat*
108. Optics in Magnetic Multilayers and Nanostructures, *Štefan Višňovský*
109. Optical Inspection of Microsystems, *edited by Wolfgang Osten*
110. Applied Microphotonics, *edited by Wes R. Jamroz, Roman Kruzelecky, and Emile I. Haddad*
111. Organic Light-Emitting Materials and Devices, *edited by Zhigang Li and Hong Meng*
112. Silicon Nanoelectronics, *edited by Shunri Oda and David Ferry*
113. Image Sensors and Signal Processor for Digital Still Cameras, *Junichi Nakamura*
114. Encyclopedic Handbook of Integrated Circuits, *edited by Kenichi Iga and Yasuo Kokubun*
115. Quantum Communications and Cryptography, *edited by Alexander V. Sergienko*
116. Optical Code Division Multiple Access: Fundamentals and Applications, *edited by Paul R. Prucnal*

Optical Inspection of Microsystems

edited by

Wolfgang Osten

CRC Press is an imprint of the
Taylor & Francis Group, an **informa** business
A TAYLOR & FRANCIS BOOK

Cover image: The digital hologram was recorded by Andrea Finizio.

CRC Press
Taylor & Francis Group
6000 Broken Sound Parkway NW, Suite 300
Boca Raton, FL 33487-2742

First issued in paperback 2019

ISBN-13: 978-0-8493-3682-9 (hbk)
ISBN-13: 978-0-367-39057-0 (pbk)
Library of Congress Card Number 2005046670

Library of Congress Cataloging-in-Publication Data

Optical inspection of microsystems / edited by Wolfgang Osten.
p. cm.
Includes bibliographical references and index.
ISBN 0-8493-3682-1 (alk. paper)
1. Quality control--Optical methods. 2. Optical detectors--Industrial applications. 3. Microelectronics. I. Osten, Wolfgang.

TS156.2O652 2006
670.42'5--dc22 2005046670

Visit the Taylor & Francis Web site at
http://www.taylorandfrancis.com

and the CRC Press Web site at
http://www.crcpress.com

for

Angelika, Luise, and Stefan

Preface

The miniaturization of complex devices such as sensors and actuators is one of the biggest challenges in modern technology. Different manufacturing technologies — for instance, the so-called LIGA technique and UV lithography — allow the realization of nonsilicon and silicon microparts with a high aspect ratio and structural dimensions in the range from nanometers to millimeters. LIGA is an acronym standing for the main steps of the process, i.e., deep x-ray lithography, electroforming, and plastic molding. These three steps make it possible to mass-produce high-quality microcomponents and microstructured parts, in particular from plastics, but also from ceramics and metals at low cost. Techniques based on UV lithography or advanced silicon etching processes (ASE) allow for direct integration of electronics with respect to the realization of advanced microelectromechanical systems (MEMS) devices. Further technologies such as laser micromachining, electrochemical milling (ECF), electrodischarge machining, and nanoimprint lithography (NIL) offer, meanwhile, an economical, high-resolution alternative to UV, VUV, and next-generation optical lithography.

Increased production output, high system performance, and product reliability and lifetime are important conditions for the trust in a new technology and deciding factors for its commercial success. Consequently, high quality standards are a must for all manufacturers. However, with increasing miniaturization, the importance of measurement and testing is rapidly growing, and therefore the need in microsystems technology for suitable measurement and testing procedures is evident. Both reliability and lifetime are strongly dependent on material properties and thermomechanical design. In comparison to conventional technologies, the situation in microsystems technology is extremely complicated. Modern microsystems (MEMS and MOEMS) and their components are characterized by high-volume integration of a variety of materials and materials combinations. This variety is needed to realize very different and variable functions such as sensor and actuator performance, signal processing, etc. Still, it is well known that the materials´ behavior in combination with new structural design cannot be easily predicted by theoretical simulations. A possible reason for wrong predictions made by FEM calculations with respect to the operational behavior of microdevices is, for instance, the lack of reliable materials data and boundary conditions in the microscale. Therefore, measurement and testing procedures are confronted with a complex set of demands. In general, the potential for the following is challenged:

- Microscopic and nanoscopic measurement and testing on wafer scale
- Fast in-line measurement of various dimensional and functional properties of highly heterogeneous hybrid systems
- Verification of system specifications including geometrical, kinematical, and thermomechanical parameters
- Fast and reliable recognition of surface and subsurface defects, with the possibility for review and repair
- Measurement of complex 3-D structures with high aspect ratio
- Determination of material properties well defined for the bulk but to be specified for microscale

Measurement and inspection techniques are required that are very fast, robust, and relatively low cost compared to the products being investigated. The reason for this demand is obvious: properties determined on much larger specimens cannot be scaled down from bulk material without any experimental verification. Further on, in microscale, materials' behavior is noticeably affected by production technology. Therefore, simple and robust methods to analyze the shape and deformation of the microcomponents are needed. Together with the knowledge of the applied load and appropriate physical models, these data can be used for the derivation of material parameters and various system properties. It is obvious that neither a single method nor a class of measurement techniques can fulfill these requirements completely. Conventional tensile test techniques (e.g., strain gauges) are unable to test specimens from submillimeter-sized regions because of their limited local resolution and partly unwanted tactile character. Other approaches, such as, for instance, microhardness measurements, do not reveal directional variations.

However, full-field optical methods provide a promising alternative to the conventional methods. The main advantages of these methods are their noncontact, nondestructive, and fieldwise working principle; fast response potential; high sensitivity and accuracy (typical displacement resolution of a few nanometers, strain values of 100 microstrain); high resolution of data points (e.g., 1000 × 1000 points for submillimeter field of view); advanced performance of the system, i.e., automatic analysis of the results; and data preprocessing in order to meet requirements of the underlying numerical or analytical model. Thus, this book offers a timely review of the research into applying optical measurement techniques for microsystems inspection. The authors give a general survey of the most important and challenging optical methods such as light scattering, scanning probe microscopy, confocal microscopy, fringe projection, grid and Moiré techniques, interference microscopy, laser Doppler vibrometry, holography, speckle metrology, and spectroscopy. Moreover, modern approaches for data acquisition and processing (for instance, digital image processing and correlation) are presented.

The editor hopes that this book will significantly push the application of optical principles for the investigation of microsystems. Thanks are due to all authors for their contributions, which give a comprehensive overview of the state of the art in the fascinating and challenging field of optical microsystems metrology. Finally, the editor is grateful for the cooperation shown by CRC Press represented by Taisuke Soda, Preethi Cholmondeley, Gerry Jaffe, and Jessica Vakili.

Wolfgang Osten
Stuttgart

About the Editor

Wolfgang Osten received his B.Sc. degree from the University of Jena, Germany, in 1979. From 1979 to 1984 he was a member of the Institute of Mechanics in Berlin working in the field of experimental stress analysis and optical metrology. In 1983 he received his Ph.D. degree from the Martin-Luther-University Halle-Wittenberg for his thesis in the field of holographic interferometry. From 1984 to 1991 he was employed at the Central Institute of Cybernetics and Information Processes in Berlin, making investigations in digital image processing and computer vision. In 1991 until 2002 he joined the Bremen Institute of Applied Beam Technology (BIAS) to establish and direct the Department Optical 3D-Metrology. Since September 2002, he has been a full professor at the University of Stuttgart and director of the Institute for Applied Optics. His research work is focused on new concepts for industrial inspection and metrology by combining modern principles of optical metrology, sensor technology, and image processing. Special attention is paid to the development of resolution enhanced technologies for the investigation of micro and nano structures.

About [illegible] Editor

Contributors

Anand Asundi
School of Mechanical and Aerospace Engineering
Nanyang Technological University
Singapore, Singapore

Petra Aswendt
Fraunhofer Institute IWU
Chemnitz, Germany

Alain Bosseboeuf
Insitut d'Electronique Fondamentale
Université Paris Sud CNRS
Orsay, France

Patrick Delobelle
Department LMA, FEMTO-ST
Université de Franche-Comté
Besançon, France

Ingrid De Wolf
IMEC VZW
Leuven, Belgium

Angela DuparrŽ
Fraunhofer Institute for Applied Optics and Precision Engineering (IOF)
Jena, Germany

Pietro Ferraro
CNR-Instituto Nazionale
di Ottica Applicata
Napoli, Italy

Cosme Furlong
Center for Holographic Studies and Laser MicroMechatronics
Mechanical Engineering Department
Worcester Polytechnic Institute
Worcester, Massachusetts (U.S.A.)

Christophe Gorecki
Département LOPMD, FEMTO-ST
Université de Franche-Comté
Besançon, France

Roland Höfling
Vialux GmbH
Chemnitz, Germany

Markus Hüttel
Information Processing
Fraunhofer Institute Manufacturing Engineering and Automation (IPA)
Stuttgart, Germany

Michal Jozwik
Département LOPMD, FEMTO-ST
Université de Franche-Comté
Besançon, France

Klaus Körner
Institut fur Technische Optik
Universität Stuttgart
Stuttgart, Germany

Bernd Michel
Micro Materials Center Berlin
Fraunhofer Institute for Reliability and Microintegration (IZM)
Berlin, Germany

Wolfgang Osten
Institut fur Technische Optik
Universität Stuttgart
Stuttgart, Germany

Sylvain Petitgrand
Fogale Nanotech
Nimes, France

Christian Rembe
Polytec GmbH
Waldbronn, Germany

Aiko Ruprecht
Institut fur Technische Optik
Universität Stuttgart
Stuttgart, Germany

Leszek Salbut
Institute of Micromechanics and Photonics
Warsaw University of Technology
Warsaw, Poland

Joanna Schmit
Veeco Instruments, Inc.
Tucson, Arizona (U.S.A.)

F. Michael Serry
Veeco Instruments, Inc.
Santa Barbara, California (U.S.A.)

Georg Siegmund
Polytec GmbH
Waldbronn, Germany

Heinrich Steger
Polytec GmbH
Waldbronn, Germany

Dietmar Vogel
Micro Materials Center Berlin
Fraunhofer Institute for Reliability and Microintegration (IZM)
Berlin, Germany

Tobias Wiesendanger
Institut fur Technische Optik
Universität Stuttgart
Stuttgart, Germany

Michael Wörtge
Polytec GmbH
Waldbronn, Germany

Huimin Xie
School of Mechanical and Production Engineering
Nanyang Technological University
Singapore, Singapore

Bing Zhao
School of Mechanical and Production Engineering
Nanyang Technological University
Singapore, Singapore

Contents

Chapter 1
Image Processing and Computer Vision for MEMS Testing 1
Markus Hüttel

Chapter 2
Image Correlation Techniques for Microsystems Inspection 55
Dietmar Vogel and Bernd Michel

Chapter 3
Light Scattering Techniques for the Inspection of Microcomponents and Microstructures 103
Angela Duparré

Chapter 4
Characterization and Measurement of Microcomponents with the Atomic Force Microscope (AFM) 121
F. Michael Serry and Joanna Schmit

Chapter 5
Optical Profiling Techniques for MEMS Measurement 145
Klaus Körner, Aiko Ruprecht, and Tobias Wiesendanger

Chapter 6
Grid and Moiré Methods for Micromeasurements 163
Anand Asundi, Bing Zhao, and Huimin Xie

Chapter 7
Grating Interferometry for In-Plane Displacement and Strain Measurement of Microcomponents 201
Leszek Salbut

Chapter 8
Interference Microscopy Techniques for Microsystem Characterization 217
Alain Bosseboeuf and Sylvain Petitgrand

Chapter 9
Measuring MEMS in Motion by Laser Doppler Vibrometry 245
Christian Rembe, Georg Siegmund, Heinrich Steger, and Michael Wörtge

Chapter 10
An Interferometric Platform for Static, Quasi-Static, and Dynamic Evaluation of Out-of-Plane Deformations of MEMS and MOEMS 293
Christophe Gorecki, Michal Jozwik, and Patrick Delobelle

Chapter 11
Optoelectronic Holography for Testing Electronic Packaging and MEMS 325
Cosme Furlong

Chapter 12
Digital Holography and Its Application in MEMS/MOEMS Inspection 351
Wolfgang Osten and Pietro Ferraro

Chapter 13
Speckle Metrology for Microsystem Inspection 427
Roland Höfling and Petra Aswendt

Chapter 14
Spectroscopic Techniques for MEMS Inspection 459
Ingrid De Wolf

Index 483

1 Image Processing and Computer Vision for MEMS Testing

Markus Hüttel

CONTENTS

1.1 Introduction 2
1.2 Classification of Tasks 2
1.3 Image Processing and Computer Vision Components 4
    1.3.1 Behavior of Light, Colors, and Filters 5
    1.3.2 Illumination 8
    1.3.3 Lens Systems 12
    1.3.4 Sensors 15
        1.3.4.1 CCD Sensors 16
        1.3.4.2 CMOS Sensors 18
        1.3.4.3 Color Sensors and Cameras 19
        1.3.4.4 Camera Types and Interfaces 20
        1.3.4.5 Frame Grabbers 21
1.4 Processing and Analysis of Image Data 22
    1.4.1 Computer Vision Process 22
    1.4.2 Image Data Preprocessing and Processing Methods 24
        1.4.2.1 Histograms 24
        1.4.2.2 Point Transformations 25
        1.4.2.3 Spatial Filtering 27
    1.4.3 Image Data Analysis Methods 31
        1.4.3.1 Spectral Operations 32
    1.4.4 Solving Measurement and Testing Tasks 36
        1.4.4.1 Finding a Test Object or Region of Interest 36
        1.4.4.2 Position Recognition 40
        1.4.4.3 Measuring Geometric Features 42
        1.4.4.4 Presence Verification 45
        1.4.4.5 Defect and Fault Detection 47
1.5 Commercial and Noncommercial Image Processing and Computer Vision Software 49
1.6 Image Processing Techniques for the Processing of Fringe Patterns in Optical Metrology 50
1.7 Conclusion 52
References 52

1.1 INTRODUCTION

Not only is there a requirement for testing electrical and dynamic behavior of MEMS, but there is also considerable demand for methods to test these systems during both the development phase and the entire manufacturing phase. With the aid of these test methods, it is possible to assess such static properties as the dimension, shape, presence, orientation, and surface characteristics of microsystems and their components. Using an optical measurement and testing technique based on image processing and computer vision, a wide range of procedures that enable such properties to be recorded rapidly and in a robust and noncontact way can be applied.

If measurement and testing means are not based on special optical procedures but rather on illumination with normal light and imaging with normal and microscopic optics, their resolution capabilities extend to only just below the micrometer range. This is due to the diffraction of light and the dimensions of imaging sensor elements in the lateral direction. Such a degree of resolution is inadequate as it is unable to cover the entire range of microsystem structure sizes — from just a few nanometers (e.g., surface roughness) to a few millimeters (e.g., external contours). In order to measure sizes in the nanometer range, special imaging measurement and testing means are required. These include interferometers, spectrometers, near-field/scanning electron/atomic-force microscopes, and specialized reconstruction and analysis processes such as fringe processing or scanning techniques, which are described in Chapters 4 and 8 through 14.

The main advantage of implementing optical testing equipment using simple light sources and normal and microscopic optics is the speed with which images can be recorded and analyzed and the fact that they can be easily integrated into the manufacturing process, thus making the error-prone removal of components from and reintroduction into the clean environment for test purposes superfluous. For this reason, despite their limited resolution capabilities, these equipment are ideally suited for testing large piece numbers, i.e., in the manufacturing process in the areas of assembly, function, and integrity testing and packaging.

Furthermore, the algorithms developed for image processing and computer vision are not only suitable for analyzing images recorded using a video camera but can also be applied to the fields of signal analysis, data analysis and reconstruction, etc.

This chapter deals with the technical aspects of illumination and image recording techniques as well as image processing and computer vision processes relevant to optical measurement and testing techniques and their implementation in typical measurement and testing tasks. Several software products that are available commercially for image processing and computer vision will also be described. However, a classification of typical measurement and testing tasks in the field of microsystem development and production is given first.

1.2 CLASSIFICATION OF TASKS

The tendency toward miniaturization in the electronics industry, which has been observed for many years, has also been affecting the field of optics and mechanics over the last few years. The origins of miniaturizing mechanical systems to micrometer dimensions can be found in semiconductor-manufacturing research and laboratories. They possess expertise in processing delicate silicone structures and the equipment required to produce them. Most of the microelectromechanical systems (MEMS) or microopticalelectromechanical systems (MOEMS) available today have been created using the combination of electronic, optical, and mechanical functional groups. Today, MEMS or MOEMS can be found in a wide range of technical devices used in everyday life. In ink jet printers, MEMS-based microinjectors transfer ink to paper. Digital mirror devices (DMDs) — a matrix of thousands of electrically adjustable micromirrors — are responsible for producing digital images in digital light projection (DLP) projectors. Sensors for measuring pressure, force, temperature, flow rates, air mass, acceleration, tilting, and many other values are constructed as MEMS and are utilized especially in the automotive industry — a mass market with high levels of safety requirements (airbag,

Electronic Stability Program, Bosch, Germany (ESP®)). However, MEMS and MOEMS are being used increasingly in many other fields such as the medical industry or biological and chemical diagnostics. "Lab on a chip" is capable of carrying out complete investigation processes for chemical, biological, or medical analyses. Microspectrometers, just a few millimeters in size, enable the construction of extremely small and low-cost analysis devices that can be used as in-flight equipment for monitoring terrestrial, biological, and climatic processes and in a wide range of technical applications in the form of handheld devices. Using micromotors, microdrives, micropumps, and microcameras, instruments can be constructed for keyhole diagnoses and surgical interventions.

Although, in comparison with microelectronic components, microsystems possess a clear three-dimensional structure (e.g., microspectrometers, electrical micromotors or gears for drives made using microinjection molding techniques), classic MEMS structures, especially sensor and mirror systems, are essentially two-dimensional. This feature is the result of constructing MEMS based on semiconductor materials and on the corresponding manufacturing processes. Another conspicuous characteristic of MEMS-based systems is the common use of hybrid constructions, where the drive and analysis electronics are located on a semiconductor chip and the actual MEMS (e.g., the sensor element) on a separate chip.

Both of these properties influence the tests realizable for MEMS using image processing and computer vision. These are essentially performed using an incident light arrangement in which image recording and illumination take place at the same angle. The transmissive light arrangement (which can be much better controlled), where the object to be tested is situated between the illumination source and the image recording system, can be used to advantage if MEMS were more three-dimensional in shape. This is increasingly becoming the case.

From the point of view of image processing and computer vision, the solutions listed here are possible for the following examples of testing tasks:

- *Locate test object or regions of interest:* Test objects need to be located if their positions vary either in relation to each another and/or in relation to the reference system of the observation area. This is often the case in systems of hybrid constructions, systems where individual components are located relatively inaccurately next to one another. Regions of interest need to be located, for example, in cases in which MEMS components are individually processed or adjusted (e.g., when measuring the cross sections of laser-trimmed resistances and capacitors) or in the case of a high-resolution measurement or test system for a measuring area that is to be determined using a small measurement range (multiscaled measuring or testing).
- *Position recognition:* The position of components needs to be recognized if they are not aligned for a test or if their positions, when installed, show degrees of freedom (e.g., resistors, diodes, or gears). Another example is when tests need to be carried out on the components themselves, such as the recognition of component coding or the measurement of geometric features.
- *Measuring geometric features:* Investigating a component from a metrological point of view shows whether production tolerances have been adhered to or not; these affect both the mechanical and electrical behavior of the object. The location of geometric features using contour or edge recognition and the calculations based on them to obtain straightness, circularity, length, enclosed surfaces, angle, distance, diameter, etc. form the principles of optical metrology.
- *Presence verification:* The monitoring of production and assembly processes often requires a purely qualitative statement regarding certain features without specific knowledge of their geometrical characteristics. For example, in production processes, tool breakage can be monitored by checking a work piece for the presence of bore holes,

grooves, etc. With assembly processes, the focus of interest is usually on the final assembled item, i.e., the presence of all components requiring assembly.

- *Fault detection:* In contrast to presence verification, in the case of fault detection, features are checked for deviations from required standards. To detect faults, for example, text recognition is used for reading identification markings on components, color checking for verifying color-coded components, and texture analysis for investigating structured surfaces for flaws (e.g., scratches).

In order to solve these examples of measuring and verification tasks based on two-dimensional image data, image processing and computer vision have a whole range of proven processing and interpretation methods available. Shading correction, the averaging of image series, spatial and morphological filtering, edge detection, pattern and geometric feature matching, segmentation, connectivity analysis, and the metrology of geometric features denote some of these methods. Other techniques, required, for example, in interferometry, spectroscopy, or holography and which also permit imaging metrology, are described in Chapters 5, 6, and 12.

1.3 IMAGE PROCESSING AND COMPUTER VISION COMPONENTS

The elementary task of digital image processing and computer vision is to record images; process, improve, and analyze the image data using appropriate algorithms; supply results derived from them; and interpret these results. Figure 1.1 shows a typical scenario for an image processing system. Images of an object illuminated by a light source are recorded by an electronic camera. After conversion by a frame grabber into digital values, the analog electrical image signals from the camera are stored in a computer. The digital representation so obtained forms the starting point for subsequent algorithmic processing steps, analyses, and conclusions, otherwise known as image processing and computer vision.

As well as requiring knowledge of the theory of digital image processing (which represents the fundamentals for the algorithms used in image processing and computer vision), it is clear that expertise in the fields of light physics, optics, electrical engineering, electronics, and computer hardware and software is also necessary. The first steps toward finding a successful image processing solution are not in the selection and implementation of suitable algorithms but rather much earlier on, i.e., in the depiction of the images. Particular attention must be given to the illumination of a scene. A type of illumination well adapted to the task in question produces images that can be analyzed using simple algorithmic methods; badly illuminated scenes may produce images from which even the most refined algorithms are unable to extract the relevant features. It is also equally important to take the dynamic behavior of imaging sensors into consideration. If illumination is too bright, sensor elements may reach saturation and produce overilluminated images that could

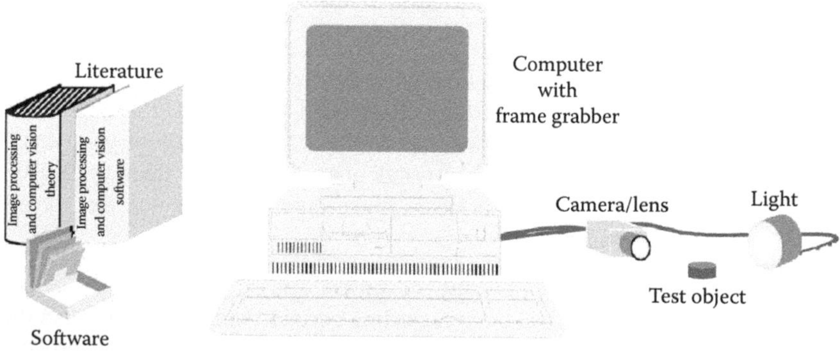

FIGURE 1.1 Components of a computer vision scenario.

lead to false conclusions. As far as metrological tasks are concerned, it is essential to understand the imaging properties of lenses. For these reasons, the following sections are concerned with the most important aspects of illumination, imaging sensors, camera technology, and the imaging properties of lenses.

1.3.1 Behavior of Light, Colors, and Filters

The range of electromagnetic waves in the *electromagnetic spectrum* (see Figure 1.2), which are of interest as far as image processing is concerned, lies between ultraviolet and near-infrared — a very small part of the entire electromagnetic spectrum. The human eye is only capable of recognizing part of the spectrum, that of visible light.

The *intensity* I (W/m^2) of a light source is defined by the quantity of photons radiated per unit of area. If the distance is increased between a diffusely illuminating spotlight source and a surface element, the intensity of the light decreases in proportion to the square of the distance. The areas of constant intensity are present as concentric spheres around the light source (see Figure 1.3). As described later on, with light collimators the conformity with this law leads to an inhomogeneous density of light, which decreases from the center towards the periphery.

Light from the sun or a bulb is seen by the human eye as white light. This is due to the fact that these light sources radiate electromagnetic waves that cover the entire spectrum of visible light. As depicted in Figure 1.4, this becomes clear when light propagates through a prism made of glass or a similar transparent material: because of the higher optical density of glass, expressed as the *refraction index* n, light waves are refracted at different degrees depending upon their wavelength λ and disintegrate into the colors of the rainbow. This is known as *dispersion*.

If light dispersed into the colors of the rainbow is merged again using a lens, white light results. This observation leads to the phenomenon of mixing colors. If lights of different colors are mixed,

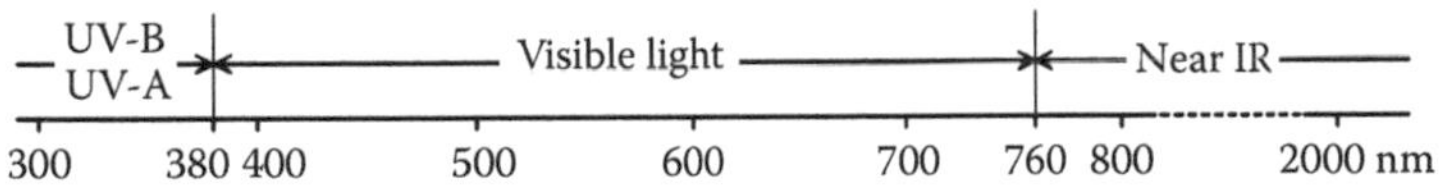

FIGURE 1.2 Electromagnetic spectrum.

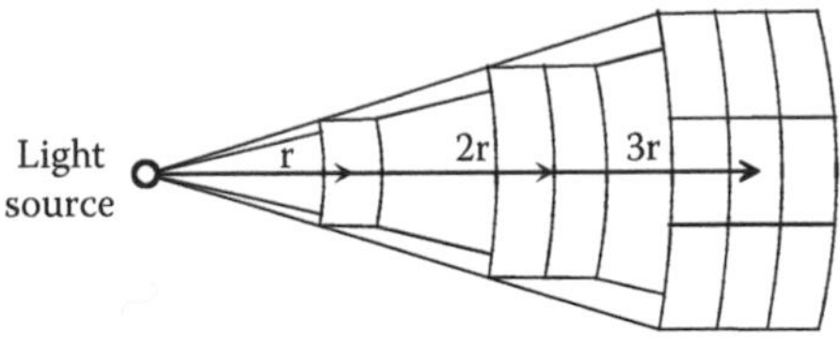

FIGURE 1.3 Illustration of the inverse square law.

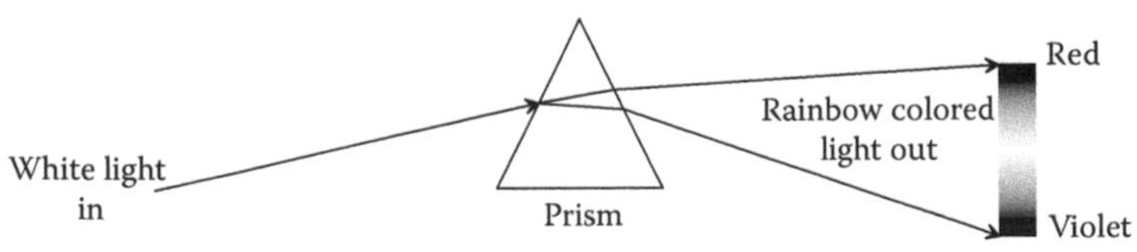

FIGURE 1.4 Refraction of white light by a prism.

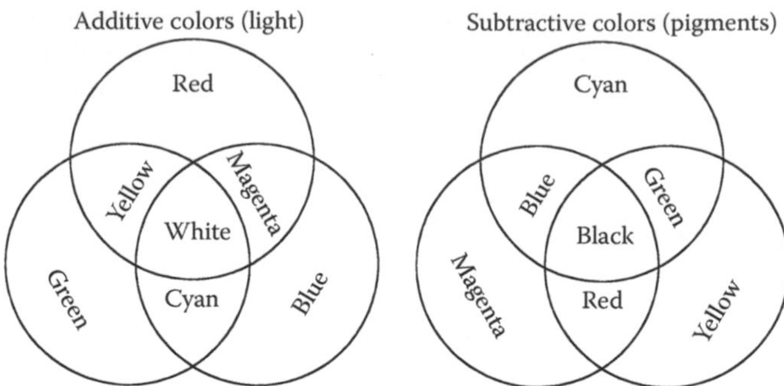

FIGURE 1.5 Additive (left) and subtractive (right) colors.

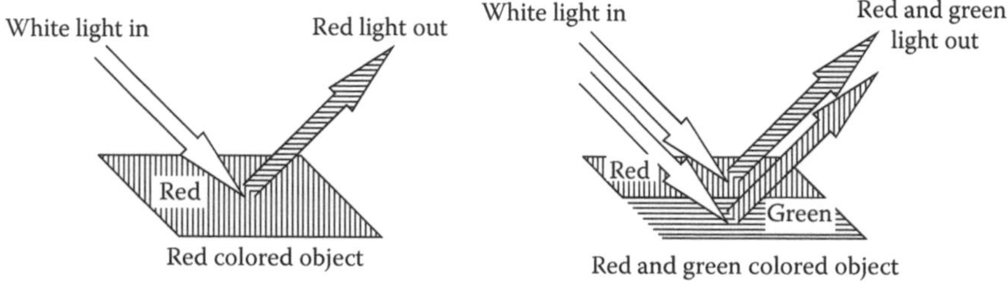

FIGURE 1.6 Response of colored objects to white light.

the colors get added together and form white. If, however, colored substances (e.g., pigments) illuminated by white light are mixed, the colors are subtracted from one another and form black, as shown in Figure 1.5.

From this arise the questions why objects appear colored if they are illuminated by white light and also what are the effects of colored lights in conjunction with colored objects.

If an object appearing red to the human eye is illuminated by white light, the electromagnetic waves of the spectrum corresponding to the color red are reflected and reach the eye. The light from all the other wavelengths is adsorbed by the object's pigments and is transformed into heat. Naturally, the same applies for objects made up of several colors, as shown in Figure 1.6.

As monochrome cameras are often used in image processing, colored light can be used advantageously to highlight colored objects. If, as shown in Figure 1.7, a red-and-green-colored

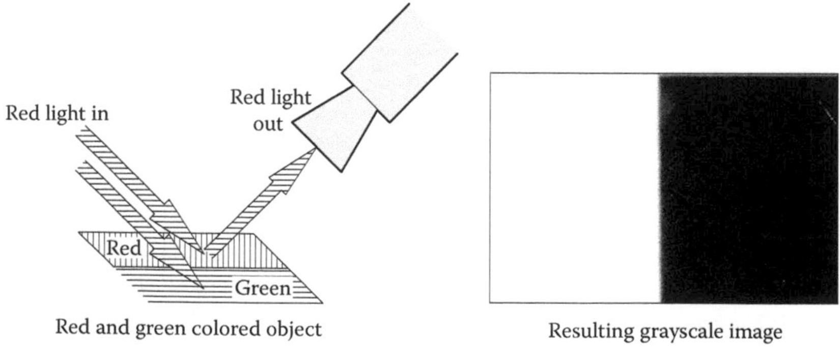

FIGURE 1.7 Response of colored objects to colored light.

object is illuminated by red light, essentially only the red-colored areas of the object reflect the light, leading to pale gray values in an image taken by a monochrome camera.

LEDs are often used as colored light sources in image processing today if the area to be illuminated is not too large. Because of their specific color types, LEDs cover the entire frequency range of visible light and the adjacent areas of near-IR and UV, induce minimal loss of warmth, and can be switched on and off very quickly. However, if intense light sources are required, sources of white light such as halogen lamps are implemented and combined with optical filters. The filters are generally made of colored glass, which selectively allow light of a certain wavelength to pass through and either adsorb or reflect all other wavelengths.

If a scene is illuminated with colored light but recorded with a monochrome camera, it often makes sense to place the filter directly in front of the camera lens rather than to filter the light from the source of white light. In this way, any stray light from the environment that does not possess the wavelength of the transmitted filter light is also filtered out and is therefore unable to reach the camera's sensors (see Figure 1.8).

In cases in which the colors of an object are irrelevant, near-infrared light sources are often used in conjunction with a filter that only transmits this light. As a result, light conditions in which the object is illuminated are almost completely independent of visible light. If the infrared light source is also monochromatic, i.e., only light from a narrow range of the spectrum is transmitted or, if the filter is constructed as a narrow band-pass one (see the following text), the color distortions of the lens cause fewer chromatic errors in the images recorded, which can be used advantageously especially when metrological tasks are concerned.

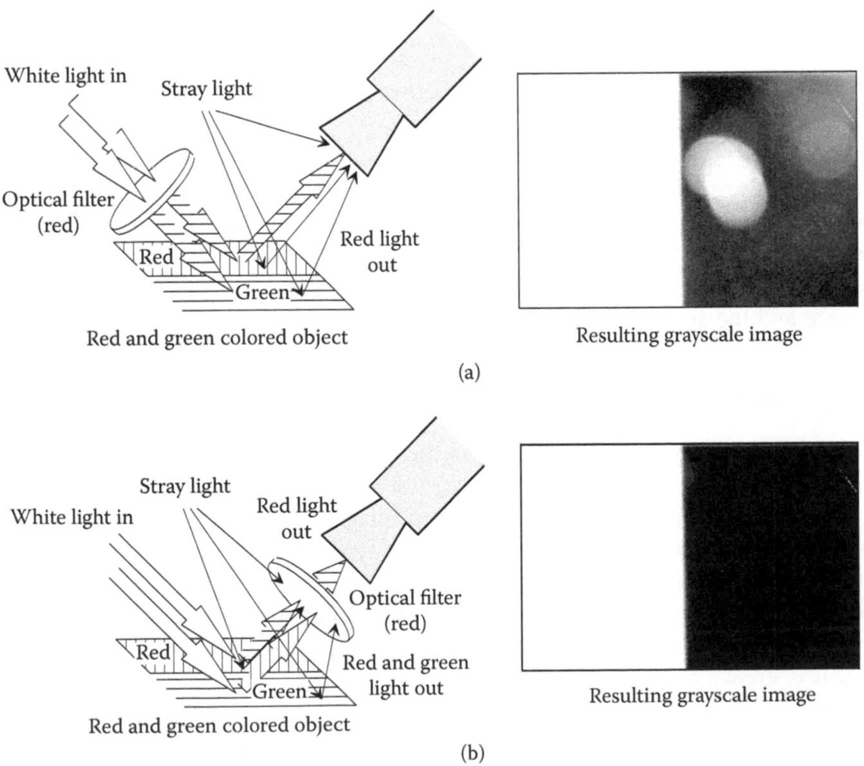

FIGURE 1.8 Effect of filter location on the resulting grayscale image.

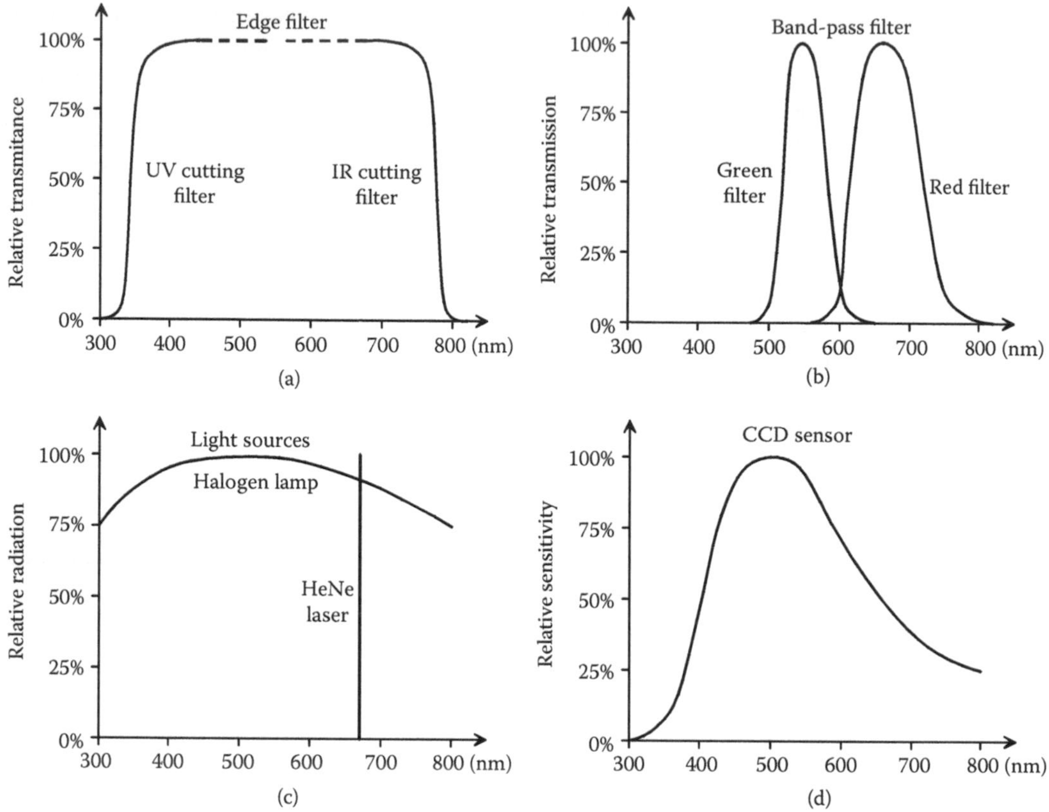

FIGURE 1.9 Spectral response plot of: (a) edge filters, (b) band-pass filters, (c) light sources, (d) CCD sensors.

As far as the transmission of light is concerned, the behavior of optical filters is characterized using spectral response plots. To do this, in general, the relative transmission of the filter material is plotted as a function of the wavelength of the electromagnetic spectrum (see Figure 1.9).

In the same way, it is also possible to characterize the spectral fractions of irradiated light from a light source and the spectral sensitivity of light-sensitive sensors.

1.3.2 Illumination

The type of illumination plays a crucial role in image processing. By using an illumination suited to the special task in question, the features of a test object requiring measurement or testing can often be better highlighted. Thus, the processing of images is simplified drastically in many cases.

Illumination can essentially be classified into two types in accordance with the arrangement of the light source, test object, and image-recording system. If the test object is situated between the light source and the image-recording system, it is known as a *transmissive light arrangement*; if the light source and the image-recording system are situated on the same side in relation to the test object, it is known as an *incident light arrangement*.

The simplest realization of a *transmissive light arrangement* is represented by a light panel (see Figure 1.10) in which the light source (usually consisting of several lamps, tube lamps, or LEDs) is positioned behind a ground-glass screen that scatters light diffusely. This arrangement is especially used to determine and measure the contours of flat, nontransparent test objects because the camera sees the shadowed image of the object. In cases in which objects are translucent or

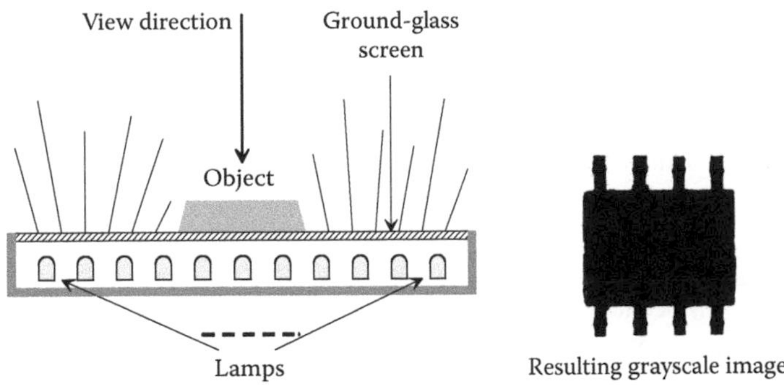

FIGURE 1.10 Transmissive light arrangement: (a) principle of a light panel, (b) silhouette of an object.

transparent, this arrangement enables internal structures to be recognized. However, the diffusely scattered light from a ground-glass screen is disadvantageous if the dimensions of test objects are particularly large in the direction of the optical axis of the camera. This is because surfaces that are almost parallel to the axis could also be illuminated, thus falsifying the true silhouette of the test object and the resulting measurement data obtained.

This disadvantage of diffusely scattered light can be reduced if an illumination system composed of a concave lens and a spotlight source (see Figure 1.11), a so-called collimator, is used instead of ground-glass illumination. When the spot light source is placed at the focal point of the lens, the light emerges as almost parallel rays. LEDs (without lens optics) are especially useful as light sources for this because light-emitting semiconductors are so small that they are almost punctiform.

Despite the advantage of parallel light rays, illuminating collimators have the disadvantage of producing inhomogeneous illumination. In accordance with the aforementioned law regarding the decrease in light intensity of a spotlight source, the light intensity of light rays passing through the lens decreases from the optical axis towards the periphery; this is known as shading or vignetting. This effect, which is disadvantageous as far as image processing is concerned, can be avoided if a homogeneously illuminated diffuse ground-glass light source, placed at the image level of a lens, is projected instead of using a spotlight source at the focal point of a collimator. If a telecentric lens (see Subsection 1.3.3) is used instead of a normal lens to project a ground-glass light source, homogeneous illumination with parallel light rays results.

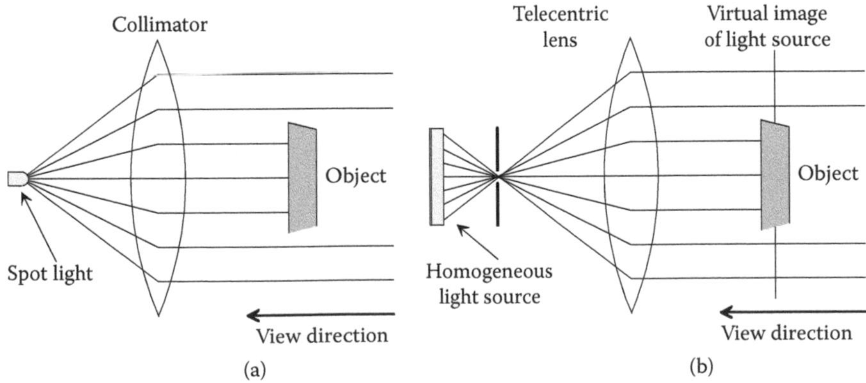

FIGURE 1.11 Transmissive light arrangement: (a) collimator, (b) projection of a homogeneous light source.

Because many MEMS or their components are based on silicone wafers, which are, according to their nature, essentially two-dimensional, the main type of illumination used in image processing to test such elements is that of incident light arrangement. In contrast to the transmissive light arrangement, by using this type of illumination the test features of the object and all other areas are equally illuminated. This results in the features of the images recorded often being more difficult to differentiate in subsequent image processing. From this point of view, a type of illumination that is suited for the task is of particular importance.

With incident light arrangements, an essential difference is made between the *dark field* and *bright field arrangement* because they highlight very different aspects of a test object.

As far as the angle of observation is concerned, with dark field arrangement (see Figure 1.12) the test objects are illuminated from an almost perpendicular angle. As a result, in an ideal situation, only a small amount of light or no light falls on the object surfaces that are parallel to the angle of incidence and thus perpendicular to the angle of observation. These fields are seen by the observer as dark areas. In contrast, all other nontransparent object details are highlighted, provided they are not positioned in the shadows of other object details. With nontransparent test objects, dark field illumination is, therefore, advantageous if raised sections of an object have to be highlighted. Another example of using dark field illumination to advantage is in detecting particles on smooth, even surfaces. Transparent test objects such as glass and plastic objects can also be examined for inclusions or edge defects because these features stand out well against a dark background, especially when the latter is matte black in color. Generally, spot and linear, diffuse and directed light sources are used for dark field illumination. In many cases, low-angle glass fibers or LED ring lights reduce the problem of shadow formation.

In contrast to the dark field arrangement, in *bright field arrangement* test objects are illuminated from nearly the angle of observation. In simple cases, the light source for bright field arrangement

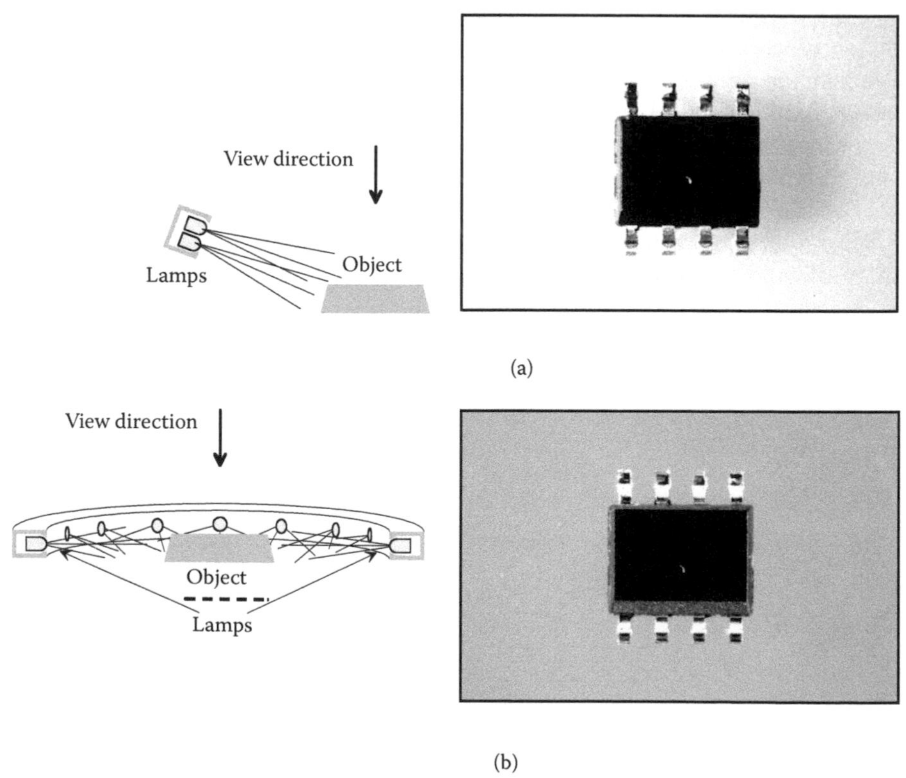

FIGURE 1.12 Principles of dark field illumination: (a) single-sided, (b) circular light.

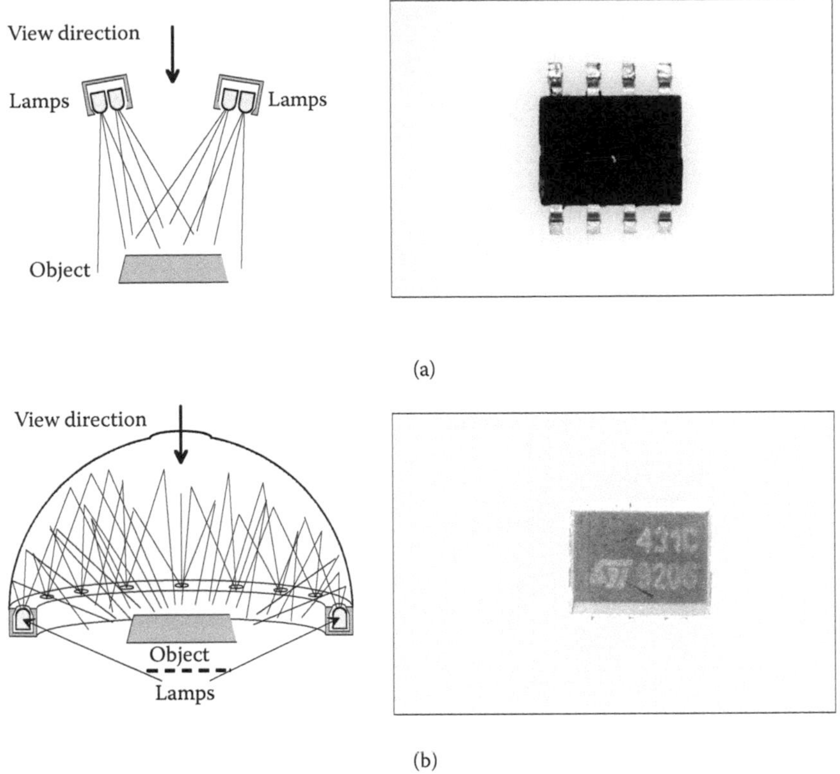

FIGURE 1.13 Bright field illumination: (a) spotlight source, (b) light dome.

is realized using either one or several spotlight sources, high-angle glass fiber lights, or LED ring lights placed around the camera lens. Using this illumination arrangement, a good differentiation of features can be obtained if object surfaces are diffusely dispersive, provided the colors of the features to be tested contrast well against their surroundings, such as in the case of dark print markings or barcodes on a pale or colored background.

The bright field arrangement described earlier becomes problematic if object surfaces are shiny or reflective. This is because the light sources may be projected into the recording camera system in the form of reflections, thus impeding or preventing reliable feature analysis. Homogeneous illumination can be achieved in these cases if all visible sides of an object are illuminated equally by a diffuse light source. The light dome (half sphere) shown in Figure 1.13 is capable of this: light sources fixed at the edge of the light dome illuminate its internal surface, which is matte and coated white.

In the case of flat, shiny surfaces, reflection-free bright field illumination can be achieved if the light is coupled coaxially to the angle of observation. As shown in Figure 1.14a, diffuse light from a homogeneously illuminated surface, such as that used in transmissive light arrangements, is deflected by 90° toward the angle of observation using a semitransparent mirror. Owing to the homogeneity of the light source and the evenness of the object surface, object areas possessing equal degrees of reflection are depicted in the camera image with the same level of brightness, making this a reliable method for imaging and analyzing features on such surfaces.

A special form of coaxial bright field illumination is utilized for telecentric lenses (see Subsection 1.3.3). As shown in Figure 1.14b light from a spotlight source is coupled to the beam path of the lens using a semitransparent mirror. Ideally, the light emitted from the lens is made up purely

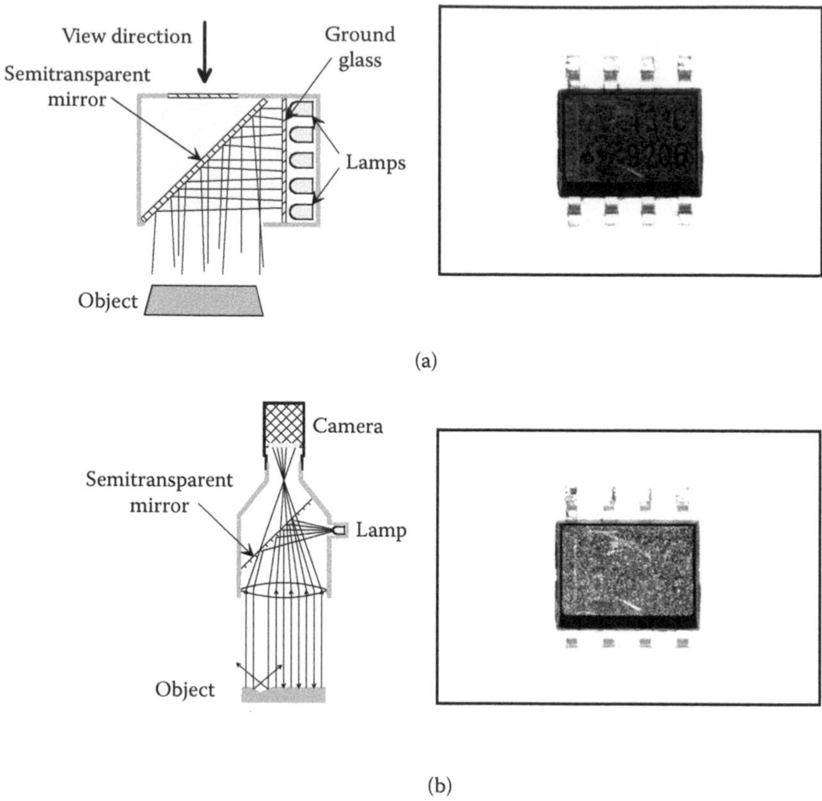

FIGURE 1.14 Bright field illumination: (a) coaxially diffuse, (b) coaxially directed.

of parallel light rays, which are only reflected towards the telecentric lens from object surfaces if those surfaces are perpendicular to the optical axis. The properties of telecentric lenses prevent the light of all other surfaces from reaching the camera, with the result that these appear as dark areas in the recorded image. This feature can be used to make the edges of flat, three-dimensional structures visible.

Besides the types of illumination described here, other forms of illumination can also implemented in image processing. Among others, these include types of structured illumination such as fringe projection and laser scanning for recording and measuring objects in three dimensions (see Chapters 5, 6, and 12).

1.3.3 Lens Systems

To project a real image on the imaging sensor of a camera, a wide range of various lenses are available. Depending on the task, normal, macro- and microscopic lenses can be used. In principle, the imaging properties of these lenses do not differ much from the imaging behavior of a single thin lens, as shown in Figure 1.15, following the imaging equation:

$$\frac{1}{f} = \frac{1}{i} + \frac{1}{o} \tag{1.1}$$

As can be seen in Figure 1.15a, the geometric construction of the image *Im* results from the object *O* as follows: a line is drawn from each object point through the center of the lens, a second

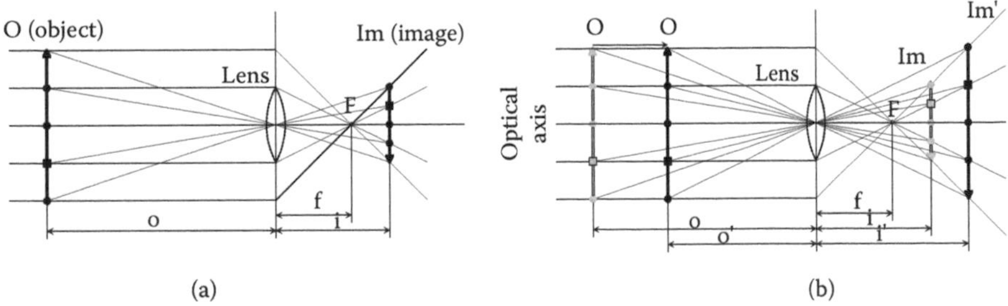

FIGURE 1.15 Imaging behavior of a thin lens: (a) geometric construction of the image of an object; (b) if an object moves toward a lens, the image moves away from the lens and gets enlarged.

line parallel to the optical axis is bisected by the principal plane of the lens, and, starting at this intersection, a third line is drawn through the focal point of the lens. The intersection of the first and third lines then gives the corresponding image point.

As shown in Figure 1.15b, it can be seen from the lens equation that both the position and the size of the image *Im* alter if the object is moved along the optical axis. In the case of lenses, in order to depict the object in focal plain, the displacement of the image *Im* needs to be corrected. This is achieved by moving the lens along the optical axis so that the focal plane (where the imaging sensors of the camera are situated) is always in the same place. Owing to the limitation of the depth of field, it can be directly deduced that objects that are particularly large in the direction of the optical axis and close to the lens cannot be completely in focus. To be more precise, this results in part of the image being in focus and surrounding areas being slightly out of focus. This may considerably impair metrological image analysis or even render it impossible.

The cause of this blurred imaging can be explained by the following observation (see Figure 1.16a). Each point of the object *O* emits light homogeneously in all directions in the form of spherical waves. Part of the light reaches the lens and this then produces an image point in the zone of sharp focus. Instead of image points, circles of confusion are formed outside the zone of sharpness, thus resulting in a blurred image.

As the geometric construction shows, light rays running parallel to the optical axis possess two characteristic features: independent of the distance between the object and the lens, they always run parallel, and when they are projected, they always pass through the focal point *F* of the lens. This results in the generation of a sharp image, independent of distance, by the use of a screen with a small aperture at the height of the focal point (focal pupil) in the beam path of the lens (see Figure 1.16b). Using this measure, a so-called *object-sided telecentric lens* is created.

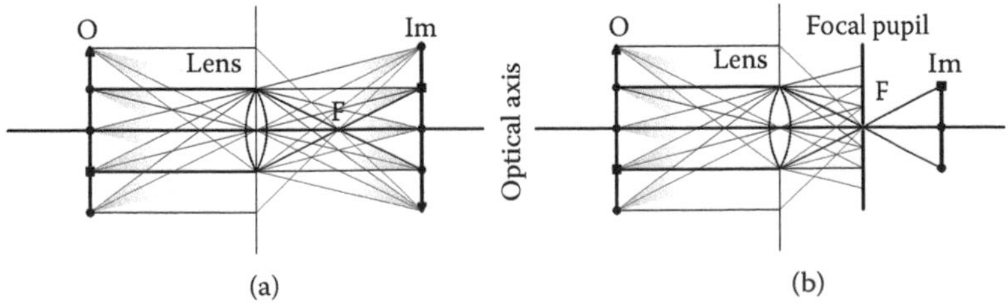

FIGURE 1.16 (a) Formation of image points from object points, (b) telecentric imaging.

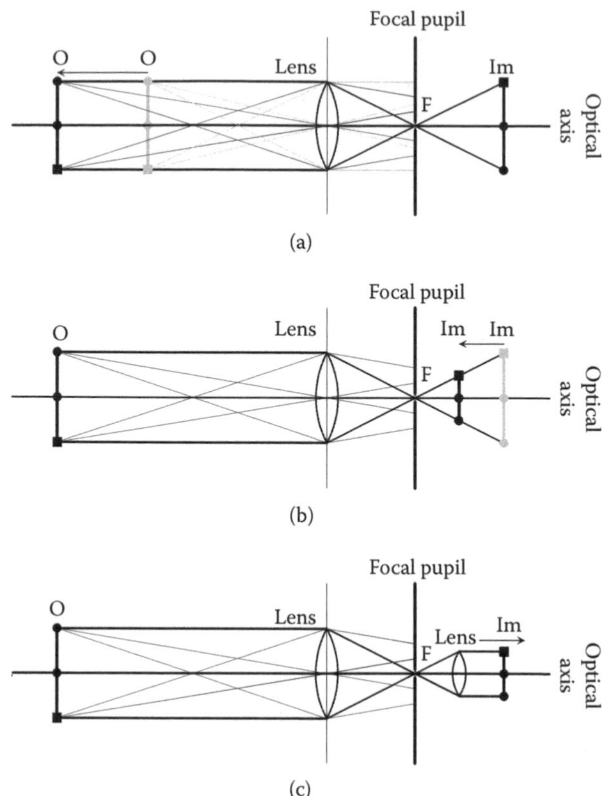

FIGURE 1.17 Behavior of one-sided and double-sided telecentric lenses: (a) within limits, object movement always results in a sharp image; (b) within limits, movement of the focal plain changes size of image but not sharpness; (c) in case of double-sided telecentric lenses, sharpness and size of image (within limits) do not change if focal plain moves along the optical axis.

As illustrated in Figure 1.17, telecentric lenses possess the following properties:

1. The field of view is equal to the size of the lens because light rays running parallel to the optical axis are not imaged outside the lens area.
2. The imaging of an object *O* (see Figure 1.17a) always results in (within limits) a sharp image *Im*, independent of the distance between the object and the lens. This is limited by the finite dimensions of the focal pupil (almost no light passes through an infinitely small point).
3. For all distances from the lens, the object is imaged with the same magnification as far as the focal plain remains fixed.
4. Owing to the focal pupil, telecentric imaging is of low light intensity.
5. The displacement of the focal plane (see Figure 1.17b) along the optical axis alters the size of an image but not its sharpness.

These properties predestine telecentric lenses to being used for tasks in which precise metrological image analysis is required.

The transition from an object-sided to a double-sided telecentric lens is achieved by introducing a second lens behind the focal pupil (see Figure 1.17c) in such a way that the focal points of the two lenses coincide. As a result, the focal rays are turned into rays that run parallel to the optical axis, and subsequently the alteration in image size, which takes place when the focal plane is moved along the optical axis, no longer occurs. Double-sided telecentric lenses can be used to advantage if imaging sensors are equipped with microlenses (one lens per pixel) to give a better light yield. In this case, light entering obliquely may lead to vignetting, thus resulting in inhomogeneous grayscale value distributions in the recorded images.

1.3.4 Sensors

Imaging sensors built into today's cameras are based on semiconductor materials. The formation of "electronic" images using such sensors utilizes the principle of the *internal photo-effect*. With this principle, light quanta (photons) possessing a minimum material-specific energy release atomic electrons from their bonded state (see Figure 1.18a). However, in the process, the freed electrons remain in the material.

The *external photo-effect* describes the process in which electrons are emitted from a material. Photomultipliers (not discussed here) are based on this effect, for example. The light-sensitive elements of imaging sensors based on semiconductor materials are constructed as photodiodes and use the depletion-layer photo-effect. The functioning principle of such diodes is explained briefly in the following text.

A silicone substrate is *p*-doped with boron atoms; the three free valence electrons of boron are insufficient electrons to bond with the four valance electrons of silicone. The substrate is then coated with a thin layer of silicone *n*-doped with fluoride. As fluoride atoms possess five free valence electrons, i.e., one electron more than necessary in order to bond with silicone — some of the free surplus electrons fill in the missing electrons sites (holes) in the *p*-substrate of the junction zone (see Figure 1.18b). Through this recombination, the *p*-substrate in the junction zone becomes negatively charged and the *n*-doped silicon layer positively charged, thus preventing additional electrons from migrating from the *n*-doped silicone to the *p*-doped silicone. The junction zone is thus transformed into a depletion layer.

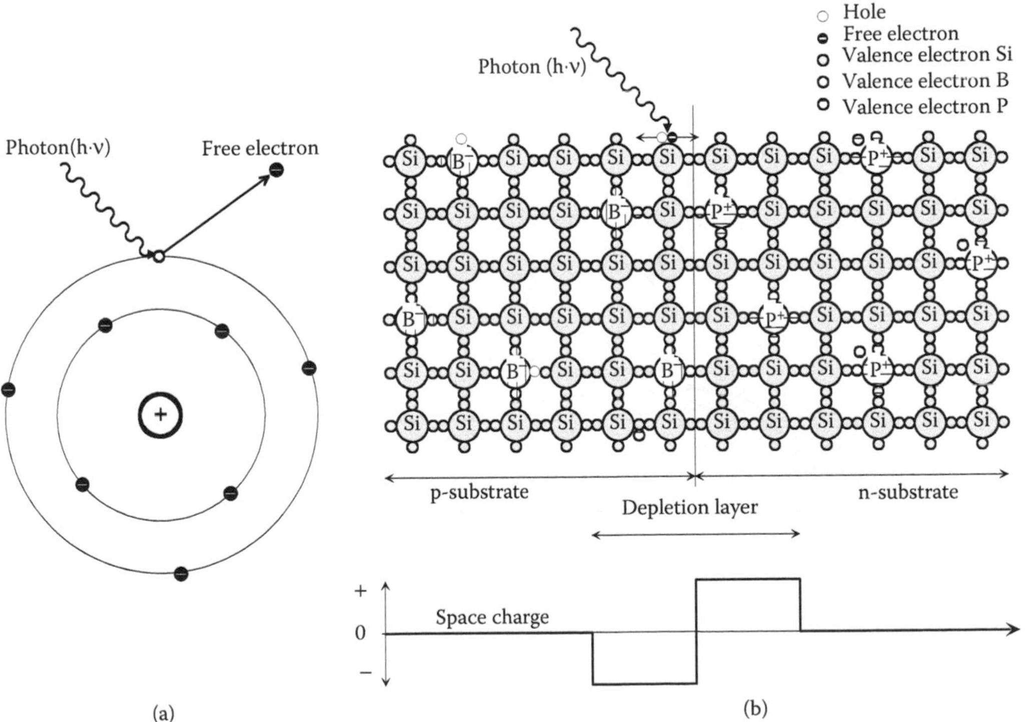

FIGURE 1.18 Illustration of the principle of the photo-effect (a) and of the photodiode (b).

If photons possessing sufficient energy when reaching the depletion layer, valence electrons are set free and electron–hole pairs are formed. Owing to the space charge at the depletion layer, holes migrate towards the *p*-substrate and the electrons towards the *n*-doped layer. A charge is formed outside the depletion layer, which can be measured as an electrode voltage. The charge-forming process continues until a balance is achieved between the charge of the depletion layer and the charge separation caused by the photons. The photodiode is then saturated, i.e., electrons set free by the photons recombine as soon as they have been released.

1.3.4.1 CCD Sensors

The first semiconductor-based imaging sensors were developed in the 1960s. These sensors utilize the effects of electrical fields to spatially separate and transport the electrical charge generated by photons from a large-surface photodiode. For this reason, these sensors are known as *charge-coupled devices* or *CCDs*.

The essential construction of an area-scanning CCD sensor is shown in Figure 1.19. A thin *n*-doped layer of silicone is applied to a *p*-doped substrate. This element thus constitutes the diode as such, and it would now be possible to measure the total number of photons reaching the sensor area in the form of a total charge. However, because of charge equalization, the spatial photonic flows of varied strengths would not result in an image.

To prevent the charge generated by the photons from being distributed over the entire diode, strongly *p*-doped depletions (channel isolations), which are insurmountable as far as the charges carried are concerned, are inserted into the substrate in a vertical direction. In this way, column-like photodiodes are formed. Another way of determining the number of photons falling on a column would be to measure the electrical charge specific to each column. *CCD line sensors* are actually constructed in this way.

However, in order to obtain a two-dimensional image, the electrical charge needs to be fixed in position along the columns (the strength of this charge varies locally because of the flow of photons, which is spatially inhomogeneous). To fix the charge, very thin strip conductors made of polycrystalline silicone are mounted horizontally on to a layer of insulating silicone dioxide. In each case, three adjacent strip conductors are used to form a line so that one of the outer conductors

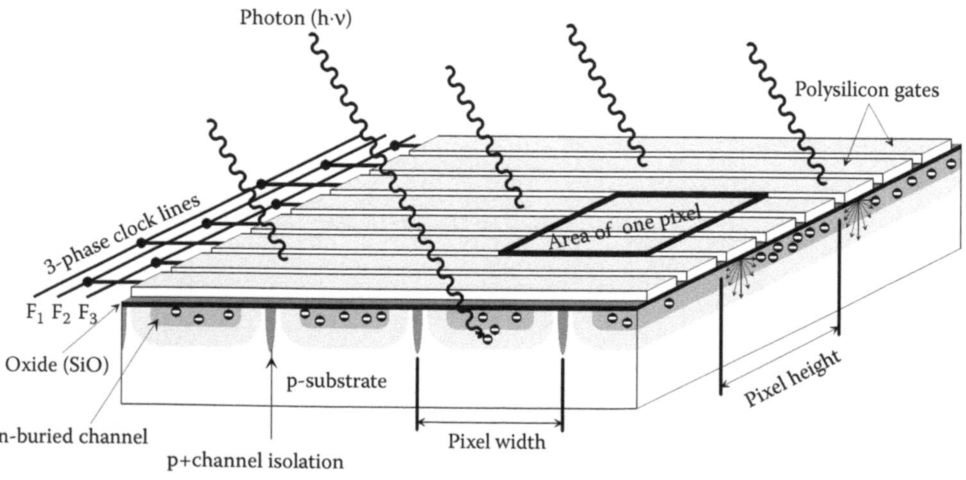

FIGURE 1.19 Essential construction of a CCD sensor.

is impinged with a negative electrical potential and the other two conductors with a positive potential while the sensor is being exposed. The charged particles are fixed in position in field valleys along the columns by the resulting electrical field.

A charge distribution is thus formed on the surface of the sensor that equates to the spatially inhomogeneous photon flow of the exposed image. The surface elements, composed of these sets of three strip conductors and also of adjacent channel isolations, are the image points of the area-scanning sensor and are known as *pixels*.

In order to read out the sensor image represented by the electrical charges, three-phase clocking (parallel clocking) is fed to the strip conductors on completion of the exposure, which induces a cyclic polarity change in the potential of the conductors. As a result, the electrical fields migrate in the direction of the columns and transport the charge bundles generated by the photons along them. After each cycle, the charge bundles from an image line reach a readout line, which is essentially constructed in the same way as a sensor column. In the readout line, using a second three-phase clocking (serial clocking), the charge bundles from one line are transported by the same mechanism into a readout cell. An amplifier connected to it then generates an electrical signal that is proportional to the charge of the corresponding pixel.

The advantage of these *full-frame CCD* sensors is that almost all the photons present on the entire sensor surface can be converted into an electrical charge, thus making it extremely light sensitive. A disadvantage is that the sensor must be darkened after exposure (a mechanical shutter fulfills this purpose in cameras), as otherwise, the recorded images could become severely smeared and falsified through further exposure during the readout process.

With *frame-transfer CCD* sensors, the need for a mechanical shutter is avoided by doubling the length of the column diodes and covering half of them with a material impervious to light. As a result, a CCD sensor that possesses not only a light-sensitive detector area but also a light-insensitive memory area is created. The electrical image information can be transmitted from the detector area to the memory area so quickly (less then 500 μsec) that almost no additional exposure takes place. The image information is then read out from the memory area with the aid of line shift registers and amplifiers. At the same time, a new image can be exposed in the detector area.

By realizing a pixel using the column diodes and three strip conductors as described earlier, a rigid grid of pixels that can only read out as a whole if three-phase clocking is used is formed on the sensor for the electrical image information. This mode of operation is known as *noninterlaced mode*.

However, for television technology, which is based on various norms and uses half frames transmitted in series because of the limited transmission capacity of radio channels, sensors working in the *interlaced mode* are required.

Full-frame- and half-frame-transfer CCD sensors fulfill these requirements; here the pixels are realized using four strip conductors and a four-phase shift clock. The number of pixels in a vertical direction equals the number of lines in a half frame. The frames are exposed using even or odd line numbers (related to the full frame) by shifting the site of the charge barriers (strip conductors with a negative potential) each time by two strip conductors. Partial overlapping of adjacent pixels in a vertical direction occurs as a result of this. This overlapping leads to the formation of mean values of the image information along the length of the lines. Known as *field integration*, this results in a poorer resolution and less sharp images. In general, interlaced full-frame- and frame-transfer CCD sensors can also be operated in the noninterlaced mode. However, in this case only images containing half the number of lines as that of a complete image are obtained.

With modern *interline-transfer CCD* sensors, also known as *IT sensors*, the number of pixels in a vertical direction equals the number of lines in a full frame. The image information from the pixels is transported in vertical direction by separate, light-insensitive shift registers. In the *interlaced mode,* by using control signals it is possible to determine whether image information from the pixels has been transported from even- or odd-numbered lines. While image information is being transported from one half frame, the second half frame can be exposed. This technology, known as *frame integration*, gives qualitatively better resolved images. IT sensors are also capable of operating in interlaced and noninterlaced *field integration* modes.

The use of the interlaced mode required for TV technology has an adverse effect as far as full-frame, frame-transfer and interline-transfer CCD sensors are concerned in cases in which fast-moving objects need to be recorded. Because of the time delay in the exposure of the half frames, distortions in images may occur, such as alterations in length in a vertical direction and the so-called comb effects (displacement of image lines) in a horizontal direction.

Modern *progressive scan CCD* sensors are constructed in a similar way to interline-transfer CCDs. In contrast, each cell of the column shift register is directly connected to each pixel in a column, with the result that images can only be transmitted in the noninterlaced mode. This mode of operation is sufficient for technical applications that are not based on TV technology and has the advantage that images of fast-moving objects are not distorted.

To enable the interlaced mode compatible with TV technology to be implemented, most progressive scan CCD sensors are equipped with a second column shift register. To transmit a recorded image in this case, all even-numbered lines are read out by one shift register and odd-numbered lines by the other shift register. For a half image, a multiplexer switches only one of the registers to the final amplifier. When the next image is recorded, the roles of the shift registers are changed over, with the result that both shift registers transmit complete frames.

Although advances made in the development of CCD sensors have considerably reduced the effect of smudging by exposing the sensor elements during image transmission, they still tend to falsify images if image areas are overexposed. Owing to the photo-effect, in the overexposed areas more electrons are generated than can be held together by electrical field depletions. The surplus electrons migrate to neighboring pixels and cause the so-called blooming effect.

1.3.4.2 CMOS Sensors

As with CCD sensors, imaging CMOS are also based on the internal photo-effect. However, in contrast to CCD sensors, CMOS sensors do not primarily utilize photonic charges to generate an image. Instead, they measure the alteration in the off-state current of the light-sensitive semiconductors (photodiodes, bipolar transistors, or MOSFET transistors) caused by the effect of the photons.

The essential construction of a CMOS sensor based on a photodiode is shown in Figure 1.20. Vertical strip conductors arranged in parallel are placed at 90° to the horizontal strip conductors

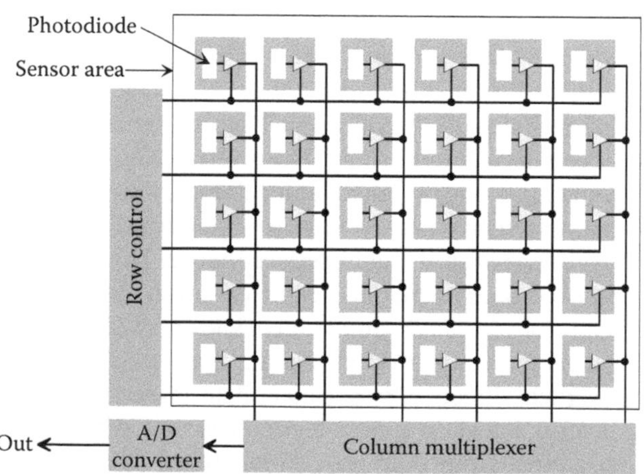

FIGURE 1.20 Principle of an imaging CMOS sensor.

(also arranged in parallel) in order to form a grid. Photodiodes are then placed near intersections in such a way so as to connect the strip conductors that are insulated from one another. Via a row access logic, all the photodiodes in one row can be biased in the nonconducting direction in order to generate an electrical current (off-state current), which is logarithmically proportional to the amount of incidental light. The current is directly transformed into voltage signals at the pixel level, which occurs at the column lines. Each time, a column multiplexer switches one of these signals to an analog-to-digital converter, which converts the voltage level into an equivalent digital representation.

An image sensor constructed in this way possesses the following advantages:

- The fact that each individual pixel can be addressed optionally enables any desired partial image to be depicted.
- Complete or partial images can be read out in the interlaced or noninterlaced mode.
- The blooming effect is avoided because the charges generated by the photons are converted into equivalent voltages at pixel level.
- The dynamic range of CMOS sensors is greater than that of CCD sensors.

However, some disadvantages are also associated with this type of sensor:

- Owing to the complexity of the pixel elements, the light-sensitive photodiode is relatively small compared to that of the total area of the sensor.
- Compared with CCD sensors, CMOS sensors are relatively insensitive because of the small photodiode.
- As the process for manufacturing these sensors is highly complex, the characteristic curves of the photodiodes are not always identical. Where necessary, the grayscale value of individual pixels may need to be interpolated from the grayscale values of adjacent pixels using secondary hardware or software.

An improvement in the sensitivity of the pixel elements can be achieved by placing a microlens in front of each photodiode. This technique is known as *lens-on-chip*.

1.3.4.3 Color Sensors and Cameras

Today's standard sensors and sensor systems generate digital RGB color images and use optical systems to split the incidental light into its elements of red, green, and blue before it is shone on to the light-sensitive sensor elements (CCD or CMOS). A principal difference is made between systems made using one or three sensors.

In the case of *single-chip color sensors* (see Figure 1.21a), red, green, and blue color filters are inserted in front of the light-sensitive sensor elements (pixels). The filters are generally arranged according to Bayer so that the number of green filters is twice the number of red or blue ones. The RGB image is obtained by interpolating the image information that has been separated into color channels. The advantage of single-chip technology is its compact size and that there are no adjusting elements. However, the relatively coarse picture elements impair image quality and may lead to Moiré effects (formation of streaks due to interference).

Three-chip color sensor systems give a much better image quality. These systems are composed of prisms and color filters (see Figure 1.21b). The incoming light is split into the three spectral fractions of red, green, and blue. Each spectral component passes through a red, green, or blue trim filter before it reaches the imaging sensor. The disadvantage of this technique is that it is very expensive, it requires more space, and components need to be adjusted extremely accurately.

In order to generate digital RGB images, more recent developments utilize the fact that longer light waves penetrate deeper into semiconductor materials than short ones, thus avoiding the need for an additional optical system to split the light into its red, green, and blue components.

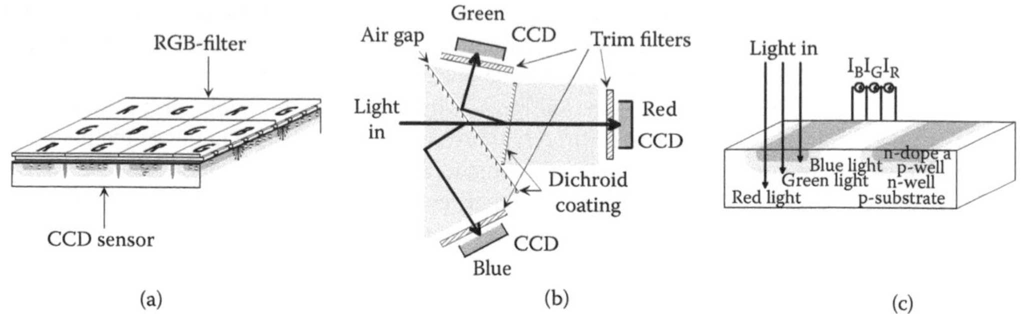

FIGURE 1.21 Principles of color sensors: (a) single-chip RGB, (b) three-chip RGB, (c) Foveon X3 color sensor.

The *Foveon X3 color sensor* (see Figure 1.21c) is based on this effect [1]. The light-sensitive elements of the sensor are made up of three photodiodes placed on top of one another, achieved using an additional pn-junction. This sensor combines the advantage of the compact size of a single-chip color sensor with the high resolution of a three-chip color sensor.

1.3.4.4 Camera Types and Interfaces

Today's camera systems can be broadly classified into the following three classes according to the way they supply their information content:

Analog cameras supply image information contained in their sensors in the form of analog electrical signals. As far as TV technology is concerned, this is a signal (a composite one) that contains both the synchronization signals for the start of images and lines and also the grayscale values of the pixels. These signals are also often supplied as single signals for technical applications.

Digital cameras contain an imaging sensor as well as at least one analog-to-digital converter that supplies image information in digital form. Additional features are also sometimes integrated into digital cameras, such as lookup tables for transforming grayscale values, hardware for interpolating RGB images from Bayer color images, and memories for the interim memorizing of images and data-transfer components or modules for various transmission protocols (e.g., Camera Link, IEEE 1394, USB).

Smart cameras possess digital camera components and processors (signal processors), program memories, and, where necessary, various interfaces such as a digital I/O, serial I/O, Ethernet, etc. In this way, image processing applications can be performed directly inside the camera. The information content supplied by the camera thus represents an interpretation of the images.

The construction used for camera interfaces to transmit image information is directly dependent upon the way the cameras supply this information.

With analog cameras, the composite video signal is supplied via an analog interface in the form of a 75-Ohm coaxial connector. In accordance with the American standard NTSC/EIA RS-170 or the European standard CCIR/BAS/FBAS, respectively, the signal level ranges between –0.4 and

+1.0 V. Voltage levels below 0 V represent synchronization signals and voltage levels above 0 V, grayscale pixel values.

If the image signal and synchronization signals are supplied separately, the analog electrical signal level of the pixels' grayscale values is approximately 1.0 V and the signal level of the synchronization signals approximately 5.0 V (TTL-compatible).

The following variations exist for digital camera interfaces:

1. Parallel transmission of image information orientated toward pixel data that uses single-ended TTL signals or signals corresponding with the specifications laid down in RS-422 or EIA-644 (LVDS = low-voltage differential signal). However, these specifications only determine the electrical aspects of the interfaces and do not cover those of transmission protocols, timing, or pin assignment. RS-422 specifies differential data transmission with a signal level of 5 V, whereas EIA-644 (LVDS) specifies differential data transmission with signal levels of 3.3 or 3.1 V.
2. Three different variations of parallel image information transmission via several serial channels defined by the Camera Link™ specification and derived from channel link technology developed by National Semiconductor. The variation known as "base" includes four serial differential LVDS data channels and one differential LVDS clock channel. The variation known as "medium" possesses twice as many, and the variation "full," three-times the number of data and clock channels. For all of these variations, in order to control camera functions (trigger, shutter, mode of operation, etc.), four additional differential LVDS camera control channels and a bidirectional RS-232 interface with differential LVDS interface are provided. As well as specifying electrical aspects, the Camera Link™ specification also defines pin assignment, transmission protocols, data widths (8, 12, and 24 bits), and the maximum possible transmission rate (340 MB/sec).
3. Serial image information transmission in accordance with the standard IEEE 1394, also known as FireWire, and serial transmission in accordance with the USB 2.0 specification. Both the IEEE 1394 standard and USB 2.0 specification define bidirectional, digital, high-speed communication via a cable that is also capable of supplying an electrical current to the connected device.

1.3.4.5 Frame Grabbers

The fundamental tasks of frame grabbers are to provide an electrical interface (analog, digital TTL, RS-422, LVDS or Camera Link™) compatible with the camera used, in order to separate synchronization and image signals (analog, composite video signal); synchronize image and line synchronization signals where necessary; digitize video signals from analog cameras if required; and transmit digital image information to the memory (RAM) of the computer used. Additional functional groups can also be added to frame grabbers. Timers generate clocking, synchronization, trigger, and reset signals for cameras. Multiplexers enable several cameras to be connected. Using lookup tables, the grayscale values of the camera pixels can be mapped onto other grayscale values. Memory modules permit the interim memorizing of images or function as FIFO (first in, first out) memories during the transmission of images to the computer's memory. High-end frame grabbers are also equipped with special signal processors or field programmable gate arrays (FPGAs) and program memories, thus enabling image processing algorithms to be performed extremely fast and without placing a strain on the computer's processor. A block diagram of the principle component of a frame grabber is shown in Figure 1.22.

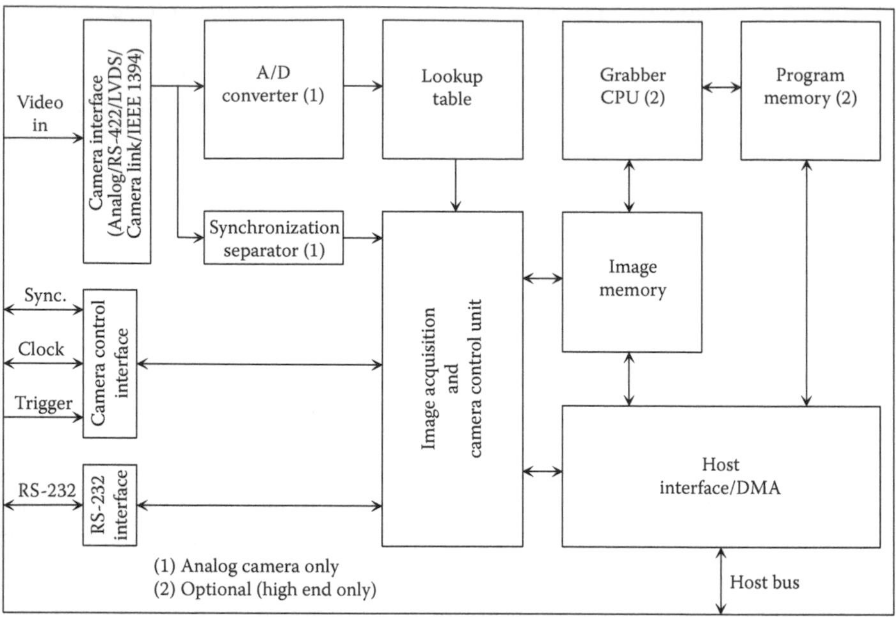

FIGURE 1.22 Principle of a frame grabber.

1.4 PROCESSING AND ANALYSIS OF IMAGE DATA

Many inspection tasks in assuring the quality of technical products are based on the interpretation of images. With visual inspections, interpretation can be carried out by people. Because of their cognitive abilities, people are able of performing a large number of highly varied tests on an object for which a precise specification is not required. However, these abilities, which to date cannot be copied, also have their disadvantages. If attentiveness decreases, faults can be overlooked. Man is only able to make an objective assessment (e.g., the assessment of a shade of color) if that shade of color exists in the form of a reference and only a comparison of the shade needs to be made. The amount of work is often time consuming and expensive.

Although computers do not possess the cognitive abilities of humans, they are capable of interpreting digital images with the aid of appropriate algorithms and programs. The advantages of computer-aided image interpretation are the reproducibility of the results, the ability to measure features, simultaneously where necessary (wafer-scale testing), and the incredibly high processing speeds, which are a prerequisite for testing mass-produced goods.

The following sections are concerned with the algorithms and methods used in today's image processing and computer vision.

1.4.1 Computer Vision Process

The central task of an image processing program for measurement and testing purposes is to interpret recorded images, supply measurement values, and make "accept/reject" decisions. The process steps are depicted in Figure 1.23 and described in detail:

1. *Image acquisition:* First, an image recording is initiated via a program instruction or an external electrical signal or trigger. The analog image signals from the image sensor are then digitized by a camera or frame grabber and transmitted to the computer's memory.
2. *Image processing:* The aim of image processing is to reduce image disturbance and highlight features for subsequent image analysis, i.e., to improve images. This is essentially

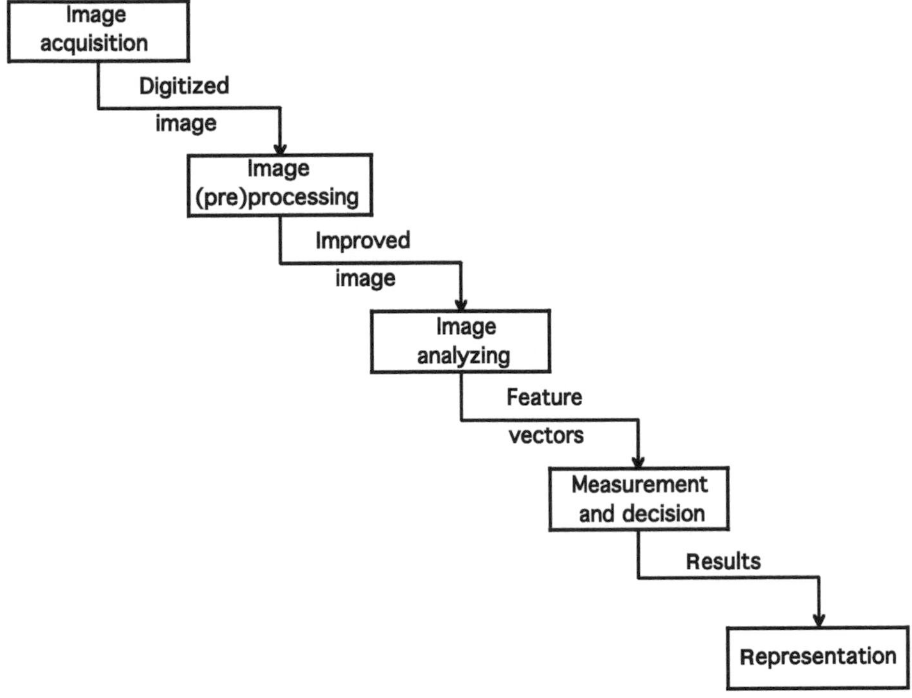

FIGURE 1.23 Stages of image processing and computer vision process.

achieved using image operations, which can be divided into two classes according to their method of function.

One class of image operations, so-called point or local operators, considers each pixel independently. Examples of this are image averaging, grayscale stretching, histogram equalization, and binarization using a threshold value.

The other class of image operations, known as neighborhood or global operators, considers not only the grayscale value of the pixel in question but also takes into account adjacent points influencing the result of the operation. Examples of this class include all filter operations such as high-pass, low-pass, and median filters; the morphological operators of erosion, dilation, opening, and closing; and gradient filtering in accordance with Roberts [2], Dewitt [3], Sobel [4], etc.

As a result, these image processing operations supply a new image that has been qualitatively improved as much as possible for subsequent image analysis. Transformed image depictions generate image operators such as the Fourier or Hough transformation; these are not described here.

3. *Image analysis:* In this process step, the information content of the improved images is examined for specific features such as intensity, contrast, edges, contours, areas, and dimensions. As these features are often limited to one segment of the images, a global analysis is not carried out but one limited to the region of interest is. The results of the analysis algorithms are feature vectors that give quantified statements about the feature concerned.

Two classes of algorithms exist with regard to test features:

One class of algorithms analyzes features associated with information regarding color or intensity. This class includes the mean grayscale value, contrast, the number of pixels of a certain color or grayscale value, and histogram analysis. The other class of algorithms is concerned with the spatial characteristics of certain features. Examples of this class of algorithms include edge detection, pattern matching, and algorithms for analyzing the connectivity of image areas (connectivity analysis) providing geometric information about the location of the main area of concern, area contents, length of perimeter and the dimensions, etc. of an object.

4. *Measurements and decision making:* Here, comparisons with set points and tolerance limits are carried out and accept/reject decisions made. With simple test features such as "mean grayscale value" or "contrast," this decision can be made directly based on image analysis features. For features that are relevant as far as metrology is concerned, the measurement values (distance, diameter, angle, etc.) must first be calculated from the analysis results before making comparisons with set points and tolerance limits. The same applies for complex test features.
5. *Presentation of results:* In the last process step, the calculated measurement values and test decisions are usually represented in numerical or graphical form. However, the presentation technique used — a domain of computer graphics — is not discussed here.

1.4.2 Image Data Preprocessing and Processing Methods

As mentioned earlier, before carrying out any feature extraction analysis, an image should be available in its best possible digital representation. To achieve this, a maximum number of disturbances need to be suppressed and the features to be analyzed must be highlighted.

Disturbances are caused by different factors. The image sensor and electronics of the camera and frame grabber cause random noise. Systematic and, therefore, predictable noise can be caused by defective pixels, inhomogeneous illumination, or dust particles on the image sensor or camera lens. Distortions caused by wrong positioning of the camera, the lens system of the camera, or by nonquadratic camera pixels are other sources of disturbance that especially falsify the optical measurement of objects if they are not taken into consideration and dealt with appropriately. Possible causes of poorly expressed features are poor contrasts, images that are too bright or too dark, and rugged feature edges, particularly in the case of binarized images.

In all cases, an attempt should be made to remove the causes of disturbance at their source. Here, the importance of illumination and image acquisition is often underestimated. If the causes of disturbance cannot be eliminated, the image processing operation described in the following text may help to improve an image. As some of these operations are based on a histogram of the image concerned, its calculation and the image features that can be read from it are described first.

1.4.2.1 Histograms

The histogram of an image is a discrete function; its functional values reproduce the frequency of the grayscale values occurring in the image. If I is the number of possible grayscale values g_i in the interval $[0, I-1]$ with $0 \leq i \leq I-1$ (e.g., $I = 256$), then g_i is counted for each possible grayscale value, the number of times it occurs in the image, and then plotted on the ordinate g_i as the functional value $h_i = H(g_i)$. The grayscale values are usually plotted on the ordinate axis in strictly monotonous increasing succession. If the image is made up of R rows and C columns, functional values no higher than $n = R \cdot C$ may occur.

Normalized histograms are often required for the calculations. These are derived from histograms by dividing the grayscale values by g_{I-1} and the functional values by n. In this way, all grayscale and functional values lie in the interval [0, 1]. The functional values then give the direct probability $p_g(g_i)$ of the grayscale value g_i reoccurring in the image.

Important features can be read immediately from the histogram of an image [5]. If the frequency of occurrence is concentrated on a small area of possible grayscale values, the image shows a low level of contrast. If the frequency of occurrence is concentrated on the left-hand or right-hand side

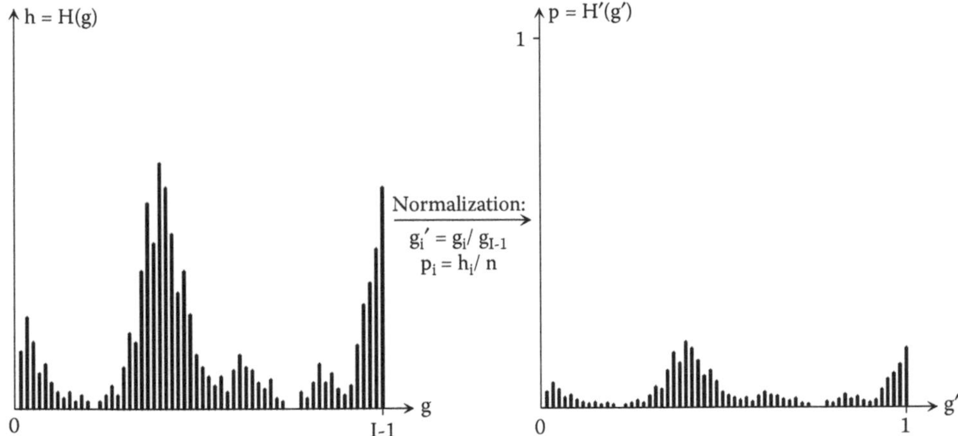

FIGURE 1.24 Histogram and normalized histogram.

of the grayscale value scale, the image is either dark or bright. An evenly distributed frequency of occurrence over the entire grayscale value scale indicates a high-contrast image.

1.4.2.2 Point Transformations

Point transformations are operations that are implemented on individual pixels without taking other pixels into account. They are used to suppress noise, increase contrast, and for segmentation purposes. The most commonly utilized operators are described in the following text.

1.4.2.2.1 Time Series Accumulation

With time series accumulation, a new destination image is calculated from several images of the same scene. In the case of the resulting image, the pixel values represent the mean value of the corresponding pixels in the original images. A prerequisite for this method is that the scene must remain unchanged during the recording of the original images.

If n original images are available, if $g_i(r,c)$ denotes the grayscale value of a pixel from ith original image and $f(r,c)$ denotes the grayscale value of the destination image at the position r,c (row, column), then the following calculation rule applies for each individual pixel:

$$f(r,c) = \frac{1}{n} \cdot \sum_{i=1}^{n} g_i(r,c) \tag{1.2}$$

By averaging the image, it is possible to reduce the stochastic noise from image sensors and electronics by a factor of $1/\sqrt{n}$. Also, fluctuations in illumination in which frequencies are similar to or higher than that of the image acquisition frequency are lessened.

1.4.2.2.2 Grayscale Stretching or Mapping

If images show a poor level of contrast or are too bright or too dark, it is possible to improve contrast or make the images darker or brighter by scaling the grayscale values in the range of interest. If N is the number of possible grayscale values g_i in the interval $[0, N-1]$ with $0 \le i \le N-1$

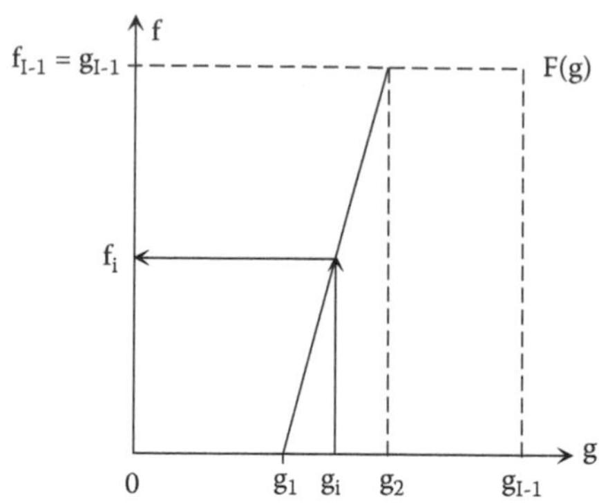

FIGURE 1.25 Transformation function for linear grayscale stretching.

and if the lowest grayscale value of interest is denoted by g_1 and the highest one denoted by g_2, with $0 \le g_1 < g_2 \le g_{I-1}$, the rule of calculation for linear grayscale stretching is as follows:

$$f_i = \begin{cases} 0, \ \text{if } g_i < g_1 \\ \dfrac{(g_i - g_1)\cdot g_{I-1}}{g_2 - g_1}, \ \text{if } g_1 \le g_i \le g_2 \\ g_{I-1}, \ \text{if } g_i > g_1 \end{cases} \tag{1.3}$$

The function $f = F(g)$ is generally realized as a lookup table in the form of a one-dimensional array. The N elements of the array are first occupied by the functional values f_i calculated according to the aforementioned rule. The grayscale values g from the original image are then transformed into the grayscale values f of the resulting image. This is achieved by addressing the ith element in the array with the original grayscale value g_i; the resulting grayscale value f_i is then read from the consigned element.

As well as linear grayscale stretching, nonlinear functions can also be utilized (mapping). If lighting conditions change, grayscale stretching has the disadvantage that the grayscale value range of interest cannot be calculated automatically. (see Figure 1.25.)

1.4.2.2.3 Histogram Equalization

An automated method for improving contrast using grayscale mapping is known as histogram equalization (see Figure 1.26). The process is based on calculating a transformation function $F(g)$ from a normalized histogram. The calculation rule for this nonlinear transformation function is as follows:

$$f_k = \sum_{i=0}^{k} P_g(g_i) \tag{1.4}$$

By using this calculation function, rarely occurring grayscale values are placed on values that are close to one another. Frequently occurring grayscale values are placed on values that are further apart from one another.

1.4.2.2.4 Binarization

Image binarization is a technique that transforms a grayscale value image into a black and white image (see Figure 1.27). To do this, a threshold value g_t is determined in the interval of possible

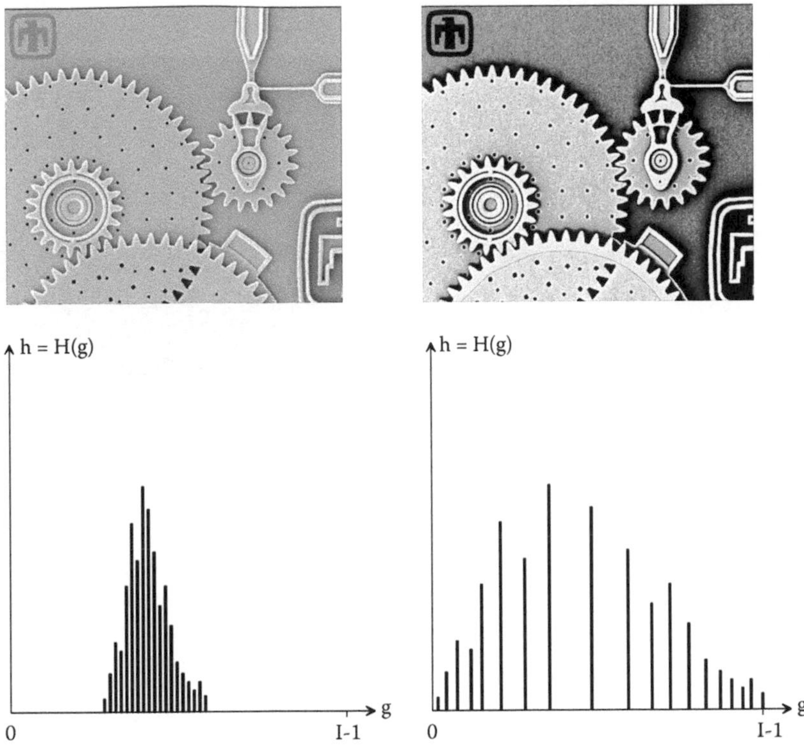

FIGURE 1.26 Grayscale image before and after histogram equalization. (Courtesy Sandia National Laboratories, SUMMiT™ Technologies, http://www.mems.sandia.gov.)

grayscale values g_i ($0 \leq i \leq I - 1$) and the following transformation carried out on each of the pixels:

$$f(r,c) = \begin{cases} 0, & \text{if} \quad g(r,c) \leq g_t \\ 1, & \text{if} \quad g(r,c) > g_t \end{cases} \tag{1.5}$$

Therefore, for each pixel, a decision is made as to whether its grayscale value is lower or equal to the threshold value. If this condition is fulfilled, the corresponding pixels are given the value 0 in the destination image. All other pixels in the destination image are allocated the value 1.

Image binarization serves to reduce image information and also to segment images based on the grayscale values of the pixels. The process of dividing up an image into sections in which the grayscale values lie below or above the threshold value is known as segmentation. Segmented images are required for such image analysis functions as binary pixel counting and connectivity analysis, which are capable of determining a wide range of features such as contours, areas, center of gravity, etc. Contours, in turn, form the basis for contour-based pattern matching.

1.4.2.3 Spatial Filtering

With spatial filtering (convolution), a new grayscale value is calculated for each pixel of the original image using a mathematical operation that takes the grayscale values of surrounding pixels into account. The neighboring points are usually defined as quadratic masks (also known as filter cores) sized 2 × 2, 3 × 3, 5 × 5, 7 × 7, etc., (see Figure 1.28a). To calculate the new grayscale values, the filter core is moved over all the pixels of the original image; a mathematical operation is then

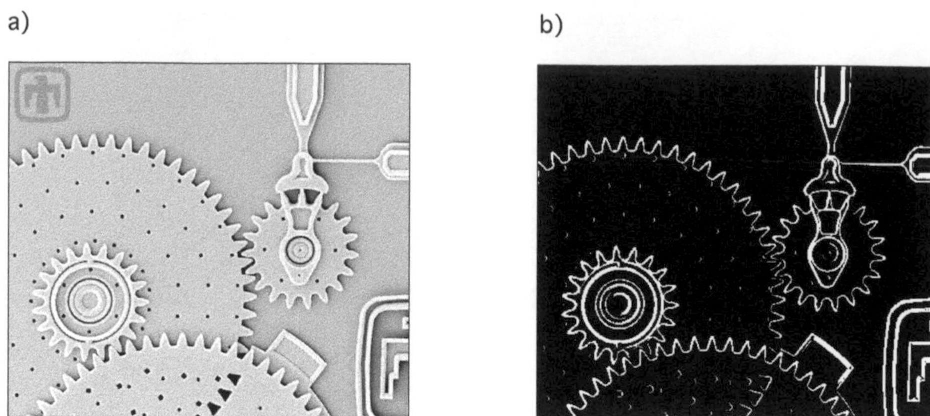

FIGURE 1.27 Original grayscale image and its binarized representation.

carried out, taking the filter core values for each pixel covered by the core elements into consideration (see Figure 1.28b). The grayscale value calculated is then placed on the destination image at the site corresponding with the central element of the mask.

1.4.2.3.1 High-Pass and Low-Pass Filters

High-pass and low-pass filters belong to the class of filters that modify localized contrasts in images while leaving image contents (object characteristics) in their original form. The filters can be realized using simple mathematical transformations.

If the grayscale values of the pixels covered by the filter core are denoted as $g_{i,j}$ and if $m = (2n + 1)$ with $n = 1,2,3,\ldots$ is the number of rows or columns in the filter core, then the rule for

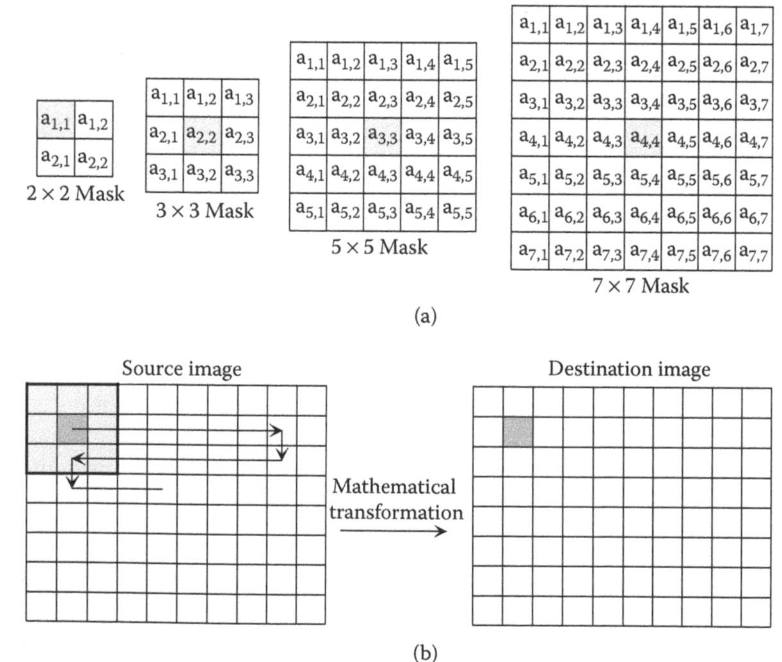

FIGURE 1.28 Spatial masks of different sizes (a) and the process of spatial filtering (b).

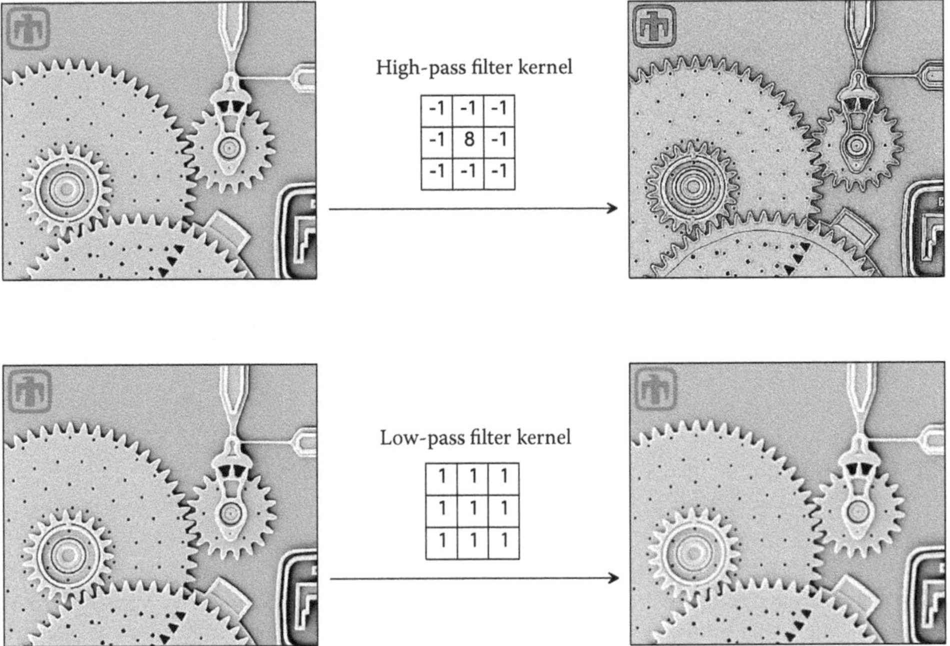

FIGURE 1.29 Grayscale image, spatial filter cores, and resulting image using a high-pass filter (top pictures) and a low-pass filter (bottom pictures).

calculating the grayscale value of the central pixel $f_{n+1,n+1}$ in the destination image is as follows:

$$f_{n+1,n+1} = \frac{1}{m^2} \cdot \sum_{i=1}^{m} \sum_{j=1}^{m} a_{i,j} \cdot g_{i,j} \tag{1.6}$$

The grayscale values of the destination image are therefore only dependent on the values $a_{i,j}$ of the filter core. Figure 1.29 shows the filter cores, the original grayscale value images and the resulting images obtained using a high-pass and a low-pass filter (mean value filter).

In the case of images processed using a high-pass filter, localized changes in grayscale values are more intense. This especially results in enhancing edges.

A low-pass filter reduces localized fluctuations in grayscale values. It is, therefore, particularly suitable for suppressing stochastic noise. However, it also blurs edges and images appear fuzzy.

The efficient calculation of high-pass and low-pass filtering is advantageous. However, the displacement of edges in filtered images may be a disadvantage, especially if the images are to be utilized for measurement tasks.

1.4.2.3.2 Median Filters

A median filter functions in a similar way to the mean value filter of a low-pass filter but does not possess the disadvantage of displacing edges. However, considerably more effort is required to calculate the grayscale values of the central pixel in the destination image. The calculation is performed according to the following rule:

1. First, all of the grayscale values $g_{i,j}$ covered by the filter core must be sorted according to size and the values entered into a one-dimensional array.
2. Then, the grayscale value of the central pixel in the destination image is replaced with that of the mean element in the array.

a)

FIGURE 1.30 (a) Grayscale image, (b) its representation using a median filter.

With filter cores possessing an even number of rows and columns, the grayscale value of the pixel in the destination image is formed using the mean value of both of the mean vector elements. Figure 1.30 shows a grayscale value image before (Figure 1.30a) and after (Figure 1.30b) using a median filter.

1.4.2.3.3 Morphological Filters

Morphological filters are operators that modify the shape of image contents by altering the grayscale values on the basis of geometric proportions to neighboring points. They can be used both for grayscale value images and for binary images. Essentially there are two types of morphological filter — erosion and dilation filters.

Erosion filters have a smoothing effect on shapes because contour roughness is removed, i.e., replaced by the background color. Small shapes caused, for example, by noise disappear. Shapes located close to one another are more clearly separated from one another. The calculation rule for erosion filters is as follows:

Binary image: The pixel in question is placed on black if its neighboring points include one or more black pixels.

Grayscale value image: The pixel in question is placed on the darkest grayscale value of its neighboring points.

Dilation filters also have a smoothing effect on shapes. In contrast to erosion, contour roughness is not removed but rather filled out, i.e., replaced by the foreground color. Shapes located close to one another may, however, merge together. The calculation rule for dilation filters is as follows:

Binary image: The pixel concerned is placed on white if its neighboring points include one or more white pixels.

Grayscale value image: The pixel concerned is replaced by the brightest grayscale value of its neighboring points.

The effect of morphological filters on a grayscale value image is illustrated in Figure 1.31. With regard to metrological analysis, it is to be noted that morphological filters alter the size of object structures.

Through the use of the filters on the images resulting from each of the complementary filters, the two morphological operators of opening and closing arise.

Opening results from using erosion and then dilation. Combined in this way, small particles are removed from images and object structures are separated more clearly from one another.

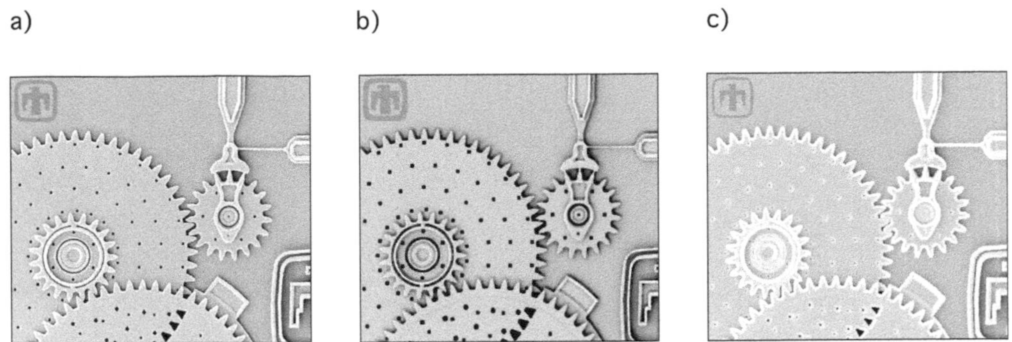

FIGURE 1.31 (a) Grayscale image, (b) filtered using erosion, (c) filtered using dilation.

FIGURE 1.32 (a) Grayscale image, (b) results of opening, (c) results from closing.

Closing is achieved by first using a dilation filter and then an erosion filter on the resulting image. By combining the filters in this sequence, small holes in object structures are filled.

The results obtained through the use of both these operators are shown in Figure 1.32. It is to be noted that both the morphological operators of opening and closing do not alter the size of object structures.

1.4.2.3.4 Gradient Filters

In contrast to high-pass, low-pass, and morphological filters, gradient filters do not generate modified images but rather new images that reproduce localized changes in grayscale values. Large grayscale value changes, such as those occurring at edges, are seen as bright pixels; homogeneous grayscale value areas appear as dark pixels. Gradient filters are therefore appropriate preprocessing methods for extracting object contours, which is a prerequisite for geometric feature-based pattern matching. The effort involved in calculating gradient images is higher than that required for high-pass and low-pass filtering because the gradients for the direction of the x and y axes are determined by two filter cores. Figure 1.33 shows an original grayscale value image, the filter cores, and the resulting images obtained using gradient transformation according to Prewitt [3] and Sobel [4].

1.4.3 Image Data Analysis Methods

Once the relevant features in the recorded images have been optimally depicted using image processing methods, the features are analyzed. As the features concerned are not usually present in the entire image, only certain segments of the image, the so-called regions of interest, are taken into consideration. Most image segments are shaped either as a rectangle, circle, annulus, circular

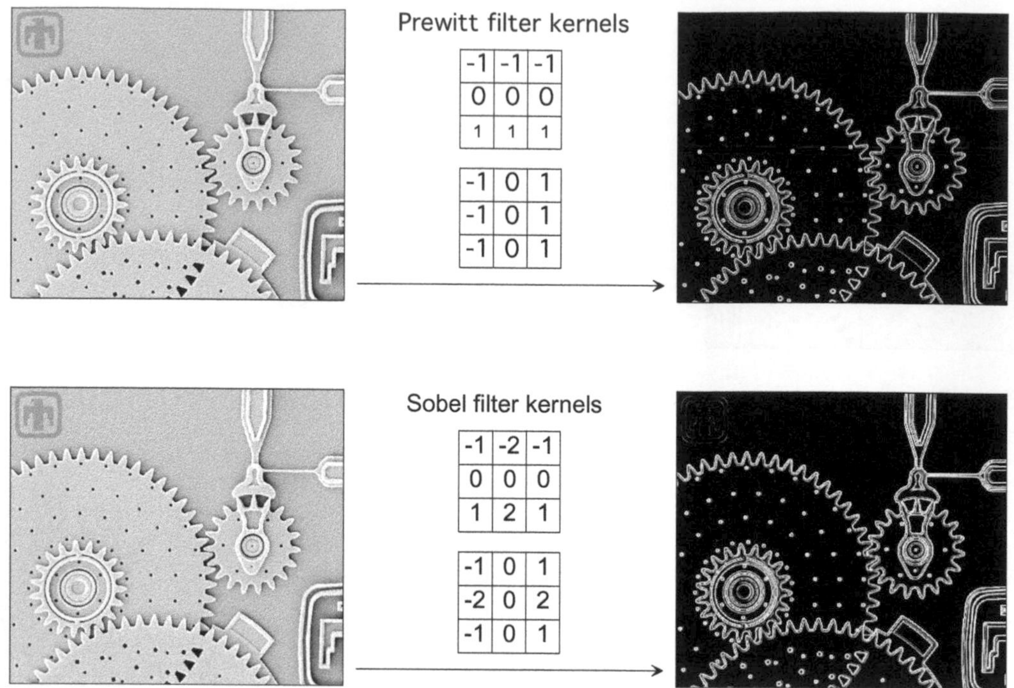

FIGURE 1.33 Grayscale image filtered according to Prewitt and Sobel.

curve, or line. By limiting the analysis to only a segment of an image, the analysis methods are not required to differentiate between the feature concerned and nonrelevant image structures such as noise. With the working methods used in image analysis, a difference is made between operations that are related to the spectral features of image segments and operations related to spatial image structures. The most important analysis methods of both these categories utilized in measurement and testing technology are described in the following text.

1.4.3.1 Spectral Operations

Spectral operations only analyze the distribution of color, grayscale values, or black and white in an image segment. Spatial image structures are not taken into account, making these methods especially suitable for testing tasks — in particular, that of checking the presence of features. The most important spectral operations are described in the following text.

1.4.3.1.1 Mean Grayscale Intensity

The simplest image analysis method of all is calculating the mean grayscale value of an image segment. This can be implemented, for example, to determine whether large areas of contamination of a different shade are present. The calculation rule for the mean grayscale value is as follows:

$$\bar{g} = \frac{1}{n} \cdot \sum_{i=1}^{n} g_i \tag{1.7}$$

where n is the number of pixels and g_i is the grayscale values of the pixels in the image segment.

1.4.3.1.2 Contrast and Contrast Ratio

Image contrast is the difference between the brightest and the darkest pixels in an image or image segment. The contrast ratio r, which is dependent on brightness, is defined as follows:

$$r = \frac{g_{max} - g_{min}}{g_{max} + g_{min}} [\%] \tag{1.8}$$

The contrast or contrast ratio is implemented, for example, in cases in which homogeneous surfaces are inspected for the presence of particles, scratches, etc.

1.4.3.1.3 Histogram Analysis

In this the histograms of two images are compared with one another. In the simplest case, this is based on the formation of differences in the frequencies of occurrence of grayscale values p_i. Histogram analysis can be used to compare images with regard to their grayscale value distribution and to reach conclusions concerning defects based on deviations in distribution. To do this, the histogram of an image segment of a defect-free feature area is checked for a match with the feature area under investigation. Deviations in the frequencies of occurrence of grayscale values may indicate the absence of features without knowing their position.

1.4.3.1.4 Binary Pixel Counting

The image analysis function of binary pixel counting is based on the same principle as histogram analysis with the difference that here the images are first binarized and then the white or black pixels are counted, what is equivalent to an area comparison.

1.4.3.1.5 Spatial Operations

Spatial operations analyze the shape characteristics of features reflected in the distribution of grayscale values in images or image segments. For this reason, they are suitable for solving both measurement and testing tasks and are used to recognize objects and positions.

1.4.3.1.6 Edge Detection

One of the most important analysis functions for performing dimensional measurements is the detection of edges in grayscale value images. As depicted in Figure 1.34, the distribution of grayscale values in a direction perpendicular to an edge is not a step function but a sigmoid one (even in the case of transmissive light arrangements). Therefore, with the aid of the grayscale value distribution, the analysis function determines the position of the real edge as being the midpoint between the

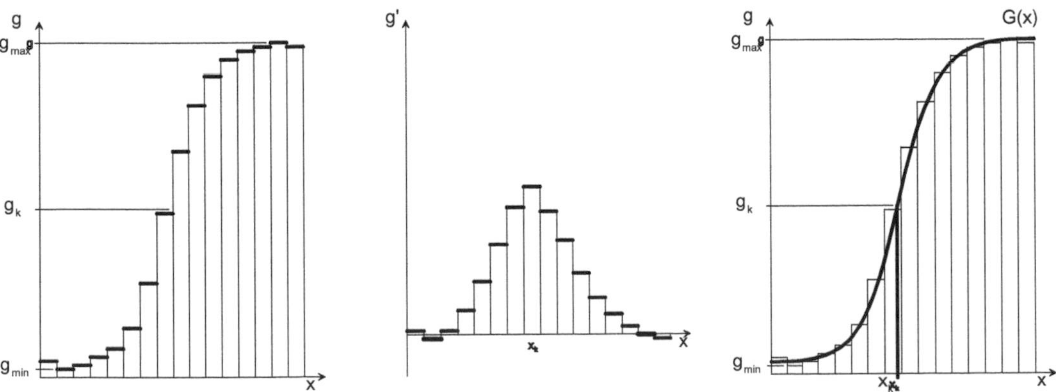

FIGURE 1.34 Edge grayscale profile, derived profile, and approximated function $G(x)$.

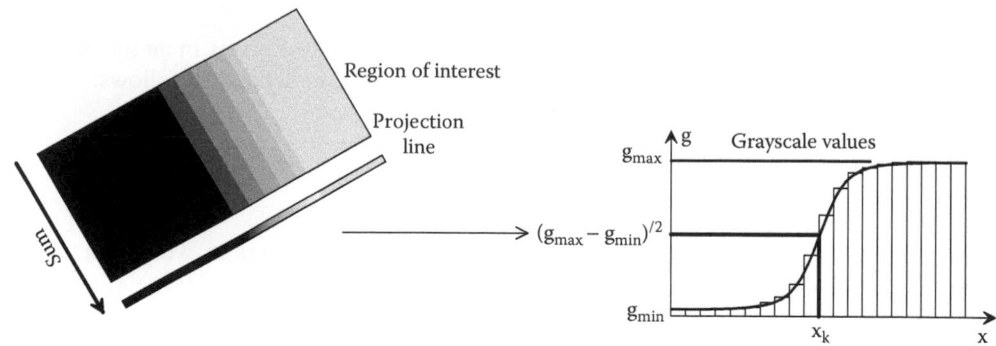

FIGURE 1.35 Edge detection with a rectangular region of interest.

brightest and darkest grayscale value in the image. The position of an edge can always be directly determined from the grayscale value distribution by locating the site of the midpoint position. However, owing to the quantization of the grayscale values through their digital representation (analog-to-digital conversion), the position determinated carries an uncertainty of ±1 pixels.

A lower degree of uncertainty can be achieved if grayscale values are recorded as the approximation points of a constant function. In this way, the midpoint transition can be determined much more accurately. Under the term "subpixel algorithm," various concepts have been described in the literature [6–10] that are based on the processes of correlation, interpolation, iteration, regression, or the observation of symmetry. Tests [10] have shown that regression processes, which calculate a model function from the *n* grayscale values around the edge area using the method of minimizing error squares, are relatively stable processes that can be rapidly calculated and possess low error rates and a high degree of reproducibility.

Model functions of regression methods are generally established in order to derive a grayscale value profile (Sobel filter) around the edge area. They may have the following appearance:

$$G'(x) = e^{-ax^2-bx-c} \tag{1.9}$$

The independent unknowns *a, b,* and *c* are calculated after taking the logarithm of the function from the parabolic equation of the discrete values

$$y_i = -ax_i^2 - bx_i - c \quad \text{with} \quad y_i = ln(G'(x_i)) \tag{1.10}$$

using the method of minimizing error squares [11]. The position of the edge is obtained from the maximum of the calculated parabola using the following equation:

$$x_k = \frac{\sum x_i}{n} \cdot \frac{n \cdot \sum \bar{x}_i^2 \cdot y_i \cdot \sum \bar{x}_i^3 - \sum \bar{x}_i \cdot y_i \cdot \left(n \cdot \sum \bar{x}_i^4 - \left(\sum \bar{x}_i^2 \right)^2 \right) - \sum y_i \cdot \sum \bar{x}_i^3 \cdot \sum \bar{x}_i^2}{2 \cdot \left(n \cdot \sum \bar{x}_i^2 \cdot y_i \cdot \sum \bar{x}_i^2 - n \cdot \sum \bar{x}_i \cdot y_i \cdot \sum \bar{x}_i^3 - \sum y_i \cdot \left(\sum \bar{x}_i^2 \right)^2 \right)}$$

$$\text{with} \quad \bar{x}_i = x_i - \frac{\sum x_i}{n}. \tag{1.11}$$

To determine the region of interest, either a rectangle or a line (the extreme case of a rectangle) lying over the edge area can be used. As shown in Figure 1.35, in the case of a rectangle, the grayscale values are projected onto a line using summation and then analyzed. Summation is similar to low-pass filtering but has the advantage of reducing small localized disturbances (noise).

1.4.3.1.7 Connectivity Analysis

Connectivity analysis, also known as blob or SRI analysis, is based on the calculation of numerous object features and characteristics from binarized images. In the case of transmissive light arrangements, they appear as black areas on a white background. The features and characteristics calculated are utilized either to assess objects directly (e.g., perimeter) or as a basis for calculating more complex analysis functions (e.g., area). Without going into detail about the algorithms involved, the most important calculable features and characteristics are listed as follows:

Contour	A list of ordered or jumbled x–y pairs of variables that describe the edge of an object.
Contour length	The length of the perimeter of an object.
Convex hull	A list of ordered or jumbled x–y pairs of variables that describe the convex edge sections of an object and their connecting pairs of lines.
Convex hull length	The length of the convex hull.
Area	The area covered by an object including or excluding the area of any holes contained within.
Compactness	The quotient of contour length and area.
Roughness	The quotient of contour length and convex hull length; this is a measure of the jaggedness of an edge.
Center of gravity	The center of gravity of an object.
Min/max coordinates	The x value of the pixels lying furthest away from the perimeter of an object either to the right or left, or the y value of the pixels lying furthest away from the perimeter of an object either to the top or bottom.
Bounding box	The rectangle encompassing the object with edges parallel to the image.
Dimensions	The quotient of the bounding box breadth and bounding box height.
Number of holes	The number of holes present within an area covered by an object.

1.4.3.1.8 Pattern Matching

The function of pattern matching is to search for image regions corresponding to a pregiven image segment. In measurement and testing technology, an image segment that contains a fault-free test feature is usually selected. With regard to the methods used for searching for appropriate image regions, a fundamental difference is made between the two following concepts:

1. Methods that are based on the calculation of similarity between the grayscale values of the fault-free image segment and the grayscale values of the test object. These methods are known as template matching.
2. Methods that are based on congruence calculations of the geometric features (edges, contours, and shapes) extracted from the fault-free image segment and from the image of the test object. These methods are also known as geometric model matching.

Template matching is the most frequently used method for determining a correlation. To achieve this, the template is gradually moved in steps of one or more pixels over the image being investigated. For each x,y position, a correlation coefficient $r_{x,y}(g,t)$ is calculated from the grayscale values $g_{i+x,j+y}$ and $t_{i,j}$ from the pixels of the image and of the template and then inserted into a resulting image.

Discreet correlation coefficients are calculated according to Reference 12 and Reference 13 using the following equation:

$$r_{x,y}(g,t)=\frac{\sum_{i=0}^{w-1}\sum_{j=0}^{h-1} g_{x+i,y+j}\cdot t_{i,j}-N\cdot\overline{g}\cdot\overline{t}}{(N-1)\cdot\sigma_g\cdot\sigma_t} \tag{1.12}$$

In the process, mean values $\overline{g}$ and $\overline{t}$ are calculated according to the formulae:

$$\overline{g}=\frac{1}{N}\cdot\sum_{i=0}^{w-1}\sum_{j=0}^{h-1} g_{x+i,y+j} \qquad \overline{t}=\frac{1}{N}\cdot\sum_{i=0}^{w-1}\sum_{j=0}^{h-1} t_{i,j}$$

and standard deviations σ_g and σ_t according to the formulae:

$$\sigma_g=\sqrt{\frac{1}{N-1}\cdot\left(\sum_{i=0}^{w-1}\sum_{j=0}^{h-1}(g_{x+i,y+j})^2-N\cdot\overline{g}^2\right)} \qquad \sigma_t=\sqrt{\frac{1}{N-1}\cdot\left(\sum_{i=0}^{w-1}\sum_{j=0}^{h-1}(t_{i,j})^2-N\cdot\overline{t}^2\right)}$$

The position of the best match can be then determined from the resulting image by carrying out a maximum search.

Correlation calculation carries the disadvantages of a relatively extensive effort required for the calculations and of rotational and scaling invariance. Rotational invariance can be overcome by correlating several rotated templates with the image. However, the calculation effort increases linearly with the number of rotated templates.

In contrast to template matching, which determines the similarity of a template with the image of the test object from all of the pixels examined, the process of geometric model matching is based on calculating a model (characteristics) from the geometric features of an fault-free object image segment and comparing it with the characteristics of the geometric features of the test object. The principal concept of geometric model matching is to select characteristics of geometric features that remain invariant in projective transformations. In this way, the method is irrespective of the viewing direction of the camera and is therefore insensitive to rotation and scaling. Details of the theories used as a basis for this method can be found in References 14, 15, and 16.

1.4.4 Solving Measurement and Testing Tasks

Now that the most important image processing and image analysis methods used in measurement and testing technology have been described, the following sections are concerned with solutions to the classes of problem mentioned at the beginning of this chapter.

1.4.4.1 Finding a Test Object or Region of Interest

Finding objects and regions of interest is a prerequisite to performing measurement and testing tasks based on the interpretation of image analysis results. As feature extraction in measurement and testing tasks is rarely carried out on the entire image contents, it is essential that the regions of interest are sensibly located with regard to the features requiring analysis.

The following points are to be taken into consideration:

- The search lines for determining an object-related coordinate system should cover the object image.
- A measuring line for determining the length of a component should run as straight as possible over the component.

- A measuring window for extracting the contour and for calculating the diameter of a bore hole should not cut into the image of the bore hole itself.

When carrying out measurement and testing tasks in practice, the following situations may arise:

- One or more identical test objects, which may be orientated in one or more directions, are present in the camera's field of view. Examples of this include products manufactured in a palletized manner, magazines loaded with test objects, or products present as identical bulk goods.
  Depending on the material and the measurement or testing task and handling mechanics involved, image acquisition may be performed using transmissive or incident light arrangements.
  One or more test objects and other objects, which may be orientated in one or more directions, are present in the camera's field of view.
  This situation arises, for example, if the bulk goods are not identical (e.g., sensors or microoptics of various sizes), where testing is only to be carried out on one type of object or if foreign particles (dirt, dust, etc.) are present in the field of view.
  Image acquisition may be performed using transmissive or incident light arrangements.
- The test object is part of a large subsystem. The position of the test object in relation to the subsystem does not have to be precise.
  This situation often occurs in the case of subsystems of hybrid construction where the microelectronic mechanical system and the electronic chip for the drive or signal analysis are separate elements, or in the case of MEMS-based sensors in which the sensor element has been assembled onto a carrier element.
  Owing to the fact that subsystems and MEMS are not generally transparent, the incident light arrangement must be used for image acquisition.
- Test objects can be processed individually. The processing site can vary in relation to the test objects.
  This situation may arise if resistors or capacitors present in hybrid subsystems are trimmed using lasers. Here, the measurement of the cutting width is important as it may lead to malfunctions if it is too narrow. The position of the cut is less important but it does denote the test feature. Another example is the markings on components where their position is less relevant in relation to component geometry.
  As a rule, the incident light arrangement must also be used here for image acquisition.
- Localization of a small test feature or test object in a relatively large, structured field of view.
  With mechanical or galvanizing processes, surfaces may possess grooves, inhomogeneities, or contamination that may impair the functionality of the surface. Information regarding such defects can be gained by carrying out roughness measurements with an atomic force microscope (AFM). However, owing to the extensive testing times involved, this device is not routinely implemented. The image processing task here is to localize possible defect areas (regions of interest) for positioning the AFM.
  Depending on the material the test object is composed of, image acquisition may be performed using transmissive or incident light arrangements.

These examples demonstrate that the quality of images may vary considerably as far as expected contrasts are concerned. With images recorded using a transmissive light arrangement, contrast is very high owing to the fact that, in ideal situations, only the silhouette of the test object is depicted. On the other hand, the contrast obtained with an incident light arrangement depends highly upon the color, amount of reflection, transparency, material structure or composition of the test object, and characteristics of the background.

For the task here of finding test objects or regions of interest, depending on contrast ratios the following image processing methods are suitable for determining solutions:

For high-contrast images:

- Edge detection
- Connectivity analysis
- Geometric pattern matching

For low-contrast images:

- Pattern matching using grayscale value correlation

The practical implementation of these methods for finding test objects and regions of interest is described in the following text with the aid of examples.

In the simplest case (see Figure 1.36), a single, essentially rectangular test object that stands out well against its background is in the camera's field of view. If test objects are also situated in a preferred orientation and are only out of line by a minimal degree of rotation, they can be located using simple edge detection in the grayscale value image. In addition to this, relative to the image coordinate system, a search line is defined in the direction of movement, and the grayscale values of the pixels below the search line are examined for an edge transition. Once this has been found, the object has also been located and other operations, such as orientation recognition, determination of the part-related coordinate system, and measurement and testing tasks, can then also be performed. If several identical test objects are present in the camera's field of view and are identically orientated, as is the case with magazined components, test objects can be located in a similar way.

The localization of test objects becomes more complicated if they are not in a preferred orientation, occur randomly in the image, when more than one test object is present, or if the test object is embedded in a larger structured environment, such as an electronic component mounted on a printed board.

If it is initially assumed that the objects stand out well against their background as far as the grayscale value image is concerned, connectivity analysis can be used for localization. To do this, the image is binarized and the bounding boxes of the objects determined. The bounding boxes, which may need to be enlarged to a certain extent, define direct image segments (regions of interest) in which position recognition, part-related coordinate system determination, and measurement and testing tasks can then be performed on separate objects (see Figure 1.37). With the aid of plausibility checks, such as area comparison, minimum and maximum object dimensions, the number of holes present in an object, etc., which can also be ascertained from the connectivity analysis, it is possible to determine whether the objects found are in fact the test objects. This is especially helpful in cases in which different objects or foreign particles (e.g., dust) are present in the camera's field of view.

If test objects do not stand out well against their background as far as their grayscale value is concerned (e.g., electronic component on a printed board) but provide sufficient contrast to enable

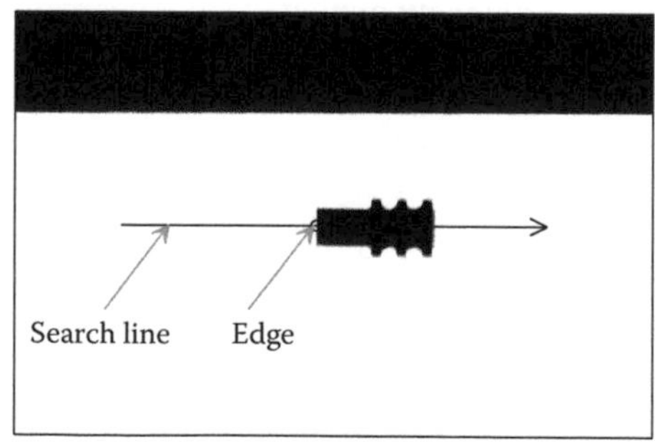

FIGURE 1.36 Simplest case of object localization using edge detection.

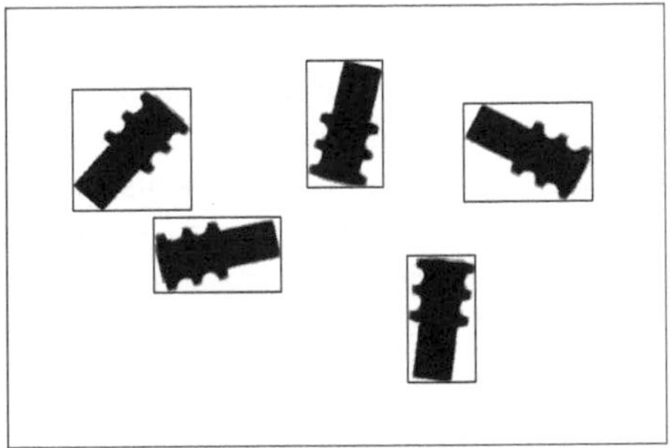

FIGURE 1.37 Detection of several (not necessarily identical) objects using bounding boxes by connectivity analysis.

distinctive edges to be generated using a gradient filter, object location can be realized using geometric pattern matching, which gives not only the position of the object but also its angle of rotation.

If contrasts are inadequate to enable distinctive edges to be generated, there is still a way of localizing test objects using pattern matching (see Figure 1.38). The following example demonstrates a scenario in which this method can be implemented.

The minimum gap widths of resistors trimmed using a laser need to be checked in a hybrid circuit mounted on a nontransparent ceramic substrate. Using a screen-printing process, the resistors are mounted onto a ceramic substrate in such a way that they partially overlap galvanized strip conductors. Owing to the manufacturing tolerances of the three process steps of the construction of the metallic conductors using electroplating, manufacture of the resistors using screen printing, and trimming of the resistors by laser cutting, the edges of the ceramic substrate cannot be utilized as a reference for determining the exact location of the laser cut. Furthermore, the position of the resistors also must be known in order to avoid false measurements.

A template containing only the image of the resistors has been shown to be suitable for recognizing their position with the aid of pattern matching using correlation calculations. Once the position of the test resistors has been located, the regions of interest are determined, which correspond to the contours of the resistors. Connectivity analysis is then carried out using a threshold value automatically adapted to the illumination conditions. This gives the edges of the laser cuts, and from these, the minimum cutting widths can be measured.

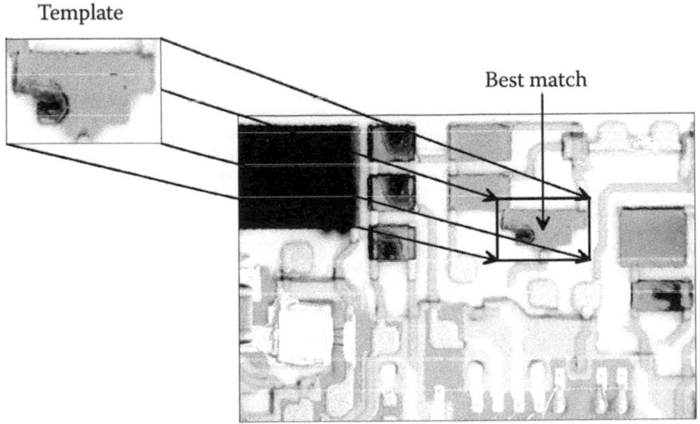

FIGURE 1.38 Detection of an object using grayscale pattern matching.

1.4.4.2 Position Recognition

In order to be able to carry out effective measurement and testing tasks on objects, either their position in relation to the camera image or the position of test features in relation to individual objects first needs to be determined. Distance measurements, such as determining the length or breadth of a test object, can only supply correct results if measurement lines are perpendicular to the external contours of the object concerned. The measurement of bore-hole diameters or verification of presence make sense only if the measurement windows required are correctly located in relation to the test object. Effective measurement and testing tasks therefore demand that measuring and testing operations on test objects be performed at the correct site and in the correct way. This can be achieved if a part-related coordinate system is determined for each test object. To do this, especially the angle of rotation α of the object in relation to the coordinate system of the image and the distance of an object feature (x_s/y_s) such as the center of gravity of an area or a defined reference point (x_r/y_r) in relation to the origin of the image coordinate system need to be determined.

As previously described in Subsection 1.3.2, the contrast ratios of recorded images may vary considerably depending on the situation. Images from transmissive light arrangements give naturally rich contrasts and appear very similar to binary images. On the other hand, the contrast present in images from incident light arrangements may vary significantly. Depending on contrast ratios, for high-contrast images, the methods of edge detection, contour-based process, and pattern matching using geometric features are suitable; for low-contrast images, the method of pattern matching using grayscale value correlation is appropriate.

If it can be assumed that, as described in the first example in Subsection 1.4.4.1, the object present in the camera's field of view is essentially rectangular in shape, located in a preferred orientation, and stands out well against its background, then recognizing the position of test objects is so simple that displacement occurs in only one direction, if at all. As described, it is possible to determine the location (x_1/y_1) of the test object by detecting an edge along a search line s_1. As shown in Figure 1.39, a part-relative coordinate system can be determined by defining two further search lines s_2 and s_3, which are perpendicular to the first search line and intersect with the contour of the test object, in such a way that they are displaced with the point of intersection (x_1/y_1) of the search line s_1.The points of intersection (x_2/y_2) and (x_3/y_3) of the search lines s_2 and s_3 then define one axis of the coordinate system. The second axis is perpendicular to the first axis at the point of intersection (x_1/y_1) of the search line s_1. The orientation of the test object, provided it is not symmetrical, is then obtained by measuring a feature-determining asymmetry.

Where test objects are present in the camera's field of view in any position, at any angle of orientation, and stand out well enough against their background to enable binarization and contour extraction using

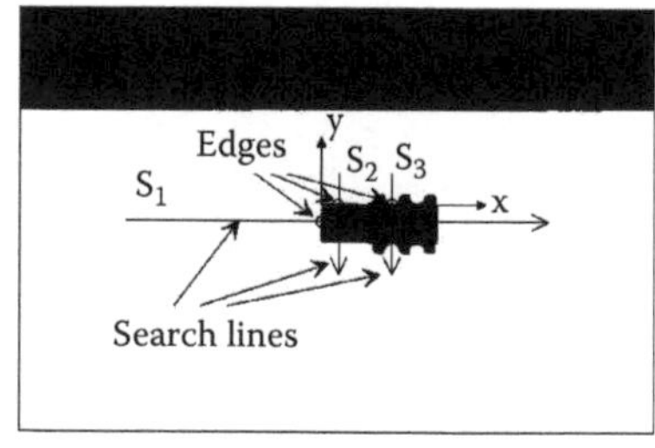

FIGURE 1.39 Recognition of position and orientation using edge detection.

connectivity analysis to be carried out, orientation recognition can be ascertained for nonsymmetrical objects by calculating the principal axis of inertia. As described in Reference 17, the following applies:

$$\tan(2\cdot\alpha) = \frac{2\cdot m_{11} - A\cdot x_s\cdot y_s}{m_{20} - m_{02} + A\cdot(y_s^2 - x_s^2)} \tag{1.13}$$

where α is the angle between the x-axis and the principal axis of inertia, and m_{mn} is the geometrical moments of inertia

$$m_{mn} = \iint_F x^m\cdot y^n\cdot dx\cdot dy \tag{1.14}$$

The area A, the coordinates x_s and y_s of the center of gravity of the area, and the geometrical moments of inertia m_{mn} of the test object can be calculated from the discreet contour data obtained using the following formulae [10]:

$$A = \frac{1}{2}\cdot\sum_{i=1}^{n}(x_i + x_{i-1})\cdot(y_i + y_{i-1})$$

$$x_s = \frac{1}{6\cdot A}\cdot\sum_{i=1}^{n}(x_i^2 + x_i\cdot x_{i-1} + x_{i-1}^2)\cdot(y_i - y_{i-1})$$

$$y_s = \frac{-1}{6\cdot A}\cdot\sum_{i=1}^{n}(x_i - x_{i-1})\cdot(y_i^2 + y_i\cdot y_{i-1} + y_{i-1}^2)$$

$$m_{11} = \frac{1}{24}\cdot\sum_{i=1}^{n}(3\cdot(x_i^2\cdot y_i + x_{i-1}^2\cdot y_{i-1}) + 2\cdot x_i\cdot x_{i-1}\cdot(y_i + y_{i-1}) + (x_i^2\cdot y_{i-1} + x_{i-1}^2\cdot y_i))\cdot(y_i - y_{i-1})$$

$$m_{20} = \frac{1}{12}\cdot\sum_{i=1}^{n}(x_i^3 + x_i^2\cdot x_{i-1} + x_i\cdot x_{i-1}^2 + x_{i-1}^3)\cdot(y_i - y_{i-1})$$

$$m_{02} = \frac{1}{12}\cdot\sum_{i=1}^{n}(x_i - x_{i-1})\cdot(y_i^3 + y_i^2\cdot y_{i-1} + y_i\cdot y_{i-1}^2 + y_{i-1}^3)$$

However, position recognition by determining a part-relative coordinate system with the angle α and the center of gravity of an area (x_s/y_s) using the method described here is limited to nonrotationally symmetrical (circles) or n-fold rotationally symmetrical (polygons) objects. In order to recognize the position of rotationally symmetrical or n-fold rotationally symmetrical objects, the method described can be extended by moments of a higher order [17] or the problem can be reduced to a contour match [10].

Another way (see Figure 1.40) of determining the position of rotationally symmetrical or n-fold rotationally symmetrical objects is to calculate the convex hull [18] of the object contours and to determine the maximum distances between the hull points with regard to all the connecting lines of neighboring hull points. From these, the greatest or smallest distance for determining the directions of the coordinate axes is selected. It is practical to place the coordinate axes through the center of gravity (x_s/y_s).

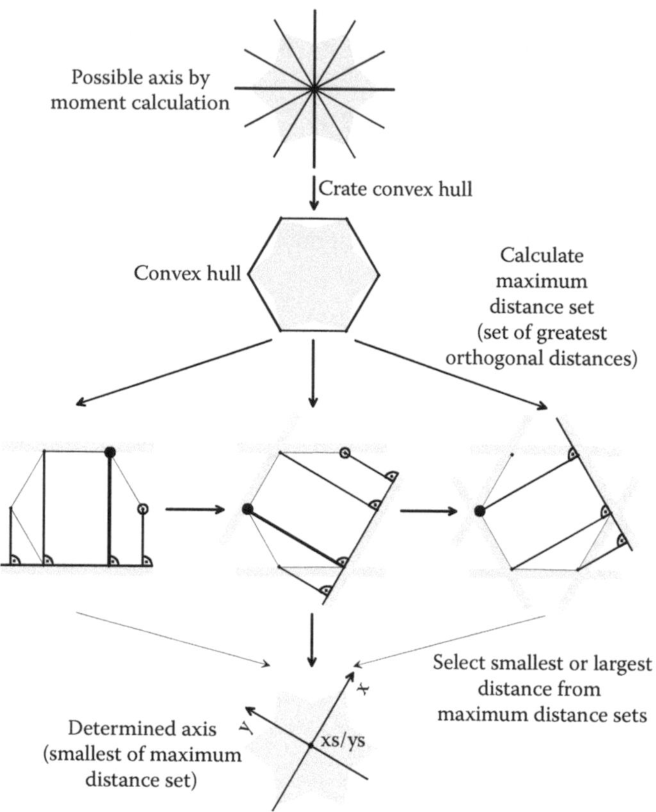

FIGURE 1.40 Position and orientation recognition using the convex hull method.

In principle, the position of objects in low-contrast images can be recognized using pattern matching. However, this is extremely time consuming and considerable calculations are required, especially in cases in which the size ratios of the depicted objects may change or any angle of orientation is present. If pattern matching cannot be avoided to achieve position recognition, the maximum possible limitations regarding scaling and rotation should be taken into account.

1.4.4.3 Measuring Geometric Features

Owing to the structure sizes of MEMS, it is only in rare cases that conventional measuring methods, such as metrology or calipers, can be implemented because the tactile character of these methods is often undesirable or the size ratios between caliper tips and structural elements are often insufficient. For this reason, as far as MEMS are concerned, the use of non-touch optical measurement technology, which is capable of performing measurements in the submicron range, is the most suitable method for determining geometric features. Just as in conventional manufacturing engineering, in the field of MEMS two-dimensional geometric features can be defined by the shapes of points, straight lines, circles, ellipses, and free-form curves. Except for free-form curves, these shapes can be derived from the general conic equation:

$$a \cdot x^2 + 2 \cdot b \cdot x \cdot y + c \cdot y^2 + 2 \cdot d \cdot x + 2 \cdot e \cdot y + f = 0 \quad (1.15)$$

To determine the unknowns *a, b, c,* etc. for a straight line with the equation

$$y = a \cdot x + b \tag{1.16}$$

two measuring points are required. For a circle with the equation

$$(x-a)^2 + (y-b)^2 = r^2 \tag{1.17}$$

three measuring points, and for an ellipse with the equation

$$\frac{(x-c)^2}{a^2} + \frac{(y-d)^2}{b^2} = 1 \tag{1.18}$$

four measuring points are required. These may, for example, be calculated accurately in the subpixel range using the edge detection method described in Subsection 1.4.3.1.6 or be obtained from previously calculated features such as best-match sites from pattern matching.

As is the case with real objects that shapes are never exactly straight or bore holes never exactly round, the calculations of shapes based on the few necessary measuring points tend to give results that deviate considerably from reality. Much better results can be obtained if the determination of shapes is based on a greater number of measurements of the real object's shape or its contour points and regression analysis subsequently carried out. This method is illustrated in the following text using the shapes of a straight line and a circle as examples [10].

Determination of regressive straight lincs

$$y = a \cdot x + b$$

from *n* measurements or contour points (x_i/y_i) using the method of minimum error squares leads to the analysis of the following:

$$\sum_{i=1}^{n} (y_i - a \cdot x_i - b)^2 \overset{!}{=} \text{Min.} \tag{1.19}$$

The zero settings of the partial derivations using both the independent variables *a* and *b* give a linear system of equations. Its solution is given by:

$$a = \frac{n \cdot \sum x_i \cdot y_i - \sum x_i \cdot \sum y_i}{n \cdot \sum x_i^2 - \left(\sum x_i\right)^2}$$

and

$$b = \frac{\sum x_i^2 \cdot \sum y_i - \sum x_i \cdot \sum x_i \cdot y_i}{n \cdot \sum x_i^2 - \left(\sum x_i\right)^2} \tag{1.20}$$

The best fit of a circle based on *n* measurements or contour points (x_i/y_i) in accordance with the method of minimizing error squares, as described in Reference 19 for example, requires a time-consuming iteration process. Provided that the measurement or contour points describe a circle and that they are distributed over most of the circle's circumference, the circle can be described [10] using the equation

$$x^2 + y^2 - 2 \cdot x \cdot x_m - 2 \cdot y \cdot y_m + c = 0 \tag{1.21}$$

For the discrete measurement or contour points (x_i/y_i), the minimizing of error squares can be formulated from this as shown:

$$\sum_{i=1}^{n} (x_i^2 + y_i^2 - 2 \cdot x_i \cdot x_M - 2 \cdot y_i \cdot y_M + c)^2 \overset{!}{=} \text{Min.} \tag{1.22}$$

The partial derivations in accordance with the variables x_M, y_M, and c give a linear equation system, which is solved using:

$$x_M = \frac{\sum x_i}{n} + \frac{\sum \bar{x}_i \cdot (\bar{x}_i^2 + \bar{y}_i^2) \cdot \sum \bar{y}_i^2 - \sum \bar{y}_i \cdot (\bar{x}_i^2 + \bar{y}_i^2) \cdot \sum \bar{x}_i \cdot \bar{y}_i}{2 \cdot \left(\sum \bar{x}_i^2 \cdot \sum \bar{y}_i^2 - \left(\sum \bar{x}_i \cdot \bar{y}_i \right)^2 \right)}$$

$$y_M = \frac{\sum y_i}{n} + \frac{\sum \bar{y}_i \cdot (\bar{x}_i^2 + \bar{y}_i^2) \cdot \sum \bar{x}_i^2 - \sum \bar{x}_i \cdot (\bar{x}_i^2 + \bar{y}_i^2) \cdot \sum \bar{x}_i \cdot \bar{y}_i}{2 \cdot \left(\sum \bar{x}_i^2 \cdot \sum \bar{y}_i^2 - \left(\sum \bar{x}_i \cdot \bar{y}_i \right)^2 \right)} \tag{1.23}$$

$$c = -\frac{\sum (x_i^2 + y_i^2)}{n} + 2 \cdot x_M \cdot \frac{\sum x_i}{n} + 2 \cdot y_M \cdot \frac{\sum y_i}{n}$$

$$\text{with} \quad \bar{x}_i = x_i - \frac{\sum x_i}{n} \quad \text{and} \quad \bar{y}_i = y_i - \frac{\sum y_i}{n}.$$

The method of best fit (of straight lines and circles) described here in accordance with the smallest error square, also known as the Gauss method, is often used in optical metrology. However, it has the disadvantage that the calculated shapes lie between the measurement or contour points; or in other words, there are always measurement or contour points to the right or left of a straight line or inside or outside the circumference of a circle (see Figure 1.41).

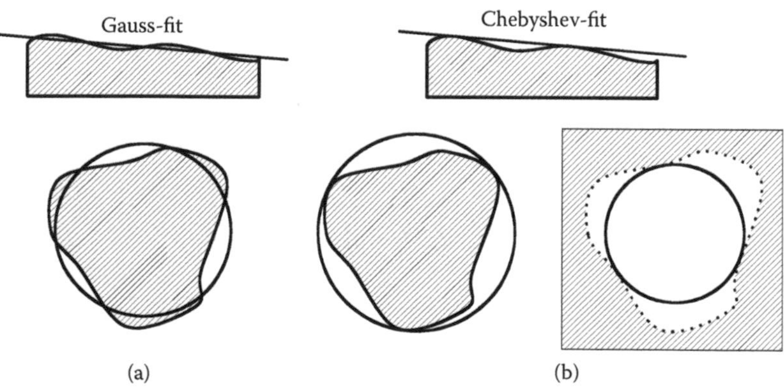

FIGURE 1.41 Best fit using the (a) Gauss and (b) Chebyshev methods.

This characteristic of the Gauss method may lead to unacceptable measurement errors, especially in cases in which mechanical components are to be inserted into one another. For example, the free cross section of the groove of a guiding channel may appear larger that it is in reality, with the result that the overdimensioned frame jams. The same problem can occur when fitting a shaft to a hub.

In order to avoid such problems, the mathematical fitting of the shapes must be performed so as to correspond as best as possible to metrological measuring means (measuring ruler, ring, or arbor). The Chebyshev best-fit methods described in Reference 20, Reference 21, and Reference 22 almost fulfill this requirement. Straight lines or surface planes are fit into the measurement or contour points so that they do not run through the material to be measured. This fitting method also enables the smallest outer circles of shafts/largest inner circles of bore holes that do not cut the material to be determined.

1.4.4.4 Presence Verification

In many automated manufacturing processes, assembly steps, and final inspections, it is often necessary to verify the presence of components or component features. The tasks associated with presence verification are as varied as the range of manufacturing processes implemented and products in existence. As it would be almost impossible to give a complete list of all the tasks involved, some of the typical areas of application of presence verification and testing tasks are mentioned in the following:

- In mechanical manufacturing, testing tasks are usually performed in association with features created as a result of mechanical processing steps. Examples of this (see Figure 14.2a and b) include the presence of bore holes and milling contours (e.g., cogs of a gearwheel), cut-ins of turning work pieces, and clips of shielded enclosures in the case of punched parts. The test results give an indication of tool wear and breakage.
- In the field of injection molding, the presence of similar features often requires checking in the same way as in mechanical manufacturing (presence of holes, contours, etc.). The causes of missing features are generally due to incorrectly set process parameters (too low or too high pressure, too low or too high temperature, etc.), the faulty composition of plastics, or mold wear.
- Etching and coating processes are utilized in the chemical industry, especially in the manufacture of MEMS and MOEMS. Testing tasks associated with these processes are concerned with checking for the absence or an overmarked presence of component features, which would indicate the use either of faulty process parameters or of incorrect concentrations of chemicals.
- Presence verification is the most commonly used testing method in automated assembly processes to ensure that the assembly has been carried out correctly. Testing tasks here are almost purely concerned with checking the presence and completeness of assembly components in their intended place. Assembly faults may be due to missing, incorrect, or defective parts in component feeding; loss of or damage to components during handling or installation; or parts falling out after their assembly.

Just as the tasks associated with presence verification are as varied as the range of manufacturing processes and products, seen from an image processing point of view, the situations that occur are equally high in number. Conditions as far as light, contrast, brightness, and shade are concerned and all the possible variations of these factors are so numerous that only fundamental procedures can be described here.

Presence verification is easiest when it can be performed using a transmissive light arrangement. In this way, for example, the presence of a bore hole can be checked by binarizing an image, defining a region of interest, and counting the number of white pixels corresponding to the cross-sectional

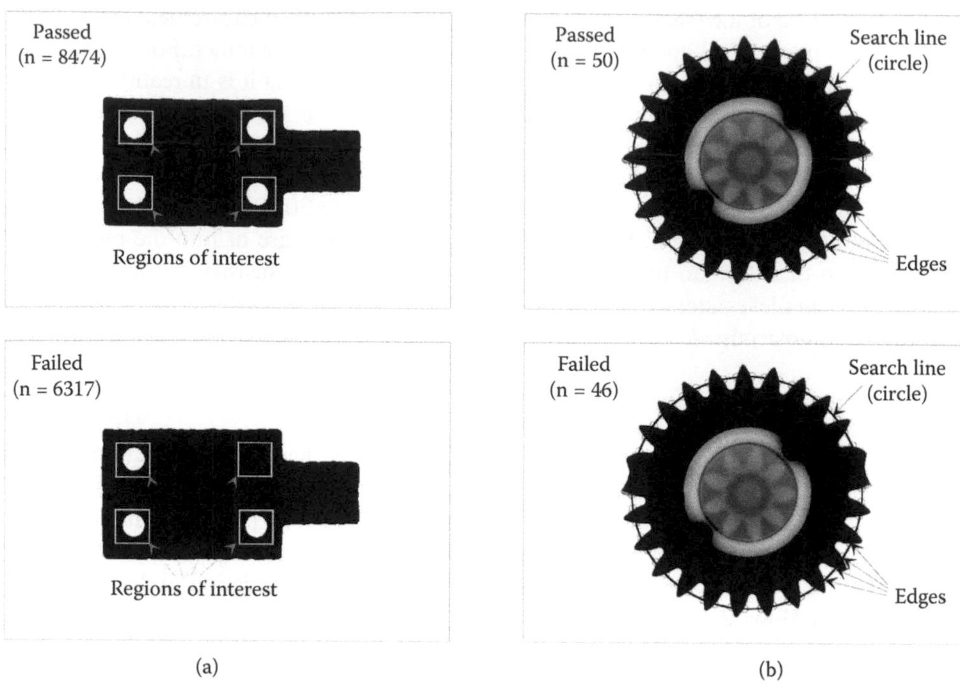

FIGURE 1.42 Examples of presence verification using (a) pixel counting, (b) edge detection.

area of the bore hole. This result can also be achieved using connectivity analysis. With a transmissive light arrangement, the presence of and number of teeth in a geared wheel can be verified using edge transitions along a circular curve laid concentrically over the midpoint of the geared wheel.

Presence verification becomes more complex and less certain in cases in which test objects or object test features need to be illuminated using incident light arrangements. Poor contrasts, light reflexes from shiny surfaces, and shadow formation may lead to faulty interpretations. For this reason, it is particularly important to ensure that test features are especially well illuminated and that the most appropriate test method is selected to deal with varying environmental conditions such as fluctuations in external light and changing surface characteristics of test objects.

If objects can be highlighted well against their background, such as in the case of bright or shiny objects present on a matte (nonshiny), dark background, presence verification can be carried out in the same way as in a transmissive light arrangement. Blind holes or other small structures in metallic or nonmetallic, diffusely shining objects can be depicted with a rich contrast and their presence checked if object surfaces are illuminated using a line-shaped or circular-shaped light source at a low angle. With grooves or channels, it may be advantageous to use a line-shaped light source. The light from a source placed at 90° to the channels strikes at a low angle causes shadows to form, thus enabling the presence of structures to be confirmed.

If test objects or backgrounds are highly structured, as is the case with assembled printed boards, richly contrasted images can often be obtained but test objects cannot be clearly separated from their background if they do not possess simple features. Here, presence verification can only be performed by using computer-intensive template matching where the template is only made from the image of the test object. The same also applies in cases in which objects only contrast slightly with their background.

Where objects requiring presence verification can be clearly differentiated from their background because of their color, it makes sense to use a color camera and color analysis.

As shown by these examples, presence verification is not limited to specific image processing or image analysis methods. The appropriate method is essentially selected according to the task in question. The use of filters to suppress noise or smooth edges may be just as helpful as connectivity analysis (which supplies numerous object features), edge detection, and other metrological processes.

1.4.4.5 Defect and Fault Detection

The field of defect and fault detection is at least as wide-ranging and varied as that of presence verification. This is due to the fact that practically all products or objects manufactured by man or machine invariably include faults. Subsequently, image processing and computer vision processes created to detect defects and faults are equally numerous and varied, and use methods used in nearly all the subfields of this discipline. The method utilized for defect and fault detection is always dependent upon the test object, the task involved, and the defects or faults to be expected or detected. It is therefore almost impossible to define a generally valid procedure for defect and fault detection. However, the fundamental aim of defect and fault detection is to identify object characteristics that deviate from requirements, although some deviations may be permitted if they are within tolerance limits.

Despite the huge variety of possible tasks required and their solutions, defect and fault detection can be divided into classes that can be solved using the same or similar image processing and computer vision methods. Examples of this include the inspection of transparent materials such as glass or optical components, of objects with regard to shape or deviations in color, of objects for breakage or the formation of cracks, or of objects for surface flaws. In assembly processes, especially in the electronics industry, among other things, installed components are checked for their correctness by recognizing markings and color codes.

With the aid of examples, the basis of methods for detecting defects and faults are described here for the classes "defects in structured surfaces" and "faults associated with component insertion."

Depending on the way they have been processed or manufactured, milled, grinded or honed surfaces, chemically produced surfaces, or the surfaces of ceramic components (provided they have not been polished) possess typical stochastic structures that may be more or less homogeneous. In image processing, these structures are known as *textures*. Defects, such as scratches, cracks, or sinkholes, are often characterized by faults in component-specific textures. As a result, defects can be detected by checking for deviations in texture, although this is no simple task.

In some cases it is possible to identify texture faults using simple image analysis methods (mean grayscale value or contrast of an image or region of interest), especially if the faults have totally different grayscale values (much brighter or much darker). In general, however, the grayscale values of textures and fault areas are usually similar or even the same, preventing simple analysis methods from supplying a useful result.

Different textures possessing identical or similar grayscale values can often be classified with the aid of the image processing method known as *texture analysis*. Depending on the principle on which the analysis is based, a difference can be made between two types:

1. Methods based on identifying differences in texture in a space domain with the aid of probabilities of grayscale values occurring in the vicinity of the pixel under investigation [23–26]
2. Methods realized using the classification of texture in a frequency domain based on Fourier analyses [27] or on a set of band-pass filters [28]

As the texture of the fault areas is not generally known, its position is determined as being the area where the texture cannot be classified. Figure 1.43 illustrates the use of texture analysis in

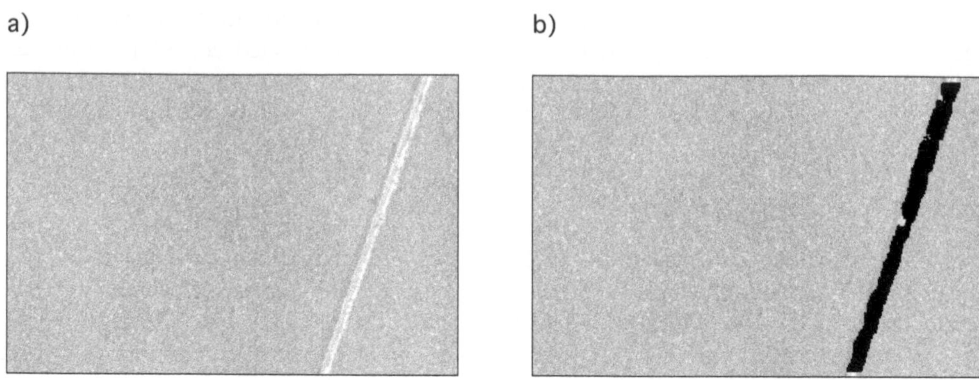

FIGURE 1.43 (a) Metallic surface with a scratch on it, (b) results of defect detection.

defect detection for recognizing a scratch on a structured metallic surface. More detailed representations and comparisons of the texture analysis method can be found in Reference 29, Reference 30, and Reference 31.

Among other things, in assembly and insertion processes, the problem arises that incorrect components are assembled or that poled parts are installed with the wrong orientation. These faults are usually due to human fault when supplying the components to pick-and-place machines or are due to picking up of components in the case of manual assembly. The detection of such faults can be achieved using image processing and computer vision methods, especially where components are marked with color or letters or numbers (see Figure 1.44), which is often the case in the electronics industry.

In order to verify these markings, the component first needs to be localized according to the methods described in Subsection 1.4.4.1 and Subsection 1.4.4.2. The identification markings can then be detected and validated. Where components are color-coded, the site of the code (diodes) is of interest for ascertaining polarity or color sequences to determine the electrical value of the part. Where components are marked with letters or numbers, optical character recognition (OCR) methods are used. The fundamental principle of these methods is based on the following steps:

- Segmentation of areas containing only one letter or number
- Improvement of the depiction of the individual letters and numbers by using various filters (low-pass, erosion, dilation, and thinning)
- Classification of the letters and numbers

Classifiers such as the hidden-Markow model, neuronal nets, and variations thereof can be implemented, as well as others that are not mentioned here. For more information, refer to the bibliography [32,33].

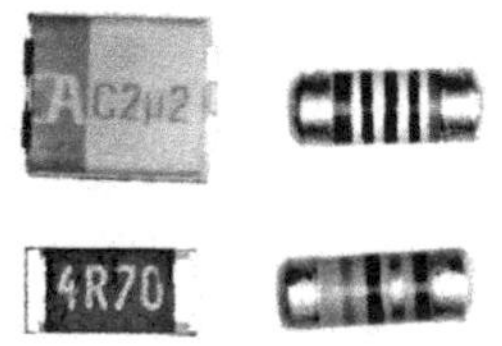

FIGURE 1.44 Examples of character-coded (left) and color-coded (right) devices.

1.5 COMMERCIAL AND NONCOMMERCIAL IMAGE PROCESSING AND COMPUTER VISION SOFTWARE

Commercial and noncommercial software products that are currently obtainable can be broadly classified with regard to their levels of abstraction into the three classes of "programming libraries," "toolboxes," and "configurable products," although the transition from one class to another may be indistinct. The herein after called *software products* are only typical examples of these classes because it would be impossible to list all the available products and manufacturers here.

Image processing and computer vision programming libraries provide elementary image processing algorithms in the form of software functions. They are implemented both in scientific fields for developing complex image analysis processes and in the field of commercial product development as a basis software. The range of function of nonspecialized libraries generally includes standard algorithms for image acquisition, image data administration, image preprocessing, image analysis, and image interpretation. Some of these products possess a considerably greater functional range and often take into account such specialized fields as texture analysis, photogrammetry, or three-dimensional image processing. The advantage of program libraries is that they are highly flexible, enabling image processing steps to be aligned and combined with one another or to be adapted to meet demands. Disadvantages include the lengthy familiarization work and programming time required, which almost exclude prototype realization work in pretesting and specification phases from being carried out, especially as these libraries hardly support graphic representations.

Examples of commercial programming library products include:

Cognex Vision Library (CVL)	Cognex Corporation http://www.cognex.com
HALCON	mvTec Software Gmbh http://www.mvtec.de
Heurisko	AEON Verlag & Studio Walter H. Dorn http://www.aeon.de
Matrox Imaging Library (MIL)	Matrox Electronic Systems Ltd. http://www.matrox.com

The following noncommercial and powerful image processing library can be downloaded in source code from Intel's Internet site:

Open Source Computer Vision Library	Intel Corporation http://www.intel.com/research/mrl/research/opencv

As far as programming knowledge and expenditure are concerned, products in the toolbox class are considerably cheaper. Image processing and computer vision toolboxes contain collections of program modules used to carry out image processing tasks such as recognizing text, bar or dot codes, and the orientation of objects or for measuring lengths, diameters, etc. With many products, toolboxes are realized as ActiveX controls so that the application programming, which is essentially limited to linking controls, can be performed in VisualBasic. As well as having image-programming functions, toolboxes often possess controls for depicting results in graphic form and also controls for interactive user operation. However, the advantage of simple programming carries the disadvantage that the application engineer is restricted to the functional range of the toolbox used.

Commercial toolbox products include:

ActiveMIL	Matrox Electronic Systems Ltd. http://www.matrox.com
Activ Vision Tools	mvTec Software Gmbh http://www.mvtec.de
Common Vision Blox	Stemmer Imaging GmbH http://www.stemmer-imaging.de
mvIMPACT	Matrix Vision GmbH http://www.matrix-vision.com
VisionPro	Cognex Corporation http://www.cognex.com

By utilizing configurable products, the user no longer needs to create programs. The scope of function of these products is usually concentrated on the field of application for which they have been constructed (two-dimensional/three-dimensional computer vision, object reconstruction, etc.) or for use in industry (metal processing, medical technology, biotechnology, etc.). Their method of functioning is based on providing the user with image processing operations that can be used in sequence on graphic images from a camera or data from another imaging system (interferometry, holography, stereography, etc.).

Examples of commercial configurable products include:

BIAS Fringe Processor	BIAS Bremer Institut für angewandte Strahltechnik GmbH http://www.fringeprocessor.de
Coake®	SAC GmbH http://www.sac-vision.net
NeuroCheck	NeuroChek GmbH http://www.neurocheck.com
Scorpion Vision Software	Tordivel Sa. http://www.scorpionvision.com
Wit	Coreco Imaging http://www.corecoimaging.com

1.6 IMAGE PROCESSING TECHNIQUES FOR THE PROCESSING OF FRINGE PATTERNS IN OPTICAL METROLOGY

Using modern optical methods such as interferometry, holography, fringe projection, and Moirè and speckle techniques, the absolute shape as well as the deformation of loaded technical components can be measured in a wide range by fringe evaluation. The quantity of primary interest is the phase of the fringes carrying all the necessary information. During the last 20 years, several techniques for the automatic and precise reconstruction of phases from fringe patterns were developed [40,42,44–46]. In this section attention is only paid for those basic concepts in which *digital image processing* is relevant:

- Fringe tracking or skeleton method [38]
- Fourier transform method [39]
- Carrier frequency method or spatial heterodyning [40]
- Phase-sampling or phase-shifting method [41]

All these methods have significant advantages and disadvantages, so the decision to use a certain method depends mainly on the special measuring problem and the boundary conditions. For simplification the discussion here is based on a modification:

$$I(x,y,t) = a(x,y,t) + b(x,y) \cdot \cos[\delta(x,y) + \varphi(x,y,t)] \tag{1.24}$$

The variables $a(x,y,t)$ and $b(x,y)$ consider the additive and multiplicative disturbances, respectively, and $\varphi(x,y,t)$ is an additionally introduced *reference phase* that categorizes the different phase-measuring techniques.

Although phase reconstruction by *fringe tracking* [34,38] is generally time consuming and suffers from the nontrivial problem of possible ambiguities resulting from the loss of directional information in the fringe formation process, it is sometimes the only alternative for fringe evaluation. Its main advantages are that it works in almost each case and it requires neither additional equipment such as phase-shifting devices nor additional manipulations in the interference field.

The *Fourier transform method* (FTM) is applied to the interferogram without any manipulation during the interferometric measurement. The digitized intensity distribution is Fourier transformed, leading to a symmetrical frequency distribution in the spatial domain. After an asymmetrical filtering including the regime around zero, the frequency distribution is transformed by the inverse Fourier transformation, resulting in a complex-valued image. On the basis of this image, the phase can be calculated using the arctan function. The disadvantage of the method is the need of individually adapted filters in the spatial frequency domain.

The most accepted techniques, however, involve calculating the phase $\delta(x,y)$ at each point, either by shifting the fringes through known phase increments (*phase-sampling method*) or by adding a substantial tilt to the wave front, causing carrier fringes and Fourier transformation of the resulting pattern (*carrier frequency method*). Both these types of *phase measurement interferometry* (PMI) can be distinguished as *phase modulation methods*: *temporal* and *spatial phase modulation techniques*. In the first case a temporal phase modulation is used [42]. This can be done by stepping the reference phase with defined phase increments and measuring the intensity in consecutive frames (*temporal phase stepping*) or by integrating the intensity while the phase shift is linearly ramped (*temporal phase shifting*). In the second case the phase is shifted spatially by adding a substantial tilt to the wave front (*spatial heterodyning*) [40] or by producing several spatially encoded phase-shifted fringe patterns (*spatial phase stepping*) by introducing at least three parallel channels into the interferometer, which simultaneously produce separate fringe patterns with the required phase shift [43]. In either case, the phase is calculated modulo 2π as the principal value. The result is a so-called *saw tooth* or *wrapped phase image*, and *phase unwrapping* has to be carried out to remove any 2π-phase discontinuities.

With respect to the successful application of time-consuming image processing algorithms to the unprocessed data or to the images that are already improved by some preprocessing, it has been proved to be useful to assess the quality of the data before they are fed to the image processing system. Some parameters that should be evaluated to test the quality of the fringes are:

- The spatial frequency distribution of the fringes over the total frame
- The fringe contrast
- The signal-to-noise ratio (e.g., the speckle index [44] or the image fidelity [45])
- The fringe continuity (analyzed, for example, with the fringe direction map [46])
- The linearity of the intensity scale (saturation effects)

Based on this evaluation, the fringe pattern can be accepted for further processing or rejected. In the case of rejection it is usually more effective to prepare the fringe pattern again under improved experimental conditions so as to expend disproportionate image processing effort.

1.7 CONCLUSION

The aspects of image processing and computer vision have only been roughly outlined here. However, they clearly show that this technology can be used to advantage to test MEMS. The main advantages include robustness, a high processing speed, the high degree of flexibility regarding adaptations to new tasks, and the noncontact testing.

The high processing speed is partly the result of more effective and highly efficient algorithms and partly due to the enormous increase in the performance of computer systems over recent years. Reduced costs brought about by the widespread availability of computers (PCs) now enable vision systems to be implemented for measurement and testing purposes in areas that would not have been considered cost-efficient a few years ago. Through the constant miniaturization of vision systems (smart cameras), it is also possible nowadays for measurement and testing systems to be integrated into manufacturing processes. Integration reduces error influences, which could occur as a result of handling if a nonintegrated solution is implemented.

The high degree of flexibility with regard to adaptations to new tasks is primarily because of the modular construction and simple programming of standard software packages for image processing and computer vision. The standardization of hardware and interfaces (Camera Link, IEEE 1394, USB, etc.) has also contributed towards this flexibility. In addition, the increasing use of vision systems has led to an ever-increasing range of vision components with the result that problem-specific solutions can be realized more and more easily.

However, the complexity of the subject of image processing and computer vision may not be ignored. As has become clear in the earlier sections, a comprehensive knowledge of illumination technology, optics, sensor and computer engineering, and software technology and algorithms is required. The ability to solve specific tasks also requires experience gained over many years because tasks are rarely identical. In fact, almost every new task demands problem-specific and often time-consuming engineering.

It is evident in many of today's applications that particular attention has been paid to solving technical challenges. Often too little thought has gone into user operation of the systems. Internal program correlations often appear unclear or the task of setting a large number of interactive parameters are too complex.

In the future, the same amount of consideration will need to be given to simple and intuitive operations by the user as that given to solving the technical problems. Self-adaptive or learning algorithms could help here significantly as they support the user in carrying out time-consuming, complicated, and obscure settings or are even capable of performing them for the user.

REFERENCES

1. Foveon, Inc., 2820 San Tomas Expressway, Santa Clara, CA 95051, http://www.foveon.com.
2. Roberts, L.G., Machine perception of three dimensional solids, in *Optical and Electro-optical Information Processing*, Tippet, J.T., Ed., MIT Press, Cambridge, MA, 1965, pp. 159–197.
3. Prewitt, J.M.S., Object enhancement and extraction, in *Picture Processing and Psychopictorics*. Lipkin, B.S. and Rosenfeld, A., Eds., Academic Press, New York, 1970.
4. Sobel, I., An isotropic 3×3 image gradient operator. In *Machine Vision for Three-Dimensional Scenes*, Freeman, H., Ed., Academic Press, Boston, 1990, pp. 376–379.
5. Gonzalez, R.C., Woods, R.E., *Digital Image Processing*, Addison-Wesley, Reading, MA, 1992.
6. Christoph, R., Bestimmung von geometrischen Größen mit Photoempfängeranordnungen, Jena, Friedrich-Schiller Universität, dissertation, 1987.
7. Seitz, P., Optical superresolution using solid state cameras and digital signal processing, *Optical Engineering*, 27(7), 535–540, 1988.
8. Nalwa, V., Binford, T., On edge detection, *IEEE Transactions on Pattern Analysis and Machine Intelligence*, 8(6), 699–704, 1986.

9. Nalwa, V.S., Pauchon, E., Edge aggregation and edge description, *Computer Vision, Graphics and Image Processing*, 40, 79–94, 1987.
10. Rauh, W., *Konturantastende und optoelektronische Koordinatenmeßgeräte für den industriellen Einsatz*, IPA-IAO Forschung und Praxis, Band 178, Springer-Verlag, Heidelberg, 1993.
11. Späth, H., *Algorithmen für multivariable Ausgleichsmodelle*, München, Wien, Oldenburg, 1974.
12. Bosch, K., *Elementare Einführung in die angewandte Statistik*, Vieweg, Wiesbaden, 1987.
13. Demand, C., Streicher-Abel, B., Waszkewitz, P., *Industrielle Bildverarbeitung*, Springer-Verlag, Heidelberg, 1998.
14. Weiss, I., Noise resistant projective and affine invariants, *IEEE Proceedings Computer Vision and Pattern Recognition*, Champaign, IL, 115–121, 1992.
15. Mundy, J.L., Zisserman, A., *Geometric Invariance in Computer Vision*, MIT Press, Cambridge, MA, 1992.
16. Rothwell, C.A., Zisserman, A., Mundy, J.L., Forsyth, D.A., Efficient model library access by projectively invariant indexing function, in *IEEE Proceedings Computer Vision and Pattern Recognition*, Los Alamitos, CA, 109–114, 1992.
17. Hu, M., Visual pattern recognition by moment invariants, *IRE-Transactions on Information Theory*, 8, 179–187, February 1962.
18. Melkman, A.V., On-line construction of the convex hull of a simple polyline, *Information Processing Letters*, 25(1), 11–12, 1987.
19. Ahn, S.J., *Least Square Orthogonal Distance Fitting of Curves and Surfaces in Space*, Springer-Verlag, Heidelberg, 2004.
20. Chebyshev Best-Fit Geometric Elements by Mathematical Programming, National Physical Laboratory, U.K., December 1992.
21. Chebyshev approximation to Data by Lines, National Physical Laboratory, U.K., January 1994.
22. Chebyshev approximation to Data by Circles, National Physical Laboratory, U.K., January 1994.
23. Haralick, R.M., Shanmugam, K., Dinstein, I., Textural features for image classification, *IEEE Transactions on Systems, Man, and Cybernetics,* 3(6), 610–621, November 1973.
24. Unser, M., Sum and difference histograms for texture classification, *IEEE Transactions on Pattern Analysis and Machine Intelligence,* 8(1), 118–125, January1986.
25. Cross, G.R., Jain, A.K., Markov random field texture models, *IEEE Transactions on Pattern Analysis and Machine Intelligence,* 5(1), 25–39, January1983.
26. Ojala, T., Pietikäinen, M., Harwood, D., A comparative study of texture measures with classification based on feature distributions, *Pattern Recognition*, 29, 51–59, 1996.
27. Weszka, J.S., Dyer, C.R., Rosenfeld, A., A comparative study of texture measures for terrain classification, *IEEE Transactions on Systems, Man, and Cybernetics*, 6, 269–285, April 1976.
28. Jain, A., Farrokhnia, F., Unsupervised texture segmentation using garbor filters, *Pattern Recognition*, 24, 1167–1186, 1991.
29. Fukunaga, K., *Introduction to Statistical Pattern Recognition*, Academic Press, New York, 1990.
30. Wagner, T., Texture analysis, in Jähne, B., Haußecker, H., Geißler, P., Eds., *Computer Vision and Applications*, Vol. 2, Academic Press, Boston, 1999.
31. Conners, R.W., Jain, A.K., A theoretical comparison of texture algorithms, *IEEE Transactions on Pattern Analysis and Machine Intelligence*, 2(3), 204–222, 1980.
32. Shunji, M., Hirobumi, N., Hiromitsu, Y., *Optical Character Recognition*, Wiley Series in Microwave and Optical Engineering, John Wiley & Sons, New York, 1999.
33. Rice, S.V., Nagy, G.L., Nartker, T.A., *Optical Character Recognition: An Illustrated Guide to the Frontier*, Kluwer Academic Publishers, Boston, 1999.
34. Osten, W., *Digital Processing and Evaluation of Interference Images*, (in German), Akademie-Verlag, Berlin, 1991.
35. Robinson, D.W., Reid, G.T., Eds., *Interferogram Analysis*, IOP Publishing, Bristol, 1993.
36. Malacara, D., Servin, M., Malacara, Z., *Interferogram Analysis for Optical Testing*, Marcel Dekker, New York, 1998.
37. Osten, W., Jüptner, W., Digital processing of fringe patterns, in Rastogi, P.K., Ed., *Handbook of Optical Metrology*, Artech House, Boston, 1997.
38. Nakadate, S., Magome, N., Honda, T., Tsujiuchi, J., Hybrid holographic interferometer for measuring three-dimensional deformations, *Optical Engineering*, 20(2), 246–252, 1981.

39. Kreis, T., Digital holographic interference-phase measurement using the Fourier-transform method, *Journal of the Optical Society of America,* 3(6), 847–855, 1986.
40. Takeda, M., Ina, H., Kobayaschi, S., Fourier-transform method of fringe pattern analysis for computer based topography and interferometry, *Journal of the Optical Society of America*, 72(1), 156–160, 1982.
41. Bruning, J.H., Herriott, D.R., Gallagher, J.E., Rosenfeld, D.P., White, A.D., Brangaccio, D.J., Digital wavefront measuring interferometer for testing optical surfaces and lenses, *Applied Optics*, 13(11), 2693–2703, 1974.
42. Creath, K., Temporal phase measurement methods, in Robinson, D.W., Reid, G.T., *Interferogram Analysis: Digital Fringe Pattern Measurement Techniques*, Institute of Physics Publishing, Bristol U.K., 1993, pp. 94–140.
43. Kujawinska, M., Spatial phase measurement methods, in Robinson, D.W., Reid, G.T., *Interferogram analysis: Digital Fringe Pattern Measurement Techniques*, Institute of Physics Publishing, Bristol U.K., 1993, pp. 141–193.
44. Crimmins, T.R., Geometric filter for speckle reduction, *Applied Optics*, 24(10), 1434–443, 1985.
45. Davila, A., Kerr, D., Kaufmann, G.H., Digital processing of electronic speckle pattern interferometry addition fringes, *Applied Optics*, 33(25), 5964–5968, 1994.
46. Osten, W., Digital processing and evaluation of fringe patterns in optical metrology and non-destructive testing, in Laermann, K.-H., Ed., *Optical Methods in Experimental Solid Mechanics*, Springer-Verlag, Wien, 2000, pp. 289–422.

2 Image Correlation Techniques for Microsystems Inspection

Dietmar Vogel and Bernd Michel

CONTENTS

2.1 Introduction 55
2.2 Deformation Measurement by Digital Image Correlation (DIC) Techniques 57
  2.2.1 Cross-Correlation Algorithms on Digitized Micrographs 57
  2.2.2 Extraction of Displacement and Strain Fields 60
  2.2.3 Determination of Derived Properties 63
  2.2.4 Capabilities and Limits 66
  2.2.5 Combining Finite Element (FE) Simulation with DIC Methods 68
2.3 Base Equipment for DIC Applications 70
  2.3.1 Components for Measurement Systems 70
  2.3.2 Requirements for High-Resolution Scanning Microscopes 72
  2.3.3 Software Tools 73
2.4 Applications of DIC Techniques to Microsystems 76
  2.4.1 Strain Analysis on Microcomponents 76
  2.4.2 Defect Detection 80
  2.4.3 Validation of Finite Element Modeling 81
  2.4.4 Measurement of Material Properties 82
  2.4.5 Microcrack Evaluation 88
  2.4.6 3-D Deformation Analysis Based on AFM Micrographs 92
  2.4.7 Determination of Residual Stresses in Microcomponents 95
2.5 Conclusions and Outlook 99
References 99

2.1 INTRODUCTION

Rapid miniaturization of microelectronics, microelectromechanical systems (MEMS), sensors, and photonics products has raised new challenges corresponding to thermo-mechanical reliability. Apart from accelerated testing of products and their components, numerical finite element analysis and deformation measurement methods are sought [1–4]. They allow an understanding of the response of components to environmental and functional thermo-mechanical loading and are part of advanced reliability studies.

The manufacturing, handling, and control of microscopic and nanoscopic objects require the quantification of their geometrical, kinematic, and mechanical properties. Although the measurement of geometrical and size data is more easily accessible by optical profilers or atomic force microscopy (AFM) and related methods, kinematic and mechanical characterization is a general problem for micro- and nanoscale objects and devices. Displacements and their derivatives are two basic mechanical properties to be measured for kinematic and mechanical description. Until now only a few methods existed that made quantified field data accessible for very microscopic regions.

Among the mechanical properties defined by material deformations we find, for example, elastic material properties like Young's modulus, Poisson's ratio, and coefficient of thermal expansion, as well as parameters describing plastic, viscoelastic, and viscoplastic material behavior. More complex material properties such as parameters in fracture mechanics, and damage and fatigue models are connected with material deformation as well. Available material data of typical materials applied in MEMS technology [5,6] is incomplete. Moreover, properties obtained by different methods and for materials from different manufacturing processes can vary significantly from each other [6]. Size effects on material properties as applied in micro- and nanotechnology are a serious issue [7–11]. Consequently, testing methods for deformation measurement on micro- and nanoscale objects are of considerable value for the future development of microsystem devices.

Different methods have been applied in the past to electronics, MEMS, and sensors. Among them, built-in stress sensors [12,13], Moiré techniques [14], strain measurement by digital image correlation (DIC) methods [15], Raman spectroscopy for stress analysis [16], and interferometric tools [17] have been involved. New developments have led to a significant downscaling of measurement resolution, making the use of higher resolution imaging for DIC [18,19] or Moiré [20] techniques possible.

At the cutting edge of nanotechnology, most published experimental studies on thermally and/or mechanically induced deformation fields are qualitative and semiquantitative analyses or make use of surface profiles depicted from AFM topography scans [21–24], using different kinds of scanning force microscope (SFM) imaging [21,25]. Among the published quantitative approaches, two techniques exist — Moiré [26–28] and DIC-based methods [18,29,30]. Both utilize load state images captured by SFMs or high-resolution scanning electron microscopes.

Published work on AFM–Moiré measurements reports single-point displacement or strain measurement [26]. Moiré fringes from crosshatched grids are taken to calculate the mentioned quantities averaging over the whole AFM scan field. Depending on the desired accuracy, one can employ, in this case, scans of moderate line number. On the other hand, the exploitation of advanced phase-shifting algorithms allows the computation of displacement maps from a couple of AFM scans shifted against each other [28]. However, this technique is not yet applicable to thermal measurements [26] and needs further development. Especially, the influence of the smallest-scan drifts during the multiple-scan capture for one measurement has to be analyzed in greater detail. Another difficulty regarding the Moiré approach is the time-consuming object grid preparation.

In contrast to Moiré measurements, correlation type measurements are based on higher-pixel-resolution AFM scans. They allow the measurement of displacements and strains with moderate spatial resolution within the AFM scan area [18,31]. DIC is the technique currently used by most researchers to measure object deformations from AFM images and seems to be a promising tool for measurements with the highest object resolution.

This chapter starts with an introduction to the testing methodology based on DIC. Underlying ideas and algorithms, capabilities and limits of the method, as well as the derivation of material properties from deformation fields are described in Section 2.2. Subsection 2.2.5 gives an introduction to the merging of experimental DIC measurements with numerical finite element analysis (FEA). Hard- and software prerequisites for DIC measurements are considered in greater detail in Section 2.3. The focus moves to necessities required for the selection of appropriate microscopic equipment and DIC analysis software. Section 2.4 gives an overview of different applications accomplished with DIC measurements. It comprises studies of stress/strain behavior on microcomponents, the investigation of failure mechanisms and defect structures, the determination of mechanical material properties from experimentally determined deformation fields, microcrack evaluation techniques, sensor deformation studies, and the determination of residual stresses.

Discussions on measurement methodology and applications refer to DIC performed exclusively on micrographs captured by different microscopes. Examples are given for results achieved from scanning electron, atomic force, and focused ion beam microscopes. The presented material is applicable to the testing and mechanical characterization of microsystems.

2.2 DEFORMATION MEASUREMENT BY DIGITAL IMAGE CORRELATION (DIC) TECHNIQUES

2.2.1 Cross-Correlation Algorithms on Digitized Micrographs

DIC methods on grayscale images were established by several research groups. Examples from different fields of applications can be found in various publications [29,32–38].

The first efforts to apply local DIC algorithms to solid-state deformation measurement were possibly inspired by the idea of transferring optical measurement principles, such as speckle techniques, to nonoptical microscopes. High-resolution scanning microscopes, such as scanning electron microscopes (SEM), AFM, and scanning near-field optical microscopes (SNOM), promised much better measurement resolution in terms of displacement determination. Moreover, the higher image resolution of these devices should make localized allocation of displacement data to particular microscopic image structures more feasible. As a consequence, inside views of the material deformation behavior of significantly smaller object areas should become possible.

Two main problems had to be tackled to achieve this. In the early 1990s, in order to extract displacement and strain fields from the comparison of load state micrographs, software algorithms had to be developed that were sufficiently reliable and fast enough, with the computing speeds available then. Much work in that direction was accomplished by Sutton and coworkers [33–35]. Another prerequisite for the successful application of DIC analysis to high-resolution imaging was the generation of suited micrographs from loaded objects. In this regard, stability problems had to be overcome for the imaging process itself, as well as for the object loading procedure. Loading stages that could be placed and operated under a microscope also had to be developed [39,40].

Early attempts to acquire high-resolution SEM images for subsequent deformation analysis were made by Davidson [36], rescanning SEM photographs. Modern SEMs allow the capture of digital images with high pixel resolution and allow correlation algorithms to be applied directly to them. This approach has been chosen by different research labs and is described in several publications [32,37,38]. Objectives for most published works were research topics in material sciences. Applications focused, for example, on fracture mechanics [36], deformation studies on the grains of loaded alloys [41], or on the material deformation behavior of loaded microdevices [29]. The authors of this chapter developed and refined different tools and equipment to apply SEM, AFM, and FIB images for deformation analysis on thermo-mechanically loaded electronics packages. This technique was established as *microDAC*, which means "micro *d*eformation *a*nalysis by means of *c*orrelation algorithms" [37].

The DIC technique, by definition, is a method of localized digital image processing. Digitized micrographs of the analyzed objects in at least two different states (e.g., before and during mechanical or thermal loading) have to be obtained by means of an appropriate imaging technique. Generally, correlation algorithms can be applied to micrographs extracted from very different sources. Digitized photographs or video sequences, as well as images from optical microscopy, scanning, electron, laser scanning, or scanning probe microscopy, are suitable for the application of digital image correlation. The basic idea of the underlying mathematical algorithms follows from the fact that images of different kinds commonly record local and unique object patterns within the more global object shape and structure. These patterns are maintained if the objects are stressed by temperature or mechanically. Figure 2.1 shows an example of images taken by SEM. Markers indicate the typical local pattern of the images. In most cases, these patterns are stable in appearance, even if a severe load is applied to the specimens. In strong plastic, viscoelastic, or viscoplastic material deformation, local patterns can be recognized after loading, i.e., they can function as a local digital marker for the correlation algorithm.

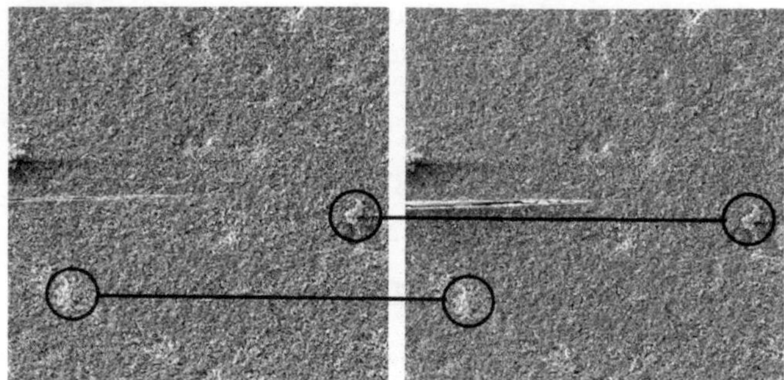

FIGURE 2.1 Appearance of local image structures (patterns) during specimen loading; SEM images of a crack in a thermoset polymer material for different crack opening displacements, image size: 315 × 315 μm, local patterns remain recognizable during loading.

The correlation approach is illustrated in Figure 2.2. Images of the object are obtained at the reference load state (1) and at a second load state (2). Both images are compared with each other using a special cross-correlation algorithm. In the image of load state 1 (Figure 2.2), rectangular search structures (kernels) are defined around predefined grid nodes (Figure 2.2, left). These grid nodes represent the coordinates of the center of the kernels. The kernels themselves act as local grayscale patterns from load state image 1 that have to be tracked, recognized, and determined by their position in the load state image 2. In the computation step, the kernel window ($n \times n$ submatrix) is displaced pixel by pixel inside the surrounding search window (search matrix) of load state image 2 to find the best-match position (Figure 2.2, right).

This position is determined by the maximum cross-correlation coefficient, which can be obtained for all possible kernel displacements within the search matrix. The computed cross-correlation coefficient, K, compares the grayscale intensity pattern of load state images 1 and 2, which have the same kernel size. A commonly used description of K is:

$$K_{i',j'} = \frac{\sum_{i=i_0}^{i_0+n-1}\sum_{j=j_0}^{j_0+n-1}\left(I_1(i,j)-M_{I_1}\right)\left(I_2(i+i',j+j')-M_{I_2}\right)}{\sqrt{\sum_{i=i_0}^{i_0+n-1}\sum_{j=j_0}^{j_0+n-1}\left(I_1(i,j)-M_{I_1}\right)^2\sum_{i=i_0}^{i_0+n-1}\sum_{j=j_0}^{j_0+n-1}\left(I_2(i+i',j+j')-M_{I_2}\right)^2}} \tag{2.1}$$

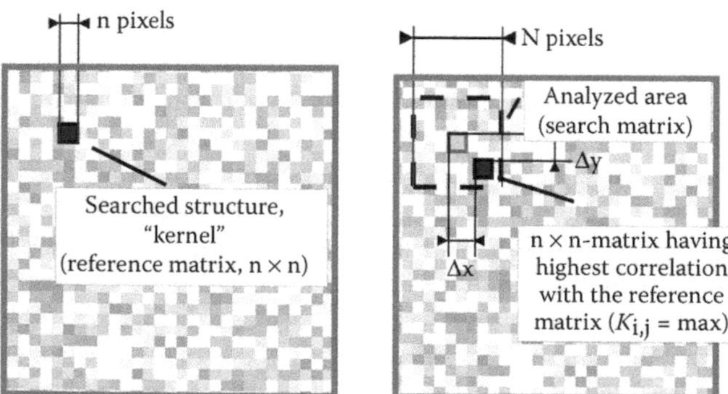

FIGURE 2.2 Displacement computation by cross-correlation algorithm on micrographs: (left) reference image at load state 1; (right) image at load state 2 (used for comparison).

$I_{1,2}$ and $M_{I_{1,2}}$ are the intensity gray values of pixels (i,j) in the load state images 1 and 2 and the average gray value over the kernel size, respectively. i′ and j′ indicate kernel displacement within the search matrix of load state image 2. Assuming quadrangle kernel and search matrix sizes, $K_{i,j}$ values have to be determined for all displacements given by $-(N-n)/2 \le i', j' \le (N-n)/2$.

The described search algorithm leads to a two-dimensional discrete field of correlation coefficients defined at integer pixel coordinates (i′, j′). The discrete field maximum is interpreted as the location where the reference matrix has to be shifted from the first to the second image to find the best matching pattern. Figure 2.3 shows an example of computed correlation coefficients inside a predefined search window, which was taken from the crack loading analysis illustrated in Figure 2.1 and Figure 2.5.

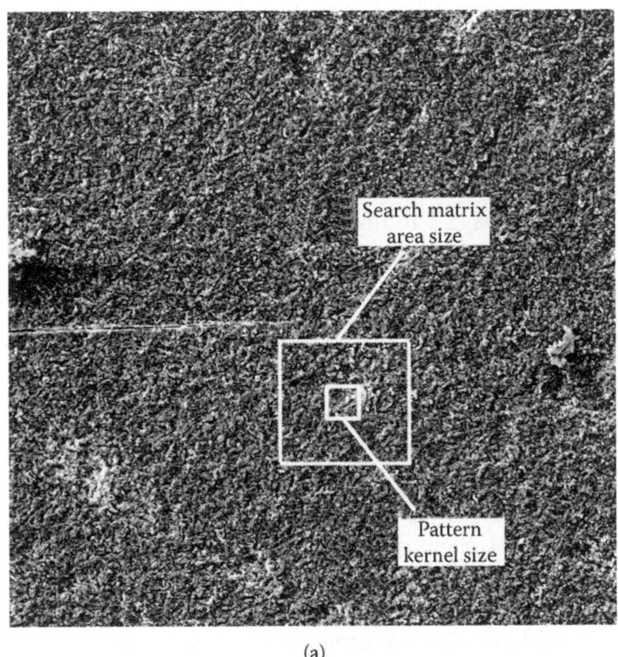

(a)

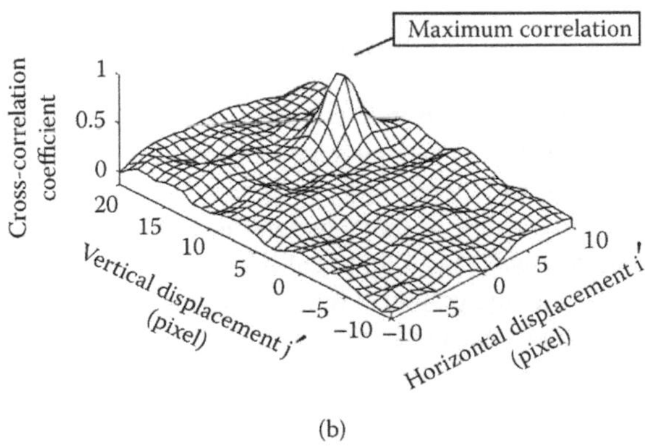

(b)

FIGURE 2.3 Discrete correlation function, $K_{i',j'}$, defined for integer i′, j′ (b), kernel size and position in the reference image, as well as applied search matrix are marked in the crack tip micrograph (a).

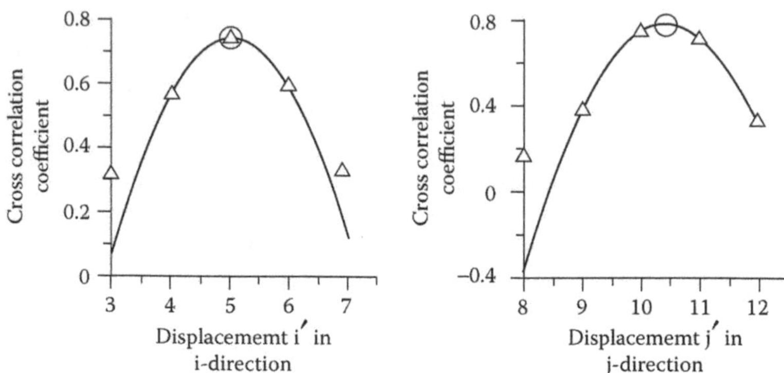

FIGURE 2.4 Principle of the parabolic subpixel algorithm: a parabolic function is fitted through three of the largest coefficients. Fitting is carried out independently in both horizontal (left) and vertical (right) directions. Calculated displacement values: i′ = 5.04, j′ = 10.44 (see circle marker).

Using this calculated location of the best matching submatrix of load state image 2, an integer value of the displacement vector is determined. In order to calculate a displacement field with high accuracy the algorithm has to be improved, i.e., a certain amount of subpixel displacement in structure displacements from image 1 to image 2 has to be determined.

Using the codes applied by different research groups, the accuracy of the cross-correlation technique is improved in a second calculation using additional subpixel algorithms. Presumably, the simplest and fastest procedure to find a value for the noninteger subpixel part of the displacement is realized in parabolic fitting. The algorithm searches for the maximum of the parabolic approximation of the discrete function of correlation coefficients in close proximity to the maximum coefficient $K_{max,discrete}$. The approximation process is illustrated in Figure 2.4.

The location of the maximum of the fitted parabolic function defines the subpixel part of the displacement. For parabolic fitting only the three largest coefficient values are taken for each coordinate direction; this saves computation time. This algorithm, implemented quite often, achieves a subpixel accuracy of about 0.1 pixels. Even so, it must be stated that it can fail considerably and introduce large systematic errors under some circumstances. The reason for this is the unfounded assumption of parabolic behavior, as well as the type of digitizing process applied to the microscope used. More advanced algorithms are more accurate, reaching subpixel accuracies of up to 0.01 to 0.02 pixels for common 8-bit depth digitizing, but demand sophisticated analysis and depend on the kind of image sources and data being treated.

2.2.2 Extraction of Displacement and Strain Fields

The result of a single two-dimensional discrete cross correlation and subsequent subpixel analysis in the surrounding region of a measuring point gives the two components of the displacement vector. Applied to a set of measuring points (e.g., to a rectangular grid of points with user-defined pitches), this method allows extraction of the complete in-plane displacement field. These results can be exported, in the simplest way, as a numerical list that can be postprocessed using standard scientific software codes. Commonly, graphical representations such as vector plots, superimposed virtual deformation grids, or color scale coded displacement plots are implemented in commercially available or in in-house software packages. Figure 2.5 shows two typical examples of graphical presentations for a correlation analysis performed on images of Figure 2.1.

Taking into account the magnification used for image capture, absolute values of horizontal and vertical in-plane displacement components u_x and u_y can be determined. Commonly, software codes for correlation analysis allow this scaling, in some cases in a convenient way by simply marking structures of known size.

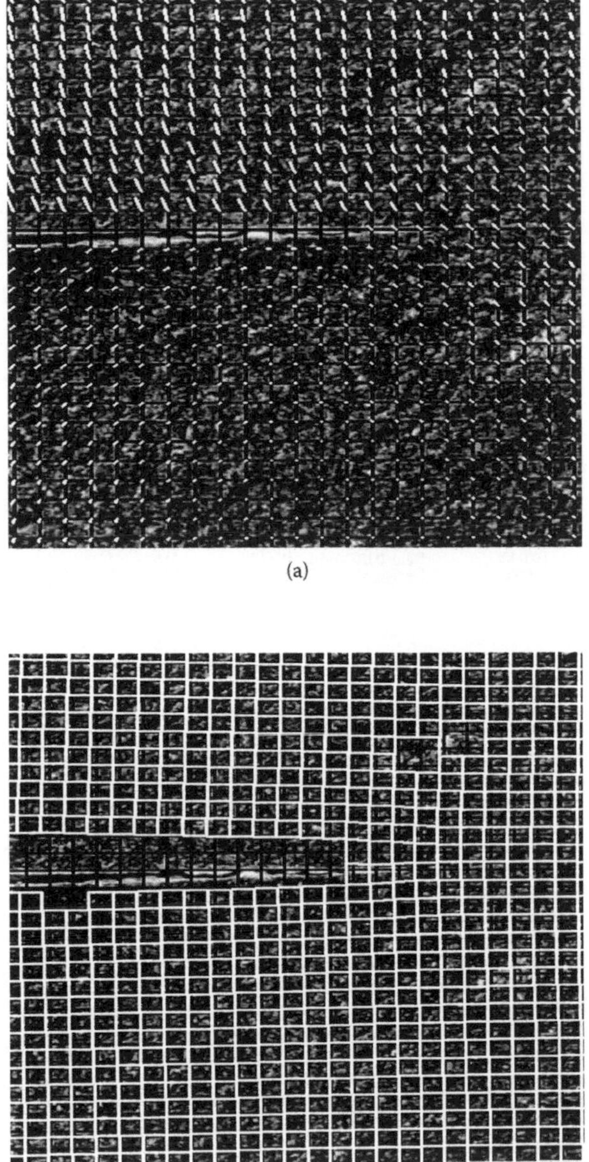

(a)

(b)

FIGURE 2.5 Digital image correlation results derived from SEM images of a crack tip, scan size: 315μm × 315 μm. (a) Image overlaid with user-defined measurement grid and vector plot; (b) image overlaid with user-defined measurement grid and deformed measurement grid. Displacements of the vector and deformed grid presentations are enlarged (by a factor of 2 and 3, respectively) with regard to image size.

Finally, taking numerical derivatives of the obtained displacement fields, $u_x(x,y)$ and $u_y(x,y)$, the in-plane strain components ε_{xx}, ε_{yy}, ε_{xy} and the local rotation angle ρ_{xy} are determined:

$$\varepsilon_{xx} = \frac{\partial u_x}{\partial x}, \varepsilon_{yy} = \frac{\partial u_y}{\partial y}, \varepsilon_{xy} = \tfrac{1}{2}\left(\frac{\partial u_x}{\partial y} + \frac{\partial u_y}{\partial x}\right), \quad \rho_{xy} = \tfrac{1}{2}\left(\frac{\partial u_x}{\partial y} - \frac{\partial u_y}{\partial x}\right) \tag{2.2}$$

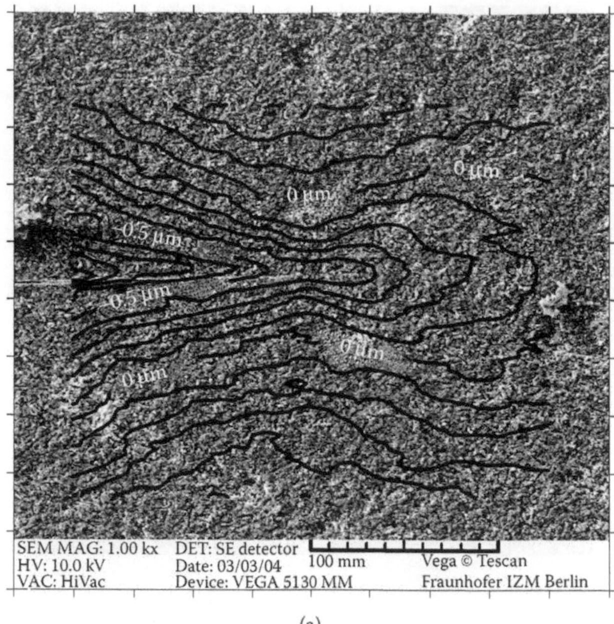

(a)

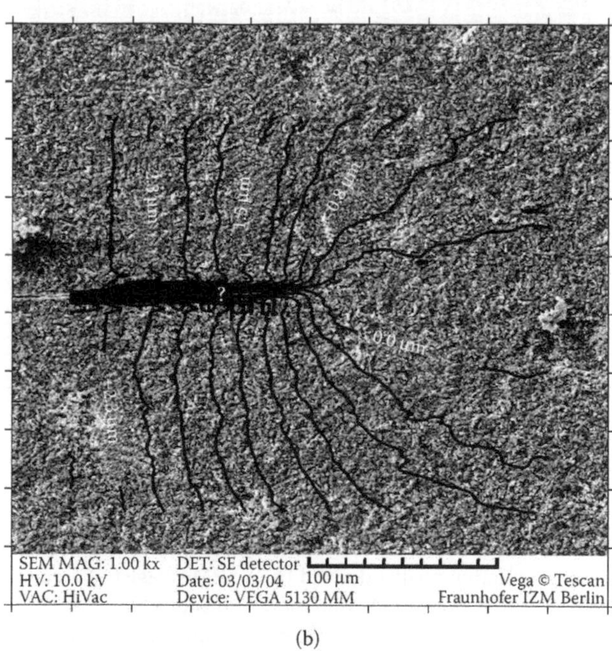

(b)

FIGURE 2.6 Displacement components, u_x (a) and u_y (b), for the crack opening example given in Figure 2.1. Contour lines are lines of equal displacement component; u_x, horizontal displacements and u_y, vertical displacements.

Derivation is included in some of the available correlation software codes or can be performed subsequently with the help of graphics software packages.

Figure 2.6 and Figure 2.7 show displacement and strain fields as contour line plots for crack opening derived from Figure 2.1 and Figure 2.5. The so-called mode I crack opening in a vertical

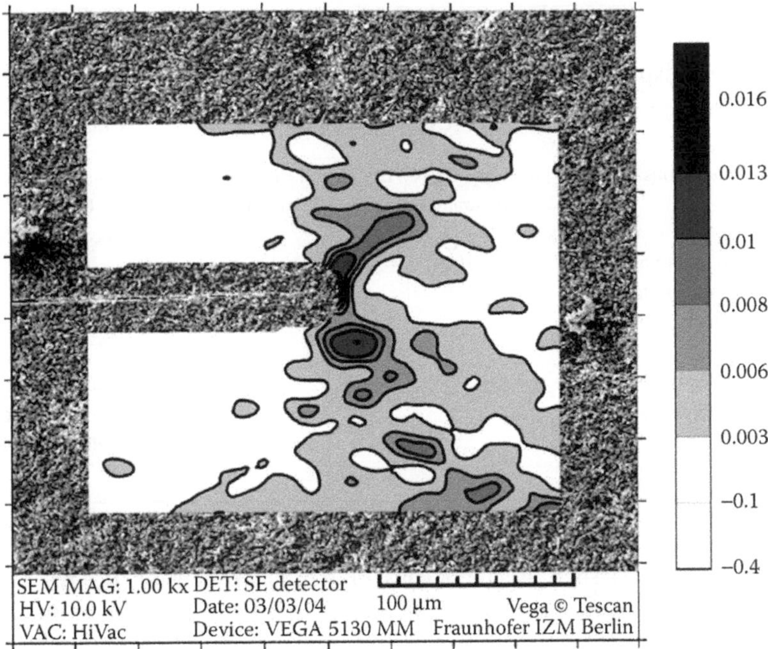

FIGURE 2.7 Strain component, ε_{yy}, for the crack opening demonstrated in Figure 2.1 and Figure 2.6. Contour lines are lines of equal vertical strain.

direction in reference to the crack boundaries is represented by the u_y displacement field. The corresponding ε_{yy} strain band, starting at the crack tip location and extending vertically is clearly seen in Figure 2.7. Local strain variations are due to material inhomogeneities.

2.2.3 Determination of Derived Properties

The quality and suitability of thermo-mechanical design, and the optimization of advanced MEMS and MEMS packaging by means of FEA, to a large extent depend on the data fitted into corresponding models. Properties for materials, as applied in MEMS and microoptoelectromechanical systems (MOEMS) applications, can differ significantly from those determined for bulk materials. These differences can be caused by changes in topology, as in that of a thin polymer layer, by the impact of manufacturing, by component-specific aging, and because of the increasing role of material surfaces and interfaces [6–10]. Consequently, material properties for microcomponents often have to be measured directly on the components to which they are applied or on adequate small-sized specimens. For this purpose, standard testing equipment cannot be used. New tools have to be developed to load small-sized specimens, and to sense load and sample load response during testing.

One of the most necessary values to measure during testing is specimen strain. In many cases, this becomes the base of the subsequent determination of mechanical material properties. Because of the sample size in microtesting, conventional extensiometers or strain gages often cannot be utilized. But correlation techniques are useful tools for measuring strains on objects of very small dimensions.

Two types of DIC measurements of material properties on a strain base can be distinguished with respect to the principal approach. In the first case, it is assumed that strain behaves homogeneously over a particular sample area of the specimen. Local strain values can be averaged over the whole sampled area, which significantly enhances measurement accuracy (see examples that follow). These measurements correspond to typical macroscopic mechanical testing with extensiometers

TABLE 2.1
Estimated Pixel Displacements for Typical Material Testing

	CTE Measurement	Uniaxial Tensile Testing
Admissible load conditions	Material: Cu (CTE = 16 ... 17 ppm K^{-1}), Maximum temperature interval: 100K	Maximum strain in direction of force: 0.5 %, Assumed Poisson ratio: $\nu = 0.25$
Micrograph size in pixel units	1024 × 1024	1024 × 1024
Single-point measurement accuracy for a DIC measurement (see Subsection 2.2.4)	0.1 pixel	0.1 pixel
Maximum mutual displacement of two micrograph patterns (located at two image edges) for the whole load interval	1.7 pixel	1.3 pixel (perpendicular to direction of force) 5.1 pixel (in direction of force)

or strain gages. Properties such as coefficients of thermal expansion (CTE), Young modulus, Poisson's ratio, or creep characteristics are determined in this way. However, it has to be ensured that the sensed (averaged) area fulfills the homogeneity requirement for deformation.

In the second case, inherently nonhomogeneous specimen deformation is caused by the specimen load. Consequently, entire spatial displacement or strain fields are used to determine material properties. Typical applications are the measurement of fracture mechanics properties such as fracture toughness and the evaluation of residual stresses by stress relaxation.

Considering in more detail the first kind of material property determination, two have been chosen and listed in Table 2.1: the measurement of coefficients of thermal expansion (CTE) and the measurement of Poisson's ratios from a uniaxial tensile testing. Reasonable material load intervals for MEMS applications are assumed. Obviously, the measurement resolution of the correlation technique (see Subsection 2.2.4) is insufficient to determine the desired material properties by a single two-point displacement measurement within one micrograph. With regard to the whole scope of load, expected maximum displacements are from only 10 up to 50 times larger than the measurement resolution.

The problem may be solved by simultaneously recording two micrographs, whose distance to each other is significantly larger than the image size [42]. So, the expected mutual displacement of two single measurement points can be increased significantly. Unfortunately, this approach is not favorable for microsystem applications if, for example, CTE values are to be determined at the cross section of the thin foils or plates or if Poisson's ratios have to be measured for fibers. The following description is another solution that has been applied to examples in Subsection 2.4.4. It utilizes the statistics of a large set of strain data points, obtained over a whole micrograph.

Under the assumption of homogeneous material expansion occurring within the micrograph area, and a dominating random error without essential systematic parts, specimen expansion can be determined by averaging a set of local strain measurements taken from different parts of the whole micrograph. However, this approach requires extensive computation because of the huge number of data points and load state images to be treated. In order to keep computation time at an acceptable level, efficient numerical correlation has to be performed. Additionally, incorrect individual data points should be identified and noted in a reliable manner, considering that about 1000 to 5000 strain values are determined for each load state image compared to a zero load reference. Advanced software tools fulfill the previously mentioned demands. Figure 2.8 shows a typical deformation field obtained during a CTE measurement between two temperatures states. This figure

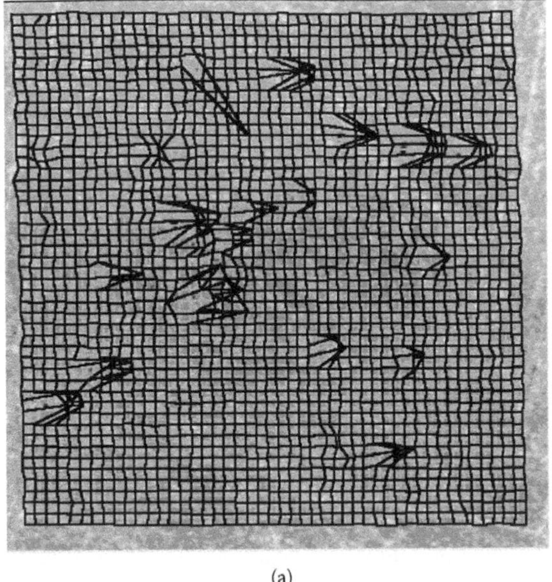

(a)

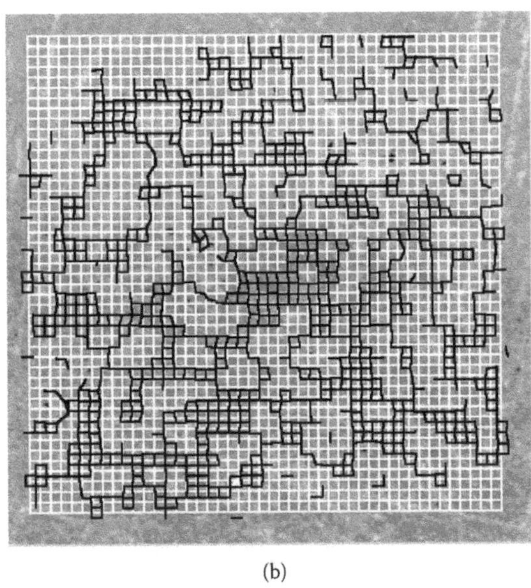

(b)

FIGURE 2.8 Deformation field (visualized by deformed virtual object grids) extracted from a CTE specimen [software mTest, courtesy of Image Instruments GmbH], temperature difference between the two image captures: 125K; (a) poor deformation field without eliminating error points, (b) original undeformed (white) and deformed (black) mesh, deformed black mesh after automatic elimination of error data points in the measurement mesh (reductant point discrimination).

illustrates the necessity and the efficiency of an implemented suppression algorithm for incorrect individual data points.

Application examples of the first kind of material properties (homogeneous strain) are demonstrated in Subsection 2.4.4, while examples of the second kind (nonhomogeneous strain) are referred to in Subsection 2.4.5.

2.2.4 Capabilities and Limits

Considering the application of DIC-based techniques to microsystem reliability issues, emphasis is put on:

- The analysis of strain–stress states under the environmental and functional load of components in order to discover weak points using exclusively deformation measurements.
- Material characterization and the determination of material properties. Strain measurements are performed on homogeneous deformation fields (e.g., the measurement of the coefficients of thermal expansion or Poisson's ratios), or more complex nonhomogeneous deformation fields (e.g., determination of fracture parameters or residual stresses).
- The validation of mechanical modeling for finite element simulations by comparing the measurement and simulation results of particular loadings.
- The detection of failure occurrences, the investigation of failure propagation due to component loading, and the tracking of failure mechanisms.
- The positioning purposes of micro- and nanocomponents with respect to each other, as well as the maintenance of component positions during loading (e.g., in optical setups).

Compared to other displacement and strain measurement techniques, such as interferometric or Moiré methods, DIC has several advantages:

- In many cases, only relatively simple low-cost hardware is required (optical measurements), or more expensive, existing microscopic tools like SEM and SFM can be utilized without any essential changes.
- Once implemented in a well-designed software code, the correlation analysis of grayscale images is user-friendly and easy to understand in measuring and postprocessing.
- For optical micrographs, in most cases, no special preparation of the objects under investigation is needed. Under these circumstances of lower magnification, demands on the experimental environment (e.g., vibration isolation, stability of environmental parameters like temperature) are lower than in interferometric-based measurements. Comparing DIC techniques with Moiré measurements, especially Moiré interferometry, DIC makes the cumbersome preparation of component-site grids unnecessary, and this makes measurements timesaving.
- The method possesses excellent downscaling capability. By using different microscopic imaging principles, very small objects can also be investigated. Therefore, the correlation analysis of grayscale images is predestinated for the qualitative and quantitative characterization of micromechanical and nanomaterial properties. Moreover, by combining different microscopes, e.g., SEM and AFM, large dynamic ranges can be realized.

The achievable accuracy of DIC displacement measurements $\delta u_{x,y}$ is defined by the object site's field-of-view or field-of-scan length, $l_{x,y}$, the amount of pixels in the micrograph in vertical (y) and horizontal (x) directions m_x and m_y, and the subpixel shift resolution, k. It is given by equation (2.3):

$$\delta u_{x,y} = (l_{x,y} / m_{x,y})\, k \tag{2.3}$$

For a given magnification, the subpixel shift resolution, k, is the main measure of the available displacement resolution. It strongly depends on the experimental conditions, the quality of the images obtained in the experiment (e.g., noise, sharpness), and the quality of the software algorithms that are used. k varies in most relevant cases between $\frac{1}{4}$ and $\frac{1}{100}$ pixels.

For displacement results of multiple point measurements obtained by statistical means (e.g., rigid body displacement calculated as the mean value of shifts measured at multiple points or linear elastic strain calculated from the slope of a regression line), resolution equivalents of 0.01 pixels and better are reached, although the correlation analysis, applied to a single measurement point, does not provide this level of accuracy. This averaging technique is used for some kinds of material property determination (see Subsection 2.2.3 and Subsection 2.4.4).

Assuming, for example, that low-resolution SEM imaging with a field of view of l_x, l_y = 100 μm, an image resolution of m_x, m_y = 1024, and a subpixel accuracy of $k = 0.1$, a measurement resolution of a single point displacement value down to 10 nm can be obtained. Generally, this displacement measurement resolution can be extended to 1 Å, if AFM scans are used with scan sizes of l_x, l_y = 500 nm, pixel arrays of 512 × 512 and if a value of $k = 0.1$ is reached.

The aforementioned consideration are based on limitations due to the image digitizing and spatially localized correlation. However, it should be pointed out that the actual measurement accuracy achieved depends on conditions of micrograph generation too. Microscope stability and imaging *reproducibility*, as well as the stability of the object's position relative to the microscope, and the stability of loading conditions can significantly influence the final resolution.

The estimation of measurement errors for strains is more sophisticated (see Equation (2.2)). Often, smoothing of displacement data in a certain neighborhood of displacement points is carried out before the computation of derivatives. Moreover, strains are computed taking into account the displacement values determined nearby. Consequently, the strain value error is a result of the kind of applied numerical algorithms, as well as data structure. For example, high displacement gradients can be "smoothed down" before calculating strains; this leads to the underestimation of strains. In order to get more reliable information about strain measurement accuracy, real object images can be deformed artificially using different spatial frequencies. When applying software codes to be tested on these images, a rough estimation of strain measurement accuracy is possible. As found for a code used by the authors [43], for moderate displacement gradients over the image (1.5 sinus waves over a 1024 × 1024 pixel image) a measurement resolution of 1×10^{-3}, in terms of strain, is realistic (for subpixel accuracies of $k = 0.1$). This value is independent of the kind of image capture, i.e., independent of image magnification.

The lateral measurement resolution is another crucial point to be examined. It is a function of the size, n, of the tracked local pattern. Nonoverlapping of neighboring searched patterns (kernels) must be provided for independent measurement points (displacement values). Therefore, the lateral resolution equals:

$$\delta l = (l_{x,y}\, n / m_{x,y}) \tag{2.4}$$

In Equation (2.4) n is the kernel pixel size (see Figure 2.2). For the aforementioned example of a typical low-resolution SEM image (l_x, l_y = 100 μm; m_x, m_y = 1024) and $n = 21$, a lateral measurement resolution of 2 μm can be found. Again, for small AFM scanned areas (l_x, l_y = 500 nm, pixel arrays of 512 × 512), this value can be significantly improved. Measured displacement values can be laterally resolved within 20 nm.

Besides the preceding resolution limits originating from the nature of the correlation approach, i.e., those defined by the correlation software, additional effects limit measurement accuracy. They are associated mostly with image acquisition and object loading, and are discussed in more detail in Subsection 2.3.2.

2.2.5 Combining Finite Element (FE) Simulation with DIC Methods

Additional values for mechanical investigations on MEMS, sensors, or other microsystems can be yielded by combining DIC deformation analysis with FEA. On one hand, DIC measurements provide surface deformation data only. Feeding DIC results into FE simulation objects allow for analysis on a three-dimensional base. On the other hand, FEA as a separate entity is a method that makes ample presumptions and simplifications in the modeling step. Principal validation of FEA findings using experiments is strongly advised. DIC measurements are useful tools for that purpose.

In order to combine DIC and FEA, some requirements have to be met. Generally, the suitability of prerequisites used for DIC measurements and FE simulations must be guaranteed. That is, the input for FEA, loading conditions, constitutive material laws, material properties, and object geometry must sufficiently describe experimental conditions. Moreover, direct comparison of data or data transfer between the two tools must take into consideration the adequacy of coordinate systems. This means that every kind of rigid body displacement and rotation must be matched before comparing data fields. A "mismatch" can be introduced if, for example, loading modules lead to rigid body movements of the object, which is not included in mechanical FE modeling during the measurement.

The following example of microcrack evaluation demonstrates possibilities of combined DIC–FEA experiments. In this example, fracture toughness values had to be determined from the crack opening displacement (COD) field, computed from AFM load state images using DIC software. Commonly, two situation are of interest:

1. Fracture mechanics parameters have to be determined from experimental COD fields, without knowing the exact load force applied to the crack. This may be the case in a complex microsystem where crack loading is not easy to deduce from loading of the known device. If material properties, such as Young's modulus, Poisson's ratio, etc. are known, FEA can yield the fracture toughness based on the experimental COD data.
2. Crack-loading forces are known, but not all material parameters for a complete FEA are available. In this case, DIC measurements can be used to compare results to FEA data in order to find a fit for the missing input material data.

In order to merge FEA and common DIC data interfaces, mutually fitting algorithms have been developed. Figure 2.9 shows the flowchart of a chosen concept [44]. The concept implies that for each measured in-plane displacement field, the best FE simulation fits are determined. In FE simulation, uncertain material properties or, e.g., crack-tip load conditions or material parameters are varied. Matching these different FE simulation results at the specimen surface with the measured displacement fields ensures that FE modeling is adequate. Now, the matched FE model can be applied to get more precise information about the mechanical behavior of the specimen, e.g., fracture mechanics parameters such as fracture toughness can be evaluated.

Before fitting the series of FEA fields with the measured DIC field, geometrical consistency between FEA and DIC has to be provided. Data exchange between DIC and FEA is achieved by importing the FE mesh to DIC software, i.e., DIC measurement points coincide with their respective FE node points. The aforementioned geometrical matching of rigid body movements is achieved by mutual derotation and rigid body translation, which makes use of the geometrical preconditions in the FE model. Geometrical matching, using best fit algorithms like least square algorithms for stepwise derotation and displacement of images, works as well [44].

This approach is general and can be applied widely to reliability issues associated with crack propagation.

Figure 2.10 illustrates the creation of common FEA or nanoDAC data interfaces. In order to obtain measurement and simulation data for the same object points, FE meshes are imported to nanoDAC correlation software. Correlation analysis is performed in the same node points as FE

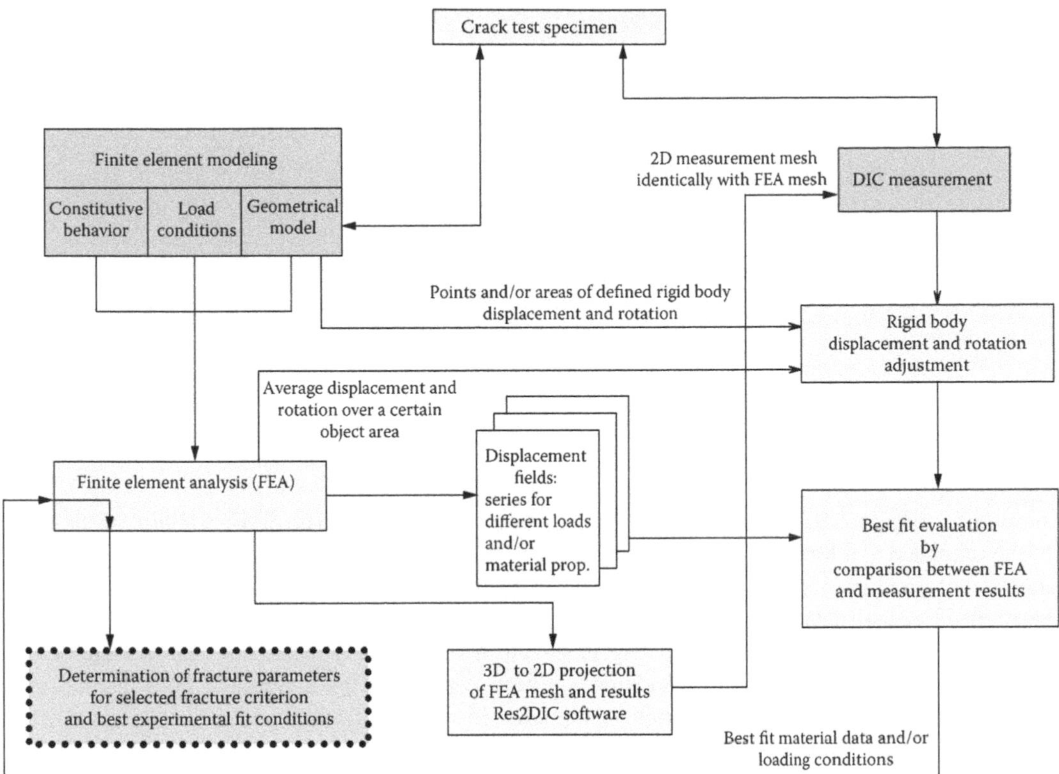

FIGURE 2.9 Common DIC measurement and FE simulation concept.

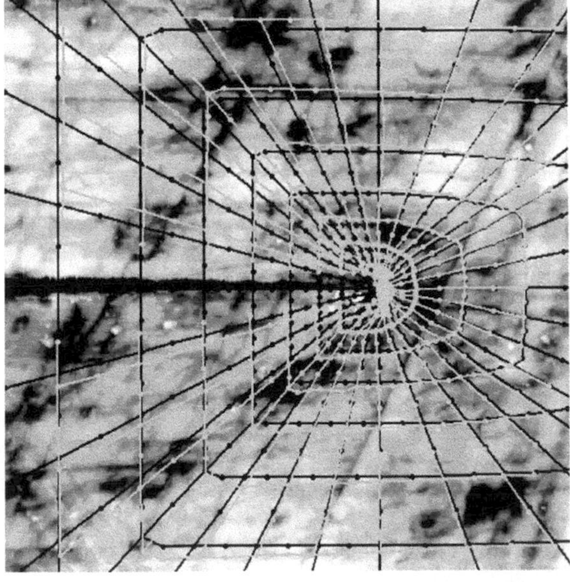

FIGURE 2.10 Displacement field measured on load state AFM scans captured around a crack tip in epoxy; dark mesh: originally imported FE mesh with measurement nodes, light mesh: deformed original mesh after crack opening (measurement result), for visibility deformations have been enlarged.

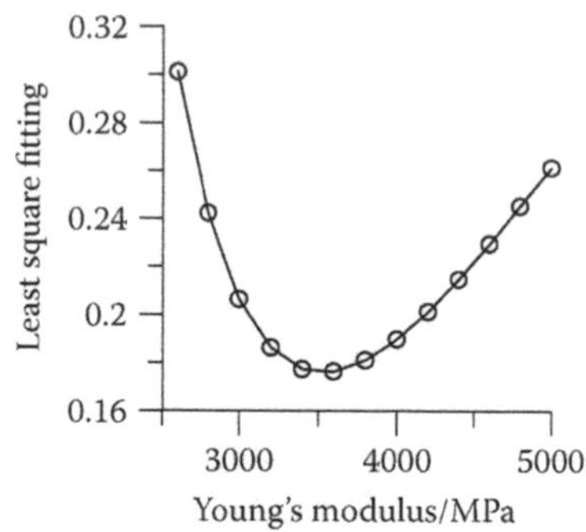

FIGURE 2.11 Comparison between FEA and DIC displacement fields of Figure 2.10 of a least-squares fit criterion; circles: comparison between a FEA–DIC pair of crack tip fields, best fitting between FE and nanoDAC at curve minimum.

simulation. Figure 2.10 contains a corresponding measurement mesh. The dark and light meshes refer to the unloaded and loaded state, respectively. Measured displacement values are exported for the FE mesh nodes and can be compared numerically with FE results directly. As a coincidence criterion between FEA and nanoDAC displacement fields, least-squares fits or correlation-type analysis, for example, are used. An example of fitting is shown in Figure 2.11, where for the measurement in Figure 2.10 an unknown Young's modulus has been determined by comparing both displacement fields. By employing the best fit Young's modulus in a final FE simulation, the correct fracture toughness value for the experimental load under investigation can be determined.

Software approaches that utilize FEA-type concepts of meshing in DIC software have to be mentioned [45]. Without aiming at combined FEA–DIC analyses described previously, these codes use FEA-type meshes to set up DIC measurement points and, later on, correlation analysis. The separate meshing of each material has specific advantages. Discontinuities in displacement fields commonly cause misinterpretations in adjacent cross-correlation analysis and can be avoided in this way. FEA-type correlation analysis is definitely restricted to single materials, which prevents error computations for the affected measurement points. Finally, different materials can lead in captured images to brightness steps at material boundaries. Localized cross correlation as applied in DIC techniques is very sensitive to such kinds of brightness steps and shows systematically shifted displacement values at the material boundary. Restricting local correlation analysis to single materials eliminates this intrusive data distortion.

2.3 BASE EQUIPMENT FOR DIC APPLICATIONS

2.3.1 Components for Measurement Systems

DIC measurement systems for microsystem applications should include the following main components:

- Appropriate microscopic equipment for image capture (e.g., optical microscopes, scanning electron microscopes, atomic force microscopes, focused ion beam systems)
- Loading modules, which can be placed inside the mentioned microscopes and which allow *in situ* object loading of the desired kind (thermally, mechanically, functionally induced stresses, or a combination of them)
- Software for DIC analysis on load state micrographs in order to extract in-plane displacement and/or strain fields; software for the computation of derived properties from obtained deformation fields

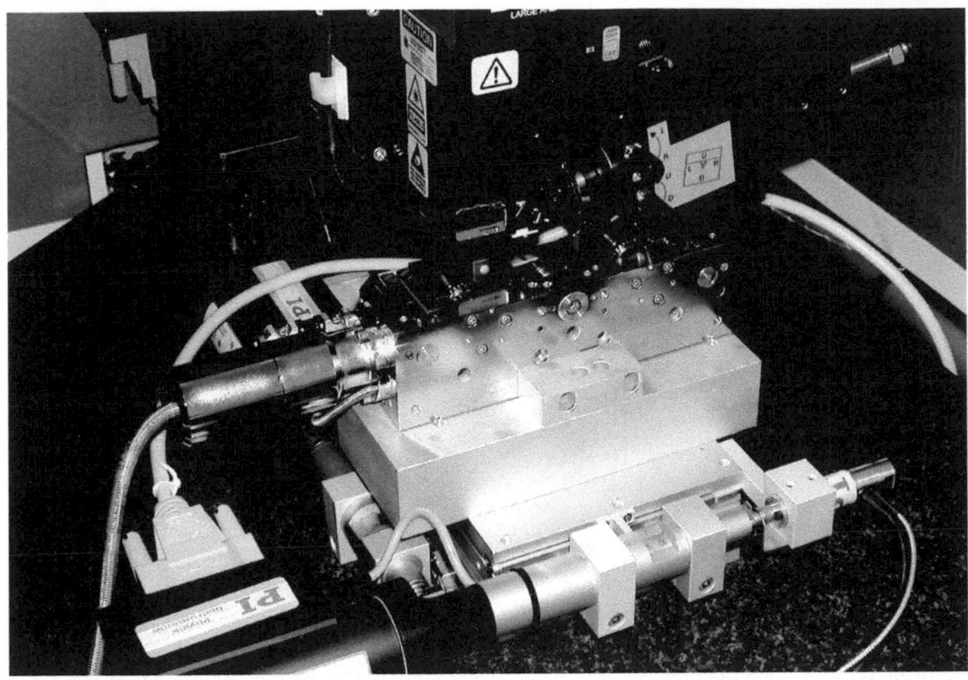

FIGURE 2.12 Three(four)-point bending module placed under an AutoProbe M5 (AFM) system to study defect mechanisms and microcrack evaluation.

In most cases, a DIC measurement system will be specified by the specific user objectives. Different components have to be assembled from different providers. The appropriate selection and mutual adaptation of these components is an important consideration when starting the system.

Figure 2.12 to Figure 2.14 give examples of loading modules integrated in high-resolution microscopes. They have been used by the Fraunhöfer IZM Berlin [46] for deformation analysis on electronics packaging structures, photonics, MEMS, and sensor devices.

The following sections give an account of some considerations referring to microscope and DIC software selection.

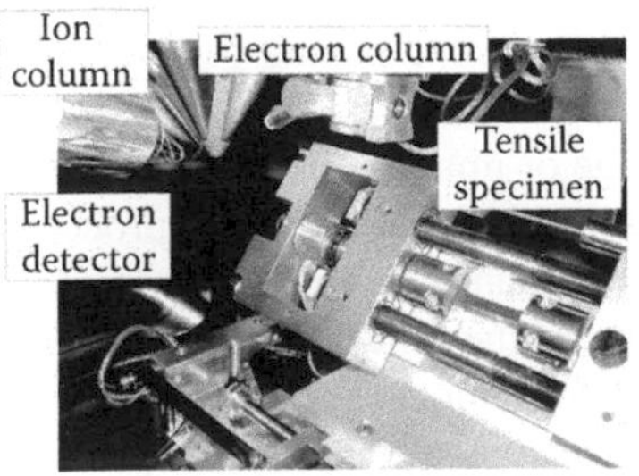

FIGURE 2.13 Inside view of a micro tensile module installed in FIB crossbeam equipment (LEO 1540XB), ion beam inclined 54° to the electron beam (perpendicular to the specimen surface).

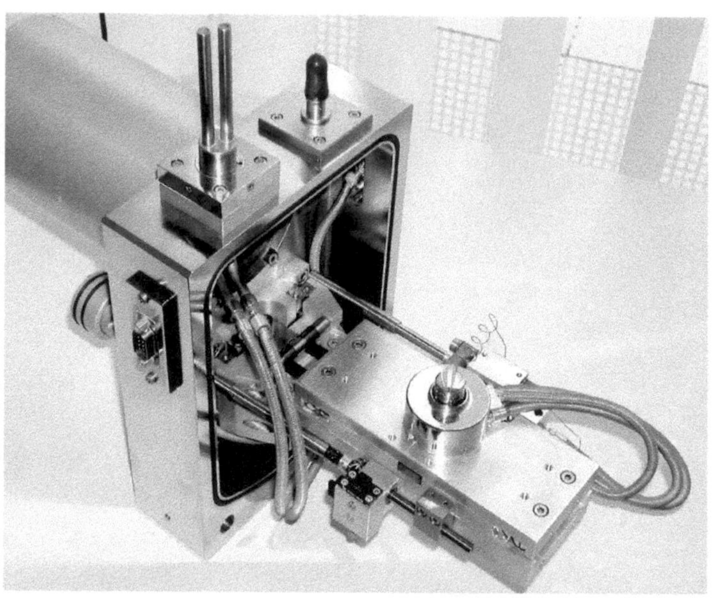

FIGURE 2.14 Contact heating/cooling device for the tensile test module shown in Figure 2.13, developed for use in SEM, AFM, and FIB; temperature range: 30 to 300°C. (Courtesy of Kammrath & Weiss GmbH.)

2.3.2 Requirements for High-Resolution Scanning Microscopes

Microscopes to be used for DIC measurements have to meet basic requirements. They must:

- Cover the range of object sizes under investigation (i.e., the respective magnification values and fields of view)
- Allow for the placement of loading stages with mounted objects inside the microscope
- Provide sufficient image resolution, depicting local image patterns suitable for local image correlation
- Allow for the creation of stable and reproducible micrographs with a low-level image distortion between subsequent images

The last condition can be a crucial demand because higher resolution microscopes utilize scanning-type imaging. Image capture requires a certain amount of time and can take even several minutes in AFM scanning. During the scan time of a load state image, imaging parameters as well as object loading conditions can drift. These drifts cause image distortions, which must be avoided. Otherwise, DIC analysis will result in superposing pseudo deformations on the real object deformation that is being studied.

In fact, time-dependent change of image distortions due to parameter drift is the actual reason for errors in DIC measurements. Drifts in the imaging process itself or in object position can be tolerated in so far as they do not change from one image to another. In the latter case, object micrographs are not a linear copy of the object surface area. But a comparison between a pair of images does not lead to pseudo deformations, because both images have been distorted in the same way. Furthermore, in most cases incremental mutual displacements are to be determined by DIC analysis, so rigid body drifts do not affect the measurement result.

When specifying a DIC-based testing system, both the scanning process drifts and objects shift due to the drift of loading conditions have to be checked separately. In terms of pixel values, drift

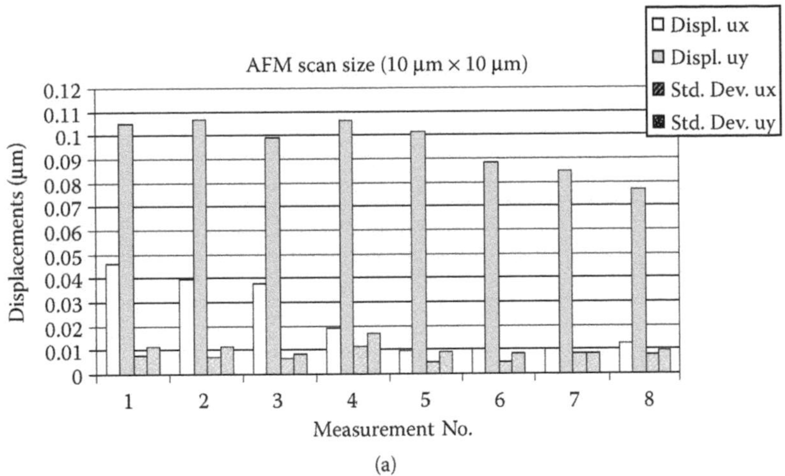

FIGURE 2.15 Evaluation of DIC measurement errors for AFM imaging, u_y, u_x – displacement in line scan and perpendicular to the line scan respectively; measurements on AutoProbe M5 with 256 × 256 pixel images; AFM phase images, comparison of scans captured over time without object loading.

errors should be comparable to or less than errors inherent in the correlation analysis applied to digitized images (see also Subsection 2.2.4). That is, in terms of pixel units, these drift-induced errors should not exceed values of 0.1 to 0.2 pixels.

Figure 2.15 shows an example of respective tests, which have been carried out for AFM imaging to consider the effects of AFM scanner stability. Subsequent scans have been captured without introducing any deformation to the object under investigation. The scans are AFM phase images (Phase Detection Microscopy [PDM]), which are material contrast images. PDM possesses the advantage of simultaneously allocating determined displacement values to material parts, e.g., in compound materials. As seen in the chart of Figure 2.15, displacements between subsequent images are rather high, up to 2 pixels for u_x (displacement perpendicular to the AFM line scan direction). DIC measurement errors with regard to object deformation are represented by u_x gradients only, characterized by the u_x standard deviation. As seen, the u_x standard deviation is much lower because of the rather stable drift of imaging conditions. These stable drifts can be caused, for example, by thermal environmental changes. The determined u_x standard deviation values range from 0.1 to 0.2 pixels, which is a satisfying value. The test results allow the conclusion that AFM imaging itself can be used for the capture of images for later DIC analysis.

2.3.3 Software Tools

The processing of images acquired from different load states comprises various steps. They can be united in one code or be split off, if features from external software codes such as data presentation tools are available and can be utilized. In some cases, DIC analysis software is merged with control software for loading modules as well. The scheme of a possible flowchart for a DIC measurement analysis is given in Figure 2.16.

Currently, different providers offer DIC analysis software for the extraction of deformation fields from grayscale images. Commonly, these software codes import externally captured images

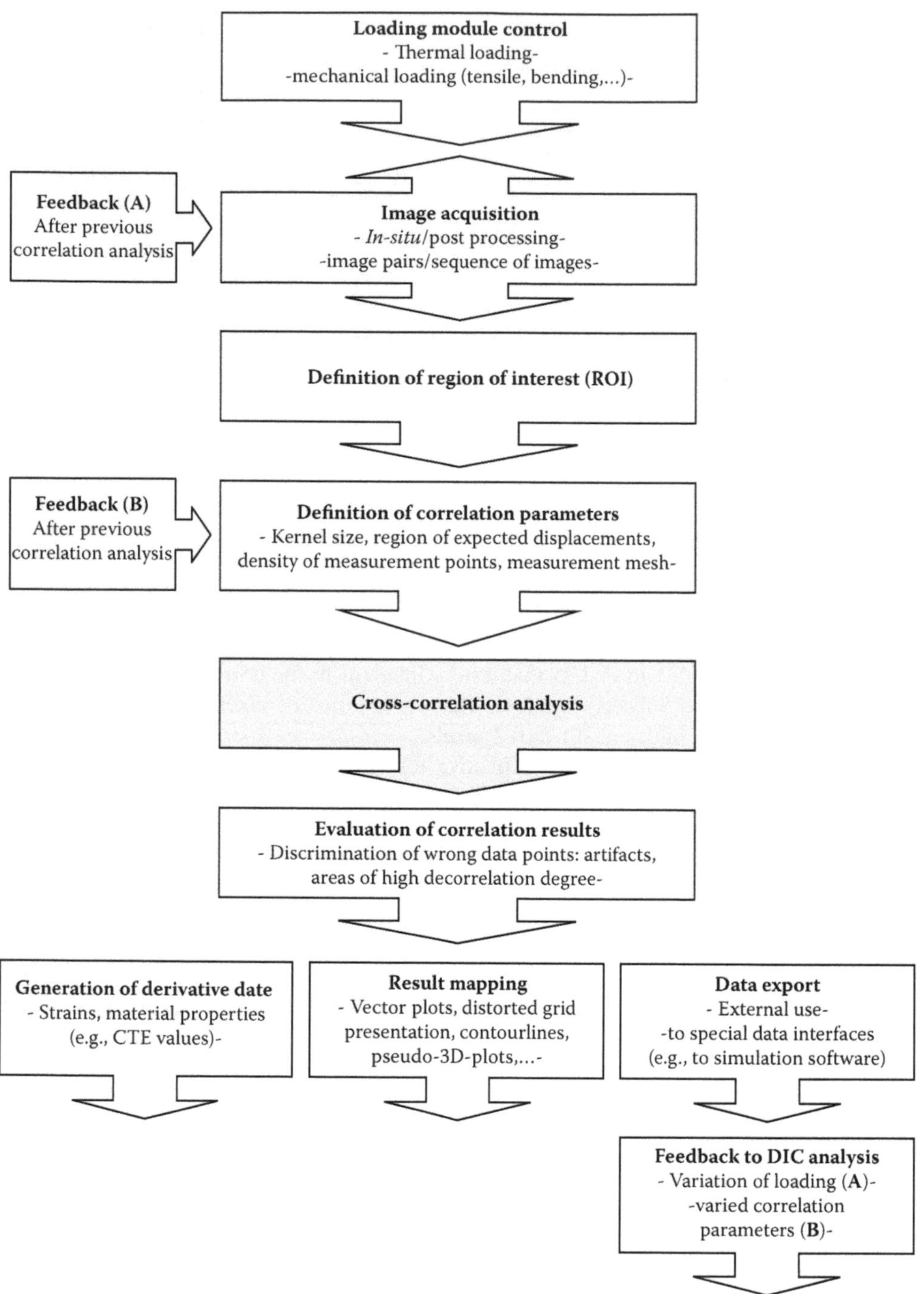

FIGURE 2.16 Principle DIC analysis scheme.

and carry out DIC processing on them. For efficient data generation from load state images, minimal software packages should comprise the following items:

1. ***Variable selection of local pattern size*** (kernel n in Figure 2.2), which is tracked by correlation algorithms. Different objects under investigation may possess different kinds of local, natural, or artificial patterns. Moreover, the imaged pattern type depends strongly on the microscopic equipment used. When analyzing software, kernel size must be tuned for existing object structures in order to obtain optimal measurement resolution.
2. ***Free generation of measurement meshes*** (e.g., for regular grids) of measurement points with a selectable amount of data points. This option is essential for adapting mesh density to the selected kernel size, as well as to the lateral resolution of measurements that is required.
3. ***Masking of object areas.*** Image parts must be masked to perform DIC analysis only where desired and to exclude areas without chance of producing reasonable results. The absence of such options can cause trouble with ample error data points, which must be eliminated using further data-processing steps.
4. ***Algorithms for automatic suppression of incorrect measurement results at single data points.*** Correlation analysis will fail for single data points, especially if some decorrelation of the local object pattern had taken place due to object loading. For that reason, automatic procedures for the elimination of incorrect data points are indispensable. Different approaches can be implemented, e.g., error discrimination by strain constraints, finding the minimum local image contrast, consideration of data neighborhoods, and by cutting off extreme data points in the whole measurement field.
5. ***Manual correction of measurement results for wrong data points.*** Automatic error screening is sometimes not perfect. In this case, single incorrect displacement vectors must be eliminated manually. Otherwise, further treatment, e.g., data smoothing and mapping, generates artifacts in final displacement and strain fields.
6. ***Scaling options for absolute object dimensions from object images.*** This feature allows for the measurement of absolute values of displacements.
7. ***Elementary displacement and strain data mapping.*** This can be realized as vector displacement plots, deformed object overlay grids, contourline maps, etc. It is necessary for a quick measurement overview, for example, as well as for the optional elimination of incorrect data points.
8. ***Data export for further treatment.***

Besides the aforelisted options, some additional features define the value of a software code for DIC analysis. Such options are:

1. ***The derivation of strain fields from originally determined displacement fields.*** Strains, as derivatives of displacements, are more sensitive to measurement noise. The kind of strain computation applied influences the quality of strain data to a large extent.
2. ***Data smoothing.*** Data smoothing in the displacement field has a major impact on derived strain fields, namely on their lateral resolution and accuracy. It can be used to fill in interpolated displacement values where error-type measurement points have been eliminated at an earlier state of treatment.
3. ***Serial image processing.*** In order to follow up entire deformation processes, series of captured load state images have to be evaluated. In many cases, parameters of DIC treatment can be kept constant for all images. So, automatic processing for a complete set of images is efficient in receiving data for a larger measurement process.
4. ***Restriction of correlation analysis to materials or parts of them.*** If kernel pattern and search areas for a kernel (see Subsection 2.2.1) are kept strictly within a defined material or part of it, there are several advantages. Wrong data points commonly obtained near areas of displacement discontinuity (e.g., at material boundaries or crack interfaces) are

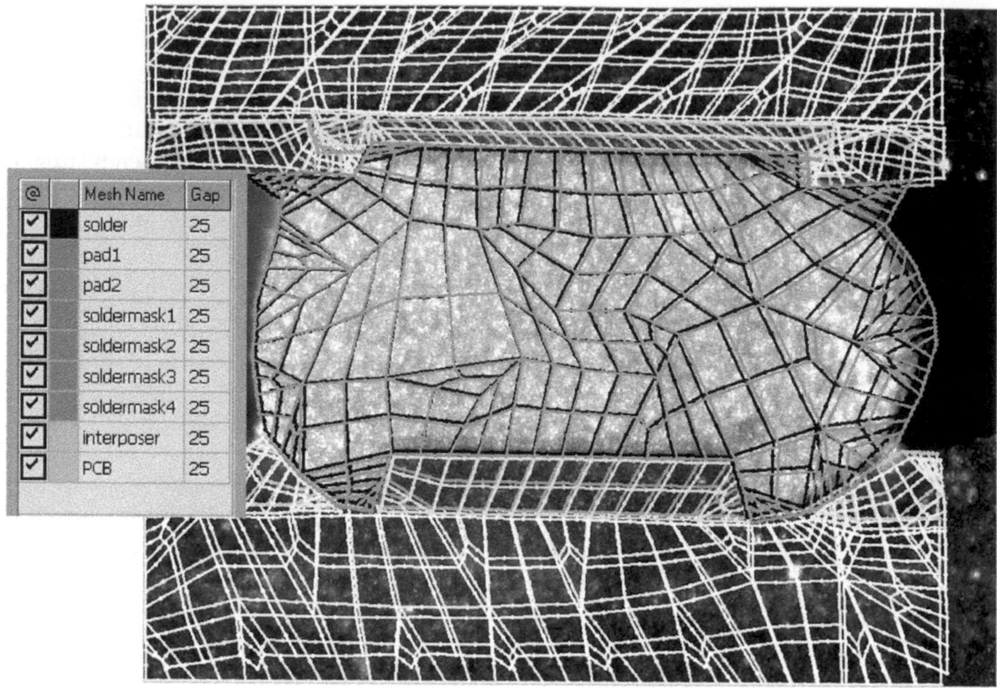

FIGURE 2.17 Deformation measurement on a solder joint, finite element (FE) compatible measurement, undeformed (bright) and deformed (dark) meshes, separate meshes are allocated to different materials. (Courtesy of Image Instruments GmbH.)

avoided that way. Additionally, drawbacks inherent in local correlation analysis, which cause systematic errors in the data field near material boundaries, are suppressed. DIC software solutions exist, which include mesh generators with material allocation. They allow for the creation of multiple measurement meshes for a single object image.

5. ***Data interfaces to finite element software.*** The mutual export and import of data between FEA and DIC software codes is a powerful tool for advanced testing approaches. It also provides better results in both FE and DIC analysis (see Section 2.25).
6. ***Software modules for the determination of derived (material) properties.*** Displacement and strain fields often are used to obtain more complex properties. Their treatment for those purposes can be implemented in DIC software. Tools have been created for the determination of coefficients of thermal expansion or Poisson's ratios (see Subsection 2.2.3).

For example, Figure 2.17 illustrates the features of a finite element type meshing for measurements. Separate measurement meshes are allocated to different materials, which results in the aforementioned advantages.

2.4 APPLICATIONS OF DIC TECHNIQUES TO MICROSYSTEMS

2.4.1 Strain Analysis on Microcomponents

Thermal and mechanical loads on microcomponents are caused by quite different functional or environmental factors. In order to figure out the weak points of devices under load, stresses and strains have to be determined and compared with critical values for a possible component failure.

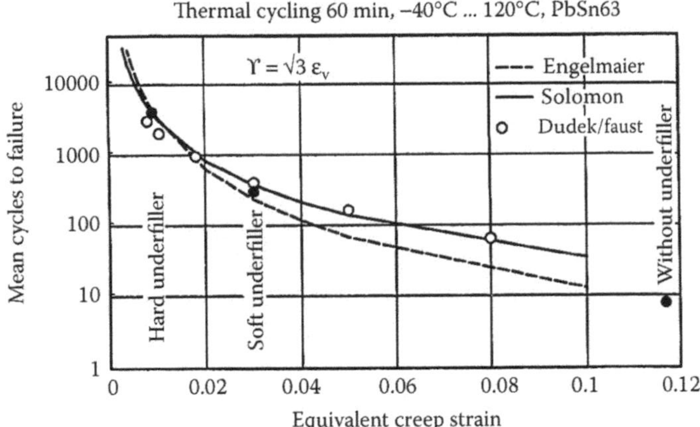

FIGURE 2.18 Dependence of mean thermal cycles to failure on accumulated equivalent creep strain in the solder. (From Dudek, R. and Michel, B., Thermomechanical reliability assessment in SM- and COB-technology by combined experimental and finite element method, in *Proc. Int. Reliability Physics Symp.*, IEEE Catalog No. 94CH3332-4, San Jose, 1994, p. 458).

For example, strain field measurements can be valuable, if creep strain is a major trigger for fatigue, as a dominant failure mechanism. In the past, such stand-alone measurements were widely performed on flip chip solder joints, which are used in the packaging of electronics, MEMS, and sensor devices. Strain measurements for thermal loads (e.g., applied in the accelerated testing of components) allows for the discovery of main mechanisms affecting the reliability of solder connections.

Solder fatigue due to thermal cycling is one of the essential concerns of mechanical reliability in flip chip technology. Accumulated solder creep strains during cycling finally result in the failure of the solder joint. The amount of acceptable strain has been determined in numerous papers [47]. The curve in Figure 2.18 gives the mean cycles to failure of a eutectic Pb–Sn solder, dependent on the accumulated equivalent creep strain. Taking this data into consideration, an accumulated creep strain of approximately 1% is acceptable to reach 2000 cycles to failure. That is, all spatial strain components contributing to the equivalent creep strain have to be kept below this value.

Flip chips (FC) mounted on organic laminates exhibit a large thermal mismatch. Thermal load can lead to high strains and stresses driven by the difference between the coefficients of thermal expansion (CTE) of silicon ($\approx$3 ppm K^{-1}) and the board material (e.g., for FR4 $\approx$ 14 ppm K^{-1}). The reliability of solder bumps against fatigue depends to a large extent on the stress–strain distribution in the whole assembly. In order to understand this, some rough principal schemes of projected global FC deformation due to cooling down the mounted components are shown in Figure 2.19. The scheme (a) means that the thermal mismatch is converted directly into an elongation difference between the substrate and the die. As a result, the outermost bumps undergo a severe shear strain. For a 10 × 10 mm^2 die and a 100 K temperature interval, a 6 to 8 % shear strain over the bump gap is possible. The scheme (b) in Figure 2.19 assumes a common elongation of both die and board, as well as an extremely stiff package with regard to bending. In this case, extremely high tension and compression stresses are likely to exist in the board and die material, respectively. Finally, the scheme (c) implies common bending of silicon and laminate material. Large shear values for outside bumps are avoided. On the other hand, peeling forces tend to strain the bump material in a direction perpendicular to the board plane. In the past, one of the main questions that arose was about what type of deformation actually existed in flip chips on organic substrates. For that purpose, DIC measurements have been used to clarify the main mechanisms of inelastic solder deformation.

To illustrate this, Figure 2.20 and Figure 2.21 show the outermost bump deformation due to heating of the flip chip component for an underfilled and a non-underfilled specimen, respectively.

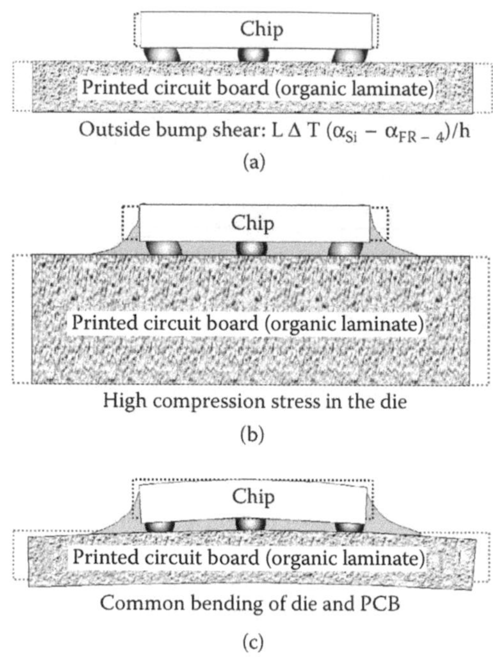

FIGURE 2.19 (a), (b), (c): Principal flip chip deformation schemes for cooling down the whole component.

Non-underfilled flip chips on FR4 have no stress redistribution layer between the board and silicon chip. As a result, the outside bump is forced to perform a significant shear movement. Already for small die sizes (see Figure 2.20 for a 10 × 10 mm^2 sized chip), high shear strains of up to 5% are introduced. In addition, bump stretching of 1.5% in the center occurs. That is, for non-underfilled

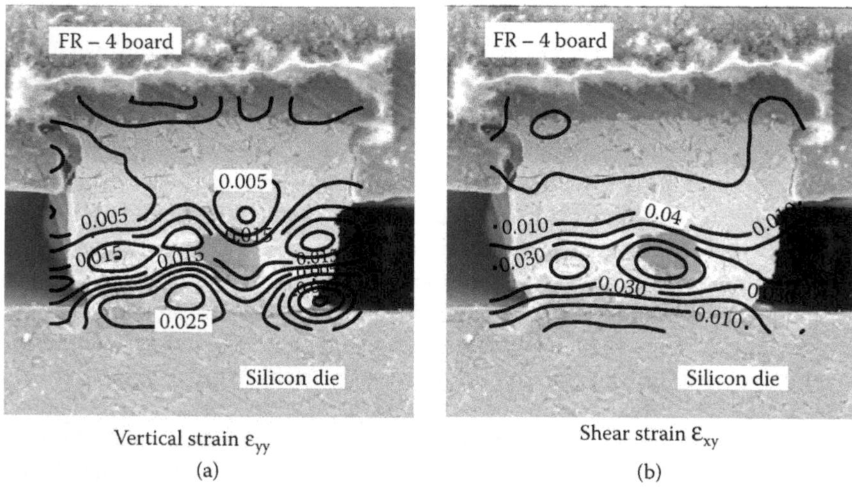

FIGURE 2.20 PbSn solder bump strain field for assembly heating from 25°C to 125°C, strain component perpendicular to board direction, ε_{yy} (a), and in-plane shear strain, ε_{xy} (b), microDAC measurement inside SEM, flip chip ***without underfill***, outermost bump.

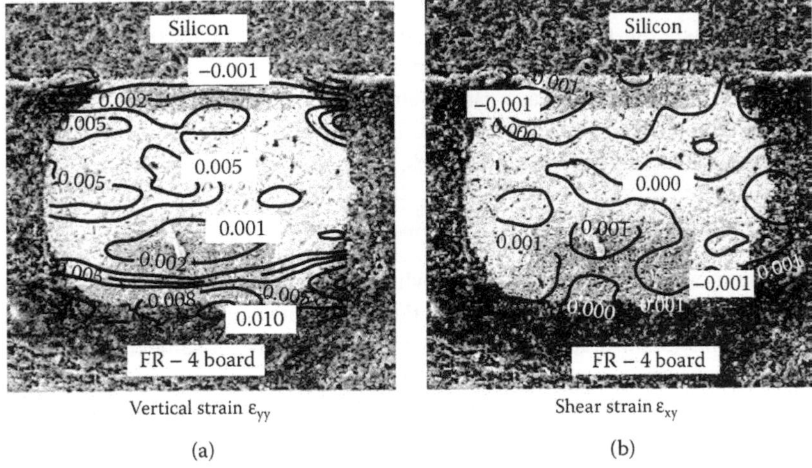

FIGURE 2.21 PbSn solder bump strain field for assembly heating from 25 to 125°C, strain component, ε_{yy}, (a) perpendicular to board and shear strain, ε_{xy}, (b) microDAC measurement inside SEM, flip chip ***with underfill***, outermost bump.

flip chips, the left deformation scheme (a) in Figure 2.19 dominates. Furthermore, slight component bending, as in Figure 2.19c, is added.

In contrast, the data in Figure 2.21 illustrates the typical strain redistribution if underfills are used. The bump in Figure 2.21 originates from a flip chip with a stiff underfill. No significant shear strain exists. Only at bump corners some shear appears (less than 0.3%). The maximum of the strain, e_{yy}, perpendicular to the board plane of approximately 0.5% in the middle of the bump, is quite moderate. Its value is twice the expansion for unrestricted solder material. The strain, e_{xx}, in board plane direction (not shown here), is of nonessential value. Consequently, underfilling makes common board-die bending dominant as in Figure 2.19c, which protects the solder from strains that are too high and consequently from early fatigue.

Figure 2.22 summarizes investigations accomplished on flip chip assemblies with varying underfill content. The epoxy formulation of the underfill, as well as the geometric assembly configuration, have been kept constant for all measurements. The figure shows the average strain

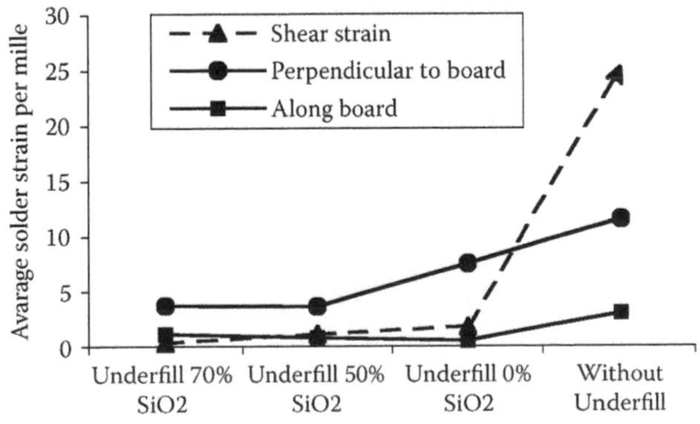

FIGURE 2.22 Dependence of average solder bump strain on filler content in epoxy underfills (corner bumps of a 10 × 10 mm² die).

values determined over the whole bump cross section. As seen, the bump shear is higher for components with a lower die-to-board interface stiffness. Nevertheless, only flip chips without any underfill exhibit remarkable shear. Even poor epoxy underfill results in a shear strain less than the bump strain, perpendicular to the board. Bump stretch along its axis increases by a factor of 2 for nonfilled epoxies compared to those that are 50 to 70% filled. This seems reasonable because of the lower CTE value of the highly filled epoxies.

2.4.2 Defect Detection

Analysis of the thermo-mechanical reliability of whole microsystem devices is required to concentrate on weak spots, where failure behavior is initiated. Powerful finite element simulation studies, including fatigue, fracture mechanics, or damage modeling, can be carried out if some *a priori* knowledge exists on major failure mechanisms and defect location. DIC-type measurements can be useful tools in providing such preliminary information. Deformation fields around component defects and imperfections reveal discontinuities if the latter are under thermal or mechanical load. An equal number of unusually high deformations may indicate weak spots that need to be examined in more detail.

Figure 2.23 gives an example of a defect occurring in the deformation field of a microscopic flip chip solder bump exposed to thermal loading. At the left bump-to-underfill interface, unusually high

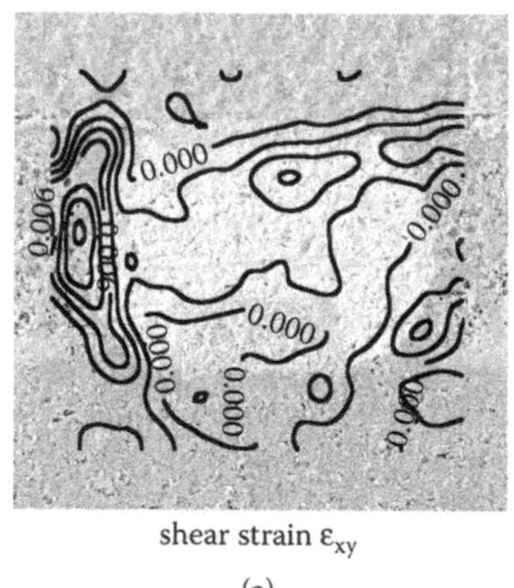

shear strain ε_{xy}

(a)

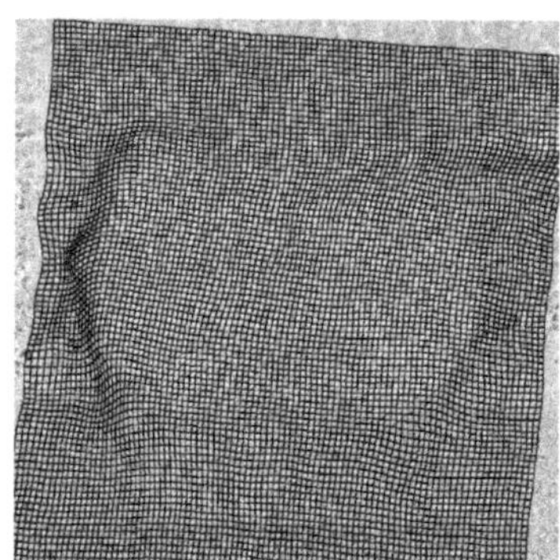

Deformed object grid

(b)

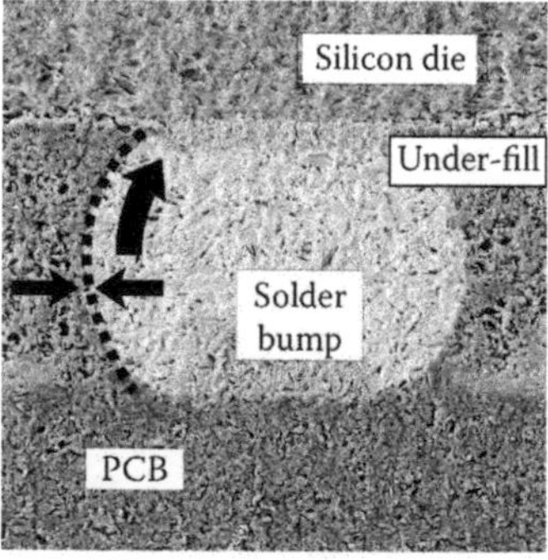

Changes at the interface

(c)

FIGURE 2.23 (a) Shear strain (ε_{xy}); (b) deformed object grid; (c) changes at the interface. PbSn solder bump strain for component heating from room temperature (over 100°C), microDAC measurement inside SEM, delamination at bump/underfill interface detected by unusually high local strains.

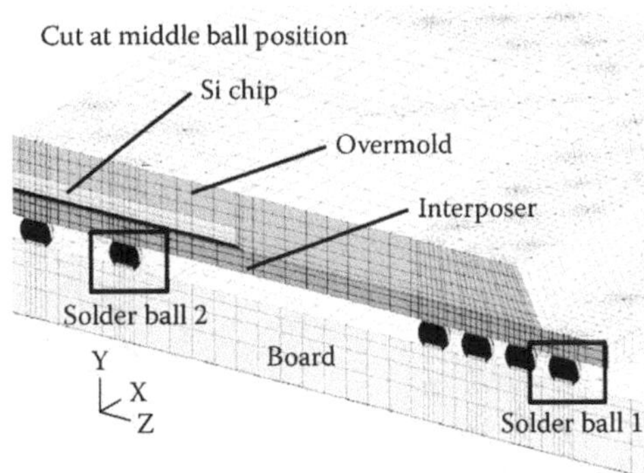

FIGURE 2.24 Finite element model for a BGA assembled on an organic board type.

shear strains appear (Figure 2.23a), which reveal the delamination of the interface. In fact, in Figure 2.23b, the corresponding deformation grid supports this result. Figure 2.23c summarizes the main changes at the interface due to heating. The bump and underfill material slide against each other and the gap between bump and underfill closes when the component is heated.

2.4.3 Validation of Finite Element Modeling

A major benefit of combining FEA and DIC is the feasibility to validate complex FE models. Devices comprising many different materials with rather complex material behavior need an experimental proof of mechanical modeling. Another obstacle comes from material properties that are not sufficiently accurate. This becomes more important as materials are applied to submicron and nanoscopic dimensions. Because of the size effects and the difficulties of material property determination, often unrealistic properties are introduced in FE models. Consequently, the validation of mechanical models by making use of accompanying measurements is strongly desired.

For example, packages such as BGAs and CSPs assembled with organic laminates can sometimes lead to unrealistic models. Subsequent prediction of failure behavior will fail if it is based on such FE modeling. There may be a way to avoid this by comparing simulated and measured deformation fields. Differences in experimental and theoretical results must be eliminated using step-by-step searching and by defining error sources. This is the classic hybrid approach for problem solution. It is illustrated by a thermal analysis performed for a BGA mounted on FR4 board material.

Figure 2.24 shows the prepared, geometrical finite element model. The assembly has been investigated for solder fatigue, which occurs due to thermal cycling. The introduced major global mismatch originates from the low CTE of the silicon die. The overmold CTE is approximately 12–13 ppm K^{-1}. The interposer and board reveal CTE values of about 14 and 12 ppm K^{-1}, respectively.

The microDAC measurements show that all measured deformations are rather moderate for "bulk" solder material. Solder strain components do not exceed values of 0.3%, as seen in the examples of Figures 2.26 and Figure 2.27. The comparison between measured and simulated displacement and strain fields exhibits reasonable agreement for "bulk" solder material (Figure 2.25 to Figure 2.27). The incline of the FE displacement contourlines in Figure 2.25 is due to the bending rotation of the outer BGA part. The respective measured displacement field has been derotated, i.e., was compensated for rigid body rotation. However, the amount of vertical displacement increment through ball height is approximately equal for both results. Deduced strain values in Figure 2.26 confirm these findings.

Considering the entire ball deformation of the outermost functional solder ball 1, no basic discrepancy between simulation and measurement can be found. In contrast, the inner thermal solder ball shows severe differences for the deformation of the solder corner areas at the interposer site. This region has been marked by an arrow in Figure 2.27. Measurements reveal a high incidence

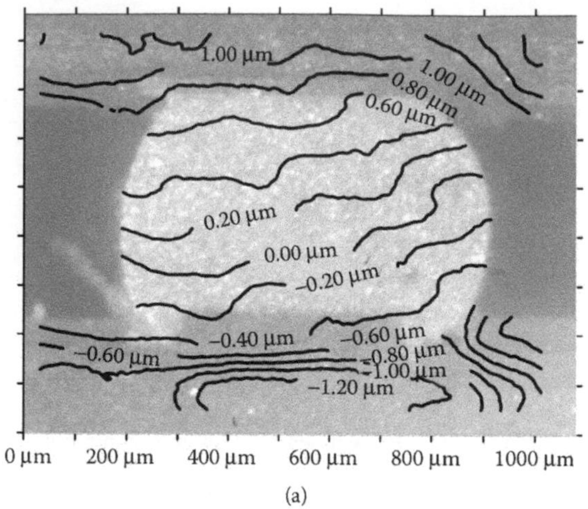

(a)

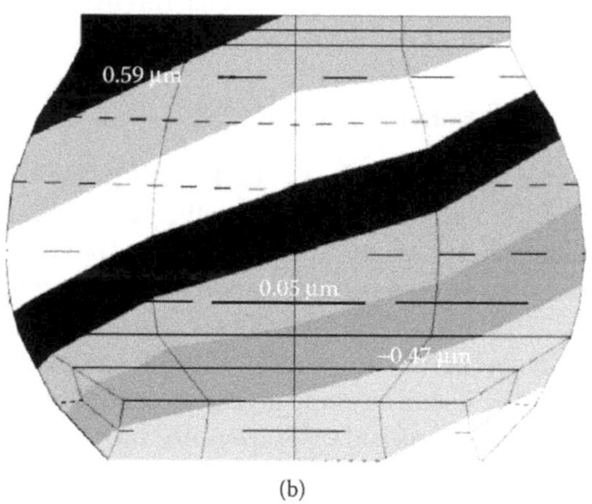

(b)

FIGURE 2.25 Vertical displacement fields, u_y, of outermost functional SnAgCu solder ball (solder ball 1 in Figure 2.24) for heating up the BGA from room temperature to 118°C; (a) DIC measurement; (b) FE simulation, incline of contourlines in the FE result due to BGA bending rotation (measurement result: derotated).

of tensile (Figure 2.27) and shear strains (Figure 2.28), whereas simulations do not. As a consequence, finite element modeling does not adequately describe the real behavior of the inner solder ball. The reason for this discrepancy could have been discovered as an incorrect geometrical modeling for FE simulations. Real thermal solder balls show significant solder overlapping with the solder mask (see Figure 2.28). The extremely high thermal expansion of the solder stop material acts as an expanding wedge, which intrudes into the solder material as the BGA is heated. Such forces lead to severe tensile and shear creep strains at the solder corners, and finally the lifespan of the solder ball is reduced due to material fatigue. Numerical material fatigue analysis, for example, the Coffin Manson approach (see Figure 2.18), should take local geometry at solder corners into account, if it is to succeed. This was what was observed by comparing finite element and measurement results for the same problem.

2.4.4 Measurement of Material Properties

This section includes the determination of material properties by DIC methods, which are based on averaging strains over a certain area in order to enhance measurement accuracy (see also

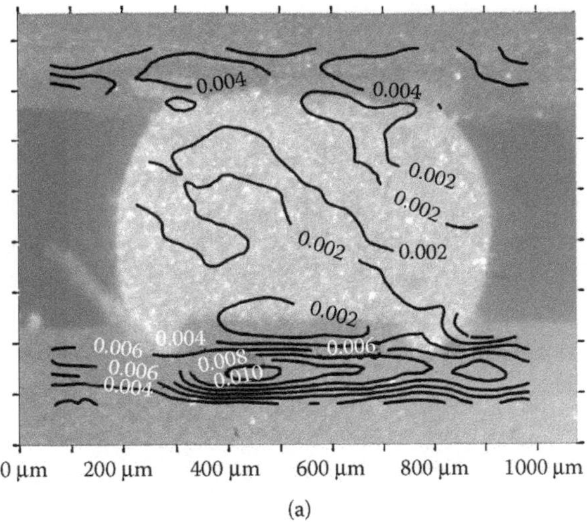

(a)

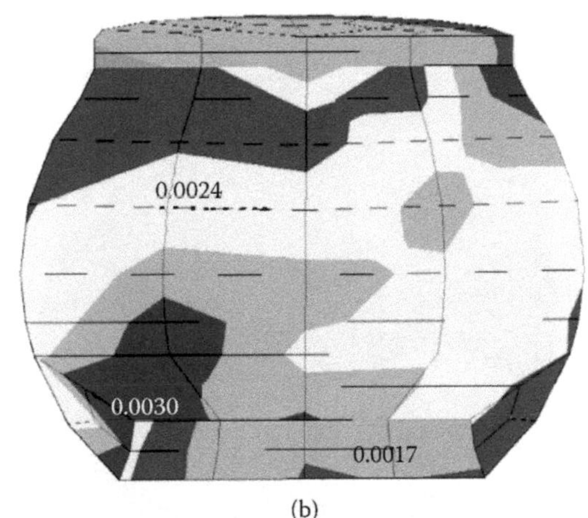

(b)

FIGURE 2.26 Vertical strain fields, ε_{yy}, of outermost functional SnAgCu solder ball (solder ball 1 in Figure 2.24) for heating up the BGA from room temperature to 118°C: (a) DIC measurement; (b) FE simulation.

Subsection 2.2.3). This modified DIC technique is illustrated for the measurement of CTE values and Poisson's ratios.

The first example corresponds to CTE values to be measured on polymers applied to the packaging of microsystems. So, appropriate selection of material properties for flip chip underfills in electronics is essential for thermo-mechanical reliability. As a rule, CTE values should be not too far from that of the applied solder material. Otherwise, very high local thermal mismatches can cause early component failure. With regard to this, it is of interest to note whether or not underfill CTEs depend on chip-to-board gap width. Furthermore, it is necessary to determine to what extent sample preparation can influence measured CTE data. Classical thermo-mechanical analysis (TMA) is often not possible with commercially available equipment or, as will be shown later, can fail for thin foils. For that reason, alternative methods are being used for these measurements.

Figure 2.29a shows a corresponding CTE measurement carried out with the help of a modified DIC tool. The analyzed underfill was a filled epoxy with 70% by weight of SiO_2. Filler content is added to match material properties such as CTE value and Young's modulus. Sample thickness was altered in the present case between 85 μm and 1000 μm.

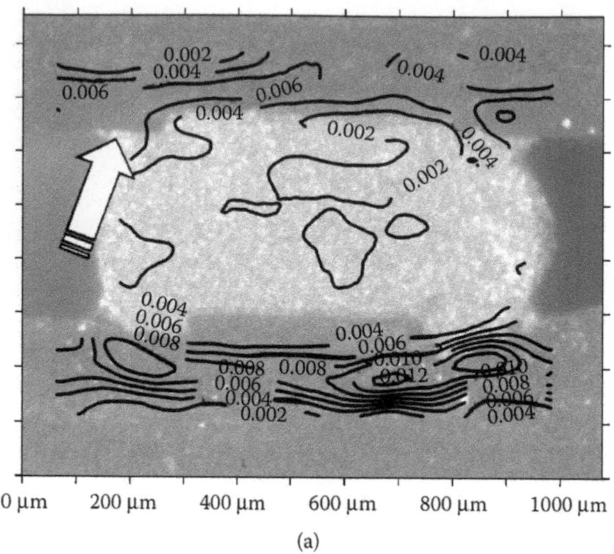

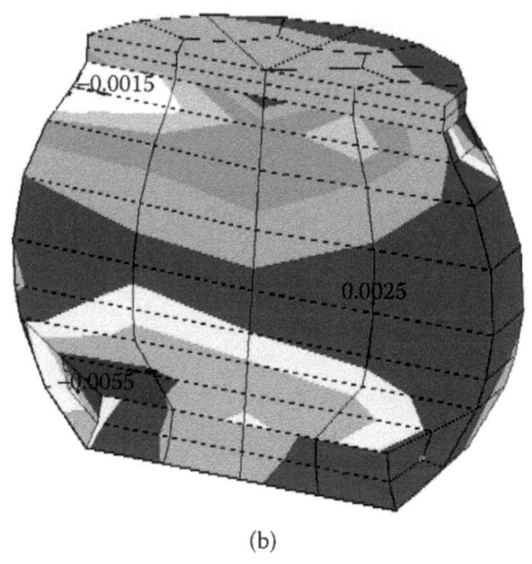

FIGURE 2.27 Vertical strain fields, ε_{yy}, of inner thermal SnAgCu solder ball (solder ball 2 in Figure 2.24) for heating up the BGA from room temperature to 119°C: (a) DIC measurement; (b) FE simulation. White arrow: discrepancy between measurement and simulation (high tensile strains in measurement, low compressive strains in simulation).

CTE values have been found from a sequence of specimen micrographs picked up with a magnification of 2.5, on a CCD camera chip. That is, CTE data has been sampled and averaged over an area of approximately 2.5 × 2.5 mm^2. Independent measurements were performed in the two in-plane directions of the material foils.

The determined CTE values as a function of foil thickness are given in Figure 2.29b. As seen from the figure, most of the CTE values exhibit slight anisotropy. The differences of the values in the x and y directions are mostly beyond measurement error. The anisotropy could be the result of local variations in filler distribution or slight changes in the epoxy constitution. Within the analyzed thickness range, no significant CTE dependence on foil thickness could have been recognized.

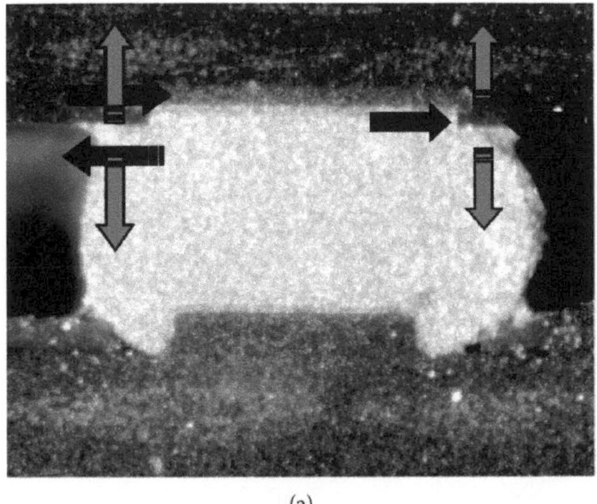

(a)

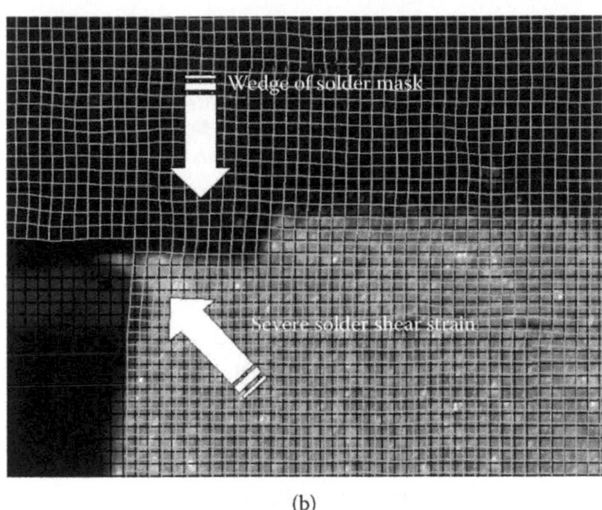

(b)

FIGURE 2.28 Thermal SnAgCu solder ball on BGA package under thermal load: (a) experimentally found main strain schemes at the solder corners on interposer site; (b) severe shear strain at the solder corner (dark mesh: undeformed solder; bright mesh: enlarged deformation due to heating).

Many of the materials applied in microtechnology are plastics. Often they are not completely homogeneous, i.e., they are either compound materials with swaying filler distribution or the material properties of the plastic alters over the specimen size as a result of frozen stresses, nonhomogeneous curing states, and so on. This is why the bending and warping of material specimens during measurement can occur. Consequently, CTE data depends on the kind of specimen and the measurement approach. As a rule, classical TMA measurements can be very uncertain, as demonstrated in the following examples.

Once again, when considering flip chip underfill materials, filler settling during sample or component preparation is a well-known phenomenon. Filler content variation over the underfill gap results in a gradient of material properties. In this case, uncritical TMA measurements can result in completely incorrect CTE values. Figure 2.30 shows schematically the measured thermal expansion of a specimen with underfill sedimentation.

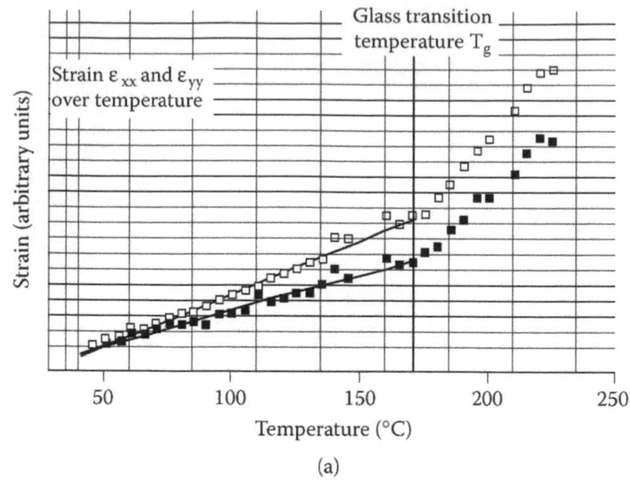

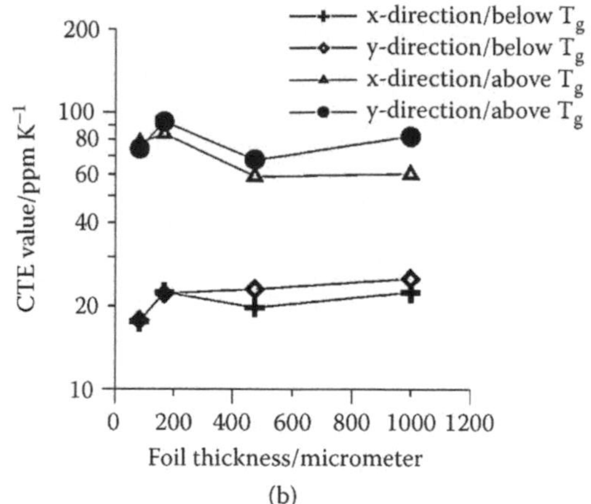

FIGURE 2.29 CTE measurement using a DIC algorithm: (a) typical strain vs. temperature curve for filled epoxy material (two perpendicular strain components); (b) dependency of CTE values on epoxy thickness.

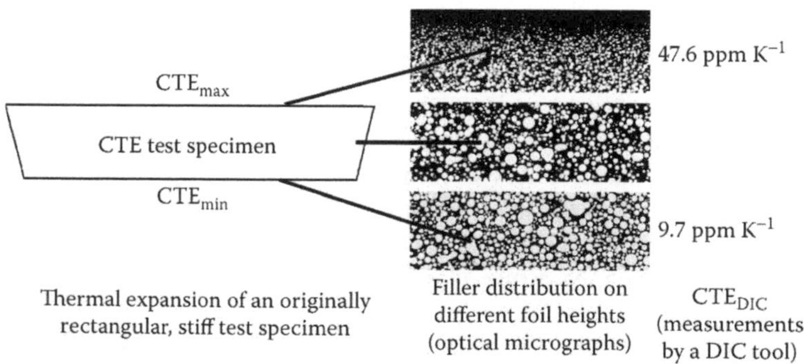

FIGURE 2.30 Thermal expansion of an underfill material foil with filler settling (low filler content at top, high filler content at bottom); CTE_{DIC} values determined at the separated underfill foil (surface measurements); CTE_{TMA} for the same specimen: 34.6 ppm K^{-1}.

For the situation illustrated by Figure 2.30, measurements with TMA and DIC on optical micrographs have been performed on the same specimen. The high degree of filler sedimentation causes a significant change in thermal expansion from the top to the bottom of the specimen. DIC measurements at both surfaces reveal a nearly 500% alteration of thermal expansion behavior.

In contrast, a classical TMA measurement has yielded a CTE value in between the DIC's top and bottom values (Figure 2.30). The specimen was clamped from the narrow sides. It was expected, in this case, that TMA should result in a value equal to the DIC value determined for the top of the foil. Obviously, a slight bending of the whole specimen has reduced this value. That is, finally an unpredictable value below CTE_{max} at the top of the specimen is recorded by TMA. The only way out of this situation is to prepare homogeneous samples or to measure with alternate tools such as the proposed DIC technique, which avoids unpredictable measurements errors.

Consequently, DIC-based CTE measurements have the advantage of evaluating CTE gradients over specimen height. This would be very helpful if, for example, a finite element stress analysis for flip chips has to be performed. Settling of underfill with respect to its consequences on material property gradients has to take into account for adequate problem modeling. It can be evaluated from optical CTE measurements on both upper and lower specimen surfaces.

Moreover, CTE measurement using the correlation technique is based on strains found simultaneously for two perpendicular directions. As a consequence, materials with anisotropic CTE behavior can be analyzed in a time-saving manner using this method. Apart from this advantage, it may be necessary to measure tiny specimens. In many cases, thermally anisotropic materials cannot be measured in both directions with sufficient specimen size for classical dilatometry. Finally, CTE values and CTE isotropy can change over specimens, which makes local two-dimensional CTE measurements indispensable.

Achievable measurement accuracy using automatically running systems has been determined with the help of materials of known CTE [48]. Figure 2.31 shows an example of strain vs. temperature curves obtained for silver and aluminum specimens. Taking into consideration Table 2.2 that has a choice of experimental CTE values determined on 1024 × 1024 pixel micrographs, the overall accuracy of a CTE measurement can be roughly estimated. Under these conditions, an accuracy of approximately ±0.5 ppm K^{-1} is possible, depending on the kind of image structuring. Consequently, the suggested method is suitable for materials with medium and high CTE values.

An example of the evaluation of Poisson's ratios is shown in Figure 2.32. Image capture from the specimen in the testing machine is carried out at predefined time intervals. For the determination

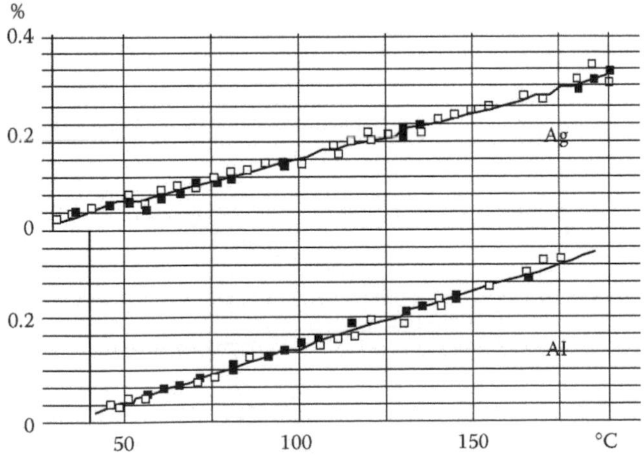

FIGURE 2.31 Average strain vs. temperature curves.

TABLE 2.2
Accuracy Estimation of the CTE Measurement Using the DIC Technique

	Al	Ag
CTE values from DIC measurement [ppm K^{-1}]	23.6	18.4
	23.7	18.4
	23.2	18.4
	23.0	19.5
		19.4
		18.7
Material property from references [ppm k^{-1}]	23.8	18.7

of Poisson values the ratio of the inclines of the linear fitting curves of the two perpendicular strains, is used. The figure demonstrates a measurement result for a filled epoxy encapsulant. Measurement was performed for standard conditions on a conventional testing machine.

2.4.5 Microcrack Evaluation

Tiny defects or cracks present in microsystems components can lead to severe crack propagation and complete failure if devices are stressed. Intrinsic stress sources, such as thermal material mismatches, and changes in environmental conditions (temperature, pressure, mechanical vibrations) can initiate fatal crack propagation. Experimental crack detection can be a crucial issue having in mind original crack sizes of about some micrometer. These cracks will open under a subcritical load of some ten nanometers or less. Their detection using DIC displacement measurements is possible if the highest resolution microscopes are utilized. Figure 2.33 shows as an estimation made by finite element simulation for a typical case. The crack opening displacement of the microcrack in the chip passivation layer is as small as 20 nm for the critical load of silicon chip bending. Figure 2.34 shows an example of potential initial cracks, which are introduced, e.g., in silicon chip edges after wafer dicing.

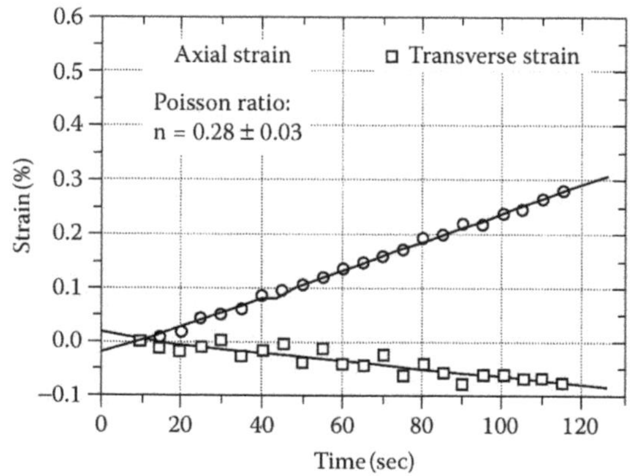

FIGURE 2.32 Poisson's ratio measurement from a modified DIC algorithm; computation of Poisson's ratio from the incline of the two strain curves vs. load time; standard dog bone specimen.

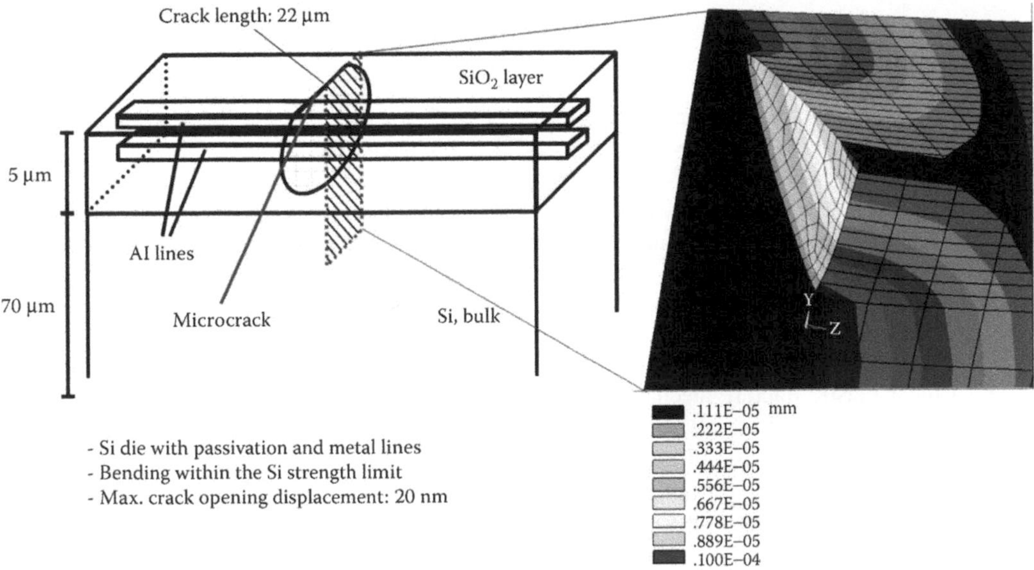

FIGURE 2.33 FEA for a microcrack opening; crack location inside a chip passivation layer.

Furthermore, the determination of real crack load in stressed components or devices can be a tough job if ample data on material laws and properties required for finite element analysis are uncertain. In this case, it could be easier to determine fracture parameters from crack opening displacements where material data is only needed for adjacent crack materials. Taking into consideration the aforementioned problems, attempts have been made to measure the parameters of simple fracture criteria such as stress intensity factors. Corresponding micrographs were captured by SFMs or high-resolution SEMs. Results for AFM migrographs follow in greater detail.

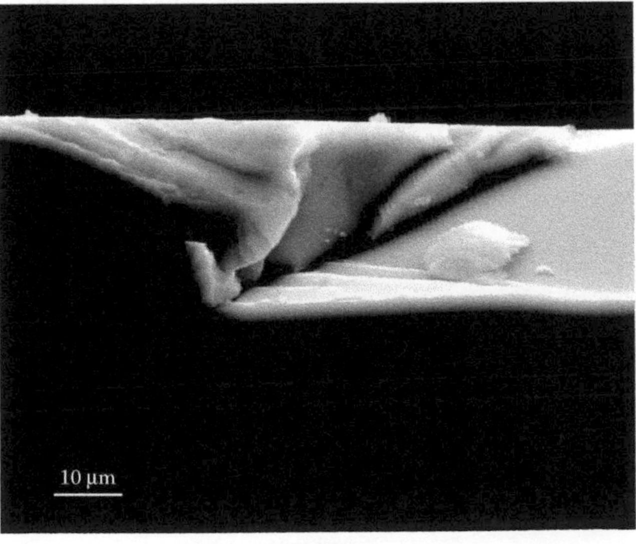

FIGURE 2.34 Example of initial cracks occurring after wafer dicing at die edges.

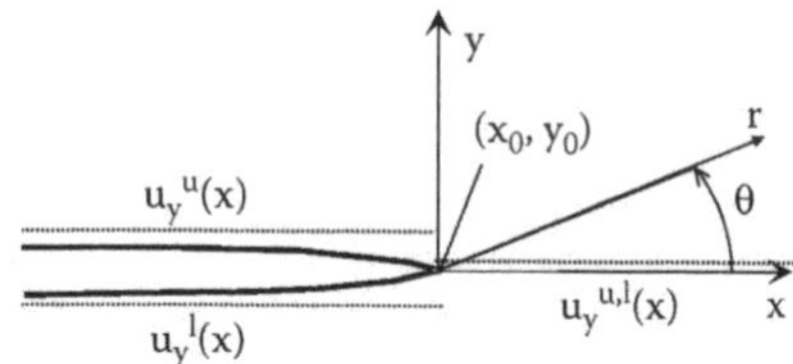

FIGURE 2.35 Scheme of crack opening displacement for infinite bulk material and mode I crack opening.

The reported DIC measurements of stress intensity factors were achieved by making the following assumptions:

- Linear elastic fracture mechanics (LEFM) can be applied within the measurement area and for the applied loads.
- The specimen consists of homogeneous material.
- Mode I and II mixing is allowed when loading the component/specimen.
- The crack is parallel to the horizontal *x*-axis.

In order to determine mode I stress intensity factor crack opening displacements, u_y^u and u_y^l have been measured along both the upper and lower crack boundaries. If determined by LEFM, they must equal the values in Equation 2.5, taking into consideration conventions made in Figure 2.35.

$$u_y^u = \frac{K_I}{2\mu}\sqrt{\frac{x}{2\pi}}(k+1), \quad u_y^l = -\frac{K_I}{2\mu}\sqrt{\frac{x}{2\pi}}(k+1) \quad \text{for } x \le 0$$

$$u_y^u = u_y^l = 0 \quad \text{for } x > 0 \tag{2.5}$$

In Equation 2.5, E is Young's modulus, ν is Poisson's ratio, K_I is the stress intensity factor, and k is a function of elastic material properties, which differs somewhat for plane stress or in the plane strain state [49]. Taking the square of the difference of upper and lower displacements, we obtain the linear function of the *x*-coordinate or 0, depending on which side of the crack tip we are:

$$\left(\frac{u_y^u - u_y^l}{2}\right)^2 = Cx \quad x \le 0$$

$$= 0 \quad x > 0 \tag{2.6}$$

Equation 2.6 does not change even if specimen rotation due to load is included. In this case, equal rotational terms are subtracted from each other. For the aforementioned formula, the crack tip is set at $x = 0$. The crack tip position on the real specimen can be found by intercepting the linear fit of the curve Cx with the *x*-coordinate axis. The incline, C, allows for the estimation of the stress intensity factor, K_I, which is a measure of the crack tip load:

$$K_I = \frac{E}{1+\nu}\frac{1}{k+1}\sqrt{2\pi C} \tag{2.7}$$

Examples of AFM images captured under mechanical load are given in Figure 2.36.

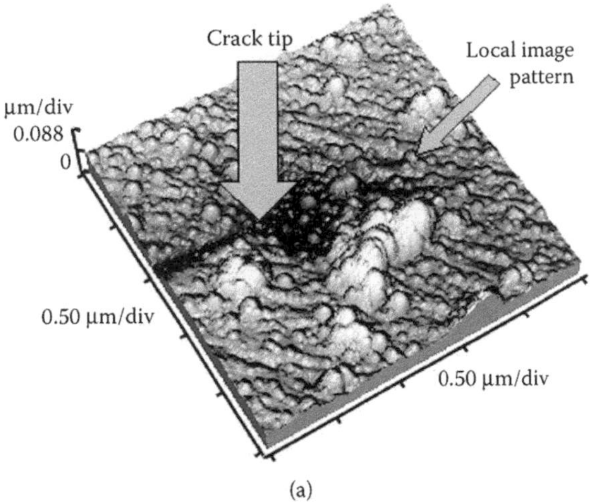

(a)

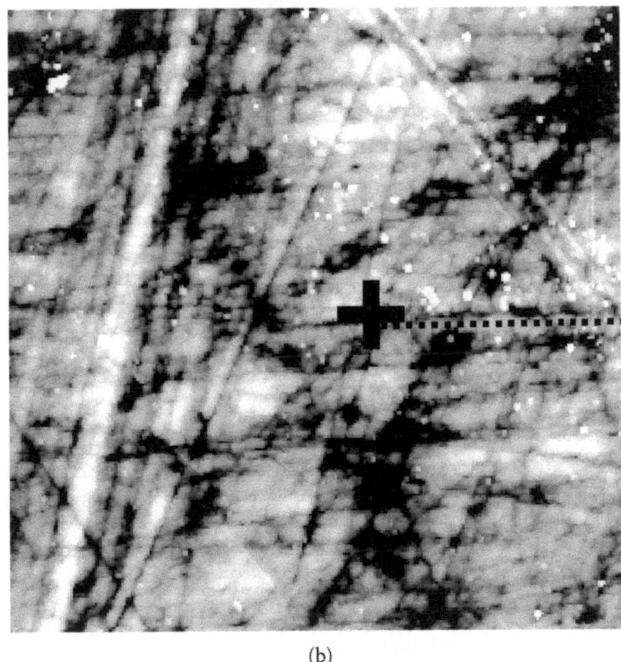
(b)

FIGURE 2.36 Examples of typical AFM scans with crack used for nanoDAC analysis: (a) AFM topography scan (size: 25×25 μm^2), and (b) crack tip position determined from the displacement fields (+) with crack boundary line (dashed line).

Figure 2.37 shows the ideal (theoretical) and measured u_y displacement fields. Maximum crack opening in the measurement area is only about 20 nm (at the area edge). The stress intensity factor was estimated from the measured displacement field as $K_I = 0.0098$ MPa m$^{1/2}$. Comparing this value with the critical stress intensity factor of the material, $K_{IC} \approx 0.6$ MPa m$^{1/2}$, it should be mentioned that the applied crack load was about $\frac{1}{60}$ of the critical value. The studied crack was characterized by a particular mode mixity. The stress intensity factor, K_{II}, for the Mode II crack

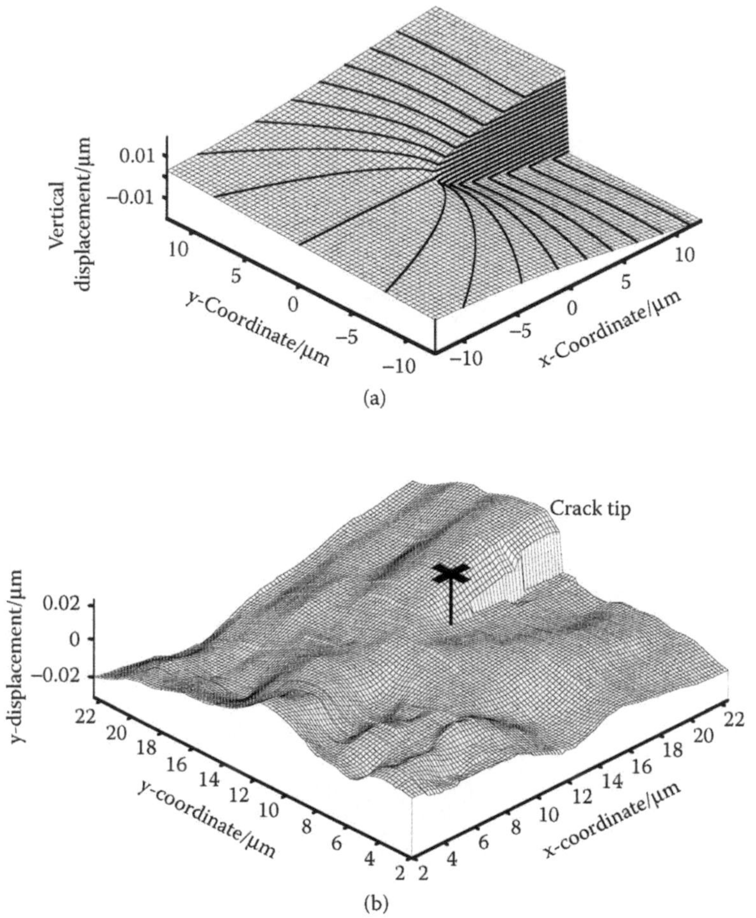

FIGURE 2.37 Displacement measurement from AFM images at the crack tip position, displacement fields, u_y (component perpendicular to the crack boundaries): (a) theoretic field from LEFM analytical solution for Mode I crack opening; (b) measured from AFM topography scans (25 μm × 25 μm image size).

opening has also been measured by making use of u_x values (displacements in crack direction) along the crack boundaries.

Figure 2.38 shows an example of incremental stress intensity factors determined from crack opening displacements computed using SFM images, and Equation 2.6 and Equation 2.7.

Materialized fracture toughness values, to a large extent, depend on material damage mechanisms in the highly stressed, submicron crack tip region. So nanoscale-modified or -filled composite materials, or nanoalloys, are developed to improve thermo-mechanical material properties. Nano-DAC deformation measurements are one tool that allows for phenomenological studies of material damage behavior in these materials.

2.4.6 3-D Deformation Analysis Based on AFM Micrographs

Thin layers used in sensor and MEMS technology undergo local stresses, remote from elastic material behavior, where permanent device alterations are feared after each load cycle. Today, the responses of submicron and nanomaterials to applied external loads (from temperature, vibrations, or chemical agents) are still not well understood. The same is true for actual failure mechanisms and damage behavior. Moreover, quite often those structures escape from simple continuum

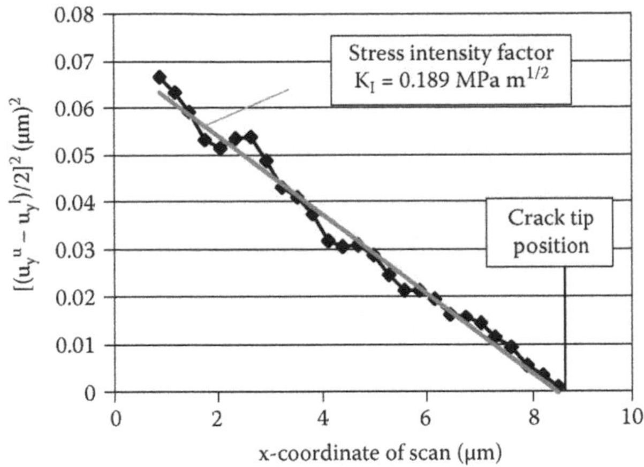

FIGURE 2.38 Determination of stress intensity factors from crack tip displacement fields, obtained by AFM imaging.

mechanics, and also from failure modeling and from numerical simulation because material properties on micro- and nanoscale are not yet known or described in available databases. Furthermore, they may change over the exploitation time. Due to these factors, efforts have been made to gain a better understanding of the material responses in submicron and nanoregions [29,50].

The way to achieve this is by combining displacement and strain measurements at micro- and nanoscale with modeling techniques based on FEA. Parameterized finite element models of MEMS are applied for faster prediction of lifespan and failure modes. The parameterization allows for the variation of model geometries and materials in order to accelerate the MEMS design process [51].

For example, sensor applications with local temperature regulation such as the gas sensor shown in Figure 2.39 and Figure 2.40 are usually thermally loaded with rapid and frequent changes in temperature. This thermal cycling and the temperature gradient over the structure imply thermal stresses and may cause the failure of components [53]. As a result, a hillock appearance and the complete layer destruction of electrode materials are observed (see Figure 2.41). In the operation mode of the reported gas sensors, thermal stresses are induced due to the activated PolySi microheater.

With *in situ* AFM measurements on this microsystem, material deformations originating from the mismatch of material properties have been investigated. The height information of the AFM topography images before and after loading is analyzed for the evaluation of movements or deformations in the *z* direction. *In situ* measurements of thermal deformations by AFM on the top of the sensor membrane have revealed a high value for the remaining deformations even after a single heat cycle (25 → 100 → 25°C). Inelastic strains remain after cooling to room temperature (Figure 2.42). They can be responsible for more severe layer deterioration if higher temperature pitch is applied.

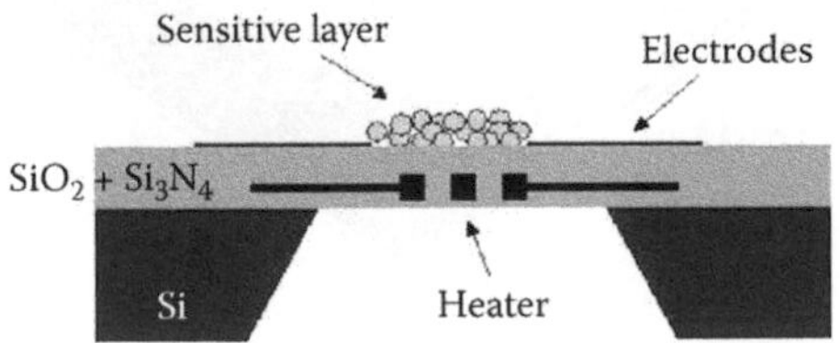

FIGURE 2.39 Layout of a gas sensor.

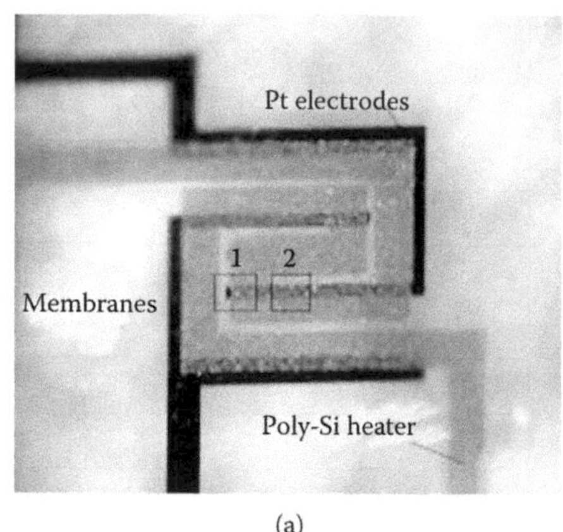

(a)

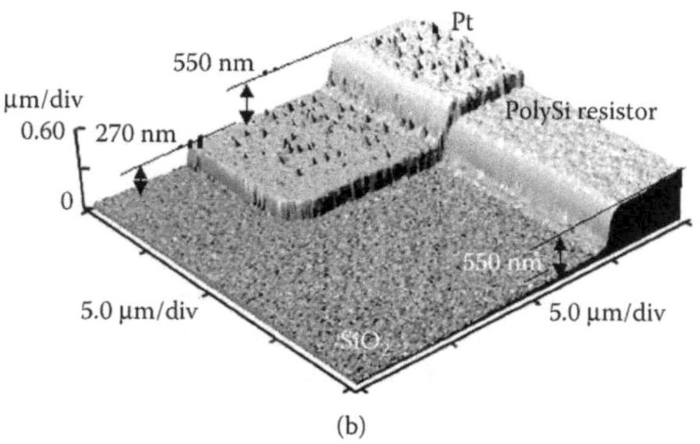

(b)

FIGURE 2.40 Gas sensor membrane: (a) microscopic image, overall membrane thickness approximately 2 μm, field of view approximately 500 μm, (b) SFM topography scan of the gas sensor depicting the Pt layer on top of the SiO_2 membrane and part of the Poly-Si heater embedded. (*Source:* Puigcorbé, J. et al., AFM analysis of high temperature degradation in MEMS elements, in *Proc. of Micro Mechanics*, Sinaia, 2002.)

Figure 2.43 shows the in-plane displacement field achieved by using DIC analysis to compare two AFM scans (captured from the same part of the membrane as in Figure 2.42). It can be seen that the platinum electrode expands toward the uncoated SiO_2 membrane base. Obviously, this is due to the significantly larger CTE value of Pt when compared with SiO_2.

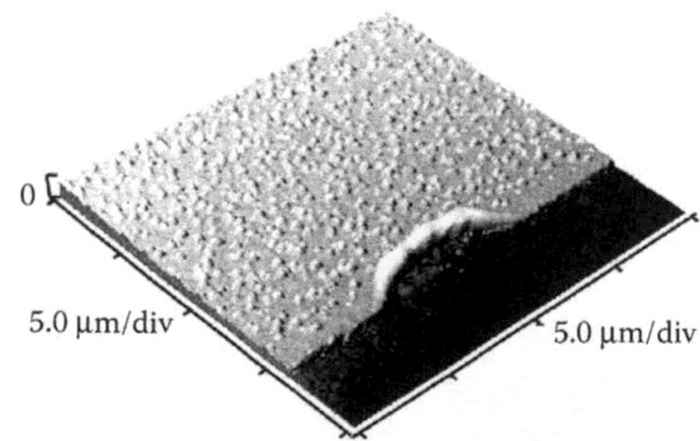

FIGURE 2.41 AFM topography scan of membrane layers after tempering at 450°C, Pt electrode destruction at edge and corners, and a hillock appearance.

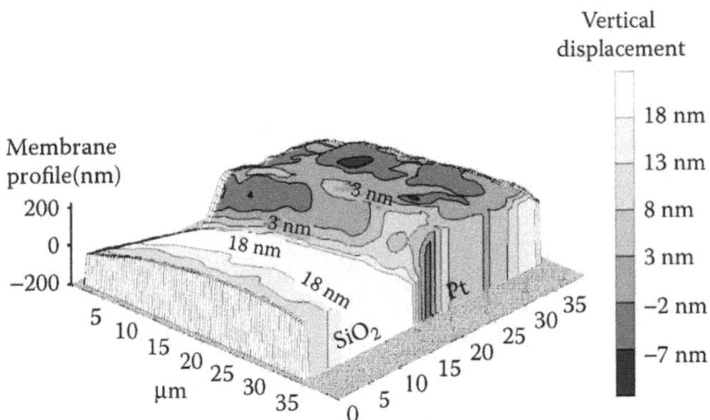

FIGURE 2.42 Residual sensor deformation after a heat cycle (AFM-based deformation measurement), 3-D plot represents part of the membrane layer profile, the coloring (gray scale) indicates the remaining vertical deformation after a heat cycle.

2.4.7 Determination of Residual Stresses in Microcomponents

In order to meet the future demands of micro- and nanotechnology, DIC and related methods have been extended by different authors to focused ion beam (FIB) equipment. The utilization of FIB-based measurements opens possibilities for the processing of microspecimens by ion milling and for the measurement of deformations on them, using the same microscope. The advantages occur in the incorporation of specimen preparation (ion milling, ion beam surface polishing, and DIC patterning), specimen loading by ion milling, and in the DIC deformation measurement in a single equipment. Some initial results are presented for the measurement of deformations caused by the release of residual stresses due to ion milling. This work aims at determining the very local residual stresses on manufactured components such as MEMS and sensors.

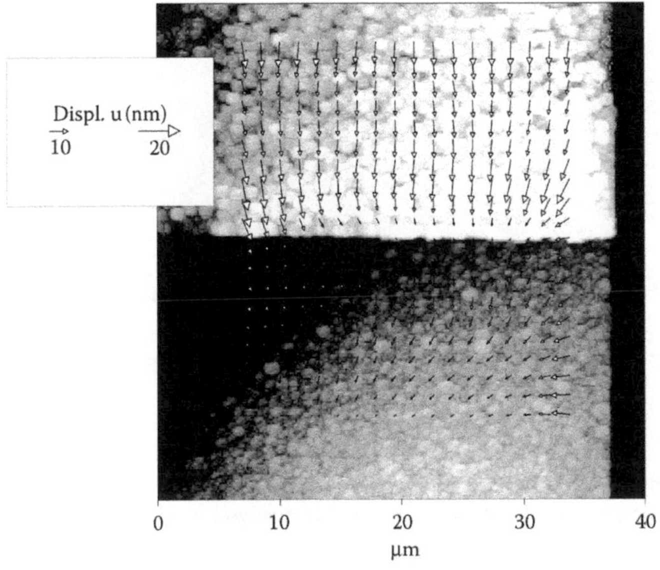

FIGURE 2.43 Displacement vectors determined by a DIC measurement, deformation caused by heating the membrane from room temperature to 100°C (upper bright material: Pt on top of SiO_2; lower dark material: SiO_2 substrate).

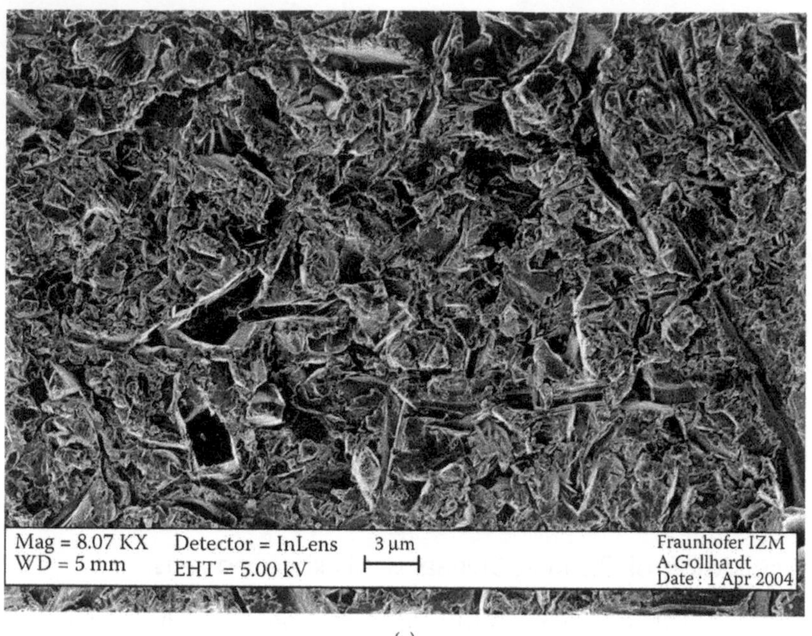

(a)

(b)

FIGURE 2.44 Copper layer on a ceramics substrate before (a) and after (b) milling a 7.2-μm-diameter hole into the layer, ion beam images, local structures for DIC analysis remain recognizable after milling even in the vicinity of the hole.

Residual stresses introduced in microsystems during manufacture severely determine their reliability. In order to study the impact of different manufacturing processes, as well as packaging, built-in structures are often used to monitor their deformation and in that way to

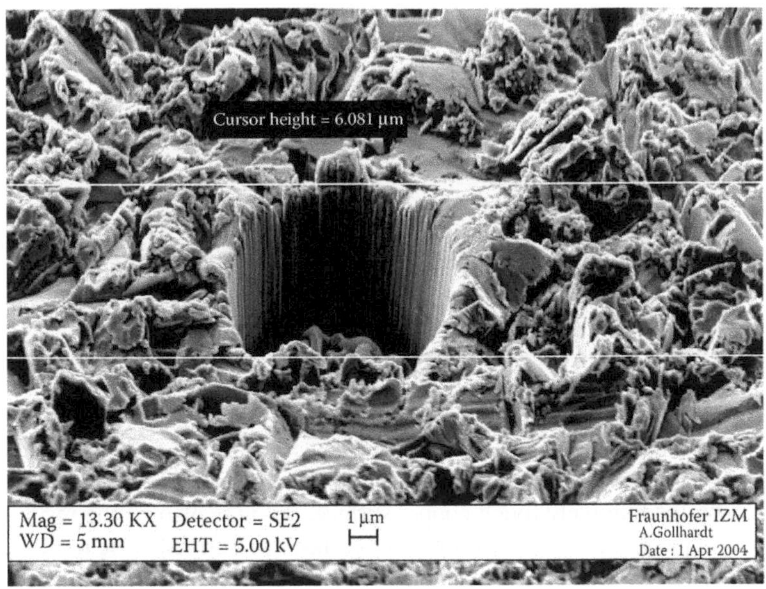

FIGURE 2.45 SEM image of the hole drilled into the Cu layer in Figure 2.44; minimum hole depth: 6.1 μm.

judge stresses [13]. The design of monitoring structures is cumbersome and problem based. In contrast, the measurement of residual stresses released by ion milling can be a more general approach. Figure 2.46 shows a stress release displacement field observed near a hole in a thin copper layer, milled with FIB equipment, while corresponding ion beam images are presented in Figure 2.44. The milled hole and copper surface can be seen on the SEM micrograph in Figure 2.45.

The stress was released on a layer of a direct copper bonding (DCB) substrate for electronics packaging. Residual stresses in DCB structures are caused by the large CTE mismatch between basic ceramics and the copper layers. Qualitatively, the measured displacement pattern coincides with the pattern expected for typical through holes or deep holes in materials with uniform residual

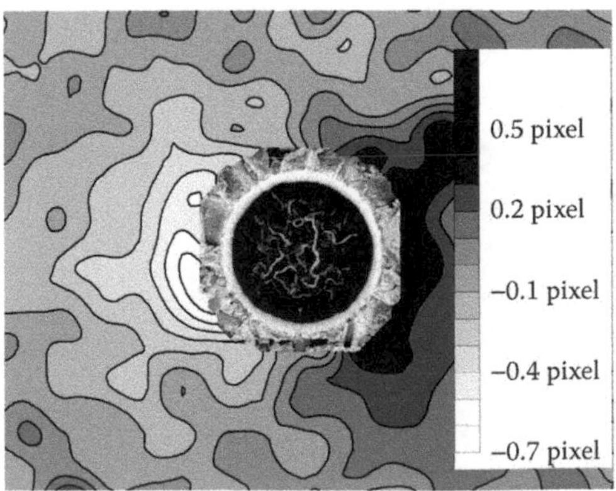

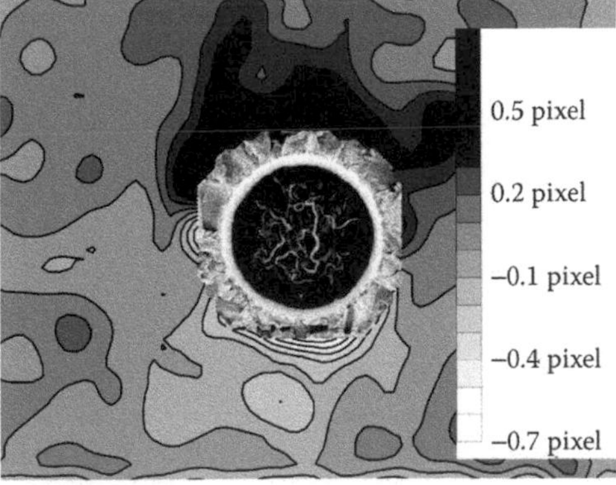

FIGURE 2.46 u_x (left) and u_y (right) displacement fields determined by DIC treatment of FIB images before and after hole milling with the ion beam; material: DCB (copper on SiO_2), hole diameter: 3.7 μm, hole depth >6 μm, contourline distance 0.1 pixel displacement.

stresses. Analytical solutions for the latter give for the displacement field, u_x and u_y,

$$u_x = -\frac{\sigma_X(1+\nu)}{2E}\left[-\frac{R_0^{\,2}}{r}\left(\frac{4}{(1+\nu)}\cos\theta + \cos(3\theta)\right) + \frac{R_0^{\,4}}{r^3}\cos(3\theta)\right] \tag{2.8}$$

$$u_y = -\frac{\sigma_X(1+\nu)}{2E}\sin\theta\left[\frac{R_0^2}{r}\left(\frac{(3-\nu)}{(1+\nu)} - 4\cos^2\theta\right) + \frac{R_0^4}{r^3}\left(-1+4\cos^2\theta\right)\right] \tag{2.9}$$

under the assumption of a homogeneous residual stress, σ_x.

In Equation 2.8 and Equation 2.9, E is Young's modulus, represents Poisson's ratio, and R_0 represents hole radius. θ and r are the coordinates of a cylindrical coordinate system with its origin in the middle of the hole. Figure 2.47 shows plots of the analytically derived field as represented in Equation 2.8 and Equation 2.9. The rather unsmooth view of the measured contour lines seems to be introduced by the relatively rough copper surface, but may also be the result of slight stress variations or due to the fact that the measurement has been performed close to the resolution limit. Future effort should be aimed at developing more appropriate schemes for milling and at finding

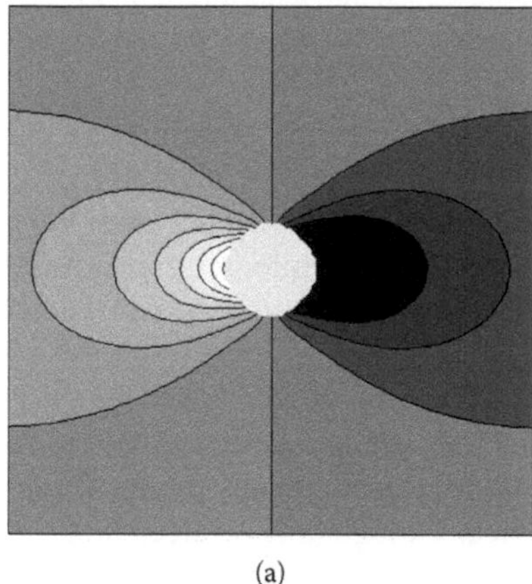

(a)

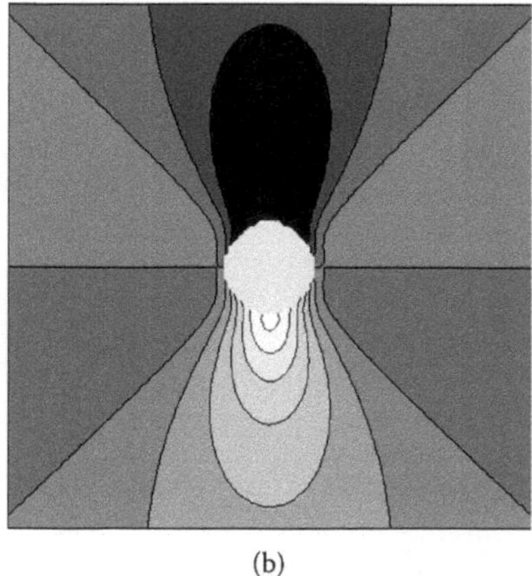

(b)

FIGURE 2.47 Typical u_x (a) and u_y (b) displacement fields determined by Equation 2.8 and Equation 2.9.

straightforward extraction algorithms of residual stress components from measurement fields utilizing FEA tools. Further details on the initial work carried out for residual stress measurements in FIB equipment can be found in Reference 54.

2.5 CONCLUSIONS AND OUTLOOK

The development of different bulk and surface micromachining techniques such as epi-poly, SCREAM, LIGA, or SU-8 processes in the past 15 years has resulted in a multitude of new MEMS and MOEMS devices. As a consequence, microsystems have become a rapidly growing market in the past few years. The speed at which new devices are introduced on the market depends to a large extent on their ability to provide *long-term* reliability under exploitative conditions. Unlike microelectronics, new approaches for mechanical device characterization and testing have to be established with regard to the specifics of MEMS technology. The DIC techniques presented in this chapter can be one of the essential methods for achieving this. The major advantages of the corresponding techniques are scalability, simplicity, software independence on imaging equipment, and compatibility with finite element simulation. As demonstrated, the DIC method can be extended to virtually all kinds of scanning microscopic imaging, thus adjusting its capability to an object's needs. The method is easy to understand and demands no additional skills apart from a knowledge of microscopy. Switching from one kind of microscopic equipment to another does not call for new software solutions because of the general approach of image comparison on the DIC base.

Until now, a variety of applications relating to microsystem characterization and testing have been demonstrated. Areas of application include defect detection, strain analysis in different kinds of interconnects (in order to figure out crucial component loading), local residual stress measurements, microcrack evaluation, and material property determination. The ongoing miniaturization of microsystem devices to nanoscale size demands new approaches and indicates the potential of this field, as illustrated by deformation studies on thin layer sensor membranes.

The future prospects of DIC application to MEMS and MOEMS will be driven by the further development of methods. There is a need to improve measurement resolution in terms of subpixel displacement accuracy. Progress in this regard would allow access to low deformation problems and to more elastic deformation states than we have at present.

The development of new loading modules that can be used inside high-resolution microscopes is another issue. One of the main challenges is to reduce loading forces and their respective displacement paths during testing. Loading forces in the micronewton range and overall specimen loading paths in the micrometer range would allow for the testing of a smart specimen prepared from MEMS components or materials, e.g., thin film separated for testing.

The combination of DIC analysis with new equipment (e.g., focused ion beam) will open new application areas. Residual stress determination by stress release due to ion milling and subsequent DIC deformation measurement seems to have good potential. Residual stresses introduced in MEMS by micromachining and MEMS packaging have to be controlled to avoid damage or functional failure. Their local measurement after particular processing steps and on the finished device are a general problem because only a few tools exist or are available in research laboratories. The application of the FIB-based method presented in this chapter demands further research and development. This applies mainly to the strategy of stress extraction from deformation fields, which assumes that certain residual stress hypotheses are required to solve an inverse mechanical problem, in addition to an effective combination of measurement and numerical simulation of the mechanical problem.

REFERENCES

1. Gad-el-Hak, M., *MEMS Handbook*, CRC Press, Boca Raton, FL, 2002.
2. Rai-Choudhury, P., *MEMS and MOEMS Technology and Applications*, SPIE Press, Bellingham, WA, 2000.

3. Freund, L.B. and Suresh, S., *Thin Film Materials*, Cambridge University Press, Cambridge, 2003.
4. Bhushan, B., *Springer Handbook of Nanotechnology*, Springer-Verlag, Berlin, 2003.
5. Alison., J., *Materials Handbook for Hybrid Microelectronics*, Artech House, Norwood, 1988.
6. Sharpe, W.N., Mechanical properties of MEMS materials, in *MEMS Handbook*, Gad-el-Hak, M., Ed., CRC Press, Boca Raton, FL, 2002, chap. 3.
7. Elsner, G., Residual stress and thermal expansion of spun-on polyimide films, *J. Appl. Polymer Sci.*, 34, 815, 1987.
8. Ogawa, H. et al., Stress-strain diagrams of microfabricated thin films, in *Proc. of Micro Materials '97*, Berlin, 1997, p. 716.
9. Willecke, R. and Ho, P.S., Study of vertical thermo-mechanical properties of polyimide thin films, in *Proc. of Micro Materials '97*, Berlin, 1997, p. 721.
10. Vogel, D. et al., Messung von Materialkennwerten an Komponenten der Mikrotechnik, *Tech. Messen.*, 2, 86, 2001.
11. Vogel, D. et al., Determination of CTE values for packaging polymers — a critical view to accustomed approaches, in *Proc. of Symp. POLY 2000*, London, 2000.
12. Jaeger, R.C. and Suhling J. C., Advances in stress test chips, in *Proc. of 1997 ASME Int. Mech. Eng. Congress and Exp.*, EEP-Vol. 22, AMD-Vol. 226, Dallas, 1997, p. 1.
13. Madou, M.J., Stress-measuring techniques, in *MEMS Handbook*, Gad-el-Hak, M., Ed., CRC Press, Boca Raton, FL, 2002, pp. 16–104.
14. Miller, M.R., Mohammed, I., Dai, X., and Ho, P.S., Study of thermal deformation in underfilled flip-chip packages using high resolution Moire interferometry, in *Proc. of MicroMat 2000*, Berlin, 2000, p. 174.
15. Vogel, D., Kühnert, R., and Michel, B., Strain Measurement in Micrometrology, in *Proc. Int. Conf. on Advanced Photonic Sensors and Applications*, Lieberman, R.A., Asundi, A., Asanuma, A., Eds., *Proc. SPIE*, Vol. 3897, Singapore, 1999, p. 224.
16. De Wolf, I. et al., High resolution stress and temperature measurements in semiconductor devices using micro-Raman spectroscopy, in *Proc. Int. Conf. on Advanced Photonic Sensors and Applications*, Lieberman, R.A., Asundi, A., Asanuma; A., Eds., *Proc. SPIE*, Vol. 3897, Singapore, 1999, p. 239.
17. Cote, K.J. and Dadkhah, M.S., Whole field displacement measurement technique using speckle interferometry, in *Proc. of Electronic Components & Technology Conference (ECTC)*, Las Vegas, 2001, p. 80.
18. Vogel, D. and Michel, B., Microcrack Evaluation for Electronics Components by AFM nanoDAC Deformation Measurement, in *Proc. IEEE-NANO*, Maui, HI, 2001, p. 309.
19. Knauss W.G., Chasiotis, I., and Huang, Y., Mechanical measurements at the micron and nanometer scales, *Mech. Mater.*, 35, 217, 2003.
20. Xie, H. et al., Focused ion beam Moiré method, *Opt. Laser. Eng.*, 40, 163, 2003.
21. Kinoshita, T., Stress singularity near the crack-tip in silicon carbide: investigation by atomic force microscopy, *Acta Materialia*, 46(11), 3963, 1998.
22. Komai, K., Minoshima, K., and Inoue, S., Fracture and fatigue behavior of single crystal silicon microelements and nanoscopic AFM damage evaluation, *Microsyst. Technol.*, 5(1), 30, 1998.
23. Marieta, C. et al., Effect of the cure temperature on the morphology of a cyanate ester resin modified with a thermoplastic: characterization by atomic force microscopy, *Eur. Polymer J.*, 36, 1445, 2000.
24. Bobji, M.S. and Bushan, B., Atomic force microscopic study of the microcracking of magnetic thin films under tension, *Scripta Materialia*, 44, 37, 2001.
25. Druffner, C.J. and Sathish, Sh., Improving atomic force microscopy with the adaptation of ultrasonic force microscopy, *Proc. of SPIE*, 4703, 105, 2002.
26. Xie, H. et al., High resolution AFM scanning Moiré method and its application to the micro-deformation in the BGA electronic package, *Microelectron. Reliab.*, 42, 1219, 2002.
27. Asundi, A. et al., Micro-Moiré methods — optical and scanning techniques, *Proc. of SPIE*, 4416, 54, 2001.
28. Asundi, A. et al., Phase shifting AFM Moiré method, *Proc. of SPIE*, 4448, 102, 2001.
29. Chasiotis, I. and Knauss, W., A new microtensile tester for the study of MEMS materials with the aid of atomic force microscopy, *Exp. Mech.*, 42(1), 51, 2002.

30. Vogel, D., Auersperg, J., and Michel, B., Characterization of electronic packaging materials and components by image correlation methods, in *Proc. of Advanced Photonic Sensors and Applications II*, Asundi, A., Osten, W., Varadan, V.K., Eds., *Proc. of SPIE*, Vol. 4596, Singapore, 2001, p. 237.
31. Vogel, D. et al., Displacement and strain field measurements for nanotechnology applications, in *Proc. IEEE-NANO*, Lau, C., Ed., Washington, 2002, p. 37.
32. Vogel, D. et al., Deformation analysis on flip chip solder interconnects by microDAC, in *Proc. of Reliability of Solders and Solder Joints Symposium* at *126th TMS Annual Meeting and Exhibition*, Orlando, 1997, p. 429.
33. Sutton, M.A. et al., Determination of displacements using an improved digital correlation method, *Image Vision Comput.*, 1(3), 133, 1983.
34. Chao, Y.J. and Sutton, M.A., Accurate measurement of two- and three-dimensional surface deformations for fracture specimens by computer vision, in *Experimental Techniques in Fracture*, Epstein, J.S., Ed., VCH, New York, 1993, p. 59.
35. Sutton, M.A. et al., Measurement of crack tip opening displacement and full-field deformations, in *Proc. of IUTAM Symp. on Advanced Optical Methods and Applications in Solid Mechanics*, Fracture of Aerospace Materials Using 2D and 3D Image Correlation Methods, 2000, p. 571.
36. Davidson, D.L., Micromechanics measurement techniques for fracture, in *Experimental Techniques in Fracture*, Epstein, J.S., Ed., VCH, New York, 1993, p. 41.
37. Vogel, D. et al., MicroDAC — a novel approach to measure in-situ deformation fields of microscopic scale, in *Proc. of ESREF'96*, Enschede, 1996, p. 1939.
38. Vogel, D. et al., High resolution deformation measurement on CSP and flip chip, in *Technical Digest of the Fourth VLSI Packaging Workshop of Japan*, Kyoto, 1998, p. 84.
39. Kühnert, R. et al., Materialmikroskopie, in *Material Mechanics, Fracture Mechanics and Micromechanics*, Winkler, Th. and Schubert, A., Eds., Chemnitz, 1999, p. 651.
40. Biery, N., de Graef, M., and Pollock, T.M., A method for measuring microstructural-scale strains using a scanning electron microscope: applications to γ-titanium aluminides, *Metall. Mater. Trans. A*, 34A, 2301, 2003.
41. Soppa, E. et al., Experimental and numerical characterization on in-plane deformation in two-phase materials, *Comput. Mater. Sci.*, 21, 261, 2001.
42. Anwander, M. et al., A laser speckle correlation method for strain measurements at elevated temperatures, in *Proc. the Symp. on Local Strain and Temperature Measurements in Non-Uniform Fields at Elevated Temperatures*, Berlin, 1996, p. 49.
43. Vogel, D, MicroDAC — a three year history of strain measurements on tiny electronics packaging structures, Materialmikroskopie, in *Material Mechanics, Fracture Mechanics and Micromechanics*, Winkler, Th. and Schubert, A., Eds., Chemnitz, 1999, p. 161.
44. Vogel, D. et al., Combining DIC techniques and finite element analysis for reliability assessment on micro and nano scale, in *Proc. of Electronics Packaging Technology Conference*, Singapore, 2003, 450.
45. Kühnert, R. et al., A new approach to micro deformation analysis by finite element based digital image correlation, in *The World of Electronic Packaging and System Integration*, Michel, B. and Aschenbrenner, R., Eds., ddp goldenbogen, 2004, p. 463.
46. Vogel, D. et al., NanoDAC — a new testing method from the IZM nanomechanics lab, in *Annual Report of Fraunhofer IZM*, Berlin, 2002, p. 54.
47. Dudek, R. and Michel, B., Thermomechanical reliability assessment in SM- and COB-technology by combined experimental and Finite Element method, in *Proc. Int. Reliability Physics Symp.*, IEEE Catalog No. 94CH3332-4, San Jose, 1994, p. 458.
48. Vogel, D. et al., Determination of packaging material properties utilizing image correlation techniques, *J. Electron. Packag.*, 124, 345, 2002.
49. Anderson, T.L., *Fracture Mechanics*, 2nd ed., CRC Press, Boca Raton, FL, 1995, p. 54.
50. Vogel, D., Jian, Ch., and de Wolf, I., Experimental validation of finite element modeling, in *Benefiting from Thermal and Mechanical Simulations in Micro-Electronics*, Zhang, G.Q., Ed., Kluwer Academic, Boston, 2000, p. 113.
51. Auersperg, J., Döring, R., and Michel, B., Gains and challenges of parameterized finite element modeling of microelectronics packages, *Micromaterials and Nanomaterials*, Vol. 1, Michel, B., Ed., Berlin, 2002, p. 26.

52. Puigcorbé, J., et al., AFM analysis of high temperature degradation in MEMS elements, in *Proc. of Micro Mechanics*, Sinaia, 2002.
53. Puigcorbé, J., et al., Thermo-mechanical analysis of micro-drop coated gas sensors, *Sensor. Actuator. A*, 97, 379, 2002.
54. Vogel, D., et al., FIB based measurements for material characterization on MEMS structures, in *Proc. Symposium on Testing, Reliability, and Application of Micro- and Nano-Material Systems III*, in *Proc. of SPIE*, Vol. 5766, San Diego, 2005.

3 Light Scattering Techniques for the Inspection of Microcomponents and Microstructures

Angela Duparré

CONTENTS

3.1 Introduction 103
3.2 Theoretical Background of Light Scattering 104
3.3 Measurement Equipment 106
3.4 Standardization of Light Scattering Methods 109
3.5 Applications for Microcomponent and Microstructure Inspection 110
3.6 Combination of Light Scattering and Profilometric Techniques 115
3.7 Conclusions and Outlook 116
References 117

3.1 INTRODUCTION

Light scattering techniques have become widely recognized as a powerful tool for the inspection of optical and nonoptical surfaces, components, and systems. This has been driven by the ever-increasing surface quality requirements; hence, the need for appropriate inspection tools. With light scattering techniques, a large variety of quality-relevant effects such as micro- and nanostructures, roughness, defects, and coating inhomogeneities can be advantageously inspected. Scattering methods are noncontact, fast, flexible, and robust. They are applicable to components with surfaces extending from large functional areas to microdimensions. This chapter is therefore intended to provide information on main scattering-measurement principles suitable for microcomponent and microstructure investigation.

Prior to the description of the measurement techniques, the brief theoretical background in Section 3.2 correlates measured light scattering signals to microroughness. This particular case of light scattering mechanism was selected for demonstration because it is widely used and has proved applicable to a large number of surface inspection problems in practice. In Section 3.3, the main types of measurement arrangements, total scattering and angle resolved scattering, are described and examples of instruments are discussed in detail. The operation wavelengths of these techniques extend from visual to both the infrared (IR) and ultraviolet (UV) spectral regions. As light scattering in the deep ultraviolet (DUV) and vacuum ultraviolet (VUV) regions has currently become a critical issue, latest developments in 193 nm and 157 nm measurement techniques are also included. Section 3.4 summarizes essential standardization aspects for total and angle resolved light scattering measurements. In Section 3.5, examples of a variety of applications involving different types of samples, wavelengths, measurement procedures, and data evaluation are given. How to advantageously combine light scattering techniques with profilometric methods is demonstrated in Section 3.6.

3.2 THEORETICAL BACKGROUND OF LIGHT SCATTERING

In this section, we introduce the basics of theoretical models of light scattering from surface microroughness [1–9]. This outline refers to the case of randomly rough structures and 2-D isotropy, because this scatter mechanism can be applied to a large variety of surfaces and is a suitable exemplary introduction to the problem of scattering models in general. In the references, further theories can be found that also include nonisotropy, deterministic surface structures and defects [3,4,8], as well as scatter effects from volume imperfections [6,10].

The microroughness of a surface can be described as a Fourier series of sinusoidal waves with different periods, amplitudes, and phases. According to the grating equation, each single grating with spacing g causes scattering at the wavelength λ into the angle Θ: $\sin\Theta = \lambda / g$. Then $f = 1/g$ represents one single spatial frequency within this grating assembly. A surface with statistical microroughness contains a large diversity of spatial frequencies, which is quantitatively described by the power spectral density (PSD) [1–4]. The PSD provides the relative strength of each roughness component as a function of spatial frequency:

$$\mathrm{PSD}(\mathbf{f}) = \lim_{L\to\infty} \frac{1}{\mathrm{L}} \left| \int_{\mathbf{L}} z(\mathbf{r}) \exp(-2\pi i \mathbf{f} \cdot \mathbf{r}) d\mathbf{r} \right|^2 \tag{3.1}$$

$z(\mathbf{r})$ is the height of the roughness profile, $\mathbf{r}$ is the position vector, and $\mathbf{f}$ is the spatial frequency vector in the x-y plane. L denotes the measured surface area **L**.

For isotropic surfaces, Equation 3.1 becomes independent of the surface direction of the spatial frequency vector $\mathbf{f}$. After transformation into polar coordinates followed by averaging over all surface directions, a function $PSD(f)$ is obtained as the 1-D representation of the isotropic 2-D PSD.

Light scattering from microirregularities can be modeled through both scalar and vector theories. Whereas scalar theories [9] are stringently restricted to near-angle scattering, vector theories [4,5] are more general and include the polarization properties of the incident and scattered light. For practical applications, the theory in Reference 5 was found particularly useful. The more complex approach in Reference 6 delivers the same results for microroughness scatter as in Reference 5. Use of the method in Reference 6 is suggested if other scatter sources such as volume imperfections or surface plasmons are to be considered in addition to surface roughness. Note that all these theories are first-order perturbation models and hence valid if the microroughness is small compared to the wavelength. In Reference 5, the scattering intensity per solid angle of a surface is given by

$$\frac{dP}{P_i d\Omega} = K(\lambda, n, \Theta, \alpha)\, PSD(f) \tag{3.2}$$

where $dP/(P_i d\Omega)$ is the differential power scattered into the direction (Θ, α) per unit solid angle $d\Omega = \sin\Theta\, d\Theta\, d\alpha$ divided by the incident power P_i. Θ and α denote the polar and azimuthal angles of scattering; n is the refractive index. The optical factor K hence contains the properties of the perfect surface (excluding the roughness) as well as the illumination and observation conditions. Backscattering as well as forward scattering can be expressed by Equation 3.2, depending on the actual conditions of illumination and observation. In this section, without loss of generality, the formulas refer to the case of normal incidence. The overall formalism, however, allows the inclusion of all other cases, including nonnormal incidence and arbitrary polarization.

It has become widespread to use the expression *angle resolved scattering* (ARS) for the term $dP/(P_i d\Omega)$, which can in turn be transformed into the also well-established scatter function BRDF/BTDF (bidirectional reflectance/transmittance distribution function) through multiplication with $\cos\Theta$:

$$\frac{dP}{P_i d\Omega} = \text{ARS} = \text{BRDF}\cos\Theta \tag{3.3}$$

and likewise for BTDF. By integration of Equation 3.3 over the forward or backward hemisphere, we obtain the total scattering (TS). TS is defined as the power P scattered into the forward or backward hemisphere divided by the incident power P_i:

$$TS = 2\pi \int_0^{\pi/2} \left(\frac{dP}{P_i d\Omega} \right) \sin\Theta \, d\Theta \tag{3.4}$$

In cases in which the correlation length, which represents the average lateral extension of the microroughness, becomes much larger than λ, simple approximations for total backscattering TS_b and total forward scattering TS_f (for transparent materials) can be derived:

$$TS_b = R_0 \left(\frac{4\pi\sigma}{\lambda} \right)^2, \; TS_f = T_0 \left(\frac{2\pi\sigma}{\lambda}(n-1) \right)^2 \tag{3.5}$$

σ is the rms roughness; R_0 and T_0 are the specular reflectance and transmittance.

If the surface considered is coated with a dielectric film or a multilayer stack, the models become more complex, because the amplitudes of the scattered fields from all interfaces add up to the overall scatter rather than just the intensities, and cross-correlation effects between the individual microroughness profiles must also be taken into account [5,11–14].

The ARS of a thin-film stack of N layers is given by [5]:

$$\frac{dP}{P_i d\Omega} = \sum_{i=0}^{N} \sum_{j=0}^{N} K_i K_j^* \, PSD_{ij}(f) \tag{3.6}$$

where K_i denotes the optical factor at the ith interface and K_j^*, the conjugate complex optical factor number at the jth interface. The optical factors now describe the properties of the ideal multilayer (refractive indices, film thicknesses), together with the illumination and observation conditions. The functions PSD_{ij} represent the power spectral densities of the interfaces. If $i \neq j$, they describe the roughness cross-correlation between two interfaces. When inspecting coated surfaces, a practical rule is that the scatter signal to a large extent results from the microstructures of the film interfaces and substrate surface. Nevertheless, volume scattering from microstructures within the film bulk may also occur and can then be described by models such as examples given in Reference 10 and Reference 15.

3.3 MEASUREMENT EQUIPMENT

The techniques for light scattering measurements of surfaces can be roughly divided into the two major categories discussed in the preceding section: ARS and TS or total integrated scattering (TIS). Details about the definition of and difference between TS and TIS are outlined in Section 3.4. For the precise measurements of microstructures and microcomponents, ARS is mostly based on goniophotometers. To measure TS or TIS, the scattering is collected over the backward and forward hemispheres by means of a Coblentz sphere [16,17] or an integrating sphere [18,19].

High-precision instrumentation for ARS measurements has been set up in a variety of laboratories [1,2,5,14,20–24]. Most such systems are typically designed for operation at 632.8 nm (He-Ne laser wavelength). The wavelength range of ARS, however, also extends into the IR and UV spectral regions [24–27]. ARS arrangements are comprehensive and flexible but rather unsuitable for fast and robust routine inspections in industry. Figure 3.1 shows the schematic principle of an ARS setup. This instrument built at the Fraunhofer Institute IOF in Jena can be operated at several laser wavelengths from 325 to 1064 nm [24,27] with a dynamic range of up to 15 orders of magnitude. The illumination system and the double goniometer constitute the central parts of the system. The sample holder and the detector are each mounted on a precision goniometer with the outer goniometer carrying the detector head, which can be revolved 360° around the sample with 0.01° resolution. The photograph in Figure 3.2 shows the inner goniometer with the hexapod for sample positioning as well as the outer goniometer with the detector arm. Another crucial part of ARS instruments in general is the illumination system. The laser radiation has to be carefully guided through a number of optical elements, such as diaphragms, polarizers, spatial filters, and mirror systems, in order to achieve high beam quality.

TS and TIS instruments based on the principle of either Coblentz spheres or integrating spheres have been developed for more than two decades in a number of laboratories [1,8,16–19,24,27, 28–30]. Figure 3.3 displays the schematic picture of a TS arrangement, which is based on a Coblentz sphere as the central element. With this equipment developed at the Fraunhofer Institute IOF, Jena

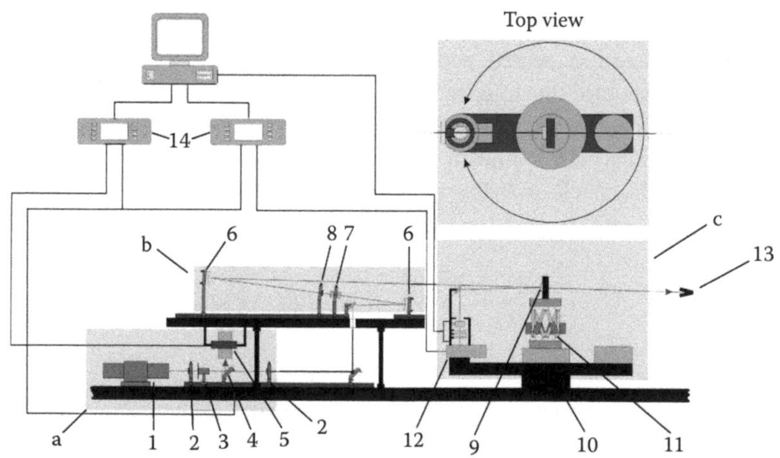

FIGURE 3.1 Schematic picture of the instrumentation for ARS measurements from 325 to 1064 nm at the Fraunhofer Institute IOF, Jena. Gray boxes: (a) light source and beam modulation, (b) beam preparation system, (c) detector arrangement. Elements: 1 — light source, 2 — diaphragm, 3 — chopper, 4 — beam splitter, 5 — reference detector, 6 — concave mirror, 7 — polarizer, 8 — pinhole, 9 — sample, 10 — double goniometer, 11 — sample positioning system with hexapod, 12 — detector system with attenuator system and detector aperture, 13 — beam stop, 14 — lock-in amplifiers.

FIGURE 3.2 Photograph of part of the ARS instrument of Figure 3.1: inner goniometer with hexapod and outer goniometer with detector arm.

[24,27], the total light scattered into the backward or forward hemisphere is detected within an angular range from 2 to 85° according to the instructions in ISO 13696 (see Reference 31 and Section 3.4). Figure 3.3 shows the Coblentz sphere in the position for backscatter (TS_b) measurement. A special arrangement allows easy change from the backscatter to the forward-scatter operation modes. This facility has been designed for TS measurements in a wide range of wavelengths

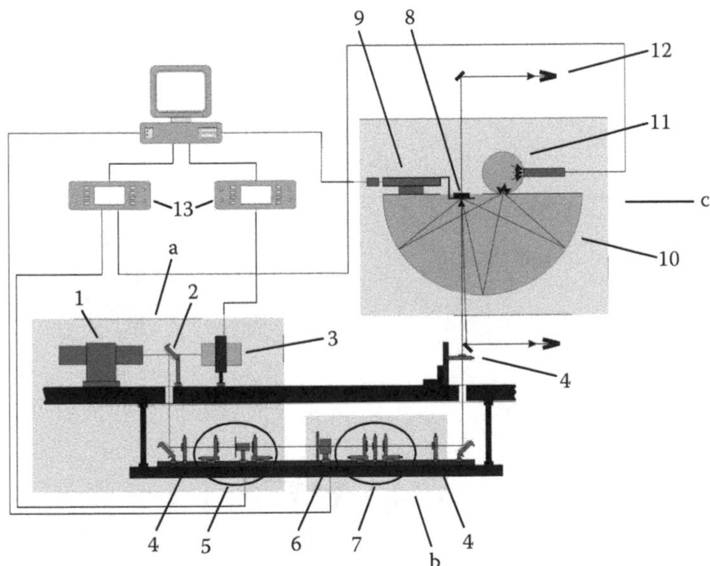

FIGURE 3.3 Schematic picture of the TS instrument (325 nm to 10.6 μm) at the Fraunhofer Institute IOF, Jena. Gray boxes: (a) light source and beam modulation, (b) beam preparation system, (c) detector arrangement. Elements: 1 — light source, 2 — beam splitter, 3 — reference detector, 4 — diaphragm, 5 — beam modulation system with chopper and lenses, 6 — attenuator, 7 — spatial filter with lenses and pinhole, 8 — sample, 9 — 2-D sample positioning system, 10 — Coblentz sphere, 11 — detector system, 12 — beam stop, 13 — lock-in amplifiers.

and is presently operated from 248 nm through 10.6 μm using several laser sources. The detector unit consists of the detector (photomultiplier in the UV, VIS, NIR, and HgCdTe-element in the IR) and a small integrating sphere for homogeneous illumination of the detector area. The laser beam is modulated by a chopper and carefully prepared in the illumination system. The incident radiation hits the sample surface at nearly 0° and the specular beam is guided back through the entrance or exit aperture of the Coblentz sphere. The positioning system enables 1-D and 2-D scanning of the sample surface, resulting in 1-D or 2-D scattering diagrams. The equipment is calibrated with an all-diffuse Lambertian standard. Through sophisticated optimization procedures, extremely low background levels (e.g., <0.05 ppm at 632.8 nm) were achieved. This is an essential precondition for precise measurements of high-quality, low-scatter microcomponents.

Apart from the wavelength range discussed in the preceding text, today's novel perspectives and demands for ultrasmooth elements and low-scatter optical components in the DUV and VUV regions for application in photolithography boost the development of scattering facilities for ever-shorter wavelengths. Therefore, a measurement system to determine TS and ARS, as well as the reflectance and transmittance at 193 and 157 nm, was developed at the Fraunhofer Institute IOF in Jena [32,33]. The schematic picture of the TS measurement module can be seen in Figure 3.4 with the Coblentz sphere positioned for TS_b measurement. The Coblentz sphere was coated with a VUV-enhanced highly reflective aluminum layer. Similar to the UV–VIS instrument described earlier, the sphere can be easily switched into the forward-scatter (TS_f) measurement position. Macro- and microcomponents for DUV and VUV applications can be investigated with high sensitivity, down to scattering levels as low as 1 ppm (10^{-6}). An excimer laser as radiation source is connected to two vacuum chambers housing the beam preparation path and the measurement arrangements. The chambers can be operated in both vacuum and nitrogen purge gas. Operation in vacuum is essential for the investigation of low-scatter samples (10 ppm and smaller) as at this low level Rayleigh scatter from gas molecules would otherwise enhance the scatter signal.

Another measurement module, based on a precision double goniometer arrangement, enables ARS, T, and R measurements at 193 and 157 nm with high resolution. This module can be inserted in the measurement chamber instead of the TS module, and the same beam preparation line is used

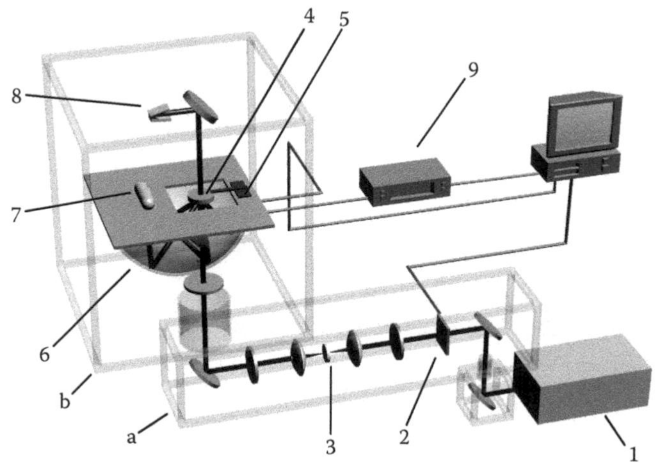

FIGURE 3.4 Schematic picture of the setup for TS measurements at 193 nm and 157 nm, Fraunhofer Institute IOF, Jena. Coblentz sphere in backscatter position. Framed boxes: (a) beam path chamber, (b) measurement chamber. Elements: 1 — excimer laser, 2 — attenuator system, 3 — beam preparation system, 4 — sample, 5 — 2-D sample positioning system, 6 — Coblentz sphere, 7 — detector system, 8 — beam stop, 9 — noise suppression system.

FIGURE 3.5 VULSTAR (system for measuring VUV Light Scatter, Transmittance and Reflectance) at the Fraunhofer Institute IOF, Jena. Measurement chamber and beam preparation chamber.

as for TS operation. The dynamic range of the ARS measurements exceeds twelve orders of magnitude, and angular resolution is better than 0.01°. The ARS, T, and R measurements can be performed as a function of the incident and observation angles and of polarization. A photograph of the entire measurement system — VULSTAR (system for measuring VUV Light Scatter, Transmittance and Reflectance) — is shown in Figure 3.5. This photograph shows the measurement chamber, which houses the Coblentz sphere or the goniometer as alternative modules, and the beam preparation chamber. The excimer laser (not seen in the picture) is connected to the rear end of the beam preparation chamber. In both measurement modules, VUV-sensitive photomultiplier tubes are employed for signal detection, which suppress fluorescent light.

By utilizing the same arrangement for transmittance and reflectance measurements as is used for ARS measurement, a consistent determination of T, R, and light scatter values becomes possible. Furthermore, gratings, being deterministic microstructures, can be advantageously inspected with respect to both the grating efficiency and scatter losses.

3.4 STANDARDIZATION OF LIGHT SCATTERING METHODS

As a result of the expanding application fields of optical measurement techniques and their impact on component technologies and quality, international standardization has gained increasing importance. This also holds for light scattering techniques. In the following, a brief summary of the main existing standard procedures for the light scattering techniques described in Section 3.3 is given.

A standard procedure for ARS measurements was defined in ASTM standard E 1392 [34]. This procedure was successfully implemented and verified in various round-robin experiments at different wavelengths [23,35–37]. Furthermore, the international working group TC 172/ SC 9/ WG 6 of the International Organization for Standardization plans activities to develop an ISO standard procedure for such measurements that meets the modified and increased demands concerning wavelength ranges, sensitivity, and flexibility.

The procedure for TIS measurements is prescribed in ASTM standard F 1048 [38]. TIS measurements collect the light scattered into the backward hemisphere. TIS is defined as the backscattered radiation divided by the total reflectance and applies only to opaque reflective surfaces. Examples of validation of this ASTM standard procedure are outlined in Reference 1 and Reference 8. The drawback of TIS, however, is that this quantity is not defined for transparent or semitransparent samples such as substrates, AR coatings, and beam splitters. Hence, TIS cannot be determined for such optical elements unless ambiguous additional assumptions are made. Therefore, the new international standard ISO 13696 [31] defines TS as the backscattered or forward-scattered radiation divided by the incident radiation. TS is hence equivalent to the scattering loss of the component. Opaque as well as transparent surfaces and coatings can be measured both in the backward and forward directions. TS and TIS can, however, be converted into one another if the reflectance of the sample is known. The ISO 13696 standard procedure was proved in an international round-robin experiment at 632.8 nm [39].

3.5 APPLICATIONS FOR MICROCOMPONENT AND MICROSTRUCTURE INSPECTION

In this section, selected results of light scattering measurements are described as typical examples from a variety of different inspection tasks, sample types, measurement procedures, and data evaluation methods.

Figure 3.6 shows TS measurements of a conventionally polished quartz surface at λ = 632.8 nm as a 3-D and 2-D (left and right) presentation of the surface mapping data. These mappings reveal enhanced scatter levels originating from microstructural features, extended scratches in particular, as well as local contaminations.

The TS measurement at λ = 632.8 nm of a dielectric microfilter arrangement for locally resolving color sensors (photograph in Figure 3.7a) is shown in Figure 3.7b. In the TS mapping, both the microcomponent pattern and scatter variations induced by microstructure inhomogeneities can be recognized.

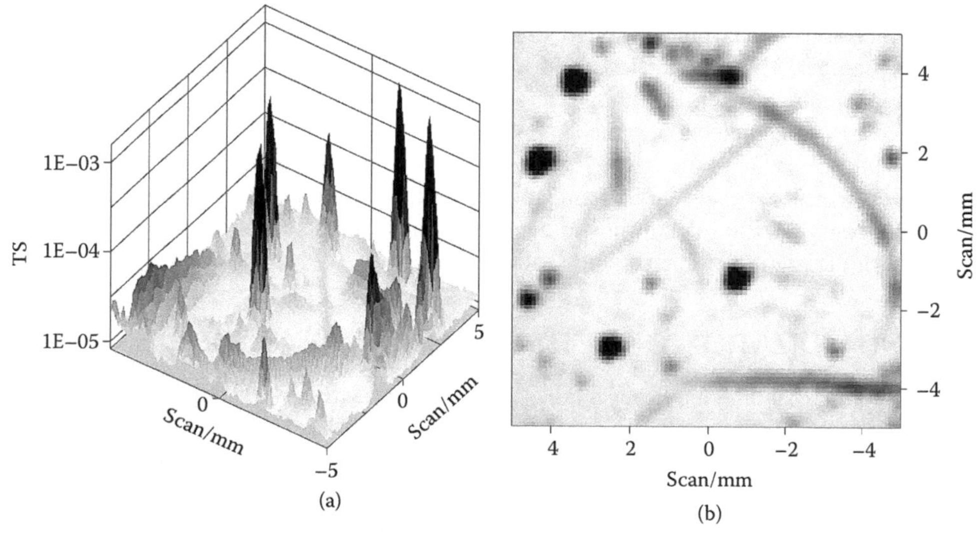

FIGURE 3.6 TS measurement of a polished quartz surface, λ = 632.8 nm. Left: 3-D, right: 2-D presentation of the surface mapping data.

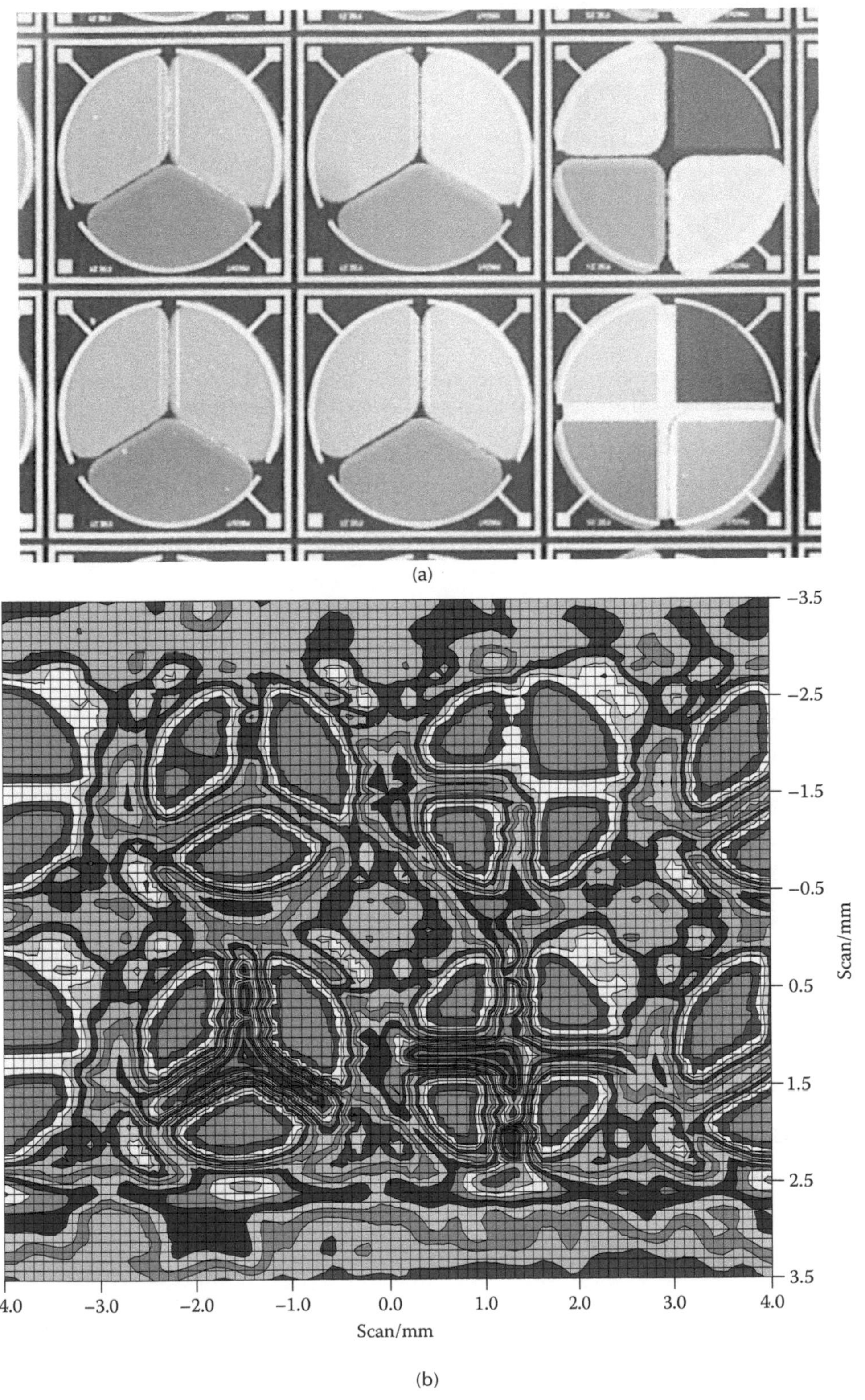

COLOR FIGURE 3.7 Photograph (a) and TS measurement at 632.8 nm (b) of a dielectric microfilter arrangement for locally resolving color sensors.

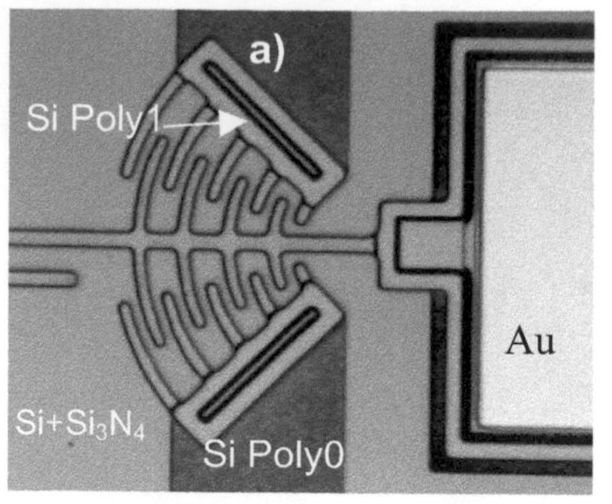

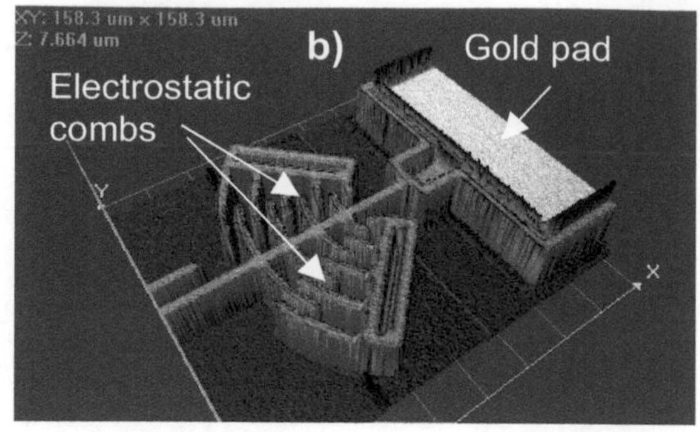

COLOR FIGURE 8.14 (a) Spectral reflectivity map and (b) 3-D profile with colors related to spectral reflectivity of an electrostatically actuated MEMS fabricated in polysilicon MUMPS technology.

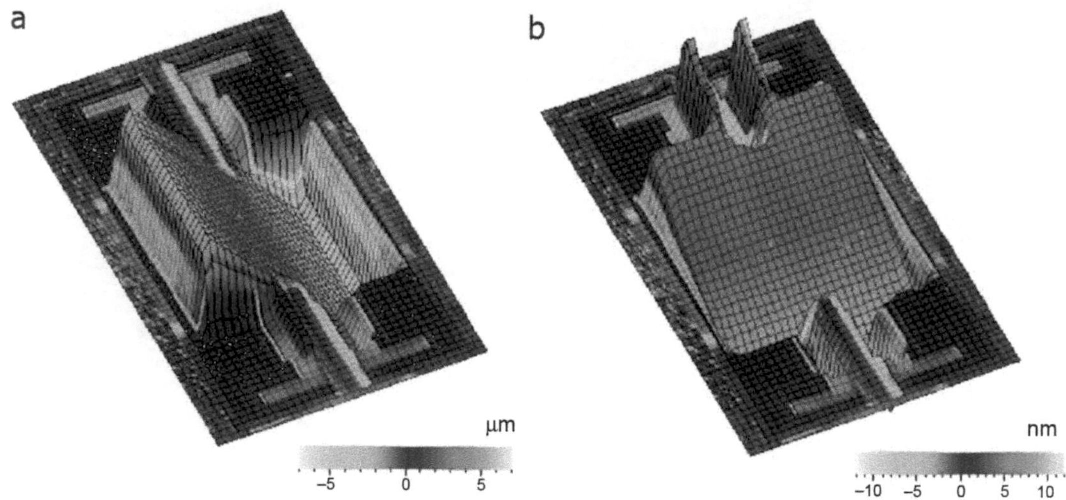

COLOR FIGURE 9.21 (a) Primary and (b) secondary mode of the micromirror as measured by scanning LDV technique.

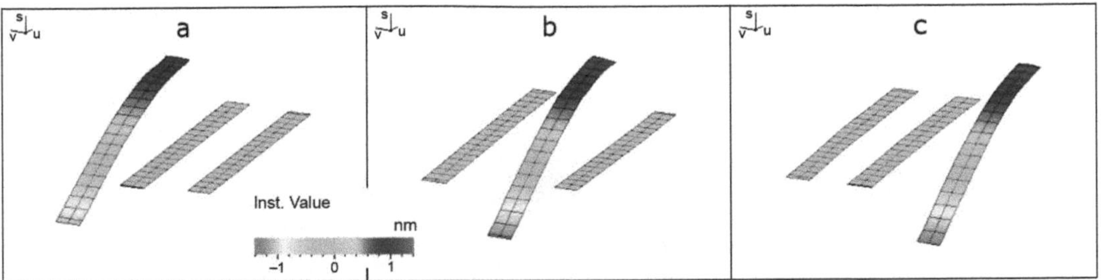

COLOR FIGURE 9.26 ODS at (a) 26.31 kHz, (b) 27.09 kHz, and(c) 27.59 kHz.

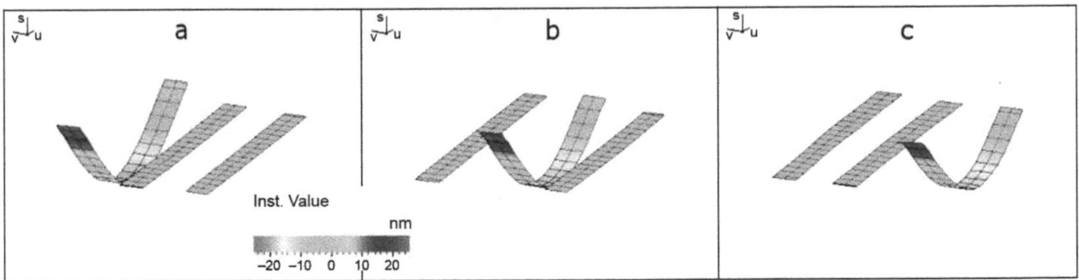

COLOR FIGURE 9.28 ODS at (a) 72.97 kHz, (b) 74.94 kHz, and (c) 76.78 kHz.

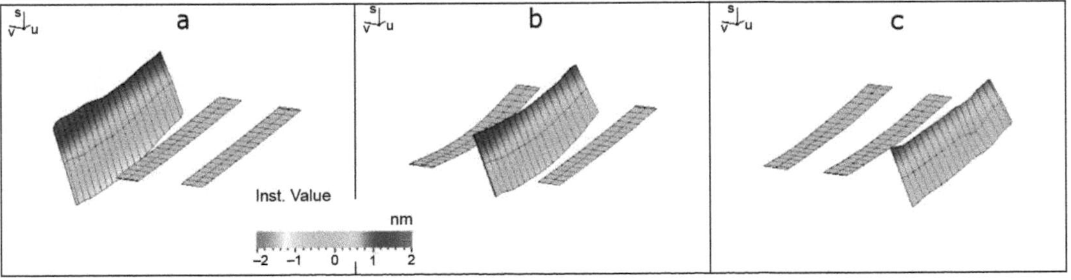

COLOR FIGURE 9.29 ODS at (a) 69.66 kHz, (b) 71.53 kHz, and (c) 73.56 kHz.

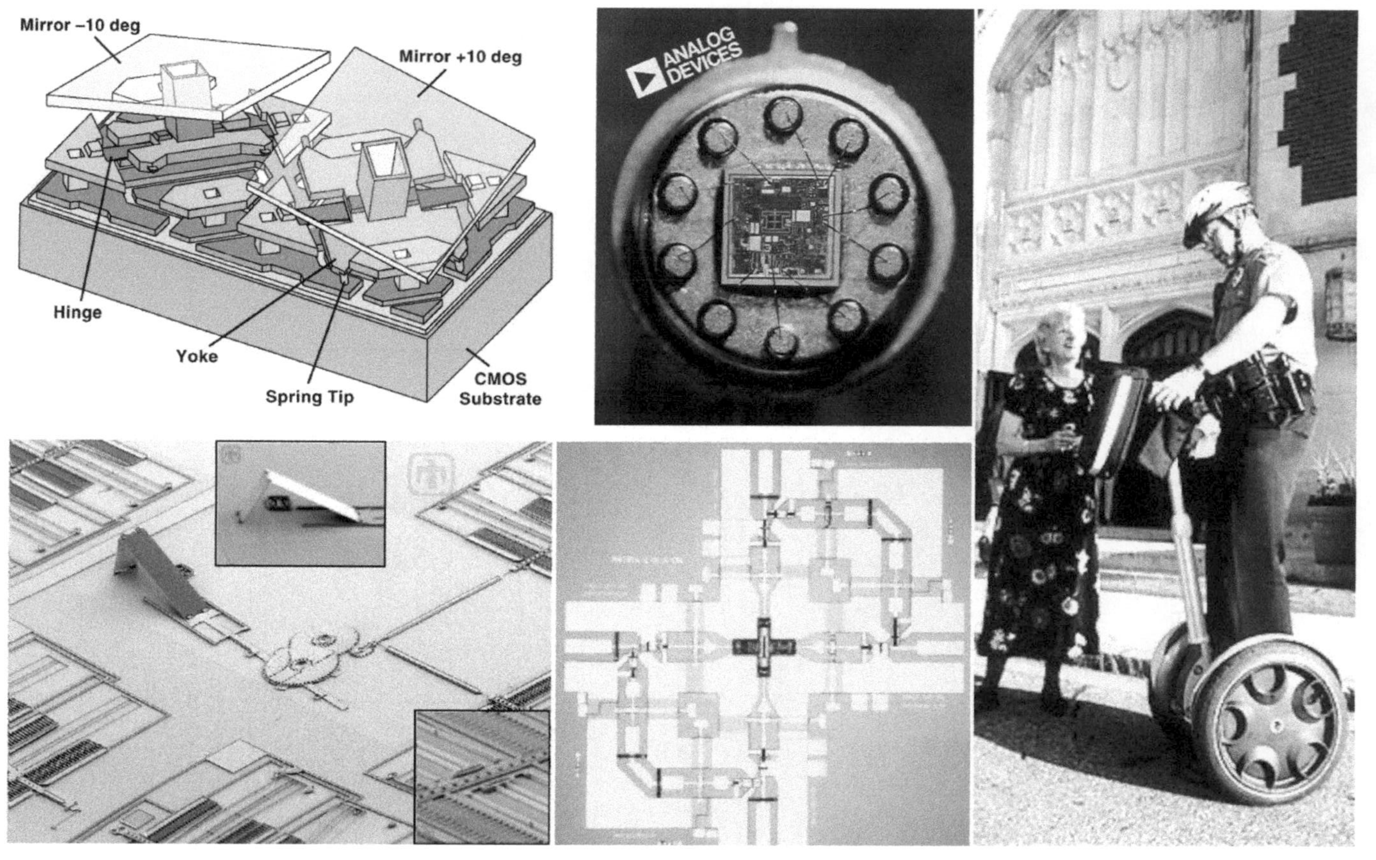

COLOR FIGURE 11.1 Representative MEMS devices and applications: (a) digital micromirror device (DMD) Courtesy of http://www.ti.com [9], (b) MEMS accelerometers used in air-bag deployment mechanisms [10], (c) Segway® Human Transporter (HI) enabled by MEMS gyroscopes [11], (d) Sandia Micromirror Device™ driven by a MEMS microengine capable to rotate at 1×10^6 rpm [12], and (e) RF MEMS switches being developed for telecommunications, radar, and automotive applications. (From Lacroix, D., Expanding the RF MEMS market — RF MEMS building blocks, *MSTNews*, 4: 34–36, 2003; Schauwecker, B., Mack, T., Strohm, K.M., Simon, W., and Luy, J.-F., RF MEMS for advanced automotive applications, *MSTNews*, 4: 36–38, 2003.)

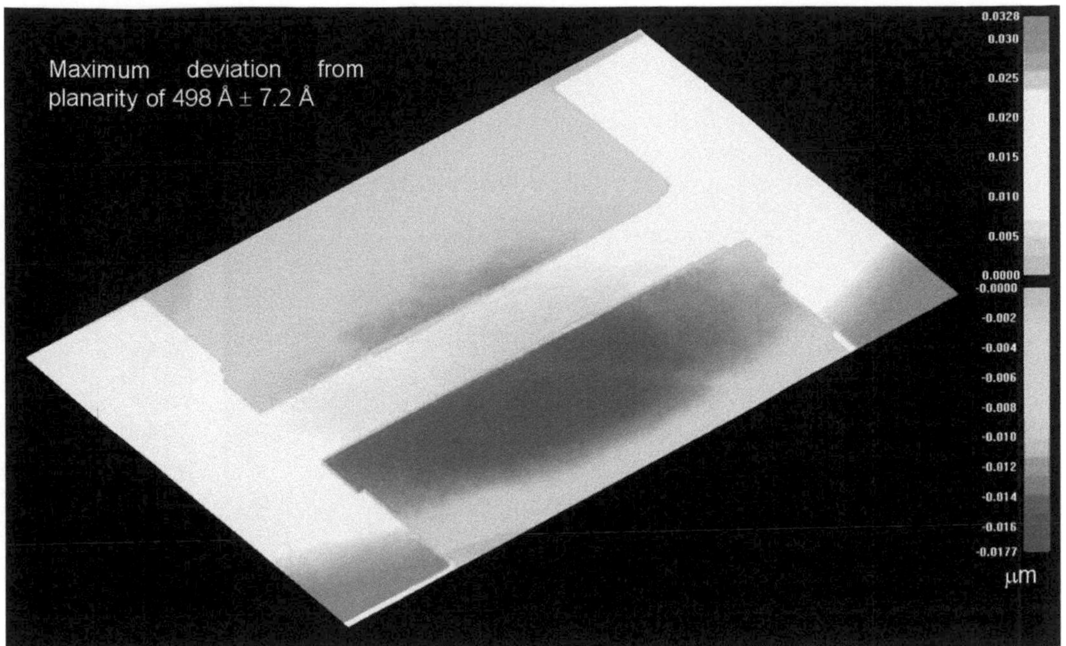

COLOR FIGURE 11.7 Deviations from planarity are obtained by difference analysis, which consists of evaluating the differences between a plane and the measured (x,y,z) data cloud shown in Figure 11.6.

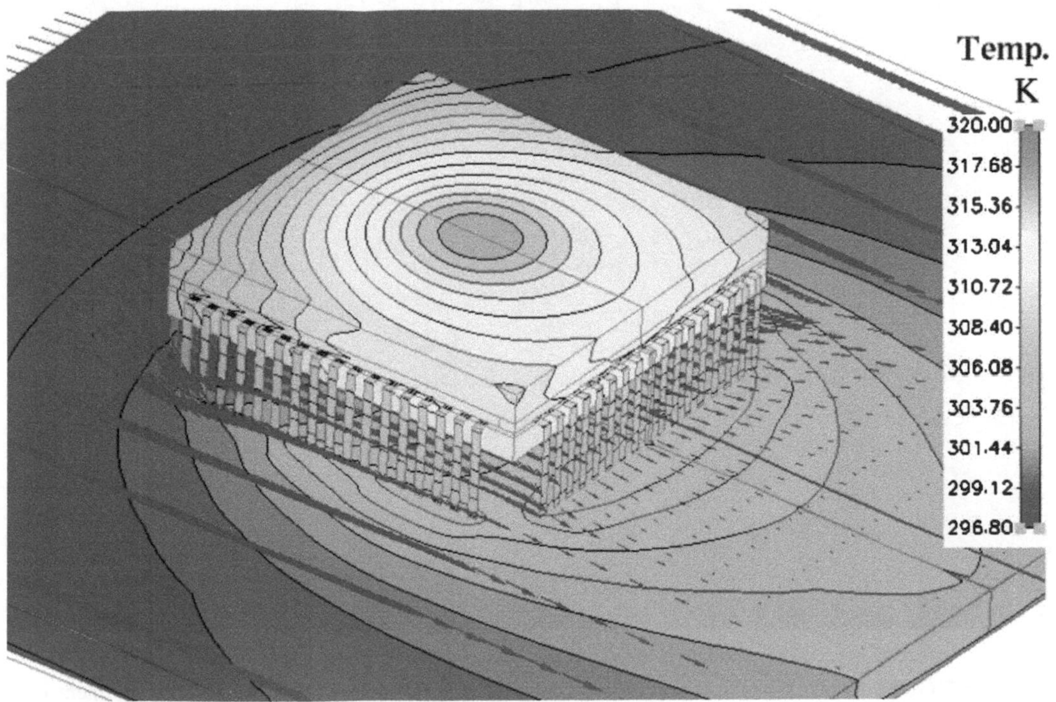

COLOR FIGURE 11.24 Computed temperature distribution for the model shown in Figure 11.23.

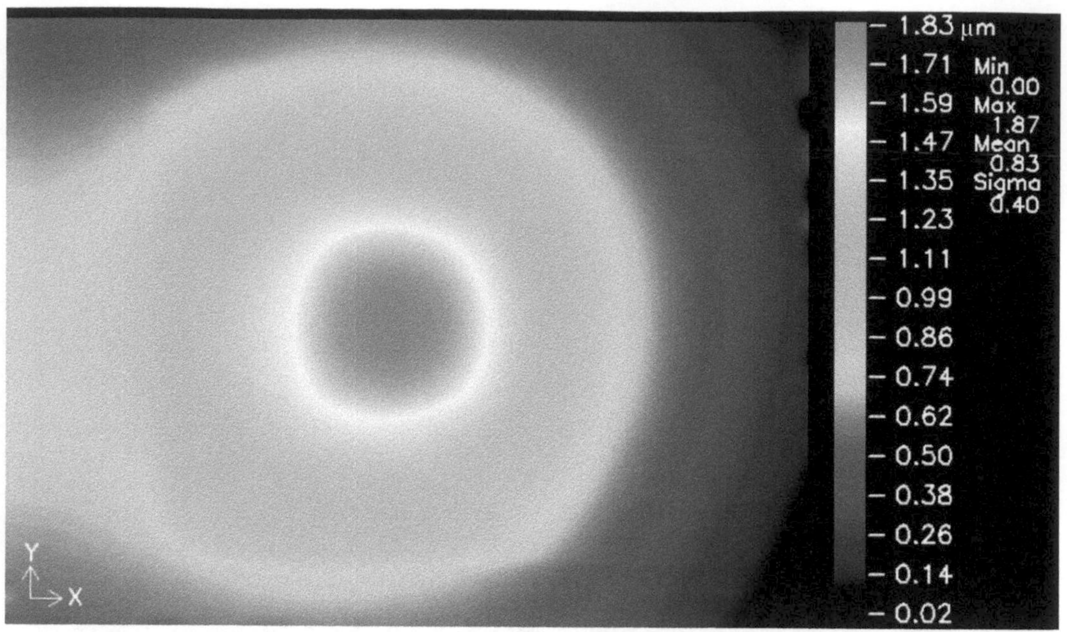

COLOR FIGURE 11.26 Thermal deformations of the SMT component based on the OEH fringe pattern of Figure 11.25.

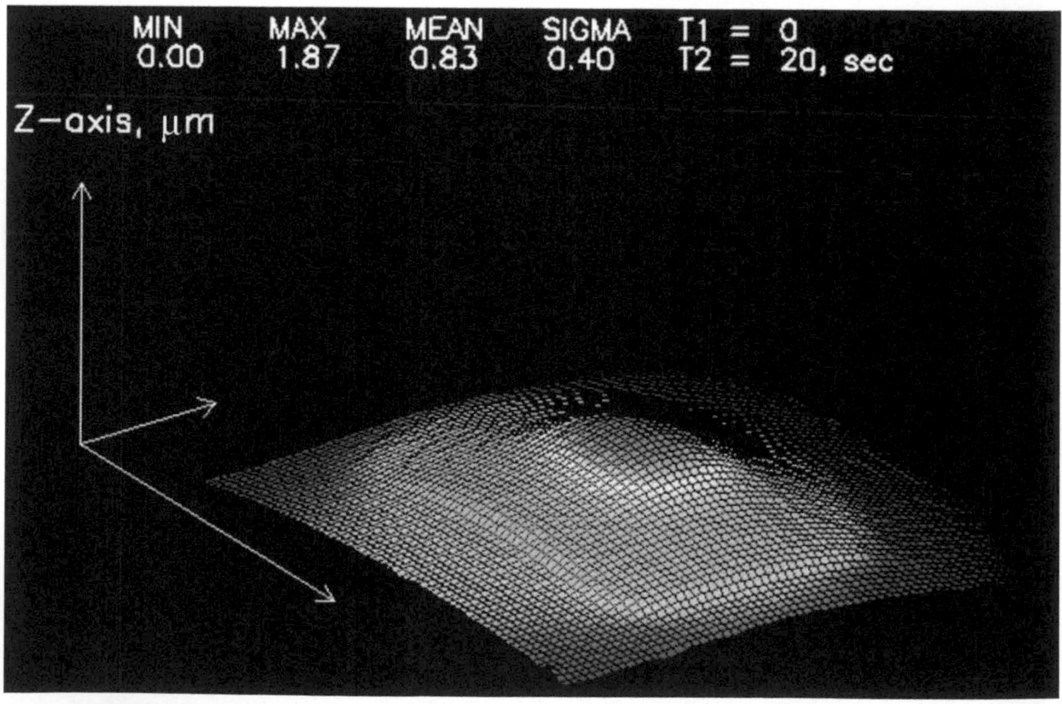

COLOR FIGURE 11.27 Wireframe representation of thermal deformations of the SMT component based on the OEH fringe pattern of Figure 11.25.

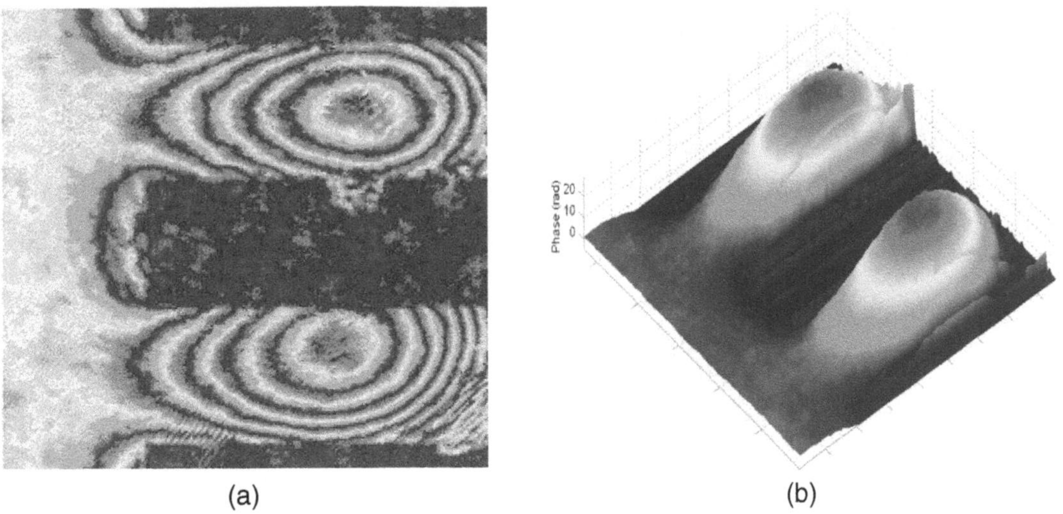

COLOR FIGURE 12.52 (a) Wrapped phase map of a severely deformed cantilever silicon MEMS, (b) unwrapped phase map showing the actual 2D structure profile.

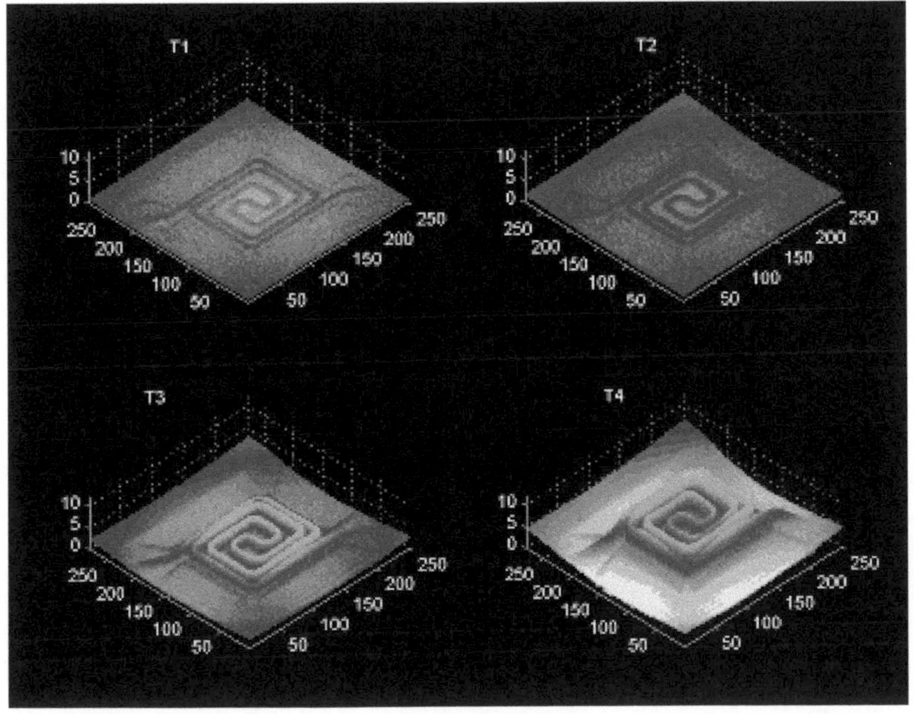

COLOR FIGURE 12.59 Out-of-plane deformation of the membrane at different temperatures: T1 = 100°C, T2 = 200°C, T3 = 400°C, T4 = 700°C (Z-axis is in radians).

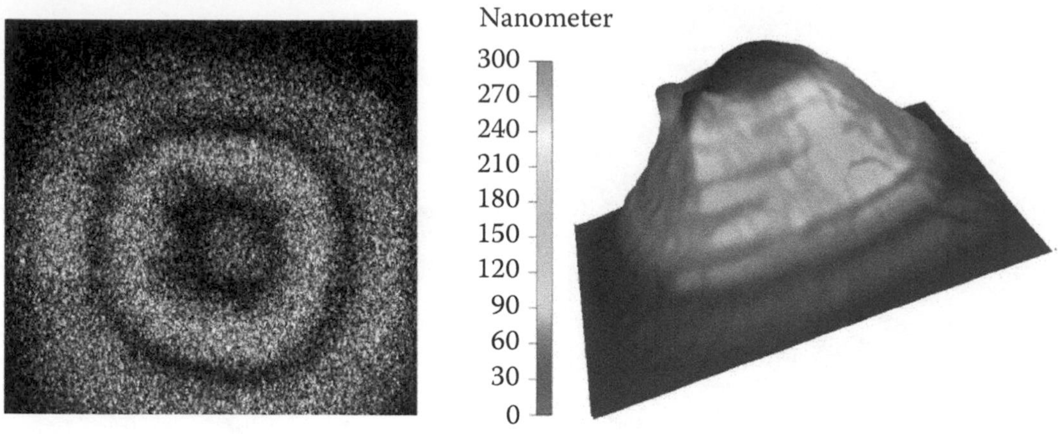

COLOR FIGURE 13.39 Static deflection of a 250-μm piezomembrane at U = 30 V: fringe pattern (left), displacement field (right).

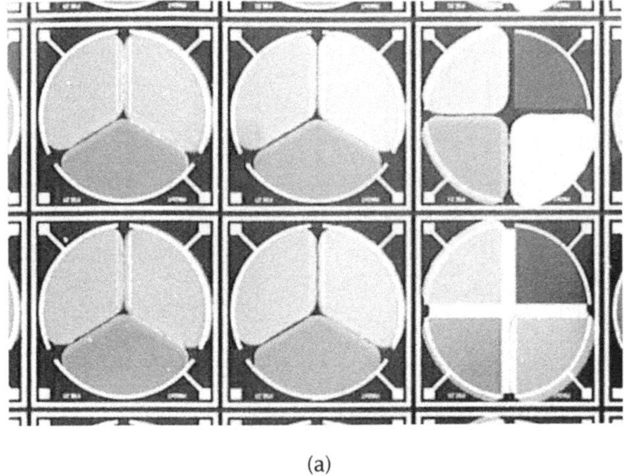

(a)

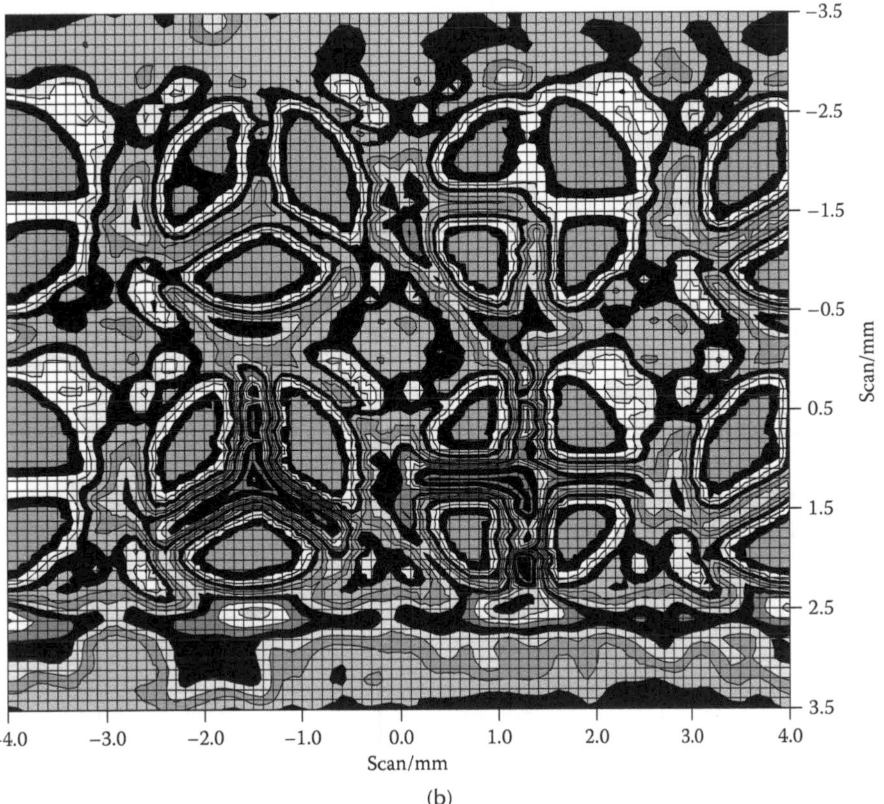

(b)

FIGURE 3.7 (See color insert following page 110.) Photograph (a) and TS measurement at 632.8 nm (b) of a dielectric microfilter arrangement for locally resolving color sensors.

In Figure 3.8, ARS was investigated for micrograting components. These high-efficiency dielectric reflection gratings for ultrashort, high-power laser application consisted of multilayer dielectric mirrors and a single corrugated top layer with 2 μm grating period and 0.84 μm grating depth. The measurements were performed at 632.8 nm in the backward hemisphere with TE-polarized incident and unpolarized detected light. The shape and magnitude of the ARS curve are determined by both the grating diffraction and the scatter loss induced by the microroughness of the thin-film layers.

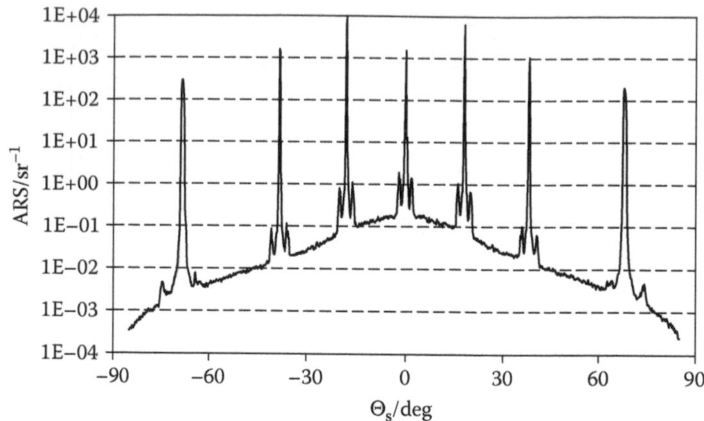

FIGURE 3.8 ARS measurement of a micrograting component, $\lambda = 632.8$ nm.

In contrast to this significantly structured ARS, the scatter curve in Figure 3.9 of a superpolished silicon wafer, as an example of a high-quality substrate for microcomponents, is smooth and appears at considerably lower levels.

Surface contaminations by particles can cause major problems in the manufacturing of microcomponents and crucially limit their optical and nonoptical functionalities. Hence, proper surface particle control is needed. Light scattering techniques have been identified as well-tailored tools for the sensitive detection of signals induced by even smallest particles in the submicron ranges [40,41]. Figure 3.10 displays TS surface mappings at $\lambda = 532$ nm of a superpolished Si wafer without (Figure 3.10a) and with particle contamination (Figure 3.10b), as well as a corresponding 1-D line scan for the contaminated wafer (Figure 3.10c). PSL spheres [41] of 0.1 μm diameter were used as test particles.

In the next example, a high-reflective fluoride multiplayer system for 193 nm was deposited on one half of a standard polished calcium fluoride substrate. Accordingly, the step in the TS mapping performed at $\lambda = 193$ nm (Figure 3.11) separates the scattering of the uncoated (right) and the coated part (left). The increased scatter of the coated surface results from the microstructure

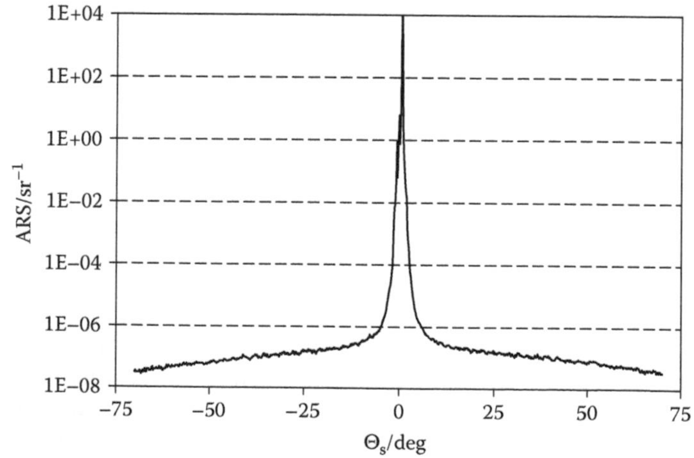

FIGURE 3.9 ARS measurement of a superpolished silicon wafer, $\lambda = 325$ nm.

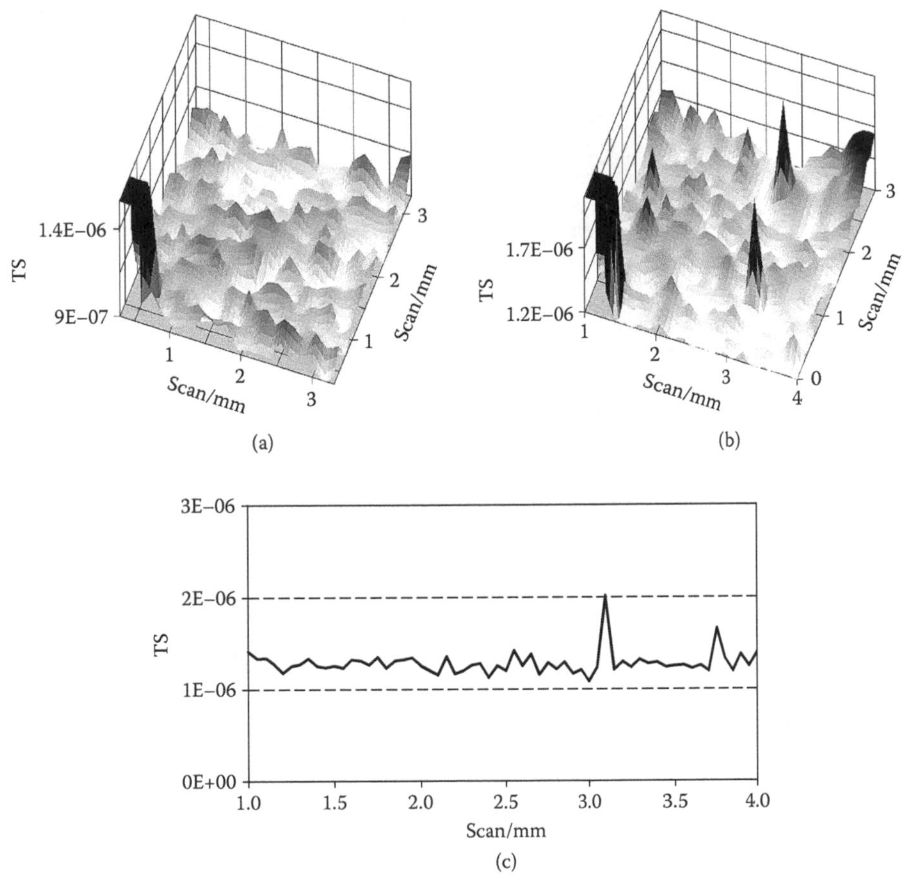

FIGURE 3.10 TS measurements (λ = 532 nm) of a superpolished silicon wafer (a) without and (b) with particle contamination. (c) 1-D line scan corresponding to the mapping in (b).

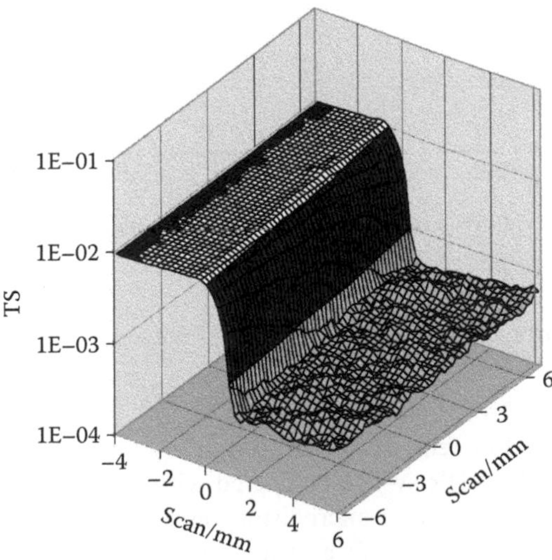

FIGURE 3.11 TS measurement (λ = 193 nm) of a fluoride high-reflective multilayer system deposited on one half of a standard polished calcium fluoride substrate. Right: uncoated, left: coated surface part.

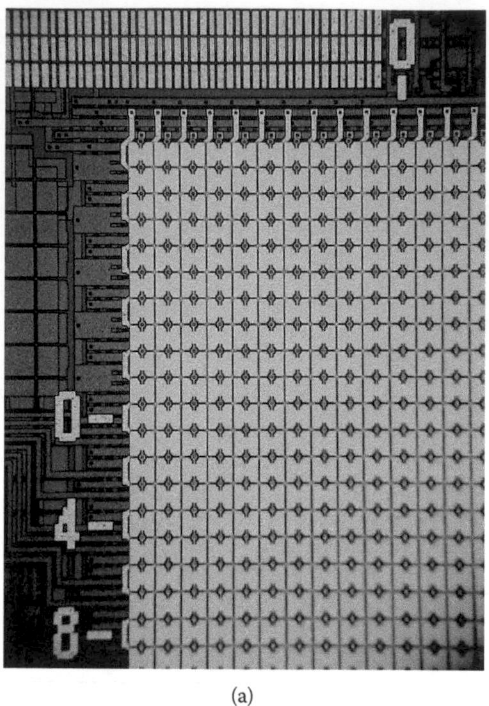

(a)

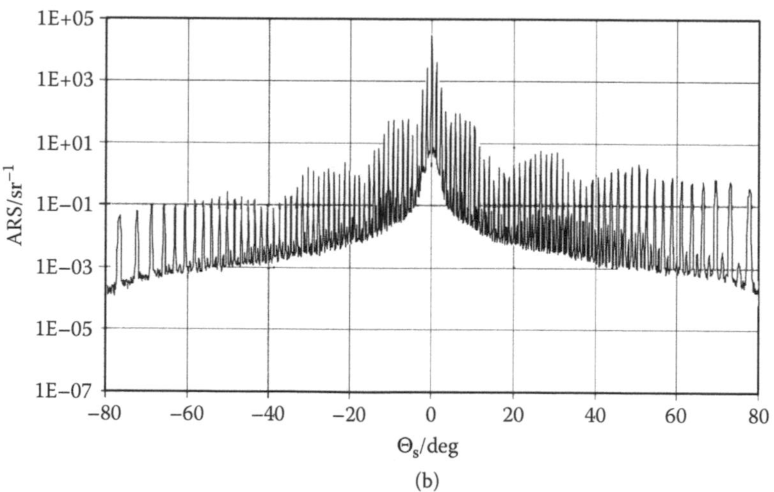

(b)

FIGURE 3.12 (a) Micromirror arrangement for DUV application, 16 mm^2 active surface area, (b) ARS measurement, $\lambda = 532$ nm.

of the film layers and related increased interface roughness, as well as from the enhanced optical factor (see Equation 3.6) of the HR system compared with the pure substrate [42].

Figure 3.12a displays a picture of a microdisplay arrangement consisting of micromirrors for DUV application. The overall active surface of the micromirror system was 16 mm^2. ARS measurements of these systems were made at 532 nm in the backscatter direction. The significant structures in the ARS-curve of Figure 3.12b are a result of both the microsystem structure and the internal microtopography of the mirrors.

3.6 COMBINATION OF LIGHT SCATTERING AND PROFILOMETRIC TECHNIQUES

Depending on the specific purpose and application, light scattering measurements can be advantageously combined with techniques that measure the profile of microstructured samples. These combination partners particularly concern, but are not limited to, methods such as atomic force microscopy (AFM), white light interferometry (WLI), mechanical profilometry, and laser scanning microscopy (LSM) [24,43–46]. In this section, we refer in particular to examples of combination with AFM. A detailed description of the AFM principle can be found in Chapter 4 of this volume.

Quite often, microstructure investigation tasks in practice require that the measurement fulfills "contradictory" demands: high vertical and lateral resolution shall be provided, and so shall results that are representative for application-relevant surface areas, which can extend to several mm^2 and cm^2. Here, light scattering techniques and AFM can ideally complement each other. An essential precondition for a proper interpretation of the obtained results, however, consists of careful consideration of the bandwidth limits of the different measurements [1,43,46]. For illustration, the spatial frequency bandwidths and vertical roughness resolution limits are depicted in Figure 3.13 for scattering measurements at wavelengths in the UV, VIS, and IR ranges, together with those for AFM and WLI.

The combination example in Figure 3.14 refers again to a superpolished Si wafer. The AFM image in Figure 3.14a provides the local rms roughness and the TS surface mapping (backscatter) in Figure 3.14b, the corresponding larger-area information. The singular peaks in the TS surface mapping originate from particle contaminations that are not registered in the local AFM scan. For calculating the rms roughness from TS using Equation 3.5 (left), the scatter level of the uncontaminated part was used. Very close roughness values were obtained from the TS and AFM measurements (0.06 and 0.07 nm). The PSD curves in Figure 3.14c were calculated from the local AFM image and from an ARS measurement at λ = 325 nm using Equation 3.2. The PSD difference within the overlapping spatial frequency range naturally occurs because the beam diameter of the ARS measurement was about 1 mm and, hence, the particle contaminations influenced the scatter signal and PSD calculation.

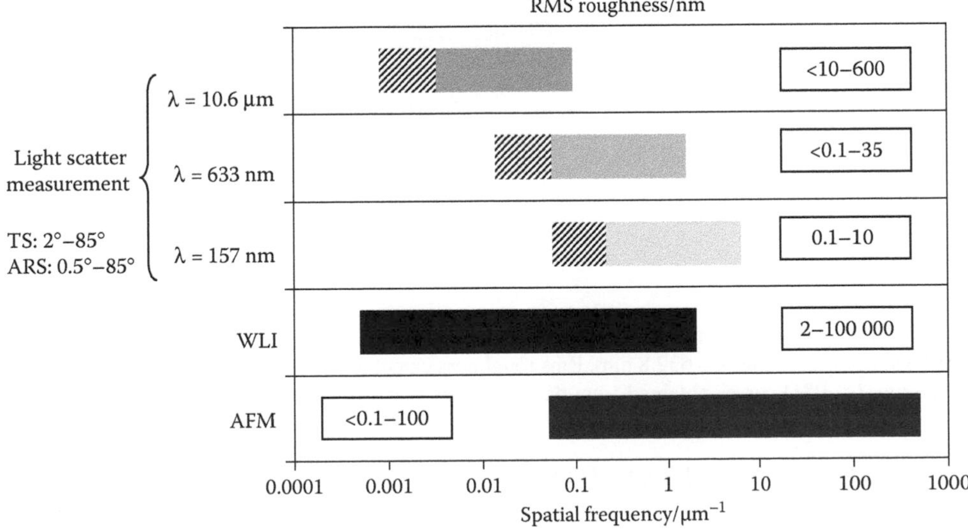

FIGURE 3.13 Spatial frequency bandwidths and vertical roughness resolution of light scattering measurements, AFM and WLI.

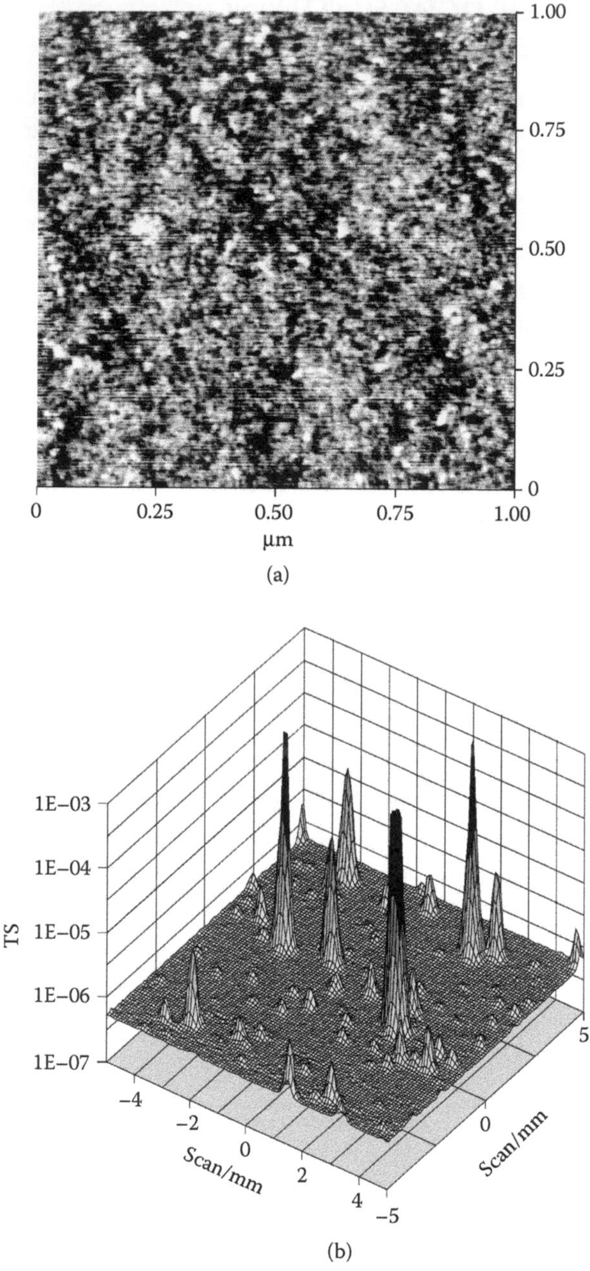

FIGURE 3.14 Investigation of a superpolished Si wafer. (a) AFM image, corresponding rms roughness: 0.07 nm, (b) TS surface mapping ($\lambda = 632.8$ nm). Rms roughness calculated from TS of the uncontaminated surface part: 0.06 nm, (c) PSD curves obtained from the AFM image and from ARS measurement at $\lambda = 325$ nm.

3.7 CONCLUSIONS AND OUTLOOK

This chapter has summarized some basic considerations on scattering measurement techniques that can be beneficially used for the nondestructive and efficient inspection of microstructures and microcomponents. A theoretical introduction provided a link between measured scatter signals and roughness data, which is applicable to manifold inspection tasks in practice.

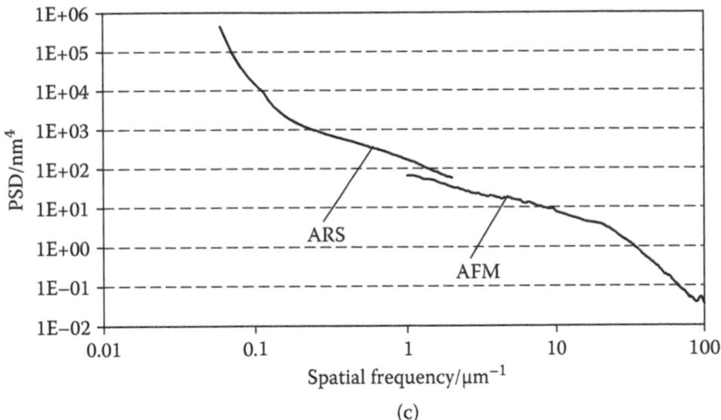

FIGURE 3.14 (Continued).

The experimental realization of today's measurement concepts was described in detail for ARS and TS equipment in the VUV-to-IR wavelength ranges. Examples of measurements that extended from superpolished surfaces to micromirrors, gratings, and thin-film coatings demonstrated the capability and flexibility of the measurement techniques.

Future work on scattering-based inspection techniques will especially include developments such as their extension to even shorter wavelengths (in particular, to 13.4 nm for inspections in EUV lithography), new concepts for particle detection down to diameters less than 50 nm, and specific methods for samples with high curvature.

REFERENCES

1. Bennett, J.M., Mattsson, L., *Introduction to Surface Roughness and Scattering*, 2nd ed., Washington, D.C.: Optical Society of America, 1999, pp. 47–70.
2. Stover, J.C., *Optical Scattering: Measurement and Analysis*, 2nd ed., Bellingham: SPIE — The International Society for Optical Engineering, 1995, pp. 3–27 and 133–175.
3. Church, E.L., Jenkinson, H.A., Zavada, J.M., Relationship between surface scattering and microtopographic features, *Opt. Eng*., 18: 125–136, 1979.
4. Elson, J.M., Bennett, J.M., Relation between the angular dependence of scattering and the statistical properties of optical surfaces. *J. Opt. Soc. Am*., 69: 31–47, 1979.
5. Bousquet, P., Flory, F., Roche, P., Scattering from multilayer thin films: theory and experiment, *J. Opt. Soc. Am*., 71: 1115–1123, 1981.
6. Elson, J.M., Theory of light scattering from a rough surface with an inhomogeneous dielectric permittivity, *Phys. Rev. B*., 30: 5460–5480, 1984.
7. Noll, R.J., Glenn, P., Mirror surface autocovariance functions and their associated visible scattering, *Appl. Opt*., 21: 1824–1838, 1982.
8. Bennett, H.E., Scattering characteristics of optical materials, *Opt. Eng*., 17: 480–488, 1978.
9. Carniglia, C.K., Scalar scattering theory for multilayer optical coatings, *Opt. Eng*., 18: 104–115, 1979.
10. Kassam, S., Duparré, A., Hehl, K., Bussemer, P., Neubert, J., Light scattering from the volume of optical thin films: theory and experiment, *Appl. Opt*., 31: 1304–1313, 1992.
11. Duparré, A., Light scattering of thin dielectric films, in Hummel, R.E., Günther, K.H., Eds., *Thin Films for Optical Coatings*, Boca Raton, FL: CRC Press, 1995, pp. 273–303.
12. Amra, C., Light scattering from multilayer optics, *J. Opt. Soc. Am. A*, 11: 197–226, 1994.
13. Duparré, A., Kassam, S., Relation between light scattering and the microstructure of optical thin films, *Appl. Opt*., 32: 5475–5480, 1993.
14. Elson, J.M., Rahn, J.P., Bennett, J.M., Light scattering from multilayer optics: comparison of theory and experiment, *Appl. Opt*., 19: 669–679, 1980.

15. Amra, C., Grèzes-Besset, C., Maure, S., Torricini, D., Light scattering from localized and random interface or bulk irregularities in multilayer optics: the inverse problem, *Proc. SPIE*, 2253: 1184–1200, 1994.
16. Duparré, A., Gliech, S., Non-contact testing of optical surfaces by multiple-wavelength light scattering measurement, *Proc. SPIE*, 3110: 566–573, 1997.
17. Rönnow, D., Veszelei, E., Design review of an instrument for spectroscopic total integrated light scattering measurements in the visible wavelength region, *Rev. Sci. Instrum.*, 65: 327–334, 1994.
18. Kienzle, O., Staub, J., Tschudi T., Description of an integrated scatter instrument for measuring scatter losses of superpolished optical surfaces, *Meas. Sci. Technol.*, 5: 747–752, 1994.
19. Rönnow, D., Roos, A., Correction factors for reflectance and transmittance measurements of scattering samples in focusing Coblentz spheres and integrating spheres, *Rev. Sci. Instrum.*, 66: 2411–2422, 1995.
20. Amra, C., Grezes-Besset, C., Roche, P., Pelletier, E., Description of a scattering apparatus: application to the problems of characterization of opaque surfaces, *Appl. Opt.*, 28: 2723–2730, 1989.
21. Neubert, J., Seifert, T., Czarnetzki, N., Weigel, T., Fully automated angle resolved scatterometer, *Proc. SPIE*, 2210: 543–552, 1994.
22. Orazio, F.D., Silva, R.M., Stockwell, W.K., Instrumentation for a variable angle scatterometer, *Proc. SPIE*, 384: 123–132, 1983.
23. Asmail, C.C., Cromer, C.L., Proctor, J.E., Hsia, J.J., Instrumentation at the National Institute of Standards and Technology for bidirectional reflectance distribution function (BRDF) measurements, *Proc. SPIE*, 2260: 52–61, 1994.
24. Duparré, A., Ferré-Borrull, J., Gliech, S., Notni, G., Steinert, J., Bennett, J.M., Surface characterization techniques for determining the root-mean-square roughness and power spectral densities of optical components, *Appl. Opt.*, 41: 154–171, 2002.
25. Amra, C., Torricini, D., Roche, P., Multiwavelength (0.45–10.6 μm) angle resolved scatterometer or how to extend the optical window, *Appl. Opt.*, 32: 5462–5474, 1993.
26. Schiff, T.F., Knighton, M.W., Wilson, D.J., Cady, F.M., Sover, J.C., Butler, J.J., Design review of a high accuracy UV to near IR scatterometer, *Proc. SPIE*, 1995: 121–130, 1993.
27. Schröder, S., Gliech, S. Duparré, A., Sensitive and flexible light scatter techniques from the VUV to IR regions, *Proc. SPIE*, 5965: 1B 1–9, 2005.
28. Detrio, J.A., Miner, S.M., Standardized total integrated scatter measurements of optical surfaces, *Opt. Eng.*, 24: 419–422, 1985.
29. Guenther, K.H., Wierer, P.G., Bennett, J.M., Surface roughness measurements of low-scatter mirrors and roughness standards, *Appl. Opt.*, 23: 3820–3836, 1984.
30. Duparré, A., Gliech, S., Quality assessment from supersmooth to rough surfaces by multiple-wavelength light scattering measurement, *Proc. SPIE*, 3141: 57–64, 1997.
31. ISO 13696, Optics and optical instruments — Lasers and laser related equipment — Test methods for radiation scattered by optical components, International Organization for Standardization, Geneva, Switzerland, 2002.
32. Gliech, S., Steinert, J., Duparré, A., Light scattering measurements of optical thin film components at 157 nm and 193 nm, *Appl. Opt.*, 41: 3224–3234, 2002.
33. Schröder, S., Gliech, S., Duparré, A., Measurement system to determine the total and angle-resolved light scattering of optical components in the deep-ultraviolet and vacuum-ultraviolet spectral regions. *Appl. Opt.* 44: 6093–6107, 2005.
34. ASTM E 1392-90, Standard practice for angle resolved optical scatter measurements on specular or diffuse surfaces, American Society for Testing and Materials, Philadelphia, 1990.
35. Leonard, T.A., Pantoliano, M., BRDF round robin, *Proc. SPIE*, 967: 226–235, 1989.
36. Leonard, T.A., Rudolph, P., BRDF round robin test of ASTM E 1392, *Proc. SPIE*, 1995: 285–293, 1993.
37. Baeumer, S., Duparré, A., Herrmann, T., Schuhmann, U., Smorenburg, C., Kirschner, V., Mattsson, L., Quinteros, T., Berglind, R., Schippel, S., SLIOS — a contribution to standard procedures in stray light measurement, *Proc. SPIE*, 3739: 414–421, 1999.
38. ASTM F 1048-87, Standard test method for measuring the effective surface roughness of optical components by total integrated scattering, American Society for Testing and Materials, Philadelphia, 1987.

39. Kadkhoda, P., Müller, A., Ristau, D., Duparré, A., Gliech, S., Schuhmann, U., Tilsch, M., Schuhmann, R., Amra, C., Deumie, C., Jolie, C., Kessler, H., Lindström, T., Ribbing, C.G., Bennett, J.M., International round-robin experiment to test the International Organization for Standardization total-scattering draft standard, *Appl. Opt.*, 39: 3321–3332, 2000.
40. Stover, J.C., Ivakhnenko, V.I., Eremin, Y.A., Use of light scatter signals to identify particle material, *Proc. SPIE*, 4449: 131–139, 2001.
41. Stover, J.C., A review of the emerging SEMI standards for particle scanners, *Proc. SPIE*, 5188: 162–168, 2003.
42. Jakobs, S., Duparré, A., Truckenbrodt, H., Interfacial roughness and related scatter in UV-optical coatings: a systematic experimental approach, *Appl. Opt.*, 37: 1180–1193, 1998.
43. Bennett, J.M., Comparison of techniques for measuring the roughness of optical surfaces. *Opt. Eng.*, 24: 380–387, 1985.
44. Amra, C., Deumie, C., Torricini, D., Roche, P., Galindo, R., Dumas, P., Salvan, F., Overlapping of roughness spectra measured in macroscopic (optical) and microscopic (AFM) bandwidths, *Proc. SPIE*, 2253: 614–630, 1994.
45. Duparré, A., Jakobs, S., Combination of surface characterization techniques for investigating optical thin-film components, *Appl. Opt.*, 35: 5052–5058, 1996.
46. Duparré, A., Notni, G., Multi-type surface and thin film characterization using light scattering, scanning force microscopy and white light interferometry, in Al-Jumaily, G.A., Ed., *Optical Metrology (SPIE Critical Reviews Series)*, Bellingham: SPIE — The International Society for Optical Engineering, 1999, pp. 213–231.

4 Characterization and Measurement of Microcomponents with the Atomic Force Microscope (AFM)

F. Michael Serry and Joanna Schmit

CONTENTS

4.1 Introduction....122
4.2 Components of AFM and Principles of AFM Operation....122
4.2.1 Probe....123
4.2.2 Scanner....123
4.2.3 Controller....124
4.2.4 Detection, Input Signal, Set Point, and Error Signal....124
4.2.5 Z Feedback Loop....125
4.3 AFM Imaging Modes....125
4.3.1 Primary AFM Imaging Modes....125
4.3.1.1 Contact Mode AFM....125
4.3.1.2 Noncontact Mode AFM....126
4.3.1.3 TappingMode AFM....126
4.3.1.4 Torsion Resonance Mode (TRMode) AFM....127
4.3.2 Secondary AFM Imaging Modes....127
4.3.2.1 Lateral Force Microscopy (LFM)....127
4.3.2.2 Phase Imaging....129
4.3.2.3 Magnetic Force Microscopy (MFM)....129
4.3.2.4 Conductive AFM (CAFM)....130
4.3.2.5 Electric Force Microscopy (EFM) and Surface Potential Imaging....131
4.3.2.6 Force Modulation Imaging....132
4.3.2.7 Scanning Capacitance Microscopy (SCM)....132
4.3.2.8 Scanning Spreading Resistance Microscopy (SSRM)....133
4.3.2.9 Tunneling AFM (TUNA)....133
4.4 AFM Nonimaging Modes....134
4.5 Applications of AFM for Microcomponent Inspection — a Case Study....134
4.6 Atomic Force Profilometer (AFP) — a Combination of AFM and Stylus Profiler....140
4.7 Optical Metrology Complementary to AFM....142
4.8 Conclusions and Outlook....142
References....143

4.1 INTRODUCTION

The atomic force microscope (AFM) is frequently used in measuring and otherwise characterizing surfaces in microsystem components, including microelectromechanical systems (MEMS) devices and magnetic recording read–write heads. The AFM is a tactile instrument that relies on direct physical contact of its mechanical microfabricated probe with the sample surface to make its measurements [1].

The AFM can generate three-dimensional maps of a surface with nanometer, and even Ångstrom, resolution. The most popular application of the AFM is in mapping surface topography and making measurements of the dimensions of individual features or of a collection of features, or statistically quantifying surface topography (e.g., surface roughness) [2].

This chapter first describes the components and principles of operation of the AFM. It then describes several different popular or new AFM imaging modes before illustrating in a case study the usefulness of the AFM in microsystems component metrology and characterization. The case study is chosen to also illustrate that the AFM is not merely a research instrument but that it is increasingly adopted as a reliable, indeed indispensable, tool in product development and industrial micromanufacturing.

The chapter also includes a section on a hybrid instrument, mainly used in industry, that combines features from a stylus profiler and an AFM. Finally, a brief description of some complementary optical measurement techniques and a comparison with the AFM leads to the concluding summary.

4.2 COMPONENTS OF AFM AND PRINCIPLES OF AFM OPERATION

The AFM is a scanning probe microscope (SPM). SPMs are a family of instruments used for studying surface properties of materials locally, on length scales that typically span a range from several tens of micrometers all the way down to the atomic level. The key components of the AFM are the probe, scanner, controller, and computer (and software). See Figure 4.1.

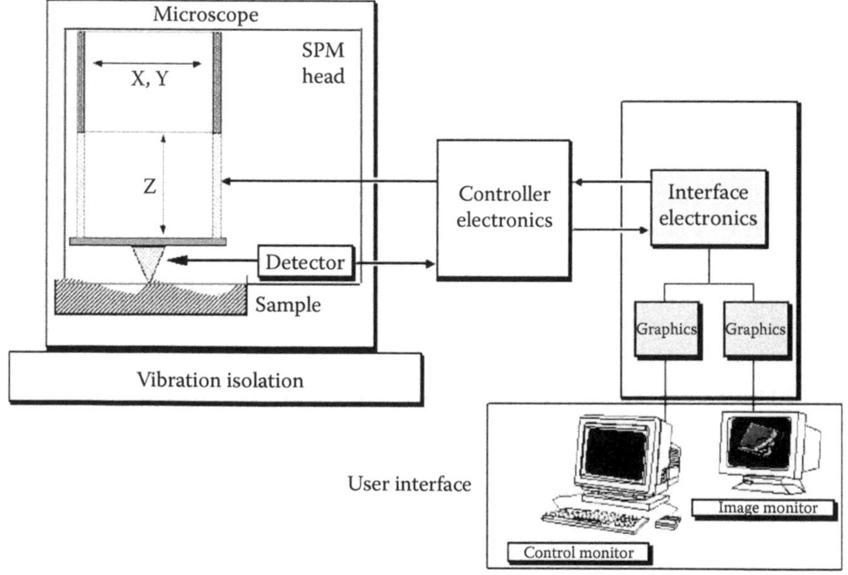

FIGURE 4.1 This generic block diagram depicts the main elements of a scanning probe microscope (SPM). The term "SPM head" usually refers to the combination of the scanner (which moves the probe or the sample) and the detector and sensors that measure the interaction between the probe tip and the sample surface.

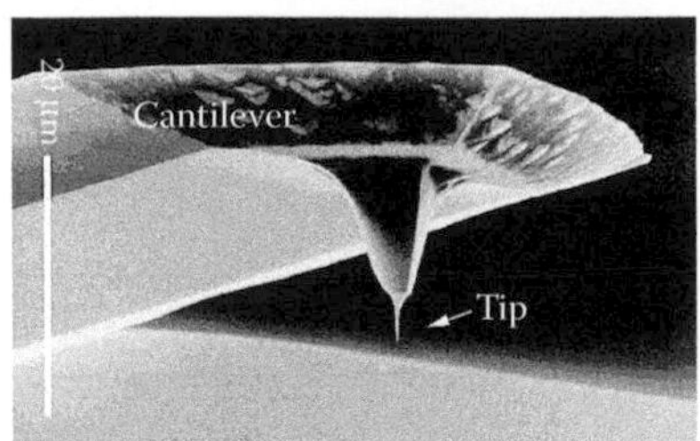

FIGURE 4.2 AFM cantilever and tip. This is a special type of AFM probe in that after batch fabrication of the tip and cantilever, the tip has been further micromachined for special measurements, such as in high-aspect-ratio trenches and holes. (Original micrograph courtesy NANOSENSORS™. Modified at Veeco Instruments, and reproduced with permission from NANOSENSORS, NanoWorld AG, Neuchatel, Switzerland.)

The probe measures the dimensions of the features on a sample surface and interrogates various qualities of the sample surface locally — that is, with an in-plane resolution that can be as small as the diameter of the probe end, typically a few nanometers.

The scanner typically has one or more piezoelectric actuating elements and controls the precise position and motion of the probe in relation to the surface, both vertically and laterally, typically with subnanometer lateral resolution. In the vertical dimension — in Z — the scanner's positioning resolution determines the AFM's resolution. Because piezoelectric actuators can be controlled to move extremely small distances, the AFM can map the Z dimension with sub-Ångstrom resolution, even with a probe that has an end radius of several nanometers.

The controller and the computer are the elements that include most of the electronics and, in most AFM designs, nearly all the software required to make the AFM work.

4.2.1 Probe

When two materials are brought very close together, subtle interactions take place at the atomic level. An AFM probe, such as the one shown in Figure 4.2, is a microfabricated cantilever with a sharp tip that comes in contact with the sample surface. This sharp tip is typically less than 5 μm long with a 2–10 nm end radius. The tip is located at the free end of a cantilever (made usually of silicon or silicon nitride) that is typically 100–500 μm long and 2–10 μm thick. Forces between the tip and the sample surface cause the cantilever to bend (cantilever deflection) as the AFM scans the sample surface. Often, the cantilever is so flexible that the tip–sample contact force is smaller than the force that holds the nearest neighbor atoms together in most solids.

4.2.2 Scanner

The (piezoelectric) scanner moves the probe relative to the sample surface (or the sample relative to the probe) as illustrated in Figure 4.1; the scanner is typically a cylindrical tube with separate portions (labeled here X, Y, and Z) that move the tube in different directions. Piezoelectric ceramics provide the kind of precise motion and positioning resolution required by the AFM scanner because these ceramics change geometry (and therefore move) when a voltage is applied. The resulting movement is related to the magnitude and polarity of the applied voltage. The size of the piezoelectric scanner determines the scan range of the AFM. The scanner raster-scans the probe (or the sample) in a defined pattern in the X and Y directions. The raster scan consists of a series of lines that define a zigzag pattern spanning a square or rectangular area (Figure 4.3). When the number of lines is large enough, then they are nearly parallel.

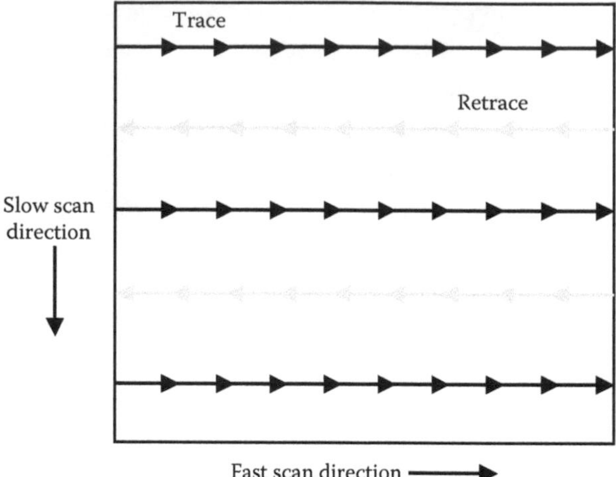

FIGURE 4.3 Raster scanning. Here, the AFM lines of scan are shown nearly parallel to each other, but in reality they form a zigzag pattern known as a raster scan. The point where a scan line to the right (trace) ends is the same point where a scan line to the left (retrace) begins. This picture shows a small portion of a scan area away from the left or the right edge; in depicting the lines as parallel, the implication is that the scan area is densely covered with numerous scan lines.

4.2.3 Controller

The architecture and details of the AFM controller vary significantly from one model (or manufacturer) to another, but AFM controllers generally comprise electronics that measure, route, and convert the signals that are exchanged between AFM system components, including those between the computer and the scanner, and between the probe-and-detector assembly and the computer, as shown in Figure 4.1.

4.2.4 Detection, Input Signal, Set Point, and Error Signal

A detector measures the cantilever deflection as the tip travels over the sample or as the sample is scanned under the tip. Most AFMs use optical techniques to detect the position (deflection) of the cantilever. In the most common scheme, shown in Figure 4.4, a laser beam bounces off the back

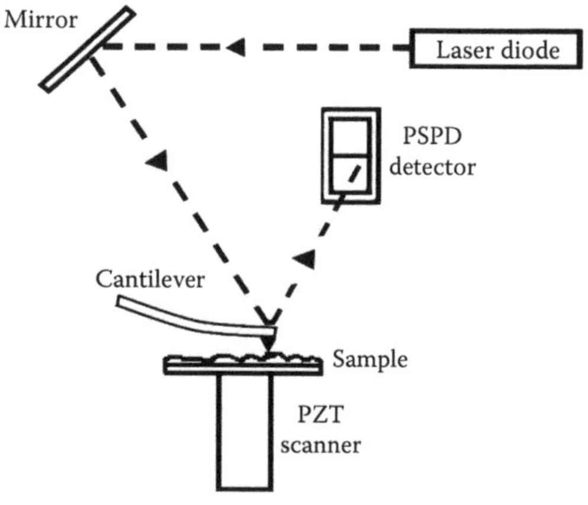

FIGURE 4.4 The AFM laser-beam-bounce detection scheme. (PZT = piezoelectric scanner.)

of the cantilever onto a position-sensitive photodetector (PSPD). As the cantilever bends, the position of the laser beam on the detector shifts. The ratio of the path length between the cantilever and the detector to the length of the cantilever is large and helps amplify the cantilever deflection signal on the detector. The setup can detect sub-Ångstrom vertical movement of the cantilever. This deflection signal, or input signal, is used to calculate another signal called the *error signal* (described in the text that follows).

During raster scanning, as the tip encounters changes in the sample topography, the cantilever moves in the Z direction, triggering a corresponding variation in the input signal. In order for the input signal to be useful for topography mapping and measurement, the AFM operator, using the AFM software, establishes a reference tip–sample force value via a software parameter commonly known as the *setpoint*. The input signal is monitored and compared to the setpoint value. The difference between these two values is referred to as the error signal, which is used in the feedback loop explained next.

4.2.5 Z Feedback Loop

As raster scanning progresses, the error signal is continuously sampled and injected (as input signal) into a feedback loop. The AFM employs a method known as *Z feedback* to ensure that the tip tracks the sample surface topography. The output of the Z feedback loop is a voltage applied to the Z portion of the piezoelectric tube actuator (Z-piezo), which is a part of the scanner. This voltage is continuously adjusted to always keep the error signal at zero; the AFM Z feedback loop tries to keep a constant deflection at the tip end of the cantilever by adjusting the distance between the sample plane and the other end of the cantilever. The voltage applied to the Z-piezo is the signal that is usually used to generate an AFM topography (or height) image; this is accomplished by plotting the voltage vs. in-plane coordinates X and Y, point by point and line by line, as raster scanning progresses.*

The AFM software is used to visualize this data in a useful, meaningful way. For example, the height level, indicated by different colors in a color scale can be adjusted to highlight features of interest. The number of data points in each line and the number of lines in each frame determine the image resolution.

4.3 AFM IMAGING MODES

4.3.1 Primary AFM Imaging Modes

There are four primary AFM modes: contact mode, noncontact mode, TappingMode, and Torsion Resonance Mode (TRMode).** From these are derived many secondary modes, which are described in the following subsections.

4.3.1.1 Contact Mode AFM

In contact mode AFM, the tip is in perpetual contact with the sample. As the scanner gently traces the tip across the sample (or the sample under the tip), the contact force causes the cantilever to bend and the Z feedback loop works to maintain a constant cantilever deflection, which corresponds to the setpoint value described in Section 4.2.

---

* The topography (height) image is usually based on the voltage applied to the Z-portion of the piezoelectric scanner. The error signal image often contains complementary information about the sample surface. For this reason, topography and error signal images are often collected simultaneously and examined side by side.

** TappingMode AFM and Torsion Resonance Mode (TRMode) AFM are patented and trademarks of Veeco Instruments.

4.3.1.2 Noncontact Mode AFM

Noncontact AFM (NC-AFM) is one of several vibrating-cantilever techniques in which the AFM cantilever is oscillating the tip in the Z-direction near the surface of a sample. The small force values in the noncontact regime and the greater stiffness of the cantilevers used for NC-AFM are both factors that make the NC-AFM signal small and, therefore, difficult to measure. In addition, NC-AFM requires that the tip stay very close to the sample surface, much closer than in TappingMode (described next). This closeness often creates conditions in which the AFM tip spends much, or all, of each oscillation cycle inside a thin liquid layer that typically covers most samples in gaseous environments, including in air. This makes imaging unstable and difficult and requires that scanning speed be significantly reduced to avoid the instability.

4.3.1.3 TappingMode AFM

TappingMode AFM, the most commonly used of all AFM modes, is a patented technique that maps topography by lightly tapping the surface with an oscillating probe tip [3]. A small piezoelectric element, using a small AC signal that it receives from the AFM controller, drives the cantilever mechanically at or near the cantilever's fundamental resonance frequency (see Figure 4.5a).

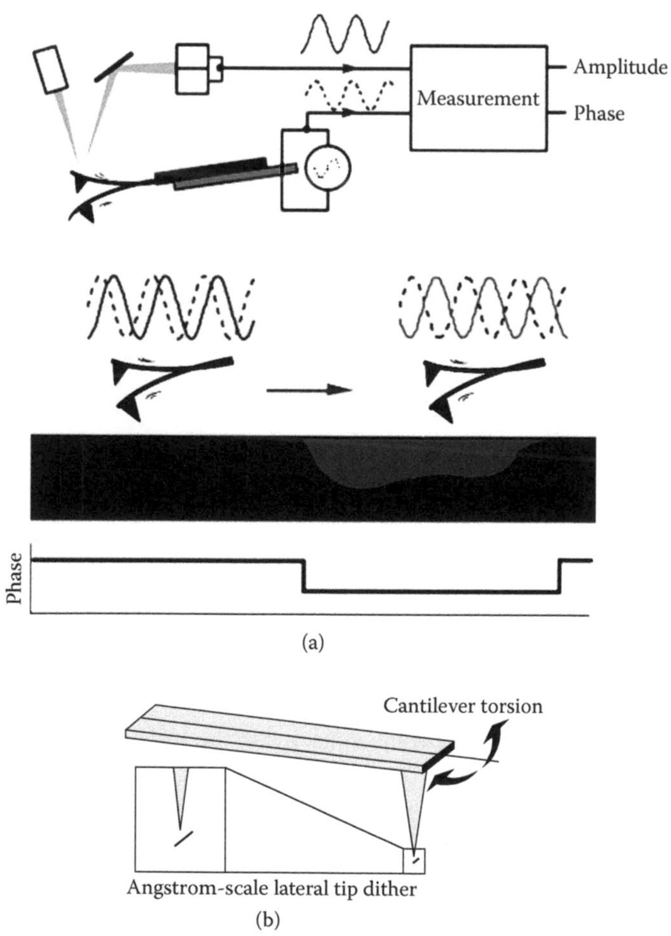

FIGURE 4.5 (a) In TappingMode AFM, the cantilever is mechanically driven (in the Z direction) at or near its fundamental resonance frequency. (b) Cantilever and tip motion in Torsion Resonance Mode (TRMode) AFM.

The cantilever's oscillation amplitude and phase change with the sample's topography and properties because the tip makes intermittent contact with the sample and loses or gains kinetic energy in the process. As the tip taps the surface, a topographic image is obtained by monitoring the changes in the amplitude and closing the Z feedback loop to minimize them. (In TappingMode AFM, the error signal described in Section 4.2 is the rms amplitude of cantilever oscillation minus the setpoint value.)

TappingMode has become an important AFM technique, as it overcomes some of the limitations of both contact and NC-AFM. By eliminating lateral shear forces that can damage soft samples and reduce image resolution, TappingMode allows routine imaging of samples that were once considered impossible to image with an AFM, especially in contact mode.

Another major advantage of TappingMode is related to the limitations that can arise owing to the thin layer of liquid that forms on most sample surfaces in an ambient imaging environment, i.e., in air and in many other gases. The amplitude of the cantilever oscillation in TappingMode is typically up to a few tens of nanometers, which ensures that the tip does not get stuck in this liquid layer. The amplitude used in NC-AFM is much smaller; as a result, in NC-AFM, the tip often gets stuck in the liquid layer unless raster scanning is performed at a very slow speed.

In general, TappingMode is much more effective than NC-AFM for imaging, but especially so for surfaces that may include large height variations of up to several micrometers. TappingMode can be performed in gases, liquids, and in vacuum. Typically, TappingMode AFM cantilevers for in-liquid operation have a lower spring constant than those used for TappingMode in air and in vacuum.

4.3.1.4 Torsion Resonance Mode (TRMode)

TRMode is a major new technique in atomic force microscopy that measures and controls dynamic lateral forces between the probe tip and sample surface. Utilizing advanced sensing hardware and electronics to characterize torsion oscillations of the cantilever, TRMode enables detailed, nanoscale examination of in-plane anisotropy and provides new perspectives in the study of material structures and properties [4]. TRMode can also be interleaved with TappingMode AFM to provide complementary lateral and vertical characterization.

In TRMode the probe oscillates along the cantilever's long axis, creating torsion, i.e., a twisting motion. This oscillation causes a dithering motion of the tip, typically a fraction of a nanometer in amplitude (see Figure 4.5b). As the probe encounters lateral forces on the sample surface, the corresponding changes in the cantilever's twisting motion are measured.

This twisting is measured by using a quadcell PSPD. Contact mode AFM uses a bicell PSPD to measure the vertical deflection of the cantilever, which indicates changes in sample topography. With a quadcell PSPD, both vertical and lateral deflections can be measured, as shown in Figure 4.6.

4.3.2 Secondary AFM Imaging Modes

The secondary imaging modes described in the following sections all derive from three of the four primary AFM imaging modes: contact mode, TappingMode, and TRMode.

4.3.2.1 Lateral Force Microscopy (LFM)

LFM is a secondary contact AFM mode that detects and maps relative differences in the frictional forces between the probe tip and the sample surface. In LFM, the scanning is always perpendicular to the long axis of the cantilever. Forces on the cantilever that are parallel to the plane of the sample surface cause twisting of the cantilever around its long axis. As in TRMode, this twisting is measured by a quadcell PSPD.

Twisting of the cantilever usually arises from either changes in surface friction or topography. Figure 4.7 shows that the tip may experience greater friction as it traverses some areas, causing the cantilever to twist more. The cantilever twists in one direction when scanning left to right (trace), and the opposite direction when scanning right to left (retrace). To separate topographic

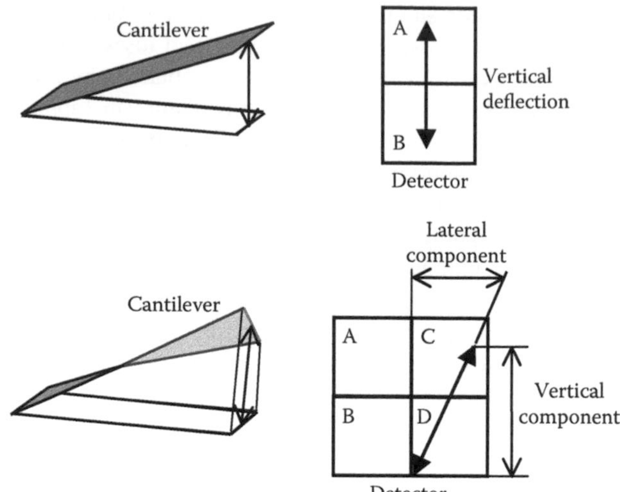

FIGURE 4.6 Quadcell PSPD used in TRMode AFM.

artifacts from real friction signals, the trace and retrace LFM signals are collected simultaneously and the trace signal is then inverted and added to the retrace signal. In this way, most topographic artifacts are removed from the LFM signal [5].

LFM applications include identifying transitions between different components in polymer blends and composites, identifying contaminants on surfaces, and delineating coverage by coatings. An additional group of applications, usually called *chemical force microscopy*, is one in which the probe tip is functionalized for specific chemical compositions (or biological species).

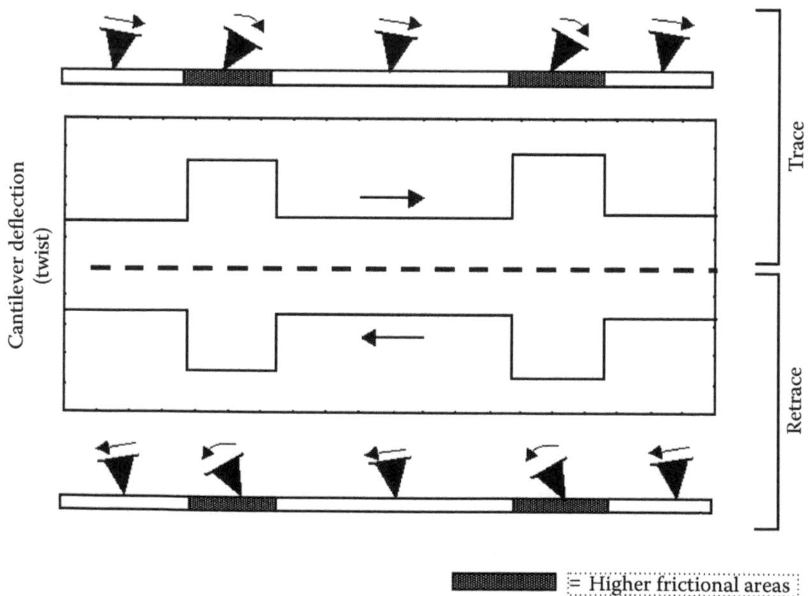

FIGURE 4.7 AFM tip lateral movement in LFM.

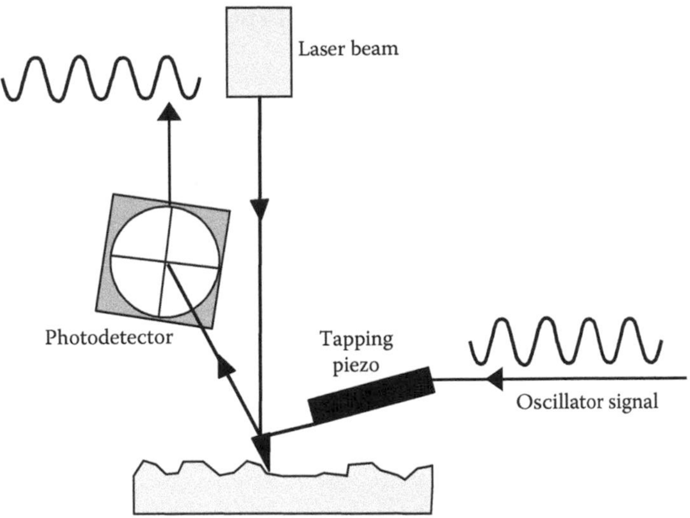

FIGURE 4.8 The phase lag (between oscillator signal and photodetector signal) varies with the properties of the sample surface. See also Figure 4.5a.

4.3.2.2 Phase Imaging

Phase imaging is derived from TappingMode and goes beyond topographic data to detect variations in composition, adhesion, friction, viscoelasticity, and other properties, including electric and magnetic. Applications for phase imaging include contaminant identification, mapping of components in composite materials, differentiating regions of high and low surface adhesion or hardness, and regions with different electrical or magnetic properties [6].

Phase imaging is the mapping of the phase lag between the periodic signal that drives the cantilever and the oscillations of the cantilever (see Figure 4.5a and Figure 4.8). Changes in the phase lag often indicate changes in the properties of the sample surface. The system's feedback loop operates in the usual manner, using changes in the cantilever's oscillation amplitude to map sample topography. The phase lag is monitored while the topographic image is being taken so that images of topography and material properties can be collected simultaneously.

The phase signal is sensitive to both short- and long-range tip–sample interactions. Short-range interactions include adhesive forces and frictional forces; long-range interactions include electric and magnetic forces.

Techniques based on phase detection, including phase imaging, are usually more sensitive than those based on amplitude detection, and this enhanced sensitivity is what makes phase imaging useful in other scanning probe techniques, including magnetic force microscopy (MFM) and electric force microscopy (EFM).

4.3.2.3 Magnetic Force Microscopy (MFM)

MFM is derived from TappingMode and maps the magnetic force gradient above the sample surface [7]. This mapping is performed through a patented two-pass technique, LiftMode.* LiftMode separately measures topography and another selected property (magnetic force, electric force, surface potential, etc.), using the topographic information to track the probe tip at a constant height (lift height) above the sample surface during the second pass (Figure 4.9).

* LiftMode is patented and a trademark of Veeco Instruments.

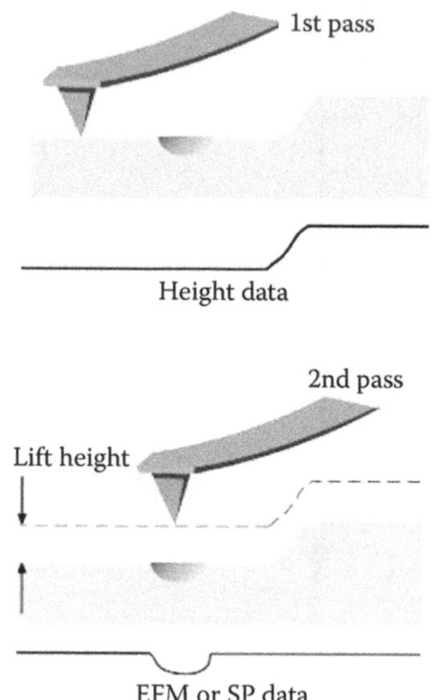

FIGURE 4.9 LiftMode AFM is used in magnetic force microscopy (MFM), electric force microscopy (EFM), and surface potential (SP) imaging. LiftMode may also be used for force modulation imaging with a negative value for lift height so that the tip is pushed into the sample surface rather than lifted above it.

The MFM probe tip is coated with a ferromagnetic thin film. While scanning, it is the magnetic field's dependence on tip–sample separation that induces changes in the cantilever's resonance frequency or phase (see Figure 4.10). MFM can be used to image both naturally occurring and deliberately written domain structures in magnetic materials. Figure 4.11 shows the TappingMode topography image and the MFM phase image of naturally occurring magnetic domains in an experimental compound.

4.3.2.4 Conductive AFM (CAFM)

CAFM is derived from contact mode AFM and characterizes conductivity variations across medium- to low-conducting and semiconducting materials. CAFM performs general-purpose measurements and usually covers a current range of pA to μA. CAFM employs a conductive probe tip; typically, a DC bias is applied to the tip, and the sample is held at ground potential. While the Z feedback signal is used to generate a normal contact AFM topography image, the current passing between the tip and the sample is measured to generate the CAFM image (Figure 4.12).

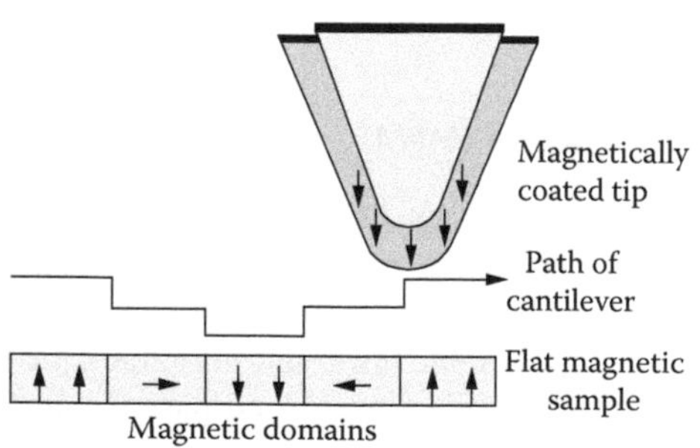

FIGURE 4.10 MFM maps the magnetic domains of the sample surface.

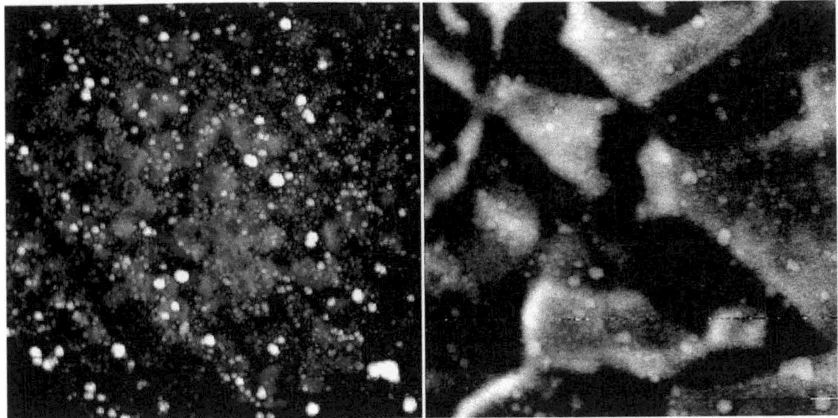

FIGURE 4.11 (Left) TappingMode topography image, (right) LiftMode MFM phase image of an experimental magnetic material. Scan area 4 μm. (Data courtesy of F.M. Serry, Veeco Instruments.)

4.3.2.5 Electric Force Microscopy (EFM) and Surface Potential Imaging

EFM is derived from TappingMode and measures electric field gradient distribution above the sample surface [8]. This measurement is performed through LiftMode. In EFM, a voltage may be applied between the tip and the sample. The cantilever's resonance frequency shift and phase change with the strength of the electric field are used to construct the EFM image. For example, locally charged domains on the sample surface are mapped in a way that is similar to how MFM maps magnetic domains (see Figure 4.13).

Surface potential imaging (also known as scanning Kelvin probe force microscopy) is a technique closely related to EFM and also uses LiftMode and TappingMode [9]. In the first pass of each LiftMode scan line, topographic data are collected in TappingMode and displayed. In the second pass, the tip lifts from the surface and scans the same line again, following the just-recorded topography for that line. In the second pass, the signal that usually drives the cantilever mechanically for TappingMode operation is diverted from the drive piezo element that oscillates the cantilever. This signal is instead applied directly to the cantilever and tip, creating an AC potential difference between the tip and the sample. If there is also a DC potential difference between the tip and the sample, the added AC signal will create a mechanical oscillation of the cantilever. The surface potential image is created by plotting the value of a DC voltage bias that the feedback loop applies to the tip. This bias is applied in order to bring the tip and the sample (at the location of the tip) to the same electrostatic potential. The system achieves this goal by monitoring the cantilever oscillations and by always trying to null those oscillations. In that sense, surface potential imaging is a nulling technique (see Figure 4.14).

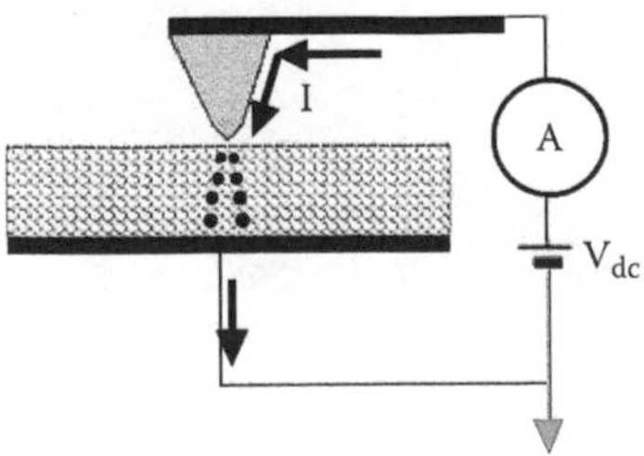

FIGURE 4.12 Conductive AFM (CAFM).

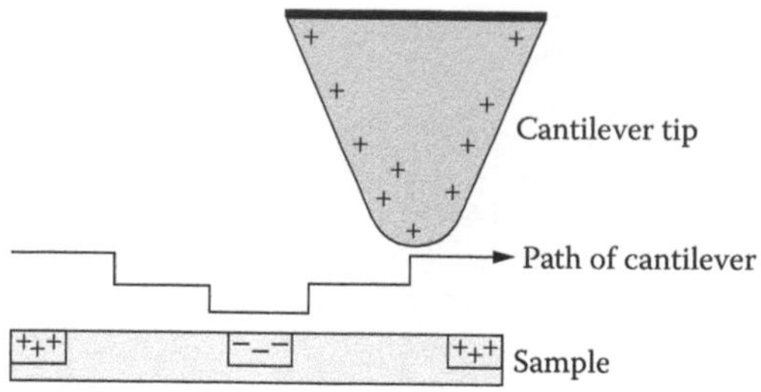

FIGURE 4.13 EFM can map the charged domains on the sample surface.

4.3.2.6 Force Modulation Imaging

Force modulation imaging is derived from contact AFM and measures the relative elasticity and stiffness of surface features [10]. It is commonly used to map the distribution of materials in a composite system. As with LFM and MFM, force modulation imaging allows simultaneous acquisition of both topographic and material properties maps.

In force modulation imaging mode, the probe tip tracks the sample topography as in normal contact AFM. In addition, a periodic mechanical signal is applied to the base of the cantilever. The amplitude of the cantilever modulation that results from this applied signal varies according to the elastic properties of the sample, as shown in Figure 4.15.

The resulting force modulation image is a map of the sample's elastic response. The frequency of the applied signal is typically a few kilohertz, which is faster than the Z feedback loop that is usually set up to track the topography. Thus, topographic information can be separated from local variations in the sample's elastic properties. Figure 4.16 shows a force modulation image.

4.3.2.7 Scanning Capacitance Microscopy (SCM)

SCM is a secondary imaging mode derived from contact AFM that maps variations in majority electrical carrier concentration (electrons or holes) across the sample surface (typically a doped semiconductor). As in EFM, an AC bias voltage is applied between the tip and sample. The tip scans across the sample surface, and changes in capacitance between the tip and the sample surface are monitored by an extremely sensitive high-frequency resonant circuit.

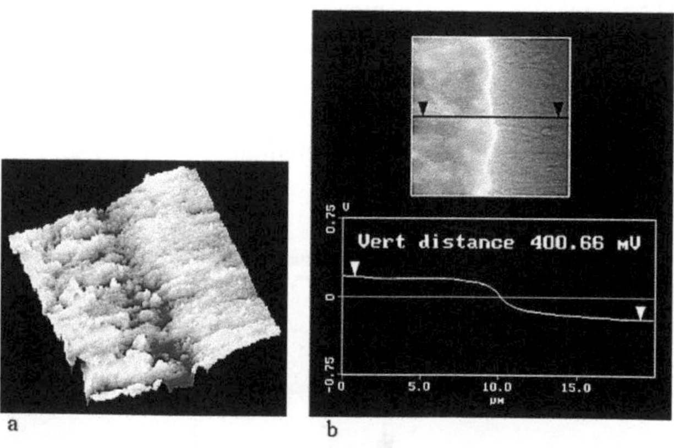

FIGURE 4.14 (a) TappingMode topography, (b) surface potential images across the boundary that separates copper from aluminum. With no external bias applied, the difference in the work function values of the two metals gives rise to a potential drop across the boundary. Scan area 20 μm. (Data courtesy of F.M. Serry, Veeco Instruments.)

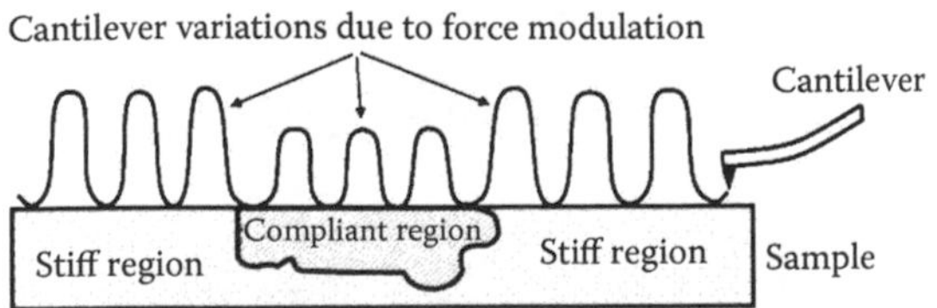

FIGURE 4.15 In force modulation imaging, the amplitude of cantilever oscillation varies according to the mechanical properties of the sample surface.

SCM is commonly used for two-dimensional profiling of dopants in semiconductor process evaluation and failure analysis. An example of the topography and SCM image of the cross section of a silicon-based transistor is shown in Figure 4.17a and b, respectively.

4.3.2.8 Scanning Spreading Resistance Microscopy (SSRM)

SSRM is a patented secondary imaging mode derived from contact AFM that maps two-dimensional carrier concentration profiles (resistance) in semiconductor materials. A conductive probe is scanned in contact mode across the sample while a DC bias is applied between the tip and sample. The resulting current between the tip and sample is measured using a logarithmic current amplifier providing a range of 10 pA to 0.1 mA.

4.3.2.9 Tunneling AFM (TUNA)

TUNA works similarly to CAFM but with higher sensitivities. TUNA measurements characterize ultralow currents (<1 pA) through the thickness of thin films. The TUNA application module can be operated in either imaging or spectroscopy mode (see Section 4.4 for AFM spectroscopy). Applications include gate dielectric development in the semiconductor industry and failure analysis of magnetoresistive heads for hard disk drives.

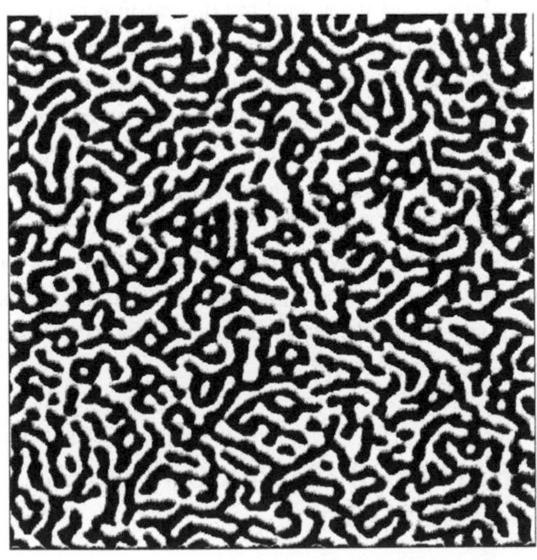

FIGURE 4.16 Force modulation image of a block copolymer identifies two phases of the material: a more glassy phase and a more rubbery phase. Scan area 1 μm. (Data courtesy of F.M. Serry, Veeco Instruments.)

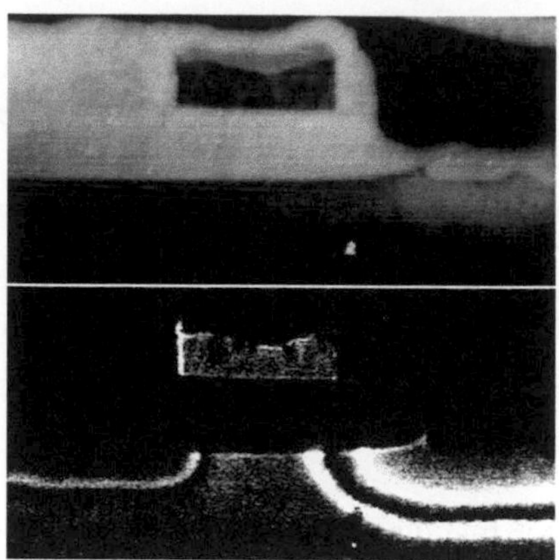

a. top b. bottom.

FIGURE 4.17 (a) Contact mode AFM topography, (b) SCM images of the cross section of a silicon-based transistor. Contrast in the SCM image outlines the distribution of majority carriers across differently doped areas. Scan area 5 μm × 2.5 μm. (Data courtesy of F.M. Serry, Veeco Instruments.)

4.4 AFM NONIMAGING MODES

The AFM is often used in a way that does not result in the images described in the previous sections. Instead of raster-scanning the sample with the probe, in many applications the AFM probe is located above the sample at a single in-plane (X,Y) coordinate, and the interaction between the AFM probe and the sample is detected and measured at that coordinate.

Sometimes a third coordinate (Z) is ramped, modulating the probe–sample separation; sometimes a different parameter (e.g., a DC bias voltage between the probe and the sample) is ramped while the separation is maintained. The probe–sample interaction is measured as a function of time and is plotted either vs. time or vs. the value of the ramped parameter. This technique is usually called *AFM spectroscopy.* A very commonly used example of AFM spectroscopy is X,Y-localized I-V characterization, such as is done on semiconductor devices.

Another example that is often used in MEMS is probe–sample force spectroscopy for investigating the underlying mechanism of stiction and wear, and for developing and evaluating possible solutions for them. Long-term mechanical reliability of microsystem elements is often investigated experimentally using cyclic loading with an AFM tip (AFM force spectroscopy). In force spectroscopy, the AFM probe and the sample approach and retract, and the strength and direction of the force between them is monitored and measured as the tip and the sample are apart, come close to contact, make contact, and press against each other with continued ramping and then separate; often the cycle repeats. (See Figure 4.18.)

Material roughness measured by AFM during different stages of testing helps to retrieve information about the fatigue behavior of the material. AFM is often used to measure lateral pitch and is the only tool to measure small structures on porous surfaces of membranes with subnanometer resolution, simultaneously in-plane and vertically.

4.5 APPLICATIONS OF AFM FOR MICROCOMPONENT INSPECTION — A CASE STUDY

The AFM is routinely used not only in research but also in product development and in failure analysis of microcomponents. Automated AFMs are now in place in numerous MEMS fabs. This section presents a case study that shows the use of the AFM as a measurement instrument for the digital micromirror

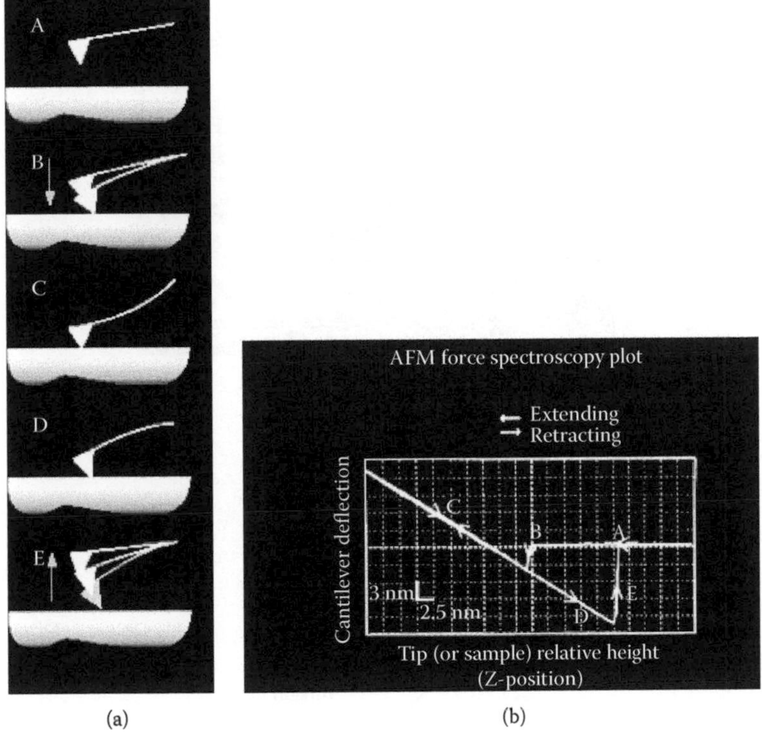

FIGURE 4.18 SPM spectroscopy, illustrated here for an AFM. At a given X,Y coordinate, tip–sample distance is changed by changing the Z-position of one or the other, and cantilever deflection is measured.

device (DMD). The DMD is incorporated in products for display applications, such as for presentations and movies. The DMD is an element in Texas Instrument's Digital Light Processing technology (DLP).*

DLP products include some of the most advanced commercially successful technology with optical MEMS at their core. The DMD (from DLP) exemplifies the value that mature MEMS are adding to products that impact people's lives and work. These products use a large array of actuated micromirrors to redirect light with high precision and color selectivity, enabling a novel projection scheme. In home entertainment systems, in movie theatres (DLP Cinema),** and in projectors used for professional presentations the DMD-based projection technology dramatically improves the visual quality of movies, still pictures, and animations.

Some of the impressive distinguishing features of DMD technology are the following:

- Number of actuated mirrors (0.5 million to 1.2 million per chip)
- Lifetime requirement (450 billion cycles of actuation)
- MEMS fully integrated with CMOS (complementary metal-oxide semiconductor) microelectronics

Some measurements (e.g., in the production environment) can be automated to various levels on semiautomated or fully automated AFMs.

Early in the development of the DMD, nonautomated AFMs were used to make measurements of device geometry, surface roughness, and other parameters on the wafer and chip level. And now, high-precision, three-dimensional measurements on MEMS and CMOS elements in the production environment are key to ensuring the high yield of this product. Various elements on the DMD (see Figure 4.19)

* DLP is a trademark of Texas Instruments Corp.

** DLP Cinema is a registered trademark of Texas Instruments Digital Light Products (DLP) division.

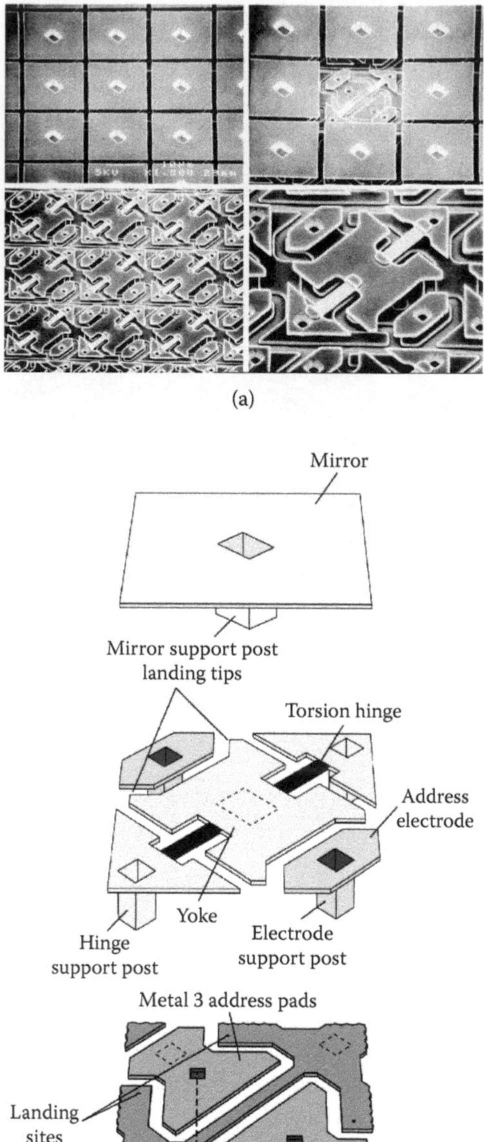

FIGURE 4.19 (a) Scanning electron microscope (SEM) micrographs of several DMDs, (b) superstructure diagrams of one pixel of DMD, reproduced with permission from Texas Instruments DLP division. An SEM micrograph, a single frozen-in-space snapshot, is severely limited for metrology, offering accuracy in at best two dimensions and from one viewing perspective only. In contrast, scalable AFM images [(c) and (d)] contain three-dimensional measures of the objects. Each image is viewable from many angles and can be processed and analyzed to extract different types of dimensional information about individual features, the entire image, or a subsection of the image. (AFM images courtesy of Pete Nagy, reproduced with permission from Veeco Instruments and Texas Instruments.)*

* All measurements on DMD are proprietary and the values undisclosed.

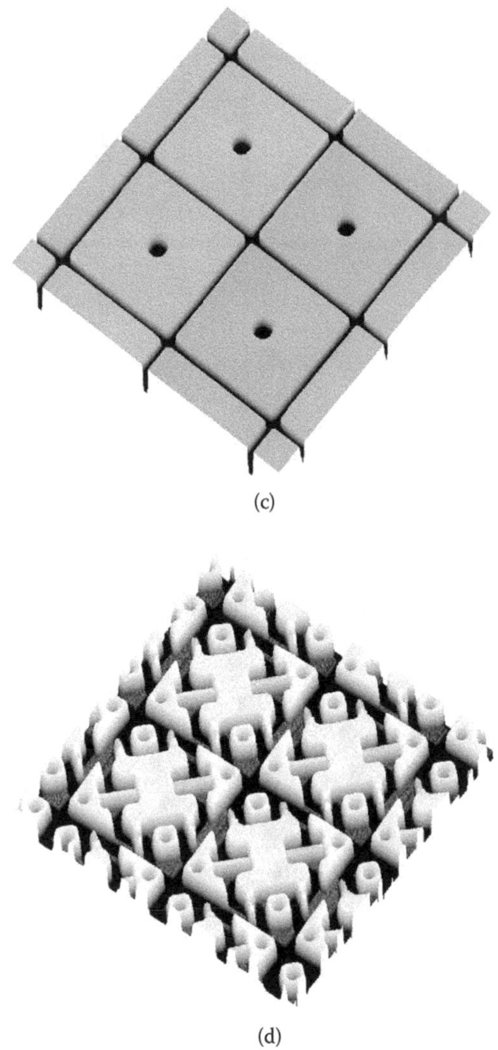

(c)

(d)

FIGURE 4.19 (Continued).

must be measured to strict dimensional requirements in any or all three dimensions (see DMD pixel superstructure in Figure 4.19b). In the figures that follow, we demonstrate the variety and scope of the measurements made on the DMD with the AFM. Among the measurements made with AFM are the following:

- Mirror planarity (Figure 4.20)
- Mirror surface smoothness (Figure 4.21)
- Mirror spacing (separation between adjacent mirrors) (Figure 4.22)
- Hinge planarity (Figure 4.23)
- Yoke planarity (Figure 4.23)
- Interlayer spacer thickness
- Metal-3 thickness (Figure 4.25)
- CMOS planarity
- In-line monitoring of planarity of spacer levels between metal deposition processes (Figure 4.25)

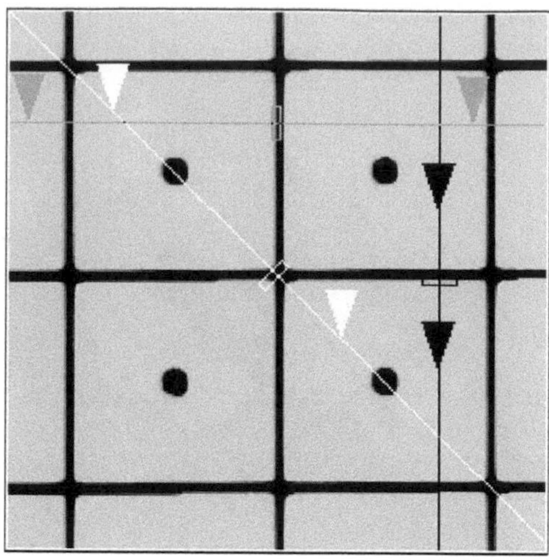

FIGURE 4.20 Mirror planarity. Top view: AFM image of four mirrors. Cross-sectional analysis quantifies departure from planarity by measuring small height variations along any direction on any segment of the image, as indicated by the three cross-sectional lines and measurement cursors in the image. The AFM measures height with sub-Ångstrom resolution. (Image courtesy of Pete Nagy. Reproduced with permission from Texas Instruments.)

In-line monitoring of the planarity of the spacer layers (which separate the three metal layers) is one of the DMD measurements made with an automated AFM. In addition, automated routines measure the spacing between the metal-3 level (drive electrode) and the torsional yoke structure, as well as between the yoke and the mirror. These are critical measurements for determination of the tilt angle of micromirrors in DMD. The images in Figure 4.20 to Figure 4.23, and Figure 4.25, were captured with an AFM.

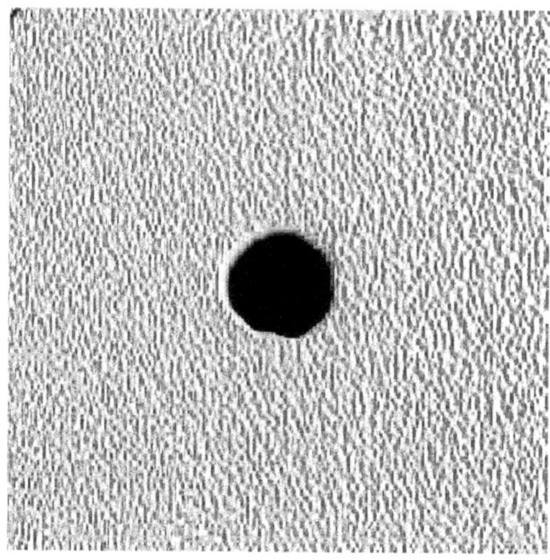

FIGURE 4.21 Mirror surface smoothness. On the whole or on any subsection of an AFM image, we can measure sub-Ångstrom-scale rms roughness, or even roughness as large as several micrometers on surfaces with peak-to-valley height variation up to several micrometers. (Image courtesy of Pete Nagy. Reproduced with permission from Texas Instruments.)

FIGURE 4.22 Mirror spacing on adjacent DMD pixels measured with "width analysis" in AFM software. The software first identifies the spacing using pattern recognition and then makes the measurement — in this case, the average value of the width of the region highlighted in white. The parameters for defining the measurement — for example, the length (L) over which to average the measurement of the spacing — are customizable. Width analysis, similar to numerous other measurement and analysis features, can be automated on many AFM models. (Image courtesy of Pete Nagy. Reproduced with permission from Texas Instruments.)

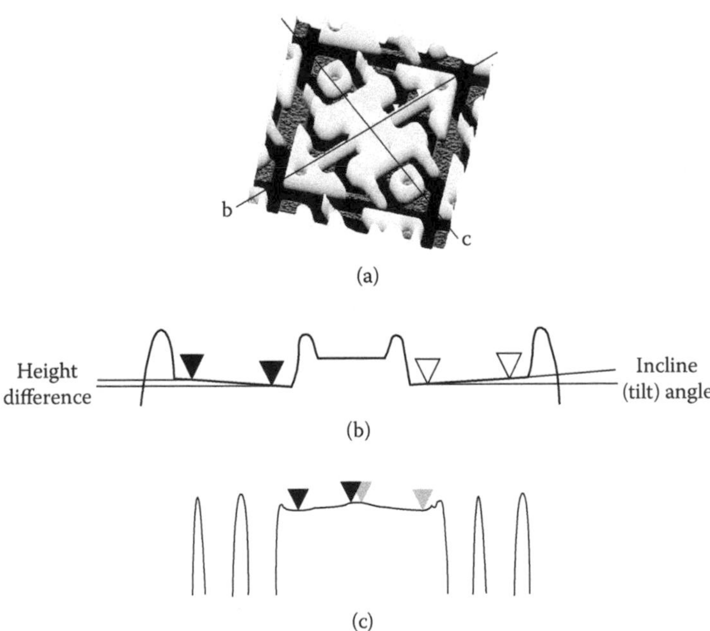

FIGURE 4.23 Hinge and yoke planarity. (a) AFM image of a DMD device, with the mirror removed to expose the support and actuation elements beneath [see Figure 4.19b]. Cross sections measure the planarity of the hinge (b) and the yoke (c). Measurements available for quantifying planarity include relative height difference and incline (tilt) angle. Steep sidewalls, especially on high-aspect-ratio structures, and even undercut structures (not shown here), can be measured using specialty AFM tips and specialized AFM systems (see Figure 4.24). (Image courtesy of Pete Nagy. Reproduced with permission from Texas Instruments.)

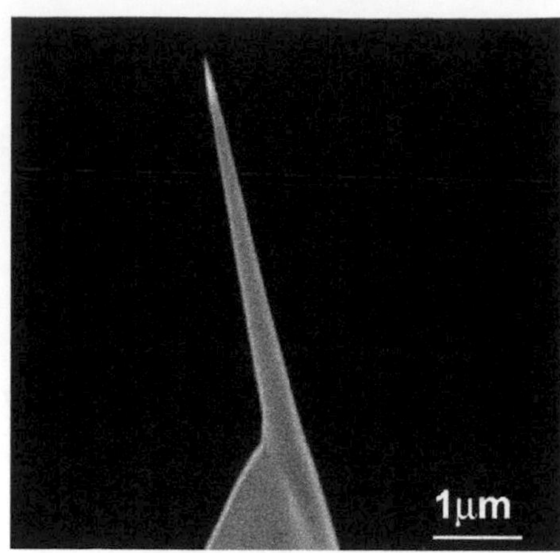

FIGURE 4.24 SEM micrograph of a specialized, high-aspect-ratio AFM tip, typically 5 to 10 nm sharp. These tips are used for making measurements on aggressive-geometry structures (including steep sidewall angles), which otherwise are often impossible to measure without breaking the wafer to expose cross sections (as is required in SEM).

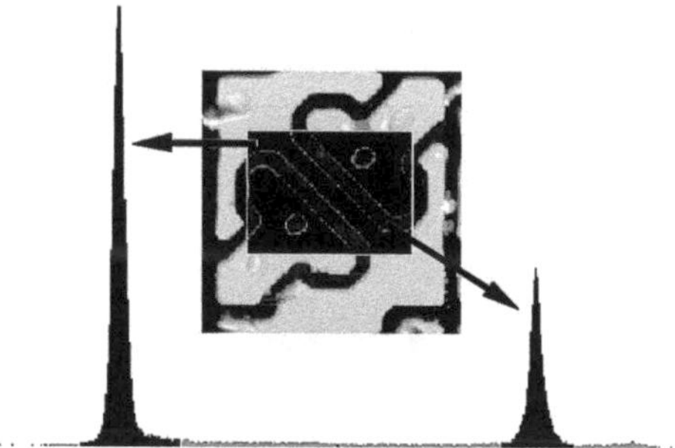

FIGURE 4.25 Measurement of CMOS metal-3 layer thickness in a DMD pixel [see Figure 4.19b]. Several methods can extract step height, including "depth analysis." The two histogram peaks correspond to height levels of metal-3 (left peak) and the underlying substrate (right peak). Among the parameters available to customize the measurement are the area (the rectangular box in the center of image), depth reference (to define the zero-height level), and percent bearing, which defines how much of the surface falls above or below a given height. To monitor and compare topographic changes at a given location in several fabrication process steps, the user can easily compare multiple histograms. (Image and data courtesy of Pete Nagy and F.M. Serry. Reproduced with permission from Texas Instruments.)

4.6 ATOMIC FORCE PROFILOMETER (AFP) — A COMBINATION OF AFM AND STYLUS PROFILER

The stylus profiler is a close predecessor of the AFM; similar to the AFM, it also incorporates a stylus that scans across the surface and is in direct contact with the sample at low force, moving up and down as it rides over the surface of the sample. The vertical motion of the stylus is converted into an electrical signal, often by using magnetic or optical sensing techniques, and produces a topographic profile along the scan. Thus, the two instruments share many similar features.

Some differences do exist between the AFM and the stylus profiler. First, in the stylus profiler the force of the tip on the sample is higher than in the AFM. The radius of curvature of the tip in the stylus profiler can be less than 0.1 μm, but this is still large compared to an AFM tip. This relatively large size limits the stylus profiler's lateral resolution as compared with an AFM. On the

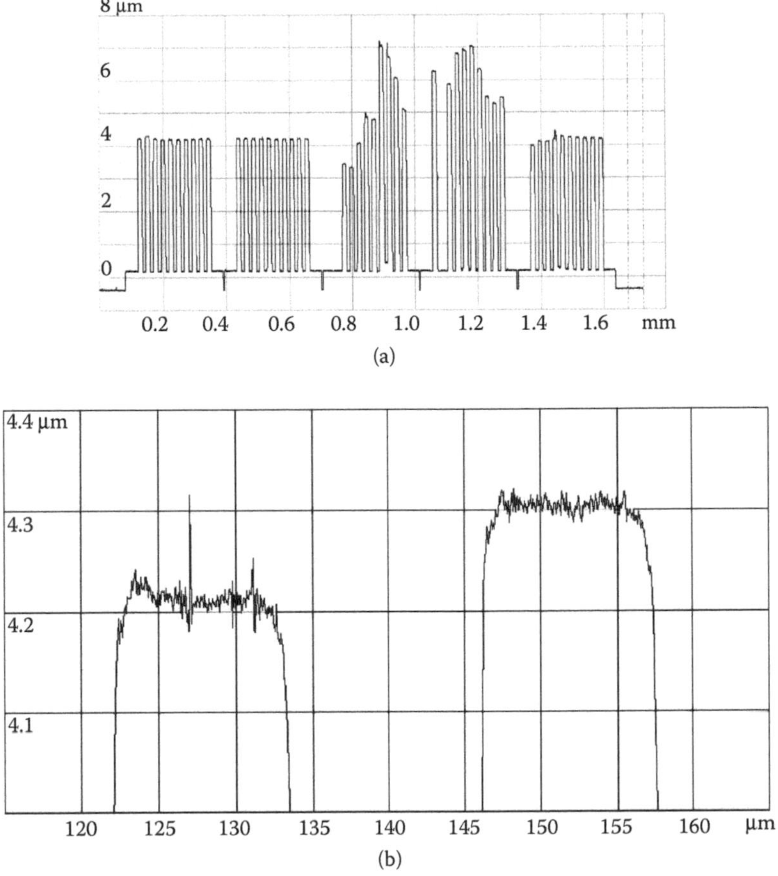

FIGURE 4.26 AFP combines advantages of the stylus profiler and AFM. (a) 1.7-mm-long profile, much longer than a typical AFM can perform, (b) zoom into part of the profile reveals the high resolution that is typical for AFM but beyond the capability of a stylus profiler. (Data courtesy of Veeco Instruments.)

other hand, the stylus profiler can measure tall features, even up to a millimeter in height, whereas an AFM's vertical range is typically less than 15 μm. In addition, the AFM provides TappingMode profiling, in which the AFM tip makes intermittent contact with the sample during raster scanning; TappingMode significantly reduces the lateral shear forces that are present in perpetual contact scanning. These differences make the AFM and the stylus profiler complementary tools for three-dimensional surface metrology.

Microsystem production processes sometimes require metrology on the chip and wafer level, including long scan profiles to measure both planarity and local roughness at high resolution. A unique combination of the AFM and stylus profiler called the AFP now exists. The development of the AFP, which combines the advantages of the stylus profiler with the sharp tip and position control of the AFM, has allowed for the scanning of continuous profiles of up to 100 mm long with a Z-range of up to 15 μm at nanometer-scale lateral and Ångstrom-scale vertical resolution. At any coordinate along a profile, the AFP can switch from profiling to imaging mode to generate three-dimensional AFM images covering a square area tens of micrometers on a side. The long profiles and images have AFM resolution and are often used to make measurements after critical process steps, including chemical–mechanical planarization and photoresist etch in semiconductor manufacturing. Figure 4.26 shows an example of an AFP profile that measures trench depth and surface roughness.

4.7 OPTICAL METROLOGY COMPLEMENTARY TO AFM

Several optical metrology techniques are commonly used for measuring and characterizing microsystems. These include optical profilometry, confocal microscopy, speckle, shearing interferometry, digital holography, and Nomarski profiling. Some of these techniques are described elsewhere in this book (see, for example, Chapters 5, 8, 12 and 13). This section briefly reviews optical profilometry (also called interference microscopy) and confocal microscopy, because they each have similarities to AFM, and they deliver complementary data to AFM data.

Confocal microscopy has recently been adapted for the measurement of microsystems. As in the optical profiler, either the sample or the objective lens is scanned vertically such that each point on the sample surface passes through the focal plane of the microscope. In the confocal microscope a very small aperture is placed in front of the detector to allow light to pass for a given point on the surface, only when this point is in focus. Similar to the AFM, confocal microscopy also relies on X-Y raster scanning.

In the optical profiler, an interference pattern of the full field of view is collected by an array sensor during a single vertical scan, rather than a point-by-point raster scan as in AFM and confocal microscopy. The two primary modes of operations are based on white light interference and monochrome light interference. White light illumination allows for fringes to be created only at the best focus for each sample point, and monochromatic illumination creates fringes with shapes representing the topography of the sample.

The speed of measurement is often important in the production environment. Because optical profilers examine the whole field of view in one vertical scan, they deliver the advantage of measurement speed over AFM and other raster-scanning techniques, including confocal microscopy.

Optical techniques in general, and optical profilometry in particular, are very well suited for measuring MEMS devices through glass packaging; this is an area that AFM cannot address. On the other hand, optical techniques may need measurement corrections when the sample's optical properties are not straightforward; this is an area in which AFM's tactile method encounters no complication.

An optical profiler equipped with stroboscopic illumination can record and measure the motion of a microsystem component. Stroboscopic illumination at a given strobe frequency is used to capture the movement of the object in a series of consecutive snapshots. The frequency or the phase of the strobe is then changed incrementally, and the captured snapshots with each new increment record the shape and deformation of the moving element at a different phase of the motion. In-plane and out-of-plane motion can be measured at the same time.

4.8 CONCLUSIONS AND OUTLOOK

The AFM is the instrument [7] of choice if nanometer and subnanometer resolution measurement and visualization of a surface are required. But the AFMs are increasingly being used to measure submicrometer features as well. AFMs are already ubiquitous in microsystem research laboratories around the world, and new applications for AFM in microsystems are on the horizon. One example involves a class of microfluidics in which surface charge patterns, produced using photolithography, attract small objects of opposite charge. Examples of these small objects are DNA and protein molecules in liquids. Measuring the resulting nanometer-scale changes in height, especially in a liquid environment, is today best possible with the AFM.

With the continuing trend toward miniaturization, going from MEMS to NEMS, more AFMs will find their way into laboratories as the only tools capable of measuring ultraminiature features. Also, with continuous improvement in automation features, fab-ready AFMs with higher measurement speeds and greater ease of use will extend their use into the production lines of microsystems, akin to what has already happened in the semiconductor integrated circuit (IC) industry.

Some further reading material on AFM is given in Reference 11 and Reference 12.

REFERENCES

1. Binnig, G., Quate, C.F., and Geber, Ch., Atomic force microscope, *Physical Review Letters,* 56(9), 930, 1986.
2. Magonov, S.N. and Whangbo, M.H., *Surface Analysis with STM and AFM: Experimental and Theoretical Aspects of Image Analysis,* Weinheim, VCH, 1996.
3. Zhong, Q., Inniss, D., Kjoller, K., and Elings, V.B., Tapping mode atomic force microscopy, *Surface Science Letters,* 290, L688, 1993.
4. Su, C., Huang, L., Nielson, P., and Kelley V., *In-situ* measurement of in-plane and out-of-plane force gradient with a torsional resonance more AFM, in *Proceedings of the 12th International Conference on Scanning Tunneling Microscopy/Spectroscopy and Related Techniques,* Eindhoven, Netherlands, 2003, 349.
5. Sundararajan S. and Bhushan. B., Topography-induced contributions to friction forces measured using an atomic force/friction force microscope, *Journal of Applied Physics,* 88(8), 4825, 2000.
6. Cleveland, J.P., Anczykowski, B., Schmid, A.E., and Elings, V.B., Energy dissipation in tapping mode atomic force microscopy, *Applied Physics Letters,* 72(20), 2613, 1998.
7. Babcock, K.L., Elings, V.B., Shi, J., Awschalon, D.D., and Douglas, M., Field-dependence of microscopic probes in magnetic force microscopy, *Applied Physics Letters,* 69(5), 705, 1996.
8. Martin, Y., Williams, C.C., and Wickramasinghe, H.K., Atomic force microscope-force mapping and profiling on a sub 100-AA scale, *Journal of Applied Physics,* 61, 4723, 1987.
9. Schmutz, P. and Frankel, G.S., Characterization of AA 2024-T3 by scanning Kelvin probe force microscopy, *Journal of the Electrochemical Society,* 145, 2285, 1998.
10. Maivald, P., Butt, H.J., Gould, S.A.C., Prater, C.B., Drake, B., Gurley, J.A., Elings, V.B., and Hansma, P.K., Using force modulation to image surface elasticities with the atomic force microscope, *Nanotechnology,* 2, 103, 1991.
11. Bhushan, B., Fuchs, H., and Hosaka, S., *Applied Scanning Probe Methods,* Springer-Verlag, New York, 2003.
12. Bhushan, B. and Liu, H., Characterization of nanomechanical and nanotribological properties of digital micromirror devices, *Nanotechnology,* 15, 1785, 2004.

5 Optical Profiling Techniques for MEMS Measurement

Klaus Körner, Aiko Ruprecht, and Tobias Wiesendanger

CONTENTS

5.1 Introduction 145
5.2 Principles of Confocal Microscopy 145
    5.2.1 Confocal Point Sensors 145
    5.2.2 Confocal Microscopes 150
    5.2.3 Measuring with Confocal Microscopes 151
    5.2.4 MEMS Measurement Applications 153
5.3 Principle of Microscopic Depth-Scanning Fringe Projection (DSFP) 156
    5.3.1 Introduction 156
    5.3.2 Intensity Model 156
    5.3.3 Experimental Realization 158
5.4 Conclusion 160
References 161

5.1 INTRODUCTION

Many different methods have been developed for the measurement of microsystems and their components. Among others, white-light interferometry, confocal microscopy, and fringe projection are well-established techniques. White-light interferometry and confocal microscopy are of increasing interest because they combine a large measurement range with a very high vertical resolution. Fringe projection is less sensitive, but it offers a higher measurement speed. It has also advantages for medium and large-sized objects in which a very high vertical resolution is not needed.

White-light interferometry and fringe projection are presented in Chapter 8 and Chapter 6, respectively. In this chapter, we will give an overview of confocal microscopy and a description of the depth-scanning fringe projection (DSFP) system.

5.2 PRINCIPLES OF CONFOCAL MICROSCOPY

5.2.1 Confocal Point Sensors

The confocal measurement principle was invented by M. Minsky in 1957 and patented by him in 1961 [1]. The aim of the invention was the minimization of stray light in the measurement of biological samples [2]. In confocal microscopes, the object is illuminated and detected pointwise. The setup of a transmissive confocal sensor is illustrated in Figure 5.1. A point-light source is imaged onto the object, and this object point is imaged onto a point detector; this is accomplished with a pinhole and a photodiode. The adjustment of the illumination and of the detection in all three dimensions is crucial. Both illumination and detection have to be coaxial and aim laterally and axially at the same point. Object points in the neighborhood of the measurement point are

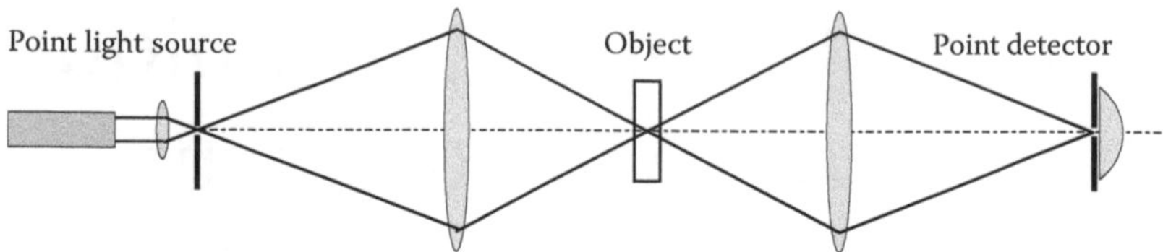

FIGURE 5.1 Confocal transmission point sensor.

imaged next to the pinhole, which blocks the light. Most stray light from points along the common optical axis is also blocked because the image of such a point is defocused on the point detector. The defocused image is larger than the point detector. Hence, most of the light does not reach the detector. In an adjusted system, the blocking of defocused light results in a strong depth discrimination.

A second possible setup that enables the measurement of reflecting surfaces is illustrated in Figure 5.2. In this case, the same imaging optics is used for illumination and detection. Therefore, the system is automatically adjusted with respect to the optical axis. Again, the point-light source is imaged onto the object surface, and this point is imaged onto a point detector. A beam splitter is used to separate the light for illumination and detection. Because light, which is reflected to the side, will not hit the lenses again, the amplitude of the confocal intensity signal varies, depending on the local gradient of the object topography. The strong depth discrimination of the confocal sensor leads to a good depth resolution. When a depth scan is performed, peak intensity is measured if the measurement point matches the object surface. The measured intensity drops rapidly when the axial distance between the object point and the measurement point becomes larger. Details of this will be discussed later.

In this chapter, the object is, for simplicity of the mathematical description, assumed to be a plane mirror perpendicular to the optical axis. When the focus of the confocal system is moved perpendicular to the plane, an intensity curve is detected, as illustrated in Figure 5.3, if the plane is at position zero. If the imaging system does not possess any aberrations, this curve can be described [3] by

$$I(w) = \left[\frac{\sin(w/2)}{w/2}\right]^2 \tag{5.1a}$$

with

$$w = \frac{8\pi}{\lambda} z \cdot \sin^2\left(\frac{\alpha}{2}\right), \tag{5.1b}$$

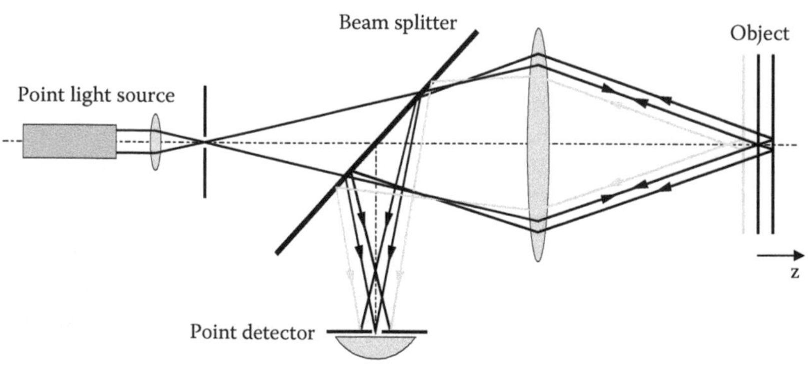

FIGURE 5.2 Confocal reflection point sensor.

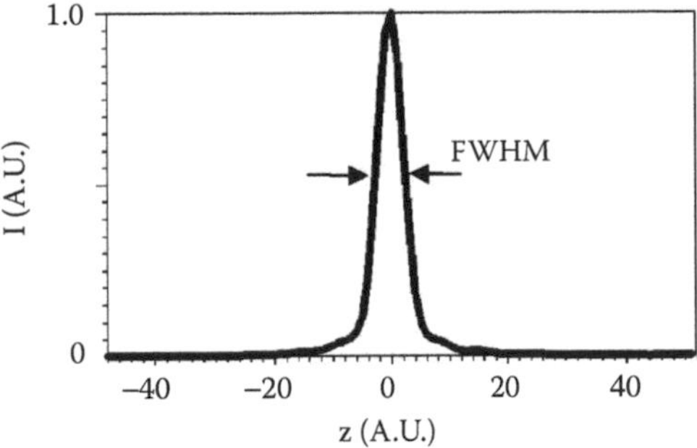

FIGURE 5.3 Confocal depth response curve.

where λ is the wavelength, z is the defocus, and α is the angle between the optical axis and the marginal ray (the *NA* angle).

The full width at half maximum (FWHM) of this confocal response curve for a system in air is given by [4,5]:

$$FWHM = \frac{0.45 \cdot \lambda}{1 - \cos\alpha}. \tag{5.2}$$

For low numerical apertures (NA < 0.5), this simplifies to [6]:

$$FWHM = \frac{0.90 \cdot \lambda}{NA^2}. \tag{5.3}$$

It can be seen that the main parameter for the width of this curve is the numerical aperture of the objective lens. This is also an important parameter if the object has inclined reflecting surfaces. A tilted mirror reflects the light sidewise and not all the light hits the objective lens again. Because a lower effective numerical aperture is used for the detection, the confocal response curve is broadened. As mentioned earlier, the signal amplitude decreases too. Reference 7 gives a rule of thumb that inclined reflecting surfaces can be measured up to a tilt angle equal to half the angle of the numerical aperture of the focusing lenses.

In a real setup, the detector is not infinitesimally small. This can also have an influence on the FWHM of the confocal response curve. In Reference 8 and Reference 9, this effect is described in detail. Up to a certain pinhole size, the FWHM is equal to the one described by Equation (5.2). Above this threshold, the FWHM increases approximately linearly with the pinhole size. In Reference 8, the pinhole size is described in optical coordinates, defined as

$$v = \frac{2\pi}{\lambda} NA \cdot r, \tag{5.4}$$

$$u = \frac{2\pi}{\lambda} NA^2 \cdot z. \tag{5.5}$$

Here, r is the radial distance perpendicular to the optical axis, z the axial position, and *NA* the numerical aperture with which the light is focused on the pinhole. Optical coordinates are useful for a mathematical description of an optical system invariant to its numerical aperture and the used wavelength. The optical system is assumed to be rotationally symmetric to the optical axis.

A less abstract approach can be undertaken by the determination of the size of the Airy disk of the focus at the detector pinhole:

$$r_{Airy} \approx 0.61 \cdot \frac{\lambda}{NA}. \tag{5.6}$$

Expressed in optical coordinates, this is

$$v_{Airy} = 2\pi \cdot \frac{r_{Airy}}{\lambda} NA \approx 2\pi \cdot 0.61 = 3.83. \tag{5.7}$$

Hence, the pinhole size can be described with coordinates normalized with respect to the radius of the Airy disk:

$$\hat{v} = \frac{r}{r_{Airy}} = \frac{v}{v_{Airy}} \approx 0.26 \cdot v. \tag{5.8}$$

The intensity distribution along the optical axis is for a (nonconfocal) focus on the object [10]:

$$I(u) = \left(\frac{\sin\left(\frac{u}{4}\right)}{\frac{u}{4}} \right)^2. \tag{5.9}$$

Therefore, the first minimum can be found at

$$u_{min} = 4\pi. \tag{5.10}$$

Accordingly, we use this axial focus size to normalize the z-axis:

$$\hat{u} = \frac{u}{u_{min}} \approx 0.080 \cdot u \tag{5.11}$$

Using the formulas in Reference 8, the FWHM is dependent on the pinhole size as plotted in Figure 5.4. The threshold up to which the FWHM is nearly constant is 0.65 times the size of the Airy disk. Because this is smaller than the size of the Airy disk, much light is lost at the pinhole. A bigger pinhole leads to a better light efficiency, but it decreases the depth resolution of the confocal system.

Because the confocal sensor is a point sensor, one has to do a lateral scan to measure a complete surface. A second effect of the use of very small pinholes is an increased lateral resolution. This is also analyzed in Reference 8 and leads to a dependency, plotted in Figure 5.5. According to the reference, the pinhole size for optimal lateral resolution at good light efficiency is 0.13 times the size of the Airy disk. This leads to a gain in lateral resolution by a factor of approximately 1.4 compared to a conventional microscope [11]. The light efficiency at this pinhole size is very low, and, therefore, a strong light source is needed.

The measured confocal intensity signal has to be analyzed relative to a common zero point. This zero point is valid for all measurement points in the field of view. Hence, the confocal measurement principle is an absolute measurement principle without ambiguities. The depth scan can be implemented in various ways. The object can be moved on a translation stage; the sensor or just a part of the imaging optics can be moved. Another approach without a mechanical depth

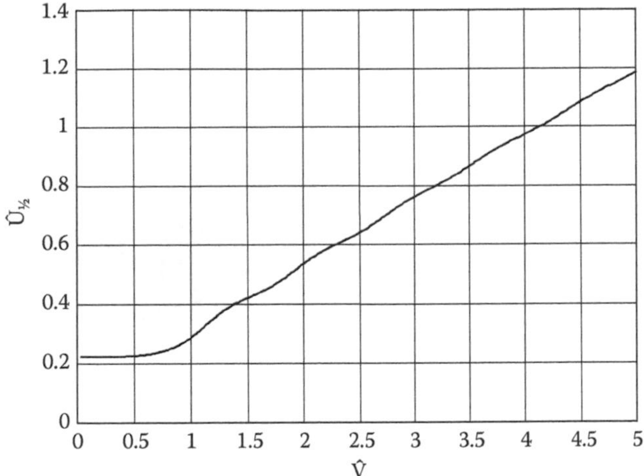

FIGURE 5.4 Signal width $\hat{U}_{1/2}$ of the confocal signal over pinhole radius $\hat{V}$.

scan is based on the application of chromatic effects by using several wavelengths or a wavelength range [12,13].

This so-called *chromatic confocal sensor principle* is illustrated in Figure 5.6. A light source with a broad spectrum is used. The lenses are designed to focus the light in different distances, depending on the wavelength. Therefore, a complete height range is measured at the same time. The confocal response curve is spectrally coded. To separate the different intensities, a spectrometer is used. The spectrally encoded confocal intensity signal can be used to calculate the topography. Therefore, the correlation between wavelength and focus position has to be calibrated first. With this wavelength–height dependence, the spectral signal can be translated to a confocal signal equivalent to a signal from a sensor with mechanical depth scan. Adaptations of the chromatic-confocal sensor principle to line- and area-measuring sensors have also been reported [14–17].

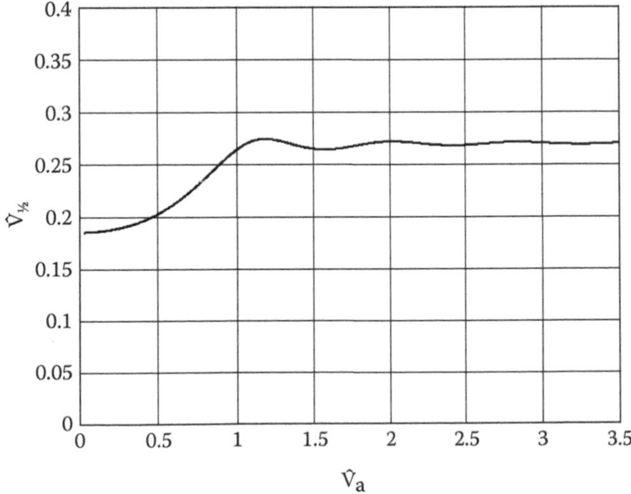

FIGURE 5.5 Half width $\hat{V}_{1/2}$ over pinhole size $\hat{V}_a$.

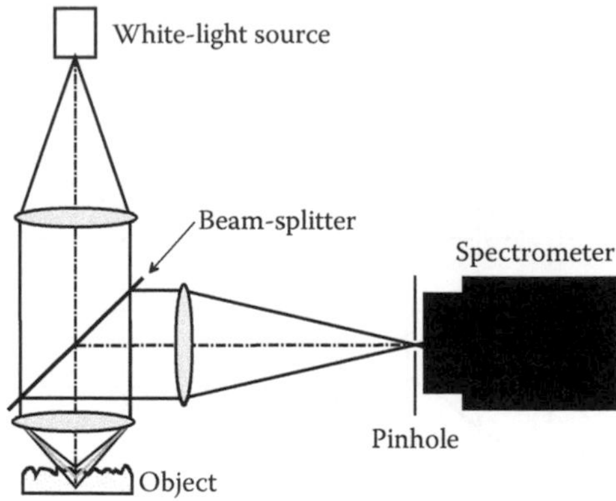

FIGURE 5.6 Chromatic confocal point sensor.

5.2.2 Confocal Microscopes

Point sensors have to scan the complete volume with one focus point. A first approach to parallelize the lateral scan by the use of several point sensors in parallel was published by Petran and Hadravsky in 1968 [18]. This setup, called a *tandem confocal microscope*, made use of a Nipkow disk (see Figure 5.7). On this disk, pinholes are distributed on a spiral path. Further developments in this approach made the system more stable and easier to adjust (e.g., [19–21]). Such a setup [22] is illustrated in Figure 5.7.

The Nipkow disk is illuminated with collimated light. Each pinhole is used as a point light source that is imaged onto the object. The reflected light is focused onto the pinholes of the Nipkow disk, which are also the confocal detection pinholes in this setup. Finally, the Nipkow disk is imaged

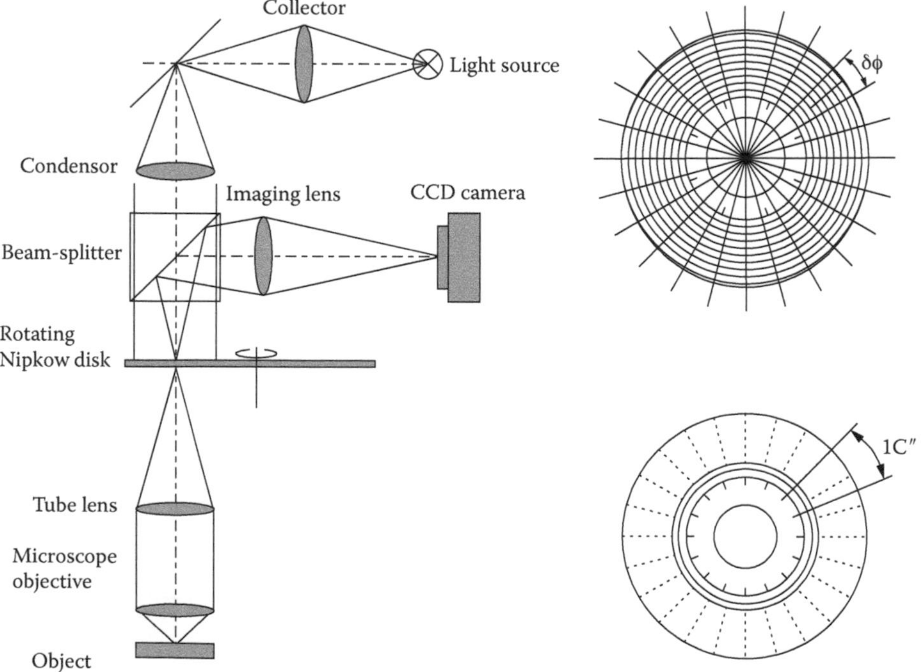

FIGURE 5.7 Confocal microscope with Nipkow disk.

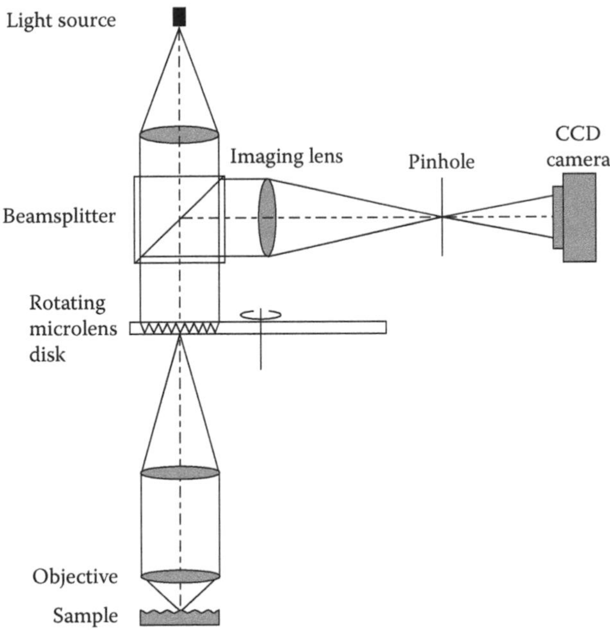

FIGURE 5.8 Confocal microscope with microlens Nipkow disk.

to a CCD camera. By rotation of the disk, the complete field of view is scanned in a very short time. The highest frame rate for a confocal image in one plane that has been reported so far, is 1 frame/msec [23]. To get the topography of the object, an additional depth scan is needed.

The pinholes on the Nipkow disk must have a distance of a multiple of their diameter in order to suppress crosstalk (e.g., [24]). A typical relation between distance and diameter is 10:1. The disk reflects much light and only a fraction, usually between one and several percent, is used for the measurement. Therefore, powerful light sources are needed and the light reflected by the disk has to be blocked with very high efficiency. A better overall light efficiency can be realized with microlenses. The microlenses can be used to focus the light on the pinholes [25] or to generate a lateral distribution of point-light sources [26]. The latter is realized in a setup illustrated in Figure 5.8 [17].

Here, the microlenses are arranged on spirals and cover the complete area of the disk. The foci of the microlenses are used as point-light sources and are imaged onto the object. As the confocal discrimination cannot be done by the microlenses alone, a central pinhole is placed in front of the camera, which can be used for all microlenses in parallel. The single pinhole has to be larger than the Airy disk to guarantee a correct image with high lateral resolution of the microlens Nipkow disk on the CCD [27]. Because the light efficiency of such a system is much higher compared to a setup using a pinhole disk, low-power sources with long lifetimes like LEDs can be used. Some topography measurements with such a setup are shown in Subsection 5.2.4.

5.2.3 Measuring with Confocal Microscopes

Confocal microscopes translate the defocus in each position into an intensity signal. The degree of defocus is dependent on the *NA* or, in other words, on the aperture angle. Therefore, it could be classified as a triangulation method. The triangulation angle is not only a solid angle around the principle ray but a complete angular range up to the aperture angle. This is an advantage with respect to shadowing effects at objects with high aspect ratios or high steps. If the light cone is partially blocked by higher parts of the object (see Figure 5.9), the effective aperture angle is reduced, but the structure is still measurable. A limiting effect is that because part of the light is

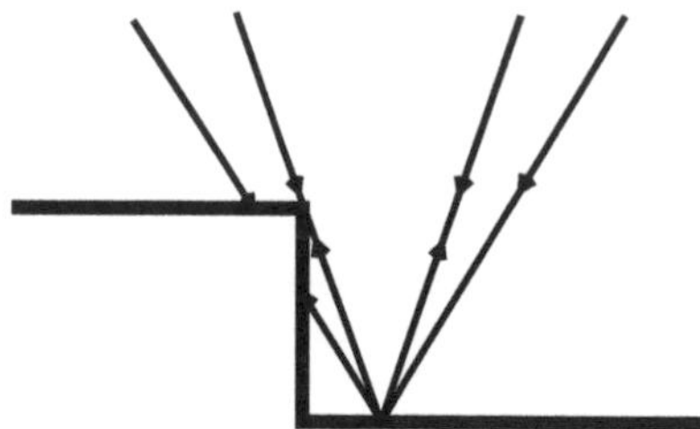

FIGURE 5.9 A light cone that is partially blocked by a high step.

blocked, the signal amplitude decreases. If too much light is blocked, the signal becomes too weak to be reliably detected. On the other hand, a high effective aperture angle is needed for a high axial resolution. This is a distinguishing feature of white-light interferometers, which can achieve high axial resolutions with low numerical apertures.

In comparison to interferometric measurement systems, confocal microscopes are much less sensitive to small vibrations. For example, a small periodic vibration that completely destroys the contrast in the interference signal causes only a slightly broader confocal signal.

The axial resolution of confocal microscopes ranges from about 5 nm with high *NA* objectives to a few microns with low *NA* objectives. Interferometric measurement systems can achieve higher axial resolutions up to the subnanometer region. Therefore, the confocal microscopy bridges the gap between interferometric systems and fringe projection.

The big advantage of chromatic-confocal sensors is that no mechanical depth scan is needed. This enables construction of compact sensors, which are very robust with respect to acceleration forces, for example, and are comparatively easy to miniaturize. Because the measurement is completely parallelized when using a white-light source, the measurement is done in all distances at the same time. This can be advantageous if the object is moving during the measurement. Absorption peaks due to a colored object surface can be a problem if the detection is not done with a high-resolution spectrometer. A limitation of the chromatic-confocal approach is the limited amount of light from white-light sources. This can restrict the measurement speed or the performance of

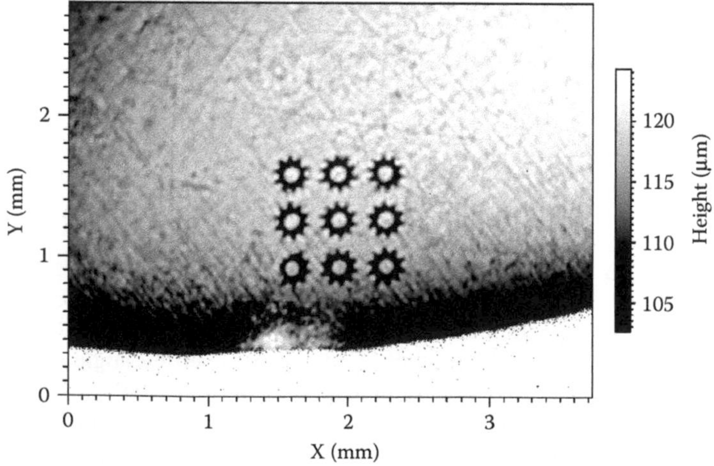

FIGURE 5.10a Topography of microcogwheels on metal background measured with 5X objective *NA* 0.15.

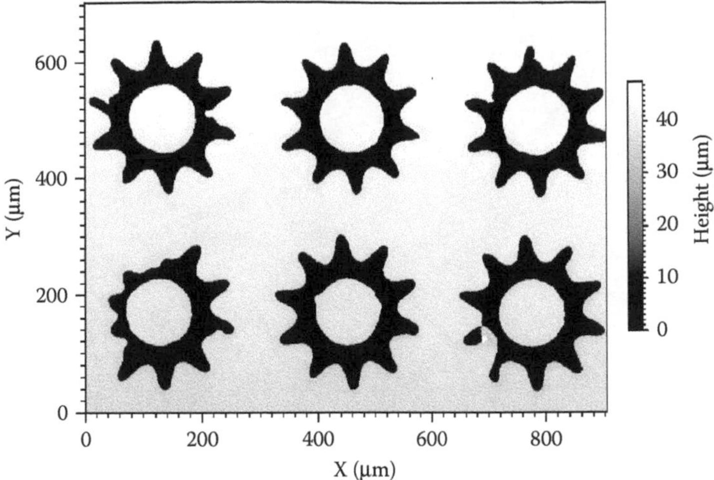

FIGURE 5.10b Topography of microcogwheels on metal background measured with 20X objective *NA* 0.4.

measurements on surfaces with low reflection coefficient. Nevertheless, chromatic-confocal point sensors with several kilohertz measurement rates are commercially available.

Because the highest intensity is measured when the image of the object point is in focus, the confocal signals in the field of view can be used to calculate an autofocus image very easily. Such autofocus images show the reflectivity of the object in great detail (see Figure 5.10d) for objects of arbitrary depth. Another possible imaging mode is the so-called *extended-focus image* where the average detected intensity in each point is displayed.

5.2.4 MEMS Measurement Applications

Figure 5.10a to Figure 5.10e show the results of measurements of an array of microcogwheel moulds on a metal surface. An objective with a magnification of 5X and a *NA* of 0.15 was used for

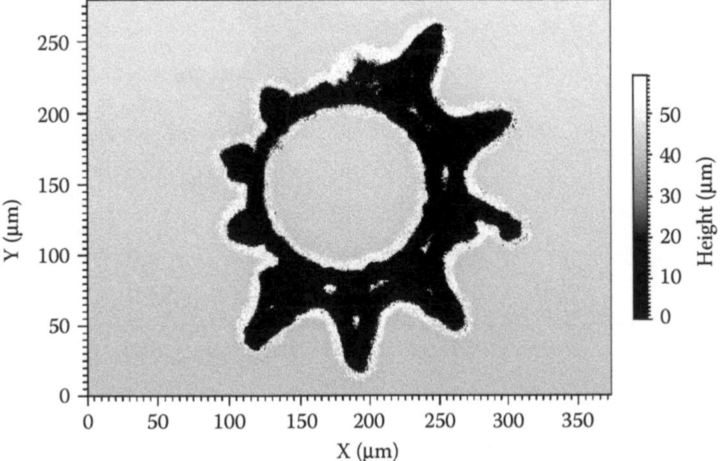

FIGURE 5.10c Topography of microcogwheels on metal background measured with 50X objective *NA* 0.5.

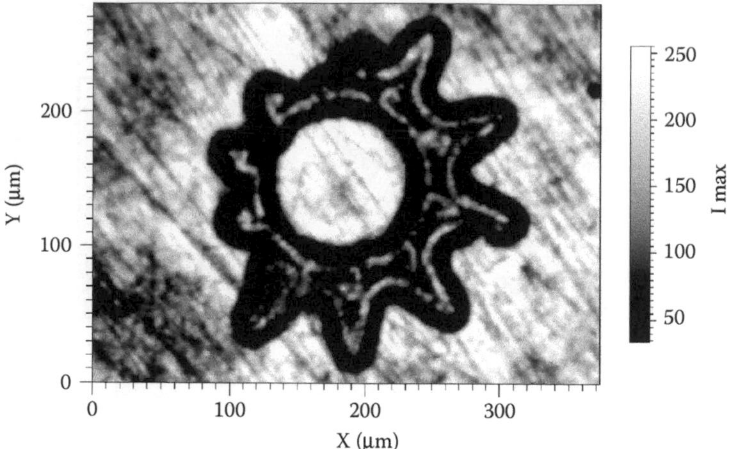

FIGURE 5.10d Intensity distribution of microcogwheels on metal background measured with 50X objective *NA* 0.5.

the measurement shown in Figure 5.10a. The field of view is relatively large (about 3.7 mm × 2.8 mm); so is the working distance of about 20 mm. The lateral and axial resolutions at this *NA*, however, amount to rather unsatisfying values. To overcome this problem, an objective with a higher numerical aperture must be used. The measurement presented in Figure 5.10b was performed using an objective with a higher magnification of 20X and a higher *NA* of 0.4. The resolution is improved in this case, but the field of view is decreased to 0.9 mm × 0.7 mm. Figure 5.10c shows the topography measured with a 50X objective (*NA* = 0.5). The autofocus image of the microcogwheels, which shows the reflectivity distribution over the surface, can be seen in Figure 5.10d. Figure 5.10e finally depicts the measured topography using an objective with 50X magnification but with an *NA* of 0.8. Comparing Figure 5.10c and Figure 5.10e, the measurement results at the steep flanges are obviously better using a higher *NA* objective (see Figure 5.10e).

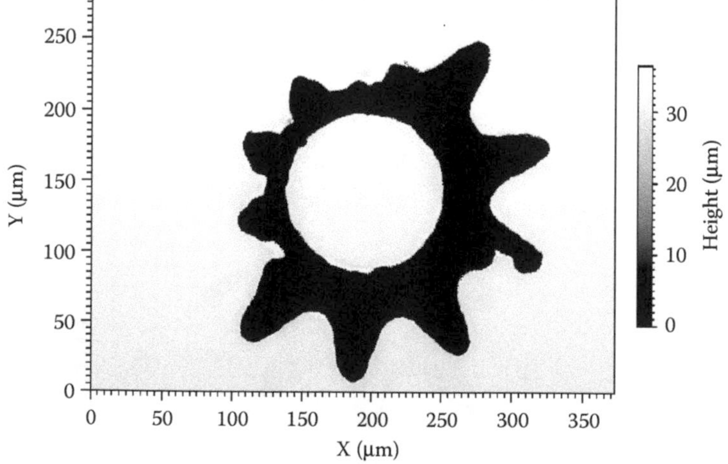

FIGURE 5.10e Topography of microcogwheels on metal background measured with 50X objective *NA* 0.8.

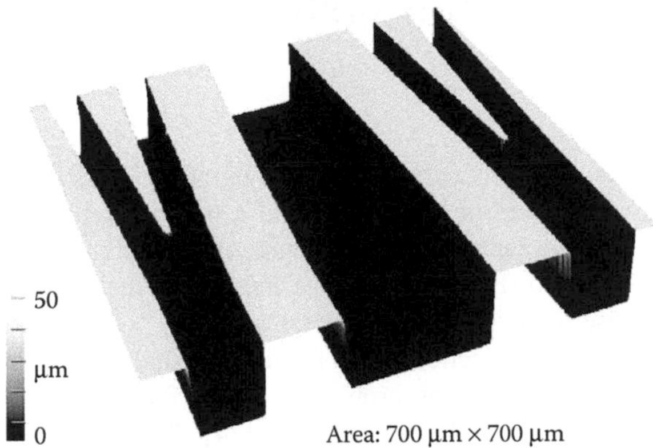

FIGURE 5.10f Topography of PMMA wave guide, measured with a commercial confocal microscope. (Nanofocus AG.)

The measurements were performed using a confocal microscope setup with a rotating microlens disk according to Figure 5.8. To show the specific influence parameters, different microscope objectives were applied for the measurements.

Figure 5.10f shows a measurement of a waveguide, carried out using a commercial confocal microscope, manufactured by Nanofocus AG, Germany (see Figure 5.11). This system makes use of a rotating Nipkow disk.

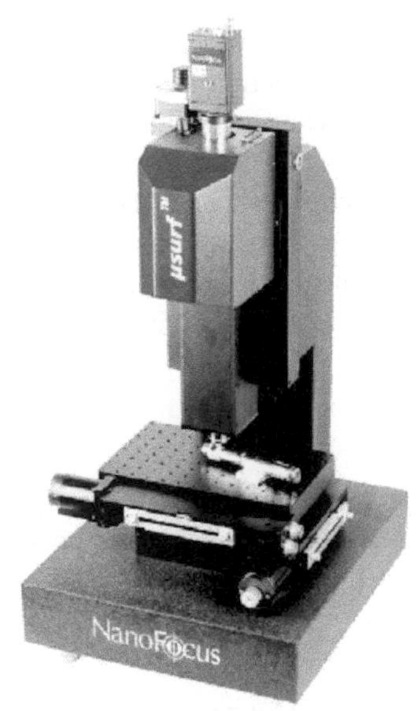

FIGURE 5.11 Commercial confocal microscope. (Nanofocus AG.)

5.3 PRINCIPLE OF MICROSCOPIC DEPTH-SCANNING FRINGE PROJECTION (DSFP)

5.3.1 INTRODUCTION

The fringe projection technique with a single frequency grating [28–31], or with projected fringes generated by a Michelson interferometer [32,33] and a monochromatic source suffers from the disadvantage that only the topography of rather cooperative objects without large steps and slopes on the surface can be measured. However, relative measurements, such as strain measurements, can be performed. For these techniques, see Chapter 6.

The application of a fringe projection setup for the measurement of surface topography with a single grating period leads to the same problem that interferometry has. The phase values are ambiguous because of the fringe identification problem [34]. The one-grating-single-period method in the commonly used triangulation setup cannot be applied to isolated surfaces or to surfaces with discontinuous height steps that are larger than the effective wavelength of triangulation. However, this problem can be solved using the Gray code technique [35] in combination with phase-shifted cosine-like fringes. Additionally, the sequential projection of more than one fringe period and hierarchical phase-unwrapping methods have been developed [34,36–39].

Furthermore, in order to obtain a large depth measurement range in the fringe projection technique, the aperture for illumination and the aperture for detection must be reduced to increase the depth of focus for both. However, a reduction in the apertures usually leads to a reduced spatial resolution due to the diffraction limit and to a reduction of the registered light intensity.

Recently, a new proposal for topography measurement on the microscopic scale to increase the application range of the fringe projection technique with only one grating [40] has been reported; this has some advantages compared to other approaches. Instead of using optics with a rather low numerical aperture, objectives with a high numerical aperture are used to reduce the depth of focus to a reasonable range. Extended objects are now illuminated and imaged with some parts that are in focus and other parts that are considerably blurred. Using the additional depth scanning of the measurement system or of the object, the surface topography is determined as in white-light interferometry, but on a coarser scale. Therefore, similar acquisition and evaluation algorithms can be used for this special type of fringe projection technique.

Other depth-scanning methods, previously published, evaluate only the contrast envelope by defocusing of a more or less blurred fringe pattern [41,42]. However, in the new method, the phase of the detected signal is also used for signal evaluation. This makes sense as the phases of the detected signals contain the information about the topography of the object because of the elongation of the observed fringes, due to the application of the triangulation principle. Using the phase information enables measurement of results with a better height resolution compared to contrast evaluation.

5.3.2 INTENSITY MODEL

The basic idea of the DSFP approach is the application of large numerical apertures together with depth-scanning data acquisition. Considering a geometric model for the setup and a grating with a cosine-like transparency function, the signal response $I(x, y)$, depending on the object (z_o) and the focus position (z), can be written as [40]:

$$I(x,y,z)=I_0\left\{1+m(x,y)\cdot\frac{J_{1,I}(\pi\rho_I)}{\pi\rho_I}\cdot\frac{J_{1,D}(\pi\rho_D)}{\pi\rho_D}\cdot\cos\left[\frac{2\pi}{\lambda_{eff}}(z-z_0)+2\pi f_x x+\varphi(x,y)\right]\right\}, \quad (5.12)$$

where I_0 is the mean intensity, $m(x,y)$ is a modulation term that describes the spatial modulation, and J_1 is a Bessel function of the first kind. The indices I and D are chosen to distinguish between

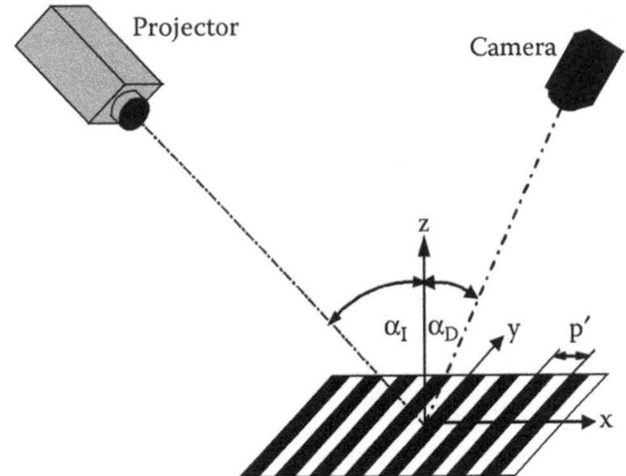

FIGURE 5.12 Coordinate system for fringe projection setup.

the illumination and the detection paths. The effective wavelength is given by [32]:

$$\lambda_{eff} = \frac{p'}{\tan(\alpha_I) + \tan(\alpha_D)}. \quad (5.13)$$

Here, p' is the fringe period in object space, α_I is the angle between the normal of the measurement plane and the illumination direction, and α_D is the angle between the normal of the measurement plane and the detection path (see Figure 5.12). The carrier frequency generated by the imaged grating of those lines parallel to the x-axis is described by the term f_x, and ρ is a generalized coordinate that is given with the following equations for illumination (I) and detection (D):

$$\rho_I = \frac{(z - z_0)}{\beta'_I p} \frac{2r_I}{f'} \quad \text{and} \quad \rho_D = \frac{(z - z_0)}{\beta'_D p} \frac{2r_D}{f'}, \quad (5.14)$$

where r denotes the aperture radius, f' the focal length of the objectives, and β' the optical magnification.

In Equation 5.14, it can be seen that the modulation of the cosine term in Equation 5.12 depends on the grating period p, the magnification β', and size of the apertures, given by the radius r of the corresponding pupil and the corresponding focal length f', equal in this case because of the symmetric optical arrangement. Furthermore, the height coordinate z_o is not only part of the Bessel functions but is also included in the phase term of the cosine function. The dependency of the phase on the height results from the triangulation arrangement. The modulation of the signal I can be used to estimate the absolute fringe order. The phase evaluation is more sensitive than the modulation evaluation and is used to interpolate the z-resolution.

The signal shape available with a depth-scanning triangulation setup is similar to that of white-light interferometry. However, there is also a great difference between the signals: in white-light interferometry, the phase at the position of maximum fringe contrast is usually approximately zero and is not a varying term. In a camera frame recorded with a DSFP setup, a strong carrier frequency is found corresponding to the fringe period of the grating. In Reference 40 a detailed description of the implemented algorithm for the DSFP-signal evaluation can be found; see also Reference 43 for fast algorithms for data reduction in modern optical 3-D profile measurement systems. In our experimental setup, a digital lock-in detection was realized. First, the signal is high-pass-filtered to remove the contribution of the mean intensity. In a second step, it is multiplied by the sine and the cosine of the carrier signal leading to a complex response signal for each pixel. The absolute value corresponds to both the contrast and the phase value and provides the object phase. The best focus position is calculated from the center of gravity of the modulation signal and is used to deliver the fringe order. The phase of this signal can be used to improve the height resolution.

5.3.3 Experimental Realization

Several experimental implementations for the depth-scanning technique are possible. Our fringe projection setup is based on a zoom stereomicroscope, MZ-12.5, from Leica with an additional motor-focus system. This setup was initially developed for standard fringe projection applications using a binary grating [44]. The grating is mounted on a computer-controlled positioning stage for the conduction of phase-shifting signal evaluation methods. The light of an external source is guided to the sensor with the help of fiber bundle optics. The exiting light is collimated and passes through the grating, which has a grating period of 178 μm. The grating is imaged onto the surface through one channel of the microscope. The other channel is used for detection. Figure 5.13 shows the experimental setup in a schematic drawing.

The whole microscope system is moved along the measurement axis, the z-axis, by means of a computer-controlled, motorized focusing system. Its z-resolution is 4 μm. The z-sensitivity of the depth-scanning sensor can be changed using another zoom objective. Thus, the field of view can easily be adapted to the desired measurement application.

A typical measured signal, as shown in Figure 5.14, was obtained by measuring with the vertical scan of the whole microscope. As already described, there is great similarity to the signals obtained by white-light interferometry.

A drawback of the chosen setup is the fact that the image is shifted laterally when the microscope is moved along the z-axis. This can be seen in Figure 5.15. This shift is a consequence of the angle between the chief rays of the observation and the movement axis of the microscope. This problem can be overcome by choosing the vertical step size in such a way that the pixel shift exactly corresponds to an integer number of pixels. This can be realized for a given triangulation angle and optical magnification. Another possibility to avoid the lateral image

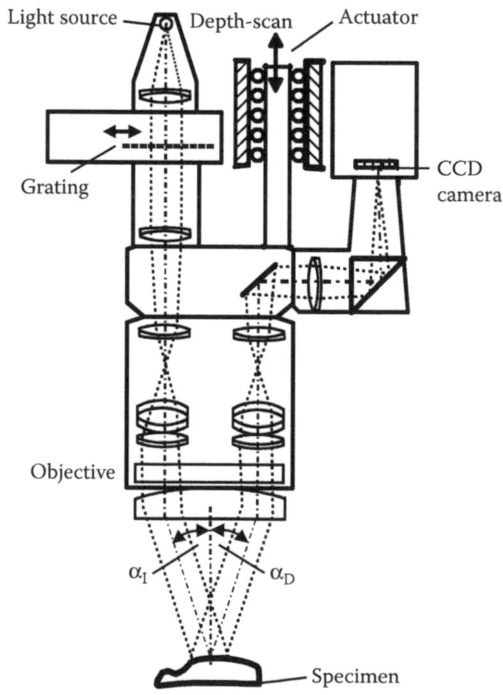

FIGURE 5.13 Experimental setup based on a zoom stereomicroscope.

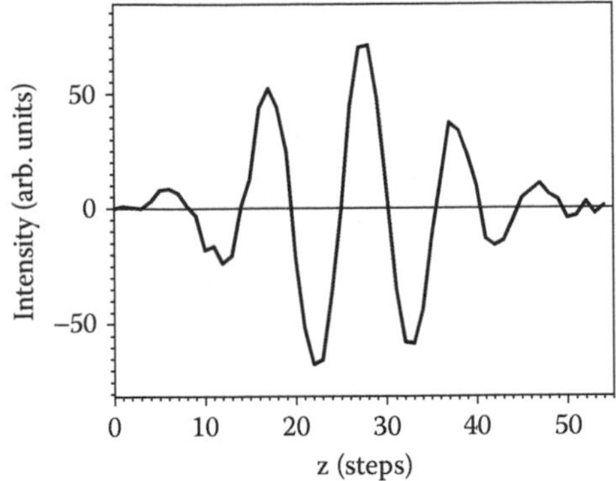

FIGURE 5.14 Typical intensity signal detected by the depth scan of the microscope body.

shifting due to depth scan is to arrange the whole triangulation setup in such a way that the direction of detection of the microscope coincides with the axis of scan movement of the steeomicroscope [45].

Figure 5.16 demonstrates a result of a measurement on a combined gearwheel. The gearwheel height is such that the depth of focus is too small to measure the object in a single shot. The scanning approach has the advantage of obtaining frames with an unlimited depth of focus. For this purpose, the maximum intensity amplitude of each pixel is stored during the data acquisition. The result is an autofocus image. Figure 5.17 shows the result from that procedure for the gear wheels from the previous measurement.

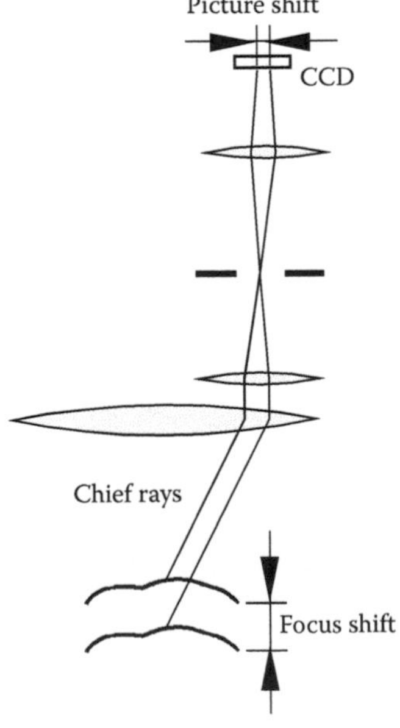

FIGURE 5.15 Picture shift due to the depth scan of the microscope body.

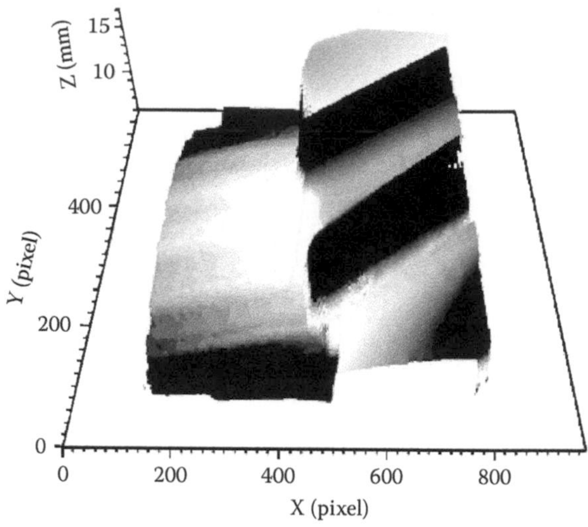

FIGURE 5.16 Three-dimensional topography of a gearwheel combination measured with the depth-scanning fringe projection (DSFP).

5.4 CONCLUSION

Confocal microscopy is a fast topography measurement technique without mechanical interaction with the measurement object, applicable to reflecting as well as to scattering MEMS surfaces. The technique is very robust with respect to vibrations and stray light. The maximum measurable surface gradient is limited by the numerical aperture of the microscope objective. The chromatic confocal technique provides single shot measurements in real time. A large variety of commercial confocal microscopes using different principles are available in the market.

The fringe projection technique, with an additional depth-scanning procedure, enables the absolute measurement of surface topographies by means of a triangulation setup using only one grating for projection. Then, modified algorithms, which are known from the white-light interferometry, can be applied efficiently to obtain the fringe order as well as the signal phase. In particular, the phase evaluation gives higher resolutions than just a simple contrast interpolation. Future work is necessary to refine this method.

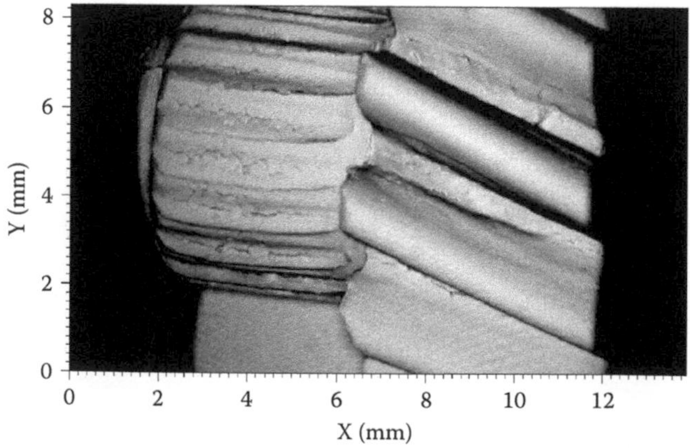

FIGURE 5.17 Autofocus image of the combined gearwheel (Figure 5.16), image rotated.

REFERENCES

1. Minsky, M., Microscopy Apparatus, U.S. Patent 3,013,467, 1961.
2. Minsky, M., Memoir on inventing the confocal scanning microscope, *Scanning* 10: 128–138, 1988.
3. Wilson, T., Confocal microscopy, in Wilson, T., Ed., *Confocal Microscopy*, San Diego: Academic Press, 1990.
4. Kino, G.S., Xiao, G.Q., Real-time scanning optical microscopes, in Wilson, T., Ed., *Confocal Microscopy*, San Diego: Academic Press, 1990.
5. Corle, T.R., Chou, C.-H., Kino, G.S., Depth response of confocal optical microscopes, *Opt. Lett.*, 11(12): 770–772, 1986.
6. Corle, T.R., Kino, G.S., *Confocal Scanning Optical Microscopy and Related Imaging Systems*, San Diego: Academic Press, 1996, p. 152.
7. Aguilar, J.F., Mendez, E.R., On the limitations of the confocal scanning optical microscope as a profilometer, *J. Mod. Opt.*, 42(9): 1785–1794, 1995.
8. Wilson, T., Carlini, A.R., Size of the detector in confocal imaging systems, *Opt. Lett.*, 12(4): 227–229, 1987.
9. Wilson, T., Optical aspects of confocal microscopy, in Wilson, T., Ed., *Confocal Microscopy*, San Diego: Academic Press, 1990.
10. Born, M., Wolf, E., *Principles of Optics*, 6th ed., Oxford: Pergamon Press, 1986.
11. Wilson, T., Sheppard, C.J.R., *Theory and Practice of Scanning Optical Microscopy*, London: Academic Press, 1984.
12. Molesini, G., Pedrini, G., Poggi, P., Quericioli, F., Focus-wavelength encoded optical profilometer, *Opt. Commun.*, 49: 229–233, 1984.
13. Browne, M.A., Akinyemi, O., Boyde, A., Confocal surface profiling utilizing chromatic aberration, *Scanning*, 14: 145–153, 1992.
14. Lin, P.C., Sun, P.-C., Zhu, L., Fainman, Y., Single-shot depth-section imaging through chromatic slit-scan confocal microscope. *Appl. Opt.*, 37: 6764–6770, 1998.
15. Tiziani, H.J., Uhde, H.-M., Three-dimensional image sensing by chromatic confocal microscopy, *Appl. Opt.*, 33(10): 1838–1843, 1994.
16. Cha, S., Lin, P.C., Zhu, L., Sun, P.-C., Fainman, Y., Nontranslational three-dimensional profilometry by chromatic confocal microscopy with dynamically configurable micromirror scanning, *Appl. Opt.*, 39(16): 2605–2613, 2000.
17. Tiziani, H.J., Wegner, M., Steudle, D., Confocal principle for macro- and microscopic surface and defect analysis, *Opt. Eng.*, 39(1): 32–39, 2000.
18. Petran, M., Hadravsky, M., Tandem-scanning reflected-light-microscope, *J. Opt. Soc. Am.*, 58: 661–664, 1968.
19. Boyde, A., Petran, M., Hadravsky, M., Tandem scanning reflected light microscopy of internal features in whole bone and tooth samples, *J. Microsc.*, 132(Pt. 1): 1–7, 1983.
20. Xiao, G.Q., Corle, T.R., Kino, G.S., Real-time confocal scanning optical microscope, *Appl. Phys. Lett.*, 53(8): 716–718, 1988.
21. Kino, G.S., Corle, T.R., Scanning optical microscopes close in on submicron scale, *IEEE Circuits and Devices*, 2: 28–31 and 34–36, 1990.
22. Jordan, M., Wegner, M., Tiziani, H.J., Highly accurate non-contact characterization of engineering surfaces using confocal microscopy, *Meas. Sci. Technol.*, 9: 1142–1151, 1998.
23. Tanaami, T., Otsuki, S., Tomosada, N., Kosugi, Y., Shimizu, A., Ishida, H., High-speed 1-frame/ms scanning confocal microscope with a microlens and Nipkow disks, *Appl. Opt.*, 41(22): 4704–4708, 2002.
24. McCabe, E.M., Fewer, D.T., Ottewill, A.C., Hewlett, S.J., Hegarty, J., Direct-view microscopy: optical sectioning strength for finite-sized, multiple-pinhole arrays. *J. Microsc.*, 184(Pt. 2): 95–105, 1996.
25. Fujita, K., Nakamura, O., Kaneko, T., Oyamada, M., Takamatsu, T., Kawata, S., Confocal multipoint multiphoton exitation microscope with microlens and pinhole arrays, *Opt. Commun.*, 174: 7–12, 2000.
26. Tiziani, H.J., Uhde, H.M., Three dimensional analysis by a microlens array confocal arrangement, *Appl. Opt.*, 33(4): 567–572, 1994.
27. Tiziani, H.J., Achi, R., Krämer, R.N., Wiegers, L., Theoretical analysis of confocal microscopy with microlenses, *Appl. Opt.*, 35(1): 120–125, 1996.

28. Meadows, D.M., Johnson, W.O., Allen, J.B., Generation of surface contours by moiré patterns, *Appl. Opt.*, 9 (4): 942–947, 1970.
29. Takasaki, H., Moire topography, *Appl. Opt.*, 9(6): 1457–1472, 1970.
30. Leonhardt, K., Droste, U., Tiziani, H.J., Microshape and rough-surface analysis by fringe projection, *Appl. Opt.*, 33: 747–748, 1994.
31. Windecker, R., Tiziani, H.J., Topometry of technical and biological objects by fringe projection, *Appl. Opt.*, 34: 3644–3650, 1995.
32. Wyant, J.C., Holographic and Moiré techniques in optical shop testing, Malacara, D., Ed., New York: John Wiley & Sons, 1978, pp. 402–406.
33. Tiziani, H.J., Automatisierung der optischen Qualitätsprüfung, *Technisches Messen tm*, 55: 481–491, 1988.
34. Osten, W., Nadeborn, W., Andrä, P., General hierarchical approach in absolute phase measurement, *Proc. SPIE*, 2860: 2–13, 1996.
35. Wahl, F.M., A Coded Light Approach for 3-Dimensional (3D) Vision, Research Report of IBM Zurich Research Laboratory RZ 1452, Log No. 52546: 1–5, 1984.
36. Gerber, J., Kühmstedt, P., Kowarschik, R., Notni, G., Schreiber, W., Three-coordinate measuring system with structured light, *Proc. SPIE*, 2342: 41–49, 1994.
37. Nadeborn, W., Andrä, P., Osten, W., A robust procedure for absolute phase measurements, *Opt. Laser Eng.*, 24: 245–260, 1996.
38. Saldner, H.O., Huntley, J.M., Profilometry using temporal phase unwrapping and a spatial light modulator-based fringe projector, *Opt. Eng.*, 36: 610–615, 1997.
39. Osten, W., Application of optical shape measurement for the nondestructive evaluation of complex objects, *Opt. Eng.*, 39: 232–243, 2000.
40. Körner, K., Windecker, R., Fleischer, M., Tiziani, H.J., One-grating projection for absolute three-dimensional profiling, *Opt. Eng.*, 40: 1653–1660, 2001.
41. Engelhardt, K., Häusler, G., Acquisition of 3-D data by focus sensing, *Appl. Opt.*, 27: 4684–4689, 1988.
42. Takeda, M., Aoki, T., Miyamoto, Y., Tanaka, H., Gu, R., Zhang, Z., Absolute three-dimensional shape measurements using coaxial and coimage optical systems and Fourier fringe analysis for focus detection, *Opt. Eng.*, 39: 61–68, 2000.
43. Fleischer, M., Windecker, R., Tiziani, H.J., Fast algorithms for data reduction in modern optical three-dimensional profile measurement systems with MMX technology, *Appl. Opt.*, 39: 1290–1298, 2000.
44. Windecker, R., Fleischer, M., Tiziani, H.J., Three-dimensional topometry with stereo microscopes, *Opt. Eng.*, 36: 3372–3377, 1997.
45. Ishihara, M., Nakazato, Y., Sasaki, H., Tonooka, M., Yamamoto, M., Otani, Y., Yoshizawa, T., Three-dimensional surface measurement using grating projection method by detecting phase and contrast, *Proc. SPIE*, 3740: 114–117, 1999.

6 Grid and Moiré Methods for Micromeasurements

Anand Asundi, Bing Zhao, and Huimin Xie

CONTENTS

6.1 Introduction 163
6.2 Grid or Grating Fabrication Methods 164
 6.2.1 Photoresist 164
 6.2.2 Moving Point Source Holographic Interferometer [8,9] 165
 6.2.3 Electron Beam Lithography [5,6] 165
 6.2.4 Focused Ion Beam (FIB) Milling [10] 166
6.3 Micro-Moiré Interferometer 167
 6.3.1 Principle 167
 6.3.2 Fiber-Optic Micro-Moiré Interferometer [12] 169
 6.3.3 Application in Microelectronic Packaging 171
 6.3.4 Conclusion 173
6.4 Moiré Methods Using High-Resolution Microscopy 174
 6.4.1 Electron Beam Moiré Method [5,6] 174
 6.4.2 AFM Moiré Method [13] 175
 6.4.3 SEM Scanning Moiré Method [14] 176
 6.4.4 FIB Moiré Method [10] 177
 6.4.5 TEM Moiré Method [15] 178
 6.4.6 Applications 179
 6.4.7 Conclusion 182
6.5 Microscopic Grid Methods 182
 6.5.1 Introduction 182
 6.5.2 Grid Line Pattern Analysis Methods with the FT Method 184
  6.5.2.1 Spatial Resolution 185
  6.5.2.2 Sensitivity 186
  6.5.2.3 Accuracy 187
 6.5.3 Grid Line Pattern Analysis Method with Phase-Shifting Method 188
 6.5.4 Grid Diffraction Method 188
 6.5.5 Applications 191
6.6 Conclusions 197
References 198

6.1 INTRODUCTION

Grid and moiré methods have been the cornerstone of deformation and strain measurements. The grid method has its foundations in the definition of normal strain as the change in length of a line segment and of shear strain as the change in the angle between two line segments originally perpendicular to each other. If this concept is extended to all points on the specimen, a grid is formed, and by comparing the grid in the undeformed and deformed states the strain at each point can be deduced from first principles. Theoretically, strain is defined as the derivative of the displacement component whereas, practically, the line segment has to have a finite length (gauge length) for measurement. Thus, to overcome strain averaging over the gauge length, there is a need for shorter gauge lengths in certain applications. Note that in case of uniform or linearly varying strains, error due to finite gauge length is zero. A shorter gauge length increases sensitivity as well as resolution but leads to an increased number of grid elements to evaluate, which can be cumbersome. Thus, the moiré method evolved. In the moiré method, the two grids (or gratings) recorded before and after loading are superposed to reveal moiré fringes that give the difference between the two grating periods. This then is directly proportional to deformation and, hence, strain can be readily deduced. The sensitivity, as with the grid method, is proportional to the pitch of the undeformed (or reference) grating. In addition, moiré methods provide full-field visualization of the deformation [1–3]. Due to diffraction effects, the coarse moiré is limited to gratings with a frequency of 100 lines/mm or to those in which the smallest pitch is 10 μm. The coarse moiré method has found a wide range of applications in both in-plane and out-of-plane deformation measurement for various engineering problems. In this chapter, we limit our discussion to in-plane deformation measurement.

Although in principle it is possible to use a finer grating pitch, it was not until the late 1980s that D. Post [4] developed moiré interferometry for highly sensitive in-plane deformation measurement. Moiré interferometry invokes the principle of interference and diffraction to create moiré fringes, which contour the displacement component similar to the coarse moiré effect. A displacement sensitivity of 25 nm/fringe has been achieved. An alternate school emerged in the early 1990s, which used the traditional coarse moiré principle of superposition of two gratings. The electron beam moiré method first developed by Kishimoto [5], and advocated by Dally and Read [6], used the scanning of the electron microscope to alias with the specimen grating that was being imaged. This was termed the *scanning moiré method.* The displacement sensitivity of the electron moiré method is 100 nm/fringe.

Grid methods have also evolved with the ability to print fine gratings on the specimen. The advantage of the grid methods is that they can directly provide strain information at a point, unlike the moiré method, in which strain has to be deduced from the moiré fringe pattern. The principle of diffraction was suggested as a means of strain determination. It was developed further into a novel optical diffraction strain sensor [7] that rivals the electrical resistance strain gauge. The ubiquitous microscope has also provided the capability to measure strain through diffraction.

In this chapter some novel advances in moiré and grid methods using various microimaging devices are exemplified with specific applications to micromeasurement.

6.2 GRID OR GRATING FABRICATION METHODS

6.2.1 PHOTORESIST

For coarse gratings (less than 300 lines/mm), a vacuum copy machine can be used to produce photoresist gratings. The specimen is coated with photoresist and prebaked. Then it is placed in the vacuum exposure machine. A master grating is placed on the specimen with photoresist. A vacuum pump is used to eliminate the gap between the specimen and the master grating. The photoresist is then exposed under an ultraviolet (UV) lamp. After developing and fixing, a photoresist grating is formed on the surface of the specimen. A typical exposure machine for producing photoresist gratings is illustrated in Figure 6.1

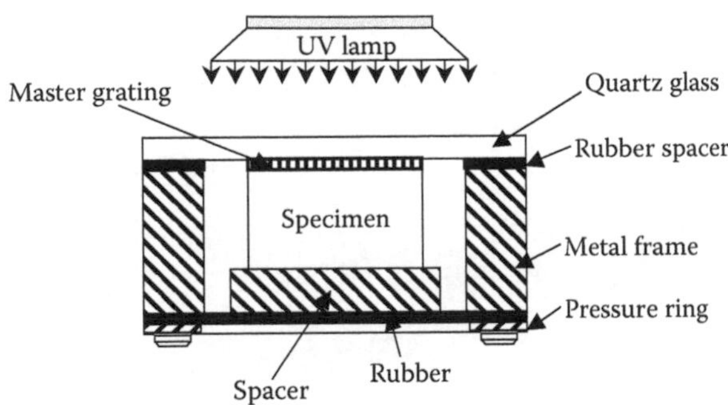

FIGURE 6.1 Schematic diagram of photoresist exposure machine.

6.2.2 Moving Point Source Holographic Interferometer [8,9]

To produce high-frequency gratings, holographic interferometry has been used. In order to eliminate speckle noise on the grating surface and to improve the quality of the grating, a moving point holographic interferometer system (as shown in Figure 6.2) is recommended for producing high-frequency gratings. In the experiment, an argon laser was the light source; the object of the fabrication grating should be coated with photoresist and exposed in the system. After developing, a photoresist grating is formed. Before a moiré test, this grating should be coated with an Al or Au layer.

6.2.3 Electron Beam Lithography [5,6]

An electron beam lithography system is set up using TOPCON SEM SX-40A equipped with a beam blanking device and a pattern generator. As displayed in Figure 6.3, under the control of the pattern generator, the electron beam is separated into blanking and unblanking parts. Suppose L_0

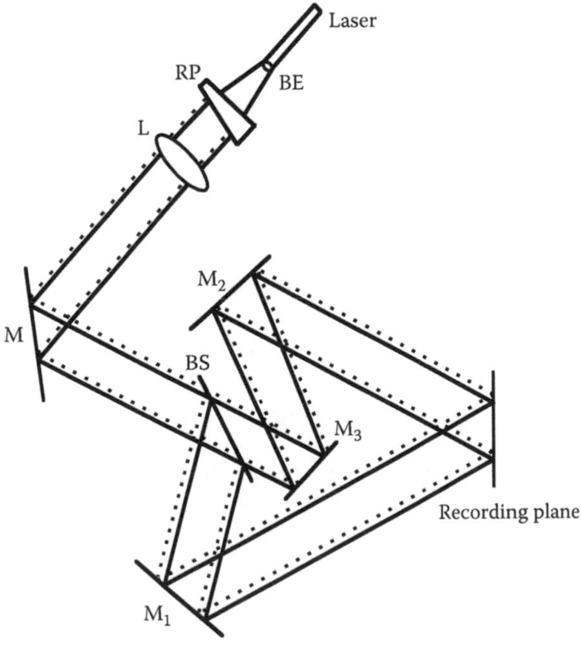

FIGURE 6.2 Schematic diagram of a moving point holographic interferometer system.

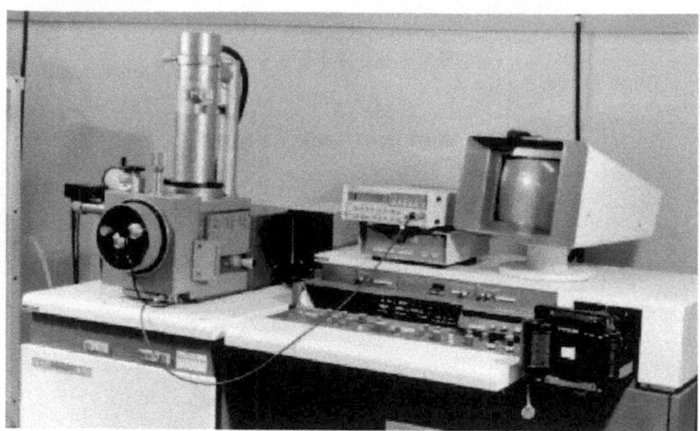

FIGURE 6.3 A photograph of the electron beam moiré system.

represents the spacing of the electron beam, the relation of the blanking number B, the unblanking number U, and the spacing of the master grid p_0 can be expressed as:

$$(U + B)\, L_0 = p_0 \tag{6.1}$$

The relation between the scanning electron microscope (SEM) magnification K and L_0 (μm) was investigated [5]. An empirical formula for this relation was obtained:

$$\log L_0 = 1.093 \log K + 2.180 \tag{6.2}$$

When a grating with pitch p is produced, the magnification K must be selected at first. Then L_0 can be calculated with Equation 6.2. From Equation 6.1, if $U = 1$, B can be derived as

$$B = \left[\frac{p_0}{L_0} - 1\right] \tag{6.3}$$

where $\left[\frac{p_0}{L_0} - 1\right]$ is the integer part of the value $\frac{p_0}{L_0} - 1$.

Once the blanking and the unblanking values are determined, the electron beam resist can be exposed. After developing and fixing, an electron beam resist grating is left on the specimen. Before a moiré test, this grating should be coated with an Al or Au layer.

6.2.4 Focused Ion Beam (FIB) Milling [10]

An FIB milling unit is essentially an SEM but with an ion instead of an electron source. When the ions hit the sample surface, not only are secondary electrons emitted but owing to their mass, ions are implanted, defects generated, and sample material removed. This sputter erosion enables local removal of material by means of focused ion beams in a direct writing mode. With the FIB milling method, a grating can be fabricated on the specimen surface. In general, a periodic line pattern can be created using the FIB control software, and then the ion beam can be used to write on the specimen surface following the designed pattern. The grating is "etched" on the surface of the specimen. A grating fabricated using FIB milling is shown in the Figure 6.4.

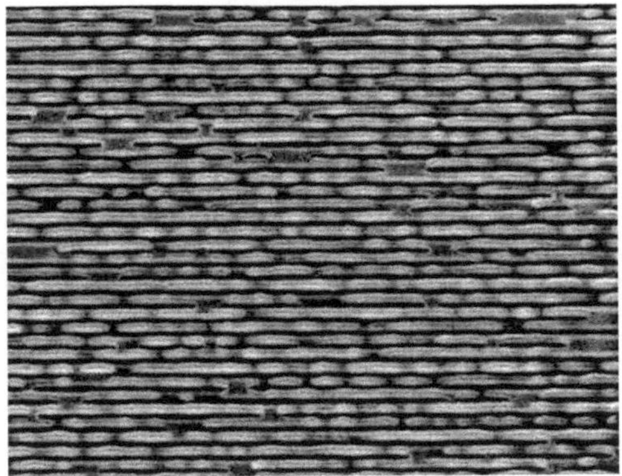

FIGURE 6.4 A 0.14-micron spacing grating on a microcantilever.

6.3 MICRO-MOIRÉ INTERFEROMETER

6.3.1 Principle

Moiré interferometry is an optical experimental technique that provides high sensitivity for in-plane deformation measurement. Moiré interferometry invokes the principles of diffraction and interference to measure deformation in the nanometer range. The basic principle of moiré interferometry is illustrated schematically in Figure 6.5.

In this method, a high-frequency (typically 1200 lines/mm) specimen grating is first adhered or printed on the specimen surface. Physically, a virtual reference grating formed by interference of two coherent beams is then superposed on the deformed specimen grating to reveal moiré fringes, which contour the displacement component perpendicular to the grating direction. The virtual reference grating has twice the frequency of the specimen grating and, thus, the sensitivity to displacement is doubled. Typically, for a specimen grating with a pitch of 1200 lines/mm, the displacement sensitivity is 0.417 μm/fringe. Optically, the concepts of diffraction and interference are used to explain the formation of the moiré fringes. First, the specimen grating diffracts the two incident beams. Because the two beams illuminate the specimen symmetrically, the −1 diffraction order of one beam and the +1 diffraction order of the symmetrical beam emerge normal to the

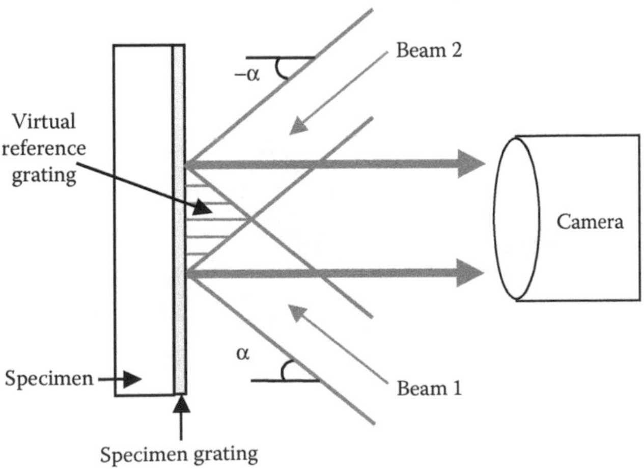

FIGURE 6.5 Moiré interferometry principles.

specimen. If the specimen grating is not deformed, the two diffracted beams emerge parallel to each other, and a null field is obtained. When the specimen (and thus, the specimen grating) deforms, the two diffracted beams emerge at an angle with respect to each other, and the resulting interference of these two fringes is the moiré pattern.

The moiré patterns depict contours of the displacement component in a direction perpendicular to the grating lines. From the fringe patterns, the displacement is obtained and the derivative of the displacement is the strain component. The displacements can be determined from the fringe orders by:

$$U = (1/f)\, N_x$$
$$V = (1/f)\, N_y \tag{6.4}$$

where N_x and N_y are the fringe orders in the U- and V-field patterns and f is the frequency of the virtual reference grating. Strains can be calculated from the measured displacement fields as:

$$\varepsilon_x = \partial U/\partial x = (1/f)\,[\partial N_x/\partial x]$$
$$\varepsilon_y = \partial V/\partial y = (1/f)\,[\partial N_y/\partial y] \tag{6.5}$$
$$\gamma_{xy} = \partial U/\partial y + \partial V/\partial x = (1/f)\,[\partial N_x/\partial y + \partial N_y/\partial x]$$

Widespread applications of moiré interferometry are found in the microelectronics industry, but other areas such as composite micromechanics and biomechanics are also finding uses for this method [11]. A high-frequency grating is replicated on the electronic package at 100°C and then cooled slowly to room temperature. Figure 6.6 shows the moiré fringe patterns of a quad flat pack (QFP) package that was subjected to an isothermal loading of $\Delta T = -75°C$. The 1200-lines/mm

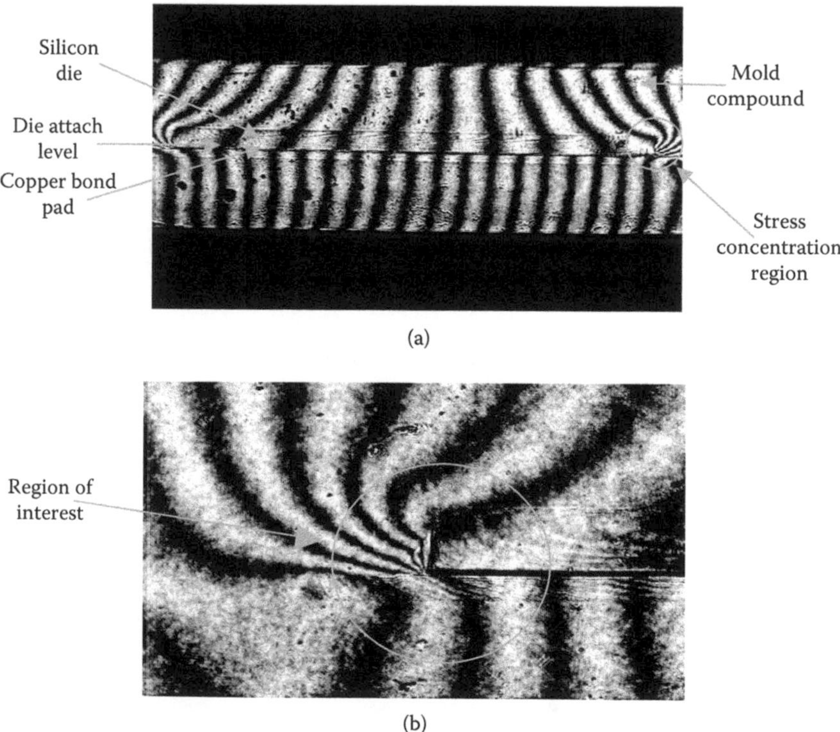

FIGURE 6.6 U-field moiré fringe patterns at the die/die–paddle interface obtained by the commercial moiré interferometer. (a) full-field, (b) zoom around the die corner.

grating that was adhered to the specimen deformed with the specimen as it cooled to room temperature. The deformation of the specimen was then recorded by a commercial moiré interferometer at room temperature. The recorded moiré fringes show the deformation that was induced by the change of temperature. Although the overall fringe patterns as shown in Figure 6.6a have good contrast and visibility, if we were to zoom into the critical interface of the die corner and copper lead frame as shown in Figure 6.6b, the limitation of the system would become apparent. Typically, the thickness of the interface is between 25 and 50 μm, and the fringes appear to be concentrated at the die corner with a large fringe gradient. However, due to poor spatial resolution of the imaging system, deformation analysis is hindered in this critical region of high strain gradient.

6.3.2 Fiber-Optic Micro-Moiré Interferometer [12]

A fiber-optic micro-moiré interferometer was developed that maintained the simplicity of the commercial moiré interferometer but provided better spatial resolution, to aid in the analysis. Figure 6.7 shows the moiré fringe patterns obtained from the micro-moiré interferometer. The limited spatial resolution of the commercial moiré interferometer limits the resolution of deformation at the critical corner of the electronic packages where cracks tend to occur. The micro-moiré interferometer, however, can resolve the critical corners. High strain concentration can be noticed at the bottom corner of the silicon die and die–paddle interface.

The photograph and schematic of the fiber-optic micro-moiré interferometer is shown in Figure 6.8. Setting up the fiber-optic micro-moiré system is easier than the previous system. However, some attention has to be paid to the nature of the light that emerges from the fibers. Details and methodologies for alignment are given in Reference 12.

An evaluation of the micro-moiré system was performed. A tensile load was applied to an aluminum alloy specimen onto which a 1200-lines/mm grating was replicated. A 3-mm gauge length strain gauge was firmly attached to the back surface of the specimen. The strain gauge and the load cell, which was attached to the loading jig, were electrically connected to a data logger, which displayed the load and strain readings. Several sets of loading were performed in the experiment. The results obtained from the strain gauge were compared with the results obtained from the micro-moiré interferometer and the theoretical values, and are shown in Figure 6.9.

During the online image acquisition of the experiment, only one fringe could be observed when the load was increased from 0 N to 80 N. With only one fringe, the displacement could not be obtained. Two fringes started to appear in the image when the load was increased from 100 N to 200 N. Thus, the strain value from the moiré fringe patterns is plotted from 100 N to 200 N.

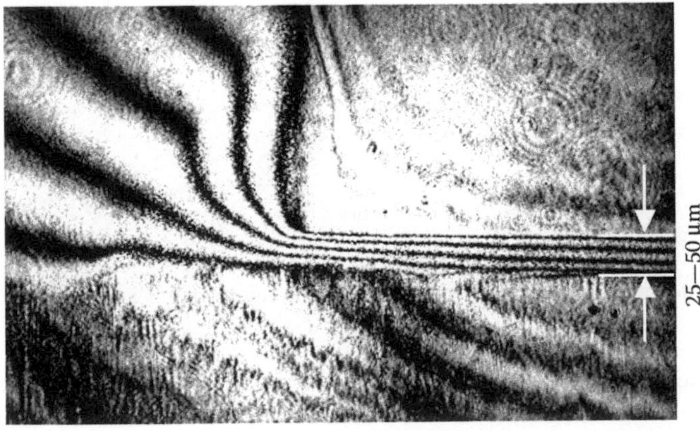

FIGURE 6.7 U-field moiré fringe patterns at the corner of the die/die–paddle interface obtained by the micro-moiré interferometer.

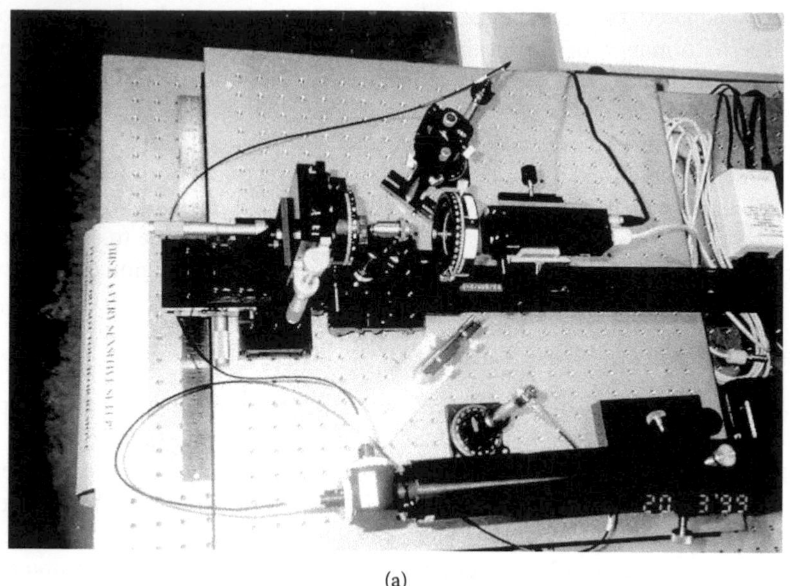

(a)

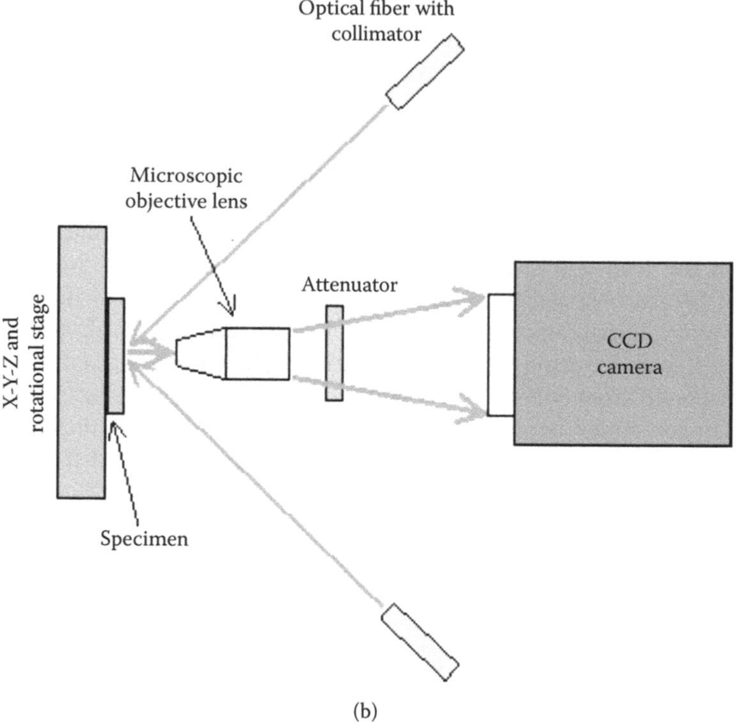

(b)

FIGURE 6.8 (a) Photograph, (b) schematic of the fiber-optic micro-moiré interferometer.

The strain gauge and the load cell were initialized to zero when a small load was applied to position the specimen for the tensile testing; the fringe pattern, however, could not be initialized. Thus, from the plot, it was found that the stress–strain graph of the fringe patterns was shifted slightly to the right when compared with that of the theoretical and strain gauge, owing to the initial loading. However, the gradient is the same as that measured by the other methods.

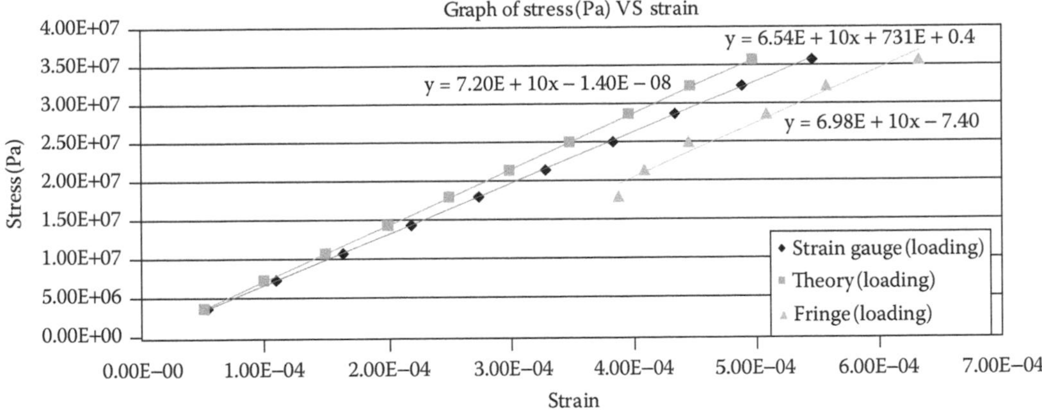

FIGURE 6.9 Evaluation of the micro-moiré interferometer.

6.3.3 Application in Microelectronic Packaging

The specimen used for this experiment was a QFP package. The specimen was initially halved and then ground to expose the internal components of the package. A uniform crossline grating on ultralow expansion (ULE) substrate with a grating frequency of 1200 lines/mm was the mold used for the grating replication process. This was to ensure that the frequency of the specimen grating did not change at the elevated temperature. Both the specimen and the grating mold were heated to an elevated temperature and kept at that temperature for at least 2 h. This was to ensure that the temperature was the same throughout the specimen and the mold, and also to make the sample relax sufficiently. Replication steps included spreading of adhesive, curing, and prying of the specimen from the mold. The grating was transferred onto the QFP package at 100°C. Immediately after the prying of the specimen from the grating mold, the specimen was cooled to room temperature (25°C). It was thus subjected to an isothermal loading of $\Delta T = -75$°C. The grating that was adhered onto the specimen would deform with the specimen during the cooling-down process. The replication process was completed when the specimen cooled down to room temperature. The deformation of the specimen was then recorded by the moiré interferometer at room temperature. The recorded moiré fringes showed the deformation that was induced by the change of temperature. The specimen was placed in the micro-moiré interferometer, and the u- and v-field deformation fringes are shown in Figure 6.10.

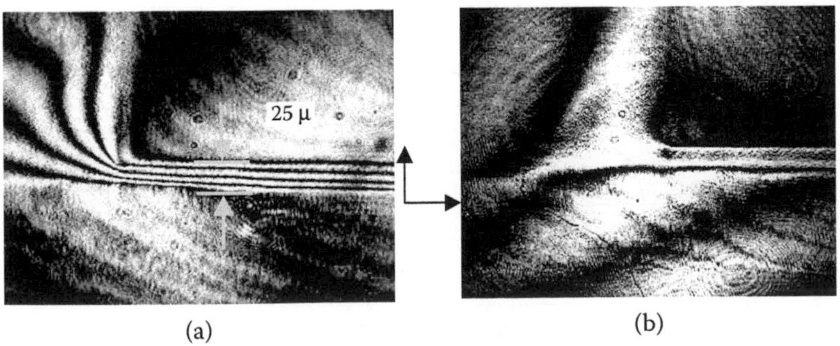

FIGURE 6.10 (a) The u-field, (b) v-field moiré fringe patterns at the corner of the die/die–paddle interface obtained by the micro-moiré interferometer.

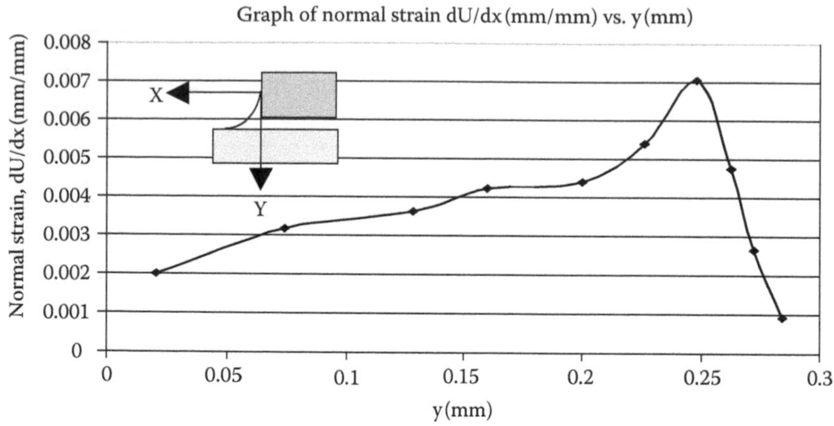

FIGURE 6.11 Normal strain dU/dx at the corner of the silicon die.

The derivatives of the displacement components that are proportional to the strain components from the critical corner to the center of the package were deduced from these fringe patterns. These derivatives were determined simply by dividing the pitch of the reference grating (0.417 μm) by the fringe spacing.

Figure 6.11 shows the distribution of dU/dx, the normal strain at the lower corner of the silicon die. It can be seen that the strain peaks at the bottom corner of the silicon die and die attachment interface and then quickly fall to zero. This can also be visualized from the concentration of fringes just ahead of the die corner. The fringes then flow in a horizontal and gradual manner into the interface of the die and the die paddle, where dU/dx is zero.

Figure 6.12 shows the distribution of the normal strain dV/dy at the lower corner of the interface along the center of the silicon die. The strain value increases steeply from the outside region of the die corner and then levels to a constant value at the center of the interface between the die and the die paddle. The graph in Figure 6.13a displays the shear strain component dU/dy from the corner of the silicon die to the center of the package along the interface. Very high values of dU/dy are obtained just outside the die corner interface and diminish gradually within the interface to the middle of the package. dV/dx is also relatively high only outside the die corner, as is evident from Figure 6.13b and decreases gradually as it approaches the die.

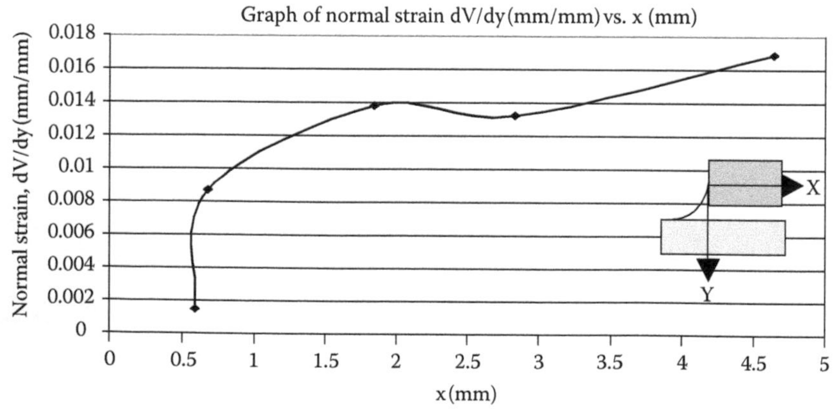

FIGURE 6.12 Normal strain dV/dy at the corner to the center of silicon die attachment interface.

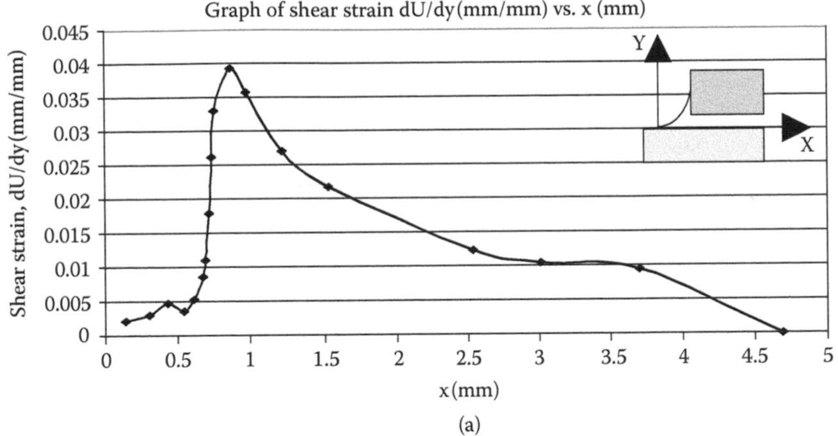

(a)

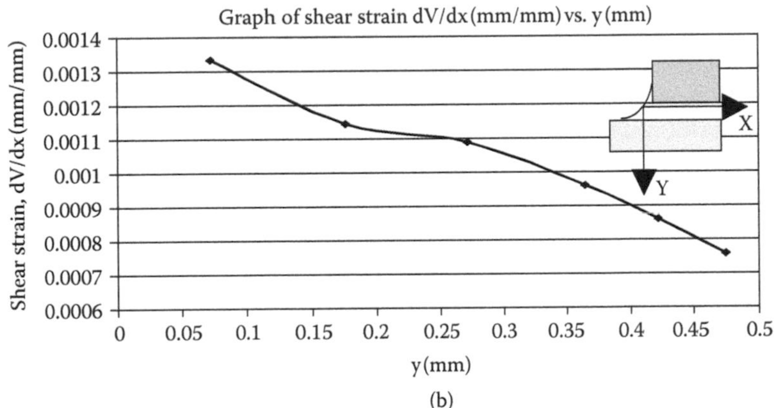

(b)

FIGURE 6.13 (a) dU/dy distribution from the corner to the center of silicon die attachment interface, (b) dV/dx distribution at the die corner.

6.3.4 Conclusion

From the experiments performed, the following are found to be the specifications of the micro-moiré interferometer:

- Displacement sensitivity of 0.417 μm per fringe order with no fringe processing
- Fringe displacement resolution of approximately 10 μm
- Imaging system resolution of 2 μm/pixel
- Strain sensitivity range of 2.71E04 to 4.17E02 with accuracy of 4.3 $\mu\varepsilon$
- Fiber-optic beam delivery system with no mirrors or optics
- Digital imaging system for online acquisition with 1/30-sec response time for capturing image
- Compact size, low weight, and flexible

Besides applications in QFP, the micro-moiré interferometer can be used to study the 75- to 100-μm solder bump in flip-chip packages. The micro-moiré interferometer has also been applied for residual stress measurement in composite material and dental biomechanics [12].

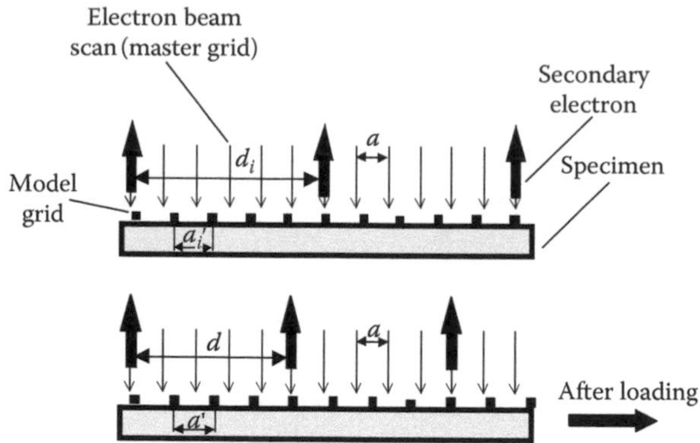

FIGURE 6.14 Principle of the electron beam moiré method.

6.4 MOIRÉ METHODS USING HIGH-RESOLUTION MICROSCOPY

In this section, some newly developed moiré methods with high-resolution microscopy are introduced that enable in-plane deformation measurement on the micro- and nanoscale. The relevant experiments are performed under a high-resolution microscope, such as the SEM, the atomic force microscope (AFM), the FIB, the transmission electron microscope (TEM), etc. These methods can be generally classified as microscopic moiré methods and retain most of the advantages of the conventional moiré method but with improved sensitivity and resolution.

6.4.1 Electron Beam Moiré Method [5,6]

The electron beam lithography system described earlier can be used for moiré pattern formation as well. First, a grating is printed on the specimen surface. The electron beam moiré is formed by the interference between the specimen grating and the patterned electron beam scan. A schematic diagram of the principle of electron beam moiré method is shown in Figure 6.14.

In the experiment, the observation field (corresponding to the selected magnification) is selected first and then the unblanking number is chosen. From Equation (6.1) to Equation (6.3) the blanking number can then be determined.

When a cross specimen grid is utilized, the u-field (displacement in x-axis) and v-field (displacement in y-axis) electron moiré fringes can be separately generated by adjusting the orientation of the specimen grating parallel or perpendicular to the grating line of the reference grating.

Using the moiré patterns corresponding to the u- or v-field, the strain components ε_x, ε_y, and γ_{xy} in the x-y plane can be measured and are expressed as:

$$\varepsilon_x = \frac{\partial u}{\partial x} = \frac{p_{ra}}{d_{xx}}$$

$$\varepsilon_y = \frac{\partial v}{\partial y} = \frac{p_{ra}}{d_{yy}} \tag{6.6}$$

$$\gamma_{xy} = \frac{\partial u}{\partial y} + \frac{\partial v}{\partial x} \approx \frac{p_{ra}}{d_{xy}} + \frac{p_{ra}}{d_{yx}}$$

where p_{ra} is the pitch of the reference grating (suppose $p_{si}^{'} = p_{ra}$, and $p_{si}^{'}$ is the pitch of the specimen grating before exerting load); d_{xx} is the spacing of adjacent u-field moiré fringes in the x-axis; d_{yy} is the spacing of adjacent v-field moiré fringes in the y-axis; and d_{xy} and d_{yx} are the spacings of adjacent u-field moiré fringes in the y-axis and adjacent v-field moiré fringes in the x-axis, respectively.

By changing the magnification of the SEM, the measured area can be enlarged or reduced easily, and the moiré patterns for different scan sizes can be obtained by adjusting the parameters of the pattern generator.

6.4.2 AFM Moiré Method [13]

The contact mode AFM operates by scanning a cantilever tip across the sample surface while monitoring the changes in cantilever deflection with a split photodiode detector. When the tip is close enough to the surface of the specimen, the atomic force between the cantilever and the surface of the substrate will lead to deflection of the cantilever. This deflection is transferred into the feedback signal. By controlling the force constant, a topographic image corresponding to the surface of the specimen can be displayed in the CRT. The measured area under the AFM is adjustable within the maximum scan size of the scanner.

The AFM scanning moiré is formed by the interference between the scanning lines in the AFM monitor (reference grating) and the specimen grating. The frequency of the reference grating f_{ra} can be defined as:

$$f_{ra} = \frac{1}{p_{ra}} = \frac{M}{L} \tag{6.7}$$

where M is the number of the scanning lines ($M = 64, 128, 256, 512, \ldots$), L is the scan size, and p_{ra} is the pitch of the reference grating.

When the area under study has two orders of moiré fringes, we have [8]:

$$\varepsilon_y = \frac{|p_s - p_{ra}|}{p_{ra}} = \frac{p_s}{L} \tag{6.8}$$

where p_s is the pitch of the specimen grating.

From the Equation 6.7 and Equation 6.8, the range of L can be derived as

$$L = p_s(M \mp 1) \tag{6.9}$$

When there are N moiré fringes in the measured area, we have

$$L = p_s\left[M \mp (N-1)\right] \tag{6.10}$$

In order to form an initial carrier moiré pattern, the number of the scanning lines M is selected first, and the scanning size L is then determined according to a specific initial number of moiré fringes, using either Equation 6.9 or Equation 6.10. After deformation, under the same parameters of scan size and number of scanning lines, the moiré pattern with the deformation can be recorded.

Using the moiré patterns before and after deformation, the real direct strain ε_y in the specimen owing to the deformation can be determined:

$$\varepsilon_y = \varepsilon_y^1 - \varepsilon_y^0 = \frac{\partial v^1}{\partial y} - \frac{\partial v^0}{\partial y} = \frac{p_{ra}}{d_{yy}^1} - \frac{p_{ra}}{d_{yy}^0} \tag{6.11}$$

where d_{yy}^0 and d_{yy}^1 are the spacing of adjacent moiré fringes in the u-field and v-field, respectively.

Similarly, when the specimen or the scanning direction is rotated 90°, the normal strain component ε_x can be measured as:

$$\varepsilon_x = \varepsilon_x^1 - \varepsilon_x^0 = \frac{\partial u^1}{\partial x} - \frac{\partial u^0}{\partial x} \tag{6.12}$$

Using the u- and v-field moiré patterns, we can also obtain:

$$\gamma_{xy} = \gamma_{xy}^1 - \gamma_{xy}^0 = \frac{\partial u^1}{\partial y} + \frac{\partial v^1}{\partial x} - \frac{\partial u^0}{\partial y} - \frac{\partial v^0}{\partial x} \tag{6.13}$$

where the superscripts 0 and 1 in the strain components represent states before and after deformation, respectively.

6.4.3 SEM Scanning Moiré Method [14]

The superposition of the specimen grating and the scanning line of the SEM monitor generate the SEM scanning moiré. In this case, the scanning line in the SEM monitor is the reference grating. When a magnification factor K is utilized, the gauge length L is magnified into length L'. Thus, the pitch p'_{rs} of the reference grating on the monitor is

$$p'_{rs} = \frac{L'}{M} \tag{6.14}$$

where M is the number of scanning lines. As we know, the magnification under an SEM is not continuously adjustable and, thus, p'_{rs} is a discontinuous function depending on the magnification.

If the pitch of the specimen grating is a multiple of the reference grating pitch, no fringes are seen. Thus,

$$p_s = \beta \frac{p'_{rs}}{K} = \beta p_{rs} \tag{6.15}$$

where $p_{rs} = p_{ra}$, = 1, 2, 3, …. When the specimen deforms, a moiré pattern with the multiplication factor can be obtained.

The displacement increment over the gauge length L is

$$\Delta v = p_{rs} \Delta N \tag{6.16}$$

where ΔN is the fringe number over the gauge length.

Thus,

$$\Delta v = \frac{L}{M} \Delta N \tag{6.17}$$

and the normal strain over a gauge length L is

$$\varepsilon_y = \frac{\Delta N}{M} \tag{6.18}$$

Using the carrier moiré method, we can calculate the normal strain due to load alone as:

$$\varepsilon_y = \varepsilon_y^1 - \varepsilon_y^0 = \frac{\Delta N_1 - \Delta N_0}{M} \tag{6.19}$$

where ΔN_0 and ΔN_1 are the fringe numbers over the gauge length before and after loading, respectively.

6.4.4 FIB Moiré Method [10]

The focused ion beam (FIB) moiré is generated by the interference between the specimen grating and the focused gallium ion beam raster scan lines. For a magnification K, the frequency of the reference grating f in the scan direction is defined as:

$$f = \frac{1}{p_r} = \frac{M}{L} \tag{6.20}$$

where M is the number of the scanning lines, L is the gauge size in the scan direction, and P_r is the pitch of the reference grating. It is observed that f changes continuously with changes in the magnification factor. For a typical commercial FEI 200 FIB system (FEI Inc.), the available number of scan lines is 221, 442, and 884, respectively. Before the experiment, the pitch of the reference grating should be calibrated by measuring the scan size of the FIB images with a magnification range from 500 to 2500. The results are shown in Figure 6.15. It illustrates that the pitch of the reference grating decreases with the magnification for a given number of scan lines.

In Equation (6.6), by replacing p_{ra} with p_r, the in-plane strain components $\varepsilon_x, \varepsilon_y$, and γ_{xy} can be determined.

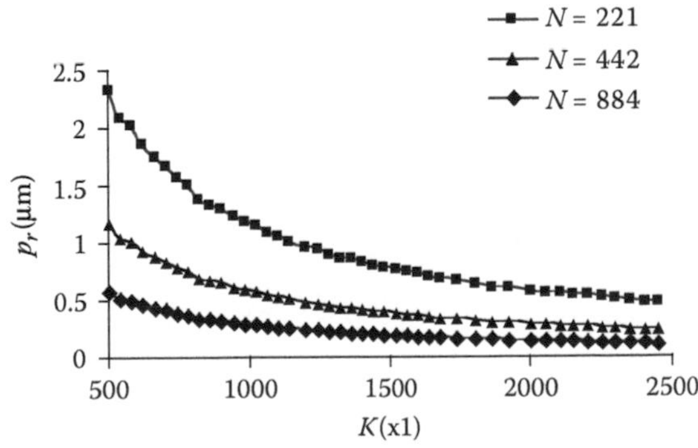

FIGURE 6.15 Relation between the pitch of the reference grating and magnification of the FIB system.

6.4.5 TEM Moiré Method [15]

In this method, nanocrystal lattices are utilized as a specimen grating. The grating frequency on a TEM image depends on the magnification of the microscope. This grating is generally used to reconstruct moiré in a 4-f Fourier system. Thus, for Fourier filtering, a frequency greater than 10 lines/mm is required to separate the diffraction orders. For TEM images with high lattice frequency, the magnification used should be as low as the recording resolution condition permits. A line grating with spacing close to that of the specimen grating is used as a reference grating. Moiré fringes are produced and reflect the differences between the lattice structure and the reference grating. The moiré fringes carry the information of the deformation of the measured object.

Because the crystal lattice lines are commonly nonorthogonal, depending on the crystal structure and observation direction under the TEM microscope, the displacements represented by the moiré fringes are the components of displacements that are normal to the reference-grating lines. As shown in Figure 6.16, the displacements in direction x and r can be obtained by adjusting the orientation of the specimen grating. u_x and u_r are expressed as:

$$u_x = N_x p_{ra} \tag{6.21}$$

$$u_r = N_r p_{ra} \tag{6.22}$$

where N_x and $N_r = 0, 1, 2, \ldots$; p_{ra} is the pitch of the reference grating divided by the magnification factor of TEM; and N_x and N_r are the fringe orders in the x- and r-direction, respectively. For an arbitrary direction n, the displacement is:

$$u_n = u_x \cos\varphi + u_r \cos(\varphi - \vartheta) = \left[N_x \cos\varphi + N_r \cos(\varphi - \theta)\right] p \tag{6.23}$$

where φ is the angle between the x-direction and the n-direction. θ is the angle between the x-direction and the r-direction.

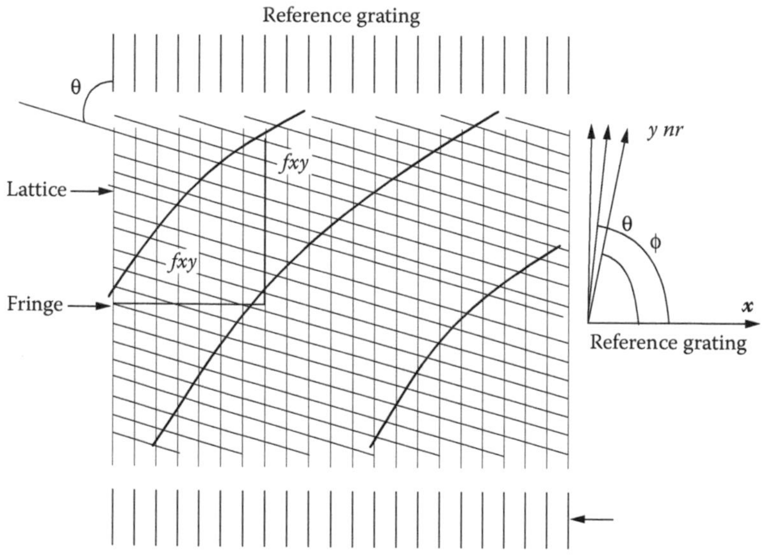

FIGURE 6.16 Schematic diagram of the TEM moiré method.

The normal strain components ε_x and ε_r can be expressed by:

$$\varepsilon_x = \frac{p_{ra}}{d_{xx}} \tag{6.24}$$

$$\varepsilon_r = \frac{p_{ra}}{d_{rr}} \tag{6.25}$$

where d_{rr} is the adjacent moiré fringe spacing in r-field moiré pattern.

The displacement component u_y in the y-direction is $u_r \sin\theta$ and thus,

$$\varepsilon_y = \frac{\partial u_y}{\partial y} = \frac{\partial u_r}{\partial y}\sin\theta \approx \frac{p_{ra}}{d_{ry}}\sin\theta \tag{6.26}$$

$$\gamma_{xy} = \frac{\partial u_y}{\partial x} + \frac{\partial u_x}{\partial y} \approx \frac{p_{ra}}{d_{rx}}\sin\theta + \frac{p_{ra}}{d_{xy}} \tag{6.27}$$

where γ_{xy} is the shear strain component.

6.4.6 Applications

A 1200-lines/mm grating was replicated onto a diced section of ball grid array (BGA) electronic package. The specimen grating was interrogated using the e-beam moiré system (Topcon SEM SX-40A, Topcon Co., Japan) to study the thermal deformation [11]. Figure 6.17a is the schematic diagram of the measured cross sections in the BGA package and Figure 6.17b shows the v-field moiré pattern at solder No. 1 of the BGA package and Figure 6.17c is the shear strain at each of the seven solders.

The AFM moiré method was applied to measure the thermal deformation in the QFP package. A schematic diagram of a measured section of the QFP package is shown in Figure 6.18a. A DI3000 SPM (Digital Instruments, U.S.) was used to generate the AFM scanning moiré under the contact mode. The AFM was set to scan a length of L = 100 μm with N = 256 scan lines. Thus, the pitch of the scanning reference grating was 0.39 μm, which corresponds to a frequency of 2564 lines/mm. With a specimen grating of 1200 lines/mm, this would give rise to carrier fringes of 164 lines/mm or in the image area of 50 μm, there would be 8.2 fringes. The AFM moiré patterns of the package at different positions in the y-axis owing to thermal deformation are shown in Figure 6.18b and Figure 6.18c. Displacement components in both the x- and y-directions are shown at the die corner.

As a demonstration of the SEM moiré, a specimen grating was fabricated on the surface of an MEMS cantilever using FIB milling. A 0.2-μm spacing line grating was "printed" on the cantilever (60 μm long, 10 μm wide, and 2 μm thick) before removing the sacrificial layer (2 μm thick). A SEM moiré pattern was formed (as shown in Figure 6.19), using 960 scan lines and a magnification of 750.

The FIB moiré method is similarly applied to measure the residual deformation of a polysilicon MEMS cantilever structure following release of the SiO_2 sacrificial layer. As before, using an FIB, a line grating with a pitch of 0.2 μm was milled on the surface of the cantilever (60 μm long, 10 μm wide, and 2 μm thick) before removal of the 2-μm thick sacrificial layer. The schematic diagram of the process is shown in Figure 6.20. An initial moiré pattern was recorded at a magnification of K = 1600 and N = 882 scan lines. The structure was then dipped into the buffered oxide etch solution (BOE solution, 40% NH_4F: 49% HF = 6:1) for 2.5 h to remove the SiO_2 sacrificial layer. After that, the cantilever was placed back into the FIB system to generate an FIB moiré with the same magnification and number of scan lines. The initial and deformed moiré patterns are shown in Figure 6.21.

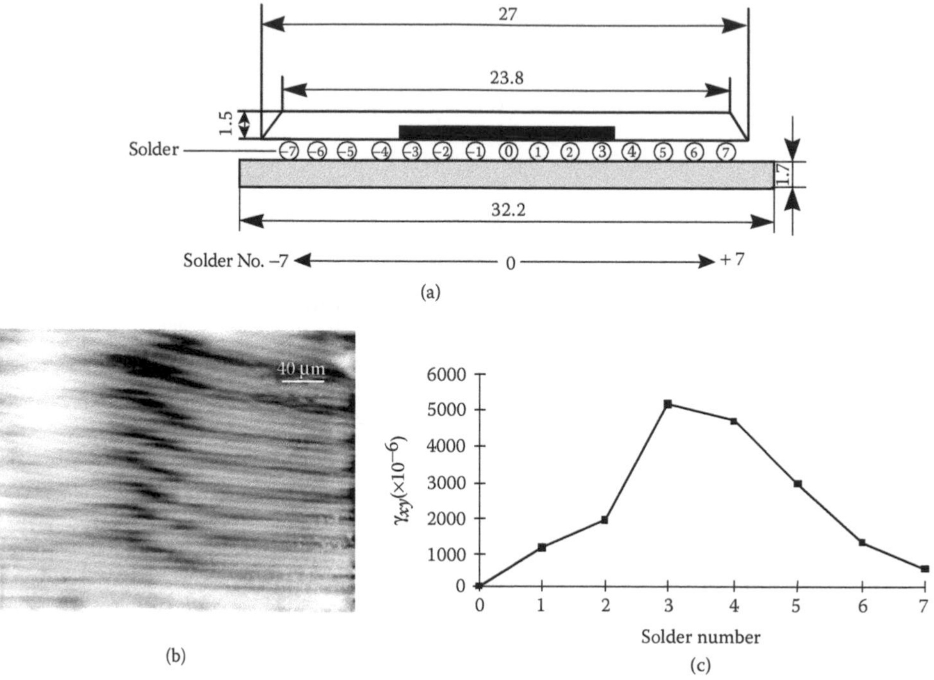

FIGURE 6.17 (a) Schematic diagram of measured section of the BGA package, (b) v-field e-beam moiré at solder 1 of a BGA package, (c) shear strain at the different solders.

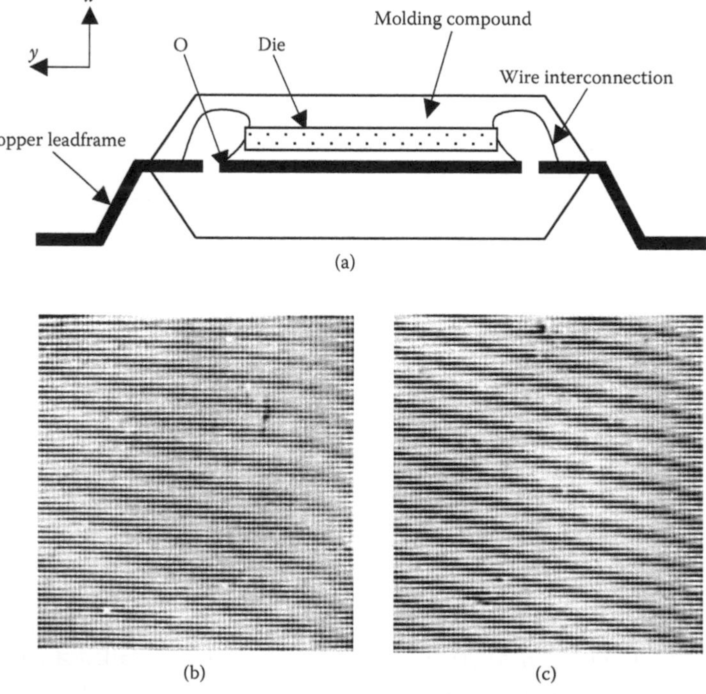

FIGURE 6.18 (a) Schematic of the QFP package, (b) u-field, (c) v-field AFM scanning moiré patterns at the die corner. Image area is 50 μm × 50 μm.

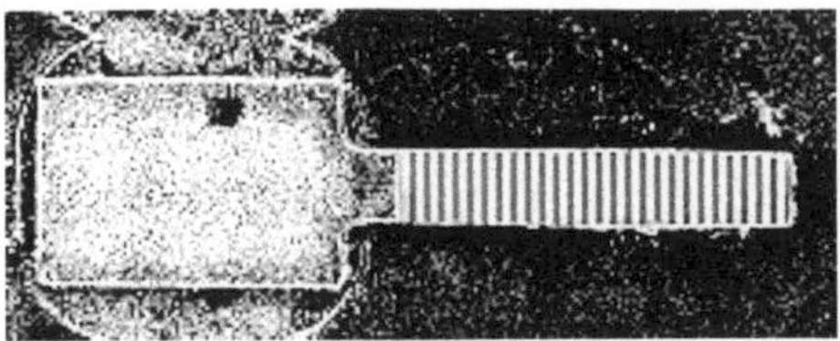

FIGURE 6.19 An SEM moiré pattern formed on a microcantilever.

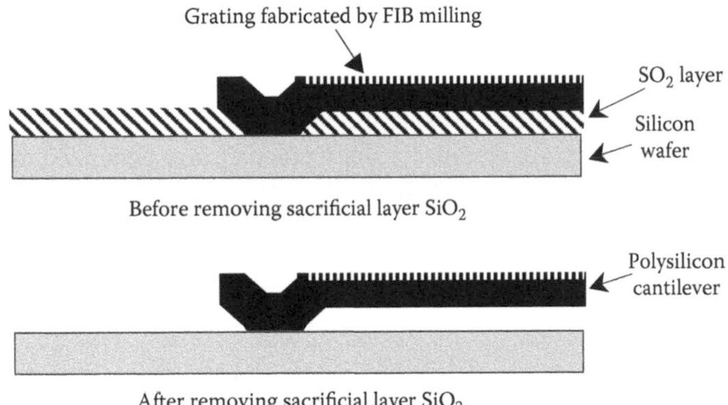

FIGURE 6.20 Schematic diagram of a polysilicon MEMS cantilever before and after removing the SiO_2 sacrificial layer.

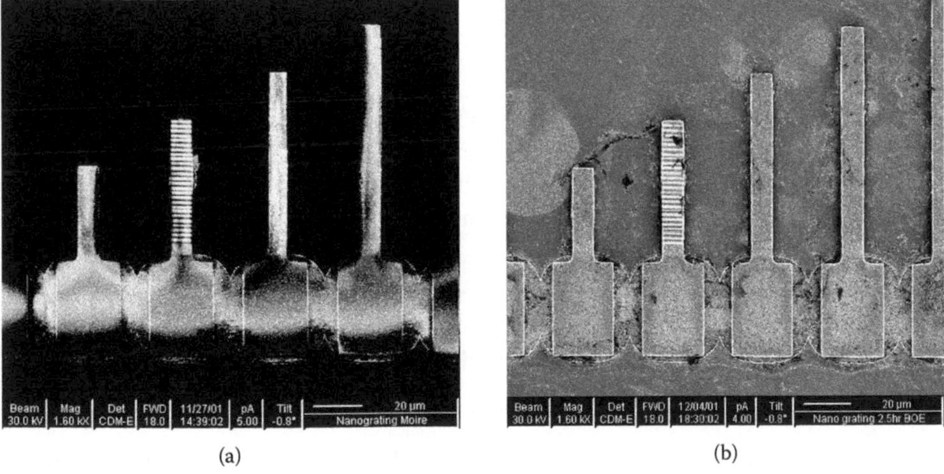

FIGURE 6.21 The FIB moiré patterns with $K = 1600$ and N = 882: (a) initial moiré pattern, (b) moiré pattern after removal of sacrificial layer.

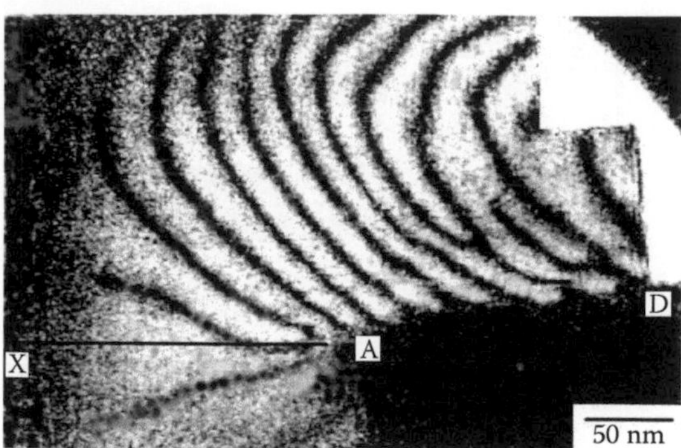

FIGURE 6.22 The moiré pattern near the crack tip.

The TEM nano-moiré was generated using the Si single crystal as the specimen. The Si (111) lattice was the specimen grating. An atomically sharp precrack was generated in the (110) direction by loading the specimen very slowly. A TEM (JEM-200cx) was used to record the lattice image, and the moiré fringes were produced by the method mentioned earlier. Figure 6.22 shows the fringe pattern near a crack tip.

6.4.7 Conclusion

A comparison of the advanced moiré methods is listed in Table 6.1. Both the spatial resolution as indicated by the measured size and the deformation sensitivity as indicated by the grating pitch are in the nanometer range without the use of phase shift or Fourier transform (FT) image processing tools. It is possible to further enhance the sensitivity using phase shifting, and thus has been demonstrated for the AFM moiré method [16] in addition to laser moiré interferometry; however, one of the drawbacks in the scanning moiré is the speed and, hence, temporal phase-shifting techniques could further reduce the data acquisition times.

6.5 MICROSCOPIC GRID METHODS

6.5.1 Introduction

In the moiré methods discussed earlier, two grids (or gratings) are needed to generate a moiré pattern. At least one of the grids (the specimen grating) has to be real, whereas the other could be real or virtual (reference grating). In all the methods discussed earlier, the reference grating was virtual — either formed by two-beam interference or by using the scanning nature of the data

TABLE 6.1
Comparison of the Advanced Moiré Methods

Method	Specimen Grating (pitch)	Measurable Size
Laser moiré interferometry	0.32 μm to 1.67 μm (For λ = 632.8 nm)	100 μm to few cm
Electron beam moiré	0.1 μm to 25 μm	100 μm to 5 mm
AFM scanning moiré	0.1 nm to 2 μm	50 nm to 100 μm
SEM (FIB) scanning moiré	0.1 μm to 4 μm	100 μm to 5 mm
TEM moiré	0.1 nm to 25 μm	50 nm to 100 μm

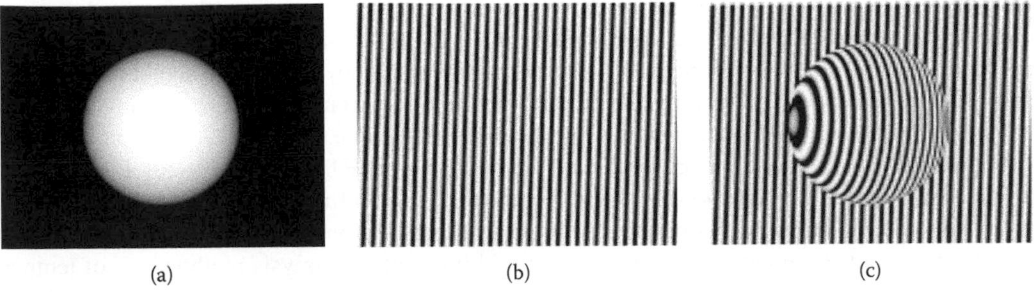

FIGURE 6.23 Simulation of grid projection for profiling: (a) ball sample, (b) projected grid image without sample, (c) projected grid image with sample.

acquisition system. The moiré fringe represents the "deformation" of the specimen grating relative to the reference grating. In the grid method, only one grid is used and no physical moiré pattern is formed. The measurement is based on either processing the grid image or analyzing the diffraction pattern from the grid. Hence, the grid method can be divided into two categories, i.e., *grid line pattern analysis method* and *diffraction method.*

The grid line pattern analysis method treats the grid line pattern as a fringe pattern. A grid has a record of the full spatial deformation information, but the deformation information appears as modulation of a carrier pattern. The moiré method automatically eliminates the carrier pattern, and thus the moiré fringe pattern directly represents the deformation information. Figure 6.23 shows an example of 3-D shape measurement of a ball using the grid projection method.

The sample ball, original projected grid line pattern, and the "deformed" grid line pattern are shown in Figure 6.23. Although the ball is symmetric, the deformed fringe pattern is not symmetric about the ball center. In other words, the "deformed" grid does not provide the desired depth information directly. We can obtain the depth information by removing the carrier grid information. Figure 6.24 shows the projection moiré fringes of the ball.

The moiré fringe pattern is symmetric about the ball center. Figure 6.23 and Figure 6.24 highlight the difference between the grid method and the moiré method. The moiré (without carrier fringes) pattern directly provides the measurand information, and the phase of the moiré fringe is

FIGURE 6.24 Projection moiré pattern for ball profiling.

proportional to the measurand. However, in the grid technique, we can obtain the measurand information only after postprocessing to remove the grid information. Another difference between the moiré method and the grid method is that the spatial frequency of the grid signal is higher than that of the moiré signal. Therefore, the spatial resolution requirements of the imaging techniques for the grid methods are higher than that for the moiré technique.

Grid line pattern analysis is generally carried out using the FT method [17–20]. Phase shifting provides higher spatial resolution than the FT method. Phase shifting using grid methods has been easily realized for the 3-D grid projection method through shifting the grid [21,22]. For in-plane displacement and strain measurement using the grid line pattern analysis method, use of temporal phase shifting is not feasible as the only grid is fixed on the sample surface. Spatial phase shifting [23] was proposed by moving the digital camera. Creath and Schmit [24,25] suggested a so-called *N*-point spatial phase-shifting technique. This technique is also suitable for grid measurement. The grid diffraction method is a special grid technique for strain measurement. Since 1956, when Bell [26] initiated the use of the "diffraction grating strain gauge," the grid diffraction strain measurement technique has advanced steadily [27–31]. The main advantage of this method is that it directly provides the strain components without having to differentiate the displacement pattern. The principle of strain measurement using grid diffraction has been well summarized in the literature.

Microscopic grid methods are finding widespread applications in scientific and industrial strain and shape measurement [32–44]. Nonoptical microscopes [32–34] such as the SEM have also been used to magnify the grid line patterns to measure the strain for microdevices and microstructures. The spatial resolution for strain measurement is high, owing to the high resolving power of the microscope. There are several examples [35–50] of strain measurement using optical microscopic grid techniques. Zhao and Asundi [35] have measured the thermal-induced residual strain, using FT analysis of a microscopic image of a specimen with a 1200-lines/mm grating. Sciammarella and Bhat [36] used a long-distance optical microscope and the FT method to measure mechanical deformation. Szanto, Dally, and Read [37] suggested a hybrid grid-moiré method using a conventional microscope and the FT technique. Conventional microscopes have higher resolving power than long-distance microscopes, but incorporating a loading frame into a microscope is not straightforward. Zhao and Asundi [38–40] have also used a conventional optical transmitting microscope for deformation measurement. They realized the grid diffraction under the conventional microscope and then developed it as a strain microscope. A technique of phase-shifting full-field grid measurement is proposed with the same microscope. Three-dimensional shape measurement using an optical microscopic projection grid technique [41–44] has also been discussed.

6.5.2 Grid Line Pattern Analysis Methods with the FT Method

The FT approach [45,46] is illustrated in Figure 6.25 and is summarized here by considering a 1-D grid as an example. The intensity of a 1-D grid line can be expressed as:

$$g(x, y) = a(x, y) + b(x, y) \times \cos\left[2\pi f_0 x + \phi(x, y)\right] \tag{6.28}$$

where the phase $\phi(x, y)$ is proportional to the required displacement $u(x, y)$:

$$\phi(x, y) = 2\pi f_0 u(x, y) \tag{6.29}$$

where f_0 is the grid frequency, and $a(x, y)$ and $b(x, y)$ are the mean and modulated intensities of the grid lines, respectively. It should be noted that the higher-order harmonics are neglected in Equation 6.28. The FT function along the sensitivity direction (x) of the grid can be expressed as:

$$G(f, y) = A(f, y) + C(f - f_0, y) + C^*(f + f_0, y) \tag{6.30}$$

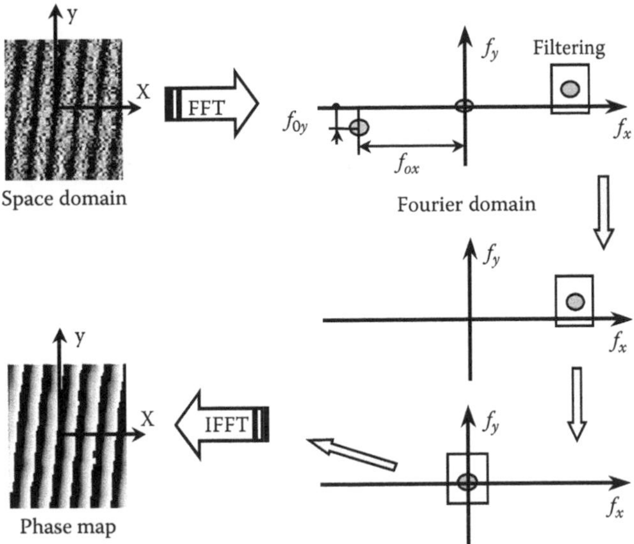

FIGURE 6.25 Principle of FT method for grid analysis.

where * denotes the complex conjugate. Because $a(x, y)$, $b(x, y)$, and $\phi(x, y)$ are slowly varying as compared to f_0, the function $G(f, y)$ will be trimodal with peaks at $-f_0$, f_0, and the origin. The function $C(f - f_0, y)$ can be isolated using a filter centered at f_0. The carrier frequency can be removed by shifting $C(f-f_0, y)$ to the origin to give $C(f, y)$. The inverse transform of $C(f, y)$ with respect to f yields:

$$C(x, y) = 0.5b(x, y) \times \exp\left[i\phi(x, y)\right] \tag{6.31}$$

from which the phase $\phi(x, y)$ can be deduced as:

$$\phi(x, y) = \tan^{-1}\left\{\mathrm{Im}\left[C(x, y)\right] / \mathrm{Re}\left[C(x, y)\right]\right\} \tag{6.32}$$

The phase thus obtained is wrapped in the range $(-\pi, \pi)$ and has to be corrected using an appropriate phase-unwrapping algorithm.

6.5.2.1 Spatial Resolution

Spatial resolution is defined as the distance between adjacent detection points [46]; more precisely, the minimum spatial unit needed to perform a measurement. It is because of this low spatial resolution that the conventional strain gauges cannot be used in a micromechanical measurement despite their high strain sensitivity and precision. This high-precision strain value is actually obtained over a sensing area of about a square millimeter. For different measurands, the spatial resolutions of the measurement are different. For example, the spatial resolution for displacement will be higher than that for strain measurement if the strain value is obtained by differentiation of the displacement data. Another important point is that the spatial resolution for 2-D displacement field measurement is not just the resolution of the test (optical) system but is also associated with the detecting system, including the post-image processing system and method, such as the filtering operation, FT, etc.

As a spatial filter, the FT method will certainly degrade the spatial resolution of the measurement system. Figure 6.26a illustrates this degradation. From the Fourier spectra of grid signals shown

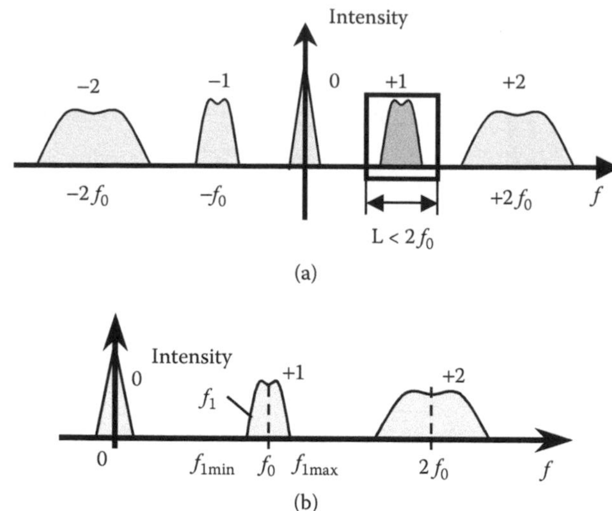

FIGURE 6.26 Relationship between filter window size, spatial resolution, and sensitivity: (a) effect of filter bandwidth on spatial resolution, (b) effect of frequency spread on the limit of measurement sensitivity.

in this figure, only the +1 spectrum is used in inverse FT to get the measured phase information. The bandwidth of the filter, L, limits the spatial resolution of the FT. All information with frequency outside this filter bandwidth is eliminated and cannot be restored. Therefore, $1/L$ is defined as the spatial resolution of FT. As L must be less than $2f_0$ (otherwise the 0 and +2 spectra are included in the filter), this means that the upper limit of the spatial resolution is $1/(2f_0) = 0.5$ pitch of the grid. Actually, a spatial resolution of 0.5 pitch cannot be reached because at least one period is needed to evaluate the information encoded in a carrier. Thus, the largest bandwidth of the filter is f_0. The practical upper limit of the spatial resolution is equal to the pitch of the grating. For a digital grating image, the grating pitch must be greater than 2 pixels. Thus, the upper limit of spatial resolution is 2 pixels. This means that subpixel interpolation is meaningless if the FT method is used.

6.5.2.2 Sensitivity

The definition of *sensitivity* for displacement measurement using the grid method is the smallest change in displacement for which the instrument shows a measurable response. This sensitivity depends not only on the ability of the measuring instrument to detect the new location of a grid point relative to its original location, but also on the grid density.

$$U = \frac{\varphi}{2\pi} P \tag{6.33}$$

Specifically, the sensitivity depends on the ability to detect the fractional fringe orders. As we can readily detect half a fringe order with the naked eye, the sensitivity should be at least equal to, if not better than, $0.5P$. In order to obtain the upper limit of sensitivity, we first evaluate the phase measurement sensitivity, which is the smallest phase change that can be detected using the FT method.

The fringe phase information is not directly reflected in the fringe spectrum image. In order to create the relationship between the fringe phase and the spatial frequency of the fringe, and to determine the sensitivity of phase measurement in the frequency domain, the local spatial frequency concept is introduced [45]. A local spatial frequency f_n for the nth spectrum is defined as

$$f_n = nf_0 + \frac{n}{2\pi}\frac{\partial\varphi(x,y)}{\partial x}, \quad n = 1,\ 2,\ \ldots \tag{6.34}$$

For $n = 1$, the phase change between two adjacent points is

$$\Delta\varphi(x,y) = 2\pi(f_1 - f_0)\Delta x \tag{6.35}$$

The minimum phase change can therefore be written as

$$\left|\Delta\varphi(x,y)\right|_{\min} = 2\pi\left|f_1 - f_0\right|_{\min}\Delta P_s \tag{6.36}$$

where ΔP_s is the pixel size in the spatial domain, f_0 is the carrier frequency and a constant, but f_1 varies in the +1 sidelobe between $f_{1\min}$ and $f_{1\max}$ (Figure 6.26b).

Theoretically, $|f_1 - f_0|_{\min}$ can approach zero and, thus, any tiny deviation of f_1 from f_0 is possible, or the sensitivity value $|\Delta\varphi(x, y)|_{\min}$ could be very small, implying very high sensitivity. However, in practice, the detectable deviation of f_1 from f_0 is limited by the pixel size. Hence, the minimum detectable change of $|f_1 - f_0|$ is the frequency increment corresponding to 1 pixel in the frequency domain, i.e.,

$$\Delta P_f = 1/(N\,\Delta P_s) \tag{6.37}$$

where N is the number of pixels in the x-direction used in the FFT to obtain the spectrum, or the number of sample points in the x-direction. The upper limit of sensitivity of phase measurement is therefore

$$\left|\Delta\varphi(x,y)\right|_{\min} = 2\pi\Delta P_f\Delta P_s = 2\pi / N \tag{6.38}$$

Therefore, the actual sensitivity of displacement for grid measurement can be estimated as

$$P / N \leq \left|\Delta U\right|_{\min} \leq 0.5P \tag{6.39}$$

Equation 6.39 is the criterion for the upper and lower limits of sensitivity without considering random errors such as noise and quantization error, etc.

6.5.2.3 Accuracy

Accuracy is defined as the closeness with which the measured reading approaches the true value of the variable being measured. Specifically, the accuracy of the displacement measurement indicates how well the displacement value detected by the FT method compares with the true displacement. Following Equation 6.33, the accuracy of displacement can be expressed as

$$\Delta U = \frac{\varphi}{2\pi}\Delta P + \frac{P}{2\pi}\Delta\varphi \tag{6.40}$$

where ΔP is the error in grid pitch and $\Delta\varphi$ is the measured phase error. The various factors that influence the displacement accuracy have been widely discussed [47–50].

6.5.3 Grid Line Pattern Analysis Method with Phase-Shifting Method

The temporal phase-shifting (TPS) method [18] is mainly used in grid profilometry. It measures the phase of a single point in an interferogram as the fringe phase is changed in a controlled way. Its main advantage is its high spatial resolution, equal to 1 pixel. No other phase detection technique can match this spatial resolution. As to its accuracy, various error sources, especially noise, have been analyzed [18,51]. For phase-shifting grid profilometry, the influence of high-order harmonics frequently becomes a source of big error and cannot be neglected. In this case, self-calibrating algorithms [52,53] are suitable to eliminate the phase error due to the harmonics and the phase shifter errors.

Spatial phase-shifting (SPS) method [54] can be used for both grid profilometry and grid displacement or strain measurement. It extracts the phase information from a single grid pattern. Its spatial resolution depends on the algorithm used. In the case of linear and uniform phase shifting, the spatial resolution of phase measurement is m pixels, where m is the grid pitch. Generally, the spatial resolution of the SPS method is better than that of the FT method but worse than that of the TPS method. The accuracy of the SPS method is generally less than that of FT, especially if the measurand varies smoothly in the field.

The minimum phase change in a SPS measurement can be written as:

$$\left|\Delta\varphi(x,y)\right|_{\min} = 2\pi\left|f_1 - f_0\right|_{\min}(m\Delta P_s) \tag{6.41}$$

The sensitivity of phase measurement in SPS method is therefore

$$\sigma_\varphi = \left|\Delta\varphi(x,y)\right|_{\min} = 2m\pi\Delta P_f\Delta P_s = 2m\pi / N \tag{6.42}$$

This equation gives an estimate of the upper limit of sensitivity of the SPS method without considering the influence of noise, etc.

6.5.4 Grid Diffraction Method

When a monochromatic collimated beam incident normal to the plane of the specimen grid (>40 lines/mm) illuminates a point on the grating, diffraction dots are observed on the screen parallel to the grating plane as shown in Figure 6.27.

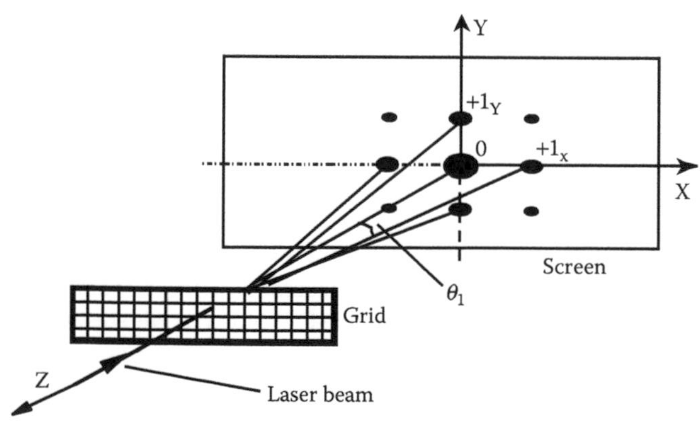

FIGURE 6.27 Diffraction pattern from a cross grid (grating).

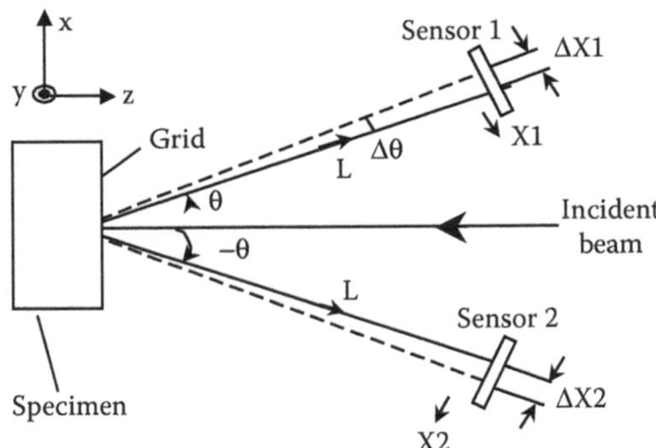

FIGURE 6.28 Principle of the optical diffraction strain sensor.

For high-frequency gratings, only the ±1 diffraction orders are observed. These beams are detected by two CCDs [27,29,31] or position sensing detectors (PSDs) [28,30] at a distance *L* from the grid. Use of two detectors eliminates error due to out-of-plane rotation of the specimen. The diffraction equation for a beam incident normal to the grating plane is

$$P \sin \theta_n = \pm n\lambda \tag{6.43}$$

where θ_n is the diffraction angle of the $\pm n$ order beams, λ is the wavelength of the laser, and P is the pitch of the grating.

Figure 6.28 illustrates the principle of strain measurement using a diffraction grid and PSD sensors. If the specimen with the diffraction grating imprinted on it undergoes a small deformation, the pitch P changes by ΔP and the diffraction angle changes to satisfy Equation 6.43. Thus,

$$P\Delta\theta_n \cos\theta_n + \Delta P \sin\theta_n = 0 \tag{6.44}$$

or

$$\Delta\theta_n = -\varepsilon_x \tan\theta_n \tag{6.45}$$

where $\varepsilon_x = \Delta P/P$ is the normal strain perpendicular to the grid lines. From Figure 6.28 the shift at detector 1 is

$$\Delta X1 = L\Delta\theta_n = L\varepsilon_x \tan\theta_n \tag{6.46}$$

For detector 2, the shift is similarly obtained by replacing θ_n with $-\theta_n$, as:

$$\Delta X2 = -L\varepsilon_x \tan\theta_n \tag{6.47}$$

Thus, from Equation 6.46 and Equation 6.47:

$$\varepsilon_x = \frac{\Delta X1}{L\tan\theta_n} = -\frac{\Delta X2}{L\tan\theta_n} = \frac{\Delta X1 - \Delta X2}{2L\tan\theta_n} \tag{6.48}$$

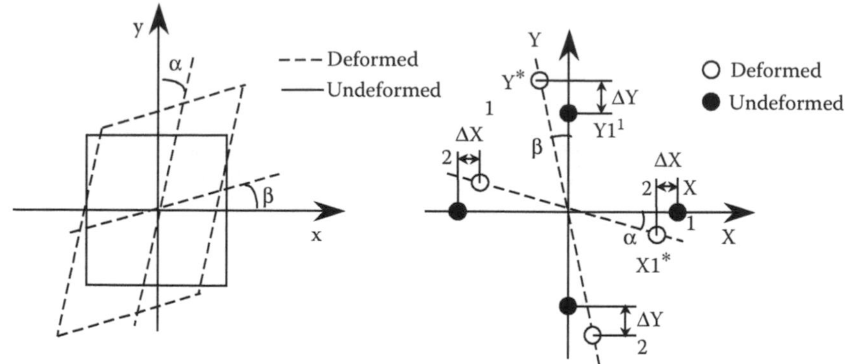

FIGURE 6.29 Two-dimensional strain measurement using optical diffraction strain sensor.

In general for a cross diffraction grating, the shifts of the diffraction spots along the x and y directions are shown in Figure 6.29. All three components of the in-plane strain, e_x, e_y, and γ_{xy}, can thus be deduced as:

$$\begin{cases} \varepsilon_y = \dfrac{\Delta Y_1 - \Delta Y_2}{2L \tan \gamma} \\ \gamma_{xy} = \dfrac{\alpha + \beta}{2} \end{cases} \tag{6.49}$$

The shift of grid diffraction pattern spots $\Delta X1$ and $\Delta X2$ are the displacement of the centroid of the +1 or −1 order diffraction spot. Thus, some image processing is required to determine the centroid of the spots. Several algorithms for centroid determination have been proposed [55,56]. A brief overview of the frequently used algorithm is provided here. An image of the grid diffraction pattern is first acquired, and then an appropriate threshold T is chosen. The threshold is defined such that any gray levels below T are set to black, and gray levels above T are set to white. The image is then raster-scanned to find a window that includes each grid point. Finally, the following equations are used within the window to calculate the grid spot centroid:

$$X = \frac{\sum_{i=i_o}^{i_f} \sum_{j=j_o}^{j_f} i(T - G_{ij})}{\sum_{i=i_o}^{i_f} \sum_{j=j_o}^{j_f} (T - G_{ij})}, \quad Y = \frac{\sum_{i=i_o}^{i_f} \sum_{j=j_o}^{j_f} j(T - G_{ij})}{\sum_{i=i_o}^{i_f} \sum_{j=j_o}^{j_f} (T - G_{ij})} \tag{6.50}$$

where i_0, i_f, j_0, and j_f define the window.

The spatial resolution for strain measurement with the grid diffraction method is generally determined by the diameter of the illuminated beam. The original laser beam diameter without any expansion is between 0.5 to 2 mm. With a lens, this beam size can be further decreased. The precision for strain measurement depends on the precision of the centroid determination and the length L between the grid and the detecting sensor. A 0.1-pixel precision for the centroid determination is attainable with current algorithms. A larger L will certainly increase the measurement sensitivity and precision, but will make the experimental setup too large.

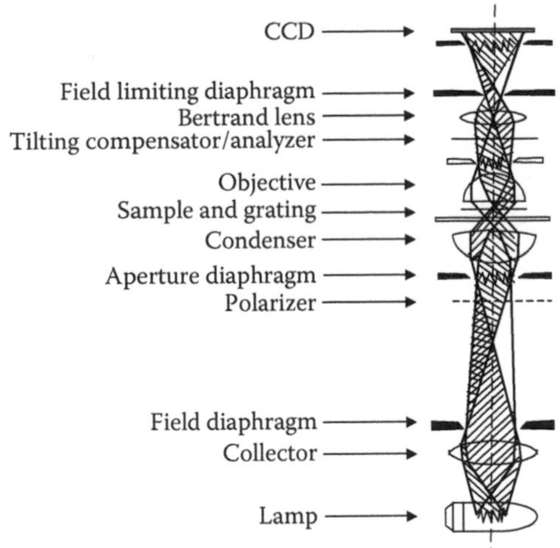

FIGURE 6.30 Schematic of a polarizing microscope.

6.5.5 Applications

Two grid methods, a grid diffraction strain method and a full-field phase-shifting method, to measure microdeformation using a conventional microscope are described.

A schematic of the optical system of a Leitz Orthoplan II Pol polarizing microscope is shown in Figure 6.30. The transmitting microscope includes a white light source, an interference filter, field and aperture diaphragms, an objective, a tilting compensator, a Bertrand lens, and a pair of eyepieces.

The specimen with the grating is placed between the condenser and the objective of the microscope. The density of the grating used ranges from 40 to 200 lines/mm. Three types of objectives, 6.3/0.2, 10/0.25, and 25/0.50 (magnification/numerical aperture), are typically used. A 10X eyepiece can be used for visual observation. The image is recorded using a CCD camera coupled to a $512 \times 512 \times 8$-bit frame grabber. Software for image capture and processing has been developed. The CCD is directly connected to the microscope eyepiece tube after removing the eyepiece. The loading system consists of a loading frame and a two-axis translation stage. The frame is fixed onto the translation stage, and the whole unit is fixed on the mechanical stage of the microscope. The specimen to be tested can be translated in two directions and can rotate with the loading frame. So, we can easily examine different areas of the specimen. The specimen must be transparent, and rotation of the gyrostat generates tensile loading to the specimen.

The principle of strain measurement described here is based on the grating diffraction properties. Using an optical microscope, the image plane, and not the FT plane, is recorded. According to the microscopic image-forming principle (Abbe's theory [57]), the transition from the object to the image involves two stages. First, the grating-like object when illuminated gives rise to a Fraunhofer diffraction pattern in the back focal plane of the objective, in which every point can be considered to be a center of a coherent secondary disturbance. The light waves that proceed from these secondary wavelets will then interfere with each other and give rise to the image of the object in the image plane of the objective. Thus, the diffraction pattern is formed in the back plane, and an image of the grating is formed at the image plane. Hence, in order to get an image of the grating diffraction pattern through a microscope, the back focal plane of the objective should be extended to the CCD sensor plane. This can be realized by using a relay or Bertrand lens. The grating diffraction in this way is a Fraunhofer diffraction and is illustrated in Figure 6.31.

The important step in the measurement is to get the grating diffraction pattern. In this test, using the Bertrand lens achieved this. This lens is generally placed between the eyepiece and the

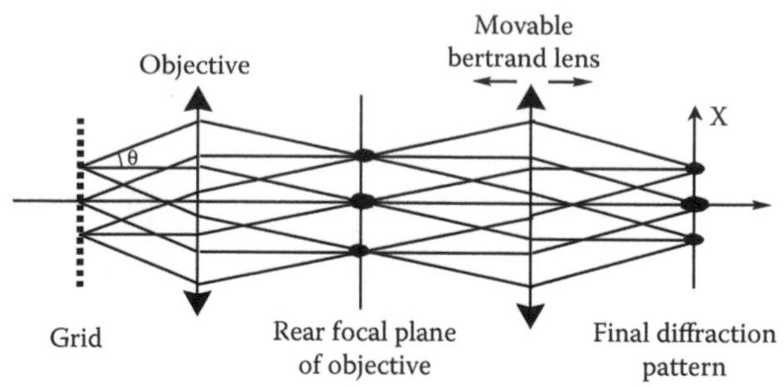

FIGURE 6.31 Bertrand lens to image the Fourier plane on to the image plane.

objective and is often used to observe the interference or polarization patterns. With this lens, one can see clearly the image formed in the back focal plane of the objective. On gradually adjusting the focusing screw that controls the up-and-down movement of this lens, the transformation from the grating image to the grating diffraction pattern is observed through the eyepiece.

After the grating is deformed, the pitch changes from P to P^*, resulting in variation in θ according to Equation 6.52. Thus, the strain along the X-direction at the illuminated point on the grating can be determined:

$$\varepsilon_{xx} = \frac{P^* - P}{P} \approx \frac{X_n^* - X_n}{X_n} \cos\theta_n \tag{6.51}$$

where X_n and X_n^* are the undeformed and deformed diffraction spot centroids of the nth diffraction order. In this experiment, the density of the grating was less than 200 lines/mm. If only the first diffraction order is considered, i.e., $n = 1$, the angle (θ_1) of the diffracted beam is small. For example, for green light, $\lambda = 500$ nm, grating pitch $P = 1/200$ mm, $\theta_1 = 5.73°$, and $\cos\theta_1 = 0.995 >> 1$. So Equation 6.51 can be simplified to:

$$\varepsilon_{xx} \approx \frac{X_1^* - X_1}{X_1} \tag{6.52}$$

If a cross grating is used instead of a linear grating, all three components of in-plane strain, e_{xx}, e_{yy}, and e_{xy}, can be determined.

There is an advantage in using a microscope to realize the grating diffraction image, i.e., the recorded image possesses good quality. This is mainly because no ground-glass screen is required, and the diffraction spots are directly projected onto the CCD sensor plane. Also, a microscope with high-quality components provides a sharp, precise optical image.

The strain sensitivity can be written as

$$\sigma_\varepsilon \approx \frac{\sqrt{2}\sigma_X}{X_1} \tag{6.53}$$

where σ_X is the sensitivity for determination of the spot centroid. One way of increasing the strain sensitivity is to decrease the value of σ_X. This approach is limited by the spatial resolution of the CCD camera and the algorithm for determining the centroid of the diffraction spot. Another way is to increase X_1, i.e., increase the distance between two spots of ±1 orders of the recorded pattern. This is generally realized by using a grating with a higher density or by increasing the distance between the grating and the image-forming screen.

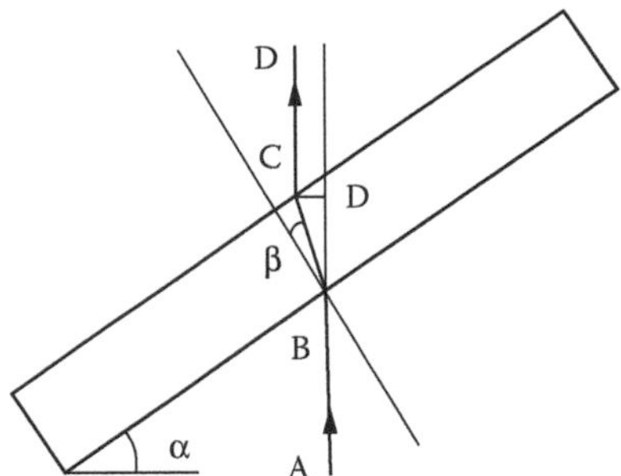

FIGURE 6.32 Shift of refracted beam when traversing a glass slab.

For strain microscope measurement, several methods can be used to increase the distance between ±1 diffraction orders. One way is to increase the distance between the grating and the CCD sensor by increasing the length of the tube. Other ways, such as using an objective with lower power or using a high-density grating, can also increase the strain measurement sensitivity. There is a limit to using high-density gratings. A high-density grating will result in a large diffraction angle and a ±1 diffracted beam cannot enter the objective. From experience, grating frequencies between 40 and 200 lines/mm have been seen to be a good compromise. For our test, the sensitivity for determination of the spot centroid is estimated to be $\sigma_X = 0.08$ pixel and X_1 cannot exceed 512/2 = 256 pixels and, thus, the strain sensitivity is 442 $\mu\varepsilon$.

In order to verify the experimental procedure and check the measurement precision, we carried out a test using the microscope translation stage to stretch a grating film specimen, which was bonded onto the stage. The specimen was a thin plastic substrate and could be easily loaded by using the micrometer-driven translation stage. The precision was not high, but the travel distance of the stage was long. We can calculate the average strain by using the measured overall displacement. Comparisons between results of the average strain and the strain measured by the diffraction method were found to be in general agreement.

An optical beam traversing an inclined optical slab will be shifted horizontally due to refraction, as illustrated in Figure 6.32. If the slab, with thickness d, is inclined at an angle of α to the horizontal, beams AB and CD represent the incident and refracted beams, respectively. The length CD is the horizontal shift due to refraction, and β is the refraction angle.

If the refractive index of the slab is n, then the length of CE is

$$|CE| = |BC|\sin(\alpha - \beta) = \frac{d}{\cos\beta}\sin(\alpha - \beta) \tag{6.54}$$

Noting that α and β are less than 90°, a nonlinear relationship between the shifted length CE and α is obtained:

$$|CE| = d\left[\sin\alpha - \frac{\sin(2\alpha)}{2\sqrt{n^2 - (\sin\alpha)^2}}\right] \tag{6.55}$$

The optical slab in Figure 6.32 can rotate around a horizontal axis that is parallel to the horizontally placed grating lines. The image of the grating will thus shift a distance CE, which corresponds to the phase shift of the grating. The equivalent phase shift is

$$\delta = 2\pi\frac{|CE|}{P*} = 2\pi\frac{d}{P*}K(\alpha) = const \times K(\alpha) \tag{6.56}$$

where P^* is the pitch after magnification by the objective, and K is a function of the rotation angle α and the refractive index n and is expressed by:

$$K(\alpha) = \sin\alpha - \frac{\sin(2\alpha)}{2\sqrt{n^2 - (\sin\alpha)^2}} \tag{6.57}$$

The phase shifter, i.e., the optical slab, is a tilting compensator in the microscope. This type of compensator is generally used for birefringence measurement in a polarizing microscope. It consists of a slab of a uniaxial crystalline medium. It is mounted so that the slab can be tilted via a graduated wheel. The crystal plate in this compensator can rotate with a precision of 0.05° and a range of ±32°.

It is difficult to know the exact value of P^* as the magnification of the Bertrand lens is not known. Thus, it is difficult to determine the correct phase-shift amount δ. Among numerous phase-shifting algorithms, the Carré [58] and Hariharan [59] algorithms do not require the value of the first δ. Thus, from the five phase-shifted fringe patterns:

$$I_i = I_a + I_b \cos(\varphi + \delta_i), \quad i = 1, 2, ..., 5 \tag{6.58}$$

where I_i represents the ith fringe pattern, I_a is the background intensity, I_b is the modulated intensity, and φ is the phase to be measured. These four parameters are functions of the designated point coordinate (X, Y). δ_i is the phase shift of the ith pattern, which can be expressed as:

$$\delta_i = (i-3)\delta_0, \quad i = 1, 2, ..., 5 \tag{6.59}$$

The phase can be derived as

$$\tan\varphi = 2\sin\delta_0 \frac{I_2 - I_4}{2I_3 - I_5 - I_1} \tag{6.60}$$

where δ_0 satisfies

$$\cos\delta_0 = \frac{I_5 - I_1}{2(I_4 - I_2)} \tag{6.61}$$

It is clear that the measured phase can be determined without knowing the phase shift δ_0, provided δ_i is proportional to δ_0. For example, in a group of δ_i, the inclined angle of the refraction slab and its corresponding phase shift δ_i is tabulated as follows:

i	1	2	3	4	5
αi(°)	−9.93	−5	0	5	9.93
δi(°)	$-2\delta_0$	$-\delta_0$	0	δ_0	$2\delta_0$

The measurement procedure is as follows:

Adjust the microscope system and record the five phase-shifted grating patterns.

Calculate initial phase field $\phi_s(X, Y)$.

Load the specimen, record the grating patterns, and obtain the final phase field $\phi_m(X, Y)$.

Unwrap the phase field to be measured $\varphi = (\varphi_m\varphi_s)$, and calculate the displacement as $u = P \times \varphi/(2\pi)$. Here, P is the real grating pitch; in Equation 6.56, and P^* indicates the pitch of the grating magnified by the objective.

The optical system and the CCD imager limit the spatial resolution of displacement. In order to detect fine grating lines with the CCD, the grating pitch must be greater than 2 pixels; thus, the CCD's resolution is defined as 2 pixels. For the CCD used, the sensor pixel size is about 10 μm. As the image formed on the CCD sensor plane is magnified by an objective, the system resolution is 20 μm divided by the objective magnification factor. Typical resolution values for three typical objectives are as follows:

Objective magnification	25	40	100
Resolution (μm)	0.80	0.50	0.20

Furthermore, the resolving power of the objective must be considered. The distance between two object points that a microscope (or objective) can just resolve is the resolution of the microscope and is given by:

$$\text{Res} = (0.61 \sim 0.82)\lambda / NA \tag{6.62}$$

where λ is the wavelength in vacuum and NA is the numerical aperture of the objective. The multiplying factor 0.61 is for incoherent light, and 0.82 is for coherent light. In the present setup, the light is partially coherent. So, the actual factor lies between 0.61 and 0.82. For white light, λ is 0.55 μm, and the resolution for the three objectives is

Objective magnification	25	40	100
NA	0.5	0.65	1.3
Resolution (μm)	0.67 ~ 0.90	0.52 ~ 0.69	0.26 ~ 0.35

The lower limits are chosen as the spatial resolution for displacement measurement, i.e., 0.90 μm, 0.69 μm, and 0.35 μm, respectively, for 25X, 40X, and 100X objectives.

Equation 6.56 is a nonlinear relation between phase-shift angle δ_i and the rotation angle of the refractive slab α_i. This relationship can be further simplified when α_i is small (less than 20°).

$$\delta_i = 2\pi \frac{d}{P} K(\alpha_i) \approx 2\pi \frac{d}{P} \frac{n-1}{n} \alpha_i \tag{6.63}$$

Thus the phase-shift error due to an error in refraction slab rotation can be written as:

$$\Delta\delta_i \approx 2\pi \frac{d}{P} \frac{n-1}{n} \Delta\alpha_i \tag{6.64}$$

The rotation of the refraction slab can be controlled to within 0.05°. For a slab with refractive index $n = 1.37859$, thickness $d = 1.52$ mm, and grating pitch multiplied by objective magnification (P^*) = 0.01 mm × 40X objective, the error is less than 0.33°. This error is negligible compared to the fringe intensity noise, which is the main error source. We assume that the standard deviation (SD) of this noise is $\sigma = 3$ gray levels. This estimate is reasonable and conservative, and can be verified by a simple test, wherein two fringe patterns in the same state but at different times are recorded and subtracted. The result obtained is a noise pattern, from which the SD of noise can be calculated. Dividing this by $\sqrt{2}$, we obtain the approximate SD of the fringe intensity noise, which is found to be less than 2.

The calculated phase error due to the noise reaches its maximum when δ_0 is 90°. In this case, we have the well-known 5-buckets error-compensating algorithm [46]. The phase error for this algorithm due to the noise is phase dependent, and its maximum SD value can be estimated as [51]:

$$\sigma_{\Delta\varphi} = \frac{1+\beta}{\eta\sqrt{N}} \frac{1}{\text{SNR}} \tag{6.65}$$

where $\beta = 1/7$, $\eta = 4\sqrt{2/35}$, and the fringe pattern number $N = 5$. The signal-to-noise ratio (SNR) = I_b/σ. If the modulated fringe intensity varies from 64 to a maximum of 128 gray levels, the corresponding SNR varies from 21.3 to 42.7. Thus, the maximum phase error due to noise is between 0.013 and 0.025 radians, or the phase error is less than 1.4 degrees. Finally, we take the SD of the phase error as $\sigma_{\Delta\varphi} = 2°$ and get the displacement measurement precision as follows:

Grating density (lines/mm)	40	80	100	150	200
Displacement precision (μm)	0.139	0.069	0.056	0.037	0.028

In order to verify the experimental procedure and check the measurement precision, a test was conducted using the microscope translation stage to stretch a grating film specimen, which was bonded onto the stage. This film (including a plastic substrate) is very thin, so it is easily loaded by using the micrometer drive, and we can precisely control the displacement. This displacement, read from the micrometer, is the overall displacement of the specimen. However, the displacement field measured with the preceding grating phase-shift method covers only a small part of the specimen length, so it is difficult to compare the localized displacement obtained with the grating phase-shift method and the overall displacement. Actually, we compare the average strain calculated on the basis of the specimen's overall displacement, i.e., micrometer displacement divided by the length of specimen and the strain calculated from the experimentally measured local displacement.

Figure 6.33 shows a test result. The first five photographs are the phase-shifted grating line patterns before loading, and the last subfigure shows the wrapped phase map. The grating density is 100 lines/mm, and a 40X objective is used. The grating line patterns are sequentially recorded

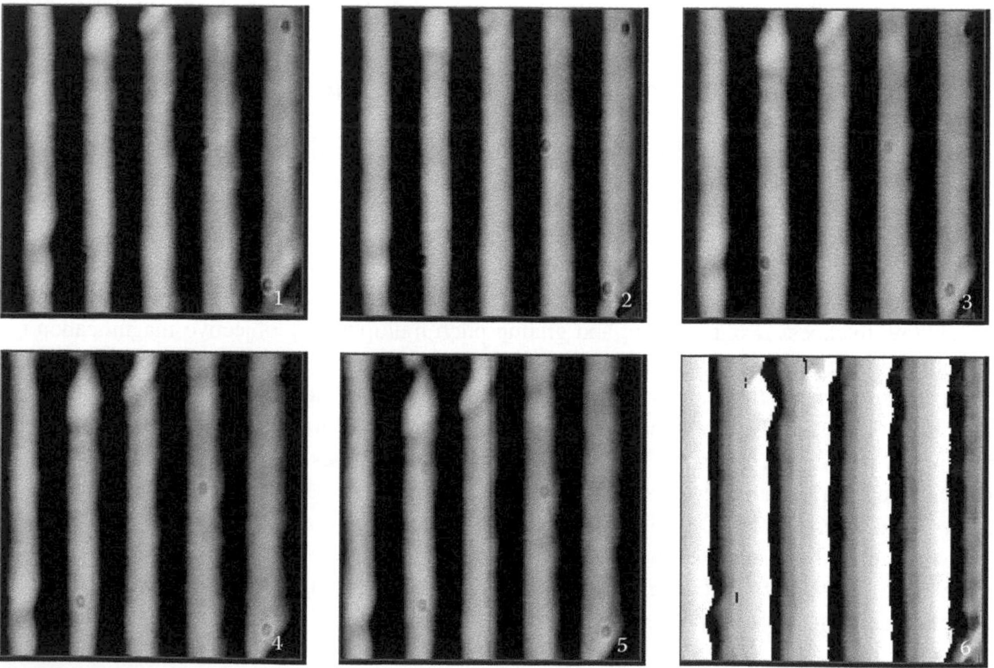

FIGURE 6.33 Phase-shifted fringe patterns and wrapped image.

FIGURE 6.34 Unwrapped displacement map.

with the CCD camera each time after turning the tilting compensator causes shift of the grating image; these grating images are stored in the extended memory of the computer. The difference among the five patterns is that the phase of the grating signal in each pattern is shifted. The phase distribution is then calculated from the five stored grating images. The corresponding unwrapped phase map representing a displacement field with an area of 50×50 μm^2 is displayed in Figure 6.34 as a gray-level map. This displacement map corresponds to a specimen with no load and is referred to as the *virtual displacement field.* After loading, the deformed gratings with added phase shifts are recorded, and the final phase (or displacement) map is obtained. The subtraction of the final and the initial phase maps gives the phase (displacement) field due to the load alone.

6.6 CONCLUSIONS

Moiré methods are a versatile set of techniques for in-plane and out-of-plane deformation. Most of the early work in moiré used visible light sources and, hence, the limit of sensitivity was half the wavelength of the light used. With this approach, moiré interferometric methods could achieve a deformation sensitivity of 0.417 μm. This was adequate for most problems. However, spatial resolution was a problem. This was resolved using the fiber-optic micro-moiré interferometer. Nanometer sensitivity and micron-level resolution provided a means to measure micron-sized interfaces and optical interconnects. However, for submicron scale resolution it was necessary to explore other imaging modalities such as the AFM and SEM. However, to increase the sensitivity, new schemes for fabricating high-frequency specimen gratings were explored. FIB milling and e-beam lithography makes these possible. In the limit, one could use the lattice structure as the specimen grating. The feasibility of these advanced moiré methods has been demonstrated.

Grid methods have also seen a resurgence with new imaging modalities, i.e., AFM and SEM. Sensitivity, accuracy, and spatial resolution are important parameters in the grid method and have been discussed. Advances in processing grids as fringe patterns using the FT method have been explained. The use of diffraction principles coupled with microscopic imaging has led to the development of novel strain sensors with high spatial resolution.

REFERENCES

1. Durelli, A.J., Riley, W.F., *Moiré Analysis of Strain*, New Jersey, Prentice Hall, 1970.
2. Chiang, F.P., Moiré methods of strain analysis, in Doyle, J.F. and Philips, J.W., Eds., *Manual on Experimental Stress Analysis*, 5th ed., Bethel, Connecticut, Society for Experimental Mechanics, 1989, chap.7.
3. Asundi, A., Photoelasticity and Moiré, in Rastogi, P.K., Ed., *Optical Measurement Techniques and Applications*, Norwood, Artech House, 1997, chap. 7.
4. Post, D., Han, B., Ifju, P., High Sensitivity Moiré, New York, Springer-Verlag, 1994.
5. Xie, H.M., Kishimoto, S., Shinya, N., Fabrication of high-frequency electron beam moiré grating using multi-deposited layer techniques, *Opt. Laser Technol.*, 32: 361–367, 2000.
6. Dally, J.W., Read, D.T., Electron-beam moiré, *Exp. Mech.*, 33: 270–277, 1993.
7. Asundi, A., Zhao, B., Optical grating diffraction method — from strain microscope to strain gauge, *Appl. Opt.*, 38: 7167–7169, 1999.
8. Post, D., Patorski, K., Ning, P., Compact grating interferometer for producing photoresist gratings with incoherent light, *Appl. Opt.*, 26: 1100–1105, 1987.
9. Post, D., Mckelvie, J., Tu, M., Fulong, D., Fabrication of holographic grating using a moving point source, *Appl. Opt.*, 28: 3494–3497, 1989.
10. Xie, H.M., Li, B., Geer, R., Focused ion beam moiré method, *Opt. Laser. Eng.*, 40: 163–177, 2003.
11. Asundi, A., Moiré interferometry for deformation measurement, *Opt. Laser. Eng.*, 11(4), 281–292, 1989.
12. Oh, K.E., Development and Applications of a Moiré Interferometric Strain Sensor, M.Eng. thesis, Nanyang Technological University, Singapore, 1999.
13. Xie, H.M., Kishimoto, S., Asundi, A., Chai, G.B., Shinya, N., In-plane deformation measurement using atomic force microscope moiré method, *J. Nanotechnol.*, 11: 24–29, 2000.
14. Kishimoto, S., Xie, H.M., Shinya, N., Electron moiré method and its application to micro-deformation measurement, *Opt. Laser. Eng.*, 34 (1): 1–14, 2000.
15. Xing, Y.M., Dai, F.L., Yang, W., An experimental study about nano-deformation field near quasi-cleavage crack tip, *Sci. China (A)*, 43: 963–968, 2000.
16. Asundi, A., Xie, H.M., Phase shifting moiré method with an atomic force microscope, *Appl. Opt.*, 40: 6193–6198, 2001.
17. Takeda, M., Ina, H., Kobayashi, S., Fourier-transform method of fringe-pattern analysis computer-based topography and interferometry, *J. Opt. Soc. Am.*, 72: 156–160, 1982.
18. Dorrio, B.V., Fernandez, J.L., Phase-evaluation methods in whole-field optical measurement techniques, *Meas. Sci. Technol.*, 10: R33–R55, 1999.
19. Morimoto, Y., Seguchi, Y., Application of moiré analysis of strain using Fourier transform, *Opt. Eng.*, 27: 650–656, 1988.
20. Morimoto, Y., Seguchi, Y., Higashi, T., Two-dimensional moiré method and grid method using Fourier transform, *Exp. Mech.*, 29: 399–404, 1989.
21. Asundi, A., Chan, C.S., Phase shifted projection grid — effect of pitch and profile, *Opt. Laser. Eng.*, 21: 31–47, 1994.
22. Su, X.Y., Zhou, W.S., Von Bally, G., Vukicevic, D., Automated phase-measuring profilometry using defocused projection of a Ronchi grating, *Opt. Commun.*, 94: 561–573, 1992.
23. Surrel, Y., Zhao, B., Simultaneous U-V displacement field measurement with a phase-shifting grid method, *Proc. SPIE*, Conference on Interferometry'94: Warsaw, Poland, 1994, 2342, pp. 66–75.
24. Creath, K., Schmit, J., N-point spatial phase measurement techniques for nondestructive testing, *Opt. Laser. Eng.*, 24: 365–379, 1996.
25. Schmit, J., Creath, K., Window function influence on phase error in phase-shifting algorithms, *Appl. Opt.*, 35(28), 5642–5649, 1996.
26. Bell, J.F., Determination of dynamic plastic strain through the use of diffraction gratings, *J. Appl. Phys.*, 27: 1109–1113 1956.
27. Bremand, F., Dupre, J.C., Lagarde, A., Non-contact and non-disturbing local strain measurement methods, I. Principle, *Eur. J. Mech. A/Solids*, 11: 349–366, 1992.
28. Asundi, A., Zhao, B., Optical strain sensor using position-sensitive detector and diffraction grating: error analysis, *Opt. Eng.*, 39: 1645–1651, 2000.
29. Ma, Y., Kurita, M., Strain measurement using high-frequency diffraction grating, *JSME Int J. Ser. A: Solid Mech. Mater. Eng.*, 36: 309–313, 1993.

30. Zhao, B., Xie, H.M., Asundi, A., Optical strain sensor using median density grating foil: rivaling the electric strain gauge, *Rev. Sci. Inst.*, 72: 1554–1558, 2001.
31. Moulder, J.C., Cardenas-Garcia, J.F., Two-dimensional strain analysis using a video optical diffractometer, *Exp. Tech.*, 5: 11–16 1993.
32. Carbonneau, X., Thollet, G., Olagnon, C., Fantozzi, G., Development of high temperature extensometric microgrids, *J. Mater. Sci. Lett.*, 16: 1101–1103, 1997.
33. Li, C.S., Orlecky, L.J., Fiducial grid for measuring microdeformation ahead of fatigue crack tip near aluminum bicrystal interface, *Exp. Mech.*, 33: 286–292, 1993.
34. Nisitant, H., Fukuda, T., Observation of local deformation and crack initiation in high-cycle torsional fatigue of carbon steel plain specimen by electron microscope, *JSME Int. J. Ser. 1: Solid Mech. Strength Mater.*, 35: 354–360, 1992.
35. Zhao, B., Asundi, A., Grid method for strain measurement in electronic packaging using optical, electronic, and atomic force microscope, *Proc. SPIE*, 3897: 260–270, 1999.
36. Sciammarella, C.A., Bhat, G., High-resolution computer aided moiré, *Proc. SPIE*, San Diego, 1991, Vol. 1554B, pp. 162–171.
37. Szanto, M., Dally, J.W., Read, D.T., Hybrid grid-moiré method for measuring strain and displacement, *Opt. Eng.*, 32: 1043–1052, 1993.
38. Zhao, B., Asundi, A., Phase-shifting grating microscope with slab-tilting method for deformation measurement, *Opt. Laser. Technol.*, 30: 431–436, 1998.
39. Zhao, B., Asundi, A., Strain microscope with grating diffraction method, *Opt. Eng.*, 38: 170–174, 1999.
40. Zhao, B., Asundi, A., Micro-measurement using grating microscope, *Sensor. Actuator. A*, 80: 256–264, 2000.
41. Korner, K., Windecker, R., Fleischer, M., Tiziani, H.J., One-grating projection for absolute three-dimensional profiling, *Opt. Eng.*, 40: 1653–1660, 2001.
42. Proll, K.P., Nivet, J.M., Voland, C., Tiziani, H.J., Enhancement of the dynamic range of the detected intensity in an optical measurement system by a three-channel technique, *Appl. Opt.*, 41: 130–135, 2001.
43. Windecker, R., Fleischer, M., Korner, K., Tiziani, H.J., Testing micro devices with fringe projection and white-light interferometry, *Opt. Laser. Eng.*, 36: 141–154, 2001.
44. Proll, K.P., Nivet, J.M., Korner, K., Tiziani, H.J., Microscopic three-dimensional topometry with ferroelectric liquid-crystal-on-silicon displays, *Appl. Opt.*, 43: 1773–1778, 2003.
45. Zhao, B., Asundi, A., Microscopic grid methods — resolution and sensitivity, *Opt. Laser. Eng.*, 36: 437–450, 2001.
46. Zhao, B., Asundi, A., Discussion on spatial resolution and sensitivity of Fourier transform fringe detection, *Opt. Eng.*, 39: 2715–2719, 2000.
47. Bone, D.J., Bachor, H.A., Sandeman, J., Fringe-pattern analysis using a 2-D Fourier transform, *Appl. Opt.*, 25: 1653–1660, 1986.
48. Roddier, C., Roddier, F., Interferogram analysis using Fourier transform techniques, *Appl. Opt.*, 26: 1668–1673, 1987.
49. Sevenhuijseen, P.J., The photonical, pure grid method, *Opt. Laser. Eng.*, 18: 173–194, 1993.
50. Kreis, T.M., Computer aided evaluation of fringe patterns, *Opt. Laser. Eng.*, 19: 221–240, 1993.
51. Surrel, Y., Additive noise effect in digital phase detection, *Appl. Opt.*, 36: 271–276, 1997.
52. Surrel, Y., Design of algorithms for phase measurements by the use of phase stepping, *Appl. Opt.*, 35: 51–60, 1996.
53. Zhao, B., Surrel, Y., Phase shifting: six-sample self-calibrating algorithm insensitive to the second harmonic in the fringe signal, *Opt. Eng.*, 34: 2821–2822, 1995.
54. Kujawinska, M., Spatial phase measurement methods, in Robinson, D.W., Reid, G.T., Eds., *Interferogram Analysis: Digital Fringe Pattern Measurement Techniques*, Philadelphia, IOP, 1993, chap. 5.
55. West, G.A.W., Clarke, T.A., A survey and examination of subpixel measurement techniques, *Proc. SPIE*, 1395: 456–463, 1990.
56. Sirkis, J.S., System response to automated grid methods, *Opt. Eng.*, 29: 1485–1487, 1990.
57. Born, M., Wolf, E., *Principles of Optics, Electromagnetic Theory of Propagation, Interference and Diffraction of Light, 6th ed.*, New York, Pergamon Press 1983.
58. Carré, P., Installation et utilisation du comparateur photoelctrique et interferentiel du bureau International des poids et mesures, *Métrologie,* 2: 13–23, 1966.
59. Hariharan, P., Oreb, B.F., Eiju, T., Digital phase-shifting interferometry: a simple error-compensating phase calculating algorithm, *Appl., Opt.*, 26: 2504–2506, 1987.

7 Grating Interferometry for In-Plane Displacement and Strain Measurement of Microcomponents

Leszek Salbut

CONTENTS

7.1 Introduction 201
7.2 Principle of Grating Interferometry 202
7.3 Waveguide Grating Interferometry 204
 7.3.1 Concept of Waveguide Grating Interferometer Head 204
 7.3.2 Modified WGI for 3-D Components of Displacement Vector Measurements 207
7.4 Measurement System 207
7.5 SG Technology 209
7.6 Exemplary Applications of WGI 210
 7.6.1 Material Constants Determination 210
 7.6.2 Polycrystalline Materials Analysis 210
 7.6.3 Semiconductor Microlaser Matrix Testing 210
 7.6.4 Electronic Packaging 212
7.7 Conclusions 214
References 215

7.1 INTRODUCTION

Various problems in experimental mechanics require measurement of very small (micro) elements or within microregions of large specimens. These include components of microoptoelectromechanical systems (MEMS/MOEMS) testing, analysis of thermal stresses in electronic packaging, crack-tip analyses in fracture mechanics, monitoring of grains behavior in polycrystalline materials, studies of joint interfaces, composite and smart materials studies, local material constants determination, etc. The experimental techniques dealing with such problems should enable determination of in-plane u and v displacement maps in fields of view smaller than 1 mm with high resolution, measurements in unstable environment, automatic analysis of experimental data, and compatibility of the results with the data required for numerical methods, i.e., finite elements method (FEM). One of the most popular optical methods, which fulfills these requirements, is grating (moiré) interferometry [1,2]. It offers real-time, whole-field mapping with submicron sensitivity and high interference fringes contrast. Recently, several versions of grating interferometers, specially adapted for micromeasurements, have been proposed [3–5]. Many of them are based on the concept of guiding the light in a waveguide (block of glass), which makes the system easy to adjust and enables work with immersion between the sample and interferometer head. The basic engineering features expected from these interferometers are low sensitivity to environmental changes (especially vibrations), portability, and low cost of the system.

7.2 PRINCIPLE OF GRATING INTERFEROMETRY

The principle of high-sensitivity grating interferometry (GI) with conjugate wavefronts is shown schematically in Figure 7.1. Two mutually coherent illuminating beams Σ_A and Σ_B impinge on the reflection-type specimen grating (SG) at angles tuned to the first and minus first diffraction order angle of the SG. The +1 diffraction order of Σ_A and the −1 order of Σ_B propagate coaxially along the grating normal. Their wavefronts $\Sigma_{A'}$ and $\Sigma_{B'}$ are no longer in-plane.

The amplitudes in the detector plane D, optically conjugate to SG, can be described as:

$$E_{+1}^{A}(x,y) \approx \exp\left\{ i\left[\frac{2\pi}{p}u(x,y) + \frac{2\pi}{\lambda}w(x,y) \right] \right\} \tag{7.1}$$

$$E_{-1}^{B}(x,y) \approx \exp\left\{ -i\left[\frac{2\pi}{p}u(x,y) - \frac{2\pi}{\lambda}w(x,y) \right] \right\} \tag{7.2}$$

where p is the spatial period of the SG grating whose lines are perpendicular to the x axis (lying in the figure plane), $u(x,y)$ is the in-plane displacement function corresponding to the departure of the grating lines from straightness, and $w(x,y)$ is the out-of-plane displacement function corresponding to the deformation of the specimen surface under load. For simplicity, the amplitude of the diffraction orders has been normalized to unity.

It can be shown that wavefront deformations caused by in-plane displacements are equal in both diffraction orders but are longitudinally reserved. On the other hand, the wavefront deformations due to out-of-plane displacements have the same value and sign in both interfering beams. Therefore, the influence of out-of-plane displacements is eliminated by the interference. Thus, assuming that out-of-plane displacements give only small variations in the slope of the wavefronts (the departure from this condition was theoretically investigated in [6]), the intensity distribution of the interferogram becomes:

$$\left| E_{+1}^{A} + E_{-1}^{B} \right| \cong 2\left\{ 1 + \cos\left[1 + \frac{4\pi}{p}u(x,y) \right] \right\} \tag{7.3}$$

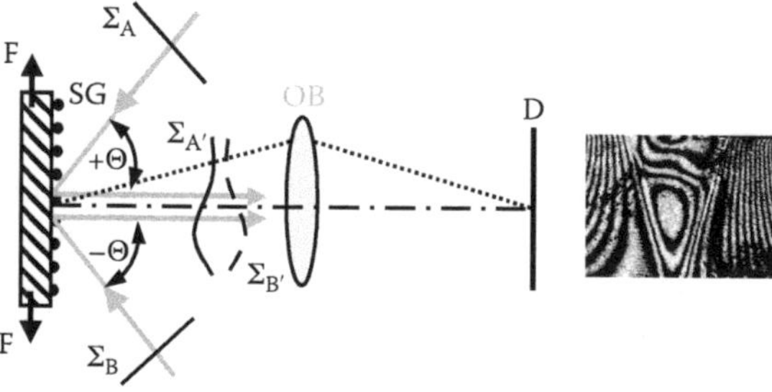

FIGURE 7.1 Schematic representation of a double illuminating beam grating interferometry for in-plane displacement studies. SG = specimen grating, OB = imaging optics, D = detector plane, Σ_A, Σ_B = wavefronts of +1 and −1 diffraction orders.

The fringes observed in the interferometer represent a contour map of in-plane displacements with half a period sensitivity. For example, when using a SG of spatial frequency 1200 l/mm, the basic sensitivity is 0.47 μm per fringe order.

When the double beam illumination system is angularly misaligned, carrier fringes are introduced into the interferogram. Equation 7.3 transforms into:

$$I(x,y) = 2\left\{1+\cos\left[2k\theta_x + \frac{4\pi}{p}u(x,y)\right]\right\} \tag{7.4}$$

where θ_x designates the angle between the diffracted orders and the grating normal $k = 2\pi/\lambda$.

For a significant gradient of out-of-plane displacements [6], function $w(x,y)$ is not completely eliminated and the intensity distribution in the image plane can be described as follows:

$$I(x,y) \cong 2\left\{1+\cos\left[\frac{4\pi}{p}u(x,y) + \frac{2\pi}{p}u_f(x,y)\right]\right\} \tag{7.5}$$

where $u_f(x,y)$ is the function of "fiction" displacement introduced due to local inclination of the SG caused by out-of-plane displacements $w(x,y)$. The relation between $u_f(x,y)$ and $w(x,y)$ is given by:

$$\frac{\partial u_f(x,y)}{\partial x} = \left[\frac{\partial w(x,y)}{\partial x}\right]^2 \tag{7.6}$$

To obtain complete strain information, a crossline SG is commonly used. A crossline grating requires two pairs of illuminating beams for its readout, as shown in Figure 7.2. This configuration provides the interferograms representing u-displacement (while illuminating A + B) and v-displacement (while illuminating C + D).

The spatial frequency f of SGs used in grating (moiré) interferometry depends on the incident angle θ and the wavelength λ of the illuminating beams. The relation between these values is given as:

$$f = \frac{1}{p} = \frac{\sin\theta}{\lambda} \tag{7.7}$$

where f is the spatial frequency of the SG.

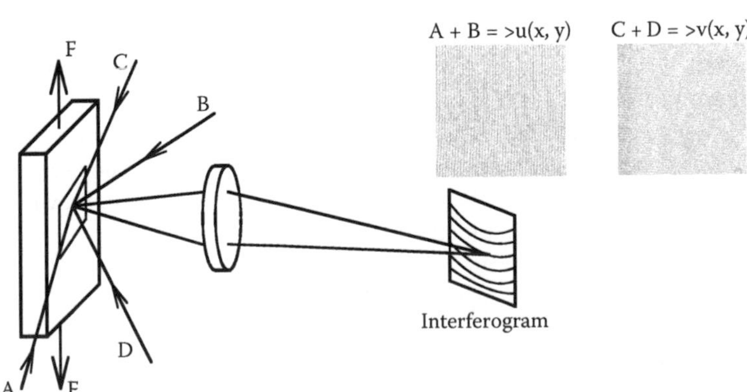

FIGURE 7.2 Scheme of four-beam GI for sequential or simultaneous u(x,y) and v(x,y) interferograms capturing.

TABLE 7.1
Conventional vs. Waveguide Realization of GI Systems (GP = Glass Plate)

Conventional	Waveguide	Sensitivity to Vibration	Sensitivity to $\Delta\lambda$
	prism; GP; immersion	High	High
mirror	mirrorized; GP; immersion	High	High
Compensating grating	GP; immersion; Compensating grating	High	Low
Compensating grating	GP; immersion; Compensating grating	Low	Low

It is clear from physical reasoning that the theoretical upper limit for the SG frequency is defined by $\theta = 90°$. For increasing the SG frequency, which allows enhancement of the sensitivity of displacement measurement, immersion fluid can be used [4]. In this case the specimen frequency is expressed as:

$$f_m = \frac{\sin\theta}{\lambda_m} = \frac{n\sin\theta}{\lambda} = nf \tag{7.8}$$

where λ_m is the wavelength of light in the medium, i.e., the frequency is increased by a factor of n (where n is the refractive index of the immersion fluid). This fact has been applied for design of various types of grating microinterferometers [1] mainly based on immersion waveguide heads [3–4].

Table 7.1 gives a comparison of various conventional [7] and waveguide realizations of grating interferometers, including their sensitivities to vibrations and achromatic light, which are substantial for the engineering applications of the method.

7.3 WAVEGUIDE GRATING INTERFEROMETRY

7.3.1 Concept of Waveguide Grating Interferometer Head

The concept of a waveguide head of grating interferometer (WGI) is shown in Figure 7.3 [3]. The illuminating beam B, introduced into the glass plate GP with parallel plane surfaces, illuminates

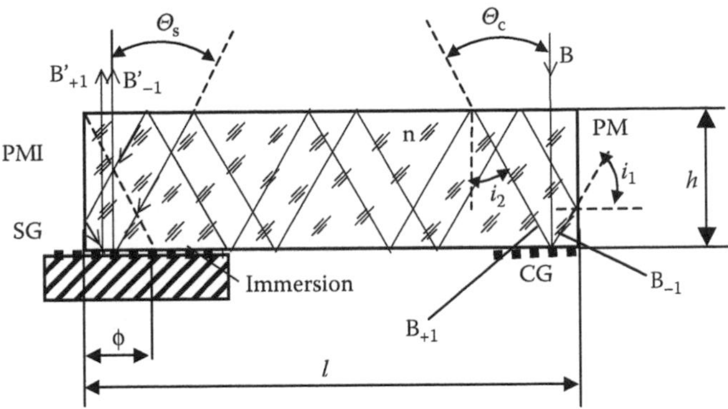

FIGURE 7.3 Scheme of the waveguide grating (moiré) interferometer. B = illuminating beam, B_{+1} and B_1 = beams diffracted on compensating grating, B'_{+1} and B'_{-1} = beams diffracted on specimen grating, CG = compensating grating, SG = specimen grating, PM and PMI = side surfaces of the plate, Θ = first diffraction order angle, i_1 and i_2 = incidence angles, l = plate length, h = plate height, n = refractive index, and ϕ = field of view.

the compensating grating CG. The plus first diffraction order beam B_{+1} is guided inside the plate and after final reflection by the surface PMI illuminates the SG. The minus first diffraction order B_{-1}, reflected by surface PM and guided by the plate, illuminates the SG directly at the same point as beam B_{-1}. In this case, the dimensions of the plate fulfill the following relation:

$$l = 2Nh\tan\Theta_c, \quad N = 1, 2, 3, \ldots \tag{7.9}$$

where l is the length of the plate, h is its height, and Θ_c is the first diffraction order angle of the CG described as:

$$\Theta_c = \sin^{-1}(\lambda/np) \tag{7.10}$$

where n is the refractive index of the glass, p is the period of diffraction gratings, and λ is the wavelength of the illuminating light.

If the SG has the same frequency as the CG, the incident angle of the beams illuminating the SG is equal to its first diffraction order angle Θ_s. In such ($\Theta_c = \Theta_s = \Theta$) a case, the diffracted beams B'_{+1} and B'_{-1} propagate along the normal to the specimen surface irrespective of the wavelength of the illuminating light (i.e., the system is achromatic). Beams B_{+1} and B_{-1} interfere, forming a fringe pattern in which information about in-plane displacements of the specimen under test is coded.

The beams B_{+1} and B_{-1} are guided by the GP due to any of the following:

- Total internal reflection inside the plate when $i_1 > \alpha_c$ and $i_2 > \alpha_c$, where α_c is critical angle described by the equation:

$$\alpha_c = \sin^{-1}(1/n) \tag{7.11}$$

- Reflection by the mirrorized surfaces of the glass plate when $i_1 < \alpha_c$ and $i_2 < \alpha_c$
- Mixed modes based on both phenomena

These modes may be referred to the diffraction angle $\Theta(\Theta = i_2 = 90° - i_1)$ and its relation to the refractive index of the waveguiding medium. The following relations give the limits for the specific guiding mode:

$$\Theta = \sin^{-1}(1/n) \tag{7.12a}$$

$$\Theta = \cos^{-1}(1/n) \tag{7.12b}$$

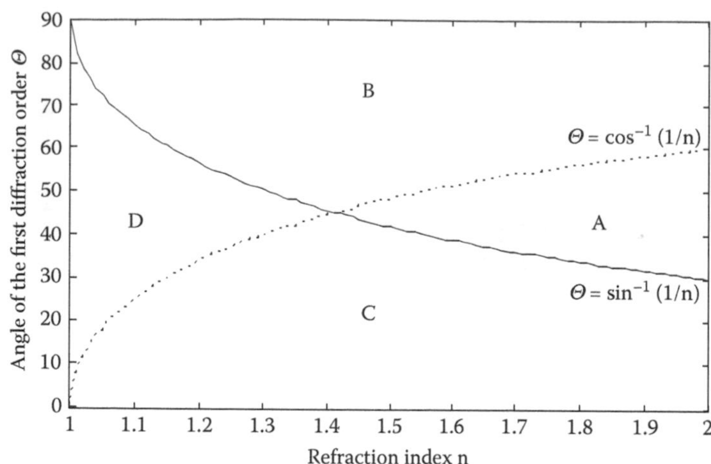

FIGURE 7.4 Diagram representing the relation between the first diffraction order angle and refraction index of the plate material.

and they are shown graphically in Figure 7.4. The areas A, B, C, and D represent the guiding modes, which are realized by the particular design of the microinterferometer head:

- Type A, based on total internal reflection at all surfaces of the plate ($\Theta > \sin^{-1}(1/n)$ and $\Theta < \cos^{-1}(1/n)$). The main advantages of the waveguide microinterferometer head type A are a very simple design and low cost as no additional surface layers are applied. However, it requires an immersion liquid between the SG and the GP (at this area total internal reflection cannot occur). The disadvantage is the relatively narrow range of grating frequencies that can be applied for a given refractive index n. For example, if glass BK7 with $n = 1.51$ is used (for = 670 nm), the allowed grating frequencies are from 1492 lines/mm to 1688 lines/mm.
- Type B, in which total internal reflection occurs only for top and bottom plate surfaces ($\Theta > \sin^{-1}(1/n)$ and $\Theta > \cos^{-1}(1/n)$). In this case, the side surfaces PM and PMI have to be mirrorized. In this interferometer the grating frequencies are not limited. Note also that this type can work properly for the gratings from area A.
- Type C, in which total internal reflection occurs only at the side surfaces of the plate ($\Theta < \sin^{-1}(1/n)$ and $\Theta < \cos^{-1}(1/n)$), so the top and bottom surfaces have to be mirrorized. In this case, low-frequency gratings can be used, especially, the most popular grating with a frequency of 1200 lines/mm. Similar to the previous case, this head also works properly for the gratings from area A. The main difference in comparison with type A and type B is the height of the plate, which is twice as much for the same field of view. Also, immersion fluid does not need to be applied, although it can increase the output beam's energy and improve the interferogram quality (it eliminates additional reflections between specimen and glass surfaces).
- Type D, based on the reflection from the mirrorized surfaces (total internal reflection cannot occur: $\Theta < \sin^{-1}(1/n)$ and $\Theta > \cos^{-1}(1/n)$). Of course, the plate with mirrorized surfaces can work for every grating frequency but the price of this waveguide plate is the highest and it can be easily damaged due to the presence of external surface layers.

An interesting adaptation of this type of microinterferometer is the case when the refractive index equals 1. In this case, the microinterferometer head is a cavity waveguide built from mirror plates. The advantage of this is that the surface layers are inside the head and that it can be easily

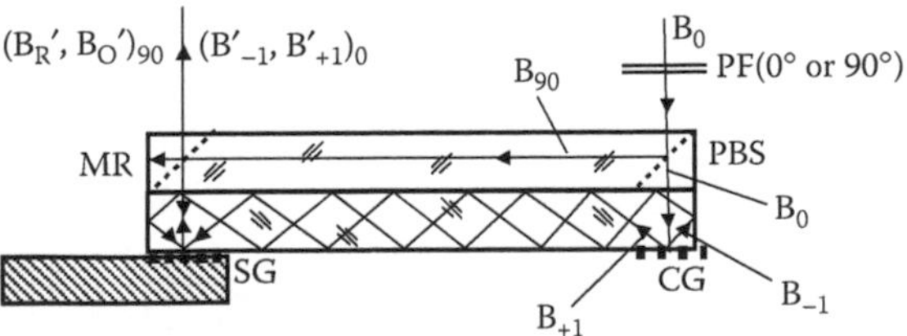

FIGURE 7.5 Configuration of WGI/TGI interferometer head for in-plane and out-of-plane displacement measurement. PF = half-wave plate, PBS = polarization beam splitter, MR = reference mirror for Twyman–Green channel, CG = compensating grating, and SG = specimen grating.

adjusted by moving one of the mirrors. A disadvantage is its relatively large dimensions in comparison with the other types of microinterferometers, especially type A and type B.

In all types of WGI head types presented, the grating CG compensates for changes in the wavelength or, if the wavelength is constant, the grating CG significantly decreases the sensitivity of the WGI to vibrations. It means that variations in the incident angle of the illuminating beam B does not cause a change in the angle between interfering beams B'_{+1} and B'_{-1}. This fact, described by Czarnek [8] in reference to the concept of the portable achromatic moiré interferometer, enables the application of the WGI head as a modification of a standard optical microscope when working in typical laboratory or workshop conditions.

7.3.2 Modified WGI for 3-D Components of Displacement Vector Measurements

The grating waveguide interferometer can be modified by adding optical elements that form the channel for interferometric shape and out-of-plane displacement measurement. This channel is based on the Twyman–Green interferometer (TGI) configuration [5]. The configuration of the WGI/TGI is shown in Figure 7.5. The coherent beam B_0, with properly oriented polarization, illuminates the interferometer glass plates. Depending on the direction of the polarization controlled by the phase plate PF, the beam is alternatively directed (by polarization beam splitter PBS) into the Twyman–Green interferometer channel — B_{90} (the upper glass plate) — or it is introduced into the waveguide grating interferometer — B_0 (described in the previous section). In the Twyman–Green configuration, the beam B_{90} is divided by the beam splitter BS into the reference (B_R) and object (B_0) beams. After reflection from the specimen and the reference mirror MR, the beams B_0' and B_R' interfere, forming an interferogram in which information about the actual shape of the specimen is coded.

7.4 MEASUREMENT SYSTEM

The scheme of the microinterferometer is shown in Figure 7.6. The light from the laser diode DL is divided into two beams by the PBS. One of these beams is collimated by lens L and illuminates the Twyman–Green microinterferometer (TGI) head. The second beam, with an orthogonal state of polarization, is introduced into the waveguide plate of the grating microinterferometer WGI. In the output, the beam splitter BS and two analyzers A are used for separation of information about in-plane displacement (beams from WGI) and out-of-plane displacement (beams from TGI). Note that in our case, due to the small field of view, the grating microinterferometer can work with a spherical wavefront of illuminating beams (errors introduced due to this fact are smaller than 10 nm). It enables miniaturization of the microinterferometer head (the beam collimator is not required) [9].

The signal from charge-coupled device (CCD) cameras is digitized by a frame-grabber and is analyzed by an automatic fringe pattern analyzer (AFPA). In the system presented, the spatial

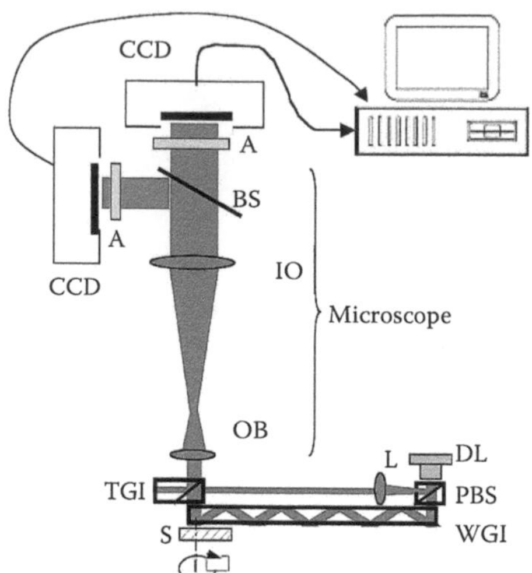

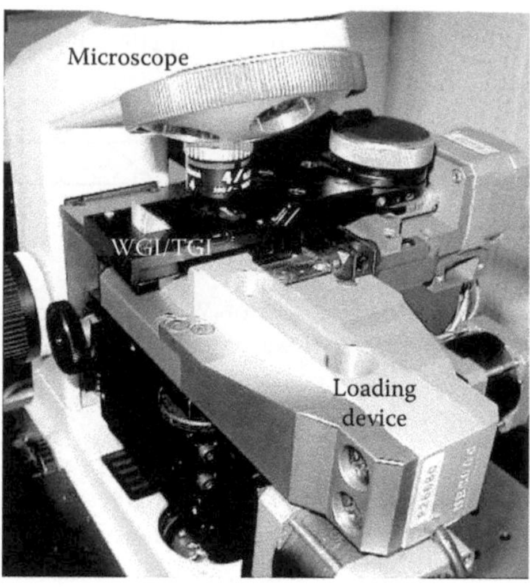

FIGURE 7.6 Scheme and photo of the waveguide grating microinterferometer (WGI) modified by Twyman–Green configuration (TGI). DL = laser diode, PBS = polarization beams splitter, L = collimating lens, S = specimen, OB and IO = optics of microscope, BS = beam splitter, and A = analyzer.

carrier phase-shifting (SCPS) method is used [10] for interferogram analysis. This method is based on the analysis of a single interferogram with a proper carrier frequency. The interferograms for u and w channels have the form:

$$I_u(x,y) = a_u(x,y) + b_u(x,y)\cos\left[\frac{4\pi}{p}u(x,y) + 2\pi f_{0x}x\right] \tag{7.13a}$$

$$I_w(x,y) = a_w(x,y) + b_w(x,y)\cos\left[\frac{2\pi}{\lambda}w(x,y) + 2\pi f_{0x}x\right] \tag{7.13b}$$

where $a(x,y)$ and $b(x,y)$ are the background and contrast functions respectively, p is the period of the SG, $u(x,y)$ and $w(x,y)$ are in-plane and out-of-plane displacements, respectively, and f_{0x} = M/4 (M is the detector resolution), the carrier frequency in the x direction. The required direction and frequency of the fringes is introduced by a proper combination of rotation of the specimen S around the axis perpendicular to its surface and the tilt of the illuminating system DL.

The interferogram is analyzed by the five-point phase-shifting algorithm modified for the assumption of constant phase derivative in the five-point sampling window, so that the phase maps are calculated according to the formula [11]:

$$\varphi(x,y) = \arctan\left[\frac{\sqrt{4\left[I(x-1,y) - I(x+1,y)\right]^2 - \left[I(x-2,y) - I(x+2,y)\right]^2}}{2I(x,y) - I(x-2,y) - I(x+2,y)}\right] \tag{7.14}$$

where $I(x + l,y)$ denotes the sequential intensities within the sampling window in the x direction ($l = -2, \ldots, +2$) and φ(x,y) is the phase equal to $(4\pi/d)u(x,y)$ or $(2\pi/\lambda)w(x,y)$.

Next, after unwrapping and scaling procedures, the displacement maps are determined.

If the in-plane displacement fields (u(x,y), v(x,y)) are known, it is possible to calculate the strain field for the specimen under load. In practice it is realized by proper differentiation of the displacement data. The in-plane strains at point i are given by the equations:

$$\varepsilon_{ix} = \frac{\partial u}{\partial x} = \frac{u_{i-m} - u_{i+m}}{2m\Delta x'} \tag{7.15a}$$

$$\varepsilon_{iy} = \frac{\partial v}{\partial y} = \frac{v_{i-m} - v_{i+m}}{2m\Delta y'} \tag{7.15b}$$

$$\gamma_{ixy} = \frac{\partial u}{\partial y} + \frac{\partial v}{\partial x} = \frac{u_{i-m} - u_{i+m}}{2m\Delta y'} + \frac{v_{i-m} - v_{i+m}}{2m\Delta x'} \tag{7.15c}$$

where $2m\Delta x'$ and $2m\Delta y'$ are the distances between the points at the sample: $\Delta x' = (1/\beta)\Delta x$ and $\Delta y' = (1/\beta)\Delta y$; β is the imaging system magnification; Δx and Δy are the distances between pixels at the CCD matrix plane; and $2m$ is the number of pixels over which differentiation is performed.

The main technical data of microinterferometry:

- Field of view: 0.18 mm × 0.12 mm ÷ 3 mm × 2 mm
- Spatial resolution: 256 × 256 or 512 × 512 pixels
- SG: 1200 l/mm (typically), cross-type
- Accuracy: 417 nm per fringe (basic), 20 nm (with AFPA) — for u measurement
  316 nm per fringe (basic), 20 nm (with AFPA) — for w measurement
- AFPA analysis: spatial carrier phase-shifting method with five-point algorithm

7.5 SG TECHNOLOGY

Grating (moiré) interferometry requires the application on a specimen of a high-frequency reflective phase type grating. Such gratings are based usually on the periodical variation of the surface relief that is produced by the molding process [1] or through direct exposure of the element covered with a photosensitive emulsion. A thin (~1 μm) photoresist layer is spun onto the specimen and then exposed with two interfering laser beams (Figure 7.7). If a simple, reflective grating is required

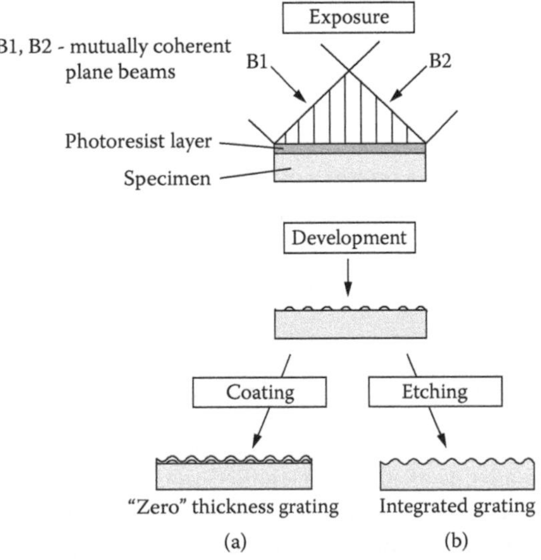

FIGURE 7.7 Stages of producing zero-thickness specimen gratings.

(e.g., for material or fracture mechanics studies), the photoresist is developed and the relief is covered (by evaporation) with a thin aluminium or gold layer (Figure 7.7a) [12]. If a grating totally integrated with, for example, a silicon element is needed (for MEMS or MOEMS studies), the development stage is followed by an etching procedure (Figure 7.7b) and the grating is produced directly in the material of the specimen [13]. These gratings are called *zero-thickness* gratings and require specimen surface finish of optical quality. Recently, special grating-preparation techniques have been proposed, including writing gratings on a coating of electron resist (e.g., polymethylmethacrylate — PMMA) or directly on PMMA microsystems housing by means of electron beams [14] or a high-power laser [15].

7.6 EXEMPLARY APPLICATIONS OF WGI

7.6.1 Material Constants Determination

The aim of this experiment was to determine Young's modulus of an inspection lot of 300-μm thick silicon wafers [9,13]. One of these wafers was cut into a few beams (Figure 7.8a). The SG with a frequency of 1000 lines/mm was integrated with the microbeam surface by electronlithography and ion-etching technology. The geometry of the beam and its loading configuration is shown in Figure 7.8b. The beam was tensile-loaded from 1 N to 2 N with the step of 0.1 N. The result of v in-plane displacement measurement in a direction parallel to the load is shown in Figure 7.8c (interferogram) and Figure 7.8d (3-D displacement map). The strain ε_y was calculated from the displacements by numerical differentiation. Next, the Young's modulus was calculated from Hook's law and the result, obtained as an average value from ten measurements, was equal to 188 ± 19 GPa.

7.6.2 Polycrystalline Materials Analysis

The local properties of materials strongly depend on their chemical and geometrical composition and the history of their manufacturing. Here we focus on the problems connected with polycrystalline materials. The published material data are based on the measurement of bulk samples with hundreds or thousands of grains. In the case of microelements it is necessary to assume, in general, that the polycrystalline material, due to crystallographic grain anisotropy, have directional properties. Knowledge about this anisotropy and possible nonhomogeneous plastic strains is very important for proper application of the material. Here, high-sensitivity grating interferometry was applied for investigation of two-phase ferritic-austenic steel polycrystalline material that was subjected to special heat treatment [3]. The cross-type diffraction grating with a frequency of 1200 lines/mm was transferred onto the steel sample and it was tensile loaded up to 0.5% plastic deformation and then unloaded. Figure 7.9 shows the in-plane displacement of u and v maps at the area where three grain borders meet and the effects of localization of high-strain regions with respect to these borders. The high quality of the output data enables analysis of the influence of different procedures of heat treatment on the local strain distributions. Specifically, the maps of shear strain of the grains can be investigated.

7.6.3 Semiconductor Microlaser Matrix Testing

The reliability of optoelectronic packages has become a subject of increasing concern for a variety of reasons, including the advent of higher integrated circuit densities, power density levels, and operating temperatures. The reliability depends, among others, on stresses caused by thermal and mechanical loadings produced in optoelectronic or electronic packages. The thermal stresses, resulting from mismatches between coefficients of thermal expansion of the materials comprising the package and the semiconductor die, can be produced during fabrication steps and during its use.

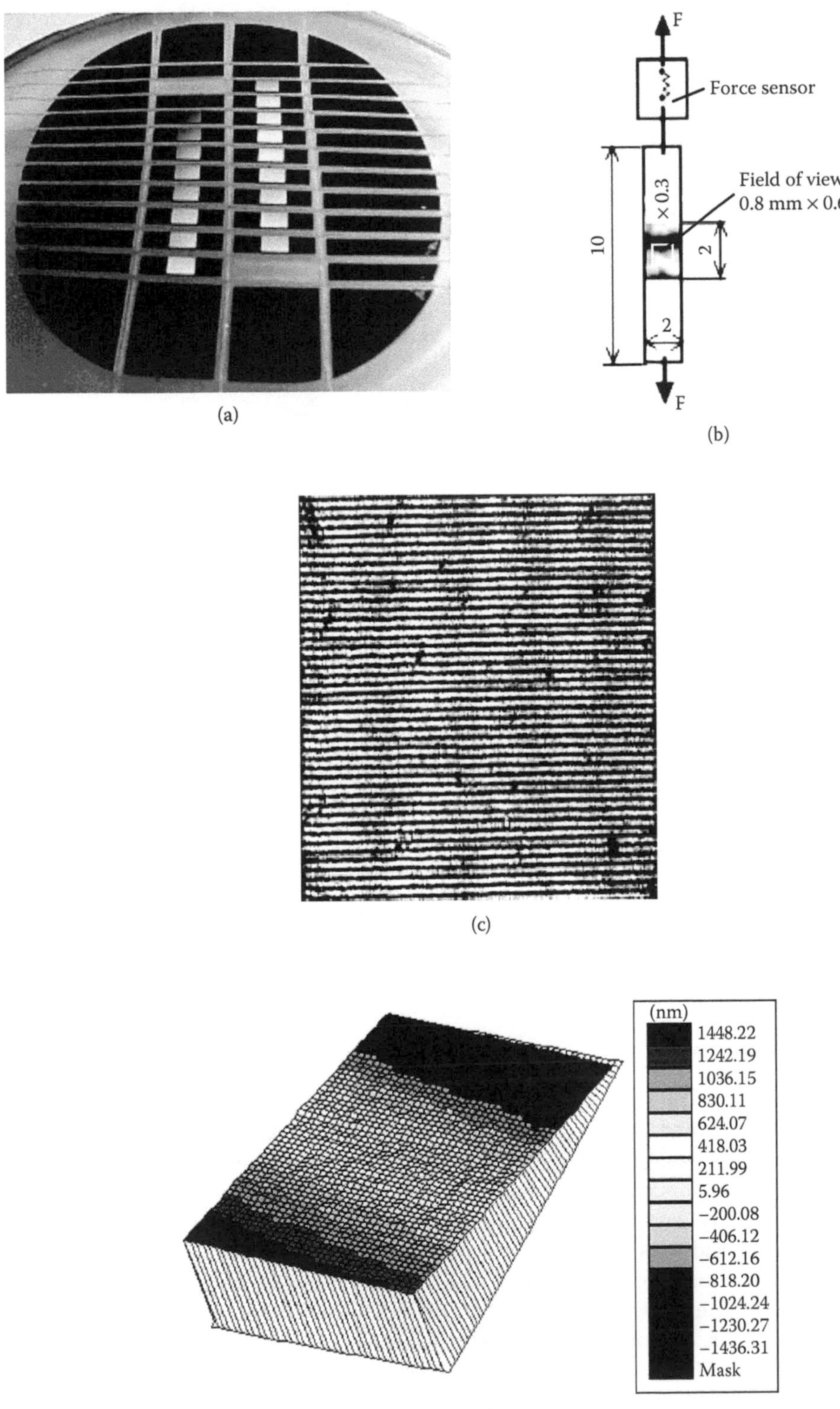

FIGURE 7.8 Silicon microbeam testing: (a) photo of a silicon wafer, (b) loading configuration and beam geometry, (c) exemplary interferogram with carrier frequency, and (d) in-plane displacement 3-D map.

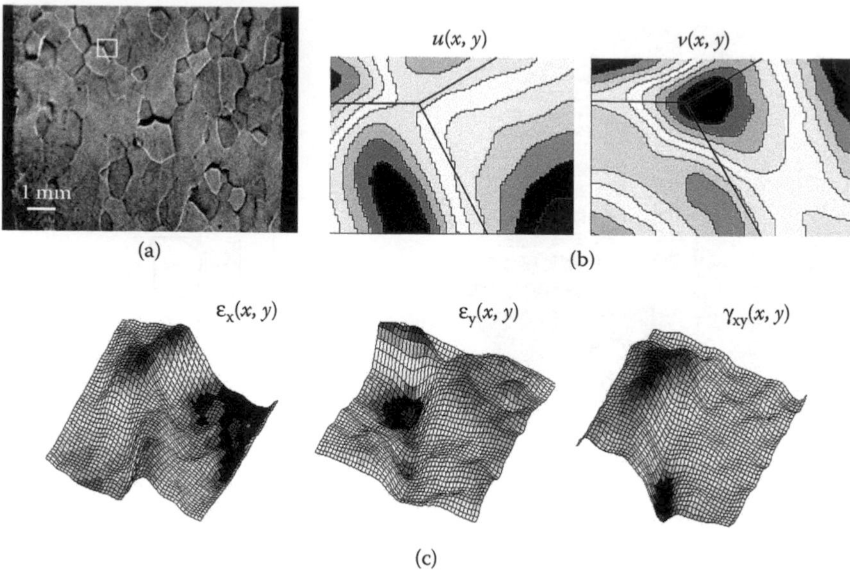

FIGURE 7.9 The analysis of local in-plane displacement or strain distribution in the microregion of two-phase steel polycrystalline material: (a) the gray-level image of microstructure, (b) contour maps representing u(x,y) and v(x,y) in-plane displacements, and (c) 3-D maps representing ε_x, ε_y, and γ_{xy} strains.

The aim of the studies performed at the microlaser matrix was the analysis of the influence of various layers (gold, dielectric, etc.) added to the silicon substrate on the strain and stress distribution in the microstructure [5]. The half-product of the matrix (Figure 7.10a) with replicated high-frequency cross-type grating was thermally loaded by coding the structure from 50°C to 20°C. The area of investigation was 1.2 × 0.8 mm^2 and refers to a section of silicon, including 3 × 3 microlasers. Figure 7.10b to Figure 7.10d present in-plane and out-of-plane displacement maps that indicate significant value variations which will result in the high strains ε_x and ε_y. The measurements were performed by a waveguide grating microinterferometer extended by TGI.

7.6.4 Electronic Packaging

The main task of mechanical analysis in electronic packaging is to study the thermal strain and stress, interfacial strength and failures, and the fatigue life of components and interconnections. These studies have a significant impact on the reliability determination of microelectronic devices and are critical in packaging design and manufacturing. Grating interferometry has been widely used for material property determination, experimental stress–strain analysis, online inspections, simulation validations, and hybrid methods in many packaging areas [16,17]. Here, the thermal strain studies performed on silicon chip UCY74S405N are presented [18]. It was a silicon chip connected with external legs by metal wires and placed inside the plastic (epoxy resin) body (Figure 7.11a). The main role of the plastic body is mechanical retention and protection of the semiconductor, protection from the environment, and the dissipation of internal heat to the outside. Due to the internal heating, the thermal stresses deform the chip elements and can lead to its destruction. During the test, the in-plane displacement and strain at the top chip surface were monitored and measured for the temperature changes from room temperature (20°C) up to operating temperature (~80°C) for 20 min. The plot of variations of the temperature in time at the central

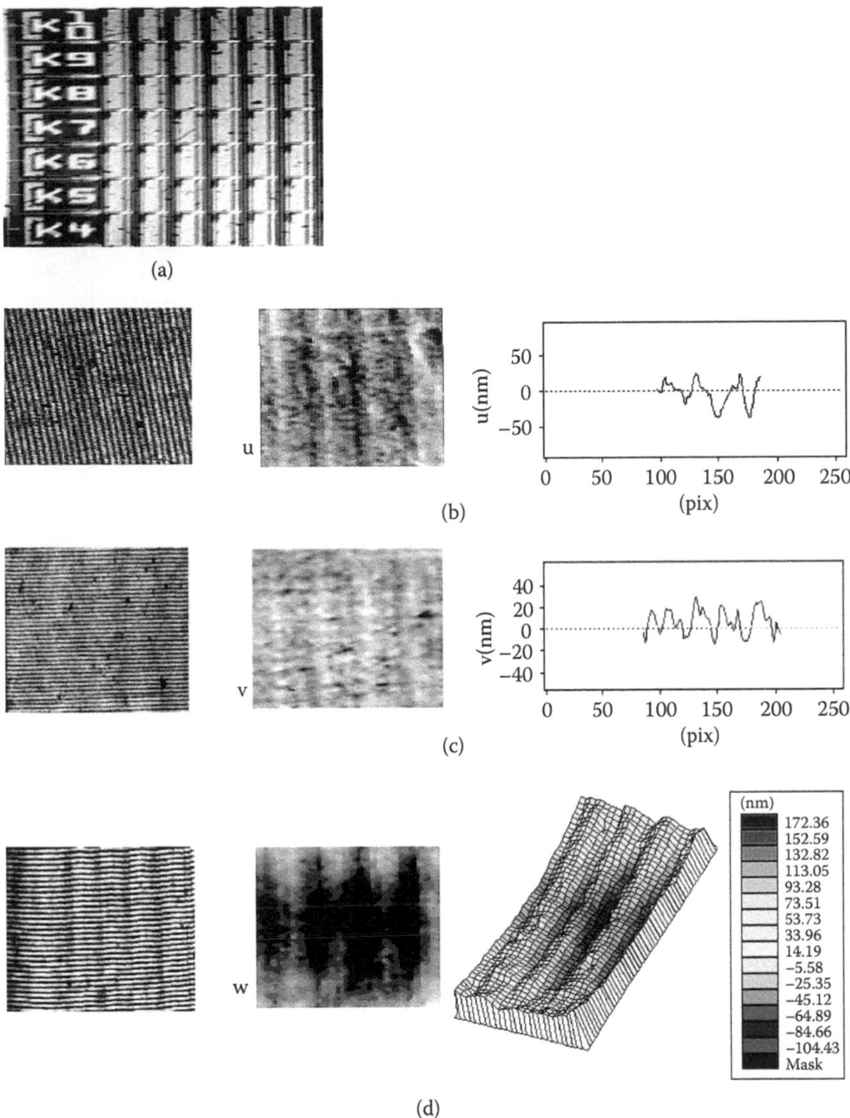

FIGURE 7.10 The behavior of half-product of semiconductor lasers matrix under thermal load: (a) overall view of the structure, (b) the gray-level map of $u(x,y)$ displacement and an exemplary horizontal profile, (c) the gray-level map of $v(x,y)$ displacement and an exemplary horizontal profile, and (d) the gray-level and 3-D maps of $w(x,y)$ displacement.

point of the chip top surface (after plugging the chip) is shown in Figure 7.11b. The temperature was stabilized after about 300 sec and was 82°C. However, the spatial distribution of the temperature at the surface was not uniform and it was additionally checked by thermovision camera. Figure 7.11c shows the temporal variations of the strains measured in the small area (about 1 mm^2) around point P. During the first 5 min (while the temperature rose), the strains increased up to approximately 0.02% for ε_x and 0.05% for ε_y. Next, during the time when the temperature is constant, both strains slightly oscillated in the range ±0.005%. This may be caused by oscillations of the heat dissipation process. This effect can be treated as a fatigue load and in the long term may decrease the life of the chip.

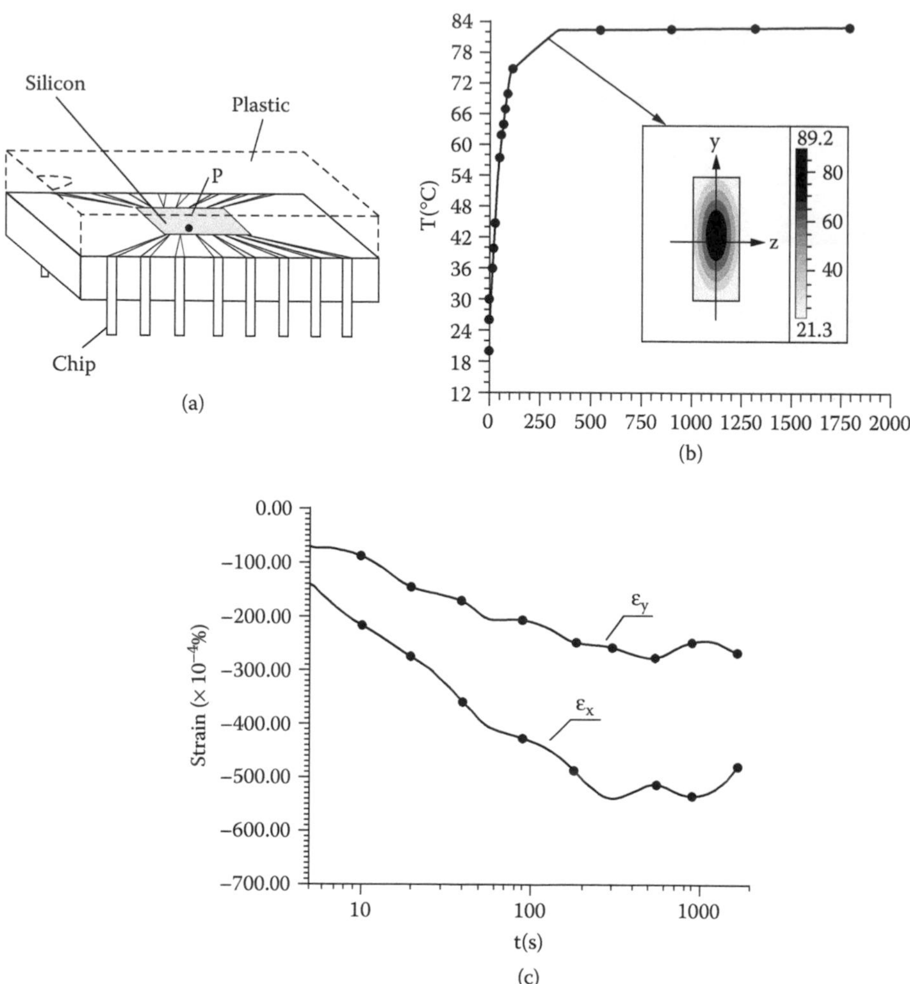

FIGURE 7.11 (a) Scheme of electronics chip, (b) temperature at the central point of the top of chip surface, and (c) temporal diagram of strains measured around the point P.

7.7 CONCLUSIONS

Analysis of microelements requires a local approach to their mechanical behavior and material constants distribution. This requires modern full-field measurement tools, among which automated grating (moiré) interferometry is one of the best-suited measurement methods. In this chapter the grating interferometry system was presented. It is based on the concept of an achromatic grating microinterferometer combined with a glass-block waveguide and is designed for work with standard optical microscopes. The various types of waveguide heads as well as their modifications for out-of-plane displacement monitoring are presented and discussed. The automated interferometric system based on these heads is described. The interferograms obtained may be analyzed by means of carrier frequency phase-shifting methods. The usefulness of grating microinterferometry is demonstrated in a few examples of in-plane displacement and strain distribution measurements.

All in all, the features of grating microinterferometry and the obtained results indicate that this technique may become a practical tool for laboratory and industrial quality assessment of MEMS, MOEMS, electronic components, and assemblies.

REFERENCES

1. Post, D., Han, B., Ifju, P., *High Sensitivity Moiré Interferometry*, Springer-Verlag, Berlin, 1994.
2. Patorski, K., *The Handbook of the Moiré Technique*, Elsevier, Oxford, 1993.
3. Salbut, L., Waveguide grating (moiré) interferometer for in-plane displacement/strain fields investigation, *Opt. Eng.*, 41: 626–631, 2002.
4. Han, B., Mikroscopic moiré interferometry, in Walker, C.A., Ed., *Handbook of Moiré Measurement*, Institute of Physics Publishing, Bristol, 2004.
5. Sabut, L., Kujawinska, M., The optical measurement station for complex testing of microelements, *Opt. Laser. Eng.*, 36: 225–240, 2001.
6. McKelvie, J., Patorski, K., Influence of the slopes of the specimen grating surface on out-of-plane displacements by moiré interferometry, *Appl. Opt.*, 27: 4603–4605, 1988.
7. Kujawinska, M., Salbut, L., Recent development in instrumentation of automated grating interferometry, *Optica Applicata,* XXV: 211–232, 1995.
8. Czarnek, R., High sensitivity moiré intererometry with compact achromatic interferometer, *Opt. Laser. Eng.*, 13: 93–101, 1990.
9. Salbut, L., Józwik, M., Gorecki, C., Lee, S.S., Waveguide microinterferometry system for microelements investigation, *Proc. SPIE,* 4400: 138–143, 2001.
10. Kujawinska, M., Spatial phase measurement methods, in *Interferogram Analysis*, Robinson, D.W., Reid, G.T., Eds., Institute of Physics, Bristol, 1993.
11. Pirga, M., Kujawinska, M., Errors in two-directional spatial-carrier phase shifting method for closed fringe pattern analysis, *Proc. SPIE,* 2860: 72–83, 1996.
12. Ifju, P., Post, D., Zero thickness specimen gratings for moiré interferometry, *Exp. Tech.*, 15(2): 45–47, 1991.
13. Jüptner, W., Kujawinska, M., Osten, W., Sabut, L., Seebacher, S., Combined measurement of silicon microbeams by grating interferometry and digital holography, *Proc. SPIE,* 3407: 348–357, 1998.
14. Dally, W., Read, D.T., Drexler, E.S., Transitioning from optical to electronic moiré, in Allison, A., Ed., *Experimental Mechanics*, Balkema, Rotterdam,1998, pp. 437–447.
15. Falldorf, C., Kopylov, C., Osten, W., Juptner, W., Digital holography and grating interferometry: a complementary approach, *Proc. SPIE,* 5457: 225–231, 2004.
16. Han, B., Guo, Y., Lim, C.K., Application of interferometric techniques to verification of numerical model for microelectronics packaging design, *Proc. EEP 10-2, Advances in Electronic Packaging*, ASME, 1995, pp. 1187–1194.
17. Guo, Y., Experimental determination of effective coefficients of thermal expansion in electronic packaging, *Proc. EEP 10-2, Advances in Electronic Packaging*, ASME, 1995, pp. 1253–1258.
18. Salbut, L., Kujawinska, M., Moire interferometry/thermovision method for electronic packaging testing, *Proc. SPIE,* 3098: 10–17, 1997.

8 Interference Microscopy Techniques for Microsystem Characterization

Alain Bosseboeuf and Sylvain Petitgrand

CONTENTS

8.1 Introduction 218
8.2 Interference Microscopes 218
8.2.1 Principle of Operation 218
8.2.2 Light Sources 219
8.2.3 Interferometers 219
8.2.4 Interference Microscopes with OPD Modulation 220
8.2.5 Interference Microscopes with Wavelength Modulation 221
8.2.6 Interference Microscopes with Direct Phase Modulation 221
8.2.7 Spectrally Resolved Interference Microscopes 221
8.3 Modeling of Two-Beam Homodyne Interference Microscopes 222
8.3.1 Two-Beam Interferometry with Monochromatic Illumination 222
8.3.2 Two-Beam Interferometry with Broadband Illumination 223
8.3.3 Two-Beam Interference Microscopy 223
8.4 Static Measurements by Interference Microscopy 224
8.4.1 Surface Profiling by Monochromatic Interference Microscopy 224
8.4.2 Surface Profiling by Low-Coherence Interferometry 225
8.5 Performance and Issues of Interference Microscopy 226
8.5.1 Edge Effects 226
8.5.2 Measurements on Heterogeneous Surfaces 226
8.5.3 Film Thickness Mapping 228
8.5.4 Spectral Reflectivity Mapping 231
8.6 Applications of Interferometric Profilometers in the MEMS Field 231
8.7 Dynamic Measurements by Interference Microscopy 233
8.7.1 Introduction 233
8.7.2 Interferometric Signal in the Dynamic Case 233
8.7.3 Vibration Measurements by Stroboscopic Interference Microscopy 235
8.7.4 Vibration Measurements by Time-Averaged Interference Microscopy 238
8.7.5 Applications of Dynamic Interference Microscopy in the MEMS Field 240
8.8 Conclusion 240
Acknowledgments 240
References 240

8.1 INTRODUCTION

Optical interferometry techniques have been used for a long time for optical components testing [1] and for deformation and vibration measurements in mechanical engineering [2]. Nevertheless, application of these techniques to the static and dynamic characterization of micro(opto)electromechanical systems (M(O)EMS) is challenging as M(O)EMS typically include high-aspect-ratio structures made of materials with various optical properties having small lateral sizes, a low surface rms roughness, and a low response time. A detection limit in the (sub)nanometer range, a lateral resolution in the (sub)micron range, a large dynamic ($>10^5$), and a high-frequency bandwidth (>1 MHz) are thus typically needed. As will be shown in this chapter, these performance standards are now commonly attained by interference microscopy techniques.

According to Richards [3], interference microscopy was initiated in Europe by Linnik, Lebedeff, and Frederikse in 1931–1935. Later, in 1945, Tolansky [4] recognized that white light interference fringes and sample scanning were useful to remove slope sign ambiguity of monochromatic interferometry. Optical profilometers based on interference microscopes began to be developed much later, from the mid-1980s. First, phase-shifting interferometry, a technique developed around 1970 for optical testing [5–7], was applied to interference microscopy for surface topography measurements with nanometer sensitivity [8,9]. Then, in 1987–1994, several authors developed profilometers based on low-coherence (white light) interferometry [10–18]. In the last 10 years, new capabilities of interference microscopes have been developed, such as out-of-plane and in-plane motion, vibration and transient measurements [19–35], micromechanical testing [36–38], thickness and group refractive index mapping of transparent films or devices [29,39–41], and spectral reflectivity map measurements [29,40].

To our knowledge, interference microscopy started to be applied to micromechanical devices profiling in 1988 [42], and it became progressively the most widely used technique for micromechanical devices and M(O)EMS profiling, testing, and reliability assessment [28,29,42–50].

In this chapter, we will review the main interference microscopy techniques used for static and dynamic characterization of micromachined surfaces and M(O)EMS. We will consider only homodyne interferometry techniques based on interferences between two light beams of equal optical frequencies. Heterodyne interferometry, in which an optical frequency shift is introduced between the two interfering beams, is discussed in Chapter 9.

Interested readers will find complementary information on interference microscopes and other examples of application in several published books and review papers [1,28,29,51–54].

8.2 INTERFERENCE MICROSCOPES

8.2.1 Principle of Operation

There are many configurations of interference microscopes. Nevertheless, they are all based on the generation, detection, and analysis of interferences between a light beam reflected on the sample surface and one reflected on a reference surface (generally, a flat mirror) (Figure 8.1). These two beams are obtained by amplitude division of the light source beam with a plate or cube beam splitter. This interference pattern (interferogram) is typically recorded with a camera that allows full field measurements but also a visualization of the interferogramy at video rate and an easy positioning of the sample.

A measurement typically consists of recording the fringe pattern for one or several values of the interferometer optical path difference (OPD) 2(h–z) (Figure 8.1) or the phase between the object and reference light beams. Then, various fringe pattern demodulation techniques are applied to extract interference phase maps, fringe amplitude maps, or other data from which a surface profile or a vibration map is extracted.

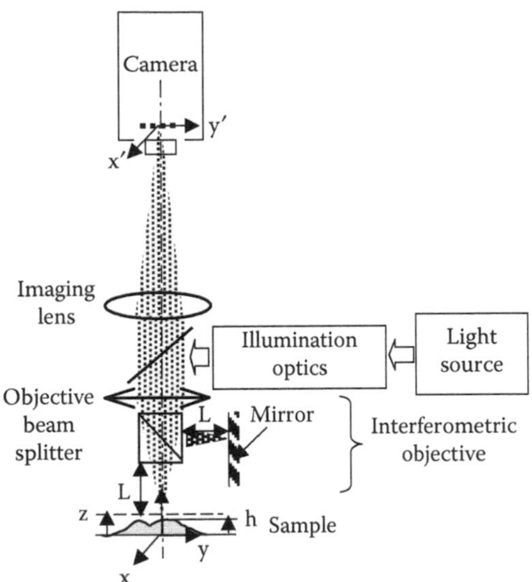

FIGURE 8.1 Principle of interference microscopy with a Michelson interferometric objective.

The generic names of this measurement technique are interference microscopy, microscopic interferometry, or interferometric homodyne profilometry/vibrometry. When a white light source is used, it is called by various other names as well: white light scanning interferometry (WLSI), scanning white light interferometry (SWLI), low-coherence interferometry, vertical scanning interferometry, correlation microscopy, coherence probe microscopy (CPM), coherence radar, or fringe peak scanning microscopy.

8.2.2 Light Sources

The choice of the light source is guided by its central wavelength, its coherence length, the light power needed, and its pulsing capabilities. Generally, visible light sources are used, although interference microscopy in the deep-UV range was demonstrated recently [55,56].

For (quasi)monochromatic interference microscopy, the light source coherence length must be larger than twice the largest height to be measured to get a good fringe contrast everywhere in the measurement area. However, highly coherent laser sources are often avoided as they tend to generate speckle and parasitic fringes, which degrade the measurements. An increasing number of interference microscopes use light-emitting diodes (LEDs), now available at low cost, with various wavelengths in the UV-visible–near-infrared range. In addition, they can easily be pulsed and have a low response time (20–50 nsec) suitable for stroboscopic measurements.

Generally speaking, as the signal-to-noise ratio of many optical detectors is a square root function of light intensity, the highest possible light power must be used. However, for point measurements on M(O)EMS, an excessive light power density can lead to unwanted static deformations [27], resonant frequency changes [57], and even permanent damage by light heating [58].

For WLSI measurements, a tungsten halogen white light source, eventually band-pass-filtered, is the most common choice, but white LEDs are an interesting alternative as they do not generate heat and can be pulsed. A light beam with an effective low-coherence length can as well be obtained by a suitable combination of two or three quasi-monochromatic light sources [59].

8.2.3 Interferometers

In interference microscopes, the sample surface acts as the second mirror of a two-beam interferometer. To minimize the optical paths in the interferometer arms and obtain a high lateral resolution,

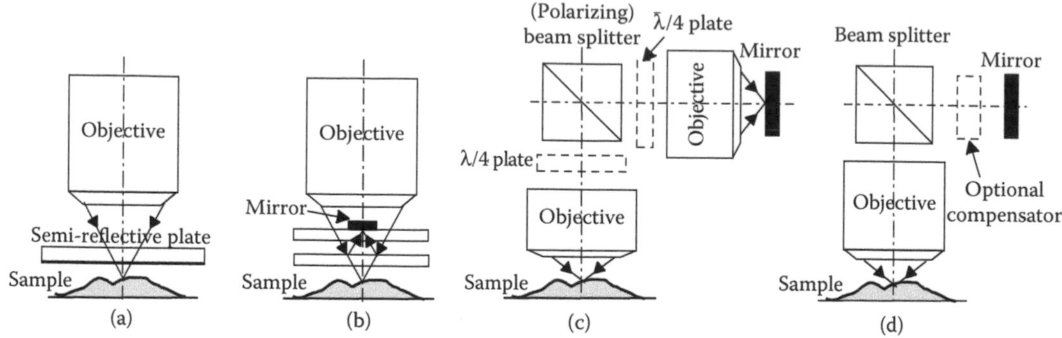

FIGURE 8.2 Common interferometric objectives: (a) Fizeau (b) Mirau, (c) Linnik, (d) "semi-Linnik."

interferometric objectives composed of a microscope objective and a miniaturized interferometer are generally used. The simplest interferometer that can be used is a beam splitter plate over the sample (Figure 8.2a). In this arrangement, recorded fringes are related to the air gap between the plate and the sample surface [47,48,51,60]. This interferometer needs a coherent light source and a compensated microscope objective to correct optical aberrations related to plate insertion. This latter requirement is common to most interferometric objectives. Michelson-type interferometric objectives (Figure 8.1) are used for low magnifications (≤ 5X). Because they are very compact, Mirau-type interferometric objectives (Figure 8.2b) are the preferred ones for intermediate magnifications (10X–50X). For high magnifications (100X–200X), Linnik interferometers are required (Figure 8.2c). Linnik interferometers need objectives with matched wave fronts and chromatic aberrations, so they are heavier, more expensive, and difficult to align. Nevertheless, they provide both the highest lateral resolution and the largest working distance. To avoid the need for matched objectives, a possible alternative is a "semi-Linnik" interferometer (Figure 8.2d), although this configuration is sensitive to wave front distortion of the reference beam if a point detector is not used [51].

In Fizeau, Michelson, and Mirau interferometers, the same objective is traversed by the object and reference beam, so chromatic aberrations and spherical aberrations of the objective have a small effect on the interference pattern [53]. Michelson and Mirau objectives are less sensitive to external disturbances than Linnik and semi-Linnik interferometers, because they have short optical paths and are lighter. Michelson, Mirau, and Linnik interferometers can be used with both coherent and low-coherence illumination. For semi-Linnik objectives, a highly coherent light source is typically needed.

Interferometric profilers can also be built with a Mach–Zehnder interferometer [52], but this interferometer is more commonly used in heterodyne vibrometers (see Chapter 9).

8.2.4 Interference Microscopes with OPD Modulation

Interferogram analysis techniques generally need a stepwise or linear scan or a sinusoidal modulation of the interferometer optical path difference (OPD) or of the interference phase [1,61–63]. In most interference microscopes, the OPD is varied by a mechanical translation of the sample or the interferometric objective with a piezoelectric transducer or equivalent. A good linearity and a resolution in the (sub)nanometer range are required. This is achieved by closed-loop control with a high-resolution displacement sensor, such as a capacitive sensor. Eventually, the scanning system can be integrated in the interferometric objective using a microactuator.

For large scans, the translation systems may generate as well some nonnegligible amount of tilt and lateral displacement. This may affect significantly the quality of measurements of M(O)EMS having fine and high-aspect-ratio patterns.

8.2.5 Interference Microscopes with Wavelength Modulation

There are various methods of varying the interference phase without mechanical scanning. One of them is wavelength modulation of the light source in an unbalanced interferometer [64–67]. For a light source with average wavelength λ_0, a variation $\Delta\lambda$ of the light source wavelength produces a phase variation $\delta\phi$ equal to:

$$\delta\phi = -\frac{4\pi\,(L-h)\,\Delta\lambda}{\lambda_0^2} \tag{8.1}$$

where L is the amount of interferometer unbalancing and h is the local height.

Accuracy and range of OPD variations that can be measured by this technique are related to the tuning range and the coherence length of the light source.

Wavelength modulation often produces a change of output intensity and fringe modulation amplitude, which must be corrected at the interferometric signal processing stage or compensated for during measurement [66].

8.2.6 Interference Microscopes with Direct Phase Modulation

The interference phase can also be varied without mechanical scanning of the sample or the interferometric objective by using polarized light beams and a phase modulator in the illumination or detection path [68] or in one of the interferometer arms. The phase modulation range is typically limited to about 2π. This technique can be used with Michelson, Linnik, semi-Linnik, or Mach–Zehnder interferometers that allow the use of a polarizing beam splitter. Additional wave plates in the illumination or detection path and in the interferometer arms are generally needed to control the polarization state of the light beams. For white light interferometry measurements, an achromatic phase shifter must be used to avoid a shift of the maximum fringe envelope function [69].

8.2.7 Spectrally Resolved Interference Microscopes

Mechanical scanning can also be avoided in a white light interference microscope by performing a spectral analysis of the detected signal. In that case, intensity variations along the chromatic axis (wavenumber $k = 2\pi/\lambda$) vary periodically with a frequency related to the optical path difference at the considered point. Spectral analysis can be performed with a spectrograph [70–72], an imaging Fourier transform spectrometer [73] (Figure 8.3), or an acoustic tunable optical filter, as well as

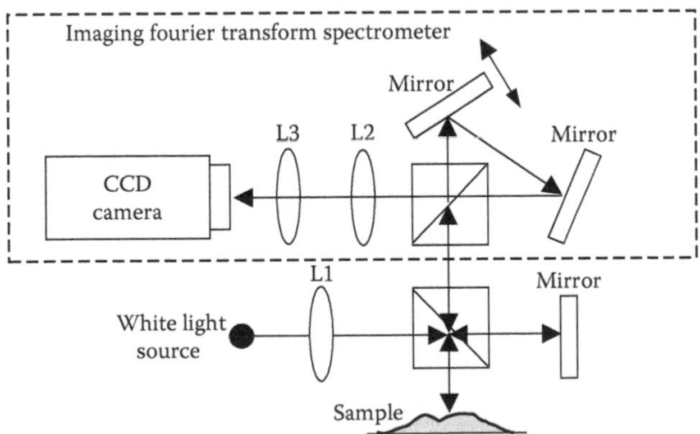

FIGURE 8.3 Spectrally resolved profilometer with an imaging Fourier transform spectrometer. L1, L3: cylindrical lenses, L2: spherical lens. (Adapted from Hart, H., Vass, D.G., and Begbie, M.L., Fast surface profiling by spectral analysis of white light interferograms with Fourier transform spectroscopy, *Appl. Opt.*, 37(10), 1764, 1998. With permission.)

by other means. In all cases, one axis of the 2-D detector is needed for the spectral (depth) information, so only profiles along a line can be measured. A resolution in the 1–5 nm range can be reached. This measurement principle can be combined with a Mirau interferometric objective to obtain a high lateral resolution [72].

8.3 MODELING OF TWO-BEAM HOMODYNE INTERFERENCE MICROSCOPES

In this section, we will first recall the classical equations describing two-beam interferometry systems with monochromatic and broadband illumination. Then, modifications that must be considered in the case of interference microscopy will be examined.

8.3.1 Two-Beam Interferometry with Monochromatic Illumination

The light intensity I_{out} at point P(x,y) at the output of a two-beam interferometer (Figure 8.4) with a monochromatic light source is given by the following well-known equation:

$$I_{out} = I_M + I_S + 2\sqrt{I_M I_S}\cos\phi \tag{8.2}$$

where I_M and I_S are the light intensities at point P of the waves reflected on the reference mirror and on the sample, respectively, and ϕ is their phase difference.

By considering propagation and reflection of these waves (Figure 8.4), it can be easily found that, for a perfect beam splitter ($l_1 = l_2$), the phase difference between these two waves at point P(x,y) is given by:

$$\phi(x,y,z,k) = 2k\left[h(x,y) - z\right] + \Delta\varphi_{MS}(x,y,k) \tag{8.3}$$

where $h(x,y)$ is the local height on the sample surface with respect to the base plane ($z = 0$), and $\Delta\varphi_{MS} = \varphi_M - \varphi_S$ is the difference of phase changes on reflection at points P′ and P″.

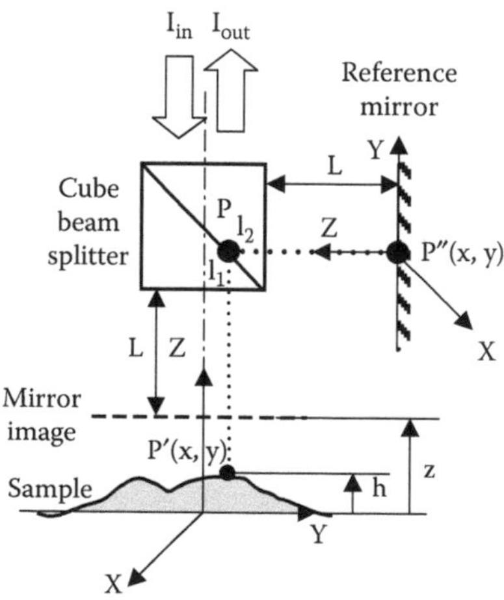

FIGURE 8.4 Two-beam Twynman–Green interferometer.

By combining Equation 8.2 and Equation 8.3, one finally finds:

$$I_{out}(x,y,z,k) = I_M(x,y,k) + I_S(x,y,k) + 2\sqrt{I_M(x,y,k)I_S(x,y,k)}$$
$$\cos\left[2k\left\{h(x,y) - z\right\} + \Delta\varphi_{MS}(x,y,k)\right] \tag{8.4}$$

The sum of the first two terms in Equation 8.4 is the background intensity, whereas the third term results from interferences between the two light waves.

Equation 8.4 is also commonly written in the following equivalent form:

$$I_{out}(x,y,z,k) = I_0(x,y,z,k)\ \left[1 + V(x,y,k)\ \cos\left[2k\left\{h(x,y) - z\right\} + \Delta\varphi_{MS}(x,y,k)\right]\right] \tag{8.5}$$

where $I_0 = I_M + I_S$ is the background intensity and $V = 2\sqrt{I_M I_S}/(I_M + I_S)$, the fringe visibility.

8.3.2 Two-Beam Interferometry with Broadband Illumination

When a two-beam interferometer is illuminated by a broadband light source, Equation 8.2 to Equation 8.5 must be modified to account for partial temporal coherence of the light source. A rigorous and detailed treatment of this problem can be found in Reference 74. The intensity at the output of the interferometer can generally be put in the following form:

$$I_{out}(x,y,z) = I_M(x,y) + I_S(x,y) + 2\sqrt{I_M(x,y)I_S(x,y)}\ E(x,y,h-z)\quad \cos\phi \tag{8.6}$$

where $E(x, y, h - z)$ is a fringe coherence envelope function and ϕ is the interference phase.

The coherence function $E(x,y,h - z)$ is maximum when the optical path difference is close to zero (i.e., $2[h(x,y)-z] = 0$) and decreases from both sides of the maximum.

If the wave number dependence of the optical constants of the beam splitter, the mirror, and the sample are neglected, the interference phase has the same expression as in the monochromatic case (Equation 8.3), except that now $k = k_m = 2\pi/\lambda_m$, where λ_m is the mean wavelength of the light spectrum.

In summary, with a broadband source, interference fringes are observed in the x,y plane with a good visibility only in sample areas in which the optical path difference of the interferometer is lower than the coherence length. More precisely, Equation 8.3 and Equation 8.6 show that, for a perfect beam splitter, the interferometric signal as a function of z is composed of fringes with period $\lambda_m/2$ shifted by $\Delta\varphi_{MS}$ inside the coherence envelope.

This is illustrated in Figure 8.5 for an experimental interferogram $I(z)$ recorded with a tungsten halogen white light source on an aluminum film evaporated on silicon.

For a more accurate modeling, dispersion effects must be considered. They induce a shift δz of the coherence envelope and an additional phase shift of the position of the interference fringes within the coherence envelope [75–78].

8.3.3 Two-Beam Interference Microscopy

Equations of two-beam interferometry given in the preceding text are suitable for the description of interferometric measurements with an interference microscope equipped with a low-aperture interferometric objective. However, to get a lateral resolution in the submicrometer range, high-numerical-aperture objectives are required. This has some important consequences, which are examined in the following text.

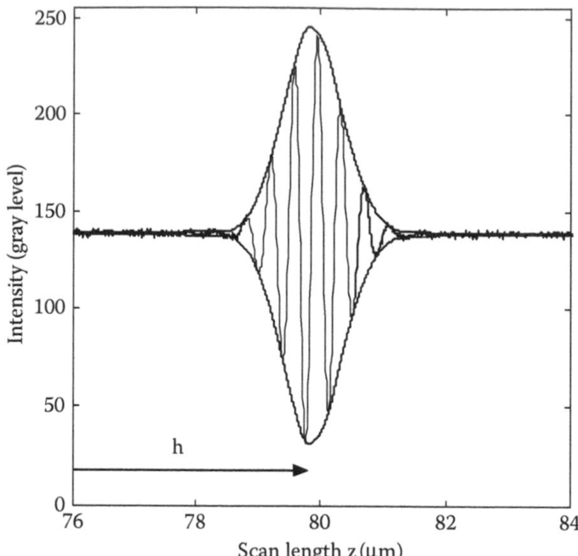

FIGURE 8.5 Interferometric signal $I(z)$ recorded by microscopic interferometry with white light illumination and a low-numerical-aperture objective on an Al/Si sample.

In a two-beam interference microscope, the reference and sample waves are actually convergent: the incidence angles on the mirror and sample surfaces vary from 0 to α_0, where α_0 is related to the numerical aperture *NA* of the objective by $NA = \sin(\alpha_0)$. This has two consequences. First, it leads to an increase in the effective fringe spacing from $\lambda/2$ to $\lambda_{eff} = \lambda/2(1 + f)$, where f is a correction factor [79–81], which varies from about 0.56% for $NA = 0.15$ to 21.5% for $NA = 0.8$ [81]. Second, it introduces a decrease of the fringe contrast with defocus and interferometer optical path difference. This limits the out-of-plane deformation range that can be measured with monochromatic illumination. These *NA* effects can be modeled by weighting and summing the contribution of each ray and each wavelength over the collection angle of the interferometric objective [79–83]. They have a complex dependence on many other parameters related to the illumination system, the light-source spectrum, the interferometric objective, and the sample position, so they are often evaluated by calibration.

8.4 STATIC MEASUREMENTS BY INTERFERENCE MICROSCOPY

In this section, we will describe the principle, the characteristics, and the issues of interference microscopy techniques for surface profiling of microstructures and for out-of-plane motion or deformation measurements.

8.4.1 Surface Profiling by Monochromatic Interference Microscopy

If we can neglect spatial variations of the background intensity I_0, of the fringe contrast *V*, and of the sample reflection phase shift φ_s, the fringe pattern $I(x,y)$ provides a contouring of the sample surface with a fringe spacing equal to $\lambda_{eff}/2$.

An example of a fringe pattern that approximately satisfies these conditions is shown in Figure 8.6, together with a line profile. It was recorded on a pressurized opaque membrane. It readily shows that without additional information, the overall sign of the topography and the sign of the local slopes cannot be determined. This is a characteristic shared by all monochromatic interferometry techniques based only on intensity measurements. In the case of Figure 8.6, the shape and sign of height variations

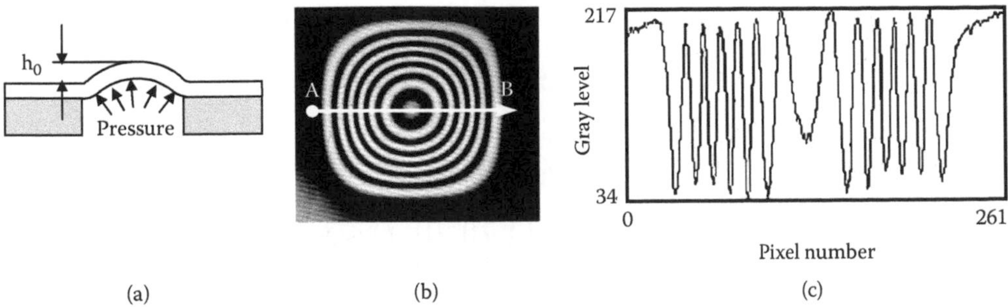

FIGURE 8.6 Interferogram recorded on a pressurized opaque square membrane with a monochromatic interference microscope: (a) schematic cross section of the deflected membrane, (b) interferogram, (c) digitized (8-bit) intensity profile along AB line. Deflection at center $h_0 \approx 1866$ nm.

can be easily guessed, so the upward central deflection h_0 can be estimated from the $I(x)$ intensity profile by fractional fringe counting, with an accuracy around $\lambda_{eff}/20$ (~30 nm).

Such an interferogram intensity map provides a visualization of out-of-plane deformations at video rate, but it depends on illumination inhomogeneities as well and on fringe contrast variations. A much more accurate measurement of the surface topography h(x,y) without slope sign ambiguity can be obtained by extracting the interference phase (see Equation 8.3). This is achieved by using phase demodulation techniques [1,61–63]. A phase map modulo 2π, the wrapped phase, is first obtained. Then, it must be unwrapped to obtain the whole phase (with an offset term) and the relative surface profile with (sub)nanometer resolution. This can be achieved only for continuous surfaces or surfaces with step heights < $\lambda/4$. This is a fundamental limitation of all standard single-wavelength interferometry techniques for microsystem profiling. Nevertheless, when only the relative out-of-plane displacement field or motion with respect to a reference state is needed, surface discontinuities are no longer an issue and accurate measurements can be performed even on stepped surfaces.

8.4.2 Surface Profiling by Low-Coherence Interferometry

As mentioned earlier, with broadband illumination, for example, a white light source, the coherence envelope of the interferometric signal $I(z)$ displays a maximum when the optical path difference is equal to zero. This is an absolute reference. Consequently, if interferometric signals are recorded as a function of z for two separate points having different heights h_1 and h_2, their envelopes will be separated by an amount $dz \approx h_2 - h_1$ (Figure 8.7). The sample surface height variations can thus be determined by finding for each point on the surface the maximum of the coherence envelope while the sample is scanned along the z axis. A 3-D surface profile can then be constructed from the relative height differences. As this process is performed independently for each point, steps, trenches, or disconnected surfaces can be measured. In addition, the signal is always recorded in the vicinity of zero optical path difference, so the range of measurable heights is only limited by the z scan range. These characteristics explain why WLSI is the most widely used technique for MEMS surface profiling.

A very large number of digital processing techniques, more or less equivalent, have been proposed to extract the position of the maximum of the coherence envelope and of the fringe phase from $I(z)$ signals. A detailed analysis of these algorithms is beyond the scope of this chapter. Some of them can be found in various references [12,15,16,18,51,72,76]. Generally, the most accurate techniques need a larger computation time, so a compromise must be found between speed and accuracy.

Figure 8.7 shows that a potentially more accurate solution is to measure the distance $\delta\phi$ between the central fringes of the interferometric signals $I(z)$ [69,76,84]. This can be achieved by identifying the fringe order within the envelope and measuring the phase of the interference fringes.

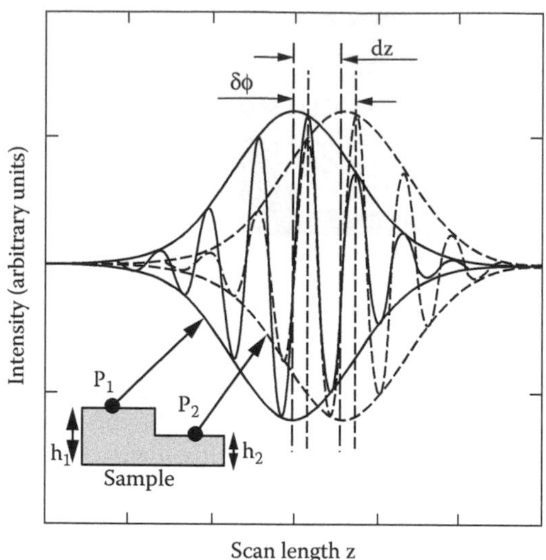

FIGURE 8.7 Principle of surface profiling by low-coherence interferometry.

A third way is to analyze the dependence of the interference phase ϕ with $k = 2\pi/\lambda$ and extract sample heights from the spatial variations of the fringe frequency $d\phi/dk$ [72,85].

8.5 PERFORMANCE AND ISSUES OF INTERFERENCE MICROSCOPY

Interference microscopy techniques are potentially able to measure surface profiles or out-of-plane deformations with a lateral resolution in the submicron range and a vertical resolution in the (sub)nanometer range. However, for highly structured and heterogeneous surfaces, as commonly encountered in the microsystem field, a certain number of issues and limitations must be known for their proper use. They will be examined in this part. Recent developments concerning the application of this technique for transparent film thickness and spectral reflectivity mapping will also be reviewed.

8.5.1 Edge Effects

The transverse resolution of interference microscopes is limited by diffraction and optical aberrations of the microscope interferometric objectives and, possibly, for low magnifications, by the camera resolution. It can be shown theoretically [53] that the different transverse resolutions of an interference microscope are similar to that of a confocal optical microscope and slightly better than a standard microscope. The reason is that the focused reference spot acts in a similar way as the pinhole in a confocal microscope, especially for high-*NA* objectives.

Nevertheless, in practice, various artifacts can be observed in measured surface topographies at the edges of patterns, such as "batwings" (Figure 8.8a) [86], an apparent broadening of the patterns [87], and invalid data. Some possible causes of these artifacts (Figure 8.8b) are diffraction effects, shadowing effects, indirect light reflection, parasitic interferences, unwanted lateral motion or tilting during sample scanning, light-source misalignment, and tilt of the interferometer or of the beam splitter with respect to the vertical optical axis.

8.5.2 Measurements on Heterogeneous Surfaces

M(O)EMS surfaces generally include patterns made with opaque and transparent materials having different optical properties. For interferometry techniques, the first consequence of material heterogeneity

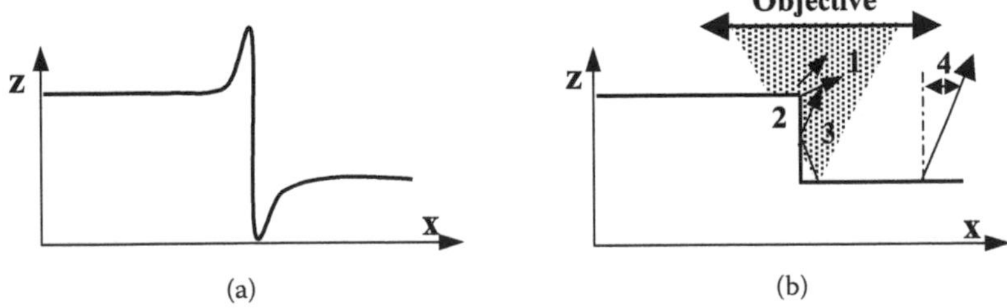

FIGURE 8.8 Two-dimensional edge effects: (a) step height profile with "batwings" artifact attributed to diffraction, (b) some origins of edge effects — (1) light diffraction, (2) shadowing, (3) light reflection, (4) lateral translation/tilting during vertical scanning.

is a spatial variation of the fringe contrast. A second and more critical issue is the material dependence of the interference phase because if it is not corrected, the true topography is not measured [88–91]. This material dependence arises from the dependence of optical constants of the sample phase change on reflection (Equation 8.3). For normal incidence, the phase change on reflection computed for an air/medium interface is:

$$\varphi_S(\theta = 0) = \tan^{-1}\left[\frac{2\kappa}{1 - n^2 - \kappa^2}\right] \tag{8.7}$$

where n and κ are the real part and the imaginary part of the complex refractive index of the sample material.

For monochromatic measurements, this phase change on reflection leads to a height offset equal to $h_{Offset}(0) = (\lambda/2)(\varphi_S(0)/2\pi)$. Figure 8.9 displays values in the visible range of $h_{Offset}(0)$ for some opaque materials used in M(O)EMS technology. They were computed from optical constants taken from Reference 92 and Reference 93. It was demonstrated [88] that for $0 < NA < 0.8$, the same formula can be used provided that $\lambda/2$ is replaced by the effective fringe spacing (see Subsection 8.3.3).

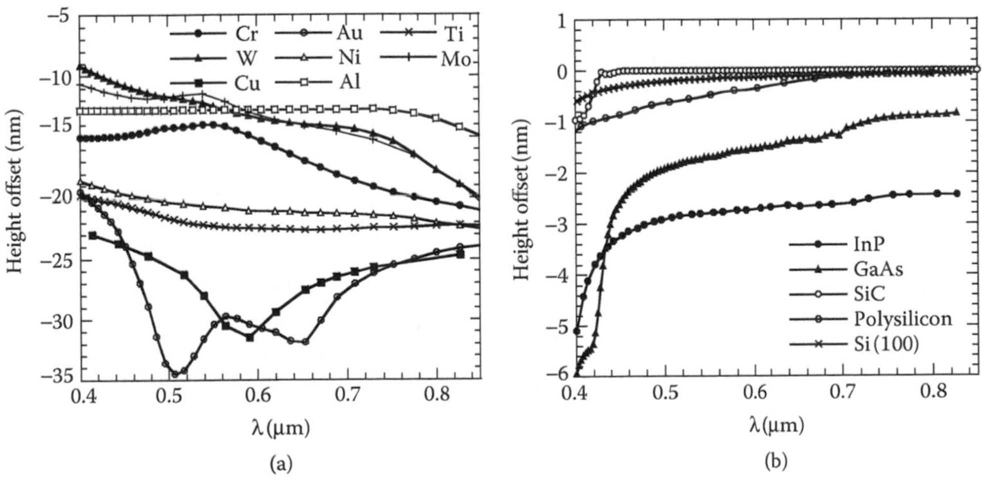

FIGURE 8.9 Normal incidence height offset as function of wavelength for some metals (a) and semiconductors (b) used in microsystem technology.

WLSI measurements based on phase demodulation are also directly affected by phase change on reflection, so the true topography is generally not measured. In addition, the dispersion of optical constants must as well be considered, so height offset corrections are more difficult to apply. WLSI measurements based only on maximum envelope detection have a low material dependence for opaque materials because the position of the coherence envelope maximum is only slightly affected by dispersion effects [75,78].

For transparent films on a substrate or for freestanding transparent films, the height offsets can be computed by using a three-medium stratified model and are in the $\pm\lambda/4$ range according to the value of the optical path in the film.

Corrections that must be applied to get the true surface topography of heterogeneous surfaces are thus far from being negligible in the microsystem field, in which patterns often have a submicron height. They can even lead to inverted steps in measured topographies. This is certainly the more serious issue of interferometry techniques for surface profiling. The true surface topography can be obtained by applying height offset corrections for each material present at the surface. This method usually requires a prior manual identification of materials and the knowledge of their optical constants. For WLSI measurements, an automated material identification is possible by computing spectral reflectivity maps that, in addition, provide real optical constants at micron scale (see Subsection 8.5.4).

Because this is a critical issue, several system modifications and other measurement procedures were proposed to allow a measurement of the real phase changes on reflection and, thus, of the true surface topography [89–91].

Let us emphasize that relative out-of-plane displacement fields can be accurately measured even on heterogeneous surfaces because height offsets can be canceled out by subtracting the initial state.

8.5.3 Film Thickness Mapping

An interesting feature of interference microscopy is its ability to perform thickness and refractive index mapping of transparent films or devices. For monochromatic light measurements, the reflection phase shifts can be exploited for this purpose, but WLSI is a more powerful technique.

For a transparent film on an opaque substrate or a transparent device (e.g., a dielectric membrane), the interferometric signal, $I(z)$, recorded by a camera pixel as a function of the scanning length, results from interferences between the reference beam and the light beam coming from the reflection on the top film surface, from the first reflection at the bottom interface and from higher-order reflections at the interface (Figure 8.10). Consequently, it is composed of several interference patterns occurring at different positions. This is illustrated in Figure 8.11 in the case of a thermal SiO_2 film on silicon.

With a low-numerical-aperture interferometric objective, angular effects can be neglected. Then for thick transparent films with an optical thickness larger than twice the coherence length, the fringe, envelope peaks are separated, and the distance between them is approximately the film optical path $n_{fg}.t_f$, where t_f is the film thickness and n_{fg} its group refractive index (Figure 8.11). Furthermore, in a first approximation, the ratio of the envelope peaks is only a function of n_{fg} and the optical constants of the substrate. Consequently, if substrate optical constants are known, both a map of the film or device thickness and its group refractive index can be measured by computing the positions and amplitudes of the maxima of the whole envelope [29]. This could be generalized to multilayer thick films.

Even when quantitative thickness measurement cannot be done, this technique remains useful in obtaining both the surface profile and the buried interface profile from the variations of the positions of the surface and interface peaks. This is illustrated in Figure 8.12 for of a 4-μm thick porous silicon film on silicon [29]. Figure 8.12 displays maps of the top surface (Figure 8.12a) and of the interface (Figure 8.12b) in a particular part of the sample. It clearly shows that the top surface

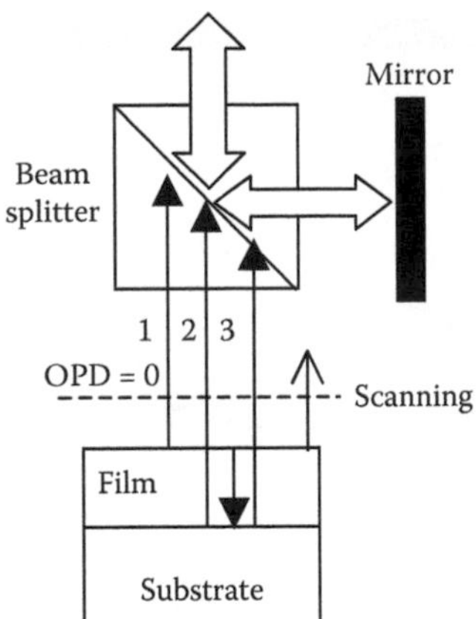

FIGURE 8.10 Principle of transparent film measurements by white light scanning interferometry (WLSI).

is flat and smooth, whereas the interface is rougher and displays a 350-nm-deep circular etching defect.

For thinner films, the envelope peaks overlap and the composite peak has still two maxima or simply an inflexion point according to the ratio R of $n_g.t_f$ to the mean interference fringe spacing. Consequently, peak maxima can no longer be used or they would lead to inaccurate measurements. Nevertheless, for some specific R values, two maxima still occur so the presence of an interface can be detected even if d < 100 nm.

Other methods can be used for thin (or thick) film mapping. The first [39] is to compute the phase of the interferometric signal, to unwrap it, and to fit its nonlinear wavenumber dependence. This technique is able to measure very thin films (down to 50 nm). However, processing can be disturbed by small phase steps in the overlapping region of the interferometric signals coming from

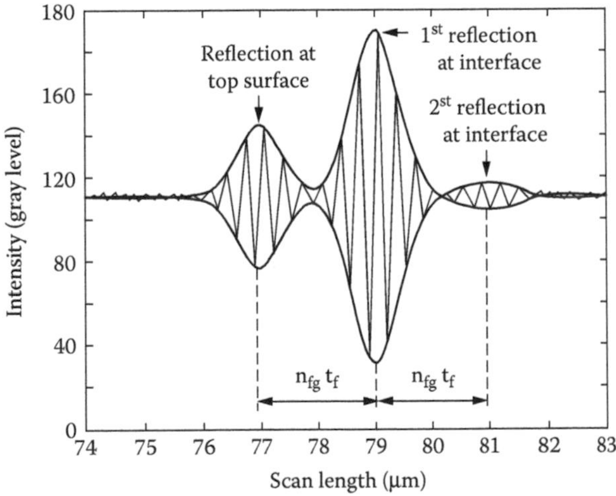

FIGURE 8.11 Interferometric signal $I(z)$ recorded on a 1.5-μm thick SiO_2/Si sample by white light scanning interferometry with a 5X Michelson interferometric objective. Fringe envelopes were computed by Fourier analysis. (From Bosseboeuf, A. and Petitgrand, S., Application of microscopy interferometry techniques in the MEMS field, in *Microsystem Engineering: Metrology and Inspection III, Proc. SPIE*, 5145, 1, 2003. With permission.)

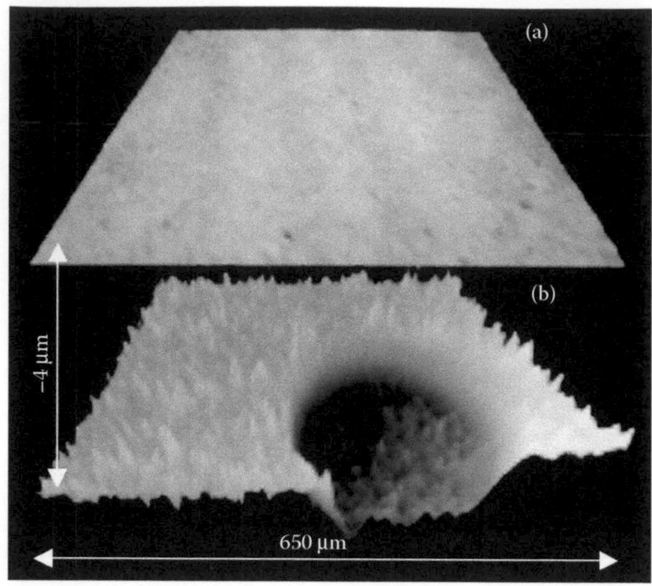

FIGURE 8.12 Three-dimensional profile of a porous silicon film on silicon: (a) top surface, (b) interface. Bottom 3-D profile was smoothed before display. (From Bosseboeuf, A. and Petitgrand, S., Application of microscopy interferometry techniques in the MEMS field, in *Microsystem Engineering: Metrology and Inspection III, Proc. SPIE*, 5145, 1, 2003. With permission.)

the film top and from the interface. An alternative method is to measure the spectral reflectance of the sample by WLSI as explained in the following section and to compute the film thickness from the distance between oscillations of the reflectance as function of $k = 2\pi/\lambda$ or from a fit of the whole normalized spectral reflectance curve (Figure 8.13).

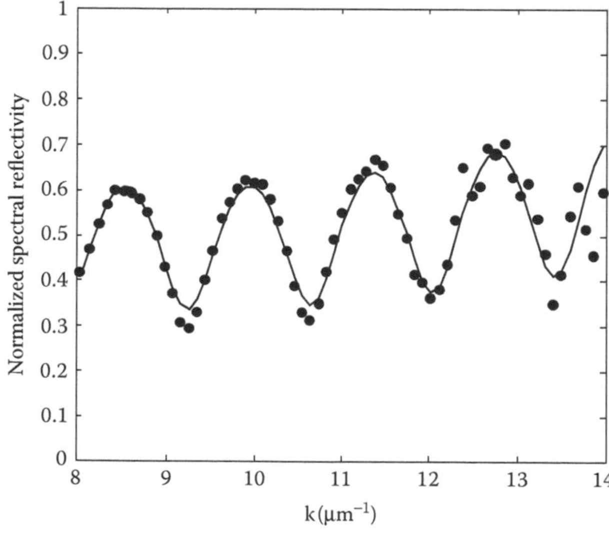

FIGURE 8.13 Normalized spectral reflectivity of a 1.53 μm thick SiO_2 film on silicon measured by white light scanning interferometry. Dots: experimental points. Solid line: fit curve.

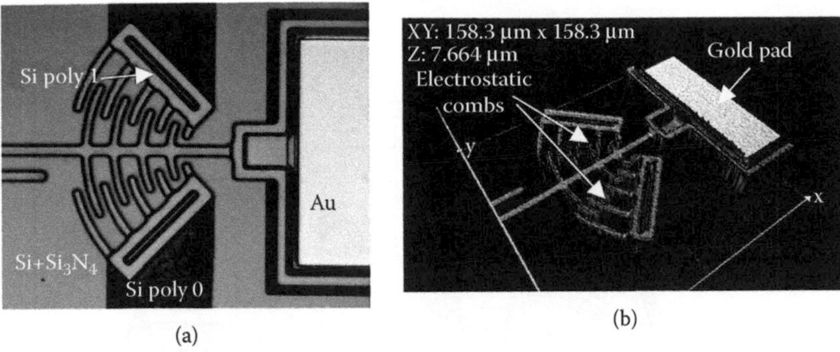

(a)

(b)

FIGURE 8.14 (See color insert following page 110.) (a) Spectral reflectivity map and (b) 3-D profile with colors related to spectral reflectivity of an electrostatically actuated MEMS fabricated in polysilicon MUMPS technology.

8.5.4 Spectral Reflectivity Mapping

When a low-coherence white light source and a not-too-wide aperture objective are used, the fringe modulation function E(z) is related to the spectrum of the light source after reflection on the sample surface, transmission back through the optical system, and detection by the camera. This spectrum can be determined at every point on the surface from the inverse Fourier transform of the envelope of the interferometric signal. Consequently, by normalizing this spectrum to that measured on a flat surface with a known reflectance spectrum, the local spectral reflectance in the visible range at each pixel can be evaluated [29]. Reflectance values measured by this technique are not as accurate as those measured with a spectrometer but are in good agreement with them. If the spectral reflectance is converted into RGB colors, it can be superimposed on the surface topography to get a colored topography in which each color corresponds to a given material and, for semitransparent materials, to different thicknesses.

This is illustrated in Figure 8.14 for an electrostatically actuated test device fabricated in MUMPS polysilicon technology. As indicated in Figure 8.14a, colors correspond to the different materials and films present at the surface. The blue color at the periphery of the high aspect ratio of the gold pattern is likely an artifact related to diffraction. The spectral reflectivity was normalized to the mean before extraction of colors. Let us emphasize that if the spectral reflectance values had been normalized to a measurement made on a sample with known reflectance in the visible range, by correcting the colors from the eye response, a 3-D topography with the true colors would have been obtained.

8.6 APPLICATIONS OF INTERFEROMETRIC PROFILOMETERS IN THE MEMS FIELD

A straightforward application of monochromatic or white light microscopic interferometry is the measurement of the full surface topography of a MEMS. As illustrated in Figure 8.14b, a full 3-D surface profile can be obtained by WLSI even for surfaces with high-aspect-ratio patterns. A side benefit of WLSI is that the maximum fringe contrast map provides a high-quality optical image with a depth of field only limited by the scanning range.

In a general way, interference microscopy techniques are useful to investigate fabrication process issues such as edge effects in electrodeposited patterns, etching nonuniformities, stiction of released parts, film delamination, etc. [29,45]. One widespread application is the measurement of out-of-plane deformations of micromechanical devices induced by internal residual stresses [19,29,37,38,43,46,47,49]. In that case, monochromatic measurements are faster and more accurate than WLSI measurements. Figure 8.15 shows an example of such measurement performed by

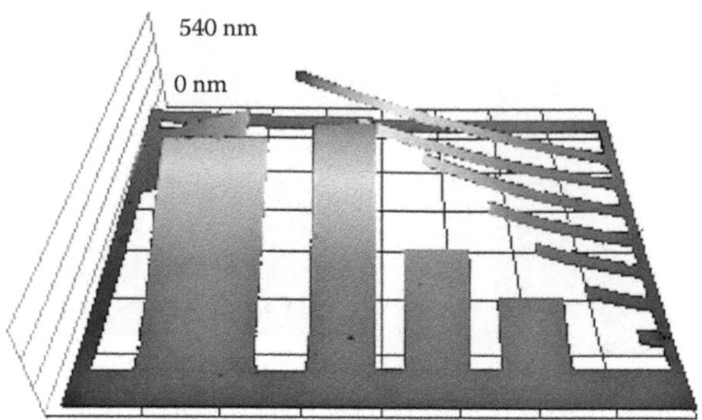

FIGURE 8.15 Three-dimensional profile of an array of silicon cantilever beams measured by phase-shifting interferometry. Sample fabricated by CNM Barcelona, Spain.

phase-shifting interferometry with automatic masking of areas not measured. It corresponds to an array of polysilicon cantilever beams bent upward by in-depth stress gradients. A closer examination showed that, as expected, the beam curvatures are parabolic and slightly dependent on the beam width because of the Poisson ratio effect.

Surface roughness is another important parameter in MEMS technology. Interference microscopy has vertical and lateral resolutions between those of mechanical profilometry and of atomic force microscopy. Some advantages of interference microscopy over these techniques are its higher measurement rate, its ability to measure a large range of rms surface roughness (from a fraction of a nanometer to several microns), and overall, to allow noncontact roughness measurements at the bottom of etched surfaces. This is particularly useful for the control of deep cavities and trenches fabricated by anisotropic chemical etching [29] or by deep reactive ion etching. For low-roughness measurement, the reference mirror roughness must be corrected by subtracting a reference measurement. A good reference can be obtained by averaging measurements recorded in several areas of a reference flat separated by a distance larger than the correlation length of its surface roughness [94].

Interference microscopy can as well be combined with external or *in situ* thermomechanical, electromechanical or optical loading/actuation for the characterization of thin films, MEMS, and actuation mechanisms. It can be used, for example, for thermal expansion coefficient evaluation from the temperature dependence of the postbuckling shape of microdevices. In our laboratory, we have applied interference microscopy to the simultaneous measurement of membrane deflection and substrate deformations during bulge tests of micromachined membranes [36]. We demonstrated that this greatly improves the accuracy of film Young's modulus and stress measurements obtained by using this technique. We have also shown with different laboratories that it can be applied to the determination of piezoelectric constants of PZT thin films from the deformations of piezo-actuated micromechanical devices.

In conclusion, the high lateral and vertical resolution of quasi-static interference microscopy measurements fulfill most of the actual requirements of microsystem inspection and testing, from the control of thin-film mechanical properties and fabrication processes up to the assessment of the final device performance. It is a mature technology and optical interferometric profilometers are now sold by more than ten companies all around the world. As detailed earlier, there are still issues that are not completely solved in a satisfactory way, notably the material dependence of measured profiles. New issues such as inspection of wafer-level-encapsulated devices, the need of faster measurements for on-wafer testing, and the higher lateral resolution required for the characterization of nanoelectromechanical systems will push for further developments. Finally, although we provided only examples of measurements on solid surfaces, it was demonstrated that interference microscopy techniques are also suitable for the characterization of liquid films and biomaterials, so it is expected that they will be used more and more in the fields of microfluidics and bio-MEMS as well.

8.7 DYNAMIC MEASUREMENTS BY INTERFERENCE MICROSCOPY

8.7.1 INTRODUCTION

Dynamic measurements are required in the MEMS field for the characterization of elastic constants, fatigue, and other properties of thin-film materials; for the characterization of motion and response time of moveable microstructures; and for the measurement of resonant frequencies, quality factor, and vibration-mode shapes of microresonators. For most devices, a detection limit in the subnanometer range and a frequency bandwidth of about 1 MHz is sufficient, but some devices need a resolution down to the picometer range and/or a frequency bandwidth up to the gigahertz range.

Identification of resonant peaks in the vibration spectrum of a microdevice might be difficult because they could have a complex geometry with many vibration modes. Even for simple devices, it is common to observe unexpected resonances that can arise from the excitation system, from the sample holder, or from various sources of vibration coupling or nonlinear effects. Vibration analysis of MEMS thus often requires vibration-mode mapping. As shown in the following text, interference microscopy techniques are particularly well suited to this type of measurement.

8.7.2 INTERFEROMETRIC SIGNAL IN THE DYNAMIC CASE

In the case of dynamic measurements, the interferometer OPD becomes a function of time. For a sinusoidal out-of-plane vibration of the sample, the OPD $2(h–z)$ becomes:

$$2(h(x,y,t)-z)=2z_0(x,y)+2a(x,y)\sin(\omega t+\phi_1(x,y)) \tag{8.8}$$

where a is the vibration amplitude; ω, its angular frequency; ϕ_1, its phase lag with respect to excitation; and z_0, the average local position during vibration.

By inserting Equation 8.8 in Equation 8.5 and developing the trigonometric term, we obtain:

$$\begin{aligned} I_{out}(t) = I_0 \Bigg\{ & 1+V(h-z)\cos\left(\frac{4\pi}{\lambda_m}z_0+\Delta\varphi_{MS}\right)\cos\left(\frac{4\pi a}{\lambda_m}\sin(\omega t+\phi_1)\right) \\ & -V(h-z)\sin\left(\frac{4\pi}{\lambda_m}z_0+\Delta\varphi_{MS}\right)\sin\left(\frac{4\pi a}{\lambda_m}\sin(\omega t+\phi_1)\right)\Bigg\} \end{aligned} \tag{8.9}$$

For low vibration amplitudes such as $a << \lambda/4\pi$, this equation simplifies to :

$$I_{out}(t)=I_0\left[1+V(z_0)\cos\left(\frac{4\pi}{\lambda_{mc}}z_0+\Delta\varphi_{MS}\right)\right]-I_0V(z_0)\sin\left(\frac{4\pi}{\lambda_{mc}}z_0+\Delta\varphi_{MS}\right)\frac{4\pi}{\lambda_{mc}}a\,\sin(\omega t+\phi_1) \tag{8.10}$$

The output intensity then consists of a constant term and a term at the excitation frequency with an amplitude proportional to a. A vibration spectrum can then be measured by scanning the excitation frequency and by using lock-in detection or a spectrum analyzer [19]. Equation 8.10

shows that to maximize the amplitude of the time-varying component, it is necessary to adjust and maintain the mean optical path difference in such a way that $4\pi z_0/\lambda_m + \Delta\varphi_{MS} = \pm(2n+1)\pi/2$, where n is an integer. This is usually achieved by adding a stabilization closed loop [19,95–97]. An alternative is the use of two balanced detectors in quadrature [95].

The lowest measurable vibration amplitude is given by [95]:

$$\delta a = \frac{\lambda_m}{4\pi}\left(\frac{2h\nu\Delta f}{\eta I_d}\right)^{1/2} \tag{8.11}$$

where I_d is the detected intensity; η, the detector quantum efficiency (<1); h, the Planck constant; ν, the light frequency; and Δf, the measurement frequency bandwidth.

In close agreement with Equation 8.11, a detection limit as low as 10^{-5} nm has been demonstrated for vibration measurements in the 1 kHz–100 kHz frequency range with a stabilized homodyne interferometer, a 2-mW He–Ne laser, and lock-in detection with a 10-sec time constant [97]. For vibration spectra of microresonators with an interference microscope, a lower time constant and light power are used and the detection limit is more typically around 10 pm. Nevertheless, with sinusoidal phase modulation, measurements up to 2 GHz with a vertical resolution of 0.3 pm and a lateral resolution of 0.5 μm have been demonstrated [26].

For large vibration amplitudes, it can be shown [21] that Equation 8.9 becomes

$$\begin{aligned}
I_{out}(t) = {} & I_0\left[1 + V(h-z)\cos\left(\frac{4\pi}{\lambda_m}z_0 + \Delta\varphi_{MS}\right)J_0\left(\frac{4\pi a}{\lambda_m}\right)\right] \\
& - 2I_0\ V(h-z)\sin\left(\frac{4\pi}{\lambda_m}z_0 + \Delta\varphi_{MS}\right)J_1\left(\frac{4\pi a}{\lambda_m}\right)\sin(2\pi ft + \phi_1) \\
& + 2I_0V(h-z)\cos\left(\frac{4\pi}{\lambda_m}z_0 + \Delta\varphi_{MS}\right)\sum_{n=1}^{\infty}J_{2n}\left(\frac{4\pi a}{\lambda_m}\right)\cos(2n(2\ \ t + \phi_1)) \\
& - 2I_0V(h-z)\ \sin(\frac{4\pi}{\lambda_{mc}}z_0 + \Delta\varphi)\sum_{n=1}^{\infty}J_{2n+1}\left(\frac{4\pi a}{\lambda_{mc}}\right)\sin((2n+1)(2\ \ ft + \phi_1))
\end{aligned} \tag{8.12}$$

where J_i are the Bessel functions of integer order i (Figure 8.16).

The time dependence of output intensity is thus no longer sinusoidal for large vibrations. Examination of Equation 8.12 shows that it is composed of a continuous term, a component at the excitation frequency, and even and odd harmonics with an amplitude proportional to Bessel functions of multiples orders of the vibration amplitude. A modulation of the fringe visibility also occurs if either a monochromatic source is not used or a large-*NA* objective is used. It can easily be shown that the number of nonzero harmonics is an increasing function of the vibration amplitude. Fitting, fringe counting, or a harmonic analysis of the signal is then required for quantitative evaluation of large vibration amplitudes. A quantitative harmonic analysis can be performed by exploiting the recurrence relations between Bessel function of integer order [98]. In any case, the vibration amplitude measurement range is limited by the frequency bandwidth

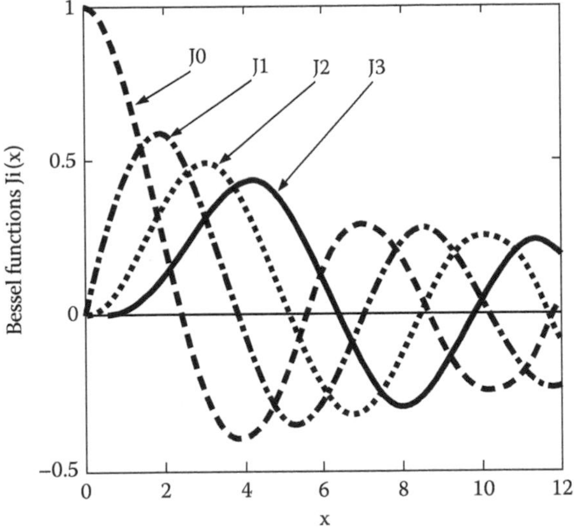

FIGURE 8.16 Shapes of Bessel functions of integer order J_i for $i = 0$ to 3.

of the detection electronics, by the depth of field of the interferometric objective, or by the coherence length of the source.

Let us emphasize that to perform point measurement with an interference microscope with a good sensitivity, the number of fringes within the detected area must preferably be less than one and, in any case, not an integer; otherwise, the sensitivity is largely degraded.

8.7.3 Vibration Measurements by Stroboscopic Interference Microscopy

In stroboscopic interference microscopy [20–22,24,25,27–32], the sample is illuminated by light pulses synchronized with the excitation signal and, eventually, with a phase delay t_0 (Figure 8.17). If the light pulse width δT is much lower than the vibration period T, the apparent motion of the sample surface is "frozen" at time t_0. Another method is to combine a four-bucket integration

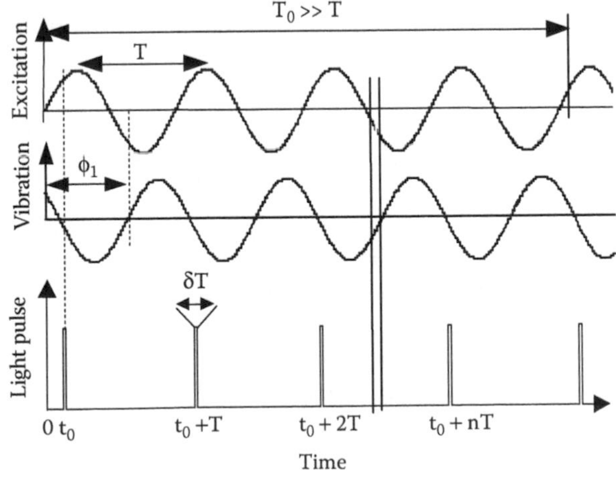

FIGURE 8.17 Time diagram of a stroboscopic measurement.

technique with synchronous detection with a fast camera [33]. In the former case, the detected intensity is then given by [21]:

$$I(x,y,h-z,t_0) = N \int_{t_0-\delta T/2}^{t_0+\delta T/2} I(x,y,h-z,t)dt$$
$$\approx N\delta T I_0 \left\{ 1 + V(z_0)\ Cos\ \left[\frac{4\pi}{\lambda_m} \left(z_0 + a\ Sin(\omega t_0 + \phi_1) \right) + \Delta\varphi_{MS} \right] \right\} \tag{8.13}$$

where $N = \text{int}(T_0/T)$ is the number of vibrating cycles during a video frame of duration T_0.

An equivalent method is to use continuous illumination and time gating with the camera. In both cases, the detected fringe pattern being constant, profilometry techniques described in Section 8.4 can then be applied to get the 3-D profile of the vibration mode at any time of the vibration cycle. The result of a single stroboscopic measurement actually provides the sum of the static profile and the vibration mode at time t_0. These two contributions can be separated by performing two stroboscopic measurements at t_0 and $t_0 + T/2$, and by computing the sum and difference of the measured profiles [30]. A drawback is that it is necessary to search for a vibration extremum before performing the actual measurements.

An automatic and powerful way to obtain a vibration-mode shape is to perform stroboscopic measurements for variable and equidistant values Δt_0 of the light pulse delay t_0 [22,30]. Then both the vibration phase and amplitude maps can be computed simultaneously by sinusoidal fitting or, better, by using algorithms similar to those used in phase-shifting interferometry [30]. This latter method, called light pulse delay shifting (LPDS) [30], can be applied to any stroboscopic measurement, even one that is not interferometric. Furthermore, it can provide at the same time the static profile. To obtain accurate measurements from this technique, it is necessary to check that the motion is sufficiently frozen at each stroboscopic measurement. A good criterion is that the fringe amplitude must be at least 30% of the fringe amplitude measured in the static case. In that case, a measurement accuracy better than 5% is obtained whatever the value of the initial light pulse phase t_0 if the duty cycle $\delta T/T$ is ≤15% [30].

Any interferogram analysis technique can be combined with stroboscopic illumination to get 3-D profiles of out-of-plane or torsion vibration-mode shapes, but with different capabilities.

By using fast Fourier transform (FFT) analysis, measurements can be performed at video rate [21]. Then, by using a repetition frequency of the light pulses slightly different from the vibration frequency to slow down the apparent motion, the 3-D shape of the microdevice during vibration can be observed in "real time" [21]. Figure 8.18a to c shows 3-D profiles of the 3 × 3 vibration mode of a membrane extracted from a sequence of profiles measured by this technique. When such measurements are performed with frequency scanning, they allow detection and investigation of vibration-mode coupling.

White light scanning stroboscopic interferometry [25] is the method to be chosen for vibration measurements on devices with rough surfaces or when the vibration amplitude is no longer negligible with respect to the depth of field of the interferometric objectives.

The most accurate measurements are obtained by phase-shifting stroboscopic interferometry. This technique is the most widely used [20,29–32] and is able to provide vibration amplitude maps with a resolution down to about 0.2 nm. An example of vibration-mode shape measurement by this latter technique is shown in Figure 8.19.

In-plane vibrations can as well be measured simultaneously by these stroboscopic techniques [31]. This is achieved by computing the fringe amplitude maps from the fringe patterns and processing them by optical flow techniques to get the relative displacement fields [22,31]. With a high-*NA* objective, a detection limit less than 0.2 nm can be reached on a textured or patterned

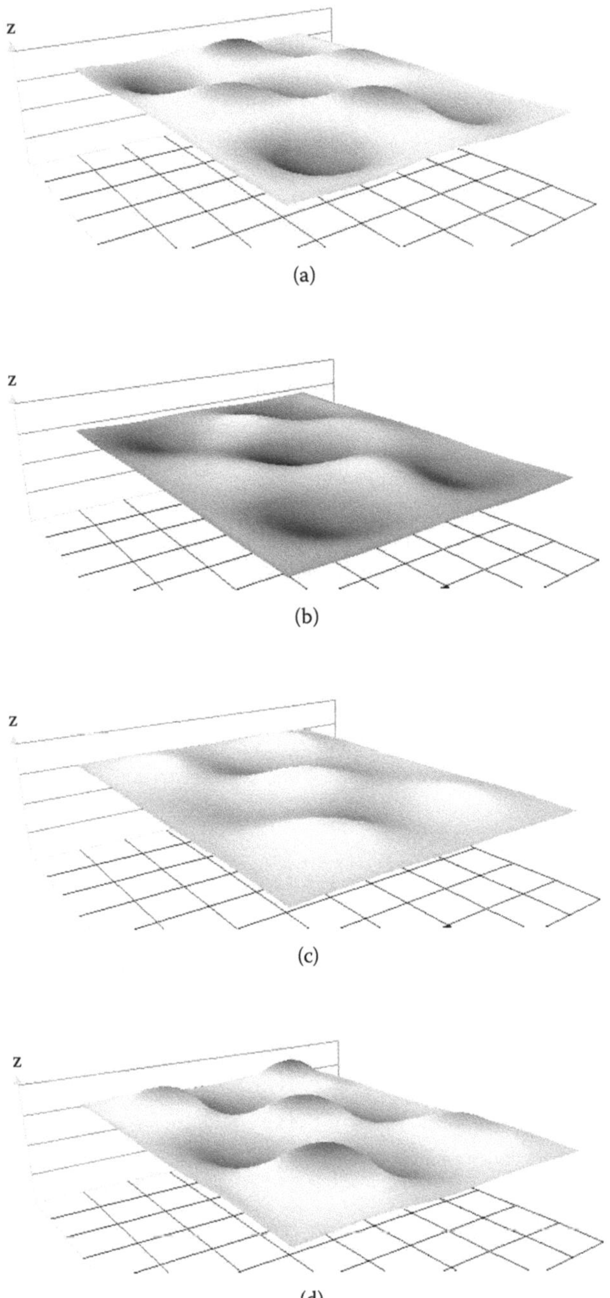

FIGURE 8.18 Time-resolved measurements by FFT stroboscopic interferometry of the 3×3 vibration-mode shape (83.21 kHz) of a rectangular gold membrane: (a) and (d): 3-D profiles at the extrema of the vibration cycle; (b), (c): 3-D profiles at intermediate times. Vibration amplitude: 108 nm. Membrane size: 1 mm × 1 mm.

surface [31]. Figure 8.20 gives an example of simultaneous in-plane and out-of-plane vibration-mode measurement by this technique in the case of a freestanding microring with a few micronwide central microbridge.

Stroboscopic interference microscopy can also be applied to repetitive transient response measurements by performing successive measurements with increasing light pulse delays. Another

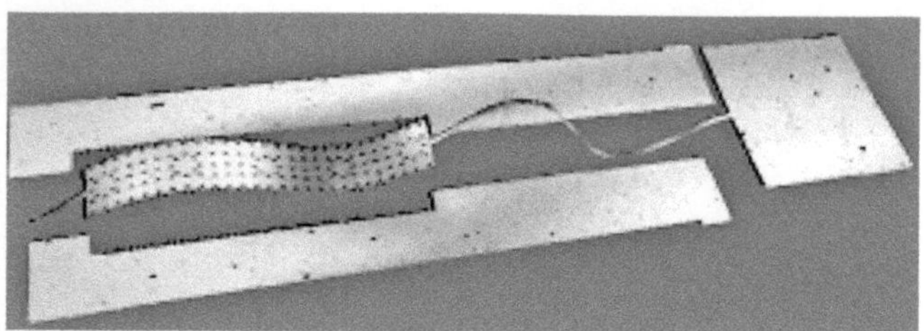

FIGURE 8.19 Measurement by phase-shifting stroboscopic interferometry of a high-order vibration mode at 507.8 kHz of an electrostatically driven test device fabricated in SOI technology. Sample fabrication: TRONIC'S Microsystems, Grenoble, France. (From Bosseboeuf, A. and Petitgrand, S., Characterization of the static and dynamic behavior of MOEMS by optical techniques: status and trends, *J. Micromech. Microeng.*, 13(4), S23, 2003. With permission.)

method is to shift the light pulse repetition rate with respect to the vibration frequency by an amount slightly larger than the video rate [27] and use single-interferogram processing techniques. This was applied to investigation of the transient response of cantilever microbeams actuated by a laser beam [27].

8.7.4 Vibration Measurements by Time-Averaged Interference Microscopy

With continuous illumination, the interference pattern recorded on a vibrating device is time-averaged by the camera. It is then described by the following equation [23]:

$$I_{averaged}(x,y,z) = I_0(x,y,z)\left[1+V(x,y,z_0)\,J_0\left(4\pi\frac{a}{\lambda_m}\right)\cos\left[4\pi z_0/\lambda_m+\Delta\varphi_{MS}(x,y)\right]\right] \quad (8.14)$$

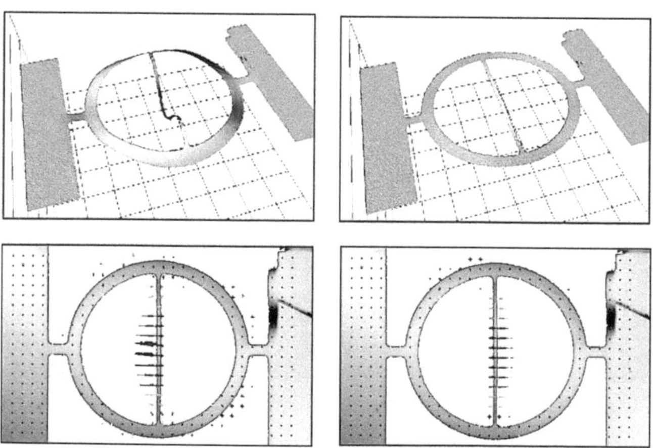

FIGURE 8.20 Simultaneous measurement of out-of-plane (top) and in-plane (bottom) vibrations at 1.365 MHz of an Al microring for a phase corresponding to the maximum (left) and minimum (right) of the out-of-plane vibration amplitude. (From Petitgrand, S. and Bosseboeuf, A., Simultaneous mapping of out-of-plane and in-plane vibrations of MEMS with (sub)nanometer resolution, *J. Micromech. Microeng.*, 14, S97, 2004. With permission.)

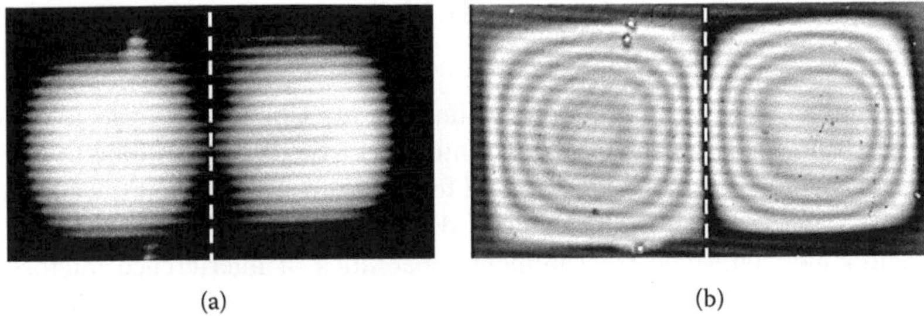

(a) (b)

FIGURE 8.21 Time-averaged interferograms recorded at 166.9 kHz (mode 21) on a 1200 μm × 600 μm × 1 μm Cr membrane: (a) maximum vibration amplitudes lower close to 110 nm, (b) maximum vibration amplitude close to 560 nm. The gray dashed line gives the location of the nodal line. Horizontal fringes are parasitic fringes.

The fringe pattern is thus similar to the static case (Equation 8.5) but with a fringe contrast in vibrating areas reduced by the factor $J_0(4\pi a/\lambda_m)$, where J_0 is the Bessel function of zero order (Figure 8.16). For a flat surface, the local mean position z_0 can be adjusted to maximize or minimize the intensity everywhere (i.e., $\cos(4\pi z_0/\lambda_m + \Delta\varphi_{MS}) = \pm 1$ for all x,y). Then, a visualization of the vibration mode is obtained. This is illustrated in Figure 8.21a in the case of vibration mode 21 of the flat rectangular Cr membrane [98]. For vibration amplitudes larger than about 100 nm, additional secondary dark "fringes" appear in places where the vibration amplitude corresponds to a zero of J_0 (Figure 8.21b). They can be used to estimate the fringe amplitude from the known values of zeros of the J_0 function (Table 8.1).

Time-averaged interference microscopy thus allows a visualization of vibration amplitude maps at video rate, whatever the vibration frequency. This unique feature is useful for a fast finding of the resonant frequencies and vibration mode of MEMS devices. Without interferogram processing, the detection limit is about 30 to 40 nm. Quantitative measurements can be performed by calibrating the fringe amplitude or by applying fringe pattern demodulation techniques such as FFT analysis [23] or phase shifting [29,32] to get the fringe amplitude maps in the static and dynamic cases. Their ratio provides a $J_0(4\pi a/\lambda)$ map. It can then be inverted to get the vibration amplitude map $a(x,y)$. This can be obtained unambiguously and automatically with a look-up table if the vibration amplitude is lower than $2.405\lambda/4\pi$, where 2.405 is the first zero of $J_0(x)$. A detection limit in the 2 to 5-nm range can then be reached. Above this limit, the resolution is ≤1 nm. This technique can as well be used for resonant peaks profiling in different areas simultaneously by using FFT processing and frequency scanning.

The main drawback of time-averaged interferometry is its sensitivity to mechanical drift and external disturbances. Fringe contrast variations related to sample drift can be avoided by controlling the sample position with a software closed loop, which maintains constant the intensity of some pixels of the interference phase in nonvibrating parts.

TABLE 8.1
Zeros of Bessel Function $J_0(x)$ and Vibration Amplitudes Corresponding to Bessel Fringes for λ = 600 nm

Zero Number	1	2	3	4	5	6	7	8	9
x	2.405	5.520	8.654	11.792	14.931	18.071	21.212	24.353	27.494
$\lambda x/4\pi$ nm (λ = 600 nm)	114.83	263.5	413.2	563.03	712.90	862.83	1012.8	1162.77	1312.74

8.7.5 Applications of Dynamic Interference Microscopy in the MEMS Field

The development of interference microscopy techniques for vibration analysis is relatively recent, but it had already been applied to various kind of microdevices [19–33]. Similar to other vibrometry techniques, it was mainly used in the MEMS field for thin-film stress and elastic constant evaluation from resonant frequencies of micromechanical devices such as cantilever microbeams, microbridges, or membranes. Full-field measurement capabilities of interference microscopy are very useful for vibration-mode profiling of more complex devices and for vibration-mode coupling studies.

In summary, interference microscopy is well suited for dynamic characterization of M(O)EMS. It allows point vibration spectra measurements with a low detection limit, but it is of interest chiefly because it allows fast and accurate mapping of vibration-mode phase and amplitude. In this respect, it is perfectly complementary with laser Doppler vibrometry, which allows fast vibration spectra measurements but remains slow for vibration-mode profiling. An increasing number of manufacturers propose stroboscopic interferometry as an option on their optical interferometric profilometers, and it is likely to become a standard technique. Some laboratories and manufacturers have begun as well to adapt their systems for measurements through the window port of a vacuum or environmental chamber [34,50]. This will allow new applications such as quality factor and resonant frequency measurements of M(O)EMS devices as function of pressure, fatigue and vibration-coupling investigations, as well as vibration measurements as a function of temperature.

8.8 CONCLUSION

Interference microscopy techniques allow both static and dynamic measurements of M(O)EMS with a sensitivity in the (sub)nanometer range and a lateral resolution in the (sub)micron range. They are particularly well suited for small-sized devices with patterns having a low surface roughness. They can be associated with other techniques such as laser Doppler vibrometry and speckle techniques to cover the whole range of characterization needs in the microsystem field, from material characterization up to the control of packaging. Actually, as illustrated in some chapters in this book, the trend is toward building measurement stations with a vacuum or an environmental chamber, which allow both electrical measurements and optical measurements by various techniques.

ACKNOWLEDGMENTS

The authors wish to thank Fogale Nanotech Co. in Nîmes (France) for providing information on their optical interferometric profilometer-vibrometers and the Centre National de la Recherche Scientifique (CNRS), the University Paris South (UPS), and the French Ministry of Research for their support. Most results presented were obtained on samples fabricated by colleagues and Ph.D. students of Institut d'Electronique Fondamentale and by co-workers from other laboratories and companies involved in the MEMS field. This chapter could not have been written without their contributions.

REFERENCES

1. Malacara, D., *Optical Shop Testing*, 2nd ed., Wiley-Interscience, New York, 1992.
2. Cloud, G., *Optical Methods of Engineering Analysis*, 1st ed., Cambridge University Press, New York, 1995.
3. Richards, O.W., One hundred years of microscopy in United States, *Centennial Yearbook of the New Microscopical Society*, 62, 1977.

4. Tolansky, S., New contributions to interferometry. Part V. New multiple beam white light interference fringes and their applications, *Phil. Mag.*, 7(36), 225, 1945.
5. Carré, P., Installation et utilisation du comparateur photoelectrique et interférentiel du bureau international des poids et des mesures, *Metrologia*, 2(13), 13, 1966.
6. Crane, R., Interference phase measurement, *Appl. Opt.*, 8, 538, 1969.
7. Bruning, J.H. et al., Digital wave front measuring interferometer for testing optical surfaces and lenses, *Appl. Opt.*, 13, 2693, 1974.
8. Wyant, J.C. et al., An optical profilometer for surface characterization of magnetic media, *ASLE Trans.*, 27(2), 101, 1984.
9. Davidson, M. et al., An application of interference microscopy to integrated circuit inspection and metrology, in *Integrated Circuit Metrology Inspection and Process Control*, Monahan, K.M., Ed., *Proc. SPIE*, 775, 233, 1987.
10. Strand, T.C. and Katzir, Y., Extended unambiguous range interferometry, *Appl. Opt.*, 26(19), 4274, 1987.
11. Lee, B.S. and Strand, T.C., Profilometry with a coherence scanning microscope, *Appl. Opt.*, 29, 3784, 1990.
12. Danielson, B.L. and Boisrobert, C.Y., Absolute optical ranging using low coherence interferometry, *Appl. Opt.*, 30, 2975, 1991.
13. Montgomery, P.C. and Fillard, J.P., Peak fringe scanning microscopy (PFSM) sub-micron 3D measurement of semiconductor components, in *Interferometry: Techniques and Analysis*, Brown, G.M., Kwon, O.Y., Kajuwinska, M., and Reid, G.T., Eds, *Proc. SPIE*, 1755, 12, 1992.
14. Chim, S. and Kino, D., Three dimensional image realization in interference microscopy, *Appl. Opt.*, 31(14), 2550, 1992.
15. Dresel, T., Haüsler, G., and Venzke, H., Three-dimensional sensing of rough surfaces by coherence radar, *Appl. Opt.*, 31(7), 919, 1992.
16. Caber, P.J., Interferometric profiler for rough surfaces, *Appl. Opt.*, 32(19), 3438, 1993.
17. Sandoz, P. and Tribillon, G., Profilometry by zero order interference fringe identification, *J. Mod. Opt.*, 40, 1691, 1993.
18. Deck, L. and de Groot, P., High speed non contact profiler based on scanning white light interferometry, *Appl. Opt.*, 33(31), 7334, 1994.
19. Bosseboeuf, A. et al., A versatile microscopic profilometer-vibrometer for static and dynamic characterization of micromechanical devices, in *Microsystems Metrology and Inspection, Proc. SPIE*, 3825, 123, 1999.
20. Hart, M. et al., Stroboscopic interferometer system for dynamic MEMS characterization, *IEEE J. MEMS*, 9(4), 409–418, 2000.
21. Petitgrand, S. et al., 3D measurement of micromechanical devices vibration mode shapes with a stroboscopic interferometric microscope, *Opt. Laser. Eng.*, 36(2), 77, 2001.
22. Freeman, D.M., Measuring motions of MEMS, *MRS Bulletin*, April 2001, 305, 2001.
23. Petitgrand, S. et al., Quantitative time-averaged microscopic interferometry for micromechanical device vibration mode characterization, in *Microsystem Engineering: Metrology and Inspection II*, Munich, June 2001, *Proc. SPIE,* 4400, 51, 2001.
24. Rembe, C. and Muller, R.S., Measurement system for full three dimensional motion characterization of MEMS, *IEEE J. MEMS*, 11(5), 479, 2002.
25. Bosseboeuf, A. et al., Three-dimensional full-field dynamical characterization of micromechanical devices by stroboscopic white light scanning interferometry in *Microsystem Engineering: Metrology and Inspection II*, Munich, Germany, *Proc. SPIE*, 4400, 36, 2001.
26. Graebner, J.E. et al., Dynamic visualisation of subangstrom high frequency surface vibrations, *Appl. Phys. Lett.*, 78(2), 159, 2001.
27. Petitgrand, S., Courbet, B., and Bosseboeuf, A., Characterization of static and dynamic optical actuation of Al microbeams by microscopic interferometry techniques, *J. Micromech. Microeng.*, 13, S113, 2003.
28. Bosseboeuf, A. and Petitgrand, S., Characterization of the static and dynamic behaviour of M(O)EMS by optical techniques: status and trends, *J. Micromech. Microeng.*, 13(4), S23, 2003.
29. Bosseboeuf, A. and Petitgrand, S., Application of microscopy interferometry techniques in the MEMS field, in *Microsystem Engineering: Metrology and Inspection III, Proc. SPIE*, 5145, 1, 2003.

30. Petitgrand, S. and Bosseboeuf A., Simultaneous mapping of phase and amplitude of MEMS vibrations by microscopic interferomery with stroboscopic illumination, in *Microscopic Engineering: Metrology and Inspection III, Proc. SPIE*, 5145, 33, 2003.
31. Petitgrand, S. and Bosseboeuf, A., Simultaneous mapping of out-of-plane and in-plane vibrations of MEMS with (sub)nanometer resolution, *J. Micromech. Microeng.*, 14, S97, 2004.
32. Salbut, L.A. et al., Active microelement testing by interferometry using time-average and quasi-stroboscopic techniques, in *Microsystem Engineering: Metrology and Inspection III, Proc SPIE*, 5145, 23–32, 2004.
33. Serio, B., Hunsinger, J.J., and Cretin, B., Stroboscopic illumination and synchronous imaging for the characterization of MEMS vibrations, *Optical Micro and Nanometrology in Manufacturing Technology*, Photonics Europe Workshop, Strasbourg, *Proc. SPIE*, 5458, 257, 2004.
34. Novak, E., Three-dimensional dynamic environmental MEMS characterization. in *Optical Micro and Nanometrology in Manufacturing Technology*, Photonics Europe Workshop, Strasbourg, *Proc. SPIE*, 5458, 1, 2004.
35. Montgomery, P.C. et al., Real time measurement of microscopic surface shape using high speed cameras with continuously scanning interference microscopy, in *Optical Micro and Nanometrology in Manufacturing Technology*, Photonics Europe Workshop, Strasbourg, *Proc. SPIE.*, 5458, 101, 2004.
36. Yahiaoui, R. et al., Automated interferometric system for bulge and blister test measurements of micromachined membranes, in *Microsystems Metrology and Inspection*, Munich, Germany, *Proc. SPIE.*, 4400, 160, 2001.
37. Wang, S.H. et al., Evaluation of microbeam deflection using interferometry, in *2nd Int. Conf on Experimental Mechanics*, *Proc. SPIE*, 4317, 262, 2001.
38. Jensen, B.D. et al., Interferometry of activated microcantilevers to determine material properties and test structure non idealities in MEMS, *IEEE J. MEMS*, 10(3), 336, 2001.
39. Kim, S.W. and Kim, G.H., Thickness-profile measurement of transparent thin-film layers by white-light scanning interferometry, *Appl. Opt.*, 38(28), 5968, 1999.
40. Abdulhalim, I., Spectroscopic interference microscopy technique for measurement of layer parameters, *Meas. Sci. Technol.*, 12, 1996, 2001.
41. Benhadou, D. et al., Buried interface characterization in optoelectronic materials by interference microscopy, *J. Modern Opt.*, 48(3), 553, 2001.
42. Lober, T.-A. et al., Characterization of the mechanisms producing bending moments in polysilicon micro-cantilever beams by interferometric deflection measurements, in *Proc. IEEE Solid State Sensor and Actuator Workshop*, Technical Digest Hilton Head Island, SC, USA, 92, 1988.
43. Boutry, M., Bosseboeuf, A., and Coffignal, G., Characterization of residual stress in metallic films on silicon with micromechanical devices, *Proc. SPIE*, 2879, 126, 1996.
44. Borenstein, J.T. et al., Characterization of membrane curvature in micromachined silicon accelerometers and gyroscopes using optical interferometry, *Proc. SPIE*, 2879, 116, 1997.
45. de Boer, M.P. et al., Role of interface properties on MEMS performances and reliability, in *Microsystems Metrology and Inspection*, Munich, Germany, *Proc SPIE*, 3825, 2, 1999.
46. Greek, S. and Chitica, N., Deflection of surface micromachined devices due to internal, homogeneous or gradient stresses, *Sensor. Actuator, A* 78, 1, 1999.
47. Brown, G.C. and Pryputniewicz, R.J., New test methodology for static and dynamic shape measurement of microelectromechanical systems, *Opt. Eng.*, 39, 127, 2000.
48. de Boer, M.P. et al., Integrated platform for testing MEMS mechanical properties at the wafer scale by the IMAP methodology, in *Mechanical Properties of Structural Films*, ASTM STP 1413, Muhlstein, C.L. and Brown, S.B., Eds., ASTM, West Conshohocken, PA, 2001.
49. O'Mahony, C. et al., Characterization of micromechanical structures using white-light interferometry, *Meas. Sci. Technol.*, 14, 1807, 2003.
50. Lafontan, X. et al., Physical and reliability issues in MEMS micro-relays with gold contacts, in *Reliability, Testing, and Characterization of MEMS/MOEMS*, Ramesham, T. Ed., *Proc. SPIE*, 4558, 11, 2001.
51. Hariharan, P., Interference microscopy, in *Optical Interferometry*, 2nd ed., Hariharan, P., Ed., Academic Press [Elsevier], Amsterdam, 143, 2003.
52. Creath, K. and Morales, A., Contact and non contact profilers, in *Optical Shop Testing*, 2nd ed., Malacara, D., Ed., John Wiley & Sons, New York, p. 687, chap. 17.

53. Corle, T.R. and Kino, G.S., *Confocal Scanning Optical Microscopy and Related Imaging Systems*, Academic Press, London, 1996.
54. Sheppard, C.J.R. and Roy, M., Low-coherence interference microscopy, in *Optical Imaging and Microscopy*, Török, P. and Kao, F.-J., Eds., Techniques and advanced systems, Springer-Verlag, Berlin, 257, 2003.
55. Erner, W. and Schwider, J., Ultraviolet interferometry with apochromatic reflection optics, *Appl. Opt.*, 38(16), 3516, 1999.
56. Montgomery, P. and Montaner, D., Deep submicron 3D surface metrology for 300 mm wafer characterization using UV coherence microscopy, *Microelectron. Eng.*, 45, 291, 1999.
57. Rao, Y.J. et al., Temperature dependence of resonant frequency in all-fibre optically addressed silicon microresonator sensors, *Electron. Lett.*, 27(11), 934, 1986.
58. Burns, D.M. and Bright V.M., Optical power induced damage in microelectromechanical mirrors, *Sensor. Actuator, A* 70(1–2), 6, 1998.
59. Wang, D.N. et al., The optimized wavelength combination of two broadband sources for white light interferometry, *J. Lightwave Technol.*, 12(5), 909, 1994.
60. Bourouina, T. et al., A new condenser microphone with a p+ silicon membrane, *Sensor. Actuator, A* 31, 149, 1992.
61. Robinson, D.W. and Reid, G.T., *Interferogram Analysis: Digital Fringe Pattern Measurement Techniques*, Institute of Physics Publishing, Bristol, 1993.
62. Dorrio, B.V. and Fernandez, J.L., Phase-evaluation methods in whole-field optical measurement techniques, *Meas. Sci. Technol.*, 10, R33, 1999.
63. Malacara, D., Servin M., and Malacarias, Z., *Interferogram Analysis for Optical Testing*, Marcel Dekker, New York, 1998.
64. Aboud-Zeid, A. and Weise, P., Interferometer with a wavelength tuned diode laser for surface profilometry, *Meas. Sci. Technol.*, 9, 105, 1998.
65. Sasaki, O., Murata, N., and Susuki, I., Sinusoidal wavelength-scanning interferometer with a superluminescent diode for step profile mesurement, *Appl. Opt.*, 39(25), 4489, 2000.
66. Hariharan, P., Phase-stepping interferometry with laser diodes: effect of changes in laser power with output wavelength, *Appl. Opt.*, 28(1), 27, 1989.
67. Kuwamura, S. and Yamaguchi, I., Wavelength scanning profilometry for real time surface shape measurement, *Appl. Opt.*, 36(19), 4473, 1997.
68. Dubois, A., Boccara, A.C., and M Lebec, M., Real-time reflectivity and topography imagery of depth resolved microscopic surfaces, *Opt. Lett.*, 24(5), 309, 1999.
69. Roy, M. et al., Geometric phase shifting for low-coherence interference microscopy, *Opt. Laser. Eng.*, 37, 631, 2002.
70. Schwider, J. and Zhou, Z., Dispersive interferometric profilometer, *Opt. Lett.*, 19, 995, 1994.
71. Calatroni, J., Sandoz, P., and Tribillon, G., Surface profiling by means of a double spectral modulation, *Appl. Opt.*, 32(1), 30, 1993.
72. Sandoz, P., Tribillon, G., and Perrin H., High resolution profilometry by using phase calculation algorithms for spectroscopic analysis of white light interferograms, *J. Mod. Opt.*, 43(4), 701, 1996.
73. Hart, H., Vass, D.G., and Begbie, M.L., Fast surface profiling by spectral analysis of white light interferograms with Fourier transform spectroscopy, *Appl. Opt.*, 37(10), 1764, 1998.
74. Born, M. and Wolf, E., *Principles of Optics*, 7th ed., Cambridge University Press, Cambridge, 1999.
75. Pförtner, A. and Schwider, J., Dispersion error in white-light Linnik interferometers and its implications for evaluation procedures, *Appl. Opt.*, 40(34), 6223, 2001.
76. de Groot, P. et al., Determination of fringe order in white light interference microscopy, *Appl. Opt.*, 41(22), 4571, 2002.
77. Doi, T., Toyoda, K., and Tanimura, Y., Effects of phase changes on reflection and their wavelength dependence in optical profilometry, *Appl. Opt.*, 36(28), 7157, 1997.
78. Harasaki, A., Schmit J., and Wyant, J.C., Offset coherence envelope position due to phase change on reflection, *Appl. Opt.*, 40(13), 2102, 2001.
79. Creath, K., Calibration of numerical aperture effects in inteferometric microscope objectives, *Appl. Opt.*, 28, 3333, 1989.
80. Sheppard C.J.R. and Larkin, K.G., Effect of numerical aperture on interfernce fringe spacing, *Appl. Opt.*, 34, 4731, 1995.

81. Dubois, A. et al., Phase measurements with wide aperture interferometers, *Appl. Opt.*, 39, 2326, 2000.
82. de Groot, P.J. and Colonna de Vega, X., Signal modelling for modern interference microscopes, in *Optical Metrology in Production Engineering*, Photonics Europe Workshop, Strasbourg, France, *Proc. SPIE*, 5457, 26–34, 2004.
83. Abdulhalim, I., Theory for double beam interference microscopes with coherence effects and verification using the Linnik microscope, *J. Mod. Opt.*, 48(2), 279, 2001.
84. Larkin, K.G., Efficient non linear algorithm for envelope detection in white light interferometry, *J. Opt. Soc. Am. A*, 13(4), 832, 1996.
85. de Groot, P. and Deck, L., Surface profiling by analysis of white light interferograms in the spatial frequency domain, *J. Mod. Opt.*, 42(2), 389, 1995.
86. Harasaki, A. and Wyant, J.C., Fringe modulation skewing effect in white light vertical scanning interferometry, *Appl. Opt.*, 39(13), 2101, 2000.
87. Montgomery, P.C. and Montaner, D., Lateral error reduction in the 3D characterization of deep MOEMS devices using white light interference microscopy, *Optical Micro- and Nanometrology in Manufacturing Technology Photonics*, Europe Workshop, Strasbourg, April 29–30, 2004, *Proc. SPIE*, 5458, 34–42, 2004.
88. Dubois, A., Effects of phase change on reflection in phase-measuring interference microscopy, *Appl. Opt.*, 43(7), 1503–1507, 2004.
89. Tsai, C.-T., Lee C.-H., and Wang, J., Deconvolution of local surface response from topography in nanometer profilometery with a dual-scan method, *Opt. Lett.*, 24(23), 1732, 1999.
90. Rogala, E.W. and Barret, H.H., Phase shifting interferometer/ellipsometer capable of measuring the complex index of refraction and the surface profile of a test surface, *J. Opt. Soc. Am. A*, 15(2), 538, 1998.
91. Feke, G.D. et al., Interferometric back focal plane ellipsometry, *Appl. Opt.*, 37(10), 1796, 1998.
92. Palik, E.D., *Handbook of Optical Constants of Solids*, Academic Press, New York, 1997.
93. Optical constants database of SOPRA spectroscopic ellipsometers, SOPRA Company, France, www.sopra-sa.com.
94. Creath, K. and Wyant, J.C., Absolute measurement of surface roughness, *Appl. Opt.*, 29(26), 3823, 1990.
95. Scruby, C.B. and Drain, L.E., *Laser Ultrasonics, Techniques and Applications*, Adam Hilger, Bristol, 1990.
96. White, R.G. and Emmony, D.C., Active feedback stabilisation of a Michelson interferometer using a flexure element, *J. Phys. E. Sci. Instrum.*, 18, 658, 1985.
97. Li, J.-F., Moses, P., and Viehland, D., Simple, high resolution interferometer for the measurement of frequency-dependent complex piezoelectric responses in ferroelectric ceramics, *Rev. Sci. Instrum.*, 66(1), 215, 1995.
98. Yahiaoui, R., Etude par microvibrométrie de films minces et de dispositifs micromécaniques; Ph.D. thesis, Université Paris XI, 2002.

9 Measuring MEMS in Motion by Laser Doppler Vibrometry

Christian Rembe, Georg Siegmund, Heinrich Steger, and Michael Wörtge

CONTENTS

9.1 Introduction.....246
9.2 Laser Doppler Effect and Its Interferometric Detection.....247
9.2.1 Laser Doppler Effect.....247
9.2.2 Shot Noise in Detection of Light.....249
9.2.3 Interferometric Detection.....250
9.2.4 Wavefront Aberrations and Laser Speckle.....250
9.3 Techniques of Laser Doppler Vibrometry.....251
9.3.1 Optical Arrangements.....251
9.3.2 Homodyne and Heterodyne Detection Techniques.....252
9.3.3 Signal Processing.....254
9.3.3.1 Fundamental Relationships of Doppler Modulation.....254
9.3.3.2 Analog Decoding Techniques.....255
9.3.3.3 Digital Demodulation by the Arctangent Phase Method.....256
9.3.4 Data Acquisition.....259
9.4 Full-Field Vibrometry.....262
9.4.1 Scanning Vibrometer.....262
9.4.2 Operating Deflection Shapes.....263
9.5 Measuring on Microscopic Structures.....263
9.5.1 Optical Arrangements.....263
9.5.2 3-D Techniques.....266
9.5.3 Ranges and Limits.....268
9.6 Resolution and Accuracy.....270
9.6.1 Noise-Limited Resolution.....270
9.6.2 Measurement Accuracy and Calibration of Laser Doppler Vibrometers.....274
9.6.2.1 General Aspects.....274
9.6.2.2 Mechanical Comparison Calibration.....275
9.6.2.3 Calibration by Means of Synthetic Doppler Signals.....276
9.6.2.4 Calibration Based on Bessel Functions.....277
9.6.2.5 Influences on Measurement Accuracy.....278
9.7 Combination with Other Techniques.....279
9.8 Examples.....282
9.8.1 Dual Mode MEMS Mirror.....282
9.8.2 Cantilever Beam Acceleration Sensor.....283

9.8.2.1 Out-of-Plane Analysis 284
9.8.2.2 In-Plane Analysis 287
9.9 Conclusion and Outlook 289
References 290

9.1 INTRODUCTION

The tremendous progress in production techniques for microelectromechanical systems (MEMS) has caused the development of reliable MEMS products in many different application fields [1–7]. Therefore, the better performance of a MEMS product compared to its competitors is nowadays the crucial factor of a commercial success, whereas in the past, the pure ability to build the device was the target of development efforts. The assurance that the product is reliable and has high performance has top priority in the design and development process and, in addition, these requirements also have to be controlled during the manufacturing process.

Optical measurement methods have gained great importance because the testing of MEMS should be nondestructive and needs to have a high spatial resolution. Many MEMS have movable mechanical structures and, therefore, the measurement should provide an analysis of the dynamics with a high resolution of motion amplitudes. In addition, nonlinear or nonrepeatable transient motions have to be detected in many cases. The measurement has to be fast if a high-volume technique is required. All these requirements can be covered with laser Doppler technique.

The laser Doppler technique was introduced in 1964 by Yeh and Cummins [8] to measure flow fields. The authors demonstrated that the frequency shift of laser light scattered at moving microscopic monodispersed polystyrene spheres can be used to detect local velocities of the water flow field. These frequency shifts are generally called Doppler shifts, named after the Austrian physicist who first proposed the phenomenon in 1842. Measurements of flow in gases [9] were demonstrated just one year after the work of Yeh and Cummins. The first motion detection of a diffuse surface using the laser Doppler technique was demonstrated by Kroeger [10], also in 1965. However, the detection of flow became known as Doppler anemometry and remained the most important application for the laser Doppler technique for the next several years. Initial papers [11,12] in the 1960s showed the detection of vibrations with homodyne interferometric techniques but were limited to small vibration amplitudes (<25 nm) or could only be used to measure a purely sinusoidal motion. This changed in the year 1970 when Eberhardt and Andrews first presented a heterodyne laser Doppler system for measurement and analysis of vibrating surfaces [13] and, thus, the laser Doppler vibrometer was born. Soon the employment of the laser Doppler technique on vibrating surfaces became known as laser Doppler vibrometry and inventions continued during the 1980s with important developments such as the fiber-based vibrometer [14] or the torsional-vibrations-measuring vibrometer [15]. Today, laser Doppler vibrometers capture applications in all kinds of vibration analysis [16–20]. Several texts and books give a detailed representation of the laser Doppler technique [21]. In this chapter, we concentrate on the methods and physical properties of laser Doppler vibrometry that are important for the characterization of dynamics in MEMS.

The Doppler frequency shifts are rather small; for example, a velocity of 10 m/sec will shift the red laser light of a helium–neon laser by about 32 MHz. This corresponds just to seven parts in 10^8 in a light frequency of 4.7×10^{14} Hz. These shifts cannot be measured directly with optical spectrometers and, therefore, the principle of heterodyning or beating two frequencies in a device having a nonlinear response is usually employed. The nonlinearity is easily realized with a photodiode, which transforms optical power into current because the oscillating electrical field contains the frequency information. The output of the detector contains a component of the difference frequency that corresponds to the heterodyne beat frequency. The instantaneous velocity signal is finally obtained from the heterodyne signal by using a frequency voltage converter. Obviously, it is also possible to demodulate the phase of the heterodyne signal instead of the frequency to acquire the instantaneous displacement information.

The major advantage of laser Doppler vibrometry is the availability of a real-time velocity or displacement signal. Together with a spectrum analyzer or an FFT analyzer, an instantaneous spectrum of the vibration can be measured. This technique allows oscillation measurements with an extremely high-frequency resolution and, therefore, enables one to measure with an extreme resolution down to oscillation amplitudes of less than 1 pm. Laser Doppler vibrometry has been proven to be superior if an instantaneous vibration spectrum has to be measured without contact with the surface. Laser Doppler vibrometry is particularly significant if the laser beam is scanned over a surface [22] to obtain the geometric distribution of vibrational shapes, the so-called operational deflection shapes (ODS).

From the beginning, laser Doppler vibrometry was especially interesting for applications in which a vibration measurement could not be obtained with an accelerometer or any other surface-contacting sensor. Typical specimens, where it is impossible or impractical to measure tactilely, are hot or rotating and, as is the case of MEMS, tiny and lightweight. Therefore, it is not surprising that laser Doppler vibrometry has gained enormous importance in today's characterization of MEMS dynamics. In the mid-1990s, several research groups started using single-point vibrometer measurements to characterize motions in micrometer-sized mechanical structures [23,24]. In the late 1990s, the first scanning vibrometer measurements on MEMS were presented [25,26], which allowed the determination of operational deflection shapes. The ODS give an understanding of excited mechanical modes in the investigated structures. In 2000, a scanning vibrometer with a microscopic lateral resolution became commercially available for the first time by Polytec [27] and made vibrometry a widespread technique for characterization of MEMS dynamics.

The laser Doppler technique can be used to measure the Doppler shift of the light scattered in the direction of the light detector. Therefore, a complete 3-D vibration measurement can be performed on macroscopic rough structures by impinging three declined vibrometer laser beams. But MEMS surfaces are usually optically flat, which means that the peak-to-peak surface-height deviation in the laser spot on the surface is much smaller than the optical wavelength. Therefore, motion can only be detected in the direction of the optical axis of the microscope optics. Hence, vibrometry needs to be combined with other techniques to obtain a full 3-D motion analysis [28].

9.2 LASER DOPPLER EFFECT AND ITS INTERFEROMETRIC DETECTION

9.2.1 Laser Doppler Effect

An interferometer illuminates a specimen with a coherent light source having a frequency ω and measures the phase of the back-reflected or the back-scattered light. The light source and the detector are usually assembled in the vibrometer measurement head and, therefore, are located in one observer's frame while the object is moving.

A vibrometer measures the motion of an object in the direction of the light ray, because the beam paths of the illumination and the detection ray have the same optical axis and polarization. Therefore, the measurement can be considered to be a one-dimensional problem and the electromagnetic wave equation [29–31] simplifies to

$$\frac{\partial^2 E(x,t)}{\partial x^2} = \frac{1}{c_m^2(x,t)} \frac{\partial E(x,t)}{\partial t} \quad \text{with } E(x=0,t) = E_o \cos(\omega t) \tag{9.1}$$

where the position of the light source is at $x = 0$. Here, $E(x,t)$ is the norm of the linear polarized electrical field vector, $c_m(x,t) = c/n(x,t)$ is the speed of light in the medium, $n(x,t)$ is the position

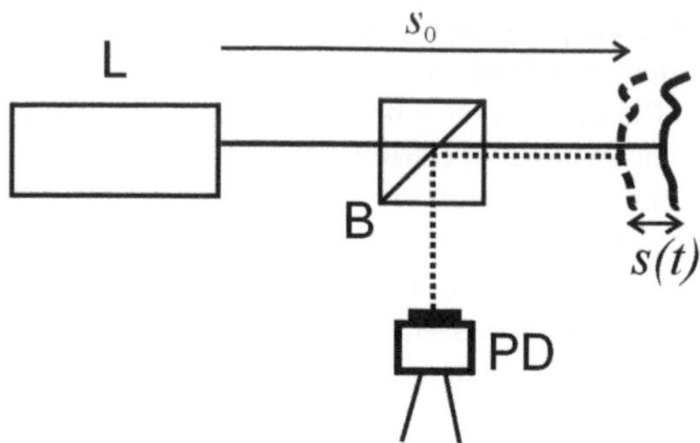

FIGURE 9.1 A displacement s produces a change of optical path length of $2s$ due to reflection. B is a beam splitter and L a laser.

and time-dependent refraction index, and c is the speed of light in vacuum. Equation 9.1 can be used to compute the electrical field $E(s(t),t)$ at the position $s_s(t)$ at the specimen. The electrical field at the fixed detector's position s_d follows from

$$\frac{\partial^2 E}{\partial x^2} = \frac{1}{c_m^2(x,t)} \frac{\partial E}{\partial t} \quad \text{and} \quad E(s(t),t). \tag{9.2}$$

A general analytic solution can be found for homogeneous media (no variation of n in space) in which the refraction index changes slowly enough so that the time dependence of n can be neglected for the duration t_m the light needs to travel from the source to the detector. Therefore, in the following, n is assumed to be a constant. Then, Equation 9.1 has a general solution $E(x,t) = f(c_m t - x) + g(c_m t + x)$ and the electrical field E_s at the specimen can be expressed as

$$E_s\left(s_s(t),t\right) = E_0 \cos\left(\omega t - k_m s_s(t)\right), \tag{9.3}$$

with $k_m = \omega/c_m = 2\pi/\lambda_m$ the wavenumber and λ_m the wavelength in the medium. Assuming a motion with $s_s(t) = s_0 + s(t)$ and $|s_0| >> |s(t)|_{\max}$ as well as $|s(t)|_{\max}/c_m << t_m$, the electrical field at the detector is approximately

$$E_m\left(s_d,t\right) = \sqrt{r}E_0 \cos\left(\omega t - 2k_m s_s\left(t - \frac{s_0}{c_m}\right) + k_m s_d\right), \tag{9.4}$$

with s_d the detector position, $\sqrt{r}E_0$ the electrical field amplitude of the reflected light, and r the reflectivity of the object. A movement of the specimen results in a total path change of two times the specimen's displacement s (see Figure 9.1).

Equation 9.4 can be simplified if $|s_s(t) - s_s(t - s_0/c_m)| << s_s(t)$ can be assumed and the approximation $s_s(t - s_0/c_m) \approx s_s(t) \approx s(t) + s_0$ is used with s_o the average distance of the laser to the specimen. Then, the electrical field at the detector can be written as

$$E_m\left(s_d,t\right) = \sqrt{r}E_0 \cos\left(\omega t - \varphi(t)\right) = \sqrt{r}E_0 \cos\left(\omega t - 2k_m s(t) - 2k_m s_0 + k_m s_d\right). \tag{9.5}$$

It should be mentioned that Equation 9.5 is an approximation in the order of v/c_m with v the instantaneous velocity of the specimen. The Doppler shift $d\varphi/dt$ can be derived using

$$\frac{d\varphi(t)}{dt} = \frac{d\left(kn\left(2s(t) + 2s_0 - s_d\right)\right)}{dt} = k\frac{d\,OPL(t)}{dt}, \tag{9.6}$$

with $k = \omega/c$ the wavenumber of the coherent light in vacuum and OPL the optical path length. Therefore, the Doppler shift is generated through an optical path length variation. The Doppler shift in a media with constant refraction index (e.g., air with $n \approx 1$) results in

$$\varphi(t) = \frac{d\varphi(t)}{dt} = 2kn\frac{ds(t)}{dt} = 2knv(t) \tag{9.7}$$

and, therefore, is proportional to the instantaneous object's velocity $v(t)$.

9.2.2 Shot Noise in Detection of Light

The photocurrent i_P resulting from the illumination by a constant-wave laser beam is not perfectly steady. The theoretical limit of the noise of photodetectors is given by the quantum character of light and current. The photocurrent can be described as Poisson process [21]

$$P_\tau(n) = e^{-\alpha_e \tau}\frac{(\alpha_e \tau)^n}{n!}, \tag{9.8}$$

with n photons in a time interval τ if $\alpha_e dt$ is the probability for the emission of an electron in an infinitesimal time interval dt. It is a Poisson process because the number n is high and each detected photon has a random detection time and gives rise to current pulses of approximately the same size. α_e can be written in terms of quantum efficiency η, light power P, and Planck's constant $h = 2\pi$ as follows:

$$\alpha_e = \frac{\eta P}{\omega} = \frac{KP}{q}. \tag{9.9}$$

K is the conversion parameter

$$K = \frac{\eta q}{\omega} \tag{9.10}$$

which describes the linear correlation of photo current and optical power $i_P = KP$. q is the charge of an electron. The Poisson process with variance

$$\sigma_n^2 = \overline{n^2} - (\overline{n})^2 = \alpha_e \tau \tag{9.11}$$

results in fluctuations of the photocurrent, which are called shot noise. The RMS (root mean square) shot noise current i_{Sh} in a bandwidth B results from Equation 9.9, Equation 9.10, and Equation 9.11 and is given by [32]

$$\overline{i_{Sh}^2} = \frac{i_P^2}{\alpha_e \tau} = \frac{q i_P}{KP}\frac{1}{\tau} i_P = q\,2B\,i_P. \tag{9.12}$$

9.2.3 Interferometric Detection

The electrical field of the measurement beam is phase-modulated through the motion of the specimen as it is expressed by Equation 9.5. The demodulation of the phase $\varphi(t)$ or its derivative $\varphi(t)$ is the task of the vibrometer optics and electronics. As optical frequencies in the range of 10^{15} Hz cannot be detected directly with a photodiode, the measurement frequency ω_1 has to be mixed down by optical beating of the measurement beam with a reference beam obtained from a coherent light source with frequency ω_2 and a fixed phase relation. Usually, the reference beam is obtained with a beam splitter from the same laser as the measurement beam and an acoustooptical frequency shifter is used to receive a frequency carrier. But, in general, it is also conceivable to obtain the reference beam from a second laser mode if a stabilized He-Ne laser is used or a second phase-locked laser source. The total field at the detector can be expressed by the sum of measurement and reference beams

$$E_d(t) = E_m \cos(\omega_1 t - \varphi(t)) + E_r \cos(\omega_2 t - \varphi_0), \tag{9.13}$$

where E_r is the amplitude of the reference beam. The current i_D of the photodiode is proportional to the power of the electrical field and, therefore, i_D is proportional to the square of the electrical field $E_d(t)$. Taking into account that the photodiode cannot be fast enough to detect frequencies of light, the remaining detector signal is

$$\begin{aligned} i_P &= \hat{K}\left[\frac{E_m^2}{2} + \frac{E_r^2}{2} + E_m E_r \cos(\omega_1 t - \omega_2 t - \varphi(t) + \varphi_0)\right] \\ &= K\left[P_m + P_r + 2\sqrt{P_m P_r}\cos(\omega_1 t - \omega_2 t - \varphi(t) + \varphi_0)\right] \end{aligned} \tag{9.14}$$

with the linear amplification $\hat{K}$, and the optical powers P_m and P_r of the reference and measurement beams, respectively. K is the already-introduced conversion parameter.

9.2.4 Wave-front Aberrations and Laser Speckle

During the derivation of Equation 9.14, we assumed that there is no spatial distribution of the phase of the electrical field vector on the specimen as well as on the detector. In reality, an angle between measurement and reference beam, wavefront aberrations of the optics, and laser speckle effects on rough specimen surfaces lead to a spatial distribution of the phase $\varphi_0 = \varphi_0(x, y)$ on the detector surface. As a result, parts of the electrical field are canceled by destructive interference of the optical field if the spot on the detector is diffraction limited or by interference of the spatially distributed currents in the detector. This is demonstrated for the case that reference and measurement waves inclined at an angle β impinge onto the detector. The spacing s_f between the interference fringes on the detector follows to

$$s_f = \frac{\lambda}{2\sin \beta/2}. \tag{9.15}$$

Only a fraction of the light power can be converted into electrical power by the detector if the detector aperture is larger than the fringe spacing s_f. The same argument is valid for wavefront aberrations that are more complex than just tilt between measurement and reference beams. If the measurement beam is scattered diffusely on a rough specimen surface, the electrical field at the detector results from the beating of a large number of sources of the same frequency but with a random phase relation. Vibrometers have usually the same effective numerical aperture NA_{eff} of the Gaussian beam for illumination and detection. The typical speckle size on the specimen is approximately $0.64\,\lambda/NA_{eff}$, which corresponds to the laser-spot diameter. The important result is that the phase φ_0 is different for every measurement point on the specimen surface if the surface is optically rough (height difference in the illuminated spot on the specimen surface is larger than half the wavelength). Assuming the phase φ_0 is the result of many contributions with a random phase relation, the vector diagram representing the sum of these contributions is a "random walk" in two dimensions. For a sufficiently large number of contributions, the probability of an intensity at the detector $P(I)$ between I and $(I + dI)$ is [21]

$$P(I)dI = e^{-I/I_0}\frac{dI}{I_0} \tag{9.16}$$

for a mean intensity I_0. There is a possibility that power of the signal is below the noise level if the intensity is too low due to a "bad speckle." In this case, a demodulation of the Doppler frequency shift is not possible. This phenomenon is known as *drop-out*.

9.3 TECHNIQUES OF LASER DOPPLER VIBROMETRY

9.3.1 Optical Arrangements

Phase noise of the laser leads to additional noise in the demodulated Doppler signal. The larger the difference between the reference and the measurement beams, the higher the noise. It is essential to have a coherent source with a low phase noise if measurements are made over a large distance. The phase noise corresponds to the line width of the light source and, therefore, to the coherence length. This requirement makes the laser the preferred source. As a nonstabilized He–Ne has at least two laser modes [33] within a frequency distance $\delta v = c/2d$, one can find a position of maximum coherence if the path length difference between measurement and reference beam is jd, $j = 1,2,3, \ldots$. Here, d denotes the cavity length of the laser. This effect is not due to the phase noise but to the beat of the two laser modes and can be understood considering that every operating laser mode produces a photocurrent with respect to Equation 9.14. The region of low coherence is defined as the object distance in which the beat of the two current signals decreases the total signal power about 20 dB. The region of low coherence has a maximum width if both laser modes operate with equal power. In this worse case, the region is ±7.5 mm around the position of the coherence minimum in which both mode signals are 180° out-of-phase. In practice, it is not easy to find these positions for a warmed-up laser due to the high sensitivity and dynamics of a laser Doppler vibrometer. The positions of low coherence can be easily determined if the system warms up. The narrow width of the single modes still makes the He–Ne laser the first choice as a light source for laser vibrometry. However, MEMS specimens are usually placed under a microscope within a well-defined distance from the laser source. Therefore, it is possible to match the reference- path length with the length of the measurement path, and a laser with a lower coherence length (e.g., a laser diode) can be used without increasing the phase noise.

The optical arrangement needs to provide the optical beat of reference and measurement beams with respect to Equation 9.14. This requirement can be realized, for example, with two well-known interferometer arrangements: the Michelson interferometer and the Mach–Zehnder interferometer [29] (see Figure 9.2).

In the Michelson interferometer, the beam coming from the laser is divided with a beam splitter into reference and measurement beams. Usually, a polarizing beam splitter is used to parcel the light efficiently to the photodetector. Maximum light can be parceled to the detector if two quarter-wave plates are placed in the reference and measurement beams, respectively.

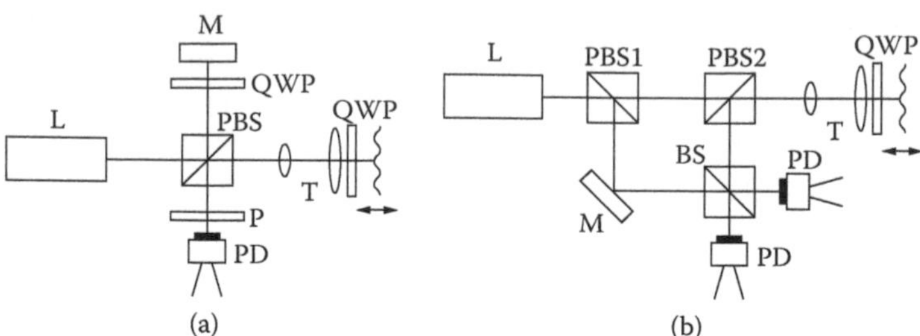

FIGURE 9.2 Optical arrangements for the classical interferometers of (a) Michelson and (b) Mach–Zehnder type. M is a mirror, BS is a beam splitter, PBS is a polarizing beam splitter, L is a laser, P is a polarizer, QWP is a quarter-wave plate, PD is a photodetector, and T is a telescope.

The quarter-wave plate rotates the beam reflected from the reference mirror or the specimen about 90° before it impinges the beam splitter a second time. The laser beams parceled to the detector are perpendicularly polarized. Therefore, a polarizer is necessary to obtain an optical beat on the detector.

The laser beam in a Mach–Zehnder interferometer is usually also divided with a polarizing beam splitter (PBS1 in Figure 9.2b). The second polarizing beam splitter (PBS2) parcels the beam to the specimen. The quarter-wave plate rotates the polarization of back-reflected light about 90°, and then the beam splitter PBS2 parcels it to the detector. Here, reference and measurement beams have the same polarization, but the beams are perpendicular. Therefore, a nonpolarizing beams splitter (BS in Figure 9.2b) is necessary to combine the two beams.

Two detectors are necessary for both interferometer types to use the full power of reference and measurement beams. The polarizer P of the Michelson interferometer in Figure 9.2a has to be a polarizing beam splitter inclined at about 45° to the two polarization directions. Then two interference signals can be detected. The second beat signal is automatically obtained with the beam splitter BS in Figure 9.2b when a Mach–Zehnder interferometer is used. As the total power of the laser beams is conserved, it is obvious that the beat signals at the two detectors PD are 180° out of phase. Therefore, the detector signals can be subtracted to receive twice the signal power. The DC components in Equation 9.14 are removed through this subtraction, which is a welcome side effect.

9.3.2 Homodyne and Heterodyne Detection Techniques

If a single laser source and an optical arrangement as shown in Figure 9.2a and Figure 9.2b is used, the light frequencies ω_1 and ω_2 are the same (i.e., $\omega_1 = \omega_2$). In this case, the detector signal of the optical beat (Equation 9.20) cannot be used directly to obtain information about the direction of the specimen motion. The motion direction can be extracted if two interference signals with a 90° phase shift are available as an I&Q signal (in-phase and quadrature).

$$i_I = I_I \cos\big(\varphi(t) + \varphi_0\big), \quad i_Q = I_Q \sin\big(\varphi(t) + \varphi_0\big) \tag{9.17}$$

The phase with sign can be computed, for example, with an arctangent operation. Homodyne vibrometers acquire the I&Q signal with a variation of the optics [34]. The variation for a Michelson interferometer is shown in Figure 9.3.

A nonpolarizing beam splitter BS is placed behind the polarizing beam splitter PBS to split the light independently from its polarization. In addition, the quarter-wave plate QWP3 in one beam behind the BS is used to delay the measurement beam with respect to the reference beam by about $\lambda/4$.

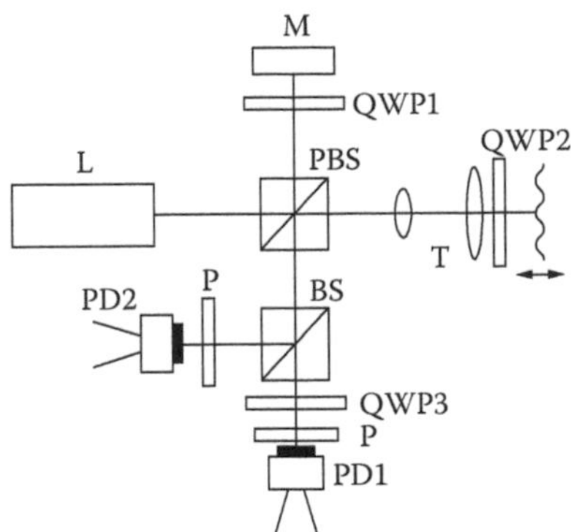

FIGURE 9.3 Optical arrangement of a homodyne vibrometer. M is a mirror, BS is a beam splitter, PBS is a polarizing beam splitter, L is a laser, P is a polarizer, QWP is a quarter-wave plate, PD is a photodetector, and T is a telescope.

Therefore, the slow and fast axes of the quarter-wave plate QWP3 are parallel to the polarizations of reference and measurement beams. The two detectors PD1 and PD2 behind the polarizers P measure the two I&Q signals in Equation 9.17.

Homodyne vibrometers suffer from some disadvantages [35]; e.g., high sensitivity against photoreceiver nonlinearity, susceptibility to electrical hum and noise, the 90° phase shift depends critically on the optical alignment. Therefore, the arrangement shown in Figure 9.4 is more widely used and is employed for the heterodyne vibrometer technique [36]. An acoustooptical modulator (Bragg cell) shifts either the reference or the measurement beam by about ω_c. Therefore, the light frequencies ω_1 and ω_2 follow the relation $|\omega_c| = |\omega_1 - \omega_2|$. Equation 9.14 can be rewritten if the Bragg cell is assembled in the optical arrangement as shown in Figure 9.4 to

$$i_S = K\left[P_m + P_r + 2\kappa\sqrt{P_m P_r}\cos\left(\omega_c t - \varphi(t) + \varphi_0\right)\right], \tag{9.18}$$

with the heterodyning efficiency $0 < \kappa < 1$ [37], which takes into account any degradation of the photocurrent due to optical distortions and misalignments of the wavefronts of the two beams. These kinds of effects are described in Subsection 9.2.4.

Equation 9.18 demonstrates that ω_c is a carrier frequency for the phase modulation generated through the optical path length alteration, which results from the specimen motion.

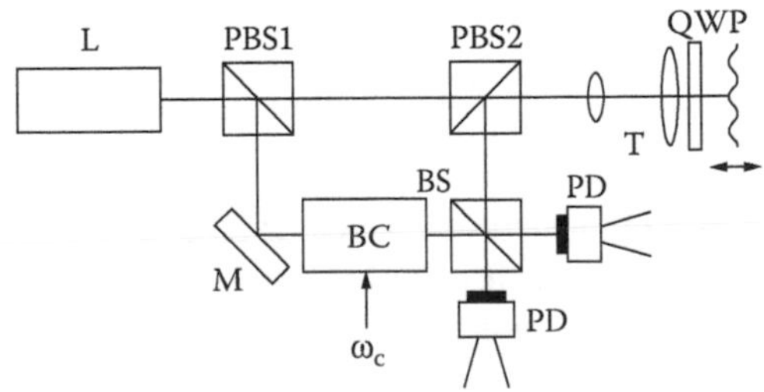

FIGURE 9.4 Optical arrangement of a heterodyne vibrometer. M is a mirror, BS is a beam splitter, PBS is a polarizing beam splitter, L is a laser, P is a polarizer, QWP is a quarter-wave plate, PD is a photodetector, BC is a Bragg cell, and T is a telescope.

9.3.3 Signal Processing

9.3.3.1 Fundamental Relationships of Doppler Modulation

The third term of Equation 9.18 represents the heterodyne Doppler signal. It is a high-frequency carrier with center frequency $f_c = \omega_c / 2\pi$, modulated phase angle $\varphi(t)$, and starting phase φ_0. This signal can carry both direction-sensitive frequency and phase modulation information resulting from target motion. When the target is moving, displacement $s(t)$ results in a phase modulation and Equation 9.7 can be rewritten as

$$\varphi(t) = \frac{4\pi \cdot s(t)}{\lambda}. \tag{9.19}$$

In case of a harmonic vibration at frequency f_{vib}, the modulated phase angle is

$$\varphi(t) = \frac{4\pi}{\lambda}\hat{s} \cdot \sin\left(2\pi f_{vib} t + \phi_s\right) \tag{9.20}$$

where $\Delta\varphi = \frac{4\pi}{\lambda}\hat{s}$ is usually called the phase deviation of the modulated carrier.

A phase modulation generates a frequency modulation at the same time. Due to the fundamental relationships $\omega(t) = d\varphi(t)/dt$ and $v(t) = ds(t)/dt$, we obtain for the corresponding frequency shift with respect to the center frequency f_c

$$\frac{\Delta\omega(t)}{2\pi} = \frac{2v(t)}{\lambda}, \tag{9.21}$$

which is commonly known as the Doppler frequency shift Δf_D.

For a harmonic vibration at frequency f_{vib}, we find again

$$\Delta f_D(t) = \frac{2\hat{v}}{\lambda}\cos\left(2\pi f_{vib} t + \phi_s\right) \tag{9.22}$$

where $\Delta f = \frac{2\hat{v}}{\lambda}$ is called the frequency deviation of the modulated carrier.

As long as, in the presence of negative velocity values, the absolute value of Δf does not exceed the center frequency f_c, i.e., $f_c > |\Delta f|$, the instantaneous frequency of the heterodyne signal correctly preserves the directional information (sign) of the velocity vector. In case of a harmonic vibration, the bandwidth of the modulated heterodyne signal is practically estimated to be limited to [38]

$$BW_{het} = 2(\Delta f + f_{vib}). \tag{9.23}$$

Consequently, f_c has to be chosen to at least $\Delta f + f_{vib}$. With the often-used center frequency 40 MHz, this condition is maintained up to a peak velocity of about 10 m/s as long as the vibration frequency does not exceed a few MHz. When measuring high-frequency vibrations on MEMS, required bandwidth and center frequency are often dominated by the vibration frequency f_{vib} rather than by the Doppler frequency Δf. This becomes clear if Equation 9.23 is written in the form

$$BW_{het} = 2 f_{vib}\left(M + 1\right) \tag{9.24}$$

where $M = \Delta f / f_{vib}$ is the so-called modulation index. For M using Equation 9.23 and Equation 9.24, we also write

$$M = \frac{\Delta f}{f_{vib}} = \frac{2\hat{v}}{\lambda \cdot f_{vib}} = \frac{4\pi}{\lambda}\hat{s} = \Delta\varphi. \tag{9.25}$$

It follows from Equation 9.24 that the carrier frequency f_c should be chosen to be at least twice the vibration frequency f_{vib} as long as $M < 1$ ($M = 1$ corresponds to an amplitude of about 50 nm at $\lambda = 633$ nm). The estimations according to Equation 9.23 and Equation 9.24 are important to understand the minimum bandwidth requirements for the RF signal path of a laser Doppler vibrometer. If the channel bandwidth is too small or the group time delay is not constant within the occupied bandwidth of the Doppler signal, harmonic distortions and frequency-dependent amplitude errors are unavoidable.

Equation 9.20 and Equation 9.22 demonstrate that the motion quantities s (displacement) and v (velocity) are encoded in phase and frequency modulation of the detector output signal, purely referenced to the laser wavelength λ. To be able to recover the time histories $s(t)$ and $v(t)$ from the modulated heterodyne signal, adequate phase and frequency demodulation techniques are utilized in the signal decoder blocks of a laser vibrometer. Theoretically it makes no difference if displacement and velocity are recovered independently by separate phase and frequency demodulators or if only one measurement quantity is recovered by a demodulator whereas the other is calculated by integration or differentiation, respectively. In the real world of analog electronics, however, all these operations cannot be performed ideally and the error terms of demodulator and subsequent differentiator or integrator blocks superimpose. Therefore, in high-end analog vibrometers, separate decoders are utilized to generate displacement and velocity signals independently.

9.3.3.2 Analog Decoding Techniques

At present, the majority of commercially available vibrometers uses analog frequency demodulators to directly convert the instantaneous Doppler frequency into a "voltage proportional to velocity" value although digital demodulation technique is more accurate and provides a lower noise level in the oscillation-amplitude spectrum. The strengths of analog demodulators are (1) a short delay time of a few microseconds between the appearance of a phase or frequency alteration at the vibrometer detector and the demodulated signal as well as (2) wide vibration frequency bandwidths up to several tens of MHz. Analog demodulation is especially important for MEMS applications if high frequencies have to be detected.

The basic concept of a heterodyne high-frequency displacement decoder is shown in Figure 9.5. This kind of an ultrasonic displacement decoder generates the reference signal by means of a

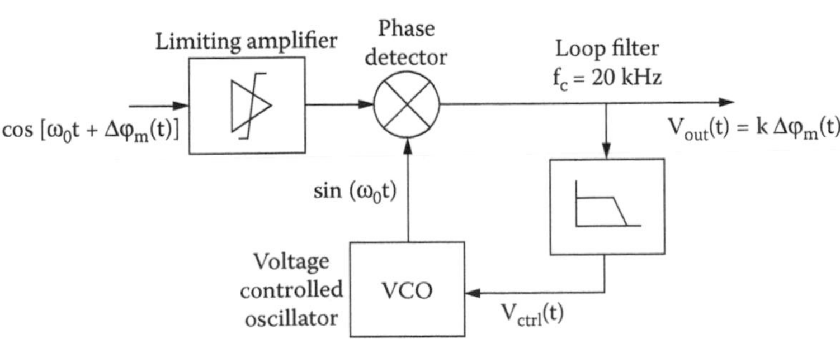

FIGURE 9.5 An analog displacement decoder for the ultrasonic frequency range.

phase-locked loop (PLL) circuit. Once the PLL is locked, the output of the voltage controlled oscillator (VCO) generates a 90° phase-shifted reference waveform of same frequency as the input signal. The AC output voltage of the phase detector is zero in this situation. This state is maintained as long as the modulation frequency of the input signal is within the bandwidth of the PLL, set by the cutoff frequency of the loop filter. At higher modulation frequencies, the loop cannot track the input signal longer as the frequency response of the control voltage $v_{ctrl}(t)$ is limited to 20 kHz. Thus, the output of the VCO serves as a stable phase reference for the input modulation. In this frequency region, the output of the phase detector is a measure of the instantaneous phase deviation at the input. Obviously, this scheme provides an inherent suppression of low-frequency modulation components, which might saturate the phase demodulator due to their large phase excursions. This feature enables the acquisition of subnanometer vibrations in the ultrasonic range even without perfect isolation of low-frequency ambient vibrations.

As the unambiguous measurement range of analog phase detectors is limited to ±90°, the corresponding displacement amplitude is limited to $\pm\lambda/4$ or about 150 nm peak-to-peak when using a He–Ne laser. However, this is sufficient for many high-frequency applications. Due to their interesting features, such demodulators are well suited for acquisition of high-frequency or fast transient displacement signals on MEMS. Also in the field of laser-based detection of ultrasound, analog phase demodulation is widely used to recover the time history of transient surface displacement or traveling sound waves [39].

9.3.3.3 Digital Demodulation by the Arctangent Phase Method

Thermal drifts in the electrical components, nonlinearities, or thermally introduced noise are the drawbacks of analog signal processing. These drawbacks can be overcome if the Doppler signal is digitized and its modulation contents are recovered by numerical methods. Almost all demodulation methods known from the analog world can nowadays be realized using digital signal processor (DSP) technology. Due to the large bandwidth of Doppler signals, high-speed A/D converters and powerful processors are absolute prerequisites for DSP-based vibrometer systems. Both standard PC platforms and tailored DSP chips are used to perform the digital high-speed processing required by this demodulation method. In all cases, signal decoding is based on the arctangent phase demodulation method, which is well known and understood from high-resolution static interferometers.

A proven standard signal decoding method for static interferometers and other displacement sensors generating an output signal in quadrature format relies on calculation of the phase angle by simple trigonometric relationships. It is also suitable for vibrometer signal decoding, provided the Doppler signal is available in the quadrature format and sufficient calculation power is available.

The key prerequisite of the arctangent method is a signal pair comprising I (in-phase) and Q (quadrature) components, whose voltage amplitudes depend on the interferometric phase angle $\varphi(t)$:

$$u_i(t) = U_i \cos\varphi(t) \tag{9.26}$$

$$u_q(t) = U_q \sin\varphi(t) \tag{9.27}$$

Such a signal combination is called an I&Q baseband signal, as there is no frequency offset present. In the baseband, a signal pair is needed to carry the complete Doppler information: whereas the absolute value of displacement s is represented by each component, its sign can only be recovered from both signals in combination.

Unlike the carrier signal whose frequency is f_c, the I&Q components are DC voltages in the case of a stationary target. If the target moves at a constant velocity v, considering Equation 9.19 and

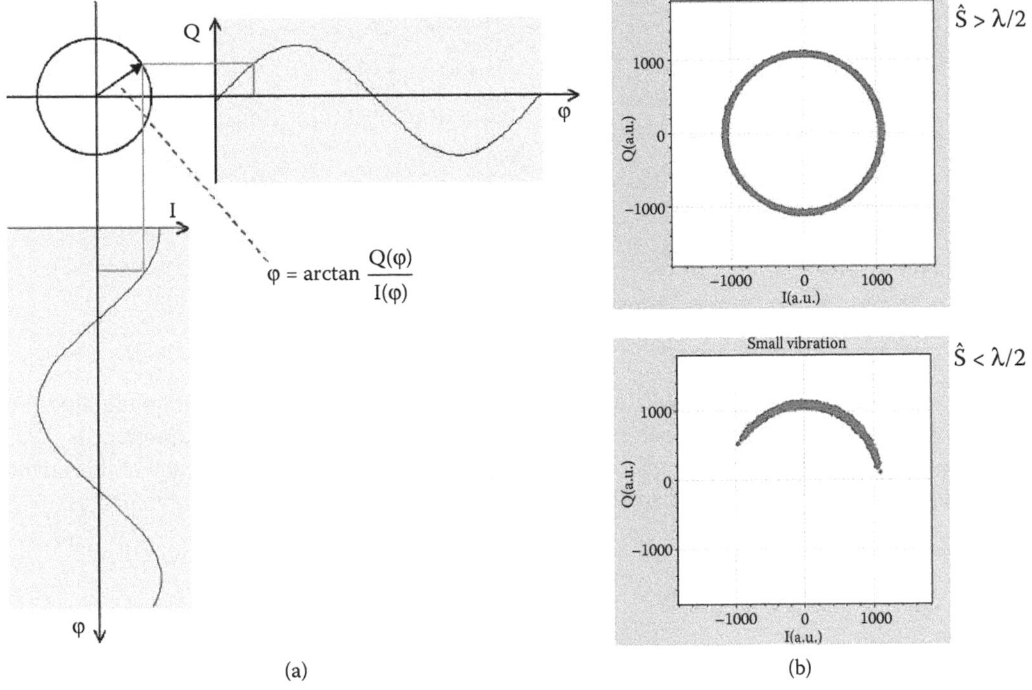

FIGURE 9.6 The I&Q signal in a vector diagram (a) and the resulting real display for different vibration amplitudes (b).

Equation 9.21, these signals become AC signals whose frequency is equivalent to the Doppler frequency shift:

$$u_i(t) = \hat{u}_i \cos\frac{4\pi \cdot s(t)}{\lambda} = \hat{u}_i \cos\left(2\pi\frac{2v}{\lambda}t\right) \tag{9.28}$$

$$u_q(t) = \hat{u}_q \sin\frac{4\pi \cdot s(t)}{\lambda} = \hat{u}_q \sin\left(2\pi\frac{2v}{\lambda}t\right) \tag{9.29}$$

In the case of an ideal interferometer, both signal amplitudes are equal, i.e., $\hat{u}_i = \hat{u}_q = \hat{u}$, no offset voltages are superimposed, and the relative phase shift is exactly 90°. The representation of such an ideal I&Q signal pair in Cartesian coordinates is shown in Figure 9.6.

With increasing phase angle $\varphi(t)$~$s(t)$, the visualizing vector describes a perfect circle centered to the coordinate origin. A quadrature signal pair can also be considered as real and imaginary parts of a complex phasor.

The I&Q format is an ideal starting point for numeric Doppler signal decoding. Recovering the object displacement $s(t)$ simply requires the calculation of the phase angle $\varphi(t)$ from the sampled instantaneous voltage values of the I&Q signals based on the trigonometric relationship $\tan\alpha = \sin\alpha/\cos\alpha$.

The inverse function provides the value of the phase angle at the sample instant t_n:

$$\varphi(t_n) = \arctan\frac{u_q(t_n)}{u_i(t_n)} + m\pi, \quad m = 0, 1, 2, \ldots \tag{9.30}$$

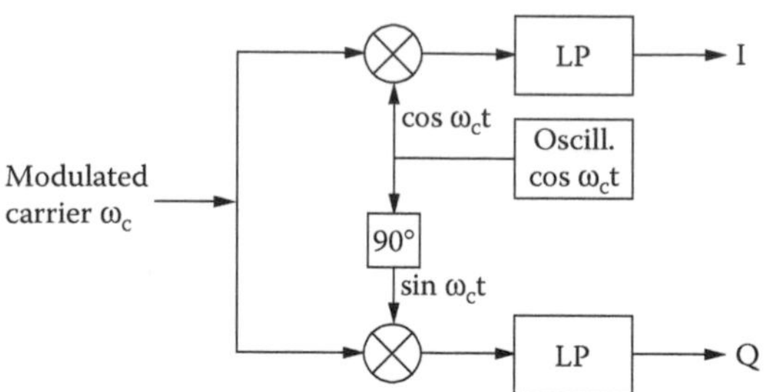

FIGURE 9.7 Conversion of a heterodyne carrier signal to I&Q baseband signals.

The ambiguity of the arctangent function can be removed by a phase unwrapping algorithm, which provides the integer number m, representing multiples of $\lambda/4$. The quotient $u_q(t)/u_i(t)$ eliminates the actual value of the signal amplitude $\hat{u}$. This is crucial for the accuracy of this method. According to Equation 9.19, the discrete displacement function $s(t_n)$ is finally obtained as

$$s(t_n) = \frac{\lambda}{4\pi}\varphi(t_n) \tag{9.31}$$

An interesting approach for demodulation of heterodyne signals by this method is down-conversion in conjunction with baseband sampling. The RF carrier itself, introduced by the Bragg cell, does not contain any Doppler information. Consequently, it can be suppressed by converting the heterodyne signal to an equivalent homodyne signal, represented by Equation 9.26 and Equation 9.27. This method combines the advantages of the heterodyne technique with the advantages of baseband sampling. An appropriate scheme for realizing the down-conversion process is shown in Figure 9.7.

The input signal is split into two fractions, each fed to a multiplier. A reference oscillator, in combination with a 90° phase shifter, provides the carrier frequency in quadrature form, fed to the other inputs of the multipliers. The outputs from the multipliers are the desired baseband signals superimposed by high-frequency components at $2f_c$, which are subsequently suppressed by the low-pass filters LP. Down-conversion according to that scheme can be performed either in the analog domain before quantization or in the digital domain.

An example of a digital decoding scheme on an intermediate-frequency (IF) level is given in Figure 9.8. It utilizes the arctangent decoding method, but in combination with a numerical down-conversion process equivalent to Figure 9.7. Prior to quantization, the original heterodyne signal is mixed down to an IF level using analog RF circuitry. A numerical oscillator (NCO) generates two data streams representing the reference frequency in I&Q format, which are then multiplied with the input data coming from the ADC. The reference frequency is matched to the IF center frequency,

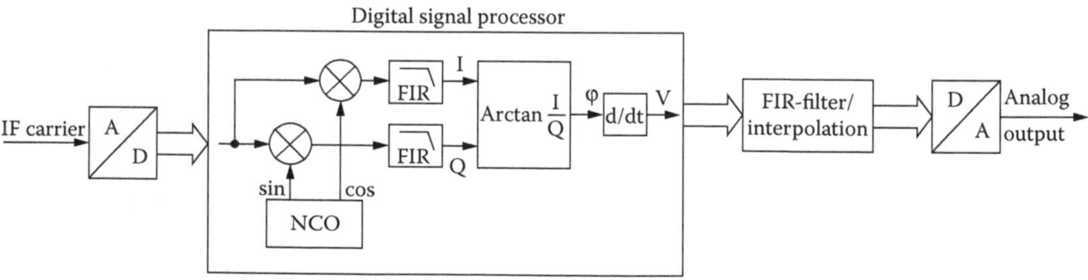

FIGURE 9.8 Signal processing scheme of an all-digital velocity decoder.

thus providing offset-free down-conversion. Obviously, numerical down-conversion does not suffer from errors introduced by analog components. Thus, the I&Q signals are inherently ideal and do not require any further correction as indispensable when using the analog method. Low-pass finite impulse response (FIR) filters define the operating bandwidth and remove high-frequency components from the I&Q signals. The arctangent calculation, based on a fast algorithm, generates the primary phase data. Instead of phase unwrapping and outputting displacement data, velocity data $v(t_n)$ are calculated by numerical differentiation. This results in a significant saving of DSP power and processing time. The remaining capacity of the DSP chip is exploited to realize digital low-pass filtering of the velocity signal with selectable cutoff frequency. Digital output data can be converted to voltage as shown, but can also be fed to subsequent processing blocks in an adequate digital format.

With an extended approach, $\varphi(t_n)$ raw phase data are unwrapped and scaled to provide an additional displacement output.

Characteristic for sequential digital data processing is a certain signal propagation delay. Current designs of DSP velocity decoders as shown in Figure 9.8 run at clock frequencies above 50 MHz, resulting in a total delay on the order of 10 μs. Compared to an analog velocity decoder with the same bandwidth, the digital one is currently slower by a factor of 3 to 4. This has to be considered in conjunction with phase-related measurements. As the delay time is constant, digital decoders are characterized by a strictly linear phase response over the full operating frequency range, which can readily be compensated during the subsequent data processing. Doppler frequency bandwidth and vibration frequency are only limited by the performance of the ADC and computation power. In terms of resolution and signal-to-noise ratio, digital demodulators are superior due to the absence of certain sources of phase noise. Displacement resolution on the order of 0.1 pm/Hz is yielded in combination with a class 2 laser optics and decoders of the outlined type. Without any doubt, DSP-based decoders represent the future of Doppler signal processing.

9.3.4 Data Acquisition

A vibration measurement system is not complete without a data acquisition and data processing system. One task is to record and process the instantaneous measurement signals, e.g., to determine frequency response functions. A second task could be the derivation of modal parameters from those frequency response functions. A common solution for these tasks is a PC-based workstation including data acquisition and signal generation boards.

To further analyze and evaluate the velocity or displacement signals of a vibrometer, these signals have to be digitized. Some advanced vibrometer systems do not require this step as they already deliver a digital velocity or displacement signal. They rather digitize the RF signal and implement the demodulation algorithm digitally as described in Subsection 9.3.3.3. This has several advantages over analog demodulation. However, it is technically very challenging and with current technology limited to frequencies below 2 MHz.

Common to both systems is the requirement that they have to observe the alias effect. One can picture an analog time signal as a series of an infinite number of samples with an infinitely short distance. To digitally process analog time signals, they must be sampled for a certain period with a sampling frequency f_S. The number of samples becomes limited, which can lead to a loss of information. Therefore, it must be made sure that the digital samples within the required accuracy provide the same information as the analog time signal. If the time signal is sampled k times, the following applies:

$$W_t = \frac{k}{f_S} \tag{9.32}$$

W_t = length of a measured time signal = time window
k = number of samples
f_S = sampling frequency

For correct digital signal processing, the sampling theorem must be fulfilled. This is guaranteed when

$$f_S > 2f_{Signal} \tag{9.33}$$

f_S = sampling frequency
f_{Signal} = maximum signal frequency

Signals with frequencies $f_{Signal} > f_S/2$ generate artificial lines at frequencies $f_{Alias} < f_S/2$. This is called the alias effect. The spectrum is falsified. To avoid infringement of the sampling theorem, signal frequencies that are higher than half the sampling frequency must be suppressed or the sampling frequency must be high enough (oversampling).

Depending on the frequency range, different analog-to-digital converter types and filter concepts are used. Delta-sigma converters apply a principle called *oversampling*. They digitize the analog signal with a very low resolution but with a very high conversion rate that is independent of the actual data rate and therefore can use a fixed low-pass filter to suppress the remaining alias problems. Through the high oversampling factor, the resolution of the digitized signal is increased. Commercially available converters have a typical sampling rate of 200 kHz at 16- or 24-bit resolution. For higher frequencies, successive approximation converters are used. Here, one has to apply low-pass filters adjusted to the sampling rate. As an alternative, one uses a fixed low-pass filter, samples at the highest rate, and reduces the data by digital filters in the data processing unit down to the desired rate.

Besides the evaluation of the recorded time data, very often vibration data are analyzed in the frequency domain for calculating frequency response, auto-spectrum, and cross-spectrum functions. Therefore, the digitized time signals are transformed using the complex discrete Fourier transform (DFT):

$$G(f_k) = \frac{1}{N}\sum_{n=0}^{N-1} g(t_n)e^{-i\frac{2\pi nk}{N}} \quad \text{forward transform} \tag{9.34}$$

$$g(t_n) = \frac{1}{N}\sum_{k=0}^{N-1} G(f_k)e^{i\frac{2\pi nk}{N}} \quad \text{inverse transform}$$

$t_n = n\Delta t$ = time corresponding to the nth time sample
$f_k = k\Delta f$ = frequency corresponding to the kth frequency component
N = number of samples
n = integer counting the time samples
k = integer counting the frequencies

When limiting the number of samples to powers of 2, one can apply a special calculation procedure known as the fast Fourier transform (FFT) algorithm, which drastically reduces the computation time by a factor of $N/\log_2 N$ [40].

The frequency response of a vibrating system is the magnitude and phase response of the system at all frequencies of interest. The frequency response function *H(f)* is defined as the Fourier transform of the impulse response *h(t)* of a system. It can be obtained from

$$H(f) = \frac{B(f)}{A(f)} \tag{9.35}$$

$A(f)$ = complex instantaneous spectrum of the input signal
$B(f)$ = complex instantaneous spectrum of the response signal

if $A(f) \neq 0$ for all f within the considered bandwidth. As all real-world signals contain noise, very often averaging methods are used to reduce the content of the undesired random noise components in the signal and to increase the data quality. Averaging can be performed in the time or frequency domain. When averaging is applied in the frequency domain, the true frequency response function H_{sys} of the measured system can be estimated by modifying Equation 9.35 in two ways.

If multiplying the top and the bottom by the complex conjugate of $A(f)$, a function known as H_1 is obtained:

$$H_1(f) = \frac{B(f)\,A^*(f)}{A(f)\,A^*(f)} = \frac{G_{AB}(f)}{G_{AA}(f)}\,. \tag{9.36}$$

In other words, the cross-spectrum normalized by the input auto-spectrum. H_1 is mainly affected by noise in the input signal (reference signal). It can be seen that with the presence of input noise, H_1 is smaller than H_{sys}.

Instead, if the complex conjugate of $B(f)$ is used, another function known as H_2 is obtained:

$$H_2(f) = \frac{B(f)\,B^*(f)}{A(f)\,B^*(f)} = \frac{G_{BB}(f)}{G_{BA}(f)} \tag{9.37}$$

which involves the cross-spectrum (from B to A) and the output auto-spectrum. H_2 is mainly affected by noise in the output signal (vibrometer signal). With the presence of output noise, H_2 is bigger than H_{sys}. Though the phase is equal for H_1 and H_2, the magnitude differs. The ratio of H_1 and H_2 equals the coherence.

Frequency response measurements require the excitation of the vibrating system with energy at all measured frequencies. The fastest way to perform the measurement is to use a broadband excitation signal that excites all frequencies simultaneously, and use FFT techniques to measure all of these frequencies at the same time. The quality of a vibration measurement depends strongly on the choice of the excitation signal. The principal criterion for the selection of the excitation signal is the purpose of the measurement. The linearity of the system also plays an important role for the selection of the excitation signal. If the device under test has nonlinearities, it is often desirable to get a linear approximation. Only certain excitation signals have the ability to average out nonlinearities. With other signals, nonlinear distortions are clearly visible in the spectrum.

Periodic chirp is an excellent choice as a broadband excitation signal for structures without any or only weak nonlinearities. It is a special form of multisine excitation signal $U_{exc}(t)$, where N frequencies, i.e., the measured frequencies, $U_{exc}(t) = U_0\, 1/\sqrt{N+1}\, \Sigma_j^N \sin(\omega_j t + \vartheta_j)$ are excited. For a given magnitude spectrum, the phase ϑ_j of these frequencies is optimized for a maximum ratio of the RMS value to the peak value of the excitation signal. As all the energy is distributed only to the selected frequencies, periodic chirp allows for fast measurements, very often without the need for averaging [41,42].

Some excitation signals generate leakage effects in the spectrum calculated by FFT, which are an additional noise source. However, a variety of signals exist that do not generate leakage effects intrinsically. Leakage errors originate from the fact that measurements have to be taken during a finite observation time T. The discrete Fourier transform then assumes that the observed signal is periodic with period T. If this condition is not met, a leakage error occurs. As a result, the discrete spectrum does not coincide with the exact one. The real energy at a frequency f is spread ("leaked") to nearby (periodic) frequencies, causing an important amplitude error. The best solution to the leakage problem is to make sure that the signal is either periodic or completely observed within the observation window. The use of time window functions offers an approximate solution to the leakage problem.

Exciting the structure with signals, which are not periodic in the observation window, another source can be small frequency deviations between the sampling rate of the excitation signal and the vibration response signal. This can be avoided by using the same base clock for all electronic signal generators and converters.

9.4 FULL-FIELD VIBROMETRY

9.4.1 Scanning Vibrometer

Taking a few boundary conditions into account, one can extend a single point LDV to a full field measurement system. By scanning the laser beam over the measurement object, one can acquire a series of single point measurements sequentially with a very high spatial resolution. However, in the analysis of the data, all points are animated simultaneously. Such a device is called a scanning laser Doppler vibrometer (SLDV). The components of an SLDV, specialized for microscopic structures (MSV), will be described in Subsection 9.5.1.

Sequential measurements can only be combined under the condition that the vibration is repeatable. In practice, this is fulfilled if the object is vibrating in a stationary state or the excitation is repeatable, which allows repeatable measurements. A vibration signal is stationary if its auto-spectrum does not change from measurement to measurement. In other words, if two successive measurements are made and the auto spectra are overlaid on one another, and if they are essentially the same (with the exception of a small amount of measurement noise), then the structure can be said to be in stationary conditions.

If the excitation is the same for all scan points, and the test object and the test setup remain unchanged during the measurements, then a sequential measurement is equivalent to a parallel one. To animate the measured vibration, one has to establish a fixed phase relation between all measurement points. This can be done by using a trigger signal for each measurement (phase-based measurement). If the trigger signal is related to a periodic process like the excitation signal of the structure (i.e., a synchronization pulse from a signal generator), one gets a well-defined phase relation between all measurement points. When calculating the complex Fourier transform from the recorded time signal of every measurement point, one obtains magnitude and phase values, which are all computed with respect to the trigger signal. By selecting the magnitude and phase values for an individual frequency and cycling the phase value φ over an entire period, one can animate the motion of the structure at this particular frequency.

$$A(\varphi)_{x,y} = \hat{A}_{x,y} \sin(\varphi + \varphi_{x,y}) \tag{9.38}$$

$\hat{A}_{x,y}$ = peak amplitude at location (x,y)
$\varphi_{x,y}$ = phase offset at location (x,y)

The true motion of the structure can be observed by animating the displacement time signals of the measurement (time-domain animation).

If there is no trigger signal available, there exists another possibility to establish a phase relation between the individual measurement points by using a reference signal, such as a signal proportional to the excitation force or a response signal from a fixed spot on the structure. In this case, one cannot use the magnitude and phase values of the Fourier transform of the response signal directly for an animation. Instead, the phase values for the animation have to be calculated from the differences between the phases of the response and the reference signal. It is not possible to synchronize the received signals for a time-domain animation with this method. Time-domain animation can only be obtained when a trigger is used for the measurement.

Scanning vibrometers are widely used for measurements on operating devices. Besides the measurement uncertainty in the individual measurement points, non-stationary conditions introduce spatial noise, which can be reduced by averaging over time and space. The additional noise is more than compensated for through the numerous advantages of an SLDV, such as non-contact, no mass load, broadband, and ease of use.

9.4.2 Operating Deflection Shapes

In general, an operating deflection shape (ODS) is any externally excited motion for two or more degrees of freedom (point and direction) on a device or structure. An ODS can be defined for a specific frequency or for a moment in time. This definition of a "shape" requires that all of the measured responses have the correct magnitude and phase relative to one another (see previous section). Scanning vibrometry is used to measure ODS.

Equation 9.38 is the mathematical representation of a time-dependent or oscillation-phase-dependent ODS. ODS analysis is the way to visualize how a product is deflecting under real-life operating conditions [43]. Not to be confused with modal analysis techniques that help to visualize the inherent resonant characteristics of a product, ODS is a very powerful tool that can solve problems related to vibrations under operating conditions. As vibrations are part of all rotating machinery, ODS analysis is very popular with people who make engines, motors, gearboxes, turbines, compressors, and other rotating components.

ODS analysis is a multichannel testing procedure used to define the system's structural deformation pattern resulting from a dynamic forcing function in time or frequency domain.

Vibration problems in structures or operating machinery often involve the excitation of structural resonances or modes of vibration. Many types of machinery and equipment can encounter severe resonance-related vibration problems during operation. In order to diagnose these problems, an animated geometric model or video image of the machine or the structure's ODS is often very helpful. In most cases, structural responses at or near a resonant (modal) frequency are "dominated" by the mode, and the ODS closely approximates the mode shape.

Modal testing (performing a modal survey) is usually done under controlled stationary (non-time-varying) conditions, using one or more exciters. Furthermore, the excitation forces and their corresponding responses are simultaneously measured. In many cases, especially with operating equipment, the measurement signals may be non-stationary (time varying) and the excitation forces cannot be measured. For these cases, different postprocessing techniques such as operational modal analysis (OMA) are required in order to extract modal parameters.

9.5 MEASURING ON MICROSCOPIC STRUCTURES

9.5.1 Optical Arrangements

A microscopic laser spot on the specimen is necessary to achieve a high lateral resolution and, therefore, a microscope objective has to be used to focus the beam on the surface. A collimated vibrometer beam should impinge the microscope objective because most objectives are infinity corrected. To estimate the focus diameter, the laser beam is approximated as a Gaussian beam [29]:

$$E_c(z,\rho)=\frac{E_0}{iz_R}\frac{w_0}{w(z)}\exp\left(-\frac{\rho^2}{w^2(z)}\right)\exp\left(-ikz-ik\frac{\rho^2}{2r(z)}+i\zeta(z)\right)e \tag{9.39}$$

where

$$w(z)=w_0\sqrt{1+\left(\frac{z}{z_R}\right)^2},\quad r(z)=z\left(1+\left(\frac{z_R}{z}\right)^2\right)$$

$$\zeta(z)=\arctan\left(\frac{z}{z_R}\right),\quad w_0=\sqrt{\frac{\lambda z_R}{\pi}}$$

with E_c the complex electric field vector, e the unit vector of the field polarization, w the beam radius, ρ the radial distance from the beam axis, r the wavefront radius, z the Guoy-phase delay, w_0 the waist radius, z_R the Rayleigh range, and λ the wavelength. The field distribution of the Gaussian beam is an approximation and is derived from the paraxial Helmholtz equation $\nabla^2 E + k^2 E = 0$ that follows from the Fourier-transformed Maxwell equations for linear, non-dispersive, homogenous, and isotropic media. A lens with focus length F transforms a beam waist w_0 at s_0 to the waist w_0' at s_0' in the following way [44]:

$$w_0' = \frac{w_0}{\sqrt{\left(1 - {s_0}/{f}\right)^2 + \left(z_R / F\right)^2}}, \qquad \frac{1}{s_0 + \frac{z_R^2}{s_0 - f}} + \frac{1}{s_0'} = \frac{1}{F} \tag{9.40}$$

The following approximation is valid in a typical microscope vibrometer system: high-input beam diameter with low beam divergence ($z_R >> F$), the waist is at the lens ($s_0 = 0$), and the image distance s_0' is approximately $s_0' = F$. Therefore, the beam waist diameter, which is the diameter of the laser spot on the sample, can be approximated by

$$d' = 2w_0' = \frac{2w_0 F}{z_R} = \frac{2\lambda F}{\pi w_0} = \frac{4\lambda F}{\pi d}, \tag{9.41}$$

where d is the laser diameter on the microscope objective (typically 3 to 5 mm). For example, a laser beam with 4 mm diameter, which is focused by a 20X objective with 10 mm focus length, has a beam diameter d' of 2 μm.

It is important to scan the laser in x and y directions over the specimen surface to determine the measurement point and to be able to collect the data to compute the operational deflection shapes. To provide a scan in x and y directions, the scanner should provide a tilt in the two directions. A vibrometer measures the instantaneous velocity through the Doppler-frequency shift of the backscattered laser light. Surfaces of silicon microdevices are usually optically flat, and light is reflected but not scattered. A microscope scanning vibrometer (MSV) can only be used to study motions along the optical axis of the imaging optics (out-of-plane). Therefore, the optics to scan a vibrometer beam over the microscopic specimen surface must be designed such that the beam impinges the surface perpendicularly. The laser beam impinges the surface perpendicularly if the center of rotation of the laser beam is in the back-sided focus (focus of the second primary focal plane) of the microscope objective. Figure 9.9 demonstrates the focusing of the laser beam and the definition of the primary planes and the positions of the front-sided and back-sided focus of the microscope objective.

Therefore, the scanner has to be placed directly in the back-sided focus, which is impractical for most objectives because the focus plane is too close to the objective optics. Usually, it is necessary to use the microscope objective as well to image the specimen on an electronic camera in order to enable the user to adjust the beam position and focus. Therefore, the scanner has to be imaged into the back-sided focus as demonstrated in Figure 9.10.

A typical x-y scanner consists of two single one-axis scanners. Piezo-based two-axes scanners are commercially available but do not offer sufficient scan angle. For high numerical-aperture lenses, the incident angle θ at the microscope objective is θ = arcsin(NA), which can be more than 30°. A design used in the Polytec system utilizes two piezo-based scanning mirrors using overlapping "virtual rotational axes" [45]. The schematic of this approach is shown in Figure 9.11.

Although this system allows higher tilt angles than x-y scanners, it is still necessary to magnify the scan angle optically, which can be performed with an imaging system that images the scanner downsized into the back-sided focus. As seen in Figure 9.11, the first piezo-bimorph-driven mirror tilts

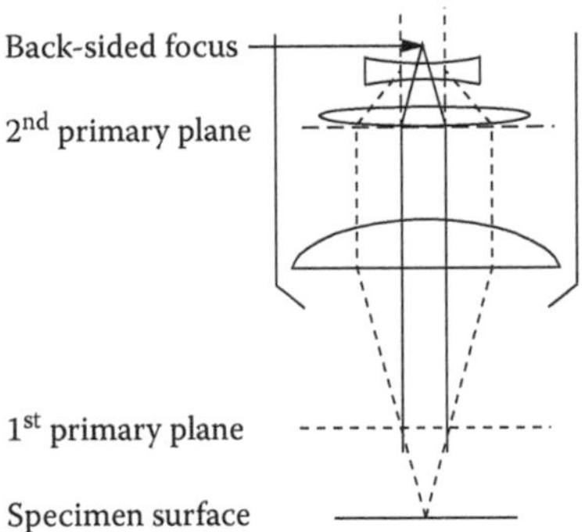

FIGURE 9.9 Definition of the 1st and the 2nd primary planes of a microscope objective.

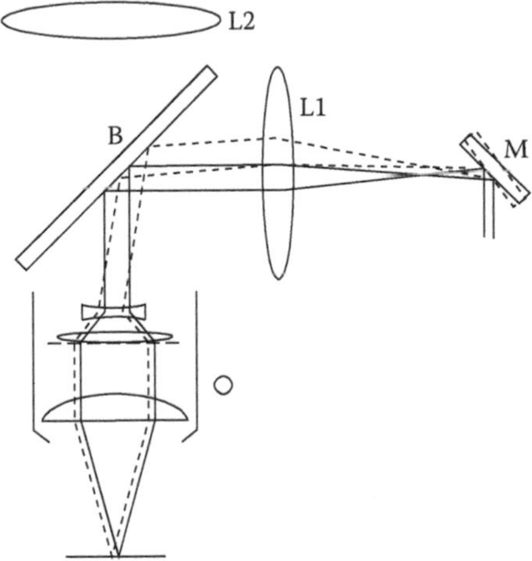

FIGURE 9.10 A microscope scanning setup. The laser beam has to enter the objective collimated and the scanner is imaged into the back-sided focal plane with lens L1. O is the objective, L2 the microscope tube lens that images the specimen on the camera chip, B a beam splitter, and M the scanning mirror.

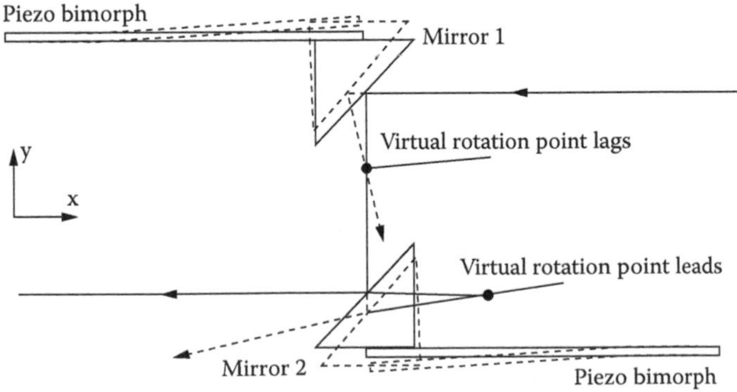

FIGURE 9.11 Scanning system using overlapping "virtual rotational points."

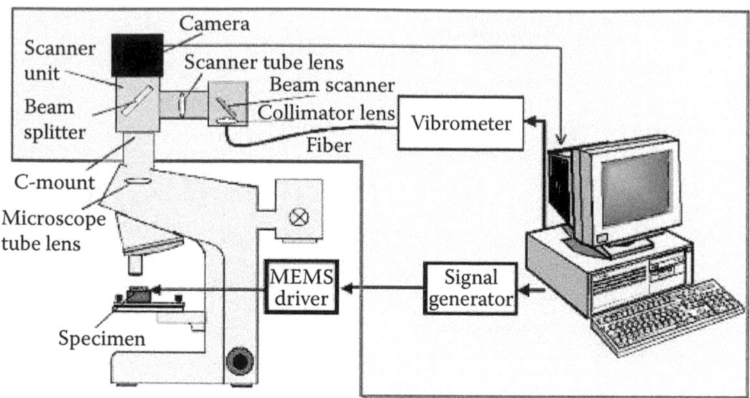

FIGURE 9.12 Microscope scanning vibrometer (MSV) system.

the beam in the plane of the paper about a center of rotation that lags behind the mirror position. The second identical assembly generates another tilt motion that appears to be centered on a point preceding the mirror. By rotating the second mirror by 90° about the *y* axis, the beam directed through the plane of the paper and in the *x* direction is controlled by mirror 1 and *y* direction is controlled by mirror 2.

Polytec has developed a vibrometer module that can be attached to the C-mount of a microscope. The laser beam of a vibrometer is coupled with a fiber into the microscope scanner unit. The schematic of the microscope scanning vibrometer (MSV) system is shown in Figure 9.12. The two piezoelectric, beam-scanning mirrors scan the laser beam over a user-defined area at user-defined scan points in the image position above the microscope C-mount. A tube lens and a collimator lens inside the MSV scanner unit focus the beam on this image position. The microscope optics image the focused laser-scan area from the C-mount position downscaled to the microscopic surface of a MEMS specimen. The laser-focus position on the device surface is visualized with a digital camera assembled on the scanner unit. The scanner unit optics transfer the image from the C-mount position to the camera chip. The scanner tube lens and the microscope tube lens image the beam scanner into the back-sided focus of the microscope objective and, therefore, the laser beam is perpendicular to the optical axis of the microscope when the beam impinges the specimen surface.

9.5.2 3-D Techniques

As outlined in Section 9.2, a vibrometer measures the Doppler-shifted laser light scattered back into the direction of the impinging beam. The only requirement is that sufficient light is scattered back. This is not the case for MEMS in general because the surface of a silicon wafer or of a thin-film layer has a very smooth surface. However, surfaces of tiny devices in precision engineering or galvanically deposited LIGA (from the German words Lithographie Galvanik and Abformung) structures do have a sufficiently rough surface to scatter enough light such that motions parallel to the surface tangent can be detected with the laser Doppler technique.

If the beam impinges with an angle α, an in-plane motion v_x will generate a velocity signal $v_1 = v_x \sin\alpha$ whereas an out-of-plane motion leads to a signal $v_1 = v_z \cos\alpha$. The velocity signal is dependent on the in-plane and out-of-plane components for a random velocity vector in the x-z plane

$$v_1 = v_x \sin\alpha + v_z \cos\alpha \tag{9.42}$$

A second beam impinging under the angle $-\alpha$ will produce a second velocity signal

$$v_2 = -v_x \sin\alpha + v_z \cos\alpha \tag{9.43}$$

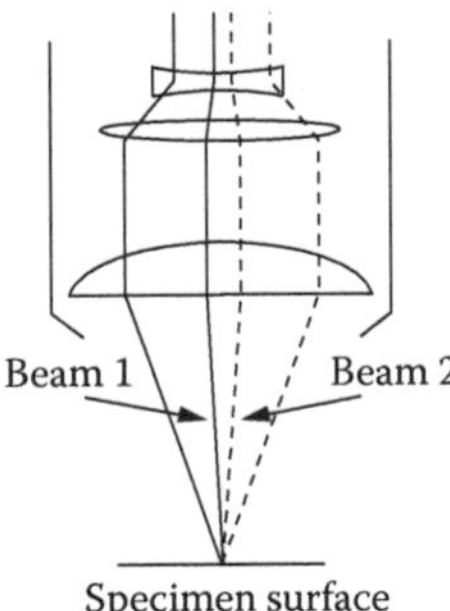

FIGURE 9.13 In-plane motion can be measured on rough surfaces if at least two vibrometer beams impinge the surface.

The schematic shown in Figure 9.13 demonstrates how these two beams can be coupled into a microscope objective to realize the arrangement.

The velocities v_x and v_z can be computed by the relations

$$v_x = \frac{v_1 - v_2}{2\sin\alpha}, \qquad v_z = \frac{v_1 + v_2}{2\cos\alpha} \tag{9.44}$$

The difference between the two vibrometer laser frequencies should be higher than the bandwidth of their detectors to avoid interference effects between the two vibrometer beams. A similar setup can be arranged with three vibrometers to obtain the v_y value. Determination of the single velocity components is a little bit more complex but follows the same idea and depends on the special arrangement of the beams. However, the high NA of the impinging beams and the needed angles of incidence to detect the in-plane motions make it impossible to realize reasonable scanning systems for 3-D microscopic vibrometer detection.

In-plane motion can also be detected interferometrically if the interference fringes are placed not on the detector but on the surface itself [46]. In this case, the system has just one detector, which does not measure the back-reflected light but rather the light scattered perpendicular to the surface as shown in Figure 9.14.

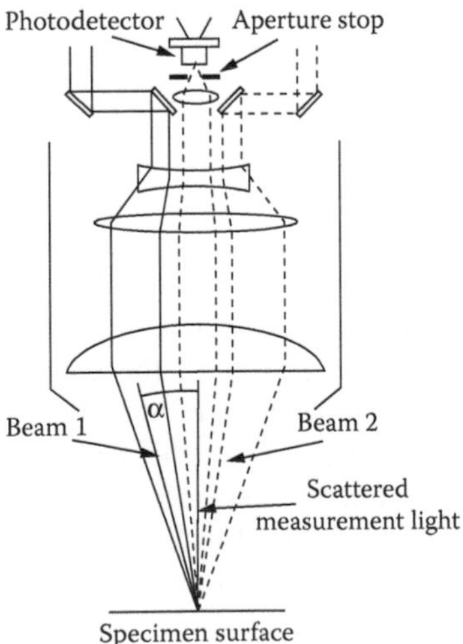

FIGURE 9.14 This arrangement provides a fringe pattern on the surface and can be used to measure in-plane motion.

A scattering point, which is moving through the fringe pattern, modulates the intensity of the scattered light, depending on its in-plane velocity and the fringe spacing

$$\Delta s = \frac{\lambda}{2\sin\alpha}, \tag{9.45}$$

with α the angle of incident of the impinging beams with respect to the optical axis defined by the photodetector. It is also possible to use a constant frequency difference between beam 1 and beam 2 to obtain the sign of the velocity. In this case, the fringe pattern would lead to an intensity modulation of the scattered light at the frequency of the heterodyne shift if a stagnant surface is measured. The frequency of intensity modulation of the scattered measurement light can be expressed by the following equation

$$\omega_i = 2\pi\frac{v_x}{\Delta s} + \omega_1 - \omega_2 = \frac{2v_x}{\lambda}\sin\alpha + \omega_c = \omega_D + \omega_c, \tag{9.46}$$

with ω_D the Doppler frequency shift. Obviously, the detector bandwidth should be designed with respect to the same design rules that were mentioned in Section 9.3 for the usual out-of-plane vibrometer. The fringes move only in in-plane direction, and therefore an in-plane motion causes a shift of the amplitude modulation frequency of the scattered light. Out-of-plane motion is not detected.

It should be mentioned here that ω_d corresponds to the sum of the Doppler frequency shifts of beam 1 and beam 2. Therefore, the arrangement shown in Figure 9.14 is widely known as the *differential Doppler technique*. We did not derive the Doppler shift in the case that the motion direction, axis of the impinging beam, and axis of the collected scattered light are not parallel, but refer to [21]. However, the argumentation of the Doppler shift is correct but not useful, because the measurement principle would be the same if the intensity pattern of the fringes were caused in a different way, e.g., with projection of a pattern. The advantage of the fringe pattern is just that the spacing Δs is very narrow.

9.5.3 Ranges and Limits

As mentioned in Subsection 9.5.2, the vibrometer can only measure if light is scattered or reflected in the direction of the impinging laser beam. Therefore, the 3-D measurement technique described in Subsection 9.5.2 can only be employed if the surface is rough. An optically flat surface (e.g., a mirror) cannot be used to detect in-plane motion with a 3-D vibrometer. The definition of optically flat is fuzzy, and peak-to-peak roughness values between $\lambda/10$ and λ can be found in the literature. Obviously, it is not the peak-to-peak roughness of the whole sample that is important for the definition of optical flatness but rather the area of the illuminated spot that is imaged on the photodetector. The diameter of this spot corresponds to the spatial coherence if the NA of the illumination is the same as the imaging NA. For microscopic measurement, this spot can have a diameter of down to the half of the wavelength as discussed in the preceding paragraph. Obviously, a surface that is macroscopically rough can appear smooth and tilted over an area with a diameter of half the wavelength. In addition, silicon-based MEMS are microscopic structures etched from a monocrystal wafer or deposited in thin-film technology. These kinds of surfaces are even macroscopically flat and do not scatter light if a beam with millimeter diameter impinges the surface. Therefore, a vibrometer cannot be used to measure in-plane motion of silicon-based microdevices. A vibrometer can only measure motions along the optical axis of the microscope imaging optics because the impinging beam is reflected. It does not change the result if the beam is reflected on a tilted surface in an angle to the optical axis because the optical path length between the illuminated diffraction-limited spot and the core of the vibrometer fiber that is a diffraction-limited pinhole is independent of path used by the ray (principle of Fermat). Only a variation in the optical path length can lead to a Doppler signal (see Section 9.2).

It is only possible to couple the portion of light into the monomode fiber of the vibrometer that accords to the electrical field mode in the fiber. In other words, the beam needs to impinge within the numerical aperture of the fiber core (NA = 0.1). The Gaussian beam has an infinitely wide spread if zero power would be the definition for the beam margin such that the full aperture of the microscope objective should be the limiting factor for the maximum tilt angle of the specimen. In practice, the coupling efficiency drops quickly so that the vibrometer is not sensitive enough to use the whole numerical aperture. The definition of the numerical aperture with $1/e^2$ opening angle of the Gaussian beam cone is a more useful way to define the maximum possible tilt of the specimen that is half the opening angle. Therefore, the maximum possible tilt ψ can be written as

$$\psi = \frac{1}{2}\arctan\frac{d'}{2f}. \tag{9.47}$$

As an example, an objective with a focal length of $f = 10$ mm and a beam diameter $d' = 4$ mm (typical values for a 20X lens with NA = 0.4) results in a maximal tilt of $\psi = 5.7°$.

The maximal out-of-plane displacement of a vibrometer is limited to a quarter of the wavelength for an ultrasonic PLL (phased locked loop), decoder. For the remaining decoders, the out-of-plane measurement range is unlimited. Therefore, the limitation results from the depth of focus of the imaging optics. The fiber core has a diameter of 4 μm and corresponds to a perfectly matched pinhole in confocal microscopes. The fiber core automatically has the diameter of the beam waist that is diffraction- limited imaged to the specimen. Therefore, the theoretical results from the considerations about confocal microscopes can be transferred to the situation in the MSV. The intensity distribution $I(z)$ at the detector in a confocal microscope with infinitesimal pinhole in dependence of the z-displacement of the specimen has been derived in [47] with nonparaxial scalar theory:

$$I(z) = I_0\left(\frac{\sin\left(knz\left(1-\cos\theta_0\right)\right)}{knz\left(1-\cos\theta_0\right)}\right)^2. \tag{9.48}$$

Here, θ_0 is the half-angle subtended by the focused beam at the objective lens, n is the index of refraction, and k is the wavenumber $k = 2\pi/\lambda$. The depth of focus d_z is commonly defined as the distance within the full width at half maximum range and results in [47]

$$\hat{d}_z = \frac{0.45\lambda}{n\left(1-\cos\theta_0\right)} \approx \frac{0.9n\lambda}{NA^2}. \tag{9.49}$$

This approximation is valid for values of NA less than ~0.5. As MEMS surfaces are usually very reflective and light is collected efficiently, and the vibrometer has a high dynamic range of 120 dB; the limitation of the maximum piston deflection goes beyond the limit demonstrated in Equation 9.49, in practice. In addition to the decreasing signal power, the laser spot diameter increases if the specimen moves out of focus. It seems to be more reasonable to define the distance z as out-of-plane motion range when $I(z)$ has its first minimum at

$$d_z = \frac{\lambda}{n\left(1-\cos\theta_0\right)} \approx \frac{2n\lambda}{NA^2}, \tag{9.50}$$

where Equation 9.50 follows from Equation 9.48 with the approximation used in Equation 9.49. Therefore, the out-of-plane, motion range of an objective lens with NA = 0.4 can be estimated to $d_z = 8$ μm using Equation 9.50. However, a measurement can be performed beyond this limit as long as the vibrometer has sufficient signal level to demodulate the Doppler shift.

The most important limit of a microscopic measurement with a laser Doppler vibrometer is the diameter of the laser spot on the specimen surface, which defines the lateral resolution of the system. The diameter can be estimated with Equation 9.41. Equation 9.41 demonstrates that the focus diameter depends on the focal length of the used microscope objective and the diameter of the collimated laser beam that impinges the pupil of the objective or, if the aperture of the objective is smaller than the beam width, the pupil diameter of the objective. Considering the confocal arrangement of the vibrometer beam optics, the lateral spatial FWHM single-point resolution results in [47]

$$d_r = 0.37\frac{\lambda}{\sin(NA)}. \tag{9.51}$$

The vibrometer beam is circularly polarized when it impinges the specimen in the focus of the microscope objective. The single-point resolution of a circularly polarized beam is even below the result for nonpolarized light (Equation 9.51). It is shown in [48] that the FWHM diameter of a focused laser light spot can be reduced about 8% if circularly polarized light is used instead of linearly polarized light. Therefore, lateral superresolution can be achieved with a microscope scanning vibrometer if the resolution is not limited by aberrations.

The maximal values for the detectable velocity $\hat{v}$ and oscillation frequency $\hat{f}_{vib}$ can be estimated with Equation 9.21 and Equation 9.23 in Subsection 9.3.3 to yield

$$\hat{f}_{vib} = f_c - \frac{2\hat{v}}{\lambda}. \tag{9.52}$$

For example, assuming a carrier frequency of $f_c = 40$ MHz and a maximal detectable velocity of 10 m/sec, the maximal theoretically possible oscillation frequency is approximately $\hat{f}_{vib} = 8.4$ MHz. An oscillation frequency of 30 MHz can have a theoretical maximal velocity of approximately $\hat{v} = 3$ m/sec. However, these values are maximal theoretical limits and other technical restrictions lead to additional limitations. A typical commercial decoder can measure, for example, 10 m/sec at frequencies up to 1.5 MHz. High-frequency PLL decoders, for example, can demodulate peak-to-peak amplitudes of maximally a quarter of the wavelength, as discussed in Subsection 9.3.3.

9.6 RESOLUTION AND ACCURACY

9.6.1 Noise-Limited Resolution

Resolution, i.e., the smallest detectable magnitude of the desired measurement quantity, is a crucial parameter in laser vibrometry. Resolution is particularly critical for measurements on small structures or for high vibration frequencies because of tiny vibration amplitudes. In the MHz region, for example, vibration amplitudes very rarely exceed a few tens of nanometers. In most cases, picometer resolution is required in order to acquire dynamic processes in the world of MEMS. Under the provision that an adequate optical arrangement is utilized in combination with state-of-the-art signal processing technology, laser interferometry can provide displacement resolution even in the subpicometer region.

In order to understand the limits of resolution in a laser vibrometer, one has to analyze its significant noise sources. As the measurement information is encoded in the instantaneous phase of the Doppler signal, the impact of noise quantities on phase angle determines the system resolution. Basically, there are three main sources of noise in the signal processing chain of a laser vibrometer: light-induced noise (shot noise), thermal noise of the detector/preamplifier combination (Johnson noise), and signal processing noise (e.g., quantization noise or phase noise of local oscillators). The ultimate limiting factor to system resolution of a laser vibrometer is the signal-to-noise ratio of the photodetector output signal. Physical limits are set here by the quantum nature of light generating shot noise, which is usually the dominant noise source (refer to Section 9.2).

According to Equation 9.14, we can write the output current of the heterodyne photodetector in the form

$$i_p(t) = K\left[P_m + P_r + 2\kappa\sqrt{P_m P_r}\cos\left(\omega_c t + \varphi(t)\right)\right] \tag{9.53}$$

ω_c = angular center frequency
$\varphi(t)$ = modulated phase angle

The detector current consists of a DC component proportional to the total optical power $P_m + P_r$ and an AC current with mean square value representing the power of the heterodyne carrier signal.

A major advantage of coherent detection is the optical amplification of the AC amplitude, expressed by the term $\sqrt{P_m P_r}$. This means that doubling the power of the reference beam yields a $\sqrt{2}$ or 3 db gain for the carrier amplitude. On the other hand, the sum power $P_m + P_r$ not only generates a DC current, but also shot noise according to Equation 9.12, represented by the mean square noise current

$$\bar{i}_{sh}^2 = 2K \cdot q \cdot B\left(P_m + P_r\right), \quad (B = \text{detector bandwidth}) \tag{9.54}$$

As mentioned in the preceding text, another noise source is thermal noise of the detector/preamplifier combination. The generated mean square noise current here is

$$\bar{i}_{th}^2 = \frac{4k \cdot T \cdot B}{R} \tag{9.55}$$

k = Boltzmann constant
T = absolute temperature
R = detector load resistance

The reference power is usually chosen such that shot noise power significantly exceeds thermal noise power under given conditions. This point is usually reached at $P_r + P_m < 1$ mW, depending on bandwidth. The system then is called shot noise limited and yields the best possible signal-to-noise ratio. Neglecting thermal noise, the signal-to-noise power ratio is consequently given by

$$SNR = \frac{P_s}{P_n} = \frac{\bar{i}_s^2}{\bar{i}_{sh}^2} = \frac{\eta \cdot \varepsilon^2 \cdot P_m \cdot P_r}{h \cdot \nu \cdot B\left(P_m + P_r\right)} \tag{9.56}$$

Furthermore, if enough reference power is available ($P_r >> P_m$), Equation 9.56 simplifies to

$$SNR = \frac{\eta \cdot \varepsilon^2 \cdot P_m}{h \cdot \nu \cdot B}, \tag{9.57}$$

i.e., the SNR of the Doppler signal is proportional to the reflected measurement beam power but independent of reference beam power.

Detectability of the desired signal in a given bandwidth is assumed, where $SNR \geq 1$.

Considering the general properties of phase modulated signals [38], the noise-induced uncertainty of the modulation phase angle can be estimated on the basis of the spectral signal-to-noise power ratio: the sinusoidal carrier signal $u_c(\varphi_c) = \hat{U}_c \sin \varphi_c$ has a maximum rise of $\frac{dU_c}{d\varphi_c} = \pm\hat{U}_c$ in the vicinity of the zero crossings. Thus, if a single noise component with peak voltage $\hat{U}_n$ is superimposed, the resulting maximum phase deviation is $\Delta\varphi_n = \frac{\hat{U}_n}{\hat{U}_c}$. Because two uncorrelated noise components below and above the carrier frequency always contribute to noise modulation, a factor of $\sqrt{2}$ has to be taken into account. Consequently, a heterodyne carrier with power $P_c >> P_n$ undergoes a peak phase deviation caused by the spectral noise power density $P_n' = P_n/1\text{Hz}$ in 1 Hz bandwidth according to

$$\Delta\varphi_n' \approx \sqrt{\frac{2P_n'}{P_c}}. \tag{9.58}$$

Using the SNR value from Equation 9.56 or Equation 9.57, we get $\Delta\varphi_n' \approx \sqrt{2}/\sqrt{SNR'}$, where SNR' is calculated for B = 1 Hz. This differential phase noise is independent of the position of the bandwidth element B = 1 Hz on the frequency axis. Consequently, phase noise or displacement noise, respectively, is white noise when caused solely by shot noise or thermal noise. Other noise sources in the subsequent signal processing chain of a vibrometer may have a nonuniform noise density distribution. In particular, *1/f* noise of amplifiers or oscillators may dominate the overall noise at low frequencies. With homodyne interferometers, amplitude noise of the laser with *1/f* characteristics may also significantly deteriorate the SNR of the Doppler signal, whereas it is in general negligible in the frequency region of heterodyne interferometers.

Noise behavior is different when analyzing the velocity signal after FM demodulation. Due to $\omega(t) = d\varphi / dt$, the frequency deviation Δf resulting from a phase deviation $\Delta\varphi$ is proportional to the modulation frequency (refer to Section 9.3). A single phase-noise component $\Delta\varphi_n'$ at frequency f_n therefore results in a frequency noise component

$$\Delta f_n' = \Delta\varphi_n' \cdot f_n \tag{9.59}$$

Again, two uncorrelated noise components at frequencies $f_c + f_n$ and $f_c - f_n$ contribute to the noise of an FM signal with center frequency f_c, and we obtain the resulting differential frequency deviation:

$$\Delta f_n' \approx f_n \sqrt{\frac{2P_n'}{P_c}} \tag{9.60}$$

The corresponding peak velocity at frequency f_n is, from Equation 9.21, $\Delta v' = \frac{\lambda}{2} \Delta f_n'$. This means that the spectral noise density in terms of velocity is proportional to frequency. For this reason, noise spectra of velocity decoders always rise with frequency, whereas noise spectra of displacement

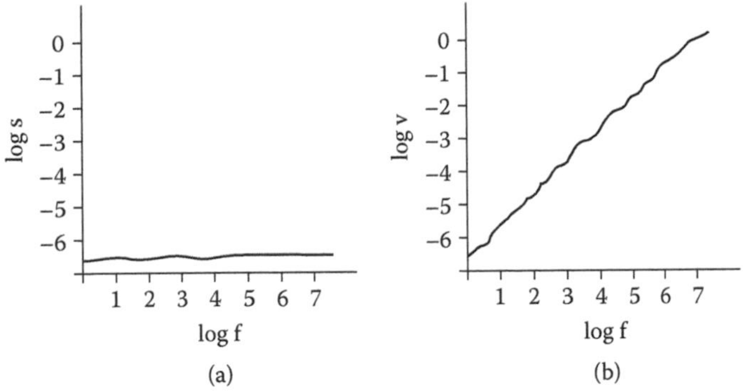

FIGURE 9.15 Spectral distribution of (a) displacement and (b) velocity noise.

decoders are flat, as shown in Figure 9.15. The spectral noise density determines the magnitude of the respective measurement quantity, which can be acquired by the laser vibrometer at 0 db SNR in a 1 Hz resolution bandwidth.

Another way to understand the limits to resolution is based on an analysis of the spectral properties of the phase-modulated heterodyne signal. When the carrier is modulated by a sinusoidal vibration, its spectrum consists of a line at the center frequency f_c and (theoretically) of an infinite number of symmetrical sidebands spaced by the vibration frequency f_s. Such a power distribution can be approximated by a set of Bessel functions of the first kind $J_n(\Delta\varphi_s)$. In the region of very small phase deviation ($\Delta\varphi_s << 1$ rad), the spectrum practically consists of only J_0 representing the carrier amplitude and J_1 representing the amplitude of the first pair of sidebands at $f_c \pm f_s$. A very good approximation here is $J_0 = 1$ and $J_1 = 0.5\Delta\varphi_s$ [38]. Figure 9.16 displays a spectrum for $\Delta\varphi_s = 0.1$ rad, corresponding to a displacement amplitude of about 5 nm at $\lambda = 633$ nm.

Obviously, if the significant sidebands are covered by noise components, the information content of the modulated signal cannot be recovered correctly. Assuming the same magnitudes for the noise component and sideband at frequency $f_c \pm f_s$, we find as an estimation for the noise-equivalent phase deviation:

$$\Delta\varphi_n' \approx \frac{2}{\sqrt{SNR'}} \tag{9.61}$$

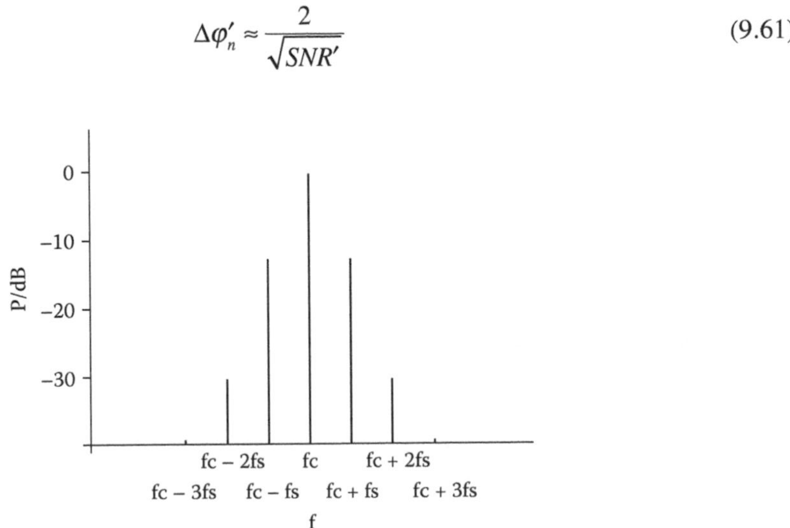

FIGURE 9.16 Power spectrum of a phase-modulated signal.

As a practical example, the theoretical noise-limited resolution shall be calculated for an interferometer powered by a 2-mW He–Ne laser. We assume a split ratio of 50:50 for reference and measurement beams, a quantum efficiency $\eta = 0.8$, and an efficiency factor $\varepsilon = 0.8$. The object may have a reflectivity of 10%, which is a good number for retroreflective film. According to Equation 9.49, we obtain for $B = 1$ Hz the spectral signal-to-noise power ratio

$$SNR = \frac{0.8 \cdot 0.8^2 \cdot 0.1\,\mathrm{mW} \cdot 1\,\mathrm{mW}}{h \cdot \nu \cdot 1\mathrm{Hz}\left(0.1\,\mathrm{mW} + 1\,\mathrm{mW}\right)} = 1.57 \cdot 10^{14}\,\mathrm{Hz}^{-1} \tag{9.62}$$

Taking this number with Equation 9.61 we get the differential noise phase deviation

$$\Delta\varphi_n' \approx 1.13 \cdot 10^{-7}\,\mathrm{rad} \cdot \mathrm{Hz}^{-\frac{1}{2}} \tag{9.63}$$

With $\Delta s = \frac{\lambda}{4\pi}\Delta\varphi$ we obtain the noise-equivalent mean square displacement $\overline{s}_n' = \frac{1}{\sqrt{2}} \cdot \frac{\lambda}{4\pi}\Delta\varphi_n' = 4.025 \cdot 10^{-15}\,\mathrm{m}/\sqrt{\mathrm{Hz}}$, or 4 femtometers per square root Hz.

In [39], a resolution limit of roughly $8 \cdot 10^{-15}$ m/$\sqrt{\mathrm{Hz}}$ was calculated for a heterodyne interferometer measuring on a mirror, but assuming a lower quantum efficiency of the detector. As mentioned, the preceding calculations of noise limited resolution are estimations of the physical limits. In reality, laser vibrometers contain additional broadband noise as well as spurious noise components, which may push the practical limits of resolution to much higher numbers.

With a commercially available class II laser vibrometer (1 mW output) in combination with digital decoding, a displacement noise floor of <50 fm/$\sqrt{\mathrm{Hz}}$ under conditions comparable to the preceding example is achieved. Spurious noise peaks, mainly caused by electronic cross talk, are not higher than 25 pm. The corresponding velocity noise of this system rises from <0.5 nm·sec^{-1}/$\sqrt{\mathrm{Hz}}$ to 0.6 μm·sec^{-1}/$\sqrt{\mathrm{Hz}}$ in the frequency range 0 to 2 MHz. In combination with an analog displacement decoder as described in Subsection 9.3.3, about the same displacement noise floor is maintained up to a maximum vibration frequency of 30 MHz. Other systems, working with about 100 mW laser power, claim a displacement resolution of <10 fm/$\sqrt{\mathrm{Hz}}$ on high-reflectance surfaces.

It should be noted here that subpicometer resolution is feasible only under the provision of sufficiently small resolution bandwidth of the subsequent signal acquisition system or using an adequate number of averages. This is usually not a problem when analyzing stationary vibrations or repetitive dynamic processes. For analysis of transient processes, however, single-shot resolution is crucial. Dependent on risetime requirements, the acquisition bandwidth is then on the order of tens of kilohertz or even megahertz. The resulting measurement resolution is calculated for a given bandwidth according to $\overline{s} = \overline{s}'\sqrt{B}$. On account of the stochastic nature of noise, one should multiply this value by a factor of five in order to get an estimation for peak-to-peak noise in a single-shot acquisition.

9.6.2 Measurement Accuracy and Calibration of Laser Doppler Vibrometers

9.6.2.1 General Aspects

Laser Doppler interferometry is the most accurate measurement technique for the motion quantities velocity and displacement because it refers them directly to the well-known wavelength of the laser source. It is therefore used in national metrology institutes (NMIs) as an absolute reference for primary measurement standards [49]. Despite this, commercially available laser vibrometers are subject to the same calibration requirements as any other transducer or measurement device.

Though the optical part is inherently accurate, a variety of factors ultimately affect the measurement uncertainty of a real vibrometer.

Basically, a laser vibrometer could be calibrated in the same way as traditional vibration transducers, i.e., by comparing to a vibration reference. Calibration procedures for vibration transducers such as accelerometers are defined in the international standard series ISO 16063 (e.g., ISO 16063-11 [50]), but at present they do not sufficiently accommodate the requirements of laser vibrometers. Depending on the level within the hierarchic traceability system, either a primary measurement standard or a reference standard is used for comparison. In both cases, mechanical vibration exciters are necessary to establish stationary vibration at the desired amplitude and frequency. A major obstacle for comparison calibration of laser vibrometers is the limited operating frequency range of calibration exciters and reference accelerometers. Though laser vibrometers are available even for the MHz region, traditional calibration equipment mainly covers the acoustic frequency range. Also, the vibration amplitude of mechanical shakers is very low compared to the measurement range limits of laser vibrometers. Shock calibration as a possible way out is very difficult to perform [51]. For these reasons, manufacturers of laser vibrometers are forced to develop alternative calibration methods as a basis for reliable specifications and calibration within the whole operating range of their products.

9.6.2.2 Mechanical Comparison Calibration

If a traceable calibration is required for a laser vibrometer, it has to be performed by comparing it to a traceable vibration reference, e.g., a reference accelerometer excited by a mechanical shaker. In NMIs, i.e., at the top level in the hierarchic traceability system, primary calibrations are performed using the national standard as a reference. At lower levels such as calibration services or manufacturer calibration laboratories, comparison calibrations are usually performed on the basis of reference standards and working standards, respectively. Common standards for calibration of traditional vibration transducers are back-to-back accelerometers. Basically, they can also be utilized for calibration of laser vibrometers, when the laser beam is steered towards their reference surface. However, this is not as simple as it appears, as a lot of error sources have to be considered with such a setup. This is mainly due to the fact that a laser interferometer acquires solely the motion vector in beam direction at the point of incidence on a surface, whereas an accelerometer always acquires a mean value on its complete reference surface. Numerous publications analyze the measurement uncertainty of such setups and give hints to achieve best performance [52,53].

Another method to calibrate laser vibrometers on a mechanical vibration exciter is one that uses a traceably calibrated laser vibrometer as a reference measurement system. With such a setup, the laser beams of the reference device and calibration object must be adjusted in such a way that both acquire exactly the same motion vector on the vibrating surface. This can be achieved by means of adjustable beam splitters and mirrors. The reference vibrometer should fulfill certain requirements regarding the signal processing method, measurement uncertainty, and long-term stability. Laser vibrometers with digital Doppler signal processing are preferred for reference measurement systems due to their inherently better accuracy and stability. In any case, a laser vibrometer used as a reference device must itself be traceably calibrated by comparing it to a higher-level standard, preferably the national measurement standard [49,51]. Under this provision, laser vibrometers can be traceably calibrated with very low measurement uncertainty in accredited calibration laboratories. Guidelines and international standards for calibration of laser vibrometers are currently in preparation.

The performance of mechanical vibration calibration is limited with respect to amplitude and frequency range by the properties of available vibration exciters. Perhaps new motion sources with higher vibration frequencies will be established in the near future, e.g., with the help of MEMS technology. At present, manufacturers of laser vibrometers are forced to seek alternative methods in order to verify and calibrate their products in the required amplitude and frequency ranges.

9.6.2.3 Calibration by Means of Synthetic Doppler Signals

The calibration method described in the following text does not use mechanical vibration as an input quantity for the device under test but uses a synthetic Doppler signal instead. Because no mechanical vibration source is involved, this method is very powerful in terms of vibration amplitude, frequency range, and accuracy. It can thus be used as a universal means for calibration and verification of laser vibrometers at the manufacturer's site. On the other hand, it is not a traceable calibration method, for the same reason.

Calibration of laser vibrometers applying synthetic Doppler signals relies on the assumption that all significant error sources are located in the electronic signal processing chain, whereas the optics are considered error-free. Consequently, the signal processing system can be calibrated and adjusted separately from the optics. As a substitute for the optics, modulated signals are generated equivalent to those coming from an ideal interferometer and fed to the input of the electronics. Figure 9.17 illustrates the principle of this calibration method. The fundamental relationships between mechanical and electrical quantities at the output of the photodetector are reflected in Equation 9.19 and Equation 9.21.

The uncertainty in the laser wavelength is less than 10^{-5} for a He–Ne laser and can be neglected in practice. Hence, the modulation parameters Δf and $\Delta\varphi$ of the Doppler signal that would occur at the output of the photodetector can be precisely calculated for any vibration amplitude and frequency. For calibration, usually pure harmonic waveforms are preferred. Depending on the required motion quantity, either frequency deviation or phase deviation of the Doppler signal are calculated for the desired calibration points according to

$$\Delta f = \frac{2\hat{v}}{\lambda} \quad \text{or} \quad \Delta\varphi = \frac{4\pi}{\lambda}\hat{s}. \tag{9.64}$$

Heterodyne or homodyne signals equivalent to original interferometer output signals are generated on basis of these modulation parameters and fed into the input section of the signal processing electronics. For calibration or frequency response testing, the resulting amplitude output of the

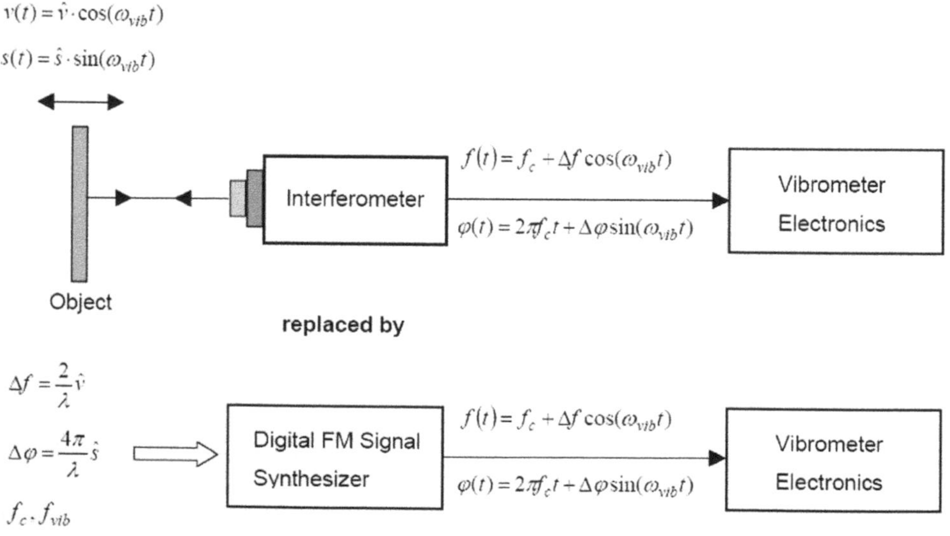

FIGURE 9.17 Principle of calibration by electronic signal substitution.

vibrometer is read and compared to the reference value. Phase response measurements can also be performed with reference to the phase of the modulating reference signal.

An indispensable prerequisite for this calibration method is an FM signal generator, capable of providing modulated RF signals of adequate modulation accuracy and spectral purity in the desired frequency range. Signals with defined phase deviation for calibration of displacement decoders can also be generated by an FM signal generator, as phase modulation and frequency modulation are coupled through the modulation index M (refer to Equation 9.25): a certain phase deviation $\Delta\varphi$ is obtained at a given vibration frequency when choosing the FM frequency deviation according to $\Delta f = \Delta\varphi \cdot f_{vib}$. As phase demodulation always relies on comparison of the instantaneous phase to a coherent reference phase, the FM signal generator should also provide the unmodulated carrier phase for reference.

The specifications of commercially available FM signal generators are often not adequate for calibration of Doppler signal decoders. Most of them are designed to generate narrow-band FM signals in a limited modulation frequency range and by far cannot cover the wide measurement range of a laser vibrometer. Also, their specified accuracy with respect to the modulation parameters Δf and $\Delta\varphi$ is often not satisfying. However, sophisticated digital technology allows the synthesis of high-precision FM signals in a wide range regarding frequency deviation and modulation frequency. Based on this technique, Doppler signal decoders can be calibrated with an uncertainty well below 1% at vibration frequencies up to hundreds of kilohertz. The limit to calibration accuracy is set by the measurement accuracy of the voltmeter rather than by the signal generator.

9.6.2.4 Calibration Based on Bessel Functions

The method described above is perfectly suited for electronic calibration of frequency and phase demodulators, but limited in vibration frequency. FM signal generators are rarely available for modulation frequencies >10 MHz. Laser vibrometers are commercially available for vibration frequencies up to 30 MHz, but basically, they can acquire vibrations even in the GHz region. This region actually becomes relevant for vibration analysis due to the latest developments in the field of RF MEMS. In the following text, a relatively simple method is described, allowing amplitude calibration of Doppler signal decoders even at very high vibration frequencies.

Fortunately, vibration amplitudes are very low at high frequencies. As outlined in Subsection 9.6.1, the corresponding Doppler signals are represented in the frequency domain by only a few equally spaced spectral lines (see Figure 9.16). The magnitude relationships between these lines are described by Bessel functions of the first kind and first order $J_n(M)$, where M is the modulation index, equivalent to the peak phase deviation (refer to Equation 9.25). The spacing of adjacent lines is equivalent to the vibration frequency f_{vib}. Some interesting properties of the Bessel functions make them very useful for calibration purposes in the region of low modulation indices. So, at certain values of M, the magnitude of particular Bessel functions is zero. The line representing J_1 at $f_c \pm f_{vib}$, for example, disappears for $M = 3.83$. These zero points can be detected very simply and accurately at any frequency. One method suggested for the primary calibration of vibration transducers according to ISO 16063-11 relies on this special property of the Bessel functions [50].

In the region M < 1, a very good approximation is $M = 2J_1$, when J_o is normalized to 1, as shown in Subsection 9.6.2. Analyzing the properties of Bessel functions in more depth, one can find a simple formula that allows to calculate any value of M on basis of the magnitudes of three adjacent spectral lines representing J_{n-1}, J_n, and J_{n+1} [54]:

$$M = \frac{2n \cdot J_n}{J_{n+1} + J_{n-1}} \tag{9.65}$$

How to benefit from this for vibrometer calibration in practice? If one is capable of establishing a mechanical vibration at the desired frequency with unknown but stationary amplitude, the spectrum of the resulting Doppler signal can be displayed on a suitable analyzer with pretty good amplitude accuracy. After selecting three adjacent lines well above the noise floor and more distant from the carrier than the line with the highest level in the spectrum, we can determine M from Equation 9.62. If, as is expected at high vibration frequencies, only J_0 and J_1 are visible, we can use the approach $M = 2J_1$. With Equation 9.19 and Equation 9.25 we finally obtain the vibration displacement amplitude $\hat{s} = \frac{M \cdot \lambda}{4\pi}$ as a calibration reference value. If a velocity reference is required, we get $\hat{v} = 2\pi f_{vib}\hat{s}$.

With the help of high-accuracy RF spectrum analyzers, amplitude calibration of both displacement and velocity decoders can be performed on the basis of the described method with an uncertainty well below 1%.

9.6.2.5 Influences on Measurement Accuracy

The metrological properties of laser vibrometers are mainly determined by the signal processing electronics, in particular by the frequency or phase demodulators. Depending on demodulation principle, measurement range, and vibration frequency bandwidth, their characteristics differ more or less from the ideal. Important specifications for evaluation of the amplitude measurement uncertainty of a laser vibrometer are the calibration accuracy at reference conditions, amplitude linearity, amplitude frequency response, and temperature coefficient of the scaling factor. For phase or time history measurements on mechanical systems, further parameters such as time delay, phase linearity, or group time delay vs. frequency play an important role.

Digital demodulators are, in terms of measurement accuracy and stability, superior to analog demodulators because they rely on mathematical methods, which do not suffer from drift, ageing, or nonlinear effects. However, even when utilizing the best demodulators, measurement errors may, in principle, also emerge due to nonideal behavior of the signal preconditioning blocks of a vibrometer. In particular, bandwidth limitation (refer to Section 9.3) may strongly affect the modulation of the Doppler signal.

Homodyne interferometers generate a quadrature signal pair in the baseband, i.e., a variable DC component is part of the information content. Special precautions must be taken here to maintain the signal integrity with respect to waveform distortions and DC accuracy. Otherwise, amplitude linearity errors, especially at very small vibration amplitudes, may occur in the measurement signal.

In most cases, the assumption of an error-free interferometer is absolutely justified; however, two optically induced errors can occur when measuring on MEMS. The first error arises for very low vibration amplitudes when back-scattered light is present, whereas the second error arises when a very high numerical aperture is used to impinge the beam with a high lateral resolution.

The first error results from the optical beat between back-scattered light and the reference beam and, as a rule of thumb, the effect has a considerable influence only on the accuracy for oscillation amplitudes smaller than the optical wavelength. Back-scattered light from all constant surfaces needs to be considered. If a very thin beam is illuminated, light that does not impinge the beam but just the static environment of the beam does contribute as an error source. If the investigated layer is very thin, light reflected at lower static layers contribute as an error source. A mathematical model that is in very good agreement with the experimental results is derived in [55]. The error can be avoided (smaller than 0.5%) if the signal light has at least 3 orders of magnitude more power than the back-reflected light. The amplitude error can be higher than 50% if the back-reflected light has more power and, consequently, the measurement is valueless. In addition, the oscillation amplitude has a strong fluctuation if the signal and back-scattered light have approximately the same amplitude. For a detailed discussion, refer to [55].

It is known from interference microscope systems (Linnik, Mirau, or Michelson type) that the fringe spacing increases with increasing numerical aperture of the interferometer-imaging optics [56]. The measured fringe spacing is in good agreement with a mathematical model derived in [57] and, therefore, this systematic error is automatically corrected in today's commercial phase-shifting interference microscopes. Obviously, a similar effect can be expected for microscopic laser Doppler measurements and, as demonstrated in [55], the measured velocity and displacement amplitudes are decreasing if measured with high-NA microscope objectives. However, it is shown in [55] that the physical approach to describe the fringe spacing in interference microscopes cannot be transferred directly to microscopic laser Doppler vibrometry. This error is smaller for microscopic LDV measurements as it is for interference microscopes and results from the Guoy phase delay of the Gaussian beam. As a rule of thumb, the amplitude error is smaller than 1% if the effective numerical aperture is NA < 0.25, which is typically the case for magnifications of the microscope objective up to 20X. The maximum amplitude error measured in [55] is 5% for a 60X objective lens with NA of 0.8.

9.7 COMBINATION WITH OTHER TECHNIQUES

Surfaces of silicon microdevices are usually optically flat, and light is reflected but not scattered. Therefore, a vibrometer can only be used to study motions along the optical axis of the imaging optics (out-of-plane). Vibrometry has to be combined with machine vision [58] techniques to realize a system with vibrometer performance that can find and measure resonances of mechanical structures, oscillating in all three dimensions [28].

In addition, vibrometry in conjunction with optical profile measurement methods [59] opens interesting measurement possibilities. The combination of interferometric profile determination techniques with 3-D motion data [60] is important to combine oscillation data with geometry data. As the MSV can be assembled on an infinity-corrected microscope with a short-coherence-length illumination, it is possible to create a white-light interference microscope by adding a precision position stage with an interference objective on one objective mount of the revolver. The white-light interferometer measures the focus position with the maximum modulation of the interference signal (correlogram) as height profile value at every pixel of the CCD sensor. The white-light technique is described in detail in Chapter 8 and, therefore, we summarize here stroboscopic video microscopy, which supplements the vibrometer about in-plane-motion measurements to a complete 3-D measurement.

The highly sensitive laser Doppler technique (with sub-picometer resolution) can be used to find all mechanical resonances rapidly without a priori information about the device if wide-band excitation is used. A frequency signal that does not fall in an FFT line is distributed over the neighboring lines, and, therefore, even a narrow vibration peak that falls between two FFT lines can be discovered.

The vibrometer laser beams and the strobe illumination have to be coupled into the beam path of the same microscope to integrate the two techniques in one setup. This can be realized by designing two beam splitters between the microscope objective and the microscope tube lens. A beam splitter is used to couple the vibrometer laser beam into the microscope beam path. The second splitter couples the strobe light source into the microscope. The vibrometer microscope optics and the Köhler illumination for the stroboscopic machine vision technique can be designed independently. The stroboscope illumination has to be designed according to the rules of Köhler to ensure high contrast and high optical resolution of microscope images. The main point of the rules of Köhler is that the specimen is placed in a pupil of the illumination beam path and, consequently, the strobe lamp is imaged to the back-sided focus of the microscope objective lens. This simplest arrangement is shown in Figure 9.18.

A stroboscopic video microscope measures in-plane motions of periodically moving structures with stroboscopic machine vision and can measure frequencies as high as 1 MHz [28]. The camera used in these kinds of detection systems usually has a CCD sensor for video frame rates and not a high-speed detector and, therefore, the stroboscopic principle has to be applied to visualize rapid motions. An LED has been proven to be a reliable solution for the light source that ensures constant

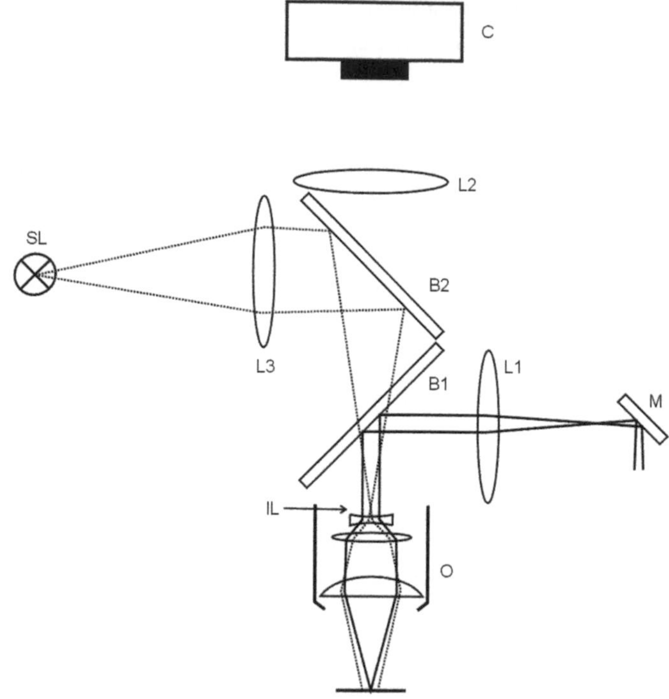

FIGURE 9.18 Optics arrangement for a combined vibrometer and machine vision measurement. L1 is the scanner lens, L2 the microscope tube lens that images the specimen on the camera chip C, lens L3 images the strobe lamp to the back-sided focus IL of the objective O, B1 and B2 are beam splitters, and M is the scanning mirror.

illumination power of the strobe pulses. The time resolution of the system is defined through the pulse width of the strobe flash because the camera is not rapid enough to capture very short events. No light is collected through the CCD sensor when the strobe light is off. Therefore, events can be recorded with a period time even shorter as the shortest possible exposure time of the camera.

The specimen-driving signal, the strobe flashes, and the camera exposure have to be accurately synchronized. The timing diagram of the strobe synchronization is shown in Figure 9.19 for an example

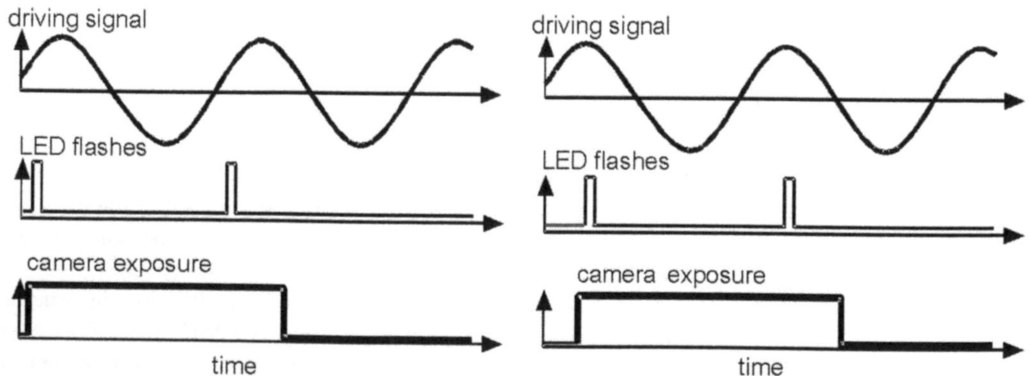

FIGURE 9.19 Timing diagram of the planar motion analyzer (PMA) signals.

of two camera shots. The shots are recorded at two different phases of the periodic excitation for the specimen.

Two LED flashes are used in Figure 9.19 within the exposure time of the camera. The number of flashes per camera shot can be used to adjust the image brightness. The time between two shots is the cycle duration of the camera framing rate. The phase delay of the strobe illumination with respect to the driving signal is adjusted by setting the duration T_{shot} between the shots to $T_{shot} = nT_{excitation} + t_{phase\,delay}$. Here, n is an integer, $T_{excitation}$ is the period length of the excitation signal, and $t_{phase\,delay}$ is the time shift that corresponds to the phase delay. The maximum frame rate F_c of the digital camera limits the shot frequency to $F_c \geq 1/T_{shot}$. The procedure demonstrated in Figure 9.19 is completed when all images necessary to derive the displacement response with image processing are captured. Short strobe pulses are necessary to freeze a rapidly moving structure. Blur is generated if the device moves a longer distance than the distance that corresponds to the diameter of a camera pixel during the strobe illumination. It is necessary to use only a few flashes per shot (best is one flash) if the device does not perform a precise periodical motion but has a jitter. In this case, blur is generated if the jitter is higher than the distance that corresponds to one pixel.

Modern video microscopy systems can automatically record frequency responses. Image sets are recorded for a number of frequencies to obtain a frequency response. Displacement vs. phase delay data is extracted for every measured frequency automatically by employing image processing techniques. Phase and amplitude are computed through a sine-function fit from the displacement vs. phase delay data for every frequency record.

In-plane shifts di and dj between image 1 (I_1) and image 2 (I_2) are computed with sub-pixel resolution by image correlation, which is discussed in detail in Chapter 3. Two images are matched if the displacement-dependent, normalized correlation coefficient

$$r_c(di,dj) = \frac{\sum_{k=1}^{K}\sum_{l=1}^{L}\left(I_1(k,l)-\overline{I_1}\right)\left(I_2(k+di,l+dj)-\overline{I_2}\right)}{\left[\sum_{k=1}^{K}\sum_{l=1}^{L}\left(I_1(k,l)-\overline{I_1}\right)^2\right]^{1/2}\left[\sum_{k=1}^{K}\sum_{l=1}^{L}\left(I_2(k+di,l+dj)-\overline{I_2}\right)^2\right]^{1/2}}, \quad (9.66)$$

at (di, dj) is a maximum:

$$\max\left(r_c(di,dj)\right). \quad (9.67)$$

Here $\overline{I}$ denotes the average intensity value of the pixels in I and ($k,l,K,L \in \mathbf{Z}$ (integer); di, $dj \in \mathbf{R}$ (real)). If r_c is maximum, the difference between the image-pattern template $I_1(k, l)$ and the shifted image $I_2(k + di, l + dj)$ is a minimum. Therefore, the displacements di and dj are the estimation parameters for an optimization algorithm that computes the maximum of r_c. The in-plane motion algorithm computes di and dj with sub-pixel resolution. The Nyquist sampling theorem can be employed to calculate a resampled image $I_r(i, j)$ ($i, j \in \mathbf{R}$), which is the key to understanding the idea of subpixel displacement computation.

The schematic of a system that combines the two methods is shown in Figure 9.20 and has been developed by Polytec [67]. The strobe light and the vibrometer beam are coupled within a confocal microscope setup.

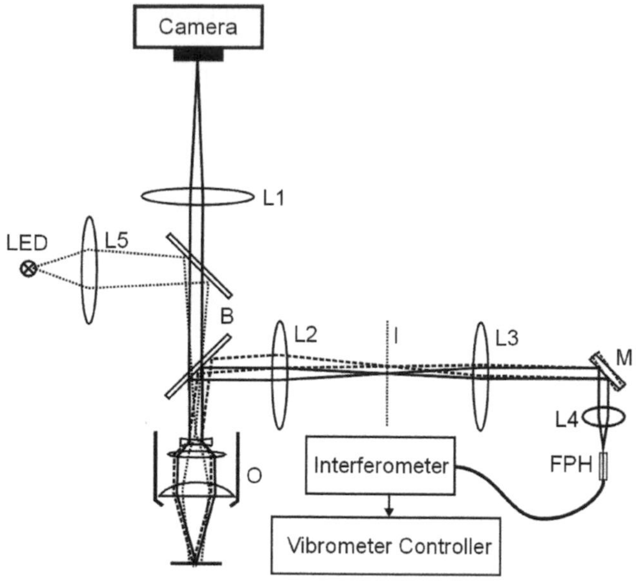

FIGURE 9.20 Schematic of a microscope-scanning setup. The laser beam is collimated at the exit pupil of the objective, and the scanner is imaged into the back-sided focal plane. FPH is a fiber that acts as a pinhole. O is the objective, L1 through L5 are lenses, B are beam splitters, I is an intermediate image, and M is the scanning mirror.

9.8 EXAMPLES

9.8.1 Dual Mode MEMS Mirror

The U.S. company Applied MEMS, Inc. [62,63] has developed a MEMS mirror device for use in optical bar code scanners. The micromirror is completely manufactured from single-crystal silicon. It is electrostatically actuated and offers a large rotation angle of at least ±14° (Table 9.1).

This mirror device was extensively tested using the microscope scanning LDV technique. The purpose of these tests was to measure the various resonant modes and to confirm that the mirror maintains optical planarity while undergoing electrostatic actuation. Modeling analysis on the mirror has shown that the single-axis mirror may be operated in a dual-axis mode, which has been confirmed by the laser measurements. Figure 9.21 shows the primary rotational mode about the x-axis and a secondary rotational mode about the y-axis.

TABLE 9.1
Design Criteria of the Micromirror Device

Criteria	Specification
Mirror size	1 mm × 1.4 mm
Die size	4.4 mm × 2.5 mm × 1.5 mm
Rotation angle	±12° operational
	±14° maximum
Resonant frequency	7585 Hz
Actuation voltage	365 V static
	~120 V dynamic
Shock tolerance	2000 g all axes

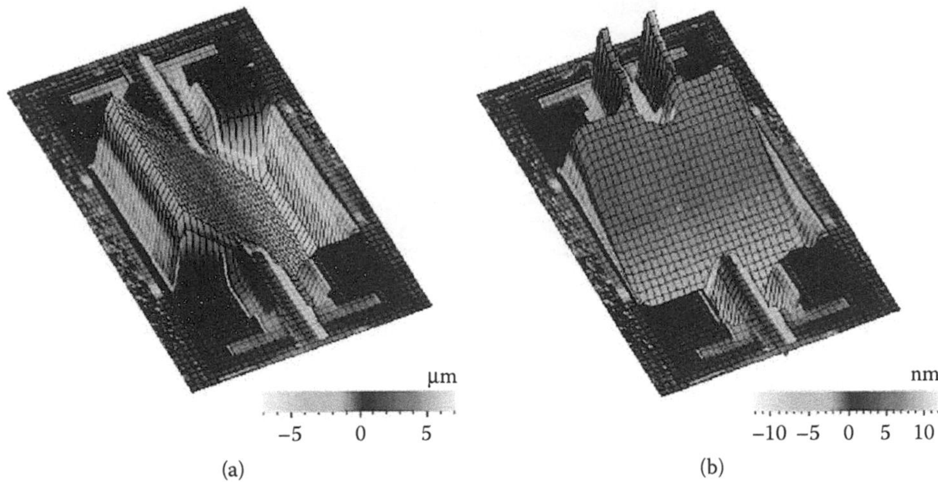

FIGURE 9.21 (See color insert following page 110.) (a) Primary and (b) secondary mode of the micromirror as measured by scanning LDV technique.

Methods to accurately measure the dynamics of microdevices under different environmental conditions are needed to get the necessary parameters for the analytical models. Laser Doppler vibrometry is a useful method to characterize and optimize device performance. The vibrometer measurements can be compared with theoretical models to improve the models for an advanced design circle. Actual vibration amplitudes and frequencies even from micron-sized elements can be obtained and entered into models for reliability and lifetime predictions. This technique is also useful in production monitoring, as defects in faulty devices will cause responses that are different from devices that are free of defects.

9.8.2 Cantilever Beam Acceleration Sensor

With the scanning ability to automatically acquire, analyze, and reconstruct complex vibration modes, microscope scanning vibrometry is the ideal technique for displaying out-of-plane deflections. A major advantage of the LDV technique is its capability to record vibration data for the full frequency spectrum of interest, i.e., it is not required to excite the structure with single sinusoidal frequencies. This allows one to obtain the broadband frequency response in a short time measurement.

However, there is one major limitation to the technique. It is difficult or impractical to measure in-plane motions because the Doppler shift is derived from a velocity vector normal to the plane of the moving surface. But it is possible to overcome this drawback using stroboscopic video microscopy as an additional and complementary technique to measure in-plane motion of periodically moving structures. This analysis is done on the basis of the broadband out-of-plane data.

The MEMS acceleration sensor displayed in Figure 9.22 has been developed for frequency selective acceleration measurements. It comprises a large number of cantilever arms of different lengths. Naturally, the resonance frequency for each lever will be different as well. When experiencing shock or vibration, only certain levers will vibrate, depending upon their resonance frequency.

The scanning LDV technique was used to measure the resonance frequencies of the individual levers and for the presentation of the out-of-plane operational deflection shape. In a second step, stroboscopic video microscopy was used to study the in-plane motion of the device. The outstanding capabilities of the SLDV for frequency response and operational deflection shape measurements in out-of-plane direction is thus expanded to full 3-D motion analysis.

The microsensor itself is placed under a standard microscope and is mechanically excited by a small piezo actuator. A broadband periodic chirp signal from a signal generator is applied to the

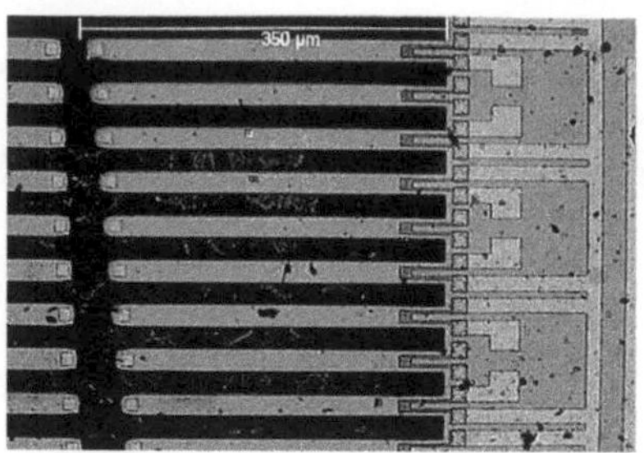

FIGURE 9.22 Video image of the micro sensor, magnification 5X.

actuator for excitation of all frequencies, with the same energy. The generator signal is looped back into the acquisition channel and thus provides the reference phase.

9.8.2.1 Out-of-Plane Analysis

Figure 9.23 shows a live video image from the CCD camera. The analysis is performed on three example cantilevers although the sample contained more than thirty. The scan point grid for each lever is defined rapidly with a freehand tool on the video image. Density is chosen to give 180 scan points on a rectangular grid (Figure 9.23). A measurement with a bandwidth of 100 kHz is then performed point by point. Figure 9.24 shows the averaged frequency response function (FRF).

The spectrum represents the average over all 180 scan points. Using a 3200 FFT-line analysis gives a frequency resolution of 31.25 Hz and a total measurement time of about 20 sec for the complete structure including complex averaging. As expected, two resonance triplets are found, one around 27 kHz and another around 75 kHz corresponding to two orders of bending modes of the three cantilevers.

First, the ODS of the 27-kHz triplet is investigated (Figure 9.25). The resonances are at 26.3 kHz, 27.1 kHz, and 27.6 kHz with similar displacement amplitudes of about ±1.5 nm. The deflection modes are displayed in Figure 9.26.

These results are quite useful when it comes to validating resonances obtained by finite element models (FEM). The resonance triplet around 75 kHz, however, reveals some new particularities (Figure 9.27).

First, the higher-order bending mode triplet is observed with similar displacement amplitudes of ±20 nm. The corresponding ODS are displayed in Figure 9.28.

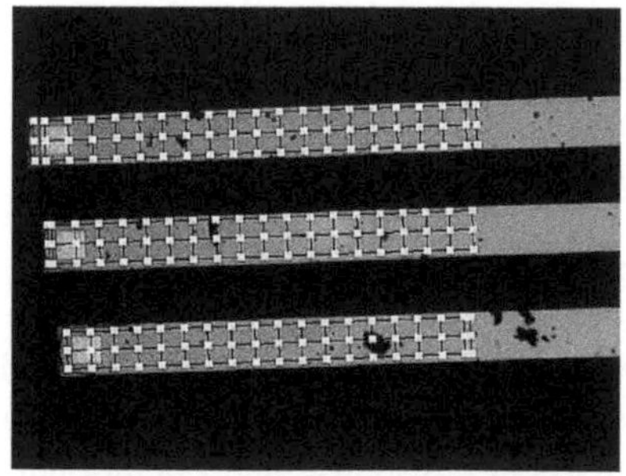

FIGURE 9.23 Live video image of cantilever structure with scan grid.

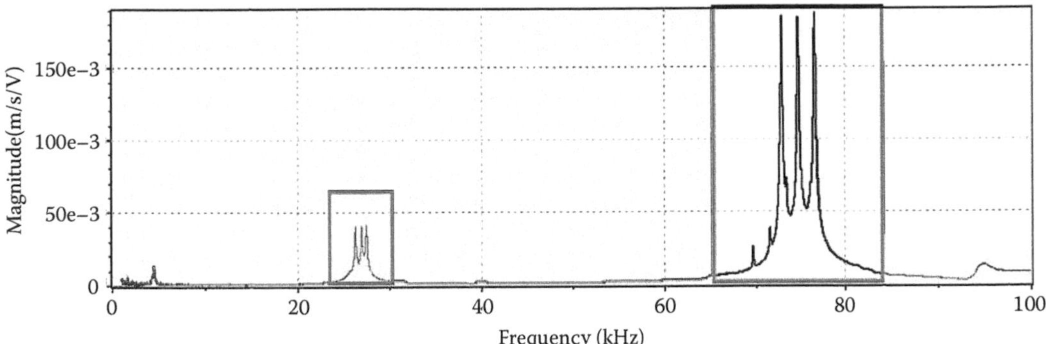

FIGURE 9.24 Averaged FRF, 100 kHz BW.

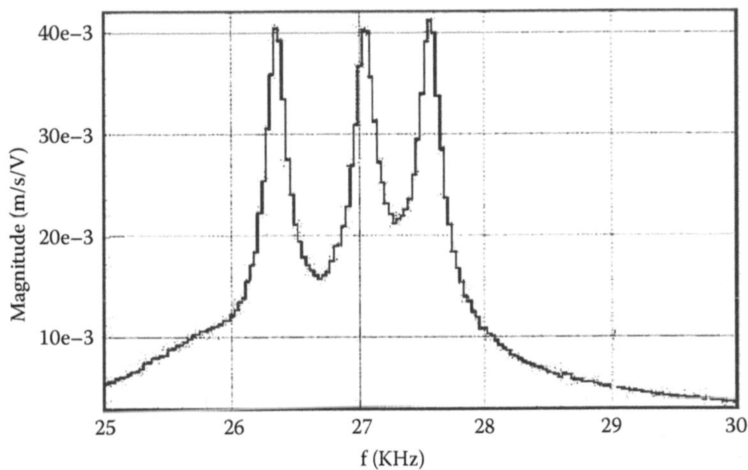

FIGURE 9.25 Enlarged view of the triplet around 27 kHz from Figure 9.24.

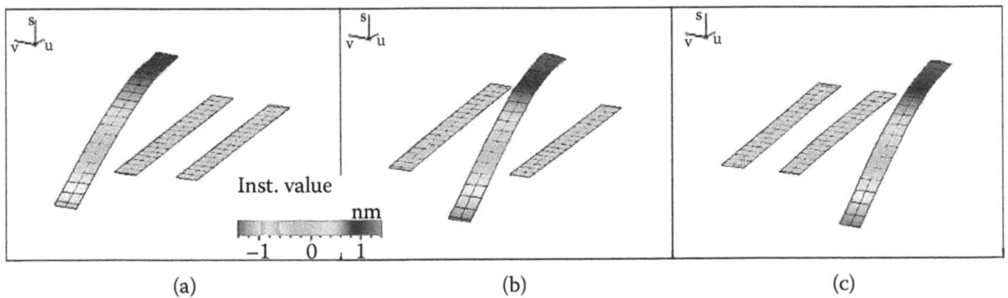

FIGURE 9.26 (See color insert.) ODS at (a) 26.31 kHz, (b) 27.09 kHz, and(c) 27.59 kHz.

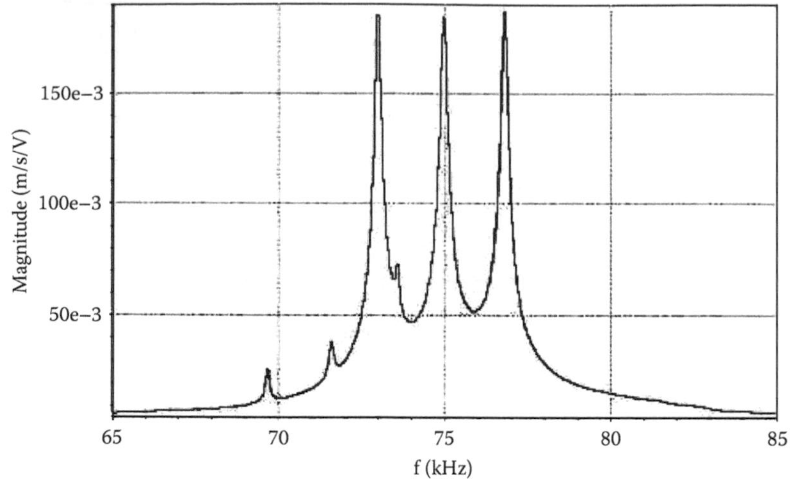

FIGURE 9.27 Enlarged view of the triplet around 75 kHz from Figure 9.24.

Apart from these higher-order bending modes, the FRF spectrum in Figure 9.27 contains a superposed third triplet of inferior amplitude. Figure 9.29 shows the resonances of this third triplet as snapshots from the 3-D ODS animation revealing tilt motion of the cantilevers at 69.7 kHz, 71.53 kHz, and 73.56 kHz. Amplitude is on the order of ±2 nm.

It can be supposed that in reality the tilting of the three cantilevers corresponds to a combined in-plane and out-of-plane mode. A rotary movement, for example, in which in- and out-of-plane motions are 90° phase-shifted would appear as a tilting in an out-of-plane-only measurement. Therefore, the 73.56-kHz resonance is chosen for an in-plane analysis using stroboscopic video microscopy.

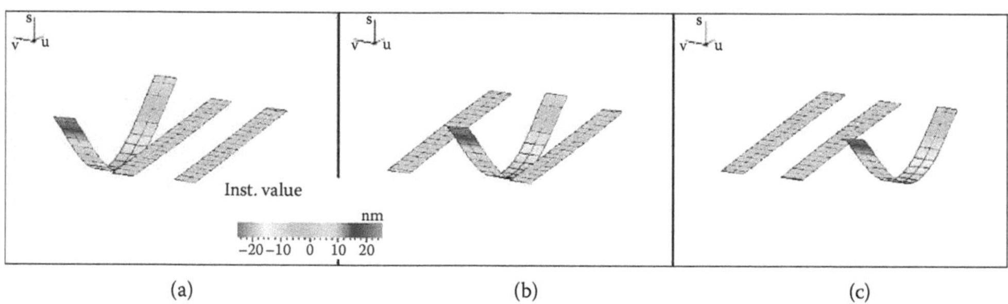

FIGURE 9.28 (See color insert.) ODS at (a) 72.97 kHz, (b) 74.94 kHz, and (c) 76.78 kHz.

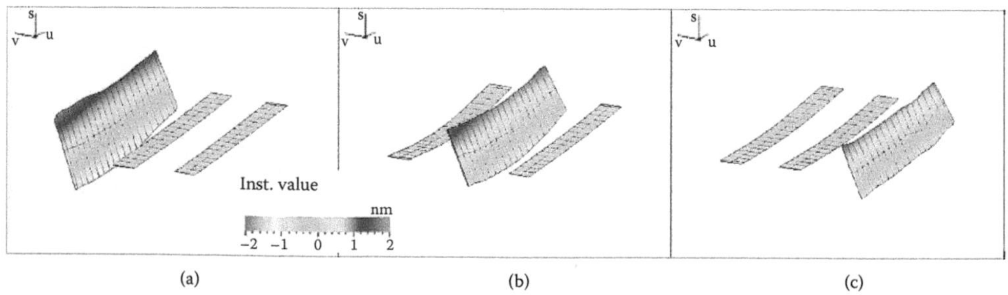

FIGURE 9.29 (See color insert.) ODS at (a) 69.66 kHz, (b) 71.53 kHz, and (c) 73.56 kHz.

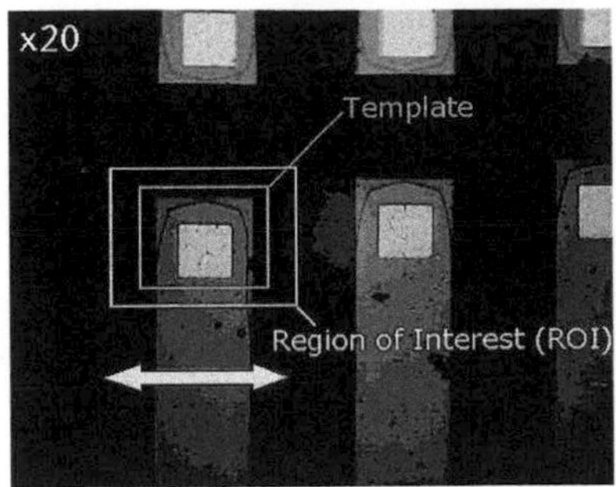

FIGURE 9.30 Image template and ROI.

9.8.2.2 In-Plane Analysis

Analysis is done by choosing a small template (inner box in Figure 9.30) of the moving structure in the image as well as a region-of-interest (ROI, outer box in Figure 9.30) in order to reduce time for calculation. Displacement (δx, δy) is obtained for each phase by image-to-image correlation algorithms (see Section 9.7). This technique has been demonstrated to determine oscillation amplitude resolutions of 5 nm [67]. Results are plotted in Bode plots (see Figure 9.31 and Figure 9.32).

The trajectory analysis (Figure 9.33) shows finally a sinusoidal in-plane motion with 6.5 μm peak-to-peak amplitude. These results are a perfect match to the previous out-of-plane analysis (± 2 nm amplitude) and show the important benefit of a hybrid system for 3-D motion analysis on MEMS structures.

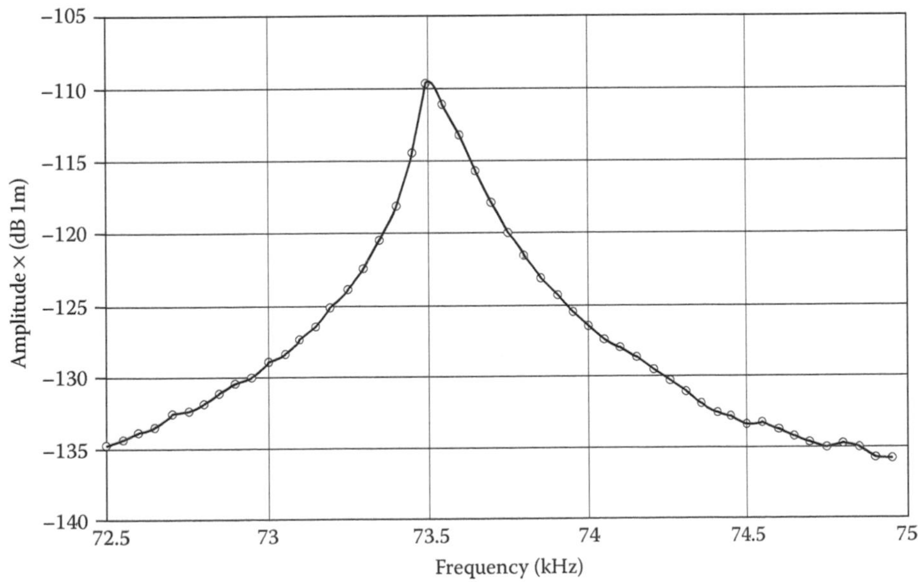

FIGURE 9.31 Bode plot of δx — amplitude.

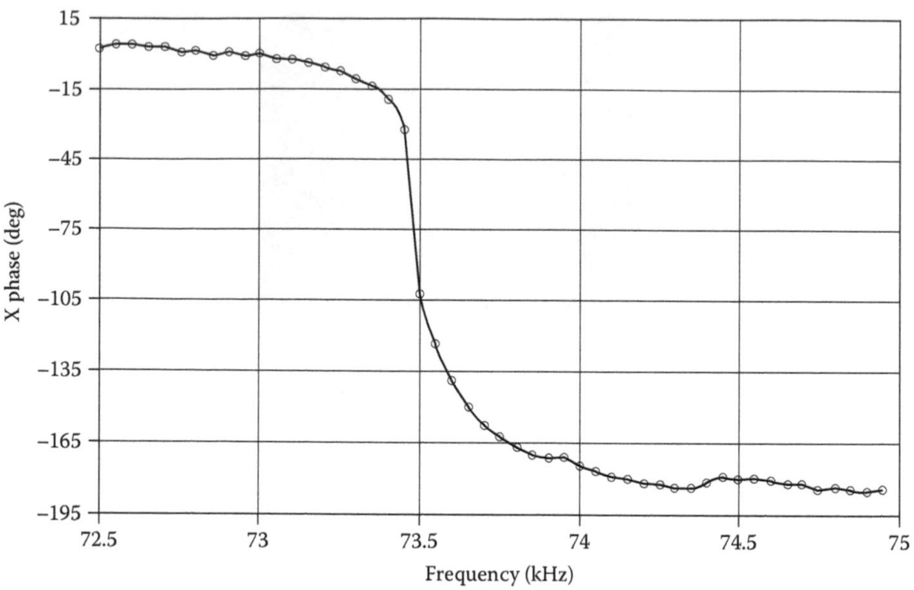

FIGURE 9.32 Bode plot of δx — phase.

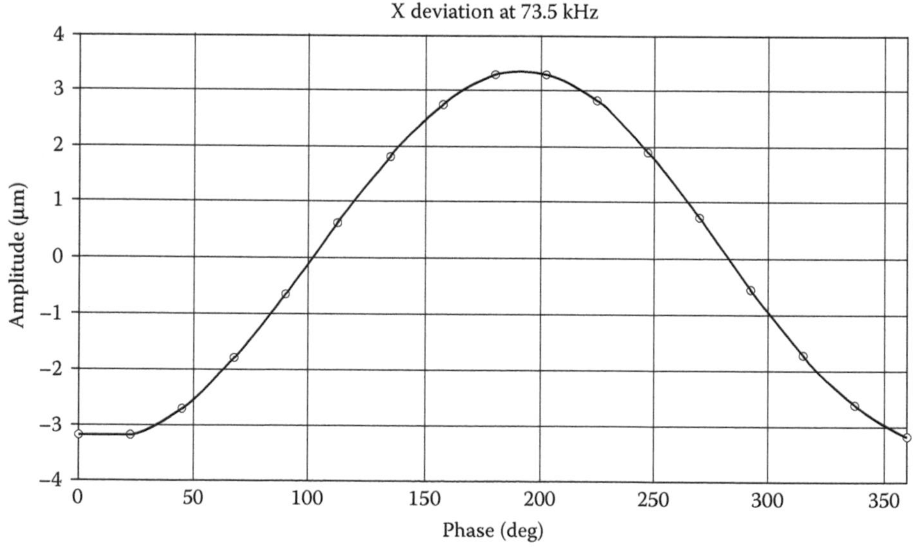

FIGURE 9.33 Trajectory plot of δx.

9.9 CONCLUSION AND OUTLOOK

Advanced testing methods for the dynamics of mechanical microdevices are necessary to develop highly reliable MEMS, which is the key for a successful introduction into the market. Mechanically testing and measuring MEMS provides feedback that helps the designer make progress in an iterative process from prototypes to fully developed products. The effectiveness of these procedures is markedly enhanced if the measurements include laser Doppler vibrometry of mechanical structures in the microsystem.

A laser Doppler vibrometer provides a real-time velocity or displacement signal of a single spot on the specimen surface. Therefore, a vibrometer can measure non-repeatable transient or chaotic processes. However, there are restrictions if a scanning measurement is performed because the surface measurement points of the operational deflection shapes need to have a defined phase relation. To do so, the specimen is excited with a periodic signal. This signal can contain a large number of harmonic frequency components. The data acquisition needs to be synchronized with the excitation or some reference signal that corresponds to the specimen oscillation. The vibrometer signals obtained on every measured spot are fast Fourier transformed and the spectrum is saved. The scan method is especially effective and accurate if the frequencies contained in the excitation signal accord with the FFT lines of the spectrum. The combination of the spatially distributed oscillations at an FFT line results in phase-dependent ODSs. The resolution for the measurement of an oscillation amplitude at a single frequency in the spectrum depends on the width of the FFT line. Very high resolution can be achieved for narrow width of the FFT lines. The noise limit is typical far below 1 pm/$\sqrt{\text{Hz}}$ without averaging at frequencies higher than 50 kHz for a well-designed vibrometer with a digital decoder.

The measurement time of a frequency spectrum corresponds to the inverse of the frequency resolution. Therefore, a spectrum with a resolution of 20 Hz is measured every 50 ms and can be displayed in real time on a monitor screen and has an amplitude resolution far below 1 nm. This feature makes laser Doppler vibrometry outstandingly suited for production monitoring and manufacturing processes such as trimming. Trimming is a procedure that is applied when the resonance frequency of a mechanical structure has to be adjusted in a postprocessing manufacturing iteration. This is usually implemented by removing material with an additional high-power laser beam.

However, laser vibrometry has some limitations. For example, it cannot be used to measure in-plane motions if the surface has a roughness less than the wavelength within the diameter of the vibrometer laser spot. Therefore, it can only be used to measure the motion component in out-of-plane direction parallel to the optical axis of the microscope objective. A vibrometer measures a displacement or a velocity signal and cannot measure the absolute position, so a vibrometer is not suited to measure the shape of a structure.

Different measurement techniques that are suited to measure surface profiles and in-plane motions are enhanced in combination with laser Doppler vibrometry. Imperfections in the MEMS device always led to mechanical coupling between in-plane and out-of-plane motions. Therefore, even in-plane resonances can be found rapidly with a laser Doppler vibrometer. Then a time-consuming stroboscopic video and image processing technique can be employed to accurately measure the in-plane frequency response at the resonance. In addition, it is very helpful to combine interference microscope measurements with vibrometer results because geometry always strongly influences the dynamic behavior. The combination of the frequency response data with surface profile data is important because design engineers need the 3-D coordinates and the oscillation data of a point in the space to perform comparisons between finite-element models and measurement data. Stroboscopic interference microscopy can be used to obtain time-dependent profile data of a harmonically excited device at a resonance found with a vibrometer measurement.

Therefore, future efforts will target the combination of profile and vibration data in a combined data set and the utilization of this data in simulation tools. By employing methods of model identification, it will be possible to extract material properties and geometric information like layer thickness from the profile and vibration data.

Commercial vibrometers can measure up to 30 MHz, which is sufficient for most of today's MEMS applications. For future RF-MEMS developments and for mobile phone SAW (surface acoustic wave) filters, frequencies up to the GHz range with picometer amplitudes need to be measured. Therefore, additional effort for future vibrometer developments is necessary to meet these demands.

REFERENCES

1. Peterson, K.E., Silicon as a mechanical material, *Proc. IEEE*, 70: 420–457, 1982.
2. Bustillo, J.M., Howe, R.T., Muller, R.S. Surface micromachining for microelectromechanical systems, *Proc. IEEE*, 86: 1552–1574, 1998.
3. Rai-Choudhury, P., Ed., *MEMS and MOEMS Technology and Applications*, Washington: SPIE Press, 2000.
4. Menz, W., Mohr, J., *Mikrosystemtechnik für Ingenieure*, 2nd ed., Weinheim: Wiley-VCH, 1997.
5. Müller-Fiedler, R., Microsystems for automotive applications, in Herbert Reichl, Ed., *Microsystem Technologies 2003*, Poing: Franzis Verlag, 2003, pp. 9–26.
6. Aigner, R., Microsystems in mobile phone applications, in Herbert Reichl, Ed., *Microsystem Technologies 2003*, Poing: Franzis Verlag, 2003, pp. 27–34.
7. Audet, S., Medical applications of microsystems, in Herbert Reichl, Ed., *Microsystem Technologies 2003*, Poing: Franzis Verlag, 2003, pp. 37–43.
8. Yeh, Y., Cummins, H.Z., Localized fluid flow measurement with an He–Ne laser spectrometer, *Appl. Phys. Lett.*, 4: 176–178, 1964.
9. Foreman, J.W., George, E.W., Lewis, R.D., Measurement of localized flow velocities in gases with a laser Doppler flow meter, *Appl. Phys. Lett.*, 7: 77–78, 1965.
10. Kroeger, R.D., Motion sensing by optical heterodyne Doppler detection from diffuse surfaces, *Proc. IEEE*, 53: 211–212, 1965.
11. Defarri, H.A., Andrews, F.A., Laser interferometric technique for measuring small-order vibration displacements, *J. Acoust. Soc. Am.*, 39: 979–980, 1966.
12. Defarri, H.A., Andrews, F.A., Vibrational displacement and mode-shape measurement by laser interferometer, *J. Acoust. Soc. Am.*, 42: 982–990, 1967.
13. Eberhardt, F.J., Andrews, F.A., Laser heterodyne system for measurement and analysis of vibration, *J. Acoust. Soc. Am.*, 48: 603–609, 1970.
14. Lewin, A.C., Kersey, A.D., Jackson, D.A., Non-contact surface vibration analysis using a monomode fibre optic interferometer incorporating an open air path, *J. Phys. E*, 18: 604–608, 1985.
15. Halliwell, N.A., Pullen, H.L., Baker, J., The laser torsional vibrometer: a new instrument, *J. Sound Vib.*, 101: 588–592, 1985.
16. Yienger, J.M., West, R.L., Mitchel, L.D., Laser-based structural imaging of an open and closed automobile door for noise control purposes, in *IMAC Proc. 11th Int. Modal Conf.*, 1983, pp. 144–149.
17. Jeong, G., Bogy, D.B., Measurements of slider-disk contacts during dynamic load-unload, *IEEE Trans. Magn.*, 27: 5073–5075, 1991.
18. Sabatier, M., Xiang, N., Laser-Doppler-Vibrometer-based anti-personnel mine detection, *Proc. IEEE Geosci. Remote Sensing Symp.*, 2001, pp. 3093–3095.
19. Frank, M., Knittel, J., Wyszynski, A., Application of laser methods to automotive seat modal testing, *Proc 2nd Int. Conf on Vibration Measurements by Laser Techniques: Advances and Applications*, University of Ancona, Italy, September 23–25, 1996, SPIE 2868, pp. 346–351.
20. Castellini, P., Miglietta, G., Revel, G.M., Scalise, L., Dynamic characterisation of teeth by laser vibrometry, *Proc. 3rd Int. Conf. on Vibration Measurements by Laser Techniques: Advances and Applications*, Ancona, Italy, June 16–19, 1998, SPIE 3411, pp. 581–590.
21. Drain, L.E., *The Laser Doppler Technique*, Chichester: John Wiley & Sons, 1980.
22. Sriram, P., Craig, J.I., Hanagud, S., Scanning laser-Doppler techniques for vibration testing, *Exp. Tech.*, 21–26, November–December 1992.
23. Burdess, J.S., Harris, B.J., Wood, D., Pitcher, R.J., A system for the dynamic characterisation of microstructures, *J. Microelectromech. Syst.*, 6: 322–328, 1997.

24. Eichner D., von Münch, W., A two-step electrochemical etch-stop to produce freestanding bulk-micromachined structures, *Sensor. Actuator A*, 60: 103–107, 1997.
25. Turner, K.L., Hartwell, P.G., MacDonald, N.C., Multi-dimensional MEMS motion characterization using laser vibrometry, Turner, K.L., *The 10th Int. Conf. on Solid-State Sensors and Actuators (Transducers '99)*, Sendai, Japan, June 7–10, 1999, Digest of Technical Papers, pp. 1144–1147.
26. Krehl, P., Engemann, S., Rembe, C., Hofer, E.P., High-speed visualization, a powerful diagnostic tool for microactuators — retrospect and prospect, *Microsyst. Technol.*, 5: 113–132, 1999.
27. *Polytec Hardware Manual*, Microscope Scanning Vibrometer MSV 300, Waldbronn, Germany: Polytec GmbH, 2002.
28. Rembe, C., Dräbenstedt, A., Heimes, F., Accurate new 3D-motion analyzer for MEMS, in *Microsystem Technologies* 2003, Herbert Reichl, Ed., Poing: Franzis Verlag, 2003, pp. 435–442.
29. Saleh, B.E.A., Teich, M.C., *Fundamentals of Photonics*, New York: John Wiley & Sons, 1991.
30. Feynman, R.P., Leighton, R.B., Sands, M., *The Feynman Lectures on Physics*, Vol. II, Reading, MA Addison Wesley, 1963.
31. Vogel, H., *Gehrtsen Physik*, 19, Auflage, Berlin: Springer-Verlag, 1997.
32. Hobbs, P.C.D., Reaching the shot noise limit, *Optics and Photonics News*, April 17–23, 1991.
33. Demtröder,W., *Laserspektroskopie*, 3, Auflage, Berlin: Springer-Verlag, 1993.
34. Strean, R.F., Mitchell, L.D., Barker, A.J., Global noise characteristics of a laser Doppler vibrometer, Part I: Theory, *Proc. 1st Int. Conf. on Vibration Measurements by Laser Techniques, Advances and Applications*, 1996, SPIE, Vol. 2868, pp. 2–11.
35. Bauer, M., Ritter, F., Siegmund, G., High-precision laser vibrometers based on digital Doppler-signal processing, *Proc. 5th Int. Conf. on Vibration Measurements by Laser Techniques, Advances and Applications*, 2002, SPIE, Vol. 4827, pp. 50–61.
36. Lewin, A., Mohr, F., Selbach, H., Heterodyn-Interferometer zur Vibrationsanalyse, *Technisches Messen*, 57: 315–362, 1990.
37. Buchhave, P., Laser Doppler Vibration Measurements Using Variable Frequency Shift, DISA Information 18, September 15–20, 1975.
38. Mäusl, R., *Analoge Modulationsverfahren*, Heidelberg: Hüthig, 1992.
39. Wagner, J.W., Optical detection of ultrasound, *Phys. Acoust.*, 19: 1990.
40. Randell, R.B., *Frequency Analysis*, 3rd ed., Copenhagen, Brüel and Kjaer, 1987.
41. Schüssler, M., Wörtge, M., Survey of excitation signals with respect to scanning vibrometer measurements, *Proc. IEEE*, 3411: 386–393, 1998.
42. Boche, H., Nahler, A., Erzeugung von Multisinussignalen mit niedrigem Crestfaktor, NTZ, 49/6, 16–19, 1996.
43. Heylen, W., Lammers, S., Sas, P., Modal Analysis Theorie and Testing, Leuven: KU, 1997.
44. Self, S.A., Focusing of spherical Gaussian beams, *Appl. Opt.*, 22: 658–661, 1983.
45. Johansmann, M., Wörtge, M., Siegmund, G., New developments in laser Doppler vibrometer optical systems and demodulation schemes for measurements on MEMS and other micro structures, *16th Brazilian Congress of Mechanical Engineering*, COBEM, Uberlandia, Minas Gerais, Brazil, November 26–30, 2001.
46. van Netten, S.M., Laser interferometer microscope for the measurement of nanometer vibrational displacements of a light-scattering microscopic object, *J. Acoust. Soc. Am.*, 83: 1667–1674, 1988.
47. Corle, T.R., Kino, G.S., *Confocal Scanning Optical Microscopy and Related Imaging Systems*, San Diego: Academic Press, 1996.
48. Dorn, R., Quabis, S., Leuchs, G., Sharper focus for a radially polarized light beam, *Phys. Rev. Lett.*, 91: 2329011–23290114, 2003.
49. Martens, H.-J.V., Current state and trends of ensuring traceability for vibration and shock measurement, *Metrologia*, 36: 357–373, 1999.
50. International Standard ISO 16063-11. Methods for the calibration of vibration and shock transducers — Part 11: primary vibration calibration by laser interferometry, Geneva: International Organization for Standardization, 1999.
51. Martens, H.-J.V., Täubner, A., Wabinski, W., Link, A., Schlaak, H.-J., Laser interferometry — tool and object in vibration and shock calibrations, *Proc. 3rd Int. Conf. on Vibration Measurements by Laser Techniques*, Ancona, Italy, 1998, SPIE Vol. 3411, pp. 195–206.

52. Martens, H.-J.V., Link, A., Schlaak, H.-J., Täubner, A., Wabinski, W., Investigations to assess the best accuracy attainable in accelerometer calibrations, *Proc. 5th Int. Conference on Vibration Measurements by Laser Techniques*, Ancona, Italy, 2002, SPIE Vol. 4827, pp. 258–276.
53. Link, A., Wabinski, W., Pohl, A., Martens, H.-J.V., Accelerometer identification using laser interferometry, *Proc. 4th Int. Conference on Vibration Measurements by Laser Techniques*, Ancona, Italy, 2000, SPIE Vol. 4072, pp. 126–136.
54. Rohde and Schwarz, Application Notes: 110-04-0289, 1994.
55. Rembe, C., Dräbenstedt, A., Additional error sources for microscopic laser-Doppler measurements, *Proc. 6th Int. Conf. on Vibration Measurements by Laser Techniques, Advances and Applications*, Ancona, Italy, June 22–25, 2004, SPIE Vol. 5503, pp. 446–457.
56. Biegen, J.F., Calibration requirements for Mirau and Linnik microscope interferometers, *Appl. Opt.*, 28: 1972–1974, 1989.
57. Tolmon, F.R., Wood, J.G., Fringe spacing in interference microscopes, *J. Sci. Instrum.*, 33: 236–238; see also note from Gates, J.W. on p. 507, 1956.
58. Rembe, C., Hofer, E.P., Tibken, B., Analysis of the dynamics in microactuators using high-speed cine photomicrography, *J. Microlelectromech. Syst.*, 10: 137–145, 2001.
59. Creath, A. Morales, Contact and noncontact profilers, in *Optical Shop Testing*, 2nd ed., Ed. Daniel Malacara, New York: John Wiley & Sons, 1992, pp. 653–686.
60. Rembe, C., Muller, R.S., Measurement system for full three-dimensional motion characterization of MEMS, *J. Microlelectromech. Syst.*, 11: 479–488, 2002.
61. *Polytec Hardware Manual*, Micro Motion Analyzer MMA 300, Waldbronn, Germany: Polytec, 2003.
62. Goldberg, H., Yu, D., Reichert, B., Speller, K., A MEMS mirror for optical scanning, *Proc. of Sensors Expo Spring* 2001, Chicago, IL, June 5–7, 2000, pp. 331–335.
63. Lawrence, E.M., Speller, K.E., MEMS characterization using laser Doppler vibrometry. reliability, testing and characterization of MEMS/MOEMS II, *SPIE Proceedings* Vol. 4980, January, 2002.
64. Oliver, A.D., Tanner, D.M., Mani, S.S., Swanson, S.E., Helgesen, K.S., Smith, N.F., Parametric monitoring for the SUMMiT V surface micromachining process, *Micromachining and Microfabrication* 2001, October 22, 2001.
65. Tanner, D.M., Owen, Jr., A.C., Rodriquez, F., Resonant frequency method for monitoring MEMS fabrication, reliability, testing and characterization of MEMS/MOEMS II, *SPIE Proceedings* Vol. 4980, January 2002, pp. 220–228.
66. Plass, R., Walraven, J.A., Tanner, D.M., Sexton, F.W., Anodic oxidation-induced delamination of the SUMMiT Polyt 0 to silicon nitride interface, reliability, testing and characterization of MEMS/MOEMS II, *SPIE Proceedings,* Vol. 4980, January 2002.
67. Rembe, C., Dräbenstedt, A., The laser-scanning confocal vibrometer microscope, *Proc. of the Int. Symp. Optical Metrology*, SPIE Vol. 5856, Munich, Germany, June 13–17, 2005, pp. 698–709.

10 An Interferometric Platform for Static, Quasi-Static, and Dynamic Evaluation of Out-of-Plane Deformations of MEMS and MOEMS

Christophe Gorecki, Michal Jozwik, and Patrick Delobelle

CONTENTS

10.1 Introduction 293
10.2 Interferometric Platform Architecture and Principle of Operation 294
10.3 Optomechanical Characterization of Membranes by "Pointwise" Deflection Method 296
10.3.1 Composition and Atomic Density of SiO_xN_y Thin Films 297
10.3.2 Mechanical Properties of SiO_xN_y Thin Films 300
10.3.3 Experimental Results 302
10.4 Mechanical Expertise of Scratch Drive Actuators via Interferometric Measurement of Out-of-Plane Microdisplacements 308
10.4.1 SDA Operation 309
10.4.2 Experimental Results 310
10.5 Dynamic Evaluation of Active MOEMS by Interferometry Using Stroboscopic Technique 314
10.5.1 Introduction 314
10.5.2 Dynamic Characterization of Active Membranes 314
10.5.3 Dynamic Characterization of Torsional Micromirrors 317
10.6 General Conclusion and Outlook 321
Acknowledgments 321
References 321

10.1 INTRODUCTION

In this chapter, we demonstrate a low-cost and multifunctional interferometric platform based on a Twyman–Green interferometer (TGI) [1], as shown in Section 10.2, for testing microelectromechanical systems/microoptoelectromechanical systems (MEMS/MOEMS). This technique may be used for relatively smooth surfaces to measure the 3-D map of out-of-plane displacements. It provides both micromechanical and material properties of buckled membranes, quasi-static actuator behavior, and vibrational analysis of microdevices. In Section 10.3, specific metrology procedures have been demonstrated to determine the residual stress of silicon membranes prestressed by silicon oxinitride (SiO_xN_y) grown by plasma-enhanced chemical vapor deposition (PECVD) [2]. The compressive stress produced by the deposition of SiO_xN_y films generates a measurable initial bending of the bimorph membrane, corresponding to an out-of-plane displacement, measured by interferometry. To measure the Young's modulus and hardness of SiO_xN_y films, we use the nanoindentation technique. The interpolation of mechanical parameters in case of compressively prestressed membranes is limited because of the absence of validated mechanical models. In order to improve this situation, the extraction of residual stress is obtained by linking the interferometry and nanoindentation data with an analytical "stress function" calculated by finite elements method (FEM). Because material properties of SiO_xN_y films depend strongly on details of the PECVD process, the relationship between the micromechanical properties, physicochemical characteristics, and, finally, the optical properties of SiO_xN_y thin films are established carefully.

The design of MEMS actuators requires precise prediction of displacements. Although simulation capabilities are advancing rapidly, many devices exhibit nonlinear effects such as electrostatic attraction or coupling of deformation modes, all of which must be accurately accounted for in the simulations. Even if the models for these effects have been developed, they are not sufficient for *a priori* prediction of actual performances. Thus, calibration and design verification require high-precision absolute measurements of displacements of structural parts. In Section 10.4, the polysilicon scratch drive actuator (SDA) is used as an example structure for demonstrating applications of interferometry for the determination of actuator performance [3]. The actuation involves contact interactions performed by the flexible polysilicon actuator plate. A better understanding of the driving and stiction mechanisms is given precisely by the length of such contact interactions that could lead to the optimization of SDA design. Until now, only in-plane characterizations have been available through microscope visualization and camera video recording allowing linear speed and driving force of actuators to be measured. However, stiction mechanisms are strongly tied to the out-of-plane actuator behavior and, particularly, the actuator/substrate contact zone. Interferometric technique measures complete deflection curves of the electrostatically actuated SDA plate. The experimental results obtained by this technique are compared with numerical data calculated by FEM.

When operating under a stroboscopic regime [4], the same interferometer allows the full-motion measurement of moving microstructures, demonstrated in the case of the response of a piezoelectric actuator and electrostatically driven torsional mirror. For vibration measurements, each of the microelements is vibrated by applying a sinusoidal voltage from a waveform generator. This generator also drives the switching electronics of the pulsed diode laser. The two channels of the waveform generator are synchronized with an adjustable phase delay, and the 3-D map of resonating modes is measured.

10.2 INTERFEROMETRIC PLATFORM ARCHITECTURE AND PRINCIPLE OF OPERATION

Chapter 8 (Section 8.2 of Chapter 8) extensively describes the general principles of interferometric microscopy and interferometric objectives in use (Figure 8.2 of Chapter 8). In our application, we use a vibration-isolated metallurgical Nikon microscope integrated with a TGI and a three-axis moveable table allowing for probing 4-inch wafers, micromanipulators, and an external light source. The optical system used to implement the principle of interferometric station is shown in Figure 10.1a, whereas Figure 10.1b represents the photograph of such an interferometric station [1]. The microscope is equipped with a set of infinity-corrected long-working-distance objectives. Table 10.1 shows the characteristics for the used microscope objectives: the magnification, the numerical aperture (N.A.), the working distance (W.D.), the field of view, and the lateral resolution in micrometers per pixel spacing. The TGI is inserted between the microscope objective and the sample to be measured and consists of a beam splitter cube of interferometric quality, coated with an antireflection coating. One of the beam splitter facets is metal overcoated, playing the role of reference mirror of the interferometer. The monochromatic light source is a temperature-stabilized and collimated laser diode operating

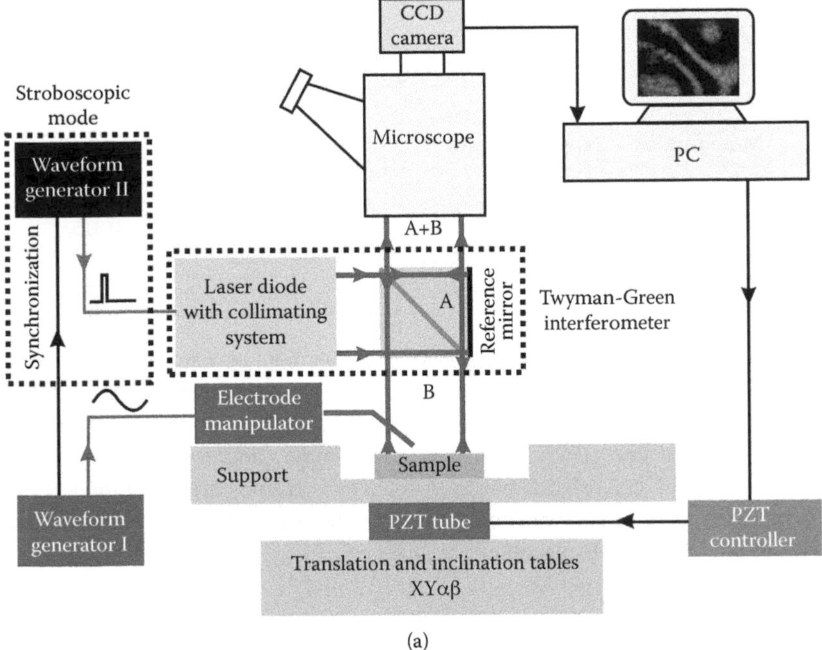

(a)

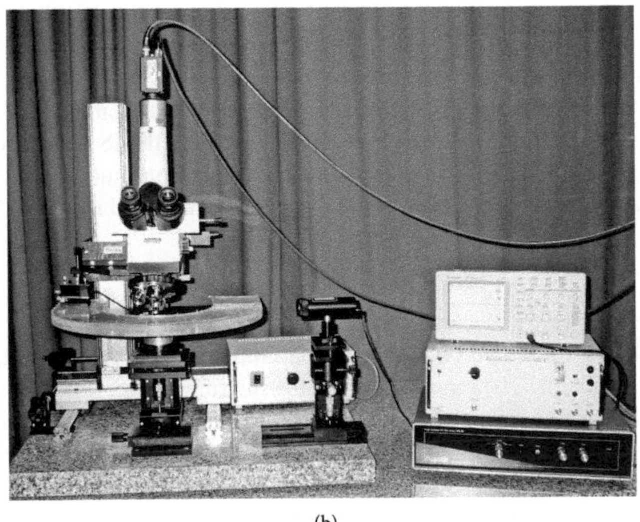

(b)

FIGURE 10.1 Architecture of TGI system: (a) schematic view, (b) photography.

at wavelength $\lambda = 670$ nm, which can be driven in pulsed or continuous-wave modes. The laser beam illuminates the beam splitter cube and is divided into two parts. Both the reference wavefront and the wavefront emanating from the sample, exiting collinearly from the beam splitter, produce an interference pattern. The intensity distribution of an interferogram in case of monochromatic illumination can be described as (see Section 8.3.2 of Chapter 8):

$$I_{out}(x, y, k) = A(x, y, k) + B(x, y, k)\cos[2kh(x, y) + \Delta\phi_{MS}(x, y, k)] \tag{10.1}$$

where $k = 2\pi/\lambda$ is a wavenumber, A is the low-frequency background intensity, B represents the local fringe visibility, and $h(x,y)$ is a wavefront function describing the out-of-plane displacement of the object.

TABLE 10.1
Microscope Objectives Used by the Interferometric System

Magnification	N.A.	W.D. (mm)	Measured Area (mm)	Resolution (μm/pixel)
5X	0.13	22.5	1.250 × 0.950	1.630
10X	0.21	20.3	0.625 × 0.475	0.815
20X	0.35	20.5	0.310 × 0.235	0.405
50X	0.45	13.8	0.125 × 0.095	0.163

The interferograms can be visualized with the microscope oculars or acquired by a 576 × 768 charge-coupled device (CCD) camera connected to a frame grabber PC card. Interferometric data are processed by an automatic fringe pattern analyzer [5] called Fringe Application 2001 (Smarttech's software).

A piezoelectric transducer (PZT) translates the sample, and five frames of interferometric intensity data are taken at $\pi/2$ relative phase shifts. These frames are combined point-by-point to calculate the phase map of the wavefront reflected by the sample (temporal phase-shift method). The phase of the out-of-plane displacement at (x,y) is:

$$\Phi\left(x,y\right)=\tan^{-1}\left[\frac{2\left(I_2(x,y)-I_4(x,y)\right)}{2I_3(x,y)-I_5(x,y)-I_1(x,y)}\right] \tag{10.2}$$

where I_1, I_2, I_3, I_4, and I_5 represent the intensity distributions of individual interferograms at (x,y) [5].

Equation 10.2 gives the phase distribution modulo 2π with a value range from $-\pi$ to $+\pi$. When the system is operating in stroboscopic mode, the diode laser is driven with a duty cycle pulse train synchronized with a sinusoidal excitation signal of vibrating MEMS to stroboscopically "freeze" this motion. This phase function is unwrapped to recover a continuous phase distribution. Once continuous $\Phi(x,y)$ is evaluated, the corresponding out-of-plane displacement distribution $h(x,y)$ is:

$$h(x,y)=\frac{\Phi(x,y)}{4\pi}\lambda \tag{10.3}$$

The deflection of the measured objects with continuous surfaces or surfaces with steps smaller than $\lambda/2$ is limited to 60 μm. The repeatability of the instrument is evaluated for objective magnification of 5X. It is obtained by measuring ten times in succession the surface of a flat mirror by using 576 × 768 resolution, and finding the rms value (root-mean-square average roughness) of the measured data. Here, the repeatability depends on the roughness of the surface and how much surface reflectivity there is. The rms value corresponding to the repeatability error is within 1.2 nm. The basic accuracy of TGI without the fringe interpolation being $\lambda/2$ per fringe, the use of phase shifting improves the accuracy, pushing it within 10 nm.

10.3 OPTOMECHANICAL CHARACTERIZATION OF MEMBRANES BY "POINTWISE" DEFLECTION METHOD

One of the significant challenges of measuring MEMS/MOEMS devices is the high-confidence measurement of material properties, crucial for optimizing micromechanical design and determining process control. Several methods have been suggested to determine simple properties such as Young's modulus, residual stress, and Poisson's ratio of the material [6–10]. The implementation of these methods is difficult in the case of thin-film materials, and the reported values of micromechanical parameters vary widely, demonstrating that the measurement accuracy is not enough to provide high-confidence testing tools. Thus, the reported values of Young's modulus for polysilicon have ranged from 120 to 201 GPa. Similarly, reported values for residual stress in polysilicon vary widely. Whereas some variability is expected, the demonstrated accuracy and resolution of these techniques are not enough to provide sufficient confidence in the data for valid process control. In addition, most of the methods cannot resolve values for residual stress below 1 MPa, even though good process control often requires keeping the stress near or below this level. Before a valid system of process control can be implemented, the demonstrated accuracy and resolution of measurement methods must be less than the expected variation due to the process.

SiO_xN_y films grown by PECVD have a wide range of applications in microelectronic technologies as insulators or passivation interlayers. Because of well-controlled refractive index and low-optical-loss characteristics, SiO_xN_y is widely used in integrated optics as a core layer for channel waveguides [10,11]. The use of PECVD films is still limited because the residual stress of deposited films is difficult to control and there is limited understanding of film properties. We have studied the material structure and composition of PECVD SiO_xN_y films by Rutherford backscattering spectroscopy (RBS), elastic recoil detection analysis (ERDA), and x-ray reflectivity (XRR). We established the correlation between these properties and the mechanical behavior of PECVD SiO_xN_y films. A special technique to determinate stress in the SiO_xN_y layer based on this membrane-bending principle was developed. The deformation of the center of the bent membrane due to the compressive stress is measured by interferometry. To extract the residual stress, both the pointwise deflection data and interferometric data are combined with micromechanical parameters obtained by means of nanoindentation test [12,13]. Because material properties of SiO_xN_y films depend strongly on the details of the deposition process and the growth conditions, the relationship between the micromechanical properties, physicochemical characteristics, and finally the optical properties of SiO_xN_y thin films are established. We compare our results with those reported in the literature.

10.3.1 Composition and Atomic Density of SiO_xN_y Thin Films

SiO_xN_y films were fabricated by a parallel plate PECVD reactor [2]. Three processing gases — silane (SiH_4), ammonia (NH_3), and nitrous oxide (N_2O) — are supplied from individual flow-control systems inside the reactor chamber. The RF-plasma-assisted reaction provides an amorphous material (SiO_xN_y) for which the characteristics depend strongly on the deposition parameters. The substrate temperature during the deposition is 350°C, the RF power is fixed at 0.11 W/cm^2, RF frequency at 150 kHz, and the pressure at 0.14 mbar for a total gas flow is fixed at 150 sccm. Under these general conditions, all SiO_xN_y samples were prepared varying only the gas-flow ratio of N_2O and NH_3. When the ratio $R = N_2O/(N_2O + NH_3)$ is varying from 0 to 1, the refractive index of the SiO_xN_y is adjusted from 1.815 to 1.469. Letters a, b, c, d, e, and f denote SiO_xN_y samples for which the deposition parameters are listed in Table 10.2 and represented in Figure 10.2. The refractive index of the samples was measured by an

TABLE 10.2
PECVD Deposition Characteristics

Sample	$R = \frac{N_2O}{N_2O + NH_3}$	N_20 (sccm)	SiH_4 (sccm)	NH_3 (sccm)	Refractive Index *n*
a	1	143	7	0	1.469
b	0.88	126	7	17	1.517
c	0.60	86	7	57	1.573
d	0.32	46	7	97	1.645
e	0.11	17	7	126	1.700
f	0	0	7	143	1.815

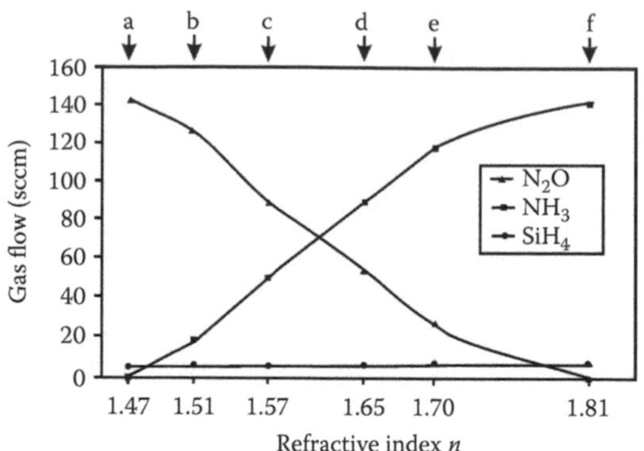

FIGURE 10.2 PECVD gas flows as a function of SiO_xN_y refractive index with sample identification codes.

ellipsometer, operating at a wavelength of 632.8 nm. The correspondence between the refractive index of SiO_xN_y and the gas-flow ratio is shown in Figure 10.3.

To define the atomic composition, samples a–f were analyzed within a Van der Graaff accelerator using an RBS detector and the ERDA technique [2]. In addition, we used an XRR system to determine electronic densities of SiO_xN_y films. Experimental data are summarized in Table 10.3, where the film thickness, the atomic percent (at %) of silicon, nitrogen, oxygen, and hydrogen, and mass density of deposited films with various refractive indexes are given. Atomic concentrations were calculated, from the layer thickness obtained in the analysis, in units of atom/cm^2. The compositional data are plotted in Figure 10.4 as a function of the refractive index of SiO_xN_y films. These data show that the quantities of incorporated nitrogen increase with increasing refractive index (0–52%), whereas the quantities of oxygen decrease with increasing refractive index (65–0%). The silicon content is relatively independent of the refractive index (29–32%). Hydrogen in particular has an important effect on the mechanical and optical properties of deposited films: his total content is a sum of N–H and Si–H bond concentrations. In the oxide-like samples, only Si–H bonds are observed; whereas in nitride-like samples, both Si–H and N–H bonds are present. In the nitride-like samples, the content of N–H bonds increases and the amount of oxide-like samples decreases when N_2O flow is increased. Because the number of N–H bonds increases significantly with the increase in refractive index whereas the concentration of Si–H bonds seems to be stable, the hydrogen atomic percentage increases rapidly with increasing refractive index (2–18.8%). Figure 10.5

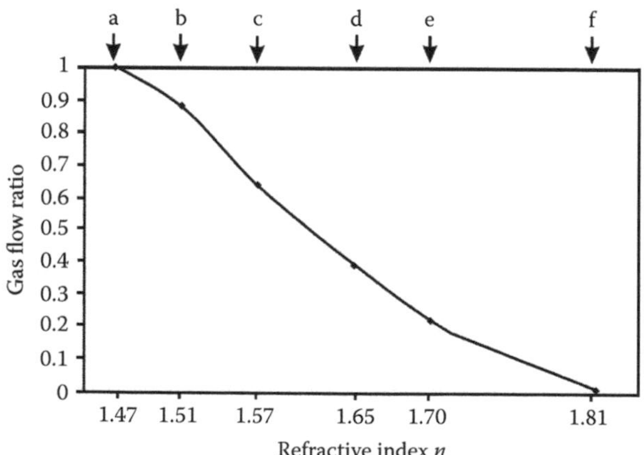

FIGURE 10.3 The dependence of SiO_xN_y refractive index vs. PECVD gas-flow ratio.

TABLE 10.3
RBS and ERDA Measurements

Sample	Thickness (nm)	*Si* (at %)	N (at %)	O (at %)	H (at %)	Mass Density (g/cm³)
a	116.8	32.4	0	65.6	2.0	2.290
b	151.6	31.7	10.9	53.8	3.6	2.440
c	115.2	29.6	19.5	44.3	6.6	2.365
d	100.6	30.1	28.0	30.0	11.9	2.435
e	107.2	30.8	35.2	17.5	16.5	2.520
f	90	29.1	52.1	0	18.8	2.710

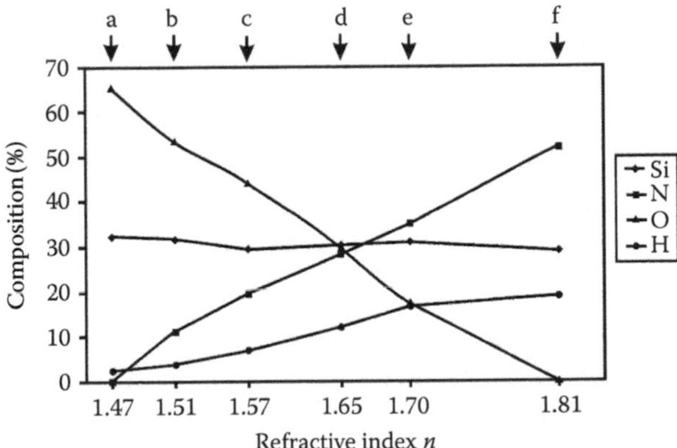

FIGURE 10.4 Chemical composition of SiO_xN_y films vs. refractive index of samples (a–f).

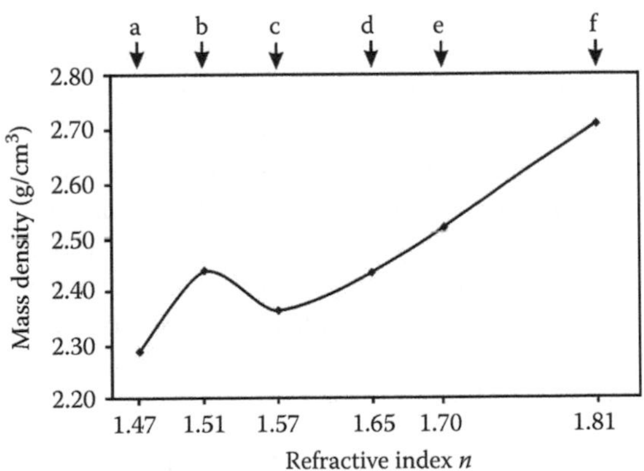

FIGURE 10.5 Atomic densities of SiO_xN_y films vs. refractive index of samples (a–f).

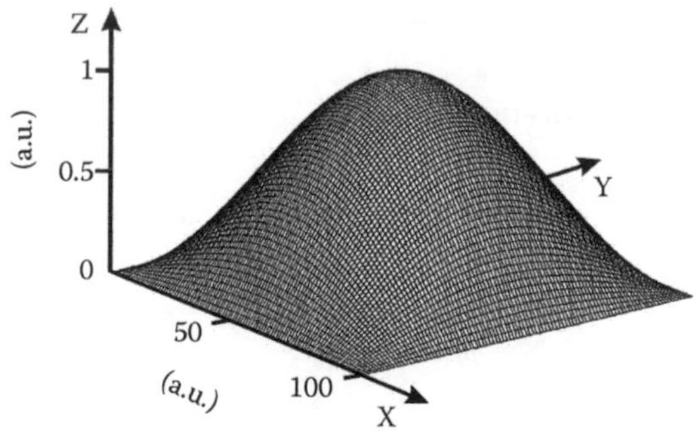

FIGURE 10.6 The shape of a square membrane operating at first mode of buckling.

displays the atomic densities as a function of refractive index. The experimental data reveal a relatively linear increase in density with increasing of refractive index, accompanied by a jump for the sample b (n = 1.517). The oxide-like film (sample a, n = 1.469) has a density slightly higher than fused silica with 2.29 g/cm^3. The nitride-like film (sample f, n = 1.815) has the higher density with 2.70 g/cm^3. It will be shown later that denser films result in more compressive stress.

10.3.2 Mechanical Properties of SiO_xN_y Thin Films

The compressive stress produced by the deposition of SiO_xN_y thin films on top of silicon membranes generates a measurable initial bending of that bimorph structure, corresponding to an out-of-plane displacement h_0 of the center of the membrane. One of the most important conditions for stress determination was the demonstration that the measured membranes were operating at first mode of buckling. In this case, the first approximation of theoretical deformation of a square membrane is [14]:

$$h = -\frac{h_0}{4}\left(1-\cos\frac{2\pi x}{a}\right)\left(1-\cos\frac{2\pi y}{a}\right) \tag{10.4}$$

where a is the size of the membrane. A more accurate second-order relation was given by Maier-Sneider et al. [15]. The shape of mode is illustrated in Figure 10.6. This shape in case of PECVD-deposited SiO_xN_y membranes are measured by interferometry, whereas the extraction of mechanical properties are obtained by nanoindentation. To measure the Young's modulus of SiO_xN_y, we used a nanoindentation technique producing elastic deformations due to the penetration of an indenter tip into the sample. The applied load P and indenter penetration depth h are recorded continuously to produce a load-displacement curve from which Young's modulus E and hardness H of the sample are calculated [12]:

$$E_r = \frac{\sqrt{\pi}}{2\beta\sqrt{A}}\frac{dP}{dh}, \quad \text{with} \quad \frac{1}{E_r} = \frac{1-\nu^2}{E} + \frac{1-\nu_i^2}{E_i} \quad \text{and} \quad H = \frac{P}{A} \tag{10.5}$$

where dP/dh is the contact stiffness measured by the continuous stiffness method, E and ν are the Young's modulus and the Poisson's ratio of the sample respectively, A is the surface of indent contact area, E_i and ν_i represents Young's modulus and Poisson ratio of the indenter, respectively, and β = 1,034 is a factor depending on the indenter geometry.

The pointwise deflection technique is performed with the nanoindenter on a square bimorph membrane where an SiO_xN_y layer is grown on top of <100> silicon substrate and subjected to the internal stress of compressive nature. It was demonstrated that the applied load P at the center of the membrane and evoked displacement h are linked by [2]:

$$\frac{P}{h+h_0} = \frac{A}{g(k)} + BC\left(\nu,\frac{h}{e},0\right)\left(h+h_0\right)^2 \tag{10.6}$$

$$\text{with} \quad k^2 = k_0^2 - 9.7\left\langle \frac{h_0}{e} - 0.8 \right\rangle^2, \; A = \frac{D_{eff}}{\alpha a^2}, \; B = \frac{E_{eff} e_{eff}}{a^2}$$

where $g(k)$ is an analytical function describing the residual stress of a square bimorph membrane, a is the size of the membrane, $\langle \; \rangle$ are Mac Cawley brackets, D_{eff} is the bending moment, E_{eff} and e_{eff} represent the effective values of Young's modulus and membrane thickness, respectively, $\alpha = 0.0056$ is a constant of the square membrane, and h_0 is the initial membrane bending measured by interferometry. C is a function of k_0 for both compressive and tensile stress calculated by the finite element method. Until the value of the limiting stress of buckling is reached, the influence of k_0 on C is negligible. For deflections until $h/e \approx 7$, C decreases with h/e, and the adjustment of FEM results by an analytic function is made for square membranes with:

$$C\left(\nu,\frac{h}{e},0\right) = \frac{6.1}{1-\nu^2}\left[1 - 0.47\left(1 - \exp - 0.168\frac{h}{e}\right)\right] \tag{10.7}$$

For bigger deflections ($h/e >> 1$), C is constant: $C = \dfrac{3.23}{1-\nu^2}$.

The effective values of Young's modulus and effective thickness defined in Equation 10.6 are:

$$E_{eff} = \frac{E_s e_s + E_f e_f}{e_s + e_f}, \; e_{eff} = e_s + e_f,$$

$$D_{eff} = \frac{E_f^2 e_f^4 + E_s^2 e_s^4 + 2E_f E_s e_f e_s\left(2e_f^2 + 2e_s^2 + 3e_f e_s\right)}{12\left(1-\nu_s^2\right)\left(E_s e_s + E_f e_f\right)}, \; \sigma_{eff} = \frac{\sigma_s e_s + \sigma_f e_f}{e_s + e_f} \tag{10.8}$$

where s and f represent the substrate and the film, respectively. σ_{eff} is the effective value of the stress.

A more complex explanation of Equation 10.8 was given in [2]. According to the nature of the residual stress, the "stress function" may be defined as the following:

- *Case of tensile stress* [2]:

$$g(k_0) = \frac{8}{k_0^2}\left[\left(\frac{K_1(k_0) - \frac{1}{k_0}}{I_1(k_0)}\right)\left(I_0(k_0)+1\right) + K_0(k_0) + Ln\frac{k_0}{2} + \gamma\right] \tag{10.9}$$

$$\text{with} \quad k_0^2 = \frac{N}{1.14 D_{eff}}\left(\frac{a}{2}\right)^2 \quad \text{and} \quad N = \sigma_f e_f + \sigma_s e_s = \sigma_{eff}(e_f + e_s)$$

where K_0/K_1 are the Bessel functions of the second kind of orders 0/1, I_0/I_1 are the modified Bessel functions of the first kind of orders 0/1, γ represents the Euler constant, and N is the applied force per unit of length.

- *Case of compressive stress* [2]:

$$g(k_0) = \frac{8}{k_0^2}\left[\frac{\frac{\pi}{2}Y_1(k_0) + \frac{1}{k_0}}{J_1(k_0)}\left(1 - J_0(k_0)\right) + \frac{\pi}{2}Y_0(k_0) - Ln\frac{k_0}{2} - \gamma\right] \tag{10.10}$$

where Y_0/Y_1 are the Bessel functions of the first kind of orders 0/1, and J_0/J_1 are the Bessel functions of the second kind of orders 0/1. The function $g(k_0)$ is plotted in Figure 10.7, where the branches correspond to three different cases of residual stress are:

(i) Tensile situation with $\sigma_{\text{eff}} > 0$ and $0 < g(k_0) < 1$
(ii) Compressive situation before the buckling (σ_{cr} is a critical stress of buckling) with $\sigma_{\text{eff}} < 0$ and $1 < g(k_0) < \infty$
(iii) Buckling regime with $\infty < g(k_0) < 0$

The calculation of these "stress functions," including the experimental data from both interferometry and nanoindentation, permits the evaluation of the effective membrane stress σ_{eff}, and then the extraction of residual stress σ_f of SiO_xN_y films (because the residual stress of silicon membranes is $\sigma_s \approx 0$).

10.3.3 Experimental Results

A series of square membranes with different optical quality of SiO_xN_y film were fabricated, as shown in Figure 10.8 [16]. First, both sides of a double-polished <100> silicon substrate were thermally oxidized at a temperature of 1050°C to grow 1-μm thick layers of oxide. A square mask aligned in

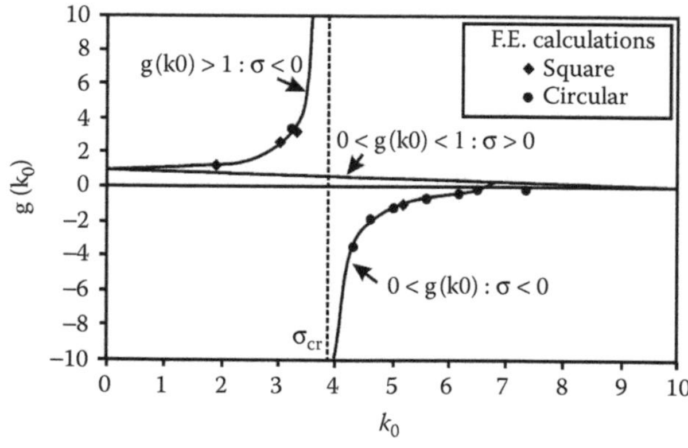

FIGURE 10.7 Plot of the function $g(k_0)$.

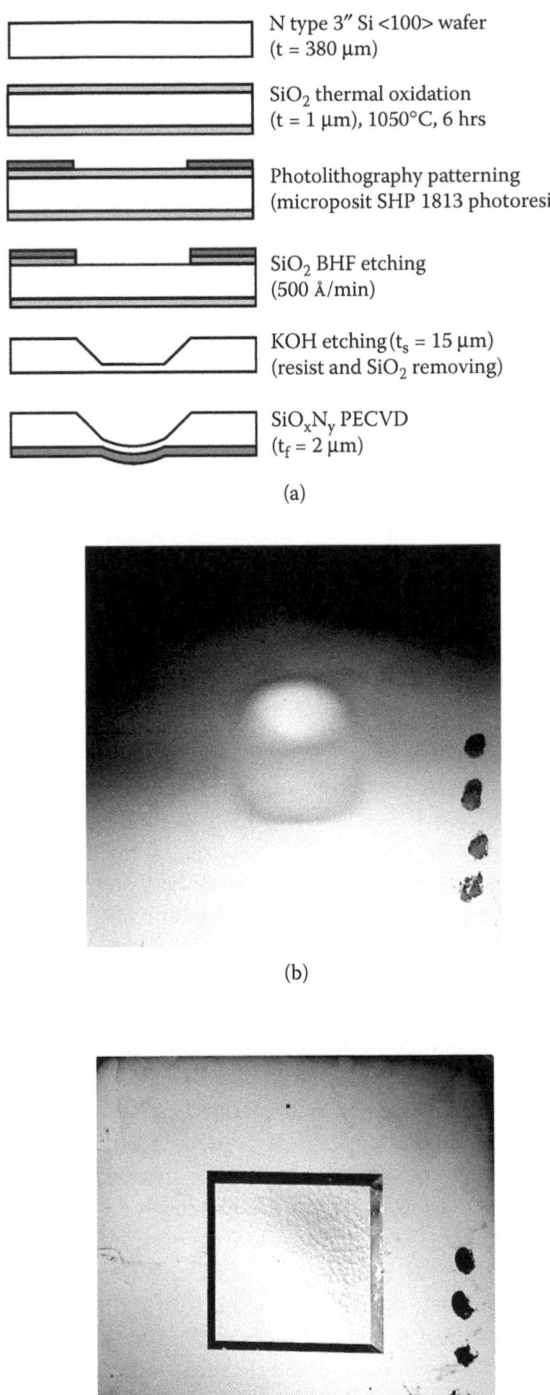

FIGURE 10.8 Fabrication of SiO_xN_y prestressed silicon membranes (a), backside (b), and topside (c) of membrane.

TABLE 10.4
The Comparison of Experimental and Theoretical Results

Series	h_0 (μm)	$g(k)$	σ_f (MPa)	E_{eff} Measured (GPa)	E_{eff} Theoretic (GPa)	$\sigma_{f\ cr}$ (MPa)
B	≈10.7+	−3.31	−119	219	122.8	−98.0
n = 1.49–1.51	17.3	−2.45	−132	115	122.8	−98.0
C	40	−1.11	−184	48	123.4	−99.3
n = 1.57–1.58	0	+5.78	−84	—	125.1	−103
	6.6+	−13.8	−106	310	125.1	−103
	36	−0.79	−237	92	125.4	−103
D	18.8	−3.62	−132	80	127.6	−107.5
n = 1.65–1.67	17.1	−3.15	−135	98	127.6	−107.5
	0*	+0.83	+16.1	—	128.2	−108.6
	11.7	−3.66	−130	173	128.2	−109.6
E	26.4	−1.26	−187	110	128.4	−109
n = 1.69–1.73	21.3+	−1.23	−189	171	128.4	−109
	26	−0.94	−227	153	129.8	−112
F	60	−0.49	−414	67	131.3	−114.2
n = 1.81–1.82	62	−0.49	−414	68	131.3	−114.2
	44	−0.48	−396	112	131.7	−115.1
	47.8+	−0.46	−452	108	131.7	−115.1
	34	−0.49	−414	132	131.7	−115.1

the <110> direction was patterned by photolithography on the topside of the wafer. Then, the oxide layer was removed using a buffered HF etch. The silicon substrate from which the oxide layer had been removed was anisotropically etched by a KOH solution, giving a 15-μm thick membrane of size 3.6 × 3.6 mm². After photoresist stripping on the topside of the wafer, the oxide layers were removed from both sides. Finally, a 2-μm-thick SiO_xN_y film was deposited by PECVD on the backside of the membrane. In the following sections, letters B, C, D, E, and F denote the range of refractive index for each set of SiO_xN_y prestressed membranes as described in Table 10.4 [2].

To measure the Young's modulus and the hardness of PECVD-deposited SiO_xN_y membranes, we used the nanoindentation technique performed by a Nano Indenter IIs on the massive part of the sample (2 μm of SiO_xN_y film on top of 380-μm-thick silicon substrate) during the well-controlled loading–unloading cycles. In this technique, a three-sided pyramidal diamond indenter is driven into the sample by applying an increasing load. It is assumed that the contact area between the indenter and the sample remains constant and moves elastically at the start of unloading. The penetration of the indenter tip into the sample is accompanied by a vibration of the tip with a 1-nm amplitude and a frequency of 45 Hz [17]. This procedure permits the measurement of the distribution of Young's modulus E and hardness H as a function of indentation depth, as shown in Figure 10.9 and Figure 10.10, respectively. Here, the maximum depth of indentation is fixed at 900 nm, and 15 indents are performed for each of five samples on an area of 300 × 300 μm². An increase in E value with increasing indentation depth is observed (Figure 10.9). This increase is faster for oxide-like samples.

Figure 10.11 shows the Young's modulus as a function of the refractive index of SiO_xN_y films, where E_f is linearly increasing with the value of the refractive index. For oxide-like films (series B, n = 1.51), the value of Young's modulus (78 GPa) is close to the modulus of thermal oxide (70 GPa); whereas for the nitride-like film (series F, n = 1.81), the Young's modulus (148 GPa) is close to the modulus of Si_3N_4 [18,19].

The behavior of hardness H (Figure 10.10), a parameter mostly related with plastic characteristics of materials, is relatively complex. Because the hardness of the silicon substrate is around

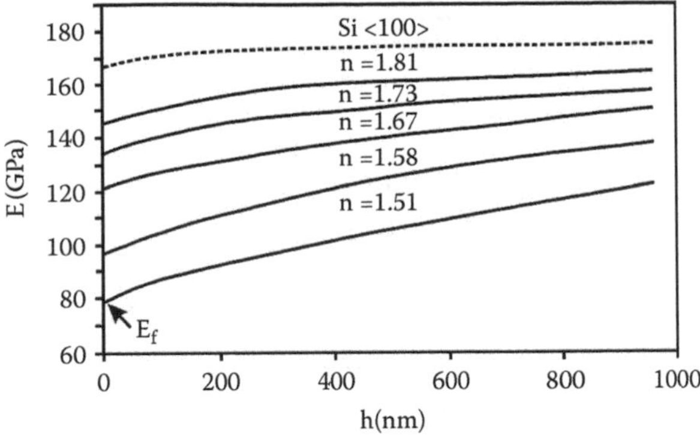

FIGURE 10.9 The distribution of Young's modulus E vs. the indentation depth.

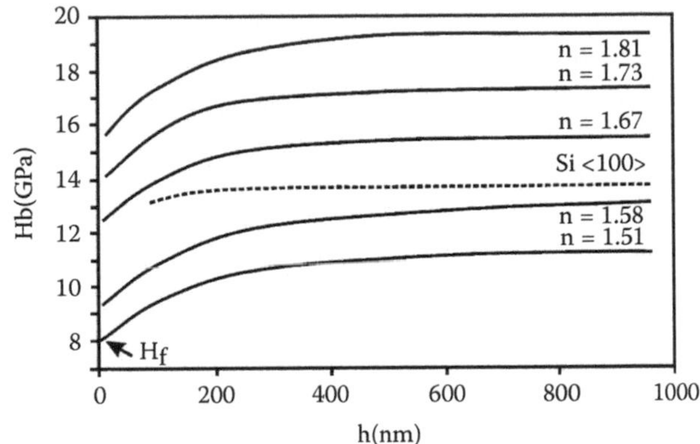

FIGURE 10.10 The distribution of hardness H vs. the indentation depth.

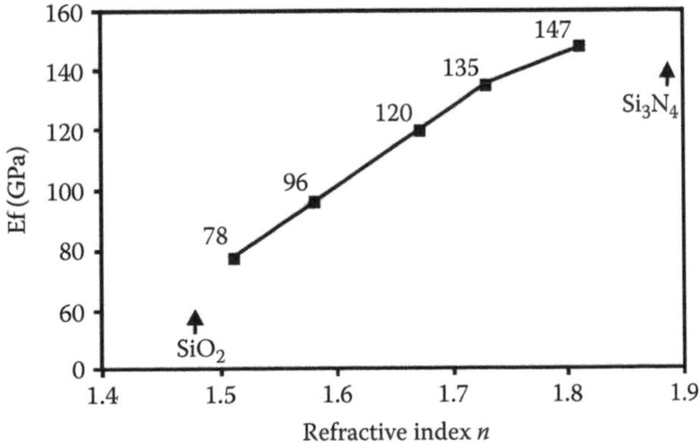

FIGURE 10.11 The Young's modulus vs. the refractive index of SiO_xN_y films.

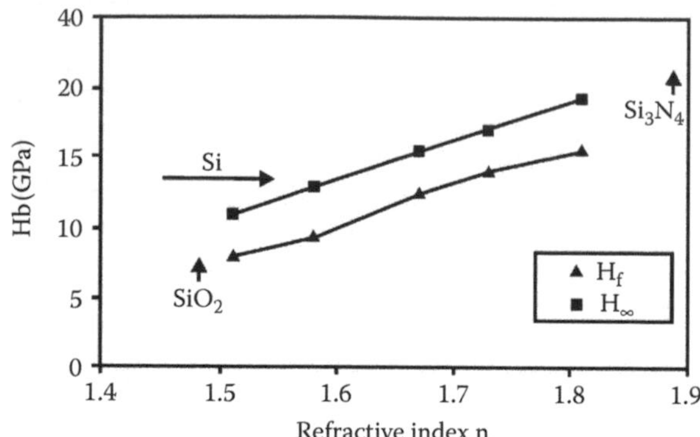

FIGURE 10.12 The hardness H_f vs. the refractive index of SiO_xN_y films.

13.5 GPa, the increase in *H* vs. the indentation depth for series B and C is in agreement with the behavior of a thin film having the hardness below that of the substrate. However, the series D, E, and F demonstrate the behavior of thin films having the hardness above that of the substrate, but the evolution of the gradient is not correct. Probably, this is due to the presence of microcracks produced by the penetration of the indenter and introducing an overvaluation of *H*. Finally, the hardness of SiO_xN_y films presents a flat region H for the indentation depth in the range from 500 to 900 nm. The hardness corresponding to this flat zone increases with the increase of refractive index. Both H_f (value of *H* for h_0) and *H* (value of *H* for h = 900 nm) increase linearly with the refractive index of SiO_xN_y films, as shown in Figure 10.12.

After mechanical characterization of SiO_xN_y films, we start measurements by interferometry. Figure 10.13a shows five fringe patterns of a half of SiO_xN_y membrane bending (series B, n = 1.51). Starting from the left side of the flat part of the wafer, we observe an increase in the number of fringes and diminution in the fringe diameter up to the smallest one on the membrane center. Figure 10.13b illustrates the phase map of out-of-plane displacements after unwrapping and Figure 10.13c, the central cross section for the profile of this membrane. These data permit access to the distribution of values of the maximum out-of-plane displacement h_0 vs. the refractive index of the SiO_xN_y films for series B, C, D, E, and F (Figure 10.14). The curve presents a minimal bending ($h_0 \approx 0$) for the refractive index n = 1.58 (series C). After that, the deflection increases with increase of the refractive index.

Finally, the distribution of residual stress σ_f calculated from Equation 10.6 vs. the refractive index of SiO_xN_y thin films is shown in Figure 10.15. This graph gives the fitting curve for the averaged experimental data value of 130 MPa. The stress is minimal for 1.58 with σ_f = 100 MPa and then, for higher refractive, becomes less compressive. We demonstrated that around this refractive index (sample C), the film density is lower. For nitride-like films (n = 1.81), the stress becomes more compressive with the increase in refractive index. Here, the stress magnitude is around 400 MPa. For the series C (n = 1.58), where $g(k_0) > 0$, the measured stress is lower than the critical stress of buckling ($\sigma_f < \sigma_f^{cr}$). This is accompanied by a very weak or zero initial deflection h_0. Other samples lead to the buckling of bimorph structures presenting $g(k_0) < 0$.

In conclusion, all membranes with SiO_xN_y thin films are of compressive nature with a minimum for n = 1.58 (series C) and a maximum for n = 1.81 (series F).

In conclusion, all tested silicon oxynitride films were in compression and have intermediate properties between silicon oxide and silicon nitride films. Table 10.5 shows an overview of measured values from the literature compared with the results of our contribution. SiO_xN_y layers are obtained from PECVD reactors operating with SiH_4–NH_3–N_2O gas mixture. We can see a clear tendency concerning the magnitude of Young's modulus. However, a direct comparison is less evident for evaluation of compressive stress magnitude because of the large number of processing parameters

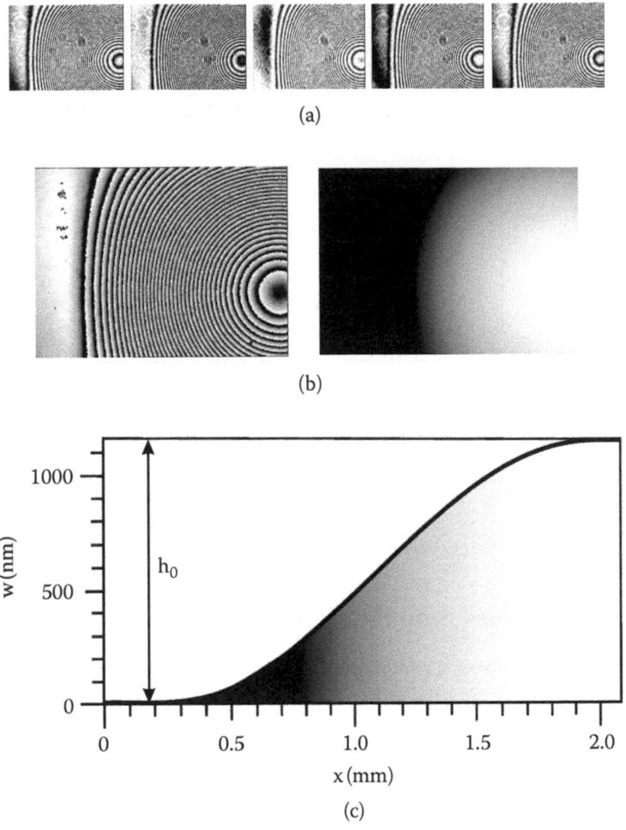

FIGURE 10.13 Initials deflections h_0: (a) interferograms, (b) phase map (left) before unwrapping, and (right) after unwrapping phase map, (c) central cross-section of phase map.

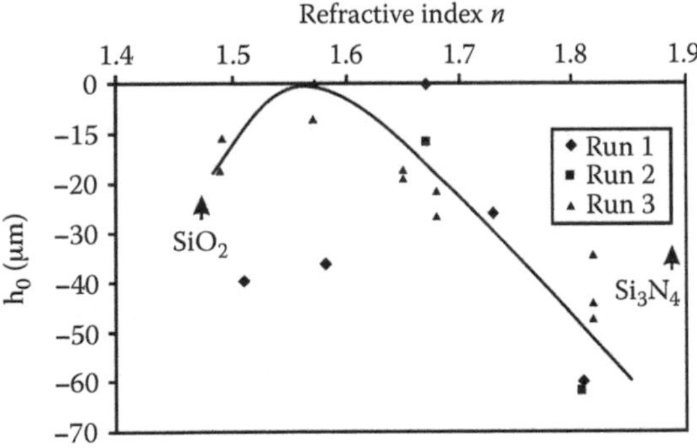

FIGURE 10.14 Initial deflections h_0 vs. the refractive index of SiO_xN_y films.

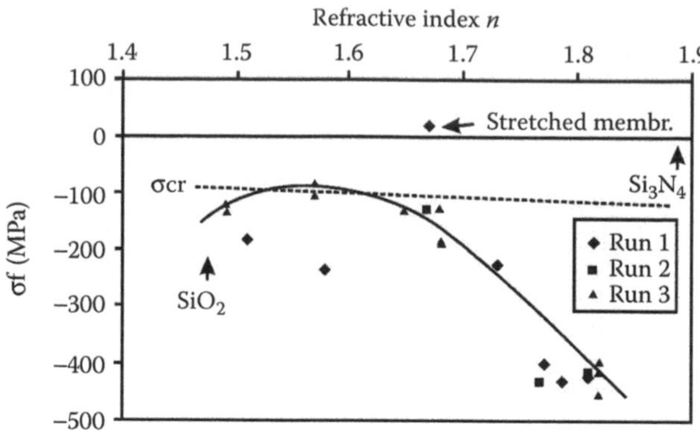

FIGURE 10.15 The distribution of residual stress vs. the refractive index of the SiO_xN_y films.

TABLE 10.5
Overview of Measured Values from the Literature

Nature of Deposition	Stress (MPa)	Young's Modulus (GPa)	Refractive Index	Pressure (mbar)	Temperature (°C)	Ref.
400 nm on Si wafer	−330 −1050	86–149	1.48–1.9	500	200–300	20
1.9 μm on Si membrane	−320	150	1.9–2.0	—	300	21
on Si wafer	−240–310	—	1.5–1.7	1.45	380	22
Si wafer	−170–780	—	1.51–1.91	0.65	300	23
100 nm on Si wafer	−180	—	1.85	3.6	220–315	24
2 μm on Si membrane	−90–400	78–147	1.51–1.81	0.14	350	This work

and the use of different PECVD reactors, although these estimates were made from measurements on membranes or on wafers. In the meantime, certain trends can be seen from the stress data, because all deposited films were found to have a compressive stress.

10.4 MECHANICAL EXPERTISE OF SCRATCH DRIVE ACTUATORS VIA INTERFEROMETRIC MEASUREMENT OF OUT-OF-PLANE MICRODISPLACEMENTS

According to scaling effects dealing with actuator downsizing, electrostatic field energy conversion can be considered as one of the most suitable driving mechanism on the micrometer scale [25]. However, the driving torque and, particularly, the driving force per unit area available from conventional silicon-based electrostatic actuators still remain limited compared to other actuation sources. The limitation of mechanical power on the micrometer scale is clearly related to the systematic use of electrostatic field interactions as an actuation source for driving conventional actuators (for example, comb drive and harmonic electrostatic actuators). Certain very useful MEMS actuators devices exploit friction. These include the scratch drive actuator (SDA) proposed by

Akiyama [26], involving frictional contact interactions. Recent investigations have already demonstrated that SDAs perform direct-drive loading characteristics, which ideally meet current actuation constraints dealing with on-chip MEMS test requirements, such as bending and tensile fracture tests [27]. Accordingly, new on-chip mechanical laboratories have been recently developed allowing systematic monitoring of MEMS manufacturing [28].

SDA creates large forces for an electrostatic drive, but the operational principle is not well understood and reliability is an issue. The step size as a function of the dynamics and magnitude of the control voltage has not been satisfactorily described. A better understanding of driving and stiction mechanisms implied in such contact interaction electrostatic actuators could lead to drawing up design optimization outlines. Until now, only in-plane characterizations were available through rough visualization techniques that allowed recording linear speed and driving force of actuators. However, stiction mechanisms are strongly tied to the out-of-plane actuator behavior and, particularly, the actuator/substrate contact zone. The proposed interferometric method measures with high-precision a 3-D map of out-of-plane displacements of the actuator plate, extracting the effective length of the plate that is in contact with the substrate as a function of electrostatic voltage. These experimental data are compared with numerical finite element simulations.

10.4.1 SDA Operation

The SDA, designed by Département LMARC, Femto-ST, Besançon, France [3], is illustrated in Figure 10.16a. It consists of a polysilicon bending plate with size of 60×60 μm^2, a bushing, and a substrate, which includes a 0.6-μm dielectric insulator. The bending plate is "stuck" to the substrate through a micrometer-height contact plot that is located on the front of the plate. These micrometer-height contact plots provide contact asymmetries, which are fundamentally needed in order to drive the actuator. The motion of an SDA has two distinct phases per cycle. The actuator begins as shown in Figure 10.16b, and a voltage is applied between the plate and the substrate. An electrostatic force causes the plate to be attracted to the substrate, and this force is largest where the plate is closest to the substrate. The elastic reaction of the plate because of this force causes it to deform and progressively stick to the substrate. Large external forces can be consequently applied on the actuator, because of the polysilicon-locking effects, which linearly depend on the electrostatic normal loading. At the fall of the electric pulse, the elastic strain energy previously stored in the plate is released. The corresponding relaxation allows a small stepping motion Δx. Elementary steps can be as small as 10 nm, taking into account a submicrometer height of the contact plot. The second phase of operation occurs as the voltage is lowered, and charge is removed from the capacitor. The plate relaxes, approaching its original state. Reapplying the voltage causes further movement. Numerous elementary cycles can be repeated, thus allowing large displacements to be made.

SDAs operate using contact interactions instead of electric field interactions that are systematically involved in conventional electrostatic actuators. Accordingly, an SDA obeys the Coulomb friction law, which means that the available driving force depends on the external normal preload applied on the bending plate through electrostatic attraction. The average electrostatic normal pressure acting on the actuator plate reaches 40 MPa, as against 10^{-2} MPa obtained using conventional parallel plate capacitors. As an example, the driving force developed by a single SDA-based actuation cell of the type in Figure 10.16 is 1500 μN, which is considerable compared to the size of the cell.

Driving force capabilities of SDA obviously suit the on-chip MEMS test requirements. Bending fracture and tensile fracture have been easily initiated within polysilicon specimens connected to a single SDA actuation cell. In addition, SDA-based actuation cells can be easily duplicated within arrays so as to provide superposition of numerous elementary driving forces. Force densities as high as 400,000 $\mu N/mm^2$ have therefore been pointed out using a millimeter-scale matrix combining several thousands of actuation cells. Moreover, speed characteristics are particularly suitable to

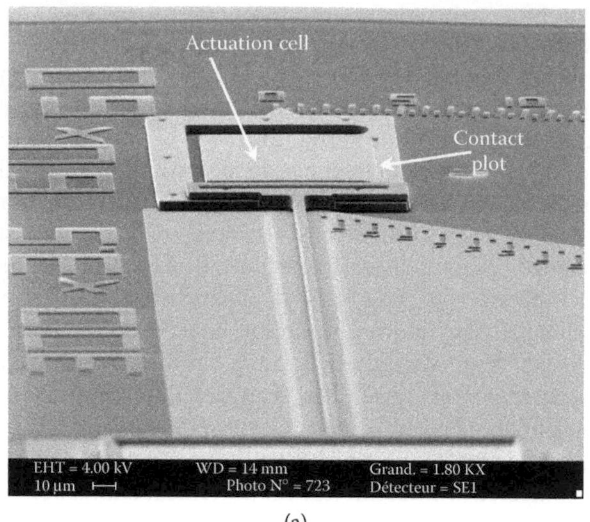

(a)

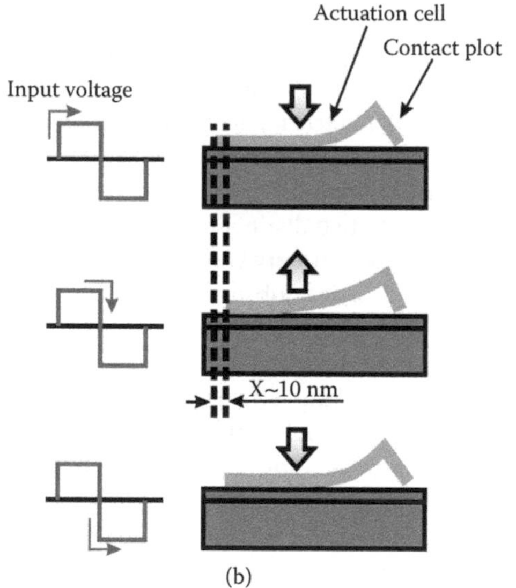

(b)

FIGURE 10.16 Elementary SDA actuation cell: (a) SEM photograph showing an actuation cell connected to a polysilicon bending specimen, (b) operating principle of a basic SDA.

on-chip test requirements. The nanometer stepping motion intrinsically leads to very low speed characteristics. However, the elementary SDA actuation cell can be driven at a high speed, depending on the input driving frequency, which can usually range from 1 Hz to several hundred kHz.

10.4.2 Experimental Results

The lifetime performances of SDA actuation cells are actually restricted because of stiction problems. In the case of the SDA mechanism, stiction includes the more complex situation of one surface (bending plate) sliding on a substrate. Thus, out-of-plane mechanical expertise given by interferometry seems to be a critical point in order to improve lifetime performances of SDA.

Figures 10.17 illustrates measuring results aiming to define the map of out-of-plane displacements of SDA structure under actuation. Thus, Figure 10.17a shows the interferogram when a ramp

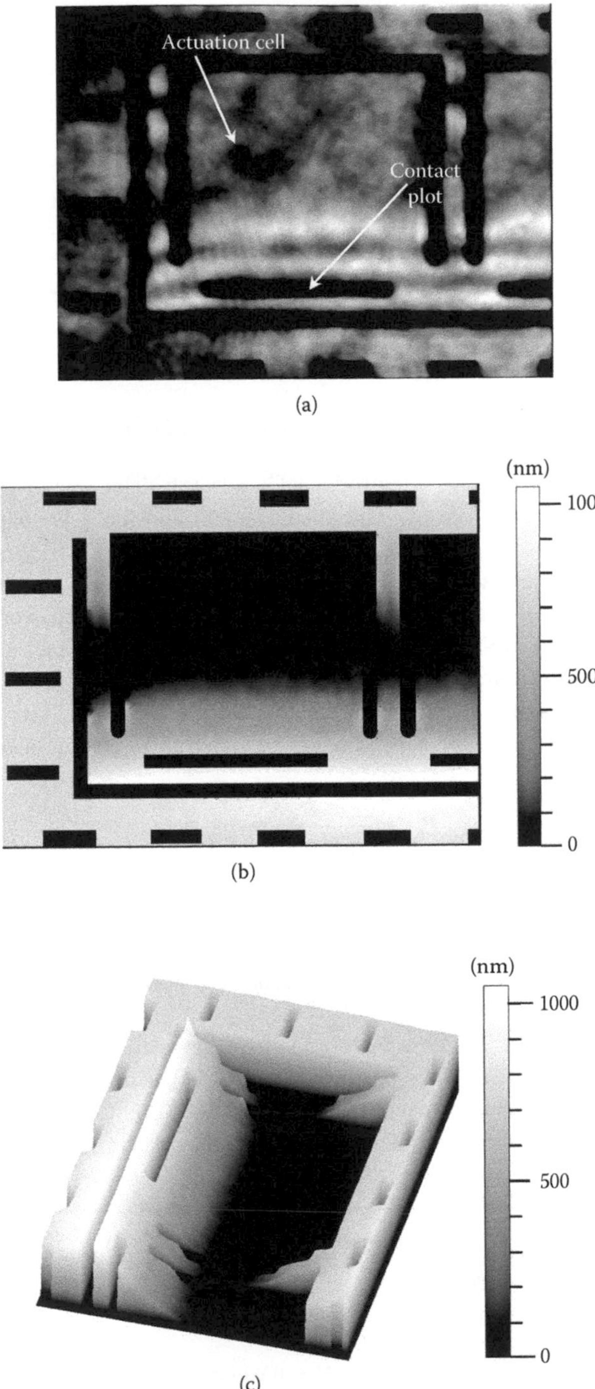

FIGURE 10.17 Experimental results for electrostatic voltage of 100 V: (a) 5 TPS interferograms, (b) top view of out-of-plane displacement, (c) 3-D shape of SDA.

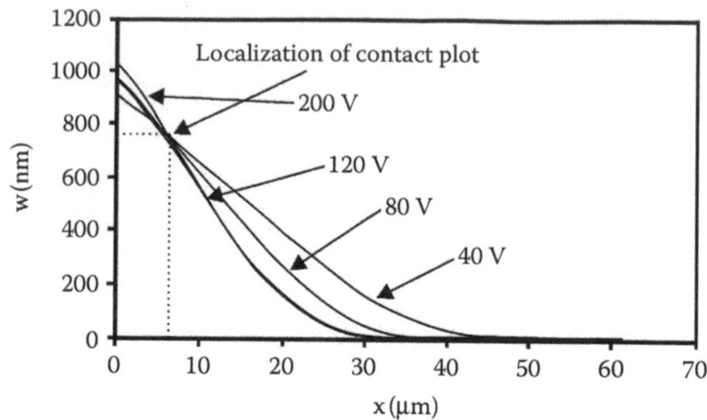

FIGURE 10.18 The deflection of bending plate deflections measured by interferometry vs. the driving voltage.

voltage of 100 V is applied to the electrodes of a scratch drive, whereas Figure 10.17b and Figure 10.17c represent data of out-of-plane displacement and the 3-D shape of deformed SDA, respectively.

Simultaneously, numerical simulations of SDA with the bending plate having width of 60 μm, length of 60μm, thickness of 2 μm, and a contact plot height of 0.75μm were performed by FEM via ANSYS software. 3-D ANSYS simulation is obtained with contact elements between the bending plate and the substrate, taking into account the increase of the bending plate stiffness during the electrostatic loading. The applied electrostatic pressure P_{elec}, depending on the air gap z between the bending plate and the substrate, is calculated through Equation 10.11:

$$P_{elec} = \frac{\varepsilon_{air}\varepsilon_{nit}^2}{2\left(\varepsilon_{air}t_{nit} + \varepsilon_{nit}z\right)^2}V^2 \tag{10.11}$$

where $(t_{nit}, \varepsilon_{nit}, z, \varepsilon_{air}, \text{ and } V)$ represent, respectively, the silicon nitride thickness, permittivity, variable air gap, air permittivity, and the driving voltage. Thus, for each driving voltage, several numerical iterations updating the electrostatic pressure are needed to achieve the plate deflection convergence.

Figure 10.18 to Figure 10.20 show the comparison between experimental data measured by interferometry and numerical simulation performed by FEM. The out-of-plane deflection of bending plate is investigated in case of four driving voltages: 40 V, 80 V, 120 V, and 200 V. Figure 10.18 shows the ANSYS data concerning the deflection of SDA under driving voltage, whereas Figure 10.19

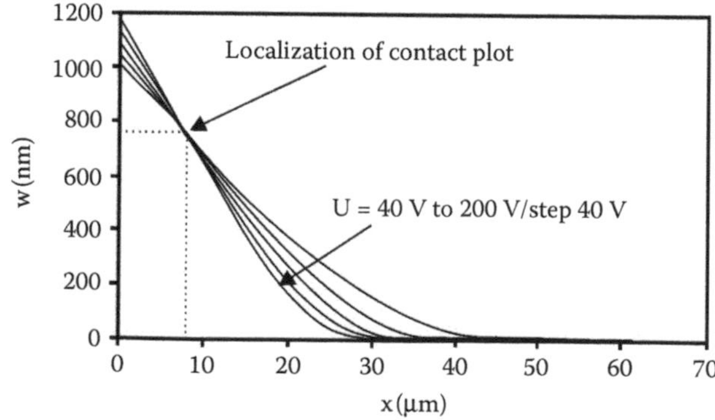

FIGURE 10.19 Deflection of bending plate calculated by ANSYS as a function of driving voltage.

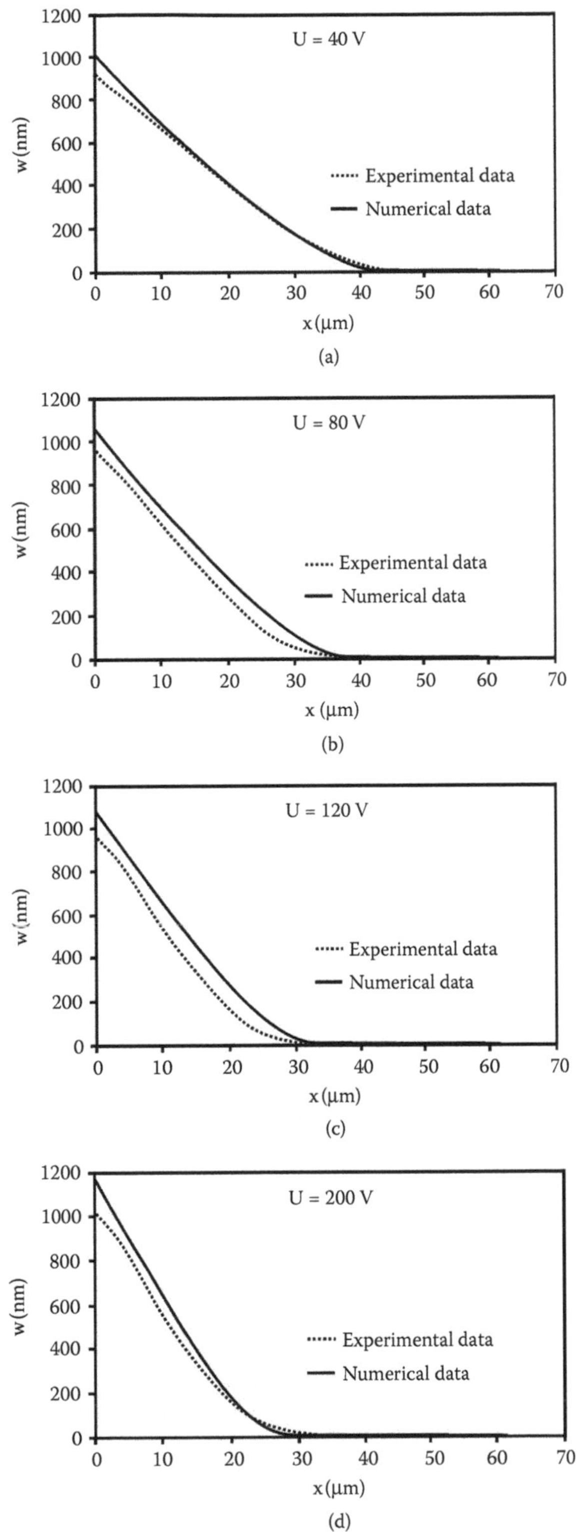

FIGURE 10.20 Comparison between experimental and numerical data of SDA out-of-plane displacement for (a) U = 40 V, (b) U = 80 V, (c) U = 120 V, and (d) U = 200 V.

represents the corresponding experimental data. We can see that the electrostatic force causes the plate to be attracted to the substrate, and this force is largest when the plate is closest to the substrate. The amplitude of deflection and the consequent plate/substrate contact length increases with the driving voltage. However, the characteristics obtained for 120 V and 200 V are quite close and deflection "saturation" effect is observed. As the plate/substrate length of contact increases, the free part of the bending plate decreases, and the bending plate stiffness rapidly increases. A direct comparison between experimental and numerical profiles of bending plate is shown in the Figure 10.20a to Figure 10.20d. We can observe a very good convergence between the experimental and numerical data. The maximal value of out-of-plane deflections is always higher in experimental cross sections. We assume that this minor difference comes from the technological process.

In conclusion, we have established a metrology procedure that is able to provide a better understanding of driving and stiction mechanisms implied in contact interactions of electrostatic actuators. The efficiency of experimental approaches is confirmed by ANSYS simulations. Future work aims at extensive usage of interferometric technique in order to improve SDA mechanical performances with regard to stiction mechanisms as well as lifetime considerations.

10.5 DYNAMIC EVALUATION OF ACTIVE MOEMS BY INTERFEROMETRY USING STROBOSCOPIC TECHNIQUE

10.5.1 INTRODUCTION

It is well known that the measurements of resonant frequencies of micromechanical devices such as microbeams, microbridges, and membranes allow one to obtain data on Young's modulus and the residual stress of materials [29–31]. However, the lack of sufficiently fast array detectors limits the method to measuring the motion of a single point or a limited number of pixels at a time. Scanning vibrometers that use a raster-scanning technique to sweep the measurement beam over a 2-D area one point at a time are available. Thus, vibrometry techniques such as deflectometry and laser Doppler vibrometry [32] were proposed for measurements of micromechanical devices. The use of such sequential techniques presents two major disadvantages of motion evaluation — that of no real-time measurement for 3-D mapping. The standard full-field interferometers are not well adapted to accurately measure MEMS devices under actuation. However, by making the assumption that the vibrations are periodic and primarily of one frequency, which are generally design goals for MEMS devices, one can use illumination strobed to match the MEMS drive frequency to effectively freeze the motion. In this case, the minimum resolvable time interval can be governed by the light-pulse width rather than by the detector integration time. Thus, the use of stroboscopic techniques permits one to adapt standard interferometry and algorithms to be employed even on these moving devices for full-field measurement or visualization of vibration modes. Among the full-field methods (with parallel acquisition and processing of the measurement data), time-averaged and stroboscopic methods can be utilized [33–36] (Section 8.7 of Chapter 8). Microscopic interferometry is used here for full-field mapping of out-of-plane or torsional vibration modes. We have illustrated the performances of the previously described probe station with TGI (Section 10.2). The system now operates in stroboscopic mode, measuring the out-of-plane motions of both piezoelectrically actuated membranes and electrostatic torsional mirrors.

10.5.2 DYNAMIC CHARACTERIZATION OF ACTIVE MEMBRANES

The object to be measured was fabricated in collaboration with Thales Research & Technology (France). This is a thin PZT transducer integrated on top of a 5-μm thick silicon membrane as shown in Figure 10.21a. The fabrication of this microactuator starts by growth of thermal SiO_2 layers on both sides of an SOI (silicon-on-insulator) wafer. A titanium layer is sputtered and oxidized

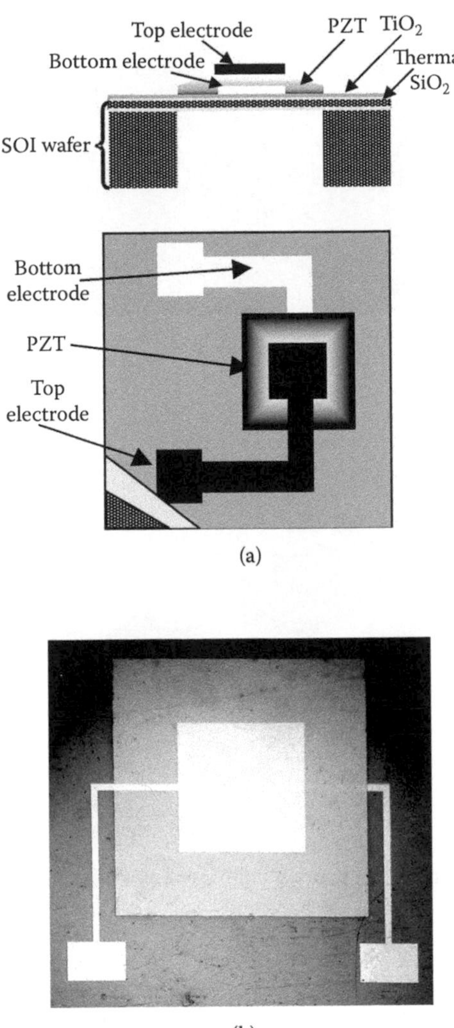

FIGURE 10.21 Micromembrane: (a) scheme of design, (b) photograph of structure 1×1 mm^2.

at 850°C to create a TiO_2 barrier. A Ti/Pt bottom electrode is sputtered onto the front side. Then, a 2-μm PZT layer using deposited using sol-gel technique and structured by a wet etch. For crystallization of amorphous PZT film, a rapid thermal annealing process is performed at 700°C. A Pt top electrode is sputtered and patterned by lift-off. The last step of the process is the backside membrane opening by deep reactive ion etching (DRIE). The size of such a square membrane is 1000×1000 μm^2, as shown in Figure 10.21b with a zoom on the PZT transducer. When a voltage is applied across PZT electrodes, the induced strain causes membrane deformation, generating a piezoelectric force. The use of PZT thin films has certain advantages, including the ability to use surface and bulk micromachining to build a complex mechanical structure that provides stable motion control at moderate voltage levels (in general, above 100 V for displacements of few micrometers). Additional advantages are features such as large force generation, high-frequency responses, and compatibility with silicon electronics. The technology process introduces an initial buckling of the membrane with the magnitude of 14.5 μm at the center. In static situations (no voltage applied), the fringe pattern and 3-D map of the membrane shape is shown in Figure 10.22 (a–b).

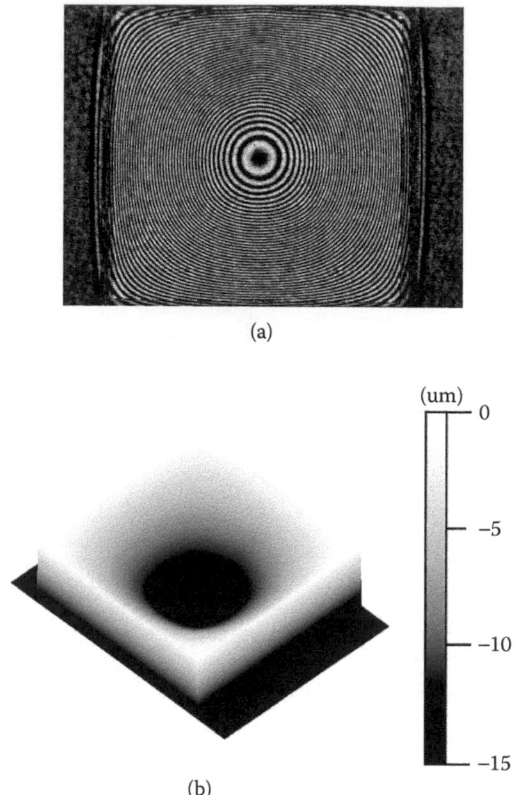

FIGURE 10.22 (a) The static interferogram, (b) the corresponding 3-D shape representation of initial deflection.

From the resonant frequencies of piezoelectric membranes, the quality factors of the resonance may be evaluated. In particular, for membranes with internal residual stress, the vibration modes of the main order may differ from that of stress-free microdevices according to the compressive or tensile nature of the stress and the stress magnitude. In this case, the vibration-mode shapes may be complex and difficult to interpret because of the lower quality factor of resonance. The objective of dynamic evaluation proposed here is to obtain a deeper understanding of the dynamic behavior of piezoelectrically actuated membranes.

The residual stress due to the deposition of piezoelectric transducer layers is still sufficiently compressive to induce a buckling of the membrane corresponding to a static out-of-plane deflection having a considerable influence to dynamic response of the membranes. 3-D vibration-mode shapes analysis is performed here by the use of a low-cost automated stroboscopic system based on the previously described TGI shown in Figure 10.1a. We show how temporal phase-shift algorithms combined with stroboscopic interferometry can be used to produce a 3-D-detailed picture of device motion and deformation from a set of captured interferograms.

When the light source works in the continuous-wave mode, the recorded interferograms present a modulation of contrast (averaging by CCD camera) [4,34]. Figure 10.23a and Figure 10.23b show two time-averaged interferograms obtained for a piezomembrane operating at two different resonance frequencies: 92.8 kHz and 172 kHz. The results of stroboscopic measurements of the vibrational modes at 92.8 kHz, 107.1 kHz, and 172 kHz are represented in Figure 10.24a to Figure 10.24c. The processing includes the subtraction of the initial static profile of the membrane from the dynamic profile obtained using a five-frame temporal phase-stepping method [5].

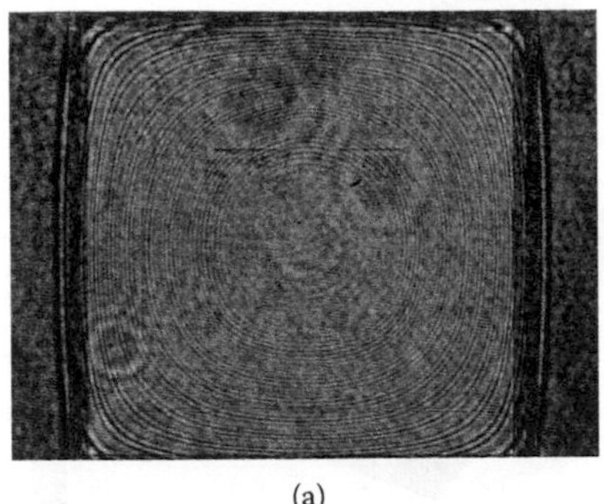

(a)

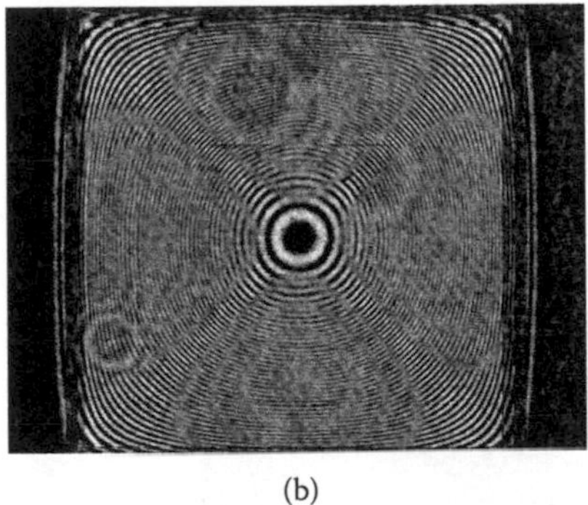

(b)

FIGURE 10.23 Two time-averaged interferograms obtained for a drive voltage of 10 V and resonance frequencies: (a) 92.8 kHz, (b) 172 kHz.

10.5.3 Dynamic Characterization of Torsional Micromirrors

Optical MEMS devices have many applications, including optical switching, bar-code scanning, light modulators, optical scanners, projection displays, variable optical attenuators, and light positioners [37–40]. The optical components, such as micromirrors, which form the building blocks of these systems, must be manufactured to very strict tolerances in order to perform their intended functions well.

We have used the interferometric platform operating in the stroboscopic regime to measure the motion and deformation of bulk-micromachined torsional micromirrors manufactured by Center of Microtechnologies, Chemnitz University of Technology (Germany). Our MOEMS has been built from two silicon wafers and a borosilicate glass plate. The main wafer contains two silicon wafers and borosilicate glass containing two independent movable mircromirrors (1.2×1.2 mm^2 and 1.5×1.5 mm^2) with electrodes, as shown in Figure 10.25. The mirror membranes are attached to the frame by 1-mm-long torsion beams with a square 30×30 μm^2 profile [41,42]. The aim of the work is determination of angular displacements at resonant frequencies for better understanding of such micromirrors and to improve the reliability of such micromechanical structures.

Table 10.6 shows the measured frequencies of torsional modes of both micromirrors, whereas Figure 10.26 represent a set of time-averaged interferograms recorded at frequencies close to the resonance frequency of second torsional mode (f_{res} = 4.57 kHz) of the 1.5×1.5-mm^2 micromirror. To measure the shape of a static mirror or a moving mirror "frozen" by strobe light, a five-frame phase-shift algorithm is used. Electrostatic force tilts the mirror about an axis defined by the torsion hinges, as shown in Figure 10.27, where a 3-D shape of moving the 1.5×1.5-mm^2 micromirror (f_{res} = 2.28 kHz) is given. The tilting motion introduces a rotation angle proportional

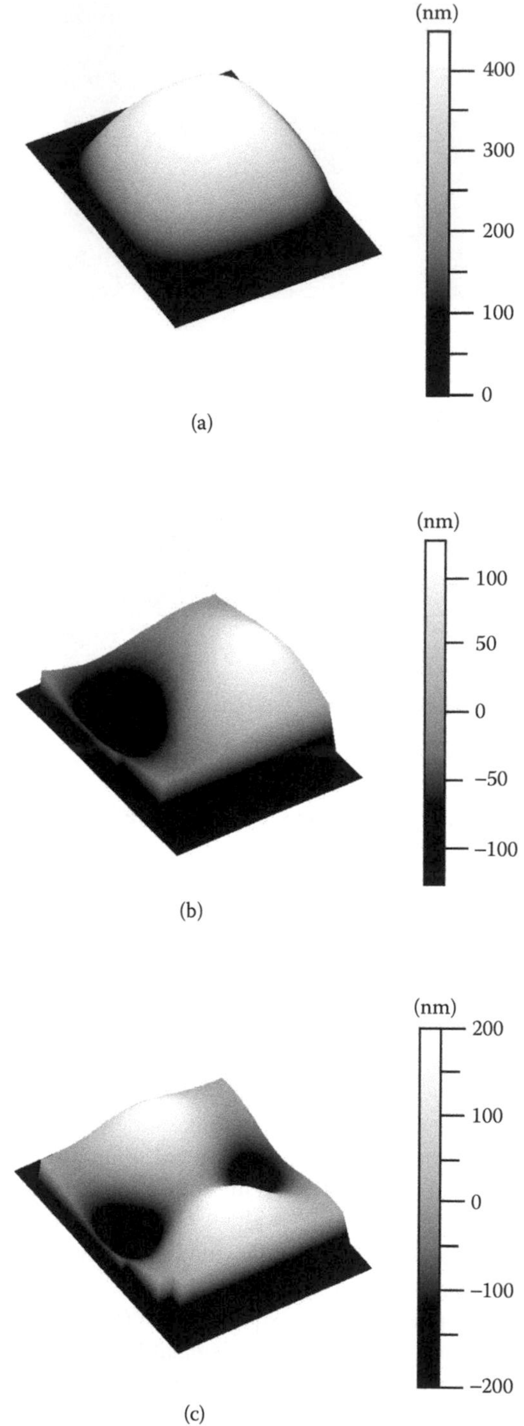

FIGURE 10.24 Stroboscopic method: shapes of the vibration modes of piezo-membrane for resonance frequencies: (a) 92.8 kHz, (b) 107.1 kHz, (c) 172 kHz (drive voltage is 10 V for all cases).

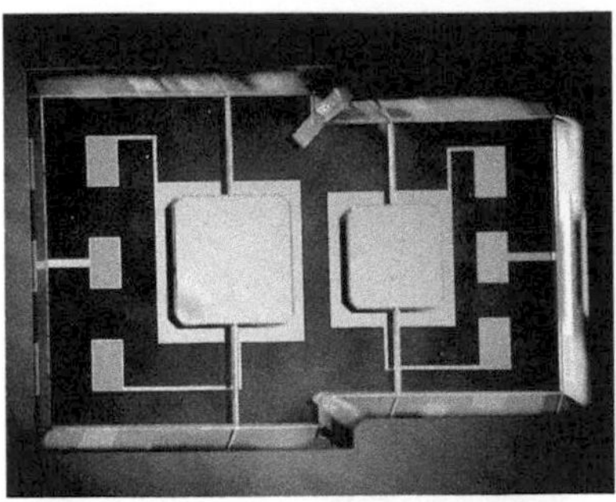

FIGURE 10.25 Micromirrors under tests.

TABLE 10.6
Resonance Frequencies of Torsional Mode of Micromirrors

Resonance Mode	Mirror 1.5 × 1.5 mm²	Mirror 1.2 × 1.2 mm²
No	f[kHz]	f[kHz]
1	2.28	3.93
2	4.57	7.86
3	—	9.22

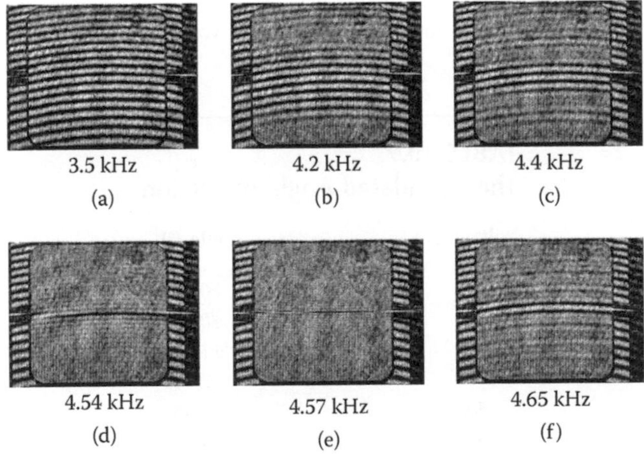

FIGURE 10.26 Time-averaged interferograms recorded at frequencies close to the resonance frequency of torsional mode (f_{res} = 4.57 kHz) of 1.5 × 1.5 mm² mirror.

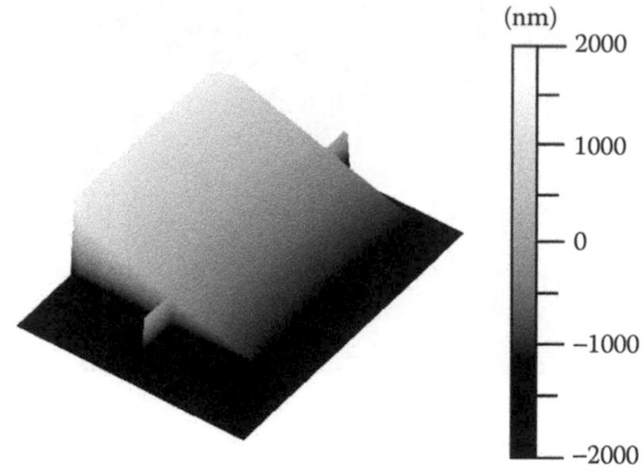

FIGURE 10.27 Torsional displacements of 1.5 × 1.5 mm² mirror for f_{res} = 2.28 kHz and U_s = 100 V.

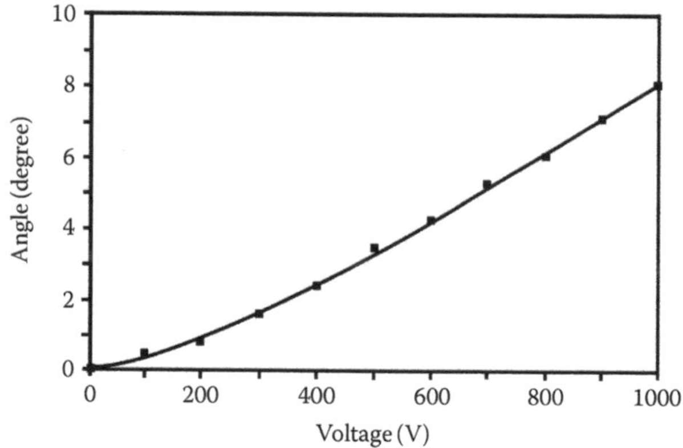

FIGURE 10.28 The calculated angle of torsion vs. the applied voltage.

TABLE 10.7
The Calculated Angle of Torsion

Voltage (V)	Angle (°)
100	±0.48
200	±0.81
300	±1.62
400	±2.41
500	±3.48
600	±4.31
700	±5.27
800	±6.09
900	±7.11
1000	±8.10

to the drive voltage. The magnitude of rotation angle as a function of applied voltage for the mirror of size of 1.5 × 1.5 mm^2 is represented in Figure 10.28 and data are collected in Table 10.7. The accuracy of measured torsional angles is within 0.05°.

In conclusion, we have demonstrated a stroboscopic phase-shifting interferometer system that is particularly useful for characterizing optical MEMS in which the control of surface deformations is critical to measure dynamic out-of-plane displacement and deformation of micromachined parts with nanometer accuracy and submicrosecond time resolution.

10.6 GENERAL CONCLUSION AND OUTLOOK

The ability to experimentally characterize static and dynamic deformation effects is crucial to the development of actuated MEMS devices and microoptical systems. The accurate metrology plays a key role in the characterization and control of critical micromachining processes. Being noncontact and capable of measurements in the nanometer to millimeter scale, optical interferometry has proved to be a versatile measurement technology. This stands as the only viable measurement technology for many current and next-generation MEMS/MOEMS applications. Results demonstrate the use of a multifunctional interferometric platform capable of measuring with nanometer resolution out-of-plane deflections of MEMS/MOEMS structures, providing both material properties and motion behavior of microdevices. The measurements allow determination of the magnitude of stress in mono- and multilayer structures in fully fabricated MEMS/MOEMS systems. Results can be used to verify or update of design and finite element simulations of MEMS behavior.

The size of the miniaturized MEMS/MOEMS structures and their high frequencies make mechanical measurement technologically difficult. As a result, there is a need for the development of optical metrology tools and methodology specifically designed for the dynamic measurement of MEMS devices.

ACKNOWLEDGMENTS

This work is supported by Growth Programme of the European Union (contract G1RD-CT-2000-00261). In particular, the authors thank the following for their active participation in this work:

- Andrei Sabac (Département LOPMD, FEMTO-ST, Besançon, France) for his help in structures' fabrication
- Thierry Dean from Thalès TRT (Orsay, France) for his help in fabrication of active membranes
- Patrice Minotti and all members of "Micromachine" group (Patrice Le Moal and Eric Joseph) from the Département LMARC, FEMTO-ST (Besançon, France) for help in ANSYS simulations and the delivery SDA samples
- Mariusz Wnuk (Warsaw University of Technology, Institute of Micromechanics and Photonics) for his help in stroboscopic measurements

REFERENCES

1. Gorecki, C., Józwik, M., Sabut, L., Multifunctional interferometric platform for on-chip testing the micromechanical properties of MEMS/MOEMS, *Proc. SPIE*, 5343: 63–69, 2004.
2. Józwik, M., Delobelle, P., Gorecki, C., Sabac, A., Nieradko, L., Meunier, C., Munnik, F., Opto-mechanical characterisation of compressively prestressed silicon-oxynitride films deposited by plasma-enhanced chemical vapour deposition on silicon membranes, accepted in *J. Thin Solid Films*, 2004.
3. Józwik, M., Gorecki, C., Le Moal, P., Joseph, E., Minotti, P., Interferometry system for out-of-place microdisplacement measurement: application to mechanical expertise of scratch drive activators, *Proc. SPIE*, 5145: 45–52, 2003.
4. Sabut, L., Patorski, K., Józwik, M., Kacperski, J., Gorecki, Ch., Jacobelli, A., Dean, T., Active micro-elements testing by interferometry using time-average and quasi-stroboscopic techniques, *Proc. SPIE*, 5145: 23–32, 2003.

5. Kujawiska, M., Automatic fringe pattern analysis, in Patorski, K., Ed., *Handbook of the Moiré Fringe Technique*, Elsevier, Amsterdam, 1993, pp. 339–410.
6. Sharpe, W.N., Jr., Yuan, B., Vaidyanathan, R., New test structures and techniques for measurement of mechanical properties of MEMS materials, *Proc. SPIE*, 2880: 78–91, 1996.
7. Ogawa, H., Suzuki, K., Kaneko, S., Nakano, Y., Ishikawa, Y., Kitahara, T., Measurements of mechanical properties of microfabricated thin films, *Proc. 10th Annu. Int. Workshop on MEMS*, Nagoya, 1997, pp. 430–435.
8. French, P.J., Sarro, P.M., Mallé, R., Fakkeldij, E.J.M., Wolffenbuttel, R.F., Optimisation of a low-stress silicon nitride process for surface-micromachining applications, *Sensor. Actuator. A*, 58: 149–157, 1997.
9. Sharpe, W.N., Jr., Yuan, B., Vaidyanathan, R., Measurements of Young's modulus, Poisson's ratio, and tensile strength of polysilicon, *Proc. 10th Annu. Int. Workshop on MEMS*, Nagoya, 1997, pp. 424–429.
10. Bezzaoui, H., Voges, E., Integrated optics combined with micromechanics on silicon, *Sensor. Actuator. A*, 29: 219–223, 1991.
11. Bonnotte, E., Gorecki, C., Toshiyoshi, H., Kawakatsu, H., Fujita, H., Wörhoff, K., Hashimoto, K., Guided-wave acousto-optic interaction with phase modulation in a ZnO thin film transducer on silicon-based integrated Mach-Zehnder interferometer, *J. Lightwave Technol.*, 17: 35–42, 1999.
12. Pharr, G.M., Oliver, W.C., On the generality of the relationship among contact stiffness, contact area, and elastic modulus during indentation, *J. Mater. Res.*, 7: 613–617, 1992.
13. Poilane, C., Delobelle, P., Lexcellent, C., Hayashi, S., Tobushi, H., Analysis of the mechanical behaviour of shape memory polymer membranes by nanoindentation, bulging and point membrane deflection tests, *Thin Solid Films*, 379: 156–165, 2000.
14. Timoshenko, S.P., Woinowsky-Krieger, S., *Theory of Plates and Shells*, Mc Graw-Hill, New York, 1959.
15. Maier-Sneider, D., Maibach, J., Obermeier, E., A new analytical solution for the load-deflection of square membranes, *J. Micromech. Syst.*, 4: 238–241, 1995.
16. Gorecki, C., Sabac, A., Józwik, M., Lee, S.S., Characterisation of internal stress of silicon oxynitride thin films fabricated by plasma-enhanced chemical vapour deposition: applications in integrated optics, *Proc. SPIE*, 4596: 9–15, 2001.
17. Oliver, W.C., Pharr, G.M., An improved technique for determining hardness and elastic modulus using load and displacement sensing indentation experiments, *J. Mater. Res.*, 7(6): 1564–1583, 1992.
18. Danaie, K., Membranes Micro-usinees par Gravure Chimique Anisotrope: Application a Caracterisation des Films Minces, Ph.D. dissertation, Université Pierre et Marie Curie (Paris VI), Paris, 2002.
19. Small, M.K., Nix, W.D., Analysis of accuracy of the bulge test in determining the mechanical properties of thin films, *J. Mater. Res.*, 7: 1553–1563, 1992.
20. Danaie, K., Bosseboeuf, A., Clerc, C., Gousset, C., Julie, G., Fabrication of UV-transparent $Si_xO_yN_z$ membranes with a low frequency PECVD reactor, *Sensor. Actuator. A*, 99: 77–81, 2002.
21. Kramer, T., Paul, O., Postbuckled micromachined square membranes under differential pressure, *J. Micromech. Microeng.*, 12: 475–478, 2002.
22. Denisse, C.M.M., Troost, K.Z., Oude-Elferink, J.B., Harbaken, F.H.P.M., Hendriks, M., Plasma-enhanced growth and composition of silicon oxynitride films, *J. Appl. Phys.*, 60(7): 2536–2542, 1986.
23. Classen, W.A.P., Pol HAJ, Th.V.D., Goemans, A.H., Kuiper, A.E.T., Characterization of silicon-oxynitride films deposited by plasma-enhanced CVD, *J. Electrochem. Soc.*, 133(7): 1458–1463, 1986.
24. Allaert, K., Van Calster, A., Loos, H., Lequesne, A., A comparison between silicon nitride films made by PCVD of N_2-SiH_4/Ar and SiH_4/He, *J. Electrochem. Soc.*, 137(7): 1763–1766, 1985.
25. Trimmer, W.S.N., Microrobots and micromechanical systems, *Sensor. Actuator.*, 19(3): 267–287, 1989.
26. Akiyama, T., Shono, K., Controlled stepwise motion in polysilicon microstructures, *J. Microelectromech. Syst.*, 3: 106–110, 1993.
27. Minotti, P., Le Moal, P., Joseph, E., Bourbon, G., Toward standard method for microelectromechanical systems material measurement through on-chip electrostatic probing of micrometer size polysilicon tensile specimens, *Jpn. J. Appl. Phys.*, 40: 120–122, 2001.
28. Minotti, P., Bourbon, G., Joseph, E., Le Moal, P., Generalized MEMS lot monitoring using fully-integrated material and structure mechanical analysis, *Sensor. Actuator. A*, 101: 220–230, 2002.

29. Ye, X.Y., Zhou, Z.Y., Yang, Y., Zhang, J.H., Yao, J., Determination of the mechanical properties of microstructures, *Sensor. Actuator. A*, 54(1–3): 750–754, 1996.
30. Tabib-Azar, M., Wong, K., Ko, W., Aging phenomena in heavily doped (p+) micromachined silicon cantilever beams, *Sensor. Actuator. A*, 33: 199–206, 1992.
31. Zhang, L.M., Uttamchandani, D., Culshaw, B., Measurement of the mechanical properties of silicon microresonators, *Sensor. Actuator. A*, 29: 79–84, 1991.
32. Burdess, J.S., Harris, A.J., Wood, D., Pitcher, R.J., Glennie, D., A system for the dynamic characterization of microstructures, *J. Microelectromech. Syst.*, 6(4): 322–328, 1997.
33. Larmer, F., Schilp, A., Funk, K., Burrer, C., Experimental characterization of dynamic micromechanical transducers, *J. Micromech. Microeng.*, 6: 177–186, 1996.
34. Petitgrand, S., Yahiaoui, R., Bosseboeuf, A., Danaie, K., Quantitative time-averaged microscopic interferometry for micromechanical device vibration mode characterization, *Proc. SPIE*, 4400: 51–60, 2001.
35. Hart, M., Conant, R.A., Lau, K.Y., Muller, R.S., Stroboscopic interferometer system for dynamic MEMS characterization, *J. Microelectromech. Syst.*, 9(4): 409–418, 2000.
36. Rembe, C., Kant, R., Muller, R.S., Optical measurement methods to study dynamic behavior in MEMS, *Proc SPIE*, 4400: 127–137, 2001.
37. Jaecklin, V.P., Linder, C., Brugger, J., de Rooij, N.F., Moret, J.M., Vuilleumier, R., Mechanical and optical properties of surface micromachined torsional mirrors in silicon, polysilicon and aluminum, *Sensor. Actuator. A*, 43: 269–275, 1994.
38. Tien, N.C., Solgaard, O., Kiang, M.H., Daneman, M., Lau, K.Y., Muller, R.S., Surface-micromachined mirrors for laser-beam positioning, *Transducers'95*, 352–355, 1995.
39. Van Kessel, P.F., Hornbeck, L.J., Meier, R.E., Douglass, M.R., A MEMS-based projection display, *Proc. IEEE*, 86: 1687–1704, 1998.
40. Urey, H., Torsional MEMS scanner design for high-resolution scanning display systems, *Proc. SPIE*, 4773: 27–37, 2002.
41. Hiller, K., Hahn, R., Kaufmann, C., Kurth, S., Kehr, K., Gessner, T., Dotzel, W., Low-temperature approaches for fabrication of high-frequency microscanners, *Proc. SPIE*, 3878: 58–66, 1999.
42. Nieradko, L., Gorecki, C., Sabac, A., Hoffmann, R., Bertz, A., Fabrication and optical packaging of an integrated Mach-Zehnder interferometer on top of a moveable micromirror, *Proc. SPIE*, 5346: 83–88, 2004.

11 Optoelectronic Holography for Testing Electronic Packaging and MEMS

Cosme Furlong

CONTENTS

11.1 Introduction 325
11.2 Overview of MEMS Fabrication Processes 327
11.2.1 Mechanical Properties of Silicon 328
11.3 Optoelectronic Holography 329
11.3.1 Optoelectronic Holography Microscope (OEHM) 329
11.3.2 Static Mode 330
11.3.2 Time-Averaged Mode 331
11.4 Representative Applications 333
11.4.1 Testing of NIST Traceable Gauges 333
11.4.2 Study and Characterization of MEMS Accelerometers 334
11.4.3 Testing at the Wafer Level 338
11.4.3.1 Inspection Procedure 341
11.4.3.2 High-Resolution Stitching 342
11.4.4 Measurement and Simulation of SMT Components 343
11.4.4.1 Computational Modeling 343
11.4.4.2 Computational and OEH Results 345
11.5 Summary 347
Acknowledgments 348
References 349

11.1 INTRODUCTION

Microelectromechanical systems (MEMS) are micron-sized electrical and mechanical structures, fabricated using very large-scale integration (VLSI) techniques adapted from those of the microelectronics industry. MEMS enable the fabrication of micromechanical structures together with the essential microelectronics circuitry used in the sensing, actuation, operating power, and control capabilities of a device [1–5]. MEMS define both the fabrication process and the devices themselves and represent one of today's most exciting areas of microelectronics activity [6]. MEMS technologies are the result of innovations from many areas of microelectronics and have evolved rapidly into a discipline of their own [6,7]. Today's MEMS combine the signal processing and computational capability of analog and digital integrated circuits with a wide range of micromechanical components. Recently, MEMS technologies have been developed and applied by a number of industries to produce components with unprecedented capabilities, which include microscale inertial sensors, pressure and temperature sensors, chemical sensors, DNA samplers, atomic force microscope

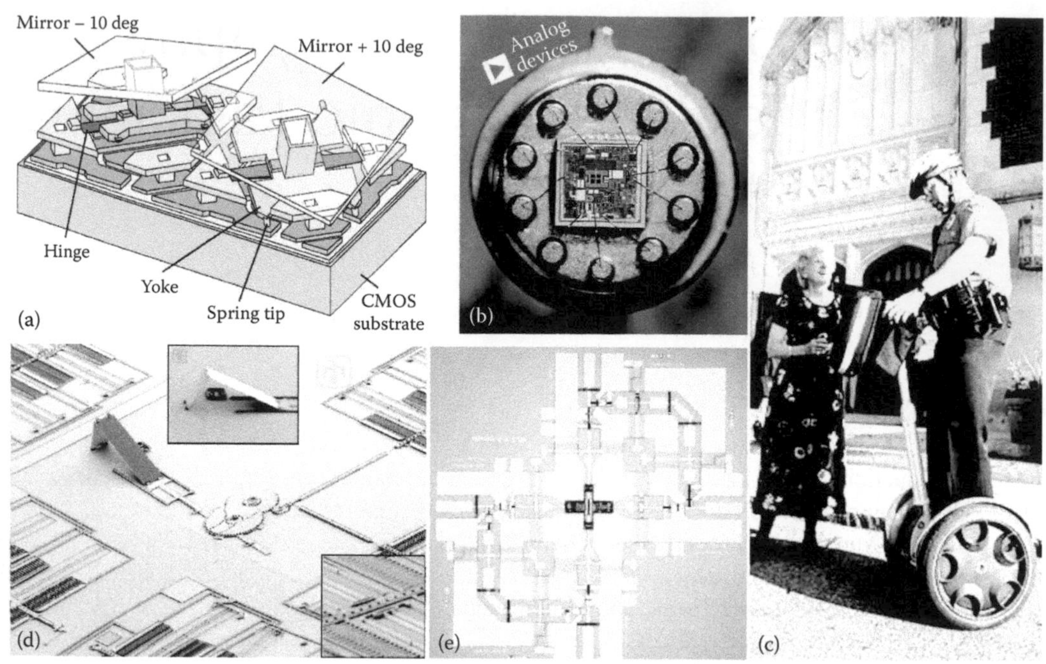

FIGURE 11.1 (See color insert following page 110.) Representative MEMS devices and applications: (a) digital micromirror device (DMD) (courtesy of http://www.ti.com [9]), (b) MEMS accelerometers used in air-bag deployment mechanisms [10], (c) Segway® Human Transporter (HI) enabled by MEMS gyroscopes [11], (d) Sandia Micromirror Device™ driven by a MEMS microengine capable to rotate at 1×10^6 rpm [12], and (e) RF MEMS switches being developed for telecommunications, radar, and automotive applications. (From Lacroix, D., Expanding the RF MEMS market — RF MEMS building blocks, *MSTNews*, 4: 34–36, 2003; Schauwecker, B., Mack, T., Strohm, K.M., Simon, W., and Luy, J.-F., RF MEMS for advanced automotive applications, *MSTNews*, 4: 36–38, 2003.)

(AFM) probe tips, microengines on a chip, RF devices, bio-MEMS, etc. [7,8]. Figure 11.1 shows representative MEMS devices and applications [9–14].

MEMS technologies have become popular for three reasons, namely, multiplicity, microelectronics, and miniaturization. Aided by photolithography and batch fabrication, large numbers of MEMS devices are fabricated simultaneously. Leveraging technologies from integrated circuit (IC) fabrication permits the direct integration of microelectronics control circuitry with micromechanical elements. MEMS can reduce the physical size and weight of sensors and actuators, making them appealing in many applications [8,15]. Finally, MEMS manufacturing facilities are not limited to the production of MEMS for only a single application. Far more agile than an automotive assembly line, MEMS foundries can create miniature sensors and actuators for every conceivable field, from aerospace to biology, to chemical spectroscopy. It is these characteristics that make MEMS an enabling technology [6–8].

As the capabilities of MEMS become more widely recognized, it is also recognized, however, that the biggest obstacle to growth of MEMS applications is the design cycle time, because it depends on tightly coordinated application of advanced design, analysis, manufacturing, and testing tools. Effective development of MEMS requires the synergism of advanced computer-aided design (CAD), computer-aided engineering (CAE), computer-aided manufacturing (CAM) and fabrication methodologies, materials science and technologies, and also of effective quantitative testing methodologies for characterizing their performance, reliability, and integrity [6–8]. Testing of MEMS includes measurements of their electrical, optical, and mechanical responses to the driving signals and environmental loading conditions. Furthermore, in order to understand the mechanics of MEMS and the materials used for their fabrication, advanced noninvasive testing methodologies, capable of measuring the shape and changes in states of deformation of MEMS packages and materials subjected to actual operating conditions, are required [6–8,15–19].

In this chapter, optoelectronic holographic microscopy (OEHM) methodologies for measuring, with submicrometer accuracy, shape and changes in states of deformation of MEMS at the different levels of electronic packaging are described.

11.2 OVERVIEW OF MEMS FABRICATION PROCESSES

MEMS structures are based upon capabilities to fabricate and micromachine 3-D structures at the microelectronics scale. These fabrication technologies can be divided into three major groups [1–5]:

1. Bulk micromachining
2. Surface micromachining
3. High-aspect-ratio electroplated structures

The first, bulk micromachining, dates back to the 1960s, when techniques for wet anisotropic etching of various forms of trenches, grooves, and membranes in silicon wafers were first developed. Advances in bulk micromachining continued rapidly through the 1970s with development of impurity-dependent etch stops, wafer-dissolution processes, and wafer fusion bonding. In the next two decades, surface micromachining, which makes use of the full lithographic capabilities of the IC processing, emerged to enable the definition of a wide range of new designs such as beams, combs, microactuators, and rotating structures. The addition of electronic circuitry to improve the sensor characteristics advanced greatly during this time also. In the 1990s, processes capable of producing high-aspect-ratio structures were developed using thick polymer molds and electroplating, and a variety of methods for merging MEMS fabrication with standard CMOS and bipolar processes were demonstrated. Figure 11.2 and Figure 11.3 illustrate general procedures used in two major fabrication technologies.

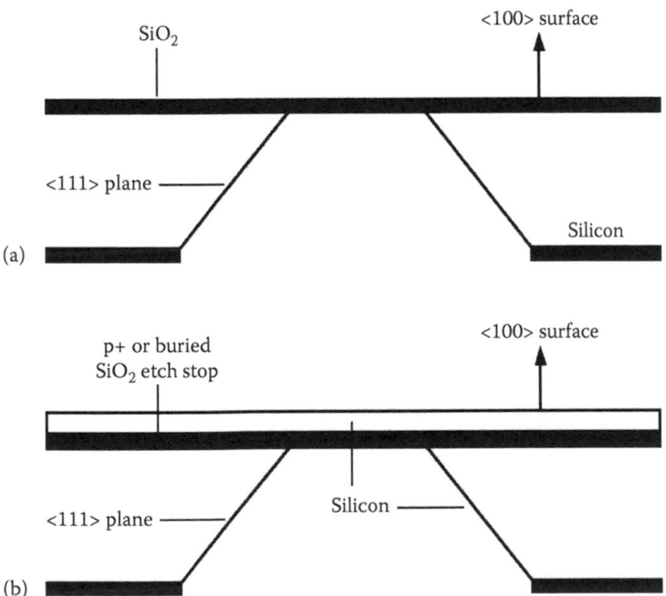

FIGURE 11.2 Diaphragms of a MEMS pressure sensor fabricated by anisotropic backside etching of a silicon wafer: (a) SiO_2 layer used as an etch stop, and (b) buried SiO_2 or p+ layer used as an etch stop to define a thin diaphragm.

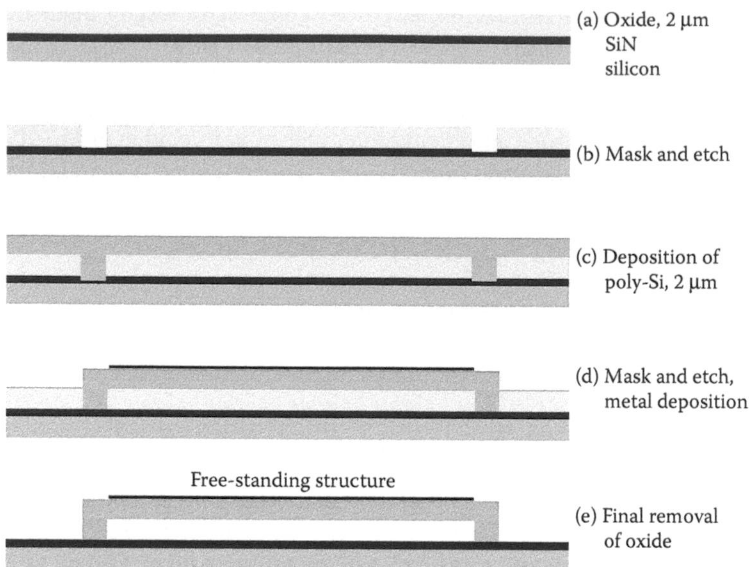

FIGURE 11.3 Surface micromachining: fabrication procedure consists of sequential deposition and removal of structural and sacrificial thin layers to define free-standing structures.

11.2.1 Mechanical Properties of Silicon

Table 11.1 shows a comparative list of representative mechanical properties of different materials used in MEMS fabrication [1–5,20,21]. Although silicon is a brittle material, its yield strength exceeds that of high-strength steel and tungsten. Its Knoop hardness is similarly large, and its elastic modulus is similar to that of iron and steel. Single crystal silicon wafers are typically large in diameter (currently up to 300 mm) and relatively thin (5 to 500 μm) and, in such configuration, they are relatively fragile. Single crystal materials tend to cleave or fracture along crystallographic planes, and breakage can

TABLE 11.1
Representative Material Properties of Selected Materials Used in MEMS

Material	Density (g/cm³)	Yield Strength (GPa)	Knoop Hardness (kg/mm²)	Elastic Modulus (GPa)	CTE (10^{-6}/°K)	Thermal Conduction (W/cm-°K)
Si	2.3	7.0	850	190	2.3	1.6
Polysilicon	2.3	1.2	—	160	2.8	—
SiO_2	2.5	8.4	820	73	0.55	0.014
Si_3N_4	3.1	14	3500	390	0.8	0.19
Diamond	3.5	53	7000	1000	1.0	20
Al	2.7	0.17	130	70	25	2.4
W	19	4.0	490	410	4.5	1.8
Steel	7.9	4.2	1500	210	12	0.97

Source: From Hsu, T.-R., *MEMS and Microsystems: Design and Manufacture*, McGraw-Hill, New York, 2002; Senturia, S.D., *Microsystem Design*, Kluwer Academic, Boston, MA, 2000; Kovacs, G.T.A., *Micromachined Transducers Sourcebook*, McGraw-Hill, New York, 1998; Elwenspoek, M. and Wiegerink, R., *Mechanical Microsystems*, Springer-Verlag, Berlin, 2001; Jaeger, R.C., *Introduction to Microelectronic Fabrication*, Prentice-Hall, Upper Saddle River, NJ, 2002; Peterson, K.E., Silicon as a mechanical material, *Proc. IEEE*, 5: 420–457, 1982; Hillsum, C. and Ning, T.H., Eds., *Properties of Silicon*, INSPEC, Institution of Electrical Engineers, New York, 1988.

often be traced to flaws in the bulk, surface, and edges of the wafers. Sections of wafers, or dies, often have microscopic defects along the edges, which can lead to premature failure of the micromachined components. On the other hand, silicon dies even as large as 20×20 mm^2 are quite robust. Equally, from Table 11.1, a high value of the Knoop hardness for silicon nitride can be observed, which is one reason nitride is an excellent protective passivation layer for finished wafers. In addition, silicon nitride is utilized to define nearly hermitic seals capable of resisting moisture penetration.

11.3 OPTOELECTRONIC HOLOGRAPHY

Optoelectronic holography (OEH) methodologies have been successfully applied to different fields of nondestructive testing (NDT) of objects [15–19,22–28]. OEH methodologies are noninvasive, remote, and capable of providing full-field-of-view qualitative and quantitative information on shape and deformation of objects subjected to a large variety of boundary conditions and loadings. The application of recent technological advances in coherent light sources, computing, imaging, and detector technologies to OEH methodologies, has dramatically increased the versatility of these methodologies. This is manifested in the development of OEH methodologies for studies of MEMS structures and components and for investigation of the micromechanics of materials used in their fabrication. To facilitate accurate characterization, the described OEH methodologies incorporate [15]:

1. Imaging systems based on high-spatial, high-frame rate, and high-digital resolution cameras
2. Light sources suitable for testing materials used in MEMS fabrication
3. Advanced nanometer resolution positioning stages
4. Computer-controlled instrumentation for loading and controlling MEMS packages under investigation
5. Analysis software capable of providing quantitative and full-field-of-view shape and deformation data in the form of 3-D point clouds
6. Routines to export data clouds into advanced CAD software packages

The capabilities of optoelectronic methodologies are illustrated with representative applications to show their use in providing indispensable information for the effective development and optimization of MEMS and nanotechnologies.

11.3.1 Optoelectronic Holography Microscope (OEHM)

Figure 11.4 depicts major components of an OEHM currently under development, specifically set up to perform high-resolution shape and deformation measurements of MEMS and for investigation of the micromechanics of materials used in their fabrication [15].

In the configuration shown in Figure 11.4, the coherent light source is a tunable laser diode (LD) with an operational wavelength centered at 640 nm, 3 mW output power, and with digital or analog modulation capabilities. The output of the LD is directed through a beam expander (BE) and through a beam splitter cube (BS). The BS splits light into the reference beam (RB) and object beam (OB). The RB is directed to the piezo-electric phase stepper (PS) and to the CCD/CMOS detector of a high-speed and high-digital- and high-spatial-resolution digital camera (CA). The OB is directed to the MEMS components of interest (CI). The reflected object beam, carrying shape and/or deformation information of the component of interest, is imaged by means of a long-working distance microscope objective (MO) and condensing lens (CL). After the CL, the OB is directed to the BS and CCD/CMOS detector of the CA. Spatial intensity distributions detected by the CA are transmitted to an image and video-processing computer (IPC). The IPC is capable of synchronizing acquisition of digital video information with data acquisition boards that control modulation of the LD and PS, instrumentation for loading the MEMS under investigation, as well as positioning of the X-Y-Z stages (XYZ).

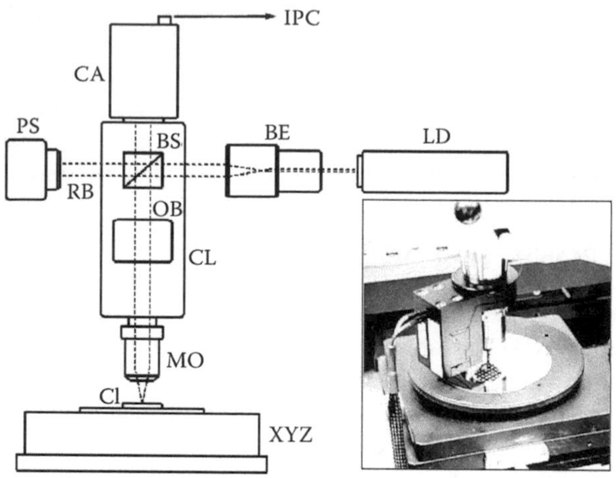

FIGURE 11.4 Optoelectronic holography microscope (OEHM) specifically set up to perform high-resolution shape and deformation measurements of MEMS materials and devices: LD is the tunable laser diode, BE is the beam expander, BS is the beam splitter, OB is the object beam, RB is the reference beam, PS is the phase stepper, CA is the high-speed, high-digital-, and high-spatial-resolution CCD/CMOS camera, IPC is the image and video processing computer, XYZ is the X-Y-Z positioner, CI is the MEMS component of interest, MO is the long-working distance microscope objective, and CL is the condensing lens.

11.3.2 Static Mode

One OEHM approach used to perform shape and deformation measurements of objects consists of acquiring and processing two sets, $I(x,y)$ and $I'(x,y)$, of phase-stepped intensity patterns, recorded before and after, respectively, the event effects to be measured. The first set of phase-stepped intensity patterns is described by [24,25]

$$I_n(x,y) = I_B(x,y) + I_M(x,y)\cos\left[\Delta\phi(x,y) + \theta_n\right], \tag{11.1}$$

where

$$I_B(x,y) = I_o(x,y) + I_r(x,y) \tag{11.2}$$

is the background irradiance, and

$$I_M(x,y) = 2\sqrt{I_o(x,y)\cdot I_r(x,y)} \tag{11.3}$$

is the modulation irradiance. In Equation 11.1 to Equation 11.3, $I_o(x,y)$ and $I_r(x,y)$ are the object and reference beam irradiances, respectively; $\Delta\phi(x,y) = \phi_o(x,y) - \phi_r(x,y)$, with $\phi_o(x,y)$ representing a random phase distribution due to light scattering from the object of interest and $\phi_r(x,y)$ representing a uniform phase distribution from a smooth reference beam wavefront; θ_n is the applied n-th phase step, the value of which is obtained during calibration procedures applied according to the specific phase stepping algorithm that is implemented; and (x,y) represents the Cartesian coordinates of the image space.

The second set of phase-stepped intensity patterns is described by [24,25]

$$I'_n(x,y) = I_B(x,y) + I_M(x,y)\cos[\Delta\phi(x,y) + \Delta\gamma(x,y) + \theta_n]\,. \tag{11.4}$$

In Equation 11.4, $\Delta\gamma(x,y)$ is the change in optical phase that occurred between acquisition of the two sets of phase-stepped intensity patterns, the value of which relates to the shape or changes in state of deformation of objects of interest. With OEHM, the two sets of phase-stepped intensity patterns are processed in the display and data modes.

In the display mode, secondary interference patterns, $Q_D(x,y)$, are generated, displayed at video rates, and modulated by a cosinusoidal function of the form

$$Q_D(x,y) = 4I_M(x,y)\ \cos[\Delta\gamma(x,y)/2] =$$

$$\sqrt{[I_1(x,y) - I_3(x,y) + I'_1(x,y) - I'_3(x,y)]^2 + [I_2(x,y) - I_4(x,y) + I'_2(x,y) - I'_4(x,y)]^2} \tag{11.5}$$

scaled to an 8-bit-resolution video image and obtained after application of four phase steps: $\theta_n = 0$, $\pi/2$, π, and $3\pi/2$. The display mode is used for adjusting, in real time, the experimental parameters for accurate OEHM investigations. Such parameters include:

1. The beam ratio $R = \text{avg}\,[I_r(x,y)]/\text{avg}\,[I_o(x,y)]$, which is important to characterize and set in order to obtain appropriate fringe visibility and also to avoid optical saturation of the CCD/CMOS array detector of the camera
2. The phase step θ_n, which is obtained by an automatic calibration procedure and used to acquire accurate phase-stepped intensity patterns, $I_n(x,y)$, based on which further processing is conducted

The data mode is used for quantitative investigations, which involve the determination of $\Delta\gamma(x,y)$, related to the shape and/or deformation of samples of interest. The discontinuous distribution $\Delta\gamma(x,y)$, modulo 2π, is determined using double-floating point arithmetic as

$$\Delta(x,y) = \tan^{-1}\left[\frac{2(I_2 - I_4)}{-I_1 + 2I_3 - I_5}\right], \tag{11.6}$$

where the (x,y) arguments have been omitted for clarity. Equation 11.6 corresponds to the implementation of the five-phase-step algorithm with $\theta_n = 0$, $\pi/2$, π, $3\pi/2$, and 2π. The application of this algorithm minimizes errors in the determination of $\Delta\gamma(x,y)$ due to possible phase-stepping miscalibration. The recovery of continuous spatial phase distributions $\Delta\gamma(x, y)$ requires the application of efficient phase-unwrapping algorithms.

11.3.2 Time-Averaged Mode

Noninvasive and full-field-of-view dynamic investigations, or modal analysis, of an object of interest can also be performed with OEHM in the dynamic data acquisition and processing mode [22,23]. In this case, it is necessary to take into consideration a time-varying fringe-locus function $\Omega_t(x,y,t)$, which is related to the periodic motion of the object under investigation [22,23,29,30]. Therefore it is possible to write

$$I(x,y,t) = I_o(x,y,t) + I_r(x,y) + 2\sqrt{I_o(x,y,t)\cdot I_r(x,y)}\cos\left[\Delta\phi(x,y,t) + \Omega_t(x,y,t)\right]. \tag{11.7}$$

Because a camera, in this case of a CCD/CMOS type, registers average intensities at the video rate characterized by the period Δt, the time-averaged intensity observed can be expressed as

$$I_t(x,y) = \langle I(x,y,t) \rangle = \frac{1}{\Delta t} \int_t^{t+\Delta t} I(x,y,t)\, dt \, . \tag{11.8}$$

Assuming invariance of I_o with respect to Δt and taking speckle statistics into account, the average sampled intensity at phase shift n that is detected can be expressed as [22,23]

$$I_{t_n}(x,y) = I_o(x,y) + I_r(x,y) + 2\sqrt{I_o(x,y)\cdot I_r(x,y)}\cos\left[\langle \Delta\phi(x,y)\rangle + \theta_n\right] M[\Omega_t(x,y)], \tag{11.9}$$

where $M[\Omega_t(x,y)]$ is known as the characteristic function determined by the temporal motion of the object. For the case of sinusoidal vibrations with a period much shorter than the video framing time of the camera utilized, the characteristic function can be expressed as

$$M[\Omega_t(x,y)] = J_o[\Omega_t(x,y)], \tag{11.10}$$

where $J_o[\Omega_t(x,y)]$ is the zero-order Bessel function of the first kind. The generation of a speckle correlation interferogram using OEHM in the dynamic mode consists in acquiring four phase-shifted speckle intensities at $\theta_n = 0,\ \pi/2,\ \pi, 3\pi/2$, which, after several mathematical operations, can be used to generate the intensity

$$Q_D(x,y) = 4\sqrt{I_o(x,y)\cdot I_r(x,y)}\ M[\Omega_t(x,y)]\, , \tag{11.11}$$

described by the function

$$Q_D(x,y) = \sqrt{(I_1 - I_3)^2 + (I_2 - I_4)^2}\, . \tag{11.12}$$

The application of Equation 11.12 to generation of dynamic speckle correlation interferograms is, as in the static mode, limited to the processing speed of the computer and video processing hardware.

Determination of $\Omega_t(x,y)$ is based on intermediate operations that require the acquisition and processing of correlation intensities of the form [22,23]

$$I_{h1}(x,y) = I_B(x,y) + I_M(x,y)\ J_o^2[|\Omega_t(x,y)|]\, , \tag{11.13}$$

$$I_{h2}(x,y) = I_B(x,y) + I_M(x,y)\ J_o^2[|\Omega_t(x,y) - \bar{B}|]\, , \tag{11.14}$$

$$I_{h3}(x,y) = I_B(x,y) + I_M(x,y)\ J_o^2[|\Omega_t(x,y) + \bar{B}|]\, , \tag{11.15}$$

where $I_B(x,y)$ and $I_M(x,y)$ are the background and modulation intensities corresponding to a dynamic speckle correlation interferogram and $\bar{B}$ is the bias function, which, for an object loaded with a sinusoidal excitation of frequency ω, takes the form

$$\bar{B}(t) = \bar{A}\sin(\omega\, t + \psi), \tag{11.16}$$

in which $\bar{A}$ is the amplitude and ψ, the bias phase. $\bar{B}$ is introduced in the optical path length by utilization of an additional *PZT* device added in the OEHM setup. However, quantitative dynamic investigations with OEHM can successfully be performed using stroboscopic illumination in combination with the static mode of data acquisition, as described earlier.

11.4 REPRESENTATIVE APPLICATIONS

11.4.1 Testing of NIST Traceable Gauges

Sensitivity, accuracy, and precision in optical metrology refer to characteristics of sensing devices or techniques related to their resolution, experimental uncertainty associated with a measurement, and capability to effectively attain similar results during repeated measurements of the quantity investigated. Measuring sensitivity is a function of the characteristics of the sensing device as well as of the optical characteristics of the imaging system utilized, whereas measuring accuracy and repeatability are functions that depend on multiple experimental parameters. These experimental parameters are divided into four groups [31]:

1. The accuracy at which the continuous spatial phase distribution is determined, which in turn depends on
   a. The accuracy of image acquisition and processing procedures
   b. The appropriate phase step calibration
   c. The appropriate algorithm to generate and evaluate continuous spatial phase distribution
2. The magnitude of the equivalent wavelength Λ and its accurate characterization, a parameter that is important when making shape measurements using multi-wavelength techniques [18]
3. The accuracy at which the metrics of the imaging system are determined, allowing characterization of optical aberrations in the imaging system
4. The accurate localization of the equivalent points of illumination and observation, allowing determination of sensitivity vectors

The characterization of sensitivity, accuracy, and precision of the optical metrology methods is important in order to demonstrate their reliability and applicability to perform high-resolution measurements, as required in MEMS and electronic packaging. Sensitivity, accuracy, and precision in OEHM are characterized with the aid of National Institute of Standards and Technology (NIST) traceable gauges. Figure 11.5 depicts one of the NIST traceable gauges specifically designed, constructed, and certified to characterize OEHM measurements [32].

Measurements of a region of interest (ROI) of a NIST traceable gauge are carried out with an OEHM setup configured with a CMOS camera containing 1024 × 1024 active pixels, capable of achieving 12-bit digitization, 25 fps at full-frame resolution, and up to 8000 fps at a reduced ROI. Measurements involve:

1. Performing flat-field calibration of the camera to achieve uniform intensity distributions
2. Setting up and optimizing integration look-up tables (LUTs) to enhance low and high intensity levels in order to achieve uniform intensity distributions by minimizing low level and saturated regions
3. Optimizing the beam ratio R
4. Performing high-resolution, automatic, phase step calibration

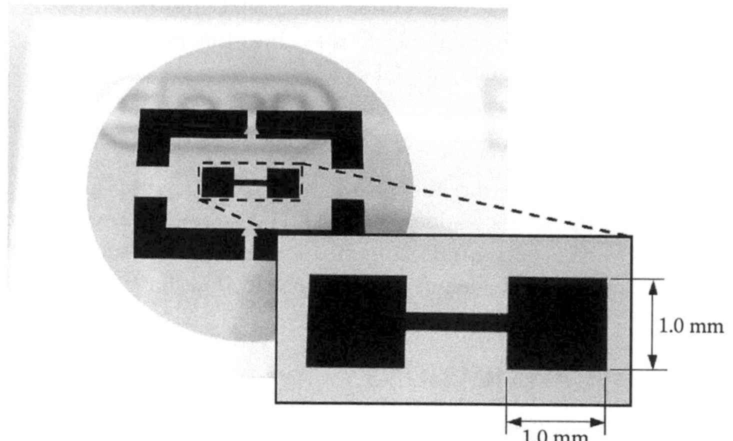

FIGURE 11.5 NIST traceable gauge utilized for characterizing sensitivity, accuracy, and repeatability of OEHM measurements. Gauges utilized contain thin Au films with certified thicknesses of 200 Å ± 2.5%, 500 Å ± 2.5%, 1 kÅ ± 2.5%, and 5 kÅ ± 2.5%. (From Veeco Instruments, Sloan DekTak calibration standards set, Veeco Metrology Group, Tucson, AZ, 2002.)

Figure 11.6 and Figure 11.7 show representative results; (x,y,z) data clouds were imported into an advanced CAD environment to perform difference analysis with respect to a plane [33–35]. Measurements and analysis of the data clouds indicate a film thickness of 498 Å ± 7.2 Å. These results are comparable to results obtained with an equivalent CCD camera.

11.4.2 Study and Characterization of MEMS Accelerometers

MEMS inertial sensors, comprising acceleration and angular rate sensors, i.e., microaccelerometers and microgyroscopes, respectively, are used in a wide range of consumer, industrial, and automotive applications. MEMS inertial sensors are typically utilized to activate safety systems, including automotive airbag deployment systems, and to implement vehicle stability systems and electronic suspensions [1–4].

Structural characterization of shape and deformation is performed on dual-axis MEMS accelerometers [36,37]. Microaccelerometers contain, on a single monolithic integrated circuit (IC), a polysilicon surface micromachined proof mass as well as signal conditioning circuitry to implement an acceleration measurement. These microaccelerometers are capable of measuring not only positive and negative accelerations up to at least ±2 g but also static acceleration forces such as gravity,

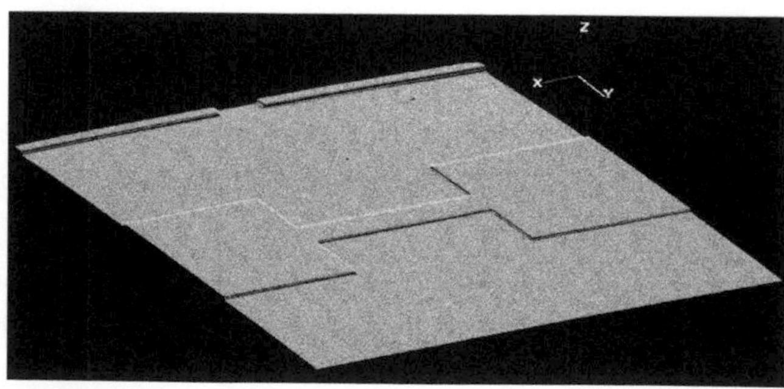

FIGURE 11.6 Measured (x,y,z) data cloud of the 498 Å ± 7.2 Å NIST traceable gauge as imported into a CAD environment.

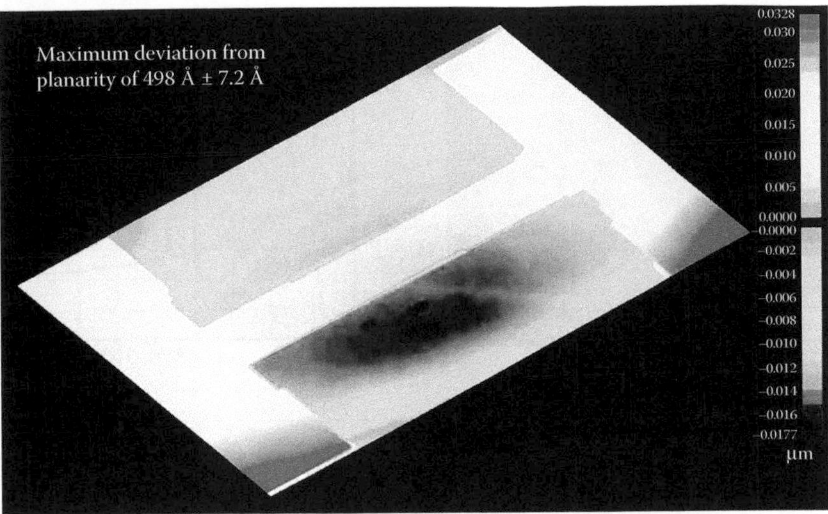

FIGURE 11.7 (See color insert.) Deviations from planarity are obtained by difference analysis, which consists of evaluating the differences between a plane and the measured (x,y,z) data cloud shown in Figure 11.6.

allowing them to be used as tilt sensors [1,36]. Table 11.2 summarizes pertinent characteristics of the microaccelerometers investigated, and Figure 11.8 and Figure 11.9 depict their different electronic packages. Figure 11.8 shows the $5.0 \times 4.5 \times 1.78$ mm^3 leadless package of the device, the electronic board set to the [10–5000] Hz bandwidth configuration, and a typical analog output response of the microaccelerometer subjected to an oscillatory driving force equivalent to 1.0 g at 1.0 kHz frequency, as observed on a spectrum analyzer. Figure 11.9 shows the die configuration of the package containing the integrated micromechanical and microelectronic components. The proof mass of the microaccelerometer has dimensions of $600.0 \times 600.0 \times 3.0$ μm^3 [37].

In an effort to understand the effects of geometry, material properties, and boundary conditions on the performance of the microaccelerometers used, finite element models (FEM) have been developed and validated with respect to the characteristics of the microaccelerometers [15,38–41]. Figure 11.10 displays an FEM and mode shape calculated at the fundamental natural frequency of 10.91 kHz for polysilicon with material properties characterized by the modulus of elasticity E = 190 GPa and mass density of 2.33 g/cm^3.

Figure 11.11 to Figure 11.16 depict typical results obtained when characterizing the shape and deformation of the microaccelerometers in two different conditions:

TABLE 11.2
Pertinent Characteristics of the Microaccelerometers Investigated

Measurement Range[a]	Sensitivity	Resonant Frequency (approx.)	Operating Voltage	Temperature Drift	Operating Temperature
±2 g	312 mV/g	10 kHz	[3.0,5.25] V	±0.5 %	[–40,85] °C

[a] 1 g = 9.81 m/sec^2

Source: Application Note, ADXL202/ADXL210 — low cost ±2 g/±10 g dual axis iMEMS® accelerometers with digital output, Analog Devices, Norwood, MA, 2004.

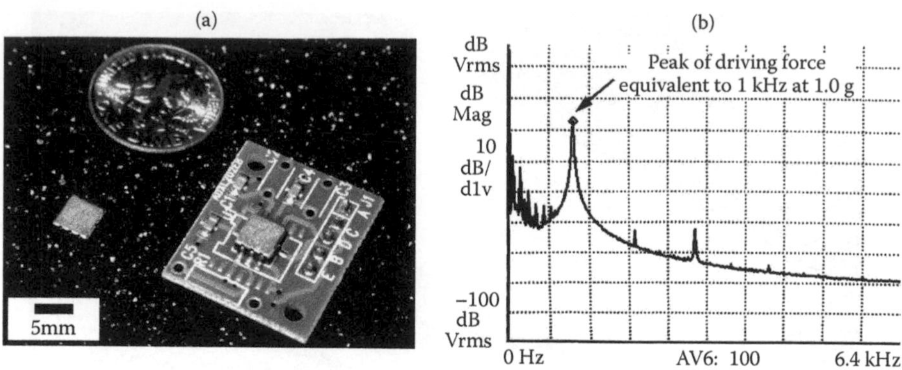

FIGURE 11.8 The microaccelerometer package and typical performance: (a) leadless package of the device and electronic board set to the [10–5000] Hz bandwidth configuration, (b) frequency spectrum of a response of the device subjected to a force equivalent to 1.0 g at 1 kHz.

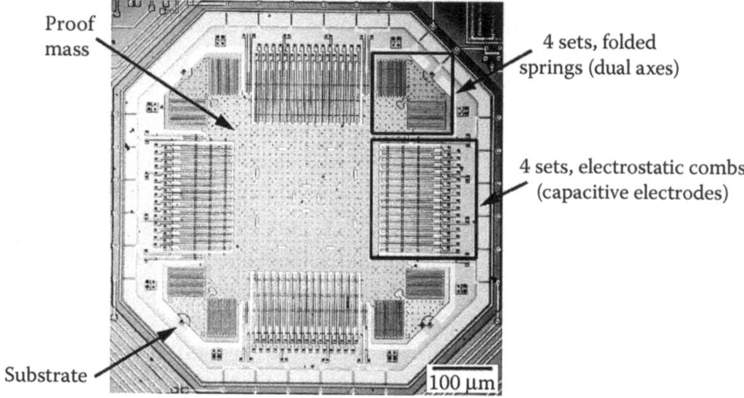

FIGURE 11.9 Electromechanical components of the MEMS accelerometer die. (From Application Note, ADXL202/ADXL210 — low cost ±2 g/±10 g dual axis iMEMS® accelerometers with digital output, Analog Devices, Norwood, MA, 2004; Kok, R., Development of a Wireless MEMS Inertial System for Health Monitoring of Structures, M.S. thesis, Worcester Polytechnic Institute, Worcester, MA, 2004.)

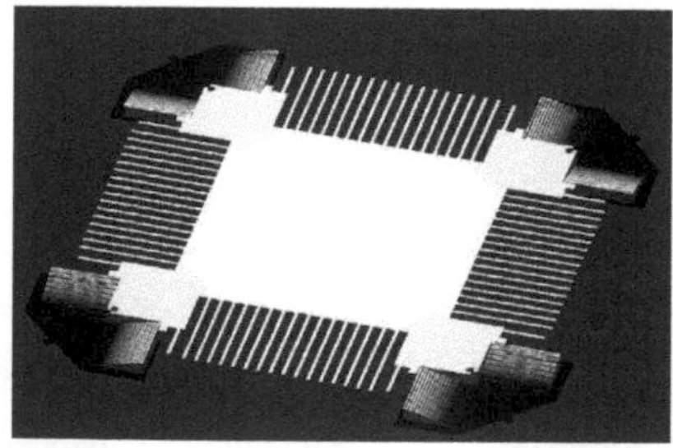

FIGURE 11.10 FEM modal analysis of the electromechanical components of the MEMS accelerometer die shown in Figure 11.9. Fundamental natural frequency calculated at 10.91 kHz. (From Furlong, C. and Pryputniewicz, R.J., Optoelectronic characterization of shape and deformation of MEMS accelerometers used in transportation applications, *Opt. Eng.*, 42(5): 1223–1231, 2003; SRAC Corporation, *COSMOS/M User's Guide v.*3.0., Los Angeles, CA, 2002.)

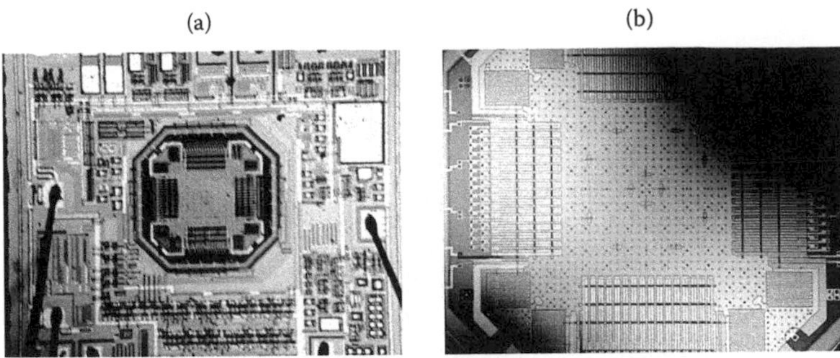

FIGURE 11.11 OEHM interferograms of the microaccelerometer subjected to the loading Case A: (a) overall view and (b) central area of the die.

1. Case A (unloaded): when an electrical potential of 0.0 Vdc is applied to power the microaccelerometer
2. Case B (loaded): when an electrical potential of 5.0 Vdc is applied to power the microaccelerometer

Figure 11.11 depicts OEHM interferograms of the microaccelerometer subjected to loading in Case A. Characterization of MEMS devices under this condition is equivalent to determining their shape without any loading. Analysis of Figure 11.11 indicates that the device has a flatness of better than 1λ. When the device is subjected to loading in Case B, out-of-plane components of deformation appear on the proof mass, the moving set of electrostatic combs, and the folded springs, as depicted in the OEHM interferograms shown in Figure 11.12.

Figure 11.13 and Figure 11.14 show higher-resolution OEHM measurements on one of the folded spring pairs of the microaccelerometer subjected to the loading in Case A and Case B, respectively. Quantitative analyses indicate that the proof mass moves 1.48 μm out of plane. Characterization of such motion and its effect on the performance of the microaccelerometer requires more detailed investigations. Quantitative results of shape and deformation, in the form of data clouds, are exported into CAD environments for further study and analysis as well as definition of CAE models, as shown in Figure 11.15.

Determination of shape and deformation on the entire microaccelerometer, although maintaining high resolution and magnification OEHM measurements, requires acquisition of data on different regions, i.e., tiles, of interest. Analyses of such regions result in individual point clouds of data that require merging, i.e., patching, in order to determine characteristics of interest on the entire microaccelerometer. In our investigations, algorithms for point cloud patching based on feature- and area-based matching are utilized [17,33,34]. Figure 11.16 shows results of point cloud patching, corresponding to shape measurements for loading in Case A, using four different regions of interest.

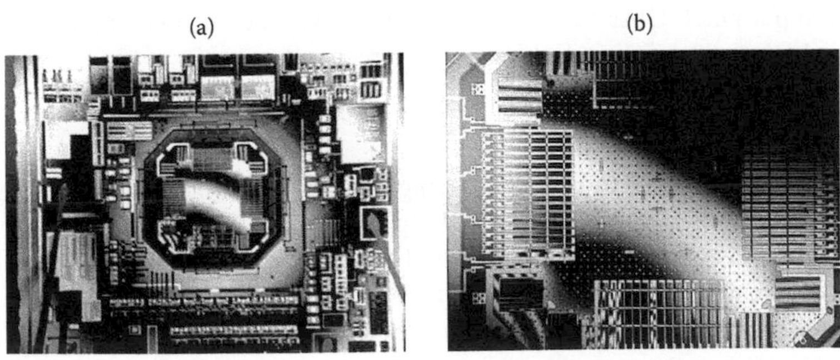

FIGURE 11.12 OEHM interferograms of the microaccelerometer subjected to the loading Case B: (a) overall view and (b) central area of the die.

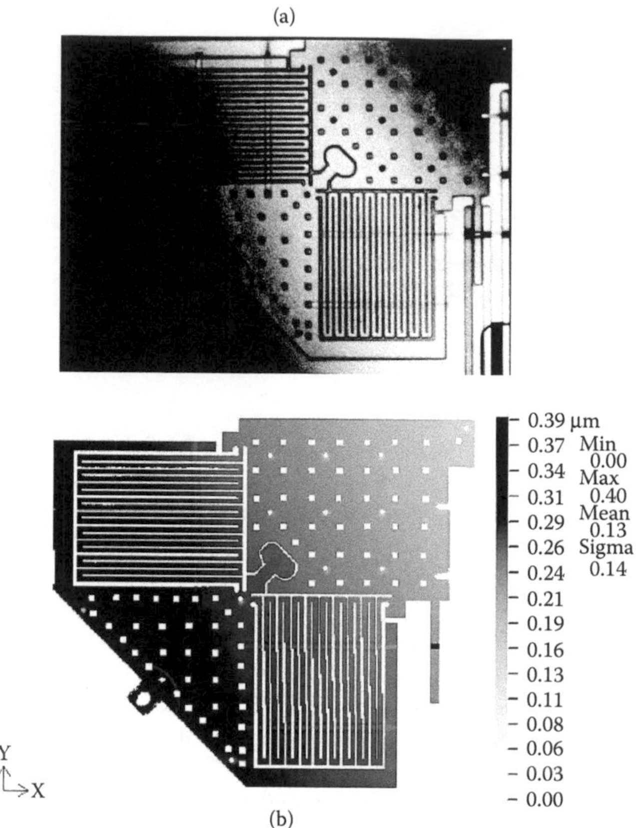

FIGURE 11.13 Quantitative OEHM analysis on one of the folded spring pairs of the microaccelerometer subjected to the loading Case A: (a) typical OEHM interferogram and (b) quantitative OEHM representation of shape.

MEMS accelerometers have very high natural frequencies, which are directly related to the design of their proof mass, folded springs, and capacitive combs. Specifically, the magnitude of the fundamental natural frequency defines the range of accelerations that can be measured with minimum errors and distortions. In order to validate and update our computational models, experimental modal analysis using OEHM methodologies was performed. Figure 11.17 shows, for the first time, the OEHM-measured fundamental natural frequency of vibration of the MEMS die considered. Frequency was observed at 10.65 kHz with dynamically induced maximum deformations on the order of 400 nm, measured with an accuracy of 5 nm. Spatial resolution shown in Figure 11.17 is 0.85 μm per pixel. Observed natural frequency deviates from computational calculations (Figure 11.10), by less than 5%, indicating good agreement.

11.4.3 Testing at the Wafer Level

Figure 11.18 shows a typical ϕ150 mm <100> single-crystal silicon wafer containing 2730 surface micromachined polysilicon MEMS devices. Taking into account that a fabrication run may contain 12 wafers, a total of 32,760 MEMS devices should be tested before dicing. Testing of these devices involves probing their response to specific electrical excitation signals. However, their detailed electromechanical responses are not repeatedly measured because of the following [15,42]:

1. Considerable addition of metrology time in the fabrication processes
2. Throughput of available metrology systems not being compatible with current and future MEMS fabrication requirements

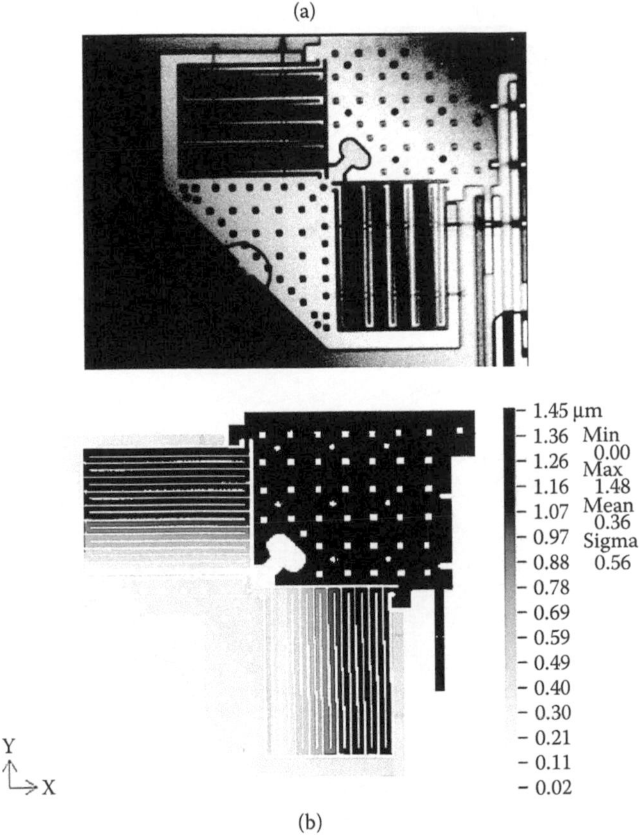

FIGURE 11.14 Quantitative OEHM analysis on one of the folded spring pairs of the microaccelerometer subjected to the loading Case B: (a) typical OEHM interferogram and (b) quantitative OEHM representation of deformation.

3. Lack of effective metrology techniques addressing the need of high-resolution measurement requirements while providing minimum inspection time
4. Compatibility of metrology techniques with probe stations for MEMS
5. Considerable efforts involved in integrating metrology with fabrication tools

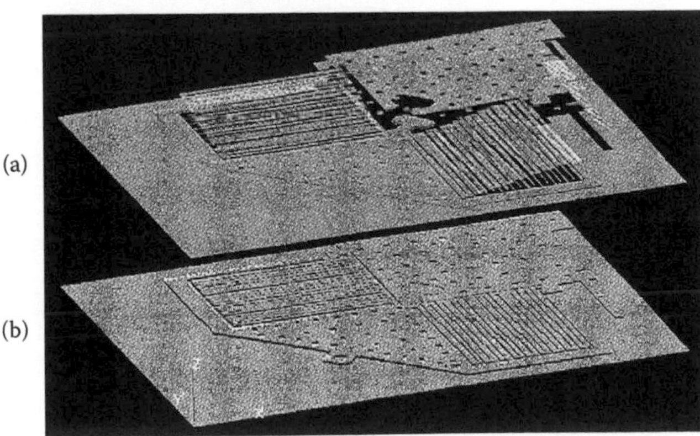

FIGURE 11.15 OEHM measurements of shape and deformation of a MEMS accelerometer imported into a CAD environment. Three-dimensional data clouds corresponding to: (a) loading Case A and (b) loading Case B.

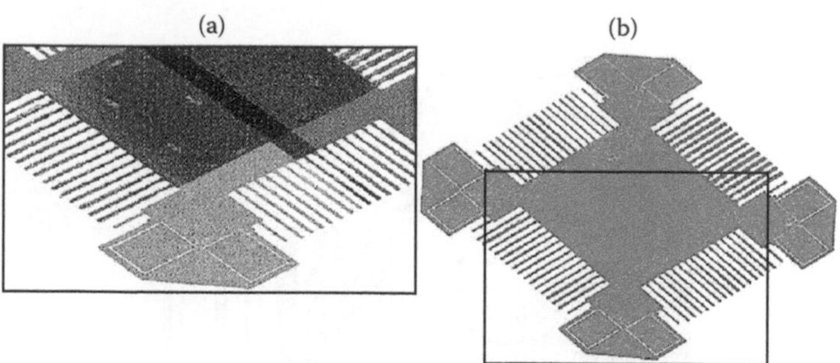

FIGURE 11.16 Point cloud patching of four regions of interest. Patched cloud contains nearly 1,000,000 data points: (a) partial view and (b) overall view.

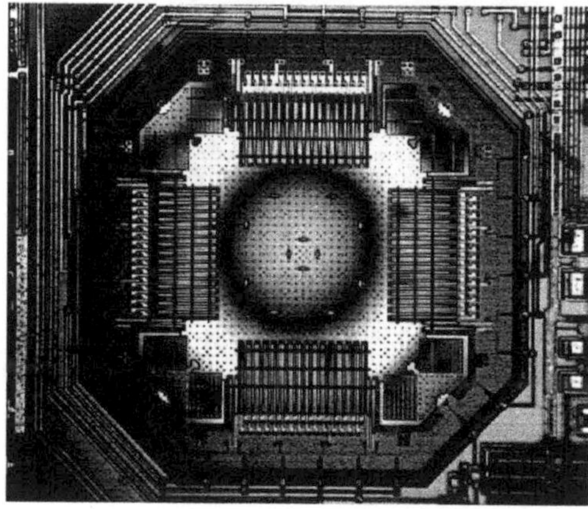

FIGURE 11.17 Fundamental natural frequency of vibration of the MEMS die shown in Figure 11.9. Frequency shown, measured with OEHM, is 10.65 kHz with dynamically induced maximum deformations on the order of 400 nm measured with an accuracy of 5 nm. Differences between computational (Figure 11.10) and measured results is less than 5%.

FIGURE 11.18 ϕ150 mm <100> single-crystal silicon wafer containing 2730 surface micromachined polysilicon MEMS. Development of effective metrology methodologies to test MEMS at the wafer level should consider minimization of added time to the fabrication processes.

In an effort to address some of these metrology constraints, we are developing OEHM methodologies for rapid inspection of MEMS at the wafer level.

11.4.3.1 Inspection Procedure

With the OEHM system described in Section 11.3, high measuring resolution and accuracy are achieved by recording multiple overlapping, high-magnification imaging data of MEMS dies [17,31]. A motorized XYZ stage of the OEHM allows positioning of the imaging system to provide a tiled coverage of the entire wafer. Each recorded imaging tile covers a different portion of a specific ROI with a slight overlapping between neighboring images. Adjacent images are stitched together based on their overlapping ROI. The overlapping edge regions are compared and aligned through correlation analysis and hill-climbing algorithms used to generate transformation matrices. These transformation matrices are applied to the images, and a correlation coefficient is calculated. The transformation that generates the highest correlation coefficient is used to combine the tiles, resulting in a single, larger image containing the union of the original imaging data. With this process, a high-resolution image of the entire wafer is constructed. Therefore, evaluation of all MEMS on the wafer can be performed prior to dicing and packaging with a higher measuring accuracy than with an individual, low-resolution image. Identification of damaged MEMS before dicing and packaging allows for earlier removal of defective MEMS, resulting in production cost savings of substantial amounts [42].

11.4.3.1.1 Correlation Analysis

Multiple images can be combined into a single contiguous image by finding the highest degree of correlation between their overlapping ROIs. The normalized correlation analysis algorithm is used to calculate the normalized correlation coefficient, i.e., the correlation coefficient *r*, of the overlapping images, defined as

$$r = \frac{\text{covariance of } A \text{ and } B}{(\text{standard deviation of } A)(\text{standard deviation of } B)}, \tag{11.17}$$

with A and B representing two image data sets. By expanding Equation 11.17, the correlation coefficient r for A and B, based on data sets containing N pixels each, becomes [17,31]

$$r = \frac{\frac{1}{N}\sum_{i=0}^{N-1}\left(A_i - \left[\sum_{k=0}^{N-1}\frac{A_k}{N}\right]\right)\left(B_i - \left[\sum_{k=0}^{N-1}\frac{B_k}{N}\right]\right)}{\sqrt{\frac{1}{N}\sum_{i=0}^{N-1}\left(A_i - \left[\sum_{k=0}^{N-1}\frac{A_k}{N}\right]\right)^2}\sqrt{\frac{1}{N}\sum_{i=0}^{N-1}\left(B_i - \left[\sum_{k=0}^{N-1}\frac{B_k}{N}\right]\right)^2}}, \tag{11.18}$$

which describes the relationship between each pair of pixels in A and B and the distances from their corresponding statistical means. For example, when $A_i - \mu_A = B_i - \mu_B$, for all *i*-th pixels, the correlation coefficient is unity. Conversely, when $A_i - \mu_A = -(B_i - \mu_B)$ for all i-th pixels, the correlation coefficient is 1.

Through a mathematical searching algorithm in the spatial overlapping domain, the transformation matrix resulting in the highest correlation coefficient for the images is found. The transformation matrix and its correlation coefficient are stored for use during the image stitching phase. The control software of the OEHM allows for the correlation analysis to be performed in parallel with image acquisition, providing stitched images of the MEMS with all available images. By running the correlation concurrently with acquisition, continuous feedback on the spatial resolution and position of motorized stages is acquired. Furthermore, by proper selection of the overlapping ROIs, computation time is decreased and stitching accuracy is increased.

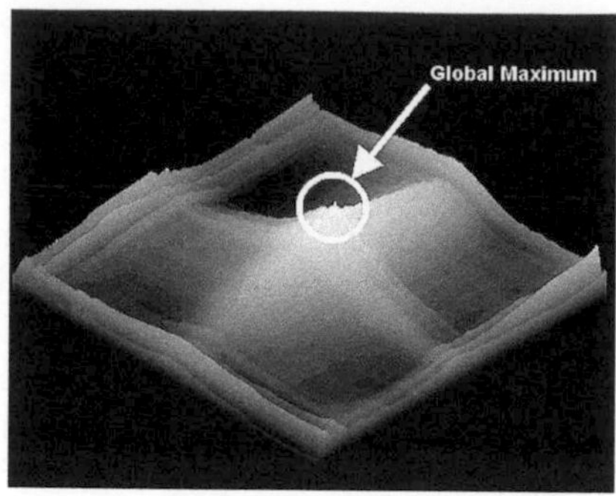

FIGURE 11.19 Three-dimensional map showing distribution of the correlation coefficient r obtained when performing sub-pixel resolution stitching of two OEHM images. Finding the maximum correlation coefficient, i.e., global maximum, defines the optimal transformation matrix and stitching conditions.

11.4.3.1.2 Hill-Climbing Searching Algorithm

The hill-climbing searching algorithm is used to find the transformation matrix resulting in the highest correlation coefficient, Equation 11.17. The starting transformation matrix is initialized with XYZ positioning information from the stages of the OEHM. Starting from this initial position, neighboring transformations are considered. The correlation coefficient, r, is found for each of the transformations. If one of these positions has a higher correlation coefficient, it is designated as the new current position. The search continues by updating the current position until no neighbor has a higher correlation coefficient. The search is terminated when the specified number of iterations is reached. The current position at the end of the search is deemed the best translation and is used for stitching the images [17,31]. Figure 11.19 shows a 3-D map distribution of the correlation coefficient, r, obtained when performing super-high-resolution, i.e., sub-pixel-resolution, stitching of two images. The developed hill-climbing searching algorithm is capable of handling an arbitrary number of degrees of freedom, which adds versatility to the OEHM when additional degrees of freedom, as required by future metrology needs, are incorporated.

11.4.3.2 High-Resolution Stitching

MEMS were inspected at the wafer level using the optical magnifications of 10X and 25X, providing spatial resolutions of 3.5 μm and 0.24 μm per pixel, respectively. Figure 11.20 shows a series of

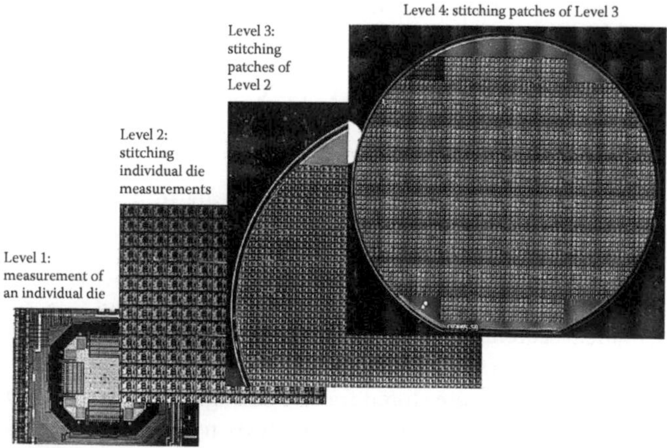

FIGURE 11.20 Inspection of an entire ϕ150 mm <100> single-crystal silicon wafer containing 2730 surface micromachined polysilicon MEMS accelerometers. Each microaccelerometer is inspected with a spatial resolution of 3.5 μm per pixel and measurements are stitched together.

overlapping images recorded at the 10X magnification and patched using the algorithms described in this section. Images were stitched with an average edge overlap of 25% and in real time, i.e., while each individual die is being inspected. Automatic calibration, which included compensation for optical distortions, determination of magnification factors, and minimization of inspection time, is incorporated in the measurements [17,31].

11.4.4 Measurement and Simulation of SMT Components

Driven by demand for delivery of more efficient devices, surface mount technology (SMT) has become the mainstream packaging technology of electronic equipment manufacturers. The significant driving force for the use of SMT is the reduced package size and cost, as well as improved board utilization as the package-to-board interconnection spacing is minimized, when compared to pinned-through-hole packages. However, placing more functions in a smaller package has resulted in higher heat densities, which require that thermal management be given a high priority in design so as to maintain system performance and reliability [28,43–45].

SMT components comprise different materials, each with a different coefficient of thermal expansion (CTE). As active components dissipate heat to their surroundings or as the ambient system temperature changes, differential thermal expansions due to different CTEs cause stresses in the interconnection structures. These stresses produce elastic and plastic strains, as well as time-dependent deformations of the joints between the relatively rigid package body and circuit board. This becomes important as the devices are thermally cycled during operation, given the differences in CTEs for the materials of package and board, and given the differences in temperature between the package, the board, and the environment. Therefore, knowledge of thermal-cycle-induced characteristics of SMT joints is important. This section describes representative results of a study using noninvasive measurements and computational simulation methodologies for evaluating thermomechanical effects in the SMT components.

The SMT components of interest are shown in Figure 11.21. More specifically, SMT packages attached to a board via gull wing leads were evaluated for thermal fields and deformations during their functional operation [28,43,44].

Thermomechanical effects in the SMT components are studied in this section with computational methodologies based on the finite differences method (FDM), the finite elements method (FEM), and an experimental methodology utilizing laser-based OEH.

11.4.4.1 Computational Modeling

To determine the magnitude of thermally induced deformations in a system, the temperature profile of the interconnected components comprising the system must be known throughout an operational cycle or system temperature excursion. In this study, the temperature profiles were evaluated by

FIGURE 11.21 SMT components of interest. SMTs are part of a motherboard of a personal computer.

FDM and FEM [39,41]. Specifically, the modeling strategy uses FEM domain discretizations to convert the geometrically based finite elements into accurate resistor networks. Then, the resistor networks are solved using a finite differences solver, which performs heat balance at each node of the model. This entails calculating a node temperature based on the resistances and the temperatures of all nodes attached to the node in question. A resistor network is obtained by converting all elements into resistances. Values of the resistances are functions of the geometry of the elements and the thermal conductivity. For example, for brick elements whose internal elements are all 90° and whose thermal conductivity is isotropic, 12 resistors, denoted by R_k, are calculated according to the equation

$$R_k = \frac{\Delta}{kA_k}, \tag{11.19}$$

where Δ is the distance between the nodes, k is the thermal conductivity of the element, and A_k, the cross-sectional area for a portion of the element associated with the interconnected nodes. This area is calculated as the product of half the width of the element normal to the heat flow in both directions for the specific resistor. If either the internal angles of a brick element are not all 90° or the thermal conductivity is anisotropic, the brick element is represented by 28 resistors, as calculated by FEM. A similar method is used for wedge-shaped elements, which is represented by 15 resistances.

Resistance R_c of a convection element, which defines convective surfaces, is calculated by the equation

$$R_c = \frac{1}{hA_c}, \tag{11.20}$$

where h is the convection heat transfer coefficient, and A_c the convective area associated with each node on the surface.

For heat to be transferred between elements of the model, they must share nodes. Because of this, in the process of calculating resistances, many resistors are generated between the same nodes. The model could be solved at this point, but the solution would be inefficient. To speed up the solution, resistors in the model that go between the same nodes are combined. If the resistances are fixed, i.e., if they are neither time nor temperature dependent, or are time and temperature dependent and reference the same dependency, they are combined into a single resistor according to the following equation:

$$R = \left(\frac{1}{R_i} + \frac{1}{R_{i+1}} + \cdots + \frac{1}{R_{i+n}} \right)^{-1}. \tag{11.21}$$

Application of Equation 11.21 takes only a few seconds on a very large model but can reduce the size of the model by a factor of 3 or 4. Resistances of surface radiation elements, R_r, are calculated from each node of the element face to a single reference node according to the equation

$$R_r = \frac{1}{K_r(T_1^2 + T_2^2)(T_1 + T_2)}, \tag{11.22}$$

where T_1 and T_2 are the absolute temperatures of the two nodes to which the radiation resistor is connected, and

$$K_r = A_r \sigma F \varepsilon. \tag{11.23}$$

In Equation 11.23, A_r defines the radiation area associated with each node of the face geometry of the specific element, σ is the Stefan-Boltzmann constant, F the view factor, and ε the emissivity. The radiation resistances are calculated in every iteration during the solution because their values depend upon the nodal temperatures. Having radiation in the network makes the model nonlinear as radiation parameters are surface temperature and/or time dependent [41].

11.4.4.2 Computational and OEH Results

Details of the component considered in this study are shown in Figure 11.22. Its FDM/FEM model [28,39,41,44], shown in Figure 11.23, displays discretization of the SMT component and board assembly used to calculate temperature distributions. Once the thermal model for a particular package is generated, it can be applied to the specific component/board environment to be evaluated. The result is a set of temperature profiles for the package and environment, as displayed in Figure 11.24.

Once the time-dependent temperature profile is known, the thermal deformations may be evaluated for all elements of the structure. A representative OEH fringe pattern, encoding thermomechanical deformations due to power-on cycle of the SMT component, is shown in Figure 11.25. Figure 11.26 indicates the maximum deformation of 1.87 μm, based on the fringe pattern of Figure 11.25, whereas Figure 11.27 displays wireframe representation of the 3-D deformations corresponding to Figure 11.26. Comparing Figure 11.24 and Figure 11.26, similarity in temperature profiles and deformation patterns, respectively, is observed. Comparison of the specific values of thermomechanical deformations obtained by FDM/FEM simulations with the deformations measured by OEH shows good correlation, as determined by experimental uncertainty.

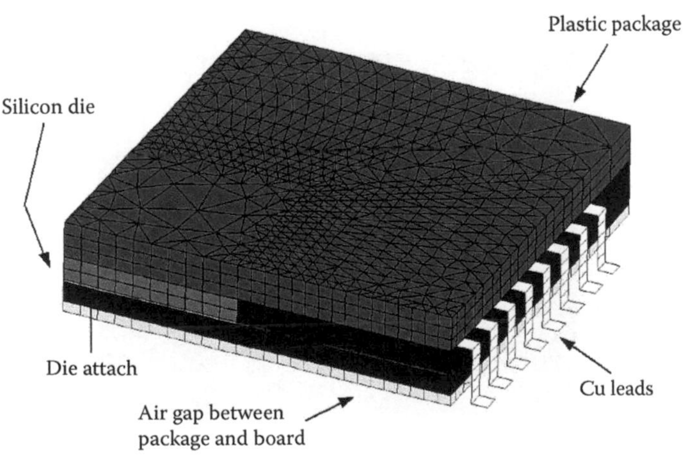

FIGURE 11.22 Details of the SMT component.

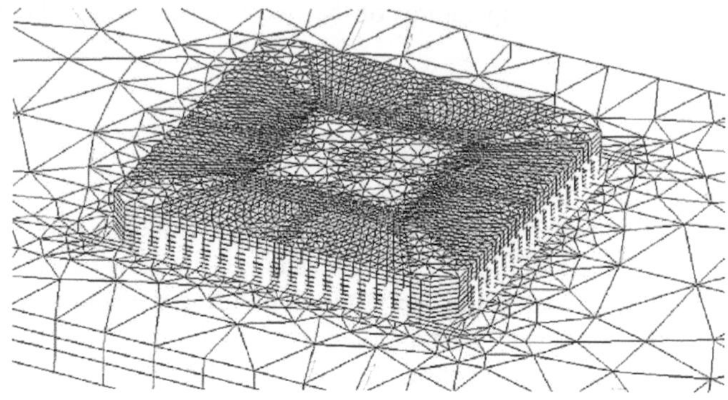

FIGURE 11.23 Computational model of the SMT component and board assembly.

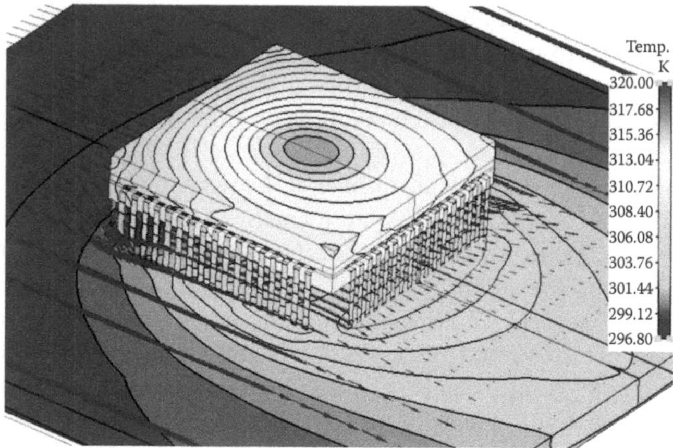

FIGURE 11.24 (See color insert.) Computed temperature distribution for the model shown in Figure 11.23.

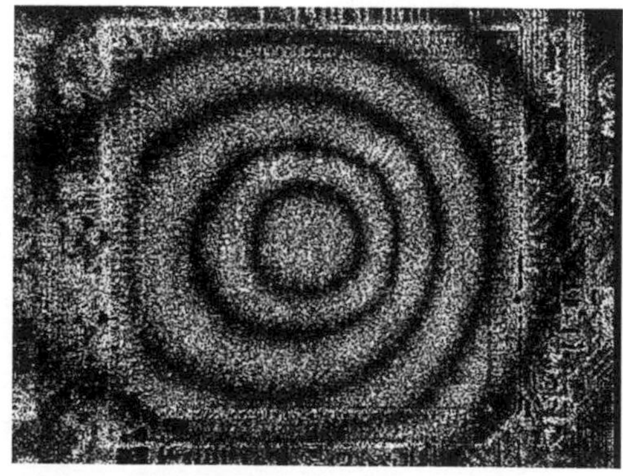

FIGURE 11.25 OEH fringe pattern displaying thermal deformations due to power-on of the SMT components shown in Figure 11.21.

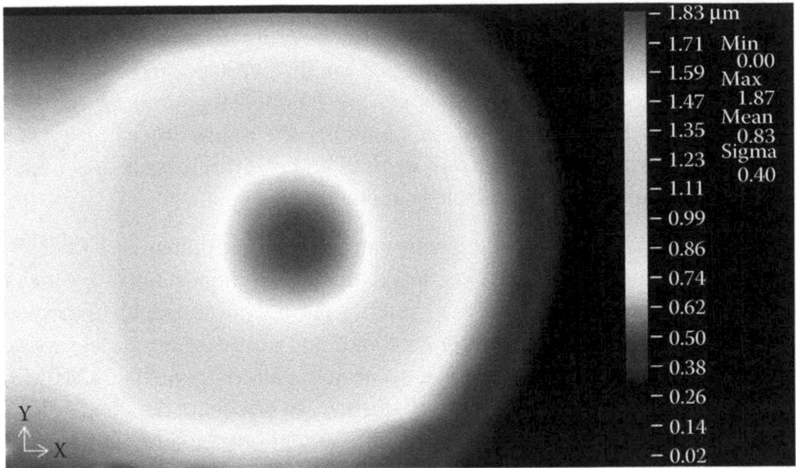

FIGURE 11.26 (See color insert.) Thermal deformations of the SMT component based on the OEH fringe pattern of Figure 11.25.

11.5 SUMMARY

OEHM methodology specifically set up to perform high-resolution shape and deformation measurements of MEMS and electronic packages has been presented. This methodology can be used either in display mode or in data mode. In the display mode, interferometric information related to shape and deformation is displayed at video frame rates, providing the capability for adjusting and setting experimental conditions. In the data mode, interferometric information related to shape and deformation is recorded as high-spatial and high-digital resolution images, which are further processed to provide quantitative 3-D information. Furthermore, the quantitative 3-D data can be conveniently exported to CAD environments and utilized for analysis and optimization of MEMS structures. Analysis of the 3-D OEHM data indicates that it is possible to achieve measurement resolutions of nearly $\lambda/500$ for the characterization of both shape and deformations.

Capabilities of the OEHM methodology were illustrated with representative applications demonstrating measurements of shape and deformations of MEMS components and electronic packaging.

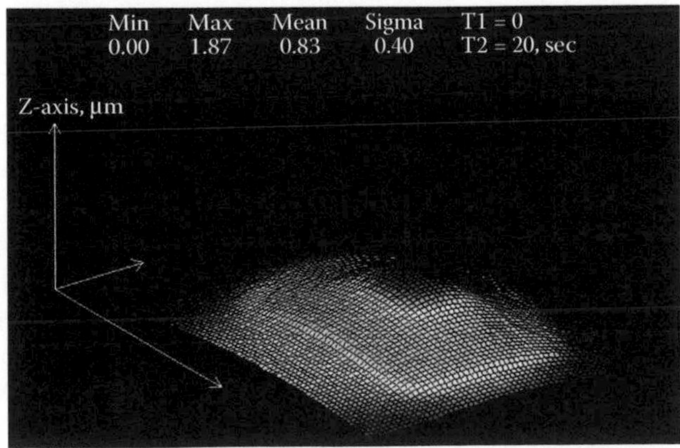

FIGURE 11.27 (See color insert.) Wireframe representation of thermal deformations of the SMT component based on the OEH fringe pattern of Figure 11.25.

OEHM measurements provide indispensable quantitative information for the effective development and optimization of MEMS structures. Furthermore, in order to provide high-spatial 3-D data, OEHM interferograms can be recorded by imaging small, overlapping regions (i.e., tiles) of the structure being investigated. Then, these tiles can be patched together, as demonstrated in this chapter, to provide detailed quantitative information over the entire object.

Testing and characterization of the MEMS devices and electronic packages impose the following challenges to quantitative optical metrology methodologies:

1. *High-speed events.* Motion of the electromechanical components of MEMS can reach very high oscillation frequencies, i.e., larger than a few hundredths of a megahertz, and future requirements call for a few gigahertz. As a consequence, high-speed control, synchronization, and data acquisition hardware and software are necessary.
2. *Depth of field.* Displacement of the electromechanical components of MEMS can reach relatively large displacements. As a consequence, an adequate depth of field, for a given optical magnification, should characterize the optical components utilized.
3. *Spatial resolution.* MEMS structures are characterized by different geometrical aspect ratios; as a consequence, very-high spatial resolutions are often required to resolve the phenomena of interest with sufficient measuring accuracy. A high spatial resolution can be achieved by increasing optical magnification; however, such an increase may limit the field of view, the depth of field, and therefore, the study of an entire electromechanical structure.

In our efforts to successfully perform testing and characterization of the MEMS devices, we are developing data acquisition and analysis algorithms and software. In addition, we are incorporating recent technological advances in coherent light sources, computing, imaging, and detection technologies into OEHM, including the following:

1. Tunable coherent light sources with analog/digital modulation capabilities to
   a. Recover the continuous fringe-locus function without the need to perform spatial phase unwrapping
   b. Improve depth of field in the measurements
   c. Enable stroboscopic illumination to investigate dynamic events
2. Long-working distance objectives to perform measurements of components characterized by high-aspect-ratio electronic packages
3. Positioning stages to perform high-magnification measurements and stitching of the measurements
4. High-speed and high-spatial- and high-digital-resolution CCD and CMOS cameras with digital control and synchronization capabilities

We are also continuing our developments of analytical, computational, and experimental techniques, particularly OEHM, to facilitate the analysis and characterization of relatively large deformations and high-speed motions in MEMS and electronic packages.

ACKNOWLEDGMENTS

The author gratefully acknowledges support by the NanoEngineering, Science, and Technology (NEST) program at the Worcester Polytechnic Institute, ME/CHSLT and contributions to the developments presented in this chapter by all members of the CHSLT-NEST laboratories, as well as resourcefulness of C. Core and B. Waterson, Analog Devices, Cambridge, MA. The author also acknowledges the following sources for illustrations: DMD picture courtesy of Texas Instruments, Inc. http://www.ti.com/; *i*MEMS inertial devices courtesy of Analog Devices, Norwood, MA 02062, http://www.analog.com/; Sandia Micromirror Device™ pictures courtesy of Sandia National Laboratories, Albuquerque, NM, http://www.sandia.gov/mstc/; RF MEMS switch picture courtesy of *MSTNews*. http://www.mstnews.de/.

REFERENCES

1. Hsu, T.-R., *MEMS and Microsystems: Design and Manufacture*, McGraw-Hill, New York, 2002.
2. Senturia, S.D., *Microsystem Design*, Kluwer Academic, Boston, MA, 2000.
3. Kovacs, G. T. A., *Micromachined Transducers Sourcebook*, McGraw-Hill, New York, 1998.
4. Elwenspoek, M. and Wiegerink, R., *Mechanical Microsystems*, Springer-Verlag, Berlin, 2001.
5. Jaeger, R.C., *Introduction to Microelectronic Fabrication*, Prentice-Hall, Upper Saddle River, NJ, 2002.
6. Pister, K.S.J., Frontiers in MEMS design, 1996 *NAE Symp. on Frontiers of Eng.*, National Academy Press, Washington, D.C., 63–66, 1997.
7. Peeters, E., Large-market applications of MEMS, 1996 *NAE Symp. on Frontiers of Eng.*, National Academy Press, Washington, D.C., 67–72, 1997.
8. Judy, J.W., Microelectromechanical Systems (MEMS): fabrication, design and applications, *J. Smart Mater. Struct.*, 10(6): 1115–1134, 2001.
9. http://www.ti.com/, Texas Instruments, DLP-DMD Discovery Website, 2005.
10. http://www.analog.com/, Analog Devices, MEMS and Sensors Website, 2005.
11. http://www.segway.com/, Segway LLC, Segway® HT Website, 2004.
12. http://www.sandia.gov/mstc/, Sandia National Laboratories (SNL), Microsystems Science, Technology and Components Website, 2005.
13. Lacroix, D., Expanding the RF MEMS market — RF MEMS building blocks, *MSTNews*, 4: 34–36, 2003.
14. Schauwecker, B., Mack, T., Strohm, K.M., Simon, W., and Luy, J.-F., RF MEMS for advanced automotive applications, *MSTNews*, 4: 36–38, 2003.
15. Furlong, C. and Pryputniewicz, R.J., Optoelectronic characterization of shape and deformation of MEMS accelerometers used in transportation applications, *Opt. Eng.*, 42(5): 1223–1231, 2003.
16. Furlong, C., Siegel, A.M., Hefti, P., and Pryputniewicz, R.J., Confocal optoelectronic holography microscope for structural characterization and optimization of MEMS, *Transducers'03*, Boston, MA, pp. 420–423, 2003.
17. Furlong, C., Ferguson, C.F., and Melson, M.J., New optoelectronic methodology for nondestructive evaluation of MEMS at the wafer level, *Proc. SPIE*, 5265: 69–78, 2003.
18. Furlong, C. and Pryputniewicz, R. J., Study and characterization of a MEMS micromirror device, *Proc. SPIE*, 5531: 54–63, 2004.
19. Furlong, C., Yokum, J.S., Phillips, C.A., and Pryputniewicz, R.J., Optoelectronic holography shape characterization of microspheres for biomedical applications, *Proc. Symp. on MEMS: Mechanics and Measurements*, Milwaukee, WI, pp. 63–66, 2002.
20. Peterson, K.E., Silicon as a mechanical material, *Proc. IEEE*, 5: 420–457, 1982.
21. Hillsum, C. and Ning, T.H., Eds., *Properties of Silicon*, INSPEC, Institution of Electrical Engineers, New York, 1988.
22. Stetson, K.A. and Brohinsky, W.R., Electro optic holography and its application to hologram interferometry, *Appl. Opt.*, 24(21): 3631–3637, 1985.
23. Stetson, K.A. and Brohinsky, W.R., Fringe-shifting technique for numerical analysis of time-average holograms of vibrating objects, *J. Opt. Soc. Am. A*, 5(9): 1472–1476, 1988.
24. Stetson, K.A., Theory and applications of electronic holography, *Proc. SEM*, Bethel, CT, pp. 294–300, 1990.
25. Pryputniewicz, R.J., *Holographic Numerical Analysis*, Worcester Polytechnic Institute, Worcester, MA, 1992.
26. Brown, G.C. and Pryputniewicz, R.J., Holographic microscope for measuring displacements of vibrating microbeams using time-average electro-optic holography, *Opt. Eng.*, 37: 1398–1405, 1998.
27. Pryputniewicz, R.J., Marinis, T.F., Hanson, D.S., and Furlong, C., New approach to development of packaging for MEMS inertial sensors, Paper No. IMECE 2001/MEMS-23906, *Am. Soc. of Mech. Eng.*, New York, 2001.
28. Furlong, C. and Pryputniewicz, R.J., Computational and experimental approach to thermal management in microelectronics and packaging, *Microelectron. Int.*, 18(1): 35–39, 2001.
29. Powell, R.L. and Stetson, K.A., Interferometric vibration analysis by wavefront reconstruction, *J. Opt. Soc. Am.*, 55(12): 1593–1598, 1965.

30. Vest, C.M., *Holographic Interferometry*, John Wiley & Sons, New York, 1979.
31. Furlong, C., Kolenovic, E., and Ferguson, C.F., Quantitative optical metrology with CMOS cameras, *Proc. SPIE*, 5532: 1–15, 2004.
32. Veeco Instruments, Sloan DekTak calibration standards set, Veeco Metrology Group, Tucson, AZ, 2002.
33. *Verdict user's guide v*8.0, SDRC Imageware, Ann Arbor, MI, 1998.
34. *Polyworks user's guide v*7.0, Innovmetric Software Inc., Sainte-Foy, Quebec, Canada, 2004.
35. Pro/Scan-Tools, *User's guide v.* 2000*i*2, Parametric Technology Corporation, Waltham, MA, 2000.
36. Application Note, ADXL202/ADXL210 — low cost ±2 g/±10 g dual axis iMEMS® accelerometers with digital output, Analog Devices, Norwood, MA, 2004.
37. Kok, R., Development of a Wireless MEMS Inertial System for Health Monitoring of Structures, M.S. thesis, Worcester Polytechnic Institute, Worcester, MA, 2004.
38. SRAC Corporation, *COSMOS/M User's Guide v.*3.0., Los Angeles, CA, 2002.
39. CFD Research Corporation, *CFD-Micromesh User's Manual*, Huntsville, AL, 2004.
40. MARC Analysis Research Corporation, *MARC: Element Library*, Palo Alto, CA, 1991.
41. Harvard Thermal, *TAS User's Manual v.*4.0, Harvard, MA, 2004.
42. Hunter, R. and Humphreys, C., Trends in 300 mm wafers: factory automation, *Semicon. Int.*, 26(6): 60–64, 2003.
43. Pryputniewicz, R.J. and Furlong, C., Novel optical-computational approach for NDT applications in microelectronics, *Proc. IX Int. Congress on Experimental Mechanics*, SEM, Bethel, CT, 2000.
44. Pryputniewicz, R.J., Rosato, D., and Furlong, C., Measurements and simulation of SMT components, *Proc. 35th Int. Symp. on Microelectronics*, pp. 151–156, 2002.
45. Janssen, G., Kole, A., and Leroux, A., Trends in cooling of electronics: the use of thermal roadmaps, *Electron. Cooling*, 10(3): 26–32, 2004.

12 Digital Holography and Its Application in MEMS/MOEMS Inspection

Wolfgang Osten and Pietro Ferraro

CONTENTS

12.1 Introduction 352
12.2 Theory and Basic Principle of Digital Holography (DH) 353
12.2.1 Digital Recording and Reconstruction of Wavefronts 353
12.2.2 Reconstruction Principles in DH 358
12.2.2.1 The Fresnel Approximation 358
12.2.2.2 Numerical Reconstruction by the Convolution Approach 360
12.2.2.3 Numerical Reconstruction by the Lensless Fourier Approach 361
12.2.2.4 Numerical Reconstruction by the Phase-Shifting Approach 362
12.2.3 Influences of Discretization 363
12.3 Digital Holographic Interferometry 364
12.3.1 Basic Principles 364
12.3.2 Holographic Displacement Measurement 366
12.3.3 Holographic Shape Measurement 368
12.3.3.1 Two-Source-Point Method 368
12.3.3.2 Two-Wavelength Method 370
12.3.4 Direct and Absolute Phase Measurement 372
12.3.5 Advantages of DH 375
12.4 Digital Holographic Microscopy (DHM) 375
12.4.1 Optical Setup in DHM 376
12.4.2 Compensation of Aberrations in DH 378
12.4.3 Removing Aberrations by Determining a Phase Mask in the Image Reconstruction Plane 378
12.4.3.1 First Method: Removing the Circular Fringe Carrier from a Recording Hologram with Phase-Shifting Method 381
12.4.3.2 Second Method: Formerly a Double-Exposure Technique 382
12.4.3.3 Third Method: Digital Adjustment by Successive Approximations 384
12.4.3.4 Comparison among the Methods and Discussion of Results 384
12.4.4 Focus Tracking in DHM 385
12.4.5 Controlling Size Independently of Distance and Wavelength 389
12.5 The Application of DH to the Investigation of Microcomponents 391
12.5.1 Experimental Prerequisites for the Investigation of Microcomponents 391
12.5.1.1 The Loading of Microcomponents 392
12.5.1.2 The Handling and Preparation of Microcomponents 394

12.5.1.3 The Observation and Evaluation of Microcomponents......................395
12.5.2 Investigation of Objects with Technical Surfaces..395
12.5.2.1 Combined Shape and Displacement Measurement of Small Objects using DH..395
12.5.2.2 The Determination of Material Parameters of Microcomponents using DH..397
12.5.3 Investigation of Microcomponents with Optical Surfaces.................................405
12.5.3.1 Testing Silicon MEMS Structures...405
12.5.3.2 Thermal Load Testing of MEMS..409
12.5.3.3 Microlens Testing..414
12.6 Conclusion...422
References...422

12.1 INTRODUCTION

There are several three-dimensional (3-D) imaging methods based on interferometry that allow the measurement of minute displacements and surface profiles. Methods such as holographic interferometry and speckle metrology can provide full-field noncontact information about coordinates, deformations, strains, stresses, and vibrations. The basic principle of these methods is the transformation of phase changes into recordable intensity changes [1]. Because of the high spatial frequency of these intensity fluctuations, the registration of a hologram requires a light-sensitive medium with adequate spatial resolution. Therefore, special photographic emulsions have dominated holographic technologies for a long period. However, the recording of holograms on electronic sensors and their numerical reconstruction is almost as old as holography itself. First ideas and implementations came up already in the 1960s and 1970s [2–4]. But only in the 1990s did the progress in high-resolution camera technology and computer hardware make it possible to record holograms directly on the charge-coupled device (CCD) target of a camera and to reconstruct the wavefront in a reasonable time [5]. But digital holography (DH) is much more than an elegant recording technique. In contrast to conventional approaches, DH allows the direct calculation of both parts of the complex wavefront, phase and intensity, by numerical solution of the diffraction problem in the computer. Several advantages for the measurement process result from this new development. Besides electronic processing and direct access to the phase, some further benefits recommend DH for the solution of numerous imaging, inspection, and measurement problems [6]. We will return to this point later. Here, only some obvious advantages will be mentioned: certain aberrations can be avoided and the remaining aberrations can be corrected numerically, different object states can be stored independently, the number of necessary holograms for quantitative evaluations can be decreased considerably, and the size of holographic sensors can be reduced drastically.

DH is well suited for the investigation of microcomponents [7]. As an interferometric method, it offers both a fieldwise and noncontact approach to very fragile components and a high sensitivity with respect to geometric quantities such as coordinates and displacements. The smallness of these objects diminishes the practical consequences of the limited space–bandwidth product (SBP) of the existing sensors, such as the restricted angle resolution of digital holograms. Nevertheless, several boundary conditions have to be observed carefully [8]. This topic will be discussed in detail. Here, we start with a description of the basic principles of DH, including the analysis of the recording process and the discussion of the main reconstruction techniques. Afterward, the interferometric methods for shape and displacement measurement are presented. Because of its importance for the quantitative evaluation process, a section is dedicated to the unwrapping of mod-2π phase distributions. Finally, the application width of DH in the investigation of microcomponents is demonstrated with several examples such as the shape inspection of a microsensor, the measurement of material parameters of basic components, residual stress, and mechanical behavior when microcomponents are subjected to thermal load.

Special attention will be given to the problems that the experimentalist has to face while using DH in either lensless and/or microscope configuration for practical evaluation and testing of microcomponents. Approaches and solutions that have been discovered and successfully applied to overcome those problems will be described and discussed, furnishing examples of real-world applications. Finally, a section will be devoted to the inspection of microlens arrays of both refractive and diffractive types and that are made of different materials.

12.2 THEORY AND BASIC PRINCIPLE OF DIGITAL HOLOGRAPHY (DH)

12.2.1 Digital Recording and Reconstruction of Wavefronts

In digital speckle pattern interferometry (DSPI) [9] (see Chapter 13), the object is focused onto the target of an electronic sensor. Thus, an image plane hologram is formed as result of interference with an inline reference wave. In contrast with DSPI, a digital hologram is recorded without imaging. The sensor target records the superposition of the reference and the object wave in the near-field region — a so-called Fresnel hologram [10,11]. The basic optical setup in DH for recording holograms is the same as in conventional holography (Figure 12.1a). A laser beam is divided into two coherent beams. One illuminates the object and causes the *object wave*. The other enters the target directly and forms the *reference wave*. On this basis, very compact solutions are possible as shown in Figure 12.1b.

For the description of the principle of digital Fresnel holography, we use a simplified version of the optical setup (Figure 12.2) [12]. The object is modeled by a plane rough surface that is located in the (x,y)-plane and illuminated by laser light. The scattered wave field forms the object wave u(x,y). The target of an electronic sensor (e.g., a CCD or a CMOS) used for recording the hologram is located in the (ξ,η)-plane at a distance d from the object. Following the basic principles of holography [1,13], the hologram $h(\xi,\eta)$ originates from the interference of the object wave $u(\xi,\eta)$ and the reference wave $r(\xi,\eta)$ in the (ξ,η)-plane:

$$h(\xi,\eta) = \left|u(\xi,\eta) + r(\xi,\eta)\right|^2 = r \cdot r^* + r \cdot u^* + u \cdot r^* + u \cdot u^* \quad (12.1)$$

The transformation of the intensity distribution into a gray-value distribution that is stored in the image memory of the computer is considered by a characteristic function t of the sensor. This function is, in general, only approximately linear:

$$T = t\left[h(\xi,\eta)\right] \quad (12.2)$$

Because the sensor has a limited spatial resolution, the spatial frequencies of the interference fringes in the hologram plane — the so-called microinterferences — have to be considered. The fringe spacing g and the spatial frequency f_x, respectively, are determined by the angle β between the object and the reference waves (Figure 12.3):

$$g = \frac{1}{f_x} = \frac{\lambda}{2\sin(\beta/2)} \quad (12.3)$$

with λ as the wavelength. If we assume that the discrete sensor has a pixel pitch (the distance between adjacent pixels) $\Delta\xi$, the sampling theorem requires at least 2 pixels per fringe for a correct reconstruction of the periodic function:

$$2\Delta\xi < \frac{1}{f_x} \quad (12.4)$$

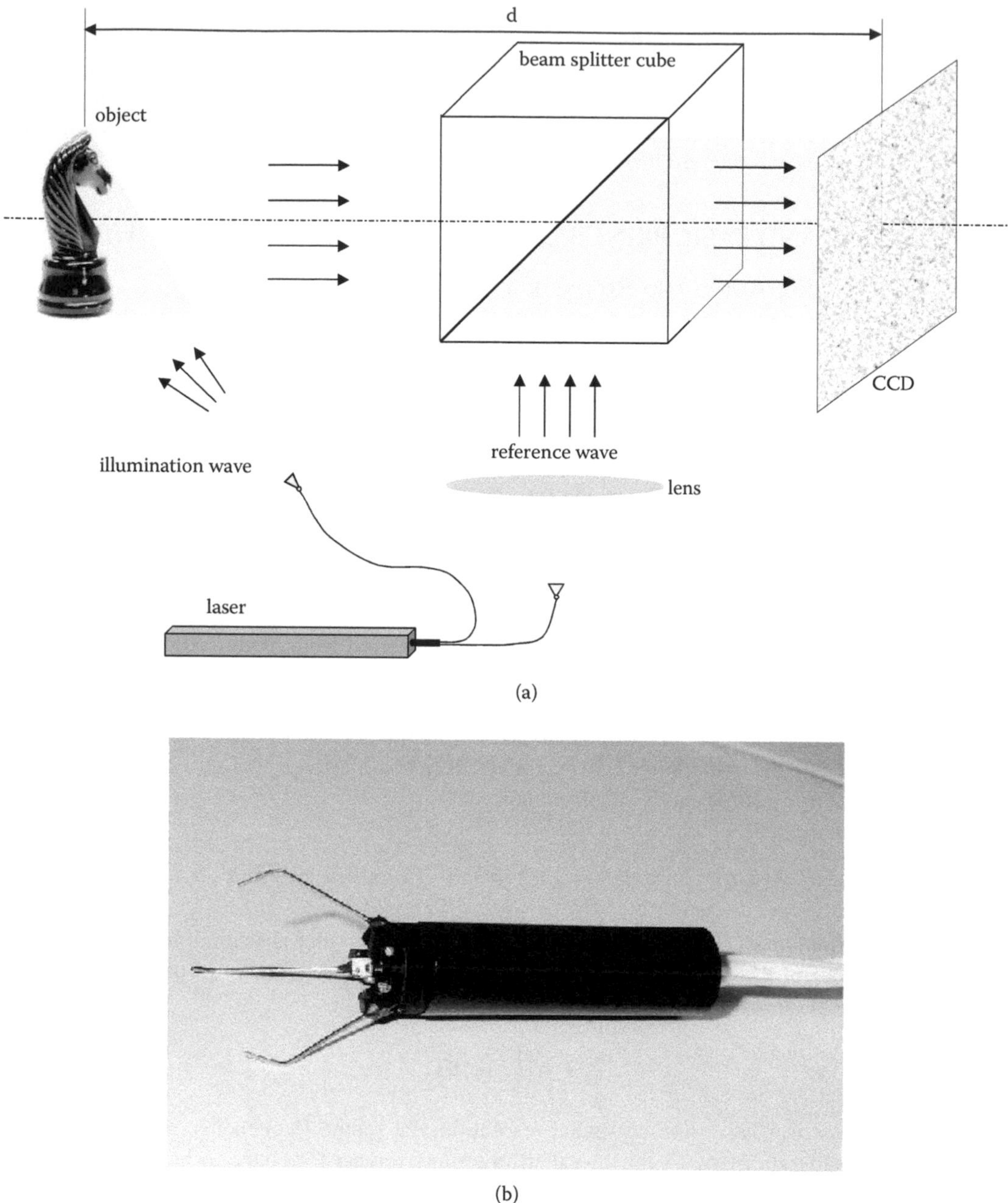

FIGURE 12.1 Optical configuration in digital holography: (a) schematic setup for recording a digital hologram onto a CCD target, (b) miniaturized holographic camera with three illumination arms. The sensor with a diameter of 9 mm includes a complete interferometer and a CCD camera. (From Kolenovic, E. et. al., Miniaturized digital holography sensor for distal three-dimensional endoscopy, *Appl. Opt.*, 42, 5167–5172, 2003.)

Consequently, we obtain for small angles β:

$$\beta < \frac{\lambda}{2\Delta\xi} \tag{12.5}$$

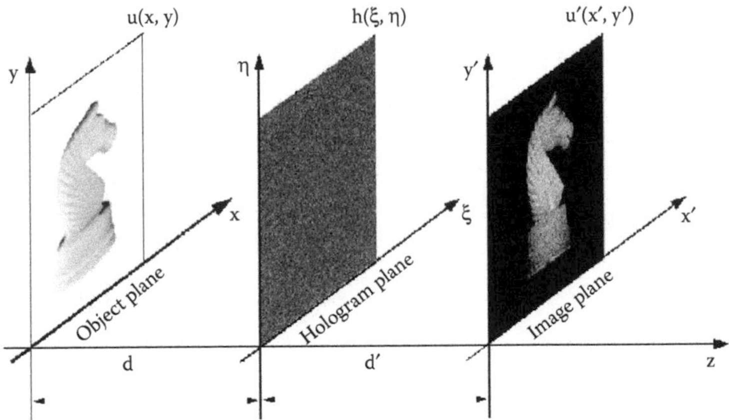

FIGURE 12.2 Schematic setup for Fresnel holography.

Modern high-resolution CCD or CMOS chips can have a pitch $\Delta\xi$ down to about 4 μm. In that case, a maximum angle between the reference and the object wave of only 4° is acceptable. One practical consequence of the restricted angle resolution in DH is a limitation of the effective object size that can be recorded "holographically" by an electronic sensor. However, this is only a technical handicap. Larger objects can be placed at a sufficient distance from the hologram or reduced optically by imaging with a negative lens [14]. A reduction in the object resolution has to be accepted in that case. Another consequence is the restricted application of off-axis setups that are used to avoid overlapping reconstructions [15]. Therefore, sometimes it is preferable to use in-line arrangements. However, in the lensless Fourier recording setup these problems becomes less serious as described in Subsection 12.2.2.3 [16].

The reconstruction is done by illuminating the hologram with a so-called reconstruction wave $c(\xi,\eta)$:

$$u'(x',y') = t\left[h(\xi,\eta)\right]\cdot c(\xi,\eta) \tag{12.6}$$

The hologram $t[h(\xi,\eta)]$ diffracts the wave $c(\xi,\eta)$ in such a way that images of the object wave are reconstructed. In general, four terms are reconstructed if the wave $u'(x',y')$ propagates in space. An assumed linear characteristic function $t(h) = \alpha h + t_0$ delivers

$$u' = T\cdot c = t(h)\cdot c \tag{12.7}$$

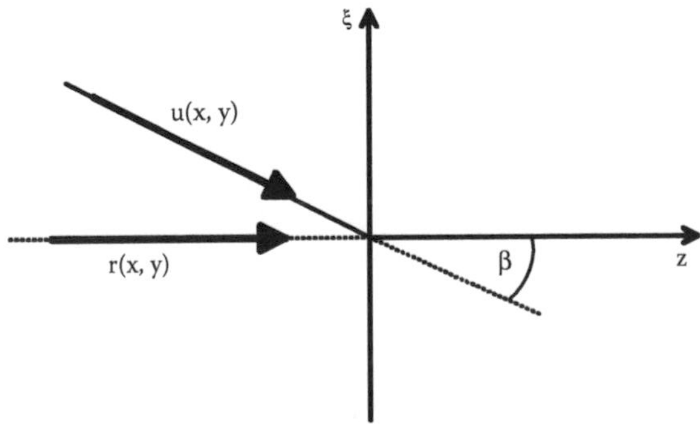

FIGURE 12.3 Interference between the object and reference wave in the hologram plane.

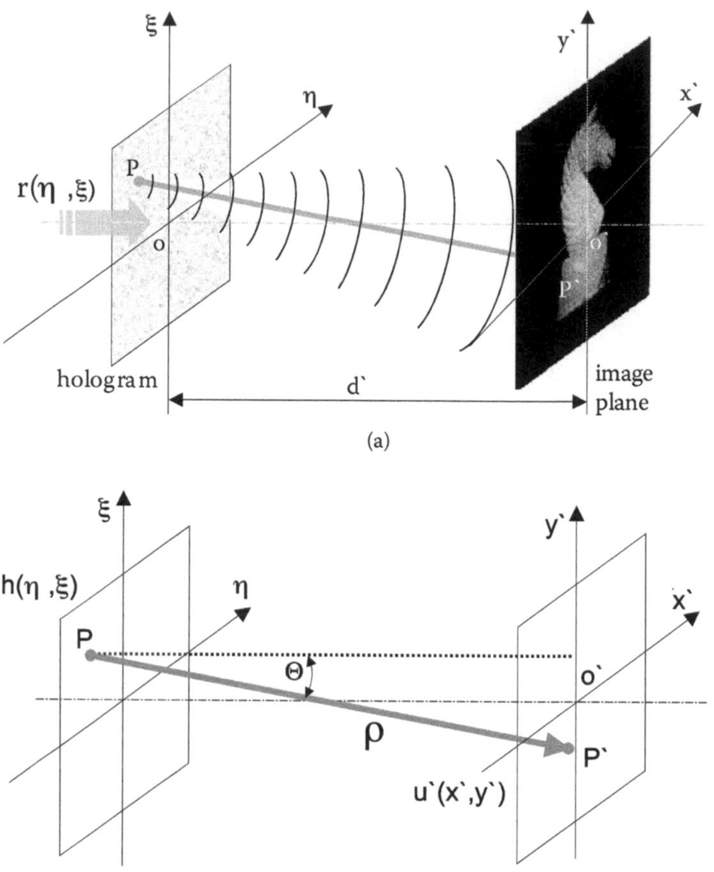

FIGURE 12.4 Reconstruction of a digital hologram: (a) principle of wavefront reconstruction, (b) light propagation by diffraction (Huygens–Fresnel principle).

and

$$u' = \alpha\left[cu^2 + cr^2 + cur^* + cru^*\right] + ct_0 \tag{12.8}$$

with two relevant image terms [cur*] and [cru*], containing the object wave and its conjugated version, respectively. The appearance of the image terms depends on the concrete shape of the reconstruction wave c. In general, the reference wave c = r or its conjugated version c = r* is applied. In the case of the conjugated reference wave, a direct or real image will be reconstructed because of a converging image wave that can be imaged on a screen at the place of the original object.

However, in DH the reconstruction of the object wave in the image plane $u'(x',y')$ is done by numerical reproduction of the physical process as shown in Figure 12.4a. The reconstruction wave, with a well-known shape equal to the reference wave $r(\xi,\eta)$, propagates through the hologram $h(\xi,\eta)$. Following Huygens' principle, each point $P(\xi,\eta)$ on the hologram acts as the origin of a spherical elementary wave. The intensity of these elementary wave is modulated by the transparency $h(\xi,\eta)$. In a given distance $d' = d$ from the hologram, a sharp real image of the object can be reconstructed as the superposition of all elementary waves. For the reconstruction of a virtual image, $d' = -d$ is used.

Consequently, the calculation of the wave field u′(x′, y′) in the image plane starts with the pointwise multiplication of the stored and transformed intensity values t[h(ξ,η)] with a numerical model of the reference wave r(ξ,η). In the case of a normally incident and monochromatic wave of unit amplitude, the reference wave can be modeled by r(ξ,η) = 1. After the multiplication, the resulting field in the hologram plane is propagated in free space. The diffracted field u′(x′, y′) at distance d′ can be found by solving the Rayleigh–Sommerfeld diffraction formula, which is also known as the Huygens–Fresnel principle [17]:

$$u'(x',y') = \frac{1}{i\lambda} \iint_{-\infty\ldots\infty} t\left[h(\xi,\eta)\right] \cdot r(\xi,\eta) \frac{\exp(ik\rho)}{\rho} \cos\theta \, d\xi d\eta \tag{12.9}$$

with

$$\rho(\xi - x', \eta - y') = \sqrt{d'^2 + (\xi - x')^2 + (\eta - y')^2} \tag{12.10}$$

as the distance between a point P′(x′, y′, z′ = d′) in the image plane and a point P(ξ,η,z = 0) in the hologram plane and

$$k = \frac{2\pi}{\lambda} \tag{12.11}$$

as the wavenumber. The obliquity factor $\cos\theta$ represents the cosine of the angle between the outward normal and the vector joining P to P′ (see Figure 12.4b). This term is given exactly by

$$\cos\theta = \frac{d'}{\rho} \tag{12.12}$$

and therefore Equation 12.9 can be rewritten

$$u'(x',y') = \frac{d'}{i\lambda} \iint_{-\infty\ldots\infty} t\left[h(\xi,\eta)\right] \cdot r(\xi,\eta) \frac{\exp(ik\rho)}{\rho^2} d\xi d\eta \tag{12.13}$$

The numerical reconstruction provides the complex amplitude of the object wavefront. Consequently, the phase distribution ϕ(x′, y′) and the intensity I(x′, y′) can be calculated directly from the reconstructed complex function u′ (x′, y′):

$$\phi(x',y') = \arctan \frac{\mathrm{Im}\left|u'(x',y')\right|}{\mathrm{Re}\left|u'(x',y')\right|} \quad \left[-\pi,\pi\right] \tag{12.14}$$

$$I(x',y') = u'(x',y') \cdot u'^{*}(x',y') \tag{12.15}$$

Figure 12.5 shows the results of the numerical reconstruction of the intensity and the phase of a chess knight recorded with a setup as shown in Figure 12.1a.

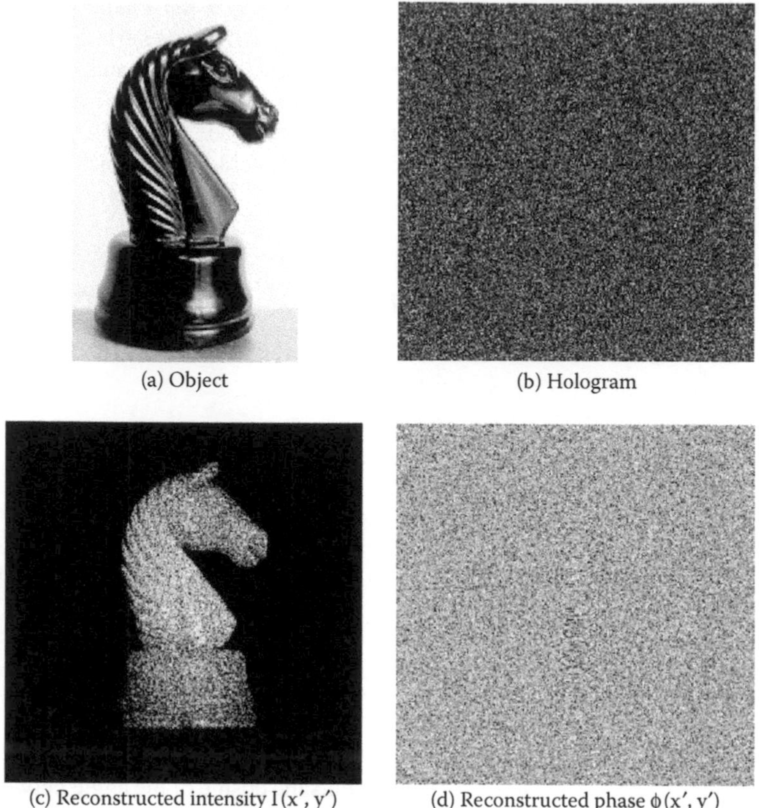

FIGURE 12.5 Reconstructed intensity and phase of a digital Fresnel hologram: (a) object, (b) hologram, (c) reconstructed intensity $I(x', y')$, (d) reconstructed phase $\phi(x', y')$.

The direct approach to the phase yields several advantages for imaging and metrology applications that will be discussed later.

12.2.2 Reconstruction Principles in DH

Different numerical reconstruction principles have been investigated: the Fresnel approximation [5], the convolution approach [3], the lensless Fourier approach [16,18], the phase-shifting approach [19], and the phase-retrieval approach [20]. In this subsection, the main techniques are briefly described.

12.2.2.1 The Fresnel Approximation

If the distance d between the object and hologram plane, and equivalently $d' = d$ between the hologram and image plane, is large compared to $(\xi - x')$ and $(\eta - y')$, then the denominator of Equation 12.13 can be replaced by d'^2, and the parameter ρ in the numerator can be approximated by a binomial expansion for the square root Equation 12.10 where only the first two terms are considered [17]:

$$\rho \approx d'\left[1 + \frac{(\xi - x')^2}{2d'^2} + \frac{(\eta - x')^2}{2d'^2}\right] \tag{12.16}$$

The resulting expression for the field at (x′,y′) becomes

$$u'(x',y') = \frac{\exp(ikd')}{i\lambda d'} \iint_{-\infty...\infty} t[h(\xi,\eta)]\cdot r(\xi,\eta)\exp\left[\frac{ik}{2d'}\left\{(\xi-x')^2+(\eta-y')^2\right\}\right]d\xi d\eta \tag{12.17}$$

Equation 12.17 is a convolution integral that can be expressed as

$$u'(x',y) = \iint_{-\infty...\infty} t[h(\xi,\eta)]\cdot r(\xi,\eta) H(\xi-x',\eta-y')d\xi d\eta \tag{12.18}$$

with the convolution kernel

$$H(x',y') = \frac{\exp(ikd')}{i\lambda d'}\exp\left[\frac{ik}{2d'}(x'^2+y'^2)\right] \tag{12.19}$$

Another notation of Equation 12.17 is found if the term $\exp\left[\frac{ik}{2d'}(x'^2+y'^2)\right]$ is taken out of the integral:

$$u'(x',y') = \frac{\exp(ikd')}{i\lambda d'} e^{i\frac{k}{2d'}(x'^2+y'^2)} \iint_{-\infty...\infty} \left\{t[h(\xi,\eta)]\cdot r(\xi,\eta)\cdot e^{i\frac{k}{2d'}(\xi^2+\eta^2)}\right\} \exp\left[-i\frac{2\pi}{\lambda d'}(\xi x'+\eta y')\right]d\xi d\eta \tag{12.20}$$

or

$$u'(x',y') = \frac{\exp(ikd')}{i\lambda d'} e^{i\frac{k}{2d'}(x'^2+y'^2)} \mathbf{FT}_{\lambda d'}\left\{t[h(\xi,\eta)]\cdot r(\xi,\eta)\cdot e^{i\frac{k}{2d'}(\xi^2+\eta^2)}\right\} \tag{12.21}$$

where $\mathbf{FT}_{\lambda d'}$ indicates the 2-D Fourier transform that has been modified by a factor $1/(\lambda d')$. Equation 12.21 makes it clear that the diffracted wave front consists of the Fourier transform **FT** of the digitally stored hologram $t[h(\xi,\eta)]$ multiplied by the reference wave $r(\xi,\eta)$ and the chirp function $\exp\{i\pi/\lambda d'(\xi^2+\eta^2)\}$. This Fourier transform is scaled by the constant factor $1/(i\lambda d')$ and a phase factor that is independent of the processed hologram.

The discrete sampling in digital processing requires the transformation of the infinite continuous integral in Equation 12.20 into a finite discrete sum. This results in the finite Fresnel transform:

$$u'(m,n) = \sum_{j=0}^{M-1}\sum_{l=0}^{N-1} t[h(j\cdot\Delta\xi, l\cdot\Delta\eta)]\cdot r(j\cdot\Delta\xi, l\cdot\Delta\eta)\exp\left[\frac{i\pi}{d'\lambda}(j^2\Delta\xi^2+l^2\Delta\eta^2)\right] \times\exp\left\{-i2\pi\left(\frac{j\cdot m}{M}+\frac{l\cdot n}{N}\right)\right\} \tag{12.22}$$

where constants and pure phase factors preceding the sums have been omitted. The main parameters are the pixel number M × N and the pixel pitches $\Delta\xi$ and $\Delta\eta$ in the two directions, which are defined by the used sensor chip.

The discrete phase distribution ϕ(m, n) of the wavefront and the discrete intensity distribution I(m,n) on the rectangular grid of M × N sample points can be calculated from the reconstructed complex function u′(m,n) by using Equation 12.14 and Equation 12.15, respectively.

Following Equation 12.22, the pixel sizes in the reconstructed image along the two directions are

$$\Delta x' = \frac{d'\lambda}{M\Delta x} \quad \text{and} \quad \Delta y' = \frac{d'\lambda}{N\Delta y} \tag{12.23}$$

In addition to the real image, a blurred virtual image, and a bright DC term, the zero-order diffracted field, is reconstructed. The DC term can effectively be eliminated by preprocessing the stored hologram [7,14], or the different terms can be separated by using the off-axis instead of the in-line technique. However, the spatial separation between the object and the reference field requires a sufficient SBP of the used CCD chip, as discussed in Section 12.2.1.

12.2.2.2 Numerical Reconstruction by the Convolution Approach

In Subsection 12.2.2.1, we have already mentioned that the connection between the image term u′(x′,y′) and the product $t[h(\xi,\eta)]\cdot r(\xi,\eta)$ can be described by a linear space-invariant system. The diffraction (Equation 12.17) is a superposition integral with the impulse response

$$H\left(x'-\xi, y'-\eta\right) = \frac{\exp\left(ikd'\right)}{i\lambda d'}\exp\left[\frac{ik}{2d'}\left\{\left(\xi-x'\right)^2+\left(\eta-y'\right)^2\right\}\right] \tag{12.24}$$

A linear space-invariant system is characterized by a transfer function G that can be calculated as the Fourier transform of the impulse response:

$$G\left(f_x, f_y\right) = \mathbf{FT}\left\{H\left(\xi,\eta,x',y'\right)\right\} \tag{12.25}$$

with the spatial frequencies f_x and f_y. Consequently, the convolution theorem can be applied, which states that the Fourier transform of the convolution $t[h(\xi,\eta)]\cdot r(\xi,\eta)$ with H is the product of the individual Fourier transforms $\mathbf{FT}\{t[h(\xi,\eta)]\cdot r(\xi,\eta)\}$ and $\mathbf{FT}\{H\}$. Thus, u′(x′,y′) can be calculated by the inverse Fourier transform of the product of the Fourier-transformed convolution partners:

$$u'\left(x',y'\right) = \mathbf{FT}^{-1}\left\{\mathbf{FT}\left[t\left(h\right)\cdot r\right]\cdot \mathbf{FT}\left[H\right]\right\} \tag{12.26}$$

$$u'\left(x',y'\right) = \left[t\left(h\right)\cdot r\right]\otimes H \tag{12.27}$$

where ⊗ is the convolution symbol. The computing effort is comparatively high: two complex multiplications and three Fourier transforms. The main difference from the Fresnel transform is the different pixel size in the reconstructed image plane [12]. The pixel size is constant independent from the reconstruction distance:

$$\Delta x' = \Delta\xi \text{ and } \Delta y' = \Delta\eta \tag{12.28}$$

12.2.2.3 Numerical Reconstruction by the Lensless Fourier Approach

In previous sections, we have already mentioned that the limited spatial resolution of the sensor restricts the utilizable numerical aperture (NA) of the solid-state sensor (i.e., CCD array) for recording the digital hologram (or equivalently, the subtended solid angle in which interfering fringes between reference and object beam can be spatially resolved by the CCD). In fact, the sampling theorem requires that the angle between the object beam and the reference beam at any point of the electronic sensor be limited in such a way that the microinterference spacing is larger than double the pixel size. In general, the angle between the reference beam and the object beam varies over the surface of the sensor, and so does the maximum spatial frequency. Thus, for most holographic setups, the full spatial bandwidth of the sensor cannot be used. However, it is very important to use the entire spatial bandwidth of the sensor, because the lateral resolution of the reconstructed image depends on a complete evaluation of all the information one can get from the sensor.

Even if the speed of digital signal processing is increasing rapidly, algorithms should be as simple and as fast to compute as possible. For the Fresnel and the convolution approaches, several fast Fourier transforms and complex multiplications are necessary. Therefore, a more effective approach, such as the subsequently described algorithm, seems promising.

The lensless Fourier approach [16] is the fastest and most suitable algorithm for small objects. The corresponding setup is shown in Figure 12.6. It allows one to choose the lateral resolution in a range, from a few microns to hundreds of microns, without any additional optics. Each point (ξ,η) on the hologram is again considered a source point of a spherical elementary wavefront (Huygens' principle). The intensity of these elementary waves is modulated by $t[h(\xi,\eta)]$ — the amplitude transmission of the hologram. The reconstruction algorithm for lensless Fourier holography

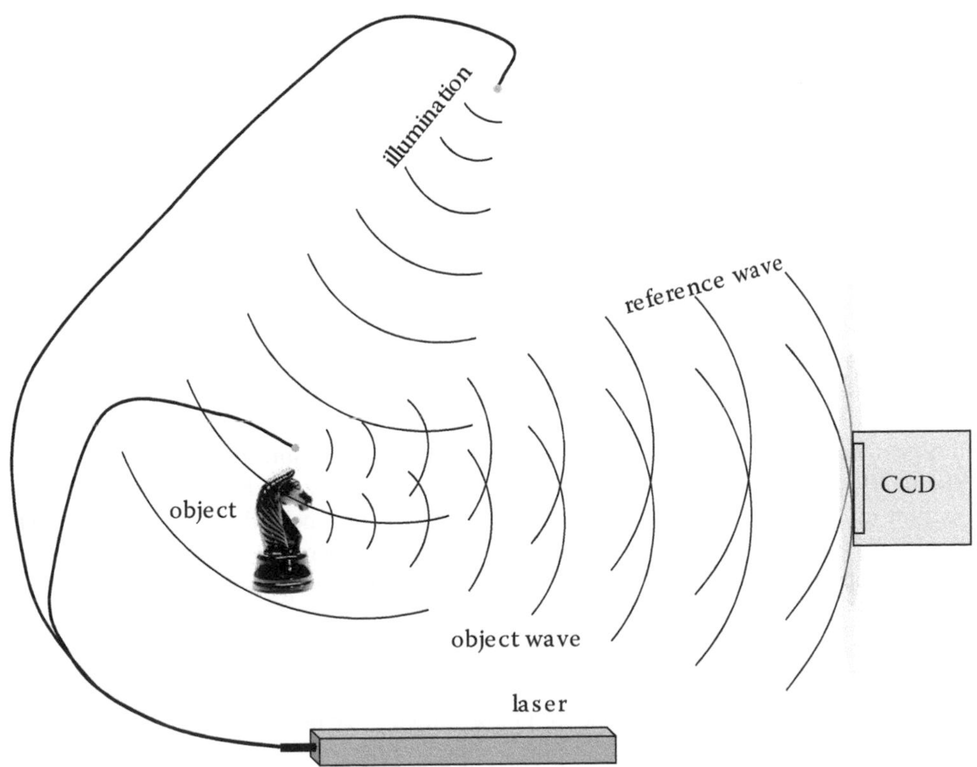

FIGURE 12.6 Scheme of a setup for digital lensless Fourier holography of diffusely reflecting objects.

is based on the Fresnel reconstruction. Here, again, u(x,y) is the object wave in the object plane, $h(\xi,\eta)$ the hologram, $r(\xi,\eta)$ the reference wave in the hologram plane, and $u'(x',y')$ the reconstructed wave field.

For the specific setup of lensless Fourier holography, a spherical reference wave is used with the origin at the same distance from the sensor as the object itself. For the case that $d' = -d$, $x = x'$, and $y = y'$, both virtual images are reconstructed and the reconstruction algorithm is then

$$u'(x,y) = \frac{\exp(\mathrm{ikd})}{\mathrm{i}\lambda \mathrm{d}} \mathrm{e}^{-\mathrm{i}\frac{\mathrm{k}}{2\mathrm{d}}(x^2+y^2)} \mathbf{FT}_{\lambda d'}\left\{t\left[h(\xi,\eta)\right]\cdot r(\xi,\eta)\cdot e^{-\mathrm{i}\frac{\mathrm{k}}{2\mathrm{d}}(\xi^2+\eta^2)}\right\} \tag{12.29}$$

where $\mathbf{FT}_{\lambda \mathrm{d}}$ is the 2-D Fourier transformation, which has been scaled by a factor $1/\lambda\mathrm{d}$. In this recording configuration, the effect of the spherical phase factor associated with the Fresnel diffraction pattern of the object is eliminated by the use of a spherical reference wave $r(\xi,\eta)$ with the same average curvature:

$$r(x,y) = const\cdot\exp\left(i\frac{\pi}{\lambda d}\left(\xi^2+\eta^2\right)\right) \tag{12.30}$$

This results in a more simple reconstruction algorithm, which can be described by

$$u'(x,y) = const\cdot\exp\left(-i\frac{\pi}{\lambda d}\left(x^2+y^2\right)\right)\mathbf{FT}_{\lambda d}\left\{h(\xi,\eta)\right\}. \tag{12.31}$$

Besides the faster reconstruction procedure (only one Fourier transform has to be computed), the Fourier algorithm uses the full SBP of the sensor chip, because it adapts the curvature of the reference wave to the curvature of the object wave.

12.2.2.4 Numerical Reconstruction by the Phase-Shifting Approach

The diameter of a CCD chip is about 10 mm, but the effective area of the sensor is usually only 4.8 × 3.6 mm. Such a small active area limits the application if the hologram is recorded in off-axis geometry, because many areas in the reconstructing plane of off-axis digital holograms are covered by the useless virtual image and zero-order beam. Therefore, in-line holography is often applied to use the full space–bandwidth (SBW) of the camera. But the reconstructed images of in-line holograms contain in this case overlapping real and virtual images and the zero-order wave. To eliminate the zero-order and the virtual image, temporal phase shifting can be used. Yamaguchi and Zhang [19] proposed a method of using four phase-shifting in-line holograms and the intensity distribution of the object wave to remove the zero-order and the virtual image. Lai et al. [21] proposed an algorithm that needs only four phase-shifting holograms.

Assume the phase shifter is placed in the reference beam, and the object wave is expressed as

$$u(x,y) = a(x,y)\exp\left[i\phi(x,y)\right]. \tag{12.32}$$

The four plane reference waves that are phase-shifted in steps of 90° can be expressed as

$$r_k(x,y) = i^k r,\quad k = 0,\ \cdots,\ 3, \tag{12.33}$$

where i represents the imaginary unit. The four in-line holograms are recorded by superposing the object beam of Equation 12.32 with each reference beam of Equation 12.33. The resulting holograms are

$$I_k = r^2 + a^2(x,y) + i^{4-k} ra(x,y)\exp\left[i\phi(x,y)\right] + i^k ra(x,y)\exp\left[-i\phi(x,y)\right]. \quad (12.34)$$

By illuminating each of the four holograms with its corresponding reference wave and adding together the resulting fields, a composite wavefront just behind the hologram is obtained:

$$u'(x',y') = \sum_{k=0}^{3} r_k \cdot I_k = 4r^2 a(x,y)\exp\left[i\phi(x,y)\right] \quad (12.35)$$

We can see that this wavefront is identical to the original object wavefront except for the coefficient $4r^2$. This means that the zero-order beam and the conjugate wave are completely removed. The remaining work is to calculate the complex wavefront in the reconstruction plane by using the diffraction formula.

12.2.3 Influences of Discretization

For the estimation of the lateral resolution in DH, three different effects related to discretization of the digitizing sensor have to be considered [16]: averaging, sampling, and the limited sensor size. The extension of the sensor field is limited. We assume a quadratic sensor with a limited active area and N × M quadratic pixels of size $\Delta\xi \times \Delta\xi$. Each pixel has a light-sensitive region with a side length $\gamma\Delta\xi = \Delta\xi_{eff}$. The quantity γ^2 is the fill factor $(0 \le \gamma \le 1)$, which indicates the active area of the pixel. The sensor averages the incident light, which has to be considered because of possible intensity fluctuations over this area. The continuous expression for the intensity $I(\xi,\eta) = t[h(\xi,\eta)]$ registered by the sensor has to be integrated over the light-sensitive area. This can be expressed mathematically by the convolution of the incident intensity $I(\xi,\eta)$ with a rectangle function [17]

$$I_1(\xi,\eta) \propto I \otimes rect_{\Delta\xi_{eff}\cdot\Delta\xi_{eff}} \quad (12.36)$$

The discrete sampling of the light field is modeled by the multiplication of the continuous-assumed hologram with the 2-D comb-function:

$$I_2(\xi,\eta) \propto I_1(\xi,\eta)\cdot comb_{\Delta\xi,\Delta\xi} \quad (12.37)$$

Finally, the limited sensor size requires that the comb function has to be truncated at the borders of the sensor. This is achieved by multiplication with a 2-D rectangle function of size $N \times \Delta\xi$:

$$I_3(\xi,\eta) \propto I_2 \cdot rect_{N\Delta\xi.N\Delta\xi} \quad (12.38)$$

The resulting consequences such as amplitude distortion, aliasing, and speckle size are discussed in Reference 16 and Reference 22.

12.3 DIGITAL HOLOGRAPHIC INTERFEROMETRY

Interferometry is the method of transforming phase differences $\delta(x,y)$ of wavefronts into observable intensity fluctuations called *interference fringes*. The holographic recording and reconstruction principle allows the comparison of an object having a rough surface with itself for two different states. We can say that a measurement method is a holographic technique if at least one of the involved wavefronts is reconstructed "holographically." The change of a part of the interferometer between two exposures, such as a displacement of the object, the application of another wavelength, or another illumination direction, causes phase differences between the interfering wavefronts. Fringes can be observed if certain boundary conditions are considered. One important boundary condition is that the microstructure of the surface remains unchanged. On this basis, both the shape of an object and the change of its shape caused by a certain load can be measured with interferometric precision [23,24]. In general, three basic techniques of holographic interferometry can be distinguished:

- The *double-exposure technique*, in which the two states of the object are stored "holographically" one after the other to be reconstructed simultaneously for interferometric comparison
- The *real-time technique*, in which one state is stored "holographically" to be reconstructed for real-time comparison with the light field scattered by the coherently illuminated object under test
- The *time-average technique*, in which numerous states of a moving or vibrating object are stored successively on a holographic sensor to be reconstructed simultaneously for the generation of an intensity distribution that is modulated by the temporal average of the sequence of phase differences

12.3.1 Basic Principles

Without loss of generality, the basic principle of holographic interferometry is described by the example of the double-exposure technique.

As we have already shown, DH allows the numerical reconstruction of the phases of both object states. For interferometric applications with respect to displacement and shape measurement, the phase difference $\delta(x,y)$ of two reconstructed wave fields $u_1'(x,y)$ and $u_2'(x,y)$ needs to be calculated. Because both phases $\phi_1(x,y)$ and $\phi_2(x,y)$ can be reconstructed independently, the algorithm can be reduced to the following simple equation:

$$\delta(x,y) = \phi_1(x,y) - \phi_2(x,y) \tag{12.39}$$

Figure 12.7 illustrates this procedure by the example of displacement measurement by DH.

Phases taken from digital holographic measurements are always mapped to values from 0 to 2π. As a consequence of the wrapped primary phases, the difference phase δ is also a mod 2π function. However, in contrast to conventional holographic techniques, the mod 2π phase can be calculated directly without additional temporal or spatial phase modulation such as phase shifting or spatial heterodyning. But phase unwrapping is still necessary in DH, because the *fringe-counting problem* remains [25] (see Subsection 12.3.4).

To derive the displacements or coordinates from the measured phase differences $\delta(x,y)$, a geometric model of the interferometer and the paths of light is needed [26]. For any object point P(x,y), the phase of the light $\phi(x,y)$ traveling from the source point Q(x,y) via P(x,y) to the observation point B(x,y) can be expressed with the wave vectors $\mathbf{k}_Q$ and $\mathbf{k}_B$ (Figure 12.8):

$$\phi = \mathbf{k}_Q\left(\mathbf{r}_P - \mathbf{r}_Q\right) + \mathbf{k}_B\left(\mathbf{r}_B - \mathbf{r}_P\right) \tag{12.40}$$

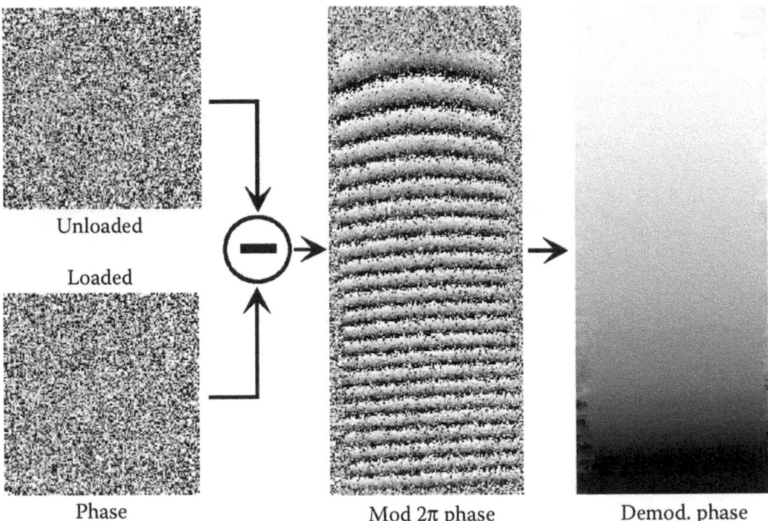

FIGURE 12.7 The principle of digital holographic interferometry demonstrated on example of a loaded small beam (size 7 mm × 2 mm, thickness 50 μm). The interference phase is the result of the subtraction of the two phases, which correspond to the digital holograms of both the object states.

with

$$\mathbf{k}_Q = \frac{2\pi}{\lambda}\mathbf{e}_Q \tag{12.41}$$

$$\mathbf{k}_B = \frac{2\pi}{\lambda}\mathbf{e}_B \tag{12.42}$$

where $\mathbf{e}_Q$ and $\mathbf{e}_B$ denote the unit vectors in the illumination and observation directions, respectively.

Based on Equation 12.40, holographic displacement analysis and contouring can be described as well. In both cases, one parameter is changed between the two exposures of the hologram. As a result, the phase ϕ registered at the position of the observer B is changed. In displacement analysis, the position of the point P, and consequently $\mathbf{r}_P$, is changed by applying a load to the object. In shape

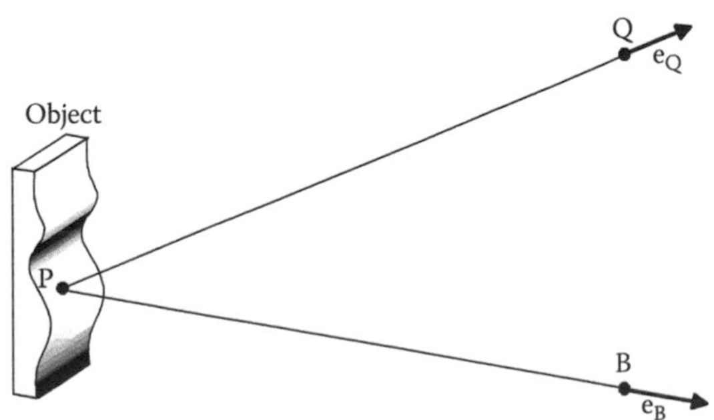

FIGURE 12.8 Geometric conditions in holography.

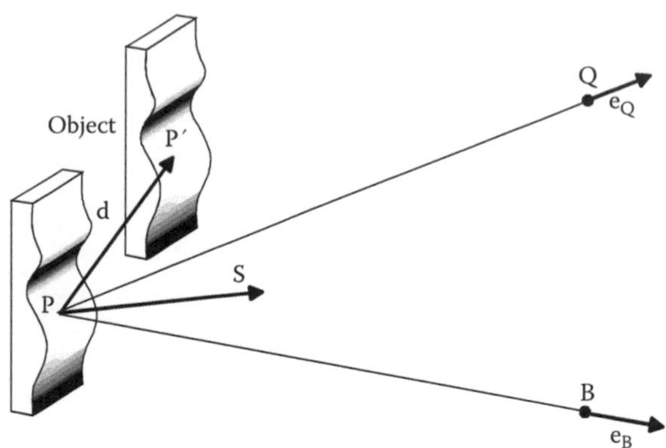

FIGURE 12.9 Geometric conditions in holographic interferometry.

measurement — also called holographic contouring — the illumination point Q, and consequently the vectors $\mathbf{e}_Q$ and $\mathbf{r}_Q$, the observation point, and consequently $\mathbf{e}_B$ and $\mathbf{r}_B$, the wavelength λ, or the refractive index n surrounding the object can be changed.

12.3.2 Holographic Displacement Measurement

Holographic interferometry provides an elegant approach to the determination of displacements of diffusely reflecting surfaces on the scale of the wavelength. The geometric conditions are shown in Figure 12.9. In a usual experimental setup, the distance between the object and the illumination point is large compared to the displacement of the object. This is also true of the distance between the object and the observation point. Thus, the relation between the phase difference $\delta(x,y) = \phi_1 - \phi_2$ registered by DH and the 3-D displacement vector $\mathbf{d}(u,v,w)$ of the point P(x,y,z) can be expressed by the basic equation of holographic interferometry [26]:

$$\delta(P) = \frac{2\pi}{\lambda} \cdot \left[\mathbf{e}_B(P) + \mathbf{e}_Q(P)\right] \cdot \mathbf{d}(P) = \frac{2\pi}{\lambda} \mathbf{S}(P) \cdot \mathbf{d}(P) \tag{12.43}$$

The double-exposure technique results with $\delta(P) = N(P) \cdot 2\pi$

$$N(P) \cdot \lambda = \mathbf{S}(P) \cdot \mathbf{d}(P) \tag{12.43a}$$

where N is the fringe number, λ is the wavelength, and **S** is the sensitivity vector. Equation 12.43a shows that the measured phase difference is equivalent to the projection of the displacement vector onto the sensitivity vector. Thus, the scalar product of the sensitivity vector and the displacement vector is constant for an interference line of the order N. Therefore, the observed interference fringes are lines of equal displacement in the direction of the sensitivity vector. This vector points in the direction of the bisector of the angle between the illumination and observation direction:

$$\mathbf{S}(P) = \left\lfloor e_B(P) + e_Q(P) \right\rfloor \tag{12.44}$$

Consequently, the usual interferometric setup is more sensitive for the out-of-plane component than for the in-plane components. For the measurement of the three displacement components $\mathbf{d}(u,v,w)$, at least three independent equations of the type of Equation 12.43 are necessary. Usually,

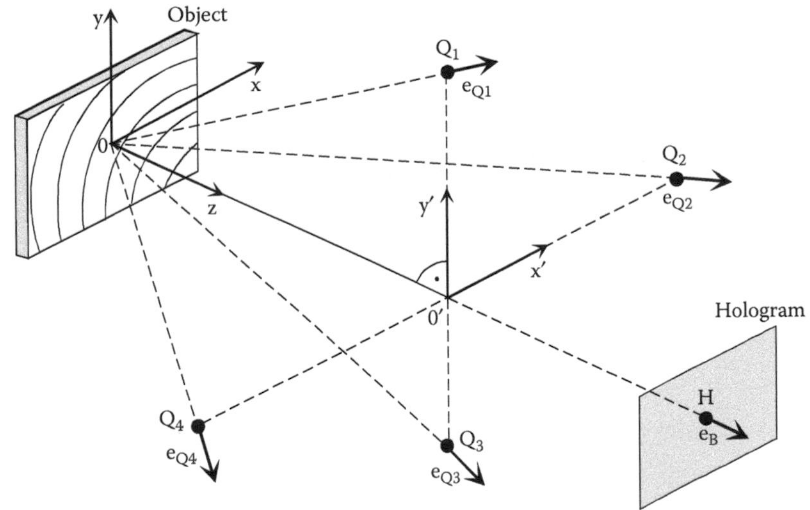

FIGURE 12.10 Scheme of a holographic interferometer with four illumination directions and one observation direction.

three or more observation and illumination directions are chosen. To avoid perspective distortions, it is more convenient to use independent illumination directions as illustrated in Figure 12.10. In this case, the following equation system can be derived:

$$N_i(P)\cdot\lambda = \left\lfloor \mathbf{e}_B(P) + \mathbf{e}_{Qi}(P) \right\rfloor \cdot \mathbf{d}(P)\,,\quad i = 1...n, n \geq 3 \tag{12.45}$$

To write in simplified terms, a matrix notation of Equation 12.45 is used:

$$\mathbf{N}\cdot\lambda = \mathbf{G}\cdot\mathbf{d}, \tag{12.46}$$

with the column vector **N** containing the n fringe numbers N_i and the (n,3)-matrix **G** containing in the i-th line the three components of the i-th sensitivity vector $\mathbf{S}_i$.

To improve the accuracy of the displacement components, overdetermined equation systems (n > 3) are often used. In our case, four illumination directions $\mathbf{e}_{Qi}$ (i = 1, …, n, n = 4) are provided. Such equation systems can be solved with the least-squares error method:

$$grad_{\mathbf{d}}\left\lfloor (\mathbf{N}\cdot\lambda - \mathbf{G}\cdot\mathbf{d})^T \cdot (\mathbf{N}\cdot\lambda - \mathbf{G}\cdot\mathbf{d}) \right\rfloor = \mathbf{0} \tag{12.47}$$

which results in the so-called normal equation system

$$\mathbf{G}^T\cdot\mathbf{N}\cdot\lambda = \mathbf{G}^T\mathbf{G}\cdot\mathbf{d}. \tag{12.48}$$

The matrix

$$\mathbf{F} = \mathbf{G}^\mathrm{T}\cdot\mathbf{G} \tag{12.49}$$

is referred to as the normal matrix with $\mathbf{G}^\mathrm{T}$ being the transpose of **G**. The design of **F** is of great importance for the construction of the interferometer and the error analysis [27,28].

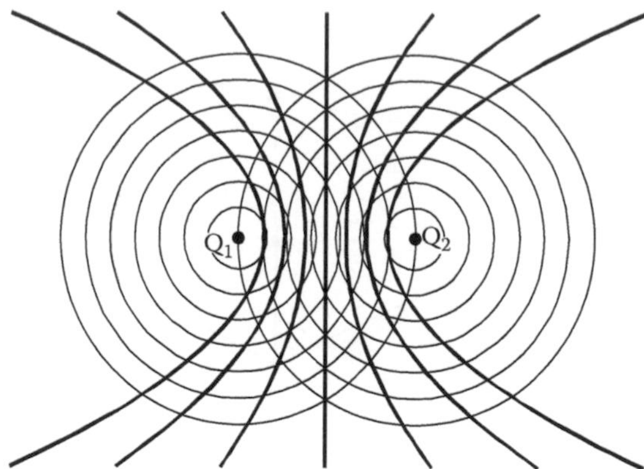

FIGURE 12.11 Surfaces of equal phase as hyperboloids of revolution resulting from the superposition of two spherical waves.

12.3.3 Holographic Shape Measurement

Coordinates and surface contours can be measured "holographically" by the generation of contour fringes that represent lines of equal phase $\delta(P)$. The conversion of lines of equal phase into lines of equal height is also based on a geometric model of the interferometer. Contour fringes can be generated, particularly by the change of one of the following parameters between the two exposures: the wavelength λ, the refractive index n, and the illumination direction $\mathbf{e}_Q$. Commonly used are the so-called two-source-point method and the two-wavelength method.

12.3.3.1 Two-Source-Point Method

This method is based on a double exposure of the holographic sensor with two different source points. Between the first and the second exposures, the source point Q_1 is slightly displaced to a new position Q_2. The reconstructed and interfering wavefronts generate a system of surfaces of equal phase with hyperbolic shape and rotational symmetry. The two source points Q_1 and Q_2 are the focal points of that system of hyperbolas (Figure 12.11). The intersecting lines of the object with the interference field form the characteristic contour lines. Their structure depends on the location of the object, its shape, and the position of the source points.

For the derivation of a relation between the phase of the object point P measured at the position of the observer and its coordinate, the light paths as shown in Figure 12.12 are considered. The refractive

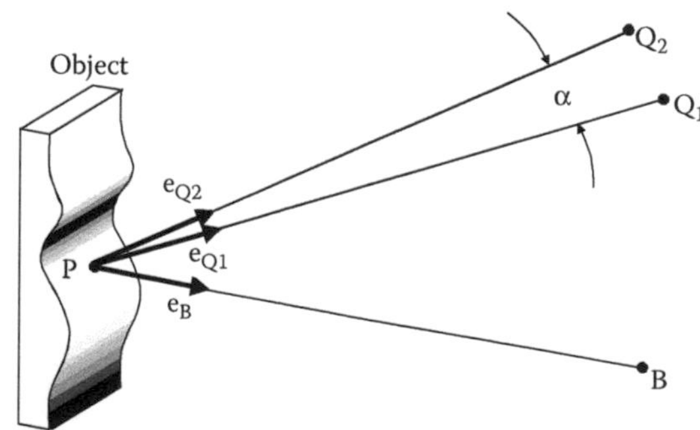

FIGURE 12.12 Light paths for the two-source-point method.

index of the surrounding medium is assumed to be n = 1. Equation 12.40 yields for the first and the second exposures with the source position Q_1 and Q_2, respectively:

$$\phi_1 = \mathbf{k}_{Q_1}\left(\mathbf{r}_P - \mathbf{r}_{Q_1}\right) + \mathbf{k}_B\left(\mathbf{r}_B - \mathbf{r}_P\right) \tag{12.50a}$$

$$\phi_2 = \mathbf{k}_{Q_2}\left(\mathbf{r}_P - \mathbf{r}_{Q_2}\right) + \mathbf{k}_B\left(\mathbf{r}_B - \mathbf{r}_P\right) \tag{12.50b}$$

Consequently, for the phase difference $\delta(P)$, it follows that

$$\delta(P) = \phi_1 - \phi_2 = \frac{2\pi}{\lambda}\left[\mathbf{e}_{Q_1}\left(\mathbf{r}_P - \mathbf{r}_{Q_1}\right) - \mathbf{e}_{Q_2}\left(\mathbf{r}_P - \mathbf{r}_{Q_2}\right)\right] \tag{12.51}$$

The condition for the surfaces of equal phase is δ = constant, i.e.,

$$\left|\mathbf{r}_P - \mathbf{r}_{Q_1}\right| - \left|\mathbf{r}_P - \mathbf{r}_{Q_2}\right| = \text{const.} \tag{12.52}$$

Equation 12.52 describes a system of hyperboloids of revolution with the focal points Q_1 and Q_2. The sensitivity vector **S** can be found with

$$\mathbf{S} = \nabla\left[\mathbf{e}_{Q_1}\left(\mathbf{r}_P - \mathbf{r}_{Q_1}\right) - \mathbf{e}_{Q_2}\left(\mathbf{r}_P - \mathbf{r}_{Q_2}\right)\right] \tag{12.53}$$

where ∇ is the Nabla operator. Consequently, the direction of high sensitivity points to the direction of the in-plane coordinates of the object (Figure 12.13).

For the situation that the illumination and observation points are located at a large distance (telecentric illumination and observation) from the object, it can be assumed that all **k**-vectors are

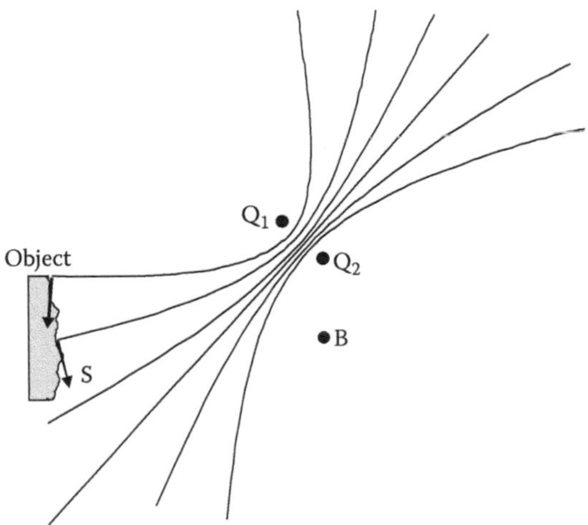

FIGURE 12.13 Position and direction of the sensitivity vector for the two-source-point method.

approximately constant with respect to changes in the position of the object point P. In this case, we can write for the sensitivity vector:

$$\mathbf{S} = \mathbf{e}_{Q_1} - \mathbf{e}_{Q_2} \tag{12.54}$$

The vector $\mathbf{e}_{Q1} - \mathbf{e}_{Q2}$, and thus the sensitivity vector, is directed perpendicular to the bisecting line of the angle α, where α is the angular distance between the sources Q_1 and Q_2 (see Figure 12.12). The amount of **S** is

$$|\mathbf{S}| = \left|\mathbf{e}_{Q_1} - \mathbf{e}_{Q_2}\right| = 2\sin\frac{\alpha}{2} \tag{12.55}$$

In the assumed setup, in which the sources are placed sufficiently far from the surface being studied, we get an approximation for the registered phase difference:

$$\begin{aligned} \delta &= N \cdot 2\pi \approx \frac{2\pi}{\lambda} 2\sin\frac{\alpha}{2} \cdot \mathbf{e}_S \cdot |r_P| \cdot \mathbf{e}_P \\ N \cdot \lambda &\approx 2\sin\frac{\alpha}{2} \cdot |r_P| \cdot \cos\left(\mathbf{e}_S, \mathbf{e}_P\right) \end{aligned} \tag{12.56}$$

In this case, the distance between two consecutive contour lines represents a height difference Δh:

$$\Delta h \approx \frac{\lambda}{2\sin\frac{\alpha}{2} \cdot \cos\left(\mathbf{e}_S, \mathbf{e}_P\right)} \tag{12.57}$$

Equation 12.57 also shows that maximum sensitivity is achieved for the in-plane coordinates. The lines obtained are contour lines of the surface relief if the planes are oriented perpendicular to the line of observation. However, such an arrangement, in which the sensitivity vector is oriented in the direction of the surface normal, causes difficulties because sections of the object are in shadow, i.e, cannot be investigated.

12.3.3.2 Two-Wavelength Method

This method is based on the consecutive illumination of the object with two different wavelengths λ_1 and λ_2. The direction of illumination and the position of the observer remain unchanged. The refractive index of the surrounding medium is again assumed to be n = 1.

For the derivation of a relation between the phase of the object point P measured at the position of the observer and its coordinate, the light paths as shown in Figure 12.14a are considered. Equation 12.40 yields for the first and the second exposures with the wavelengths λ_1 and λ_2, respectively:

$$\phi_1 = \mathbf{k}_{Q_1}\left(\mathbf{r}_P - \mathbf{r}_Q\right) + \mathbf{k}_{B_1}\left(\mathbf{r}_B - \mathbf{r}_P\right) \tag{12.58a}$$

$$\phi_2 = \mathbf{k}_{Q_2}\left(\mathbf{r}_P - \mathbf{r}_Q\right) + \mathbf{k}_{B_2}\left(\mathbf{r}_B - \mathbf{r}_P\right) \tag{12.58b}$$

Consequently, for the phase difference that follows,

$$\delta(P) = \phi_1 - \phi_2 = \left(\mathbf{k}_{Q_1} - \mathbf{k}_{Q_2}\right)\cdot\left(r_P - r_Q\right) + \left(\mathbf{k}_{B_1} - \mathbf{k}_{B_2}\right)\cdot\left(\mathbf{r}_B - \mathbf{r}_P\right) \tag{12.59}$$

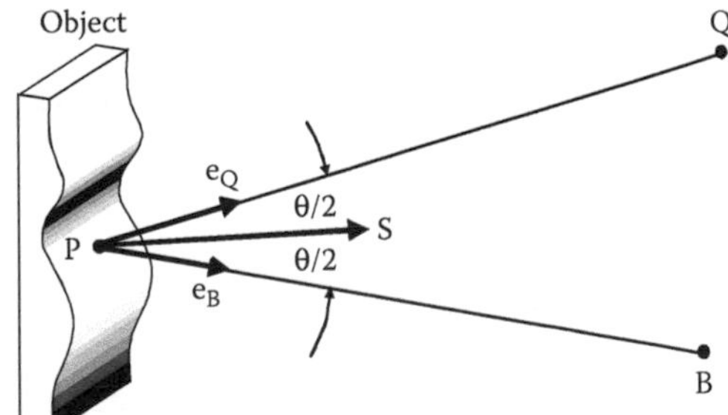

(a) Light paths for the two-wavelength-method

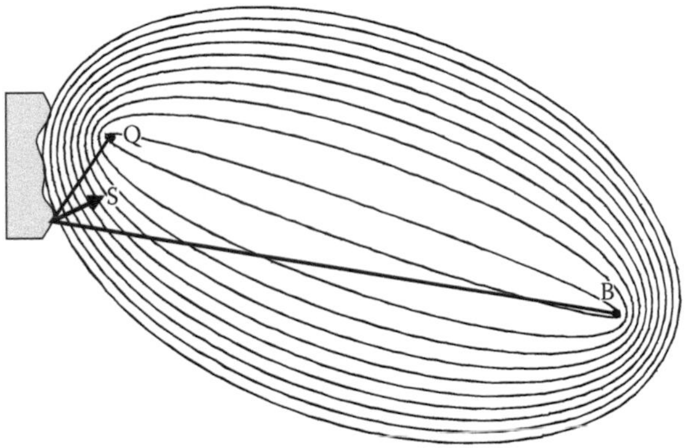

(b) Surface of equal phase for the two-wavelength-method

FIGURE 12.14 (a) Light paths for the two–wavelength method, (b) surfaces of equal phase for the two-wavelength method.

Taking into account the propagation vectors

$$\mathbf{k}_{Q_1} = \frac{2\pi}{\lambda_1}\mathbf{e}_Q(P),\ \mathbf{k}_{Q_2} = \frac{2\pi}{\lambda_2}\mathbf{e}_Q(P)$$
$$\mathbf{k}_{B_1} = \frac{2\pi}{\lambda_1}\mathbf{e}_B(P),\ \mathbf{k}_{B_2} = \frac{2\pi}{\lambda_2}\mathbf{e}_B(P) \tag{12.60}$$

it follows for the phase difference

$$\delta(P) = 2\pi\left(\frac{1}{\lambda_1} - \frac{1}{\lambda_2}\right)\cdot\left[\mathbf{e}_Q(P)(\mathbf{r}_P - \mathbf{r}_Q) + \mathbf{e}_B(P)(\mathbf{r}_B - \mathbf{r}_P)\right] \tag{12.61}$$

The condition for the surfaces of equal phase is δ = constant, i.e.,

$$|\mathbf{r}_P - \mathbf{r}_Q| + |\mathbf{r}_B - \mathbf{r}_P| = \text{const.} \tag{12.62}$$

Equation 12.62 describes a system of ellipsoids of revolution with the focal points Q and B (see Figure 12.14b). The sensitivity vector **S** can be found with

$$\mathbf{S} = \nabla\left[\mathbf{e}_Q(\mathbf{r}_P - \mathbf{r}_Q) + \mathbf{e}_B(\mathbf{r}_B - \mathbf{r}_P)\right] \tag{12.63}$$

where ∇ is the Nabla operator. If we assume that the light source Q and the observation point B are located far away from the surface, the ellipsoids degenerate to planes. The unit vectors $\mathbf{e}_Q$ and $\mathbf{e}_B$ can be regarded as approximately constant with respect to changes of the position P on the surface of the object. At infinity, it is

$$\mathbf{S} = \left(\mathbf{e}_Q + \mathbf{e}_B\right) = 2\cos\frac{\theta}{2}\cdot\mathbf{e}_S \tag{12.64}$$

where θ is the angle between the observation and illumination directions and $\mathbf{e}_S$ is the unit direction vector of **S**.

In the assumed setup, we obtain an approximation for the registered phase difference

$$\begin{aligned}\delta &= N\cdot 2\pi \approx \frac{4\pi}{\Lambda}\cos\frac{\theta}{2}\mathbf{e}_S\cdot\left|r_P\right|\cdot\mathbf{e}_P \\ N\cdot\Lambda &\approx 2\cos\frac{\theta}{2}\cdot\left|r_P\right|\cdot\cos\left(\mathbf{e}_S,\mathbf{e}_P\right)\end{aligned} \tag{12.65}$$

with

$$\Lambda = \frac{\lambda_1\cdot\lambda_2}{\lambda_2-\lambda_1} \approx \frac{\lambda^2}{\Delta\lambda} \tag{12.66}$$

as the synthetic wavelength. Consequently, the distance between two consecutive contour lines represents, in our approximation, a height difference Δz with respect to the object of

$$\Delta z \approx \frac{\lambda^2}{2\Delta\lambda} = \frac{\Lambda}{2} \tag{12.67}$$

Unlike the two-source-point method, the two-wavelength method is sensitive with respect to the out-of-plane coordinate. Therefore, a combination of both methods yields an approach to 3-D coordinate measurement on an object surface [29]. It should be emphasized that the application of the principles of DH to the two-wavelength method has the advantage that each hologram can be independently stored and reconstructed with its corresponding wavelength. This results in a great decrease of aberrations and also makes it possible to use larger wavelength differences for the generation of shorter synthetic wavelengths [30].

12.3.4 Direct and Absolute Phase Measurement

Because of the fundamental sinusoidal nature of the wave functions used in interferometry, the phase term δ to be measured is wrapped upon itself with a repeat distance of 2π. In DH, the origin of that problem lies in the storage and reconstruction of the hologram. The complex wavefronts $u_i'(x,y)$ are reconstructed numerically as a kind of interference process in which the phases $\phi_i(x,y)$ appear as the argument of a harmonic function such as the cosine. Because the cosine is an even and periodic function for the interference phase, $\delta(x,y)$ holds

$$\cos(\delta) = \cos(s\cdot\delta + N\cdot 2\pi),\ s \in [-1,1],\ N \in Z \tag{12.68}$$

Consequently, the determined phase distribution remains undefined to an additive integer multiple of 2π and to the sign s. Each inversion of expressions such as Equation 12.68 contains an inverse trigonometric function. However, all inverse trigonometric functions are expressed by the arctan function, and this function has its principal value between $-\pi/2$ and $+\pi/2$ (see Equation 12.14).

This uncertainty has an important practical consequence. When the phase angle δ extends beyond 2π to $N \cdot 2\pi$, the absolute phase value $N \cdot 2\pi$ can only be reconstructed if N can be determined. This process, called *unwrapping* or *demodulation*, leads to the major difficulties and limitations of all phase-measuring procedures. The key to phase unwrapping is the detection of the real 2π phase discontinuities. An important condition for their reliable identification is that neighboring phase samples meet the relation

$$-\pi \leq \Delta_i \delta(x, y) < \pi \quad \text{with} \quad \Delta_i \delta(x, y) = \delta(x, y) - \delta(x-1, y) \tag{12.69}$$

If the object to be analyzed is not simply connected, i.e., isolated objects or shaded regions occur, and the data are noisy, then the spatial unwrapping procedure can fail. The reason is the violation of the spatial neighborhood condition given in Equation 12.69.

There are many approaches to unwrapping complex-shaped objects such as the watch plate shown in Figure 12.15a, which contains many edges and bores [31]. Unwrapping algorithms that replace the spatial neighborhood by a temporal neighborhood are much more robust in the case of complex objects. Such algorithms are known by the names *temporal unwrapping, absolute phase measurement,* or *hierarchical demodulation* [32–36]. The objective of these algorithms consists in the extension of the 2π-unambiguity range over the whole measurement region by the generation of an adapted synthetic wavelength. However, only one coarse synthetic wavelength without mod 2π jumps reduces the accuracy of the measurement process considerably. Therefore, a sequence of gradually refined synthetic wavelengths is applied. This temporal and hierarchical unwrapping procedure avoids the sensitive spatial neighborhood and unwraps every pixel separately along the temporal axis.

The hierarchical multiwavelength technique proposed in Reference 36 and Reference 37 meets this requirement and reduces the number of demodulation steps systematically by a careful selection of synthetic wavelengths. The wavelengths $\lambda_{i,j}$ are chosen in a way that the corresponding synthetic wavelength

$$\Lambda_{i,j}^{k} = \frac{\lambda_i \cdot \lambda_j}{(\lambda_i - \lambda_j)} \tag{12.70}$$

considers the phase noise ε of the used evaluation technique. In every successive step (k + 1), it must be ensured that the corresponding synthetic wavelength $\Lambda^{(k+1)}$ of the still-ambiguous solution, including its tolerance, is longer than the remaining tolerance region $\{\Lambda^{(k)} \cdot 4\varepsilon\}$ of the previous step (k) because of the phase noise:

$$\Lambda^{(k+1)} \cdot (1 - 4\varepsilon) \geq \Lambda^{(k)} \cdot 4\varepsilon \tag{12.71}$$

Usually, only four steps are sufficient for the reliable and precise demodulation of complex mod 2π phase maps. Figure 12.15 shows four steps with the corresponding synthetic wavelengths in the case of the demodulation of the contour phase map of the watch plate.

Consequently, the absolute phase can be reconstructed very effectively and in a straightforward manner by the combination of DH and multiwavelength contouring. This quantity is the basis for

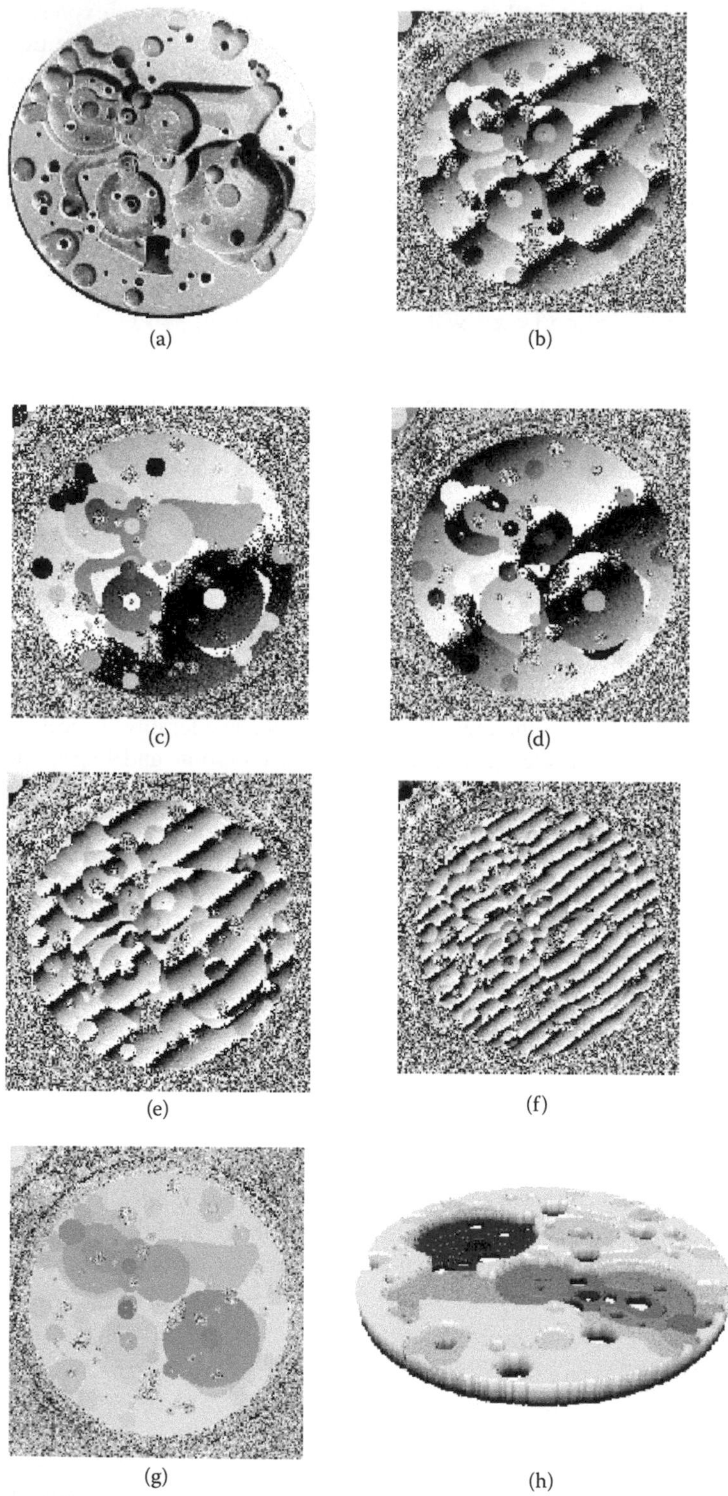

FIGURE 12.15 Hierarchical unwrapping with a 4-step multiwavelength algorithm. (a) Watch plate (diameter 20 mm), (b) reconstructed mod 2π contour fringes, (c) step 1 of 4: 0.4 mm, (d) step 2 of 4: 0.19 mm, (e) step 3 of 4: 0.049 mm, (f) step 4 of 4: = 0.025 mm, (g) demodulation result, (h) pseudo-3-D plot of the shape of the watch plate.

the calculation of the wanted shape and displacement of the object. For the transformation of the phase data into Cartesian coordinates, a calibration of the experimental setup has to be performed. The purpose of this procedure is the transformation of the loci of equal phase into the loci of equal height. Here, the curved shape of the generated surfaces of equal phase (hyperboloids or ellipsoids of revolution) has to be considered. For the numerical correction of the distorted phase map, a model-based calibration technique is preferred that is based on the knowledge of several parameters of the setup and the evaluation of an appropriate calibration object such as a flat plate [38].

12.3.5 Advantages of DH

Besides electronic processing and direct access to the phase, some further advantages recommend DH for several applications such as metrology and microscopy [6]:

- The availability of the independently reconstructed phase distributions of each individual state of the object and interferometer, respectively, offers the possibility of recording a series of digital holograms with increased load amplitude. In the evaluation process, the convenient states can be compared interferometrically. Furthermore, a series of digital holograms with increasing load can be applied to unwrap the mod 2π phase temporally [31]. In this method, the total object deformation is subdivided into many measurement steps in which the phase differences are smaller than 2π. By adding up those intermediate results, the total phase change can be obtained without any further unwrapping.
- The independent recording and reconstruction of all states also gives a new degree of freedom for optical shape measurement. In case of multiwavelength contouring, each hologram can be stored and reconstructed independently with its corresponding wavelength. This results in a drastic decrease of aberrations and makes it possible to use larger wavelength differences for the generation of shorter synthetic wavelengths [30].
- Because all states of the inspected object can be stored and evaluated independently, only seven digital holograms are necessary to measure the 3-D displacement field and 3-D shape of an object under test: one hologram for each illumination direction before and after the loading, respectively, and one hologram with a different wavelength (or a different source point of illumination), which can interfere with one of the other holograms for two-wavelength contouring (or two-source-point contouring) for shape measurement. If four illumination directions are used, nine holograms are necessary [39].
- DH is a versatile tool for the solution of numerous inspection and measurement problems because of the possibility of miniaturizing complex holographic setups [40] and of using the method for remote and comparative metrology [41].

12.4 DIGITAL HOLOGRAPHIC MICROSCOPY (DHM)

Conventional imaging optical microscopes suffer from reduced depths of focus [42] because of the high NAs of the lenses and the high magnification ratios, so that the analysis of 3-D objects requires mechanical motions along the optical axis for incremental scanning of the experimental volume of the object under investigation. These latter limitations can be surmounted if a holographic method is used. In fact, in holography, the object beam relayed by the microscope lens interferes with a reference beam on a holographic plate. The object wavefront can be optically reconstructed, and the entire 3-D volume can be reimaged by using one image acquisition. With the successive developments in digital holography (DH), wherein the hologram information is directly recorded in a CCD camera, the optical reconstruction is replaced by a computer reconstruction [2–4]. In the beginning, all the works about DH concerned amplitude contrast imaging, for which only the modulus of the numerically reconstructed optical field is considered. Then, results were obtained for quantitative information in holographic interferometry [5]. The first example of a numerically reconstructed phase-contrast image was reported by Cuche et al. [43], who used a modified reconstruction algorithm, including a multiplication of the digital hologram by a digital replica of the reference wave. Phase data carry unique information on morphology, which can be used for 3-D presentation of the microscopic object. Successively, various optical methods have been implemented to extract both the phase and the amplitude information from the fringe patterns recorded by the CCD camera [44]. In fact, as the phase and the amplitude of an optical signal are known, the DH techniques allow simulation of the reconstruction of the optical beam by using discrete forms of the Kirchhoff–Fresnel propagation equations [10,11].

We will show here that DH is an appropriate tool for retrieving the phase distribution of the object wave field for quantitative phase-contrast imaging in microscopy, meaning that the reconstructed phase distribution can be directly used for metrological applications, in particular, for surface profilometry and holographic interferometry.

It is an important feature of DH that a 3-D description of the specimen can be obtained digitally with only one image acquisition, permitting simple and compact setups that could give quantitative information about objects' shapes. Actually, there are several 3-D imaging interferometric methods that allow the measurement of minute displacements and surface profiles. Methods such as holographic interferometry, fringe projection, and speckle metrology can provide full-field noncontact information about coordinates, deformation, strains, stresses, and vibrations. However, an important advantage of DH, in comparison with interference microscopy, is that the phase aberrations can be corrected digitally [11,44–48].

In interference microscopy, this problem is solved experimentally by inserting into the setup the same microscope objective (MO) in the reference arm, at an equal distance from the exit of the interferometer. This arrangement, called a Linnick interferometer, requires that if any change has to be made in the object arm, then the same change must be precisely reproduced in the reference arm in such a way that the interference occurs between similarly deformed wavefronts. As a consequence, the experimental configuration requires a very high degree of precision. Another option is the Mirau interferometer; however, it is difficult to achieve high-resolution imaging with this technique, because a miniaturized interferometer must be inserted between the sample and the MO. It is important to point out that, in contrast to conventional approaches, DH allows the direct calculation of the complex wavefront, in phase and intensity, by the numerical solution of the diffraction problem in the computer.

In comparison with standard interference microscopy techniques, which also provide amplitude and phase contrast, the advantage of DH is that the recording of only one hologram is necessary, whereas the recording of four or more interferograms is required with interference microscopy. Moreover, no displacement or movement of optical elements is needed. As a consequence, both the acquisition time and the sensitivity to thermal and mechanical stability are reduced.

12.4.1 Optical Setup in DHM

Usually, two different experimental optical configurations are adopted in DHM: Michelson interferometer (or Twyman–Green) and Mach–Zehnder (MZ).

The MZ configuration essentially has the advantage of being more flexible because the aperture of the beams can be set up separately by using two separate beam expanders and collimators for the reference and the object beams, respectively. That advantage is obtained at the expense of a more complex setup.

In Figure 12.16 is shown a typical experimental recording holographic setup, based on the MZ configuration. Preferably, a laser source emitting a TEM00 Gaussian beam and a linearly polarized beam is used. The laser beam is divided by the polarizing beam splitter in the two beams: the reference and the object beam. The polarizing optical components allow adjustment of the intensity ratio between the object and the reference beams and setting of the same polarization to obtain the maximum contrast to the interference fringes recorded at the CCD array.

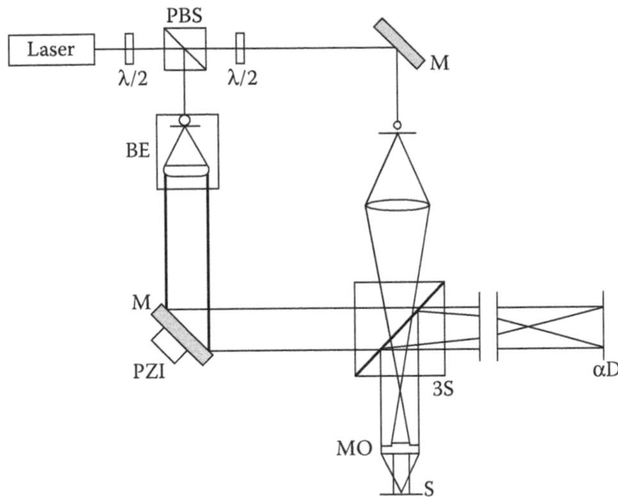

FIGURE 12.16 The experimental setup for recording digital holograms; BS — beam splitter; M — mirror; MO — microscope objective; BE — beam.

The reference beam is expanded and collimated to obtain a plane wavefront. The mirror along the path of the reference beam can be tilted to adjust the angle of incidence. In this manner, it is possible to switch the DH setup from an in-line to a variable-angle off-axis configuration.

In the same way, the object beam is expanded and an additional lens is used to obtain a plane wavefront beam at the exit aperture of the MO. The MO is, in general, interchangeable, as occurs in the standard optical microscope and is chosen as a function of the magnification required. An interesting alternative to the classical MO multilens tube is an aspheric MO lens. Typical aspheric lenses available have focal lengths and NAs of f = 4.51 mm and NA = 0.55; f = 8.00 mm and NA = 0.50; f = 15.36 mm, NA = 0.16, respectively. The lenses are MOs equivalent, respectively, to different magnification powers (i.e., 35X, 20X, and 10X). The advantage in using aspheric MO lenses is the possibility of having a more compact optical setup, because they have a reduced longitudinal size with respect to the classical MO tube. One more optical configuration for DH in microscope configuration makes use of the long-distance MO [49], even if in this case obtainable magnifications are lower than those achievable by a standard MO.

Typical detectors for visible laser sources are CCD detectors, or CMOS. A large number of arrays are available having sensitive elements ranging in number from 512 × 512 to 4096 × 4096. The single sensitive elements (pixel) have sizes ranging from 3 to 10 μm. It is preferable to have square pixels that give the same reconstruction pixel along the direction of the reconstructed image plane if equal numbers of rows and columns are used in the reconstruction process. Sometimes, in the reference beam, a mirror is mounted on a PZT actuator to apply the phase-shifting method in order to avoid the problem of twin image and eliminate the zeroth diffraction order [14] in the case of in-line recoding configuration. Otherwise, if the off-axis setup is adopted, the phase shifter element is not required as the zeroth diffraction order can be eliminated by a high-pass-filter operation applied on the recorded hologram.

It has been reported that DH can be a very useful tool for noncontact inspection and characterization of microelectromechanical systems (MEMS) and microoptoelectromechanical systems (MOEMS) [39,49,50–55].

12.4.2 Compensation of Aberrations in DH

Complete compensation of aberrations is of fundamental importance when quantitative phase determination is necessary for microscopic metrological applications, as is required, for example, in MEMS inspection and characterization.

In fact, the MO introduces a strong curvature on the object wavefront. The high curvature of the object beam envelops the quantitative phase information hindering extraction of the correct phase map related to the object under investigation.

Wavefront distortion is a common problem in interferometric microscopy. It is solved by introducing the same curvature, using the necessary optical components, in both the interfering wavefronts. For example, the interferometer with Mirau configuration allows to compensate for the curvature.

Different methods can be applied to remove the disturbing phase factor introduced by the MO. One method consists of finding a correction phase mask calculated by the knowledge of the optical parameters in the experimental setup. The mask is used in the reconstruction image plane to correct the phase-contrast image [56]. Another method allows to calculate a phase mask directly in the plane of the hologram (CCD array). Then the mask is propagated and reconstructed at the image plane where the correction has to be performed [44].

As we will show, it is interesting to note that the general problem of phase correction in DH is, for some aspects, in strict analogy with the problem of removing the carrier employed in fringe analysis, when the fast Fourier transform (FFT) methods are used [57]. Moreover, similarity between one method and double-exposure holographic interferometry can be found. In fact, it is well known that in the case of holographic interferometry, aberrations are absent because they disappear in the subtraction operation performed to get holographic interferograms from two single holograms.

In this section, approaches that can be adopted to remove the defocusing phase curvature in the reconstructed phase are described through experimental examples.

To illustrate the compensating process, the original inherent phase of the object beam, consisting of a parabolic phase factor superimposed on the characteristic phase distribution $\Delta\phi(x, y)$ of the object wavefront, has to be considered. The parabolic phase factor accounts for the wavefront curvature introduced by the imaging lens, the MO. Because of the high focal power of the MO, the recorded hologram at the CCD plane array appears basically as an ensemble of circular fringes, slightly disturbed by the characteristic object phase distribution. In Figure 12.17a, a typical digital hologram obtained by DHM is shown. The object is a silicon waveguide, 10 μm wide and 6 μm high with a rib defined by an etch 0.85 μm deep. From this hologram, circular fringes are clearly visible.

Whereas amplitude numerical reconstruction at distance **d** provides a good contrast image (Figure 12.17b), the presence of the high curvature in the reconstructed phase wavefront hinders the phase information content of the reconstructed phase (Figure 12.17c).

12.4.3 Removing Aberrations by Determining a Phase Mask in the Image Reconstruction Plane

The phase can be corrected by means of a phase factor introduced into the holographic numerical reconstruction process and determined on the basis of the geometrical parameters of the optical imaging setup at the reconstruction plane.

As demonstrated in Reference 56, the knowledge of several parameters of the optical set up is necessary in this approach. The parameters are the focal lengths of the optical components used to image the object, the lens–object and lens–CCD distances, and the reconstruction distance **d**. Analytical expressions of the equations for different experimental configurations have been derived for different typical optical configurations, both in-line and off-axis.

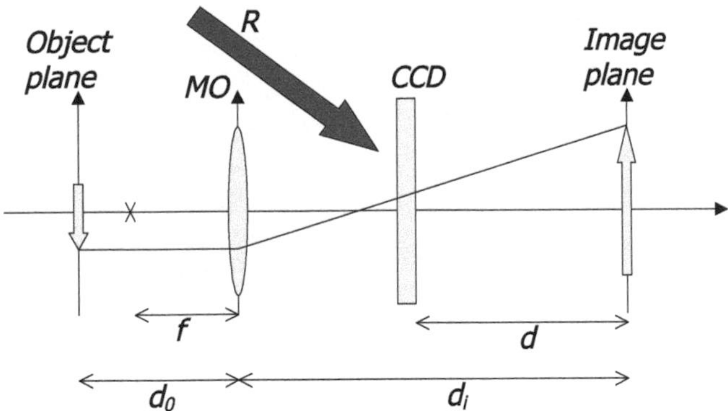

FIGURE 12.17 Configuration for holographic microscopy.

Referring to Figure 12.17, it is easy to get the following relations for the imaging system:

$$\frac{1}{D} = d_i\left(1 + \frac{d_0}{d_i}\right); \qquad \frac{1}{f} = \frac{1}{d_i} + \frac{1}{d_0} \tag{12.72}$$

where d_0 is the object–lens separation distance, d_i is the distance between the MO lens and the image reconstructed plane, and f is the MO focal length.

From Equation 12.72, it is clear that the phase parabolic curvature on the reconstructed wavefront at the image plane is given by

$$\Phi\left(x, y\right) = \frac{2\pi}{\lambda}\left(\frac{x^2 + y^2}{2D}\right) \tag{12.73}$$

where, by combining the two relations in Equation 12.72, $D = d_i - f$ is obtained.

The method essentially consists of multiplying in the image reconstruction plane, the numerically reconstructed complex wave field $b(x', y'; d')$ by a phase mask with a sign opposite to that of the phase curvature $\phi(x, y)$ of Equation 12.73.

Moreover, the phase mask can be modified to take into account the phase introduced with wave vector components k_x and k_y of the reference wavefront in off-axis configuration given by

$$\phi_R\left(x, y\right) = \frac{2\pi}{\lambda}\left(k_x x + k_y y\right) \tag{12.74}$$

Eventually, the phase mask can contain a correcting phase factor to also compensate for other aberrations such as, for example, spherical aberrations [46]. In this manner, the phase-contrast image of the object will be free of the parabolic phase factor (defocusing) as well as other aberration terms.

Obviously, the main difficulty in the preceding approach is to measure the concerned distances with high accuracy. Consequently, the employed compensation phase factor will remove the undesired spherical phase factor only partially. In fact, as has been reported, fine digital adjusting is necessary to completely remove the curvature before using the phase-contrast imaging as quantitative information [56].

An alternative way to proceed consists of calculating or estimating the unwanted wavefront curvature in the hologram plane (the plane of the CCD array) instead of that in the reconstruction plane [44]. Once the parabolic phase correction factor has been calculated in the hologram plane,

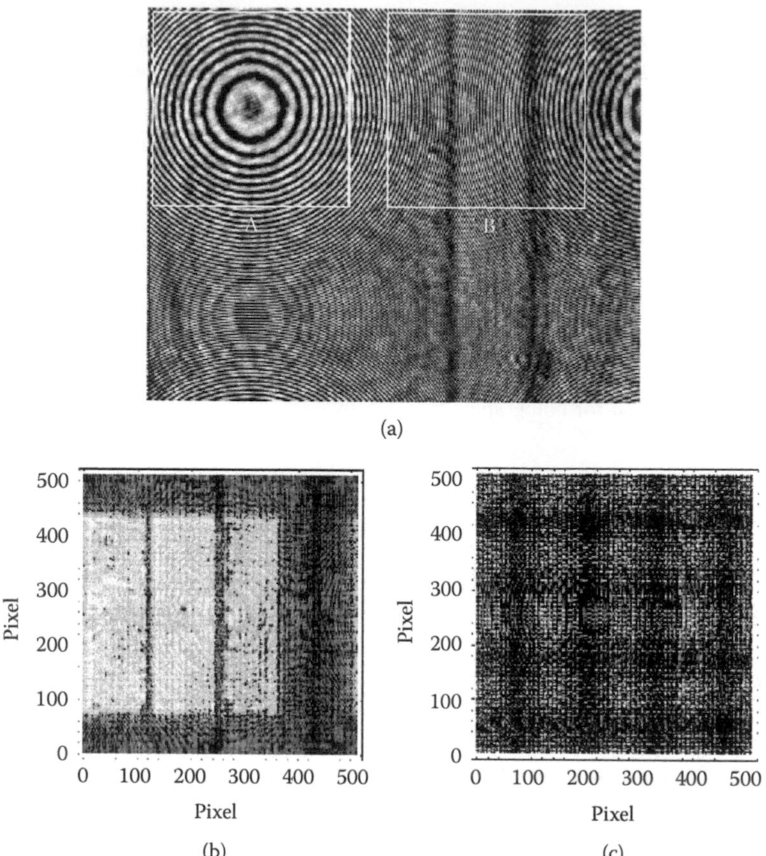

FIGURE 12.18 (a) Digital hologram of a silicon optical waveguide on a silicon substrate. A — portion of the hologram used to correct the disturbing curvature in the reconstructed phase, B — portion of the hologram of the object under investigation: the waveguide; (b) amplitude-contrast image of the waveguide; (c) phase-contrast image of the waveguide in the presence of disturbing curvature. (From Ref. 44.)

compensation of the curvature at all reconstruction planes can be easily obtained, because the correction factor is numerically reconstructed in exactly the same way as is made for the hologram. This last approach consists of essentially three steps: (1) from the four phase-shifted holograms, the phase correction wavefront is calculated using one of the three methods described below; (2) the numerical holographic phase reconstruction at distance d is performed on both the hologram and correction wavefronts; (3) the two reconstructed phase fields are subtracted at the reconstruction plane at distance d to obtain the contrast phase image of the object with compensation of the inherent curvature.

Three methods can be used to accomplish the first step. One way is to use the same hologram to recover the curvature correction factor. In fact, if there is a portion of the hologram area, as shown in Figure 12.18a, in which the circular fringes are not affected by the phase content of the object under investigation; then the phase correction factor can be calculated in that area and then extrapolated to the entire hologram. This approach is very similar to a method applied to FFT for phase demodulation of a single-image fringe pattern [57], eliminating the fringe carrier. A second method can be applied if the object is made of a plane mirror-like surface. For example, in the specific case of MEMS, microstructures realized on a silicon surface can be regarded as a good reflecting surface. In this case, the correction wavefront can be evaluated by a sort of holographic double-exposure procedure. A first hologram of an area with the microstructures is recorded.

Then, simply translating the sample transversely, with respect to the optical axis, a second hologram of a flat area is recorded. The phase obtained by this last hologram is itself the curvature correction wavefront. A third approach is to generate a synthetic phase distribution similar to those causing circular fringes at the CCD plane, by observing the fringe pattern. Fine but tedious digital adjusting can be performed to obtain the correct phase factor at the hologram plane. Independent of the method adopted for evaluating the correction phase wavefront, the aforementioned steps (2) and (3) can be implemented to remove the unwanted curvature at any reconstruction plane.

In the following subsections the three methods for compensating the inherent curvature introduced by the imaging MO lens will be illustrated in depth. Results obtainable in different cases of application for MEMS structures are discussed and compared. All holograms shown in the following section have been recorded with the same experimental apparatus, depicted in Figure 12.16.

12.4.3.1 First Method: Removing the Circular Fringe Carrier from a Recording Hologram with Phase-Shifting Method

The hologram of a silicon waveguide is shown in Figure 12.18a. The hologram has 1280 × 1024 pixels. In Figure 12.18a are also indicated two areas from the same hologram of size 512 × 512 pixels. The area A, in the upper left corner, has only circular fringes. In fact, the surface of the silicon substrate acts as a plane mirror surface and the fringe pattern is only because of the curvature introduced by the imaging MO. The area B, outlined in the upper right side, contains information about the object under investigation superimposed on the circular fringes because of the defocus due to the imaging lens on the object wavefront. Four different phase-shifted holograms with intensities I_1, I_2, I_3, and I_4 were recorded with phase shifts of 0, π/2, π, and 3π/2, respectively. The holograms were combined together to extract the complex amplitude of the object wave at the hologram plane:

$$U_{CCD}(x, y) = I_1 - I_3 - i(I_2 - I_4) \tag{12.75}$$

The phase distribution $\phi(x, y)$ at the hologram plane is then obtained:

$$\phi(x, y) = Arc\tan\left(\frac{I_1 - I_3}{I_2 - I_4}\right) \tag{12.76}$$

In Figure 12.19a and Figure 12.19b are shown the phase modulus 2π corresponding to the two areas A and B of the hologram, respectively. Assuming that the phase shown in Figure 12.19a is only because of the wavefront curvature introduced by imaging MO, its phase can be recovered by performing a nonlinear fit of the unwrapped phase relative to portion A (shown in Figure 12.20a). The correction wave front phase $\Phi_{corr}(x, y)$ can be obtained by the simple expression

$$\phi_{corr}(x, y) = \frac{2\pi}{\lambda}\left(\frac{x^2 + y^2}{2R}\right) \tag{12.77}$$

where R is the curvature radius of the correction wavefront phase, i.e., its defocus radius extrapolated to region B of the fitted sphere calculated in the region A (Figure 12.20b).

The complex field of the object can be reconstructed at distance **d**. Also, the correction wavefront phase $\phi_{corr}(x, y)$ is reconstructed at distance **d** to obtain the correction phase $\phi_{corr}(\nu, \mu)$ at the reconstructed image plane. The phase-contrast image of the object, compensated for the unwanted curvature, can be simply obtained by subtracting the phases of the two reconstructed fields

$$\phi_o(\nu, \mu) = Arg\left(\frac{O(\nu, \mu)}{\phi_{corr}(\nu, \mu)}\right) \tag{12.78}$$

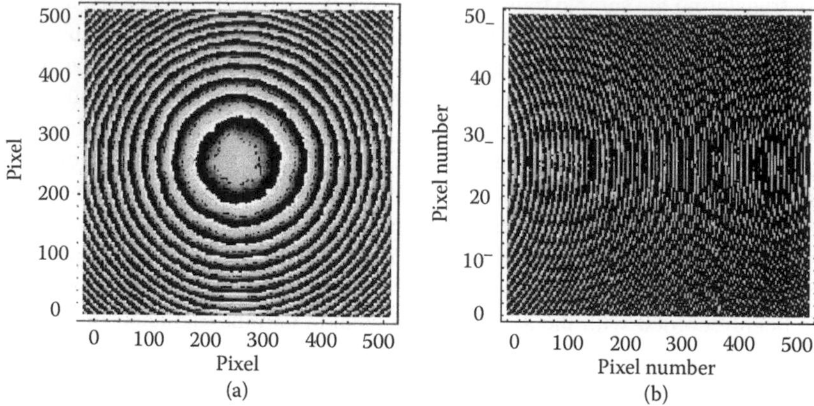

FIGURE 12.19 (a) Wrapped phase modulus 2π of the portion A of the hologram of Figure 12.18; (b) phase contrast wrapped modulus 2π of the portion B of the hologram of Figure 12.17a. [From Ref. 44.]

The corrected reconstructed phase is shown in Figure 12.21. However, it should be noted that curvature has not been completely removed, which could be attributed to additional aberrations present in the MO. In fact, the correction phase factor was calculated by the fitting procedure of a simply parabolic phase term. Anyway, if additional aberrations are present besides the defocus one, the correction method fails because the correction phase wave front has been calculated in the subaperture A of the imaging system, whereas the compensation by means the fitting operation was applied in the subaperture B.

12.4.3.2 Second Method: Formerly a Double-Exposure Technique

A second method is a kind of double-exposure digital holographic interferometry. The first exposure is made of the object under investigation, whereas the second one is made of a flat reference surface in proximity to the object. This is a typical case encountered, for example, when MEMS structures are inspected, because the micromachined parts have been realized on a flat silicon substrate. In that

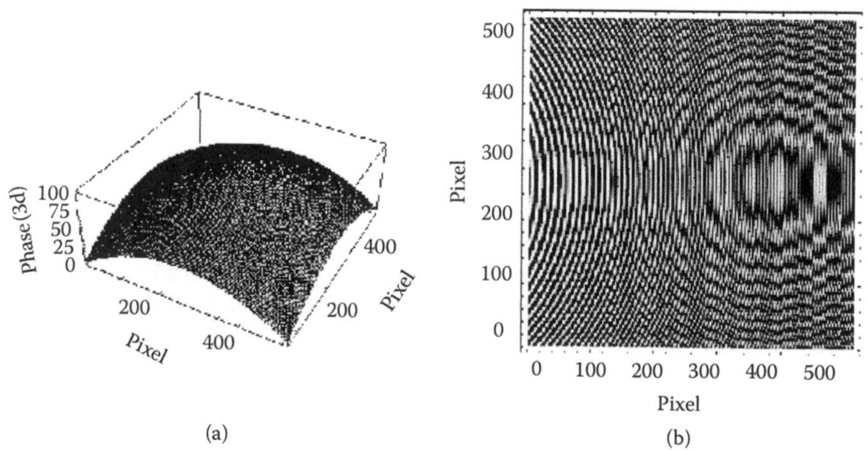

FIGURE 12.20 (a) Unwrapped phase of the portion A of the hologram of Figure 12.18a; (b) phase correction wavefront calculated by extrapolating to the region B, the fitted sphere of region A. (From Ref. 44.)

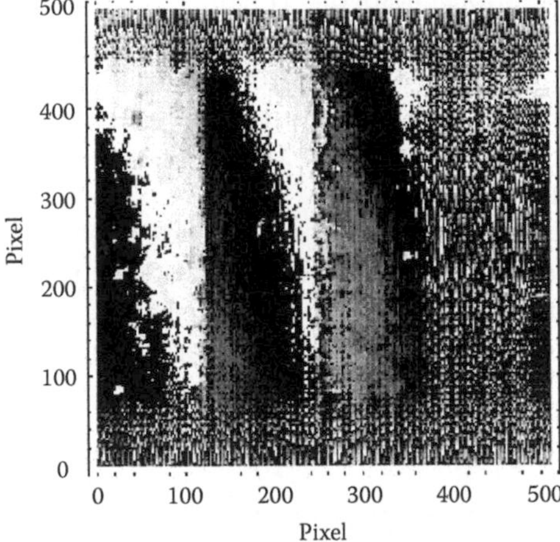

FIGURE 12.21 Phase-contrast image of the silicon waveguide where the curvature has been corrected by subtracting the compensating wavefront as calculated from the portion A of the hologram of Figure 12.18. It can be noted that the curvature has not been completely removed. (From Ref. 44.)

case, the area around the micromachined structure offers a very good plane surface that can be used as reference. A recorded hologram of two polysilicon cantilevers realized on a silicon substrate and obtained utilizing a silicon oxide sacrificial layer is shown in Figure 12.22a. The two micromachined beams are 20 μm wide, and 130 μm and 60 μm long, respectively. If the wafer is slightly

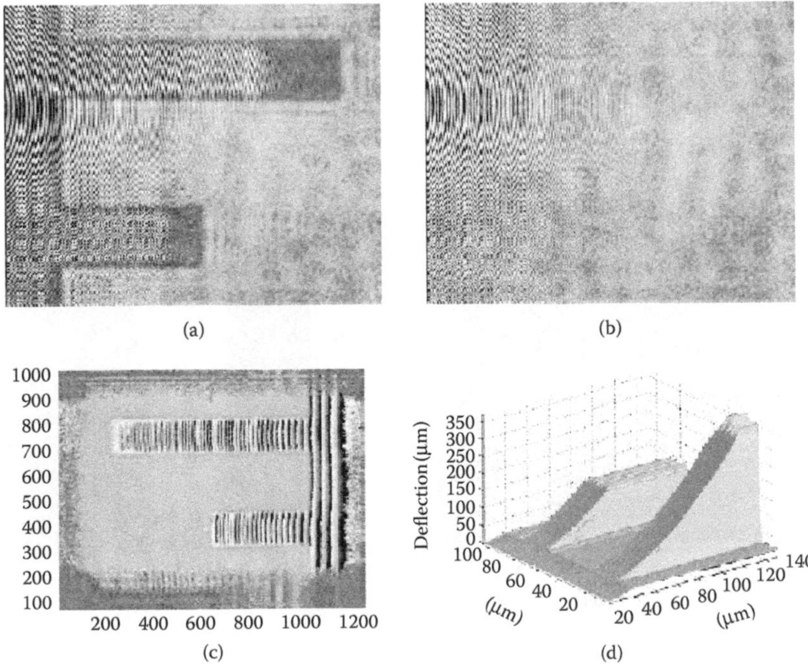

FIGURE 12.22 (a) A hologram of the polysilicon microcantilever beams; (b) a hologram recorded on a reference surface in the proximity of the micromachined beams; (c) a phase map image, wrapped mod 2π, of the polysilicon beams reconstructed at distance d = 100 mm; (d) an unwrapped phase with out-of-plane deformation and dimensions expressed in microns. (From Ref. 44.)

translated transversally, an additional hologram (Figure 12.22b), relating to the surface's substrate, can be recorded and used as reference.

Both holograms are numerically reconstructed at a distance d = 100 mm, where it is possible to calculate the phase difference to obtain the correct phase map of the object cleared of the wavefront curvature introduced by the imaging MO. In Figure 12.22c and Figure 12.22d, wrapped and unwrapped phases are shown, respectively.

This method is easy to apply but at the expense of a double image recording. Furthermore, the phase map could be affected by some local errors as a function of local imperfections on the reference surface. Moreover, the translation operation can also introduce systematic errors if, for example, some amount of tilt is introduced during the translation.

12.4.3.3 Third Method: Digital Adjustment by Successive Approximations

A third method can be applied to obtain phase-contrast imaging with compensation of the inherent wavefront phase curvature; it is called digital adjusting in published work [56]. Nevertheless, it differs from the other methods, because to eliminate the undesired wavefront curvature, we first find the correcting wavefront in the hologram plane (thus, not directly at the reconstructed plane); next, the phases of the numerically reconstructed wavefronts, relating to the recorded hologram and correcting wavefront, are evaluated at distance d where the two phases were subtracted from each other to obtain the correct phase map compensated for the unwanted curvature.

The correcting wavefront at the hologram plane can be found in a limited number of attempts, by estimating simply and approximately, with the naked eye, the center and number of circular fringes in the recorded hologram. An example of the result is shown in Figure 12.23.

12.4.3.4 Comparison among the Methods and Discussion of Results

From the description of the three methods, it is obvious that each method has some advantages and drawbacks, and selection of the appropriate method depends on the specific application. The first method requires that a portion of the recorded hologram be free of the required object phase

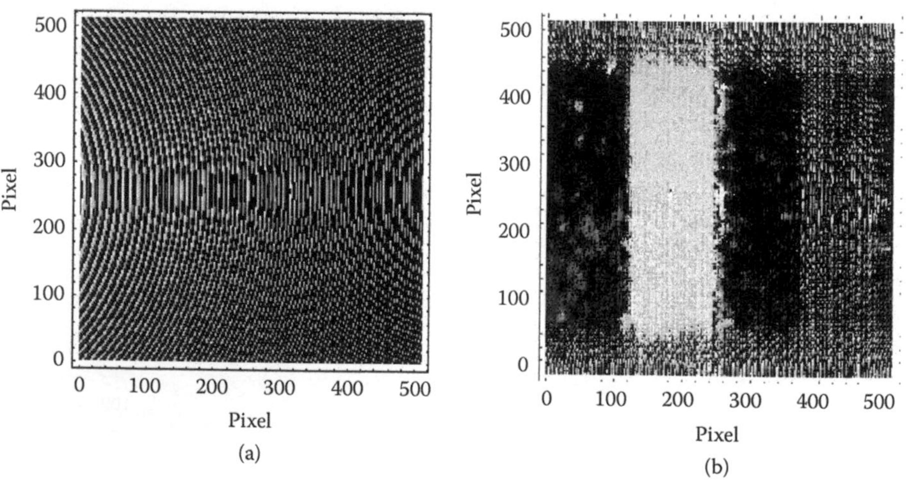

FIGURE 12.23 (a) Wrapped phase modulus 2π of the correction wavefront calculated at the CCD plane by manual digital adjusting procedure, (b) phase-contrast image of the silicon waveguide in which the inherent curvature has been completely compensated. (From Ref. 44.)

information, implying that some resolution and field of view in the useful aperture is sacrificed and lost, because one portion of the area serves as a reference surface. Moreover, inside the useful aperture of the recorded hologram, it is necessary that one portion of the object surface be flat and behave as a fairly good mirror. In addition, from the reported results, it is also clear that some aberrations introduced by the lens can affect the final results. But, the great advantage of the method lies in the fact it requires a single hologram recording. The second method, using a kind of double-exposure hologram, is unaffected by the presence of additional aberration, giving remarkably good results. Besides the need of double hologram recording, it requires a translation operation between the two recording operations. Furthermore, it is essential to have a mirror-like surface close to the object under investigation. It has been proved to be very valuable for inspection of silicon MEMS. The third method, based on manual digital adjusting, can give good results, at the price of a cumbersome procedure, in such cases where the preceding two methods cannot be applied. The interesting aspect of this approach is that the correction of the curvature is made not only to a single reconstruction plane but also directly, and just once, at the hologram plane. In this way, curvature correction can be obtained for all reconstruction distances by one of the three described methods.

12.4.4 Focus Tracking in DHM

In DHM, the image plane for best focus can determined by the *a priori* knowledge of different parameters, such as the focal length of the MO, distance between the object and the MO, and distance between the plane of the hologram and the MO. In a real-world situation, however, it may be difficult to know or measure those parameters. Because DH has the advantage of numerical reconstruction, the focus can be found by performing numerical reconstructions at different distances, analogous to mechanical translation of the MO in conventional microscopy. This numerical method can be time consuming and impractical when a long sequence of holograms has been recorded with the image plane varying unpredictably. The tedious and cumbersome search for new focal planes becomes intolerable, especially if there is the need to visualize the phenomena in anything approximating real time. Because in DH microscopy an imaging short focal lens is required, when the sample experiences even very small displacements along the optical axis, a very large variation occurs for the distance to the imaging plane, and the focus can be missing. Displacement of the object may occur for different reasons. One unavoidable situation is encountered in the case of quasi-real-time thermal characterization of silicon MEMS structures, when the temperature is changed. In this case, the distance used in the numerical reconstruction process has to be changed to get an in-focus amplitude or phase-contrast image. In fact, temperature changes cause thermal expansion of the test object and/or its mechanical holder that may not be predictable. However small the longitudinal displacement, even a few microns cause large changes of the actual reconstruction distance as illustrated in Figure 12.24. In fact, if the distance from the lens to the

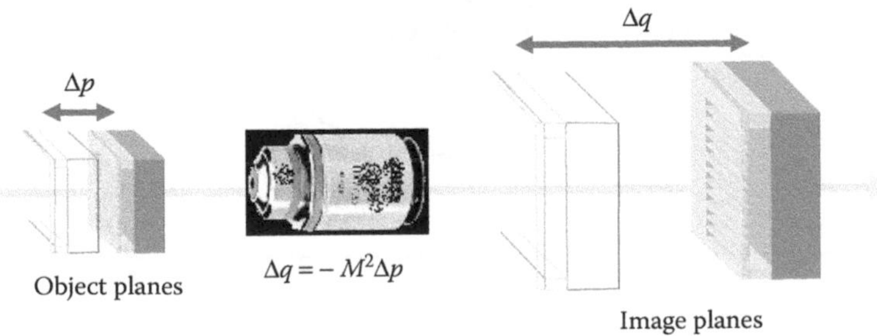

FIGURE 12.24 Microscopic axial displacements of the sample cause large variations in the reconstruction distance.

object is p, q is the distance of the image plane from the lens, and f is the focal length, then any displacement Δp of the sample results in a longitudinal shift of the imaging plane in front of the CCD, given by

$$\Delta q = -M^2 \Delta p \tag{12.79}$$

where M = q/p is the magnification. From Equation 12.79, for example, it can be shown that a displacement of 10 μm translates the image plane by Δp = 16 mm if M = 40. An approach to detecting, in real time, the axial displacement of the sample relies on measuring the phase shift of the hologram fringes. From the phone shift signal, it is possible to find the incremental change of the distance to be used in the numerical reconstruction.

Of course, the preceding procedure works correctly only if the optical beam illuminating the object is collimated. In fact, any axial displacement of the object causes a pure phase shift in the fringe pattern of the hologram only in that configuration.

That phase-shift signal can be recorded in a small flat portion of the object and, consequently, it is possible to determine the displacement Δp:

$$\Delta p = \frac{\Delta\phi(t)}{4\pi}\lambda \tag{12.80}$$

where $\Delta\phi(t)$ is the phase shift detected at time t.

Here, an example is described for testing silicon MEMS structures having out-of-plane deformation because of the residual stress induced by the microfabrication process: a cantilever (50 μm × 50 μm) and bridges 10 μm wide. The silicon wafer was mounted on a metal plate and was held by a vacuum chuck system. The metal plate was mounted on a translation stage close to the MO while the sample was heated in the range 23 to 120°C.

A first hologram was recorded before raising the temperature. The numerical reconstruction for a well-focused image was found at an initial distance of 100 mm with an estimated magnification of M = 40. The phase shift of the fringes was detected in real time by measuring the average intensity change in a group of 4 × 4. The inset of Figure 12.25 shows the recorded intensity signal while the sample was heated. The signal had 3149 points sampled at a rate of 12.5 points/sec. For every

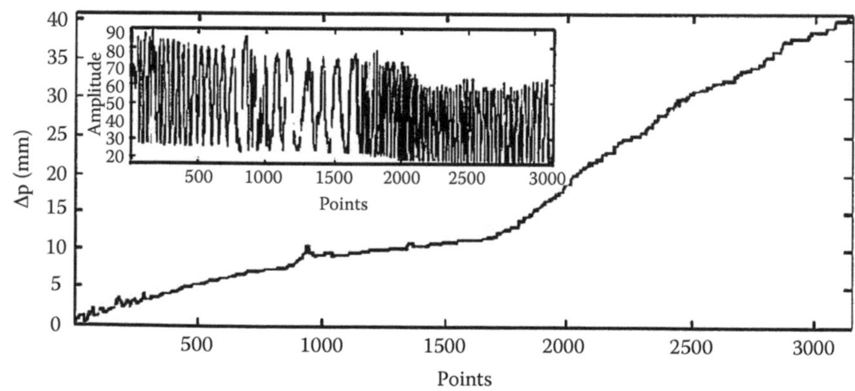

FIGURE 12.25 Displacement of the sample measured in real time by analyzing the phase shift (inset) of hologram fringes. (From Ref. 62.)

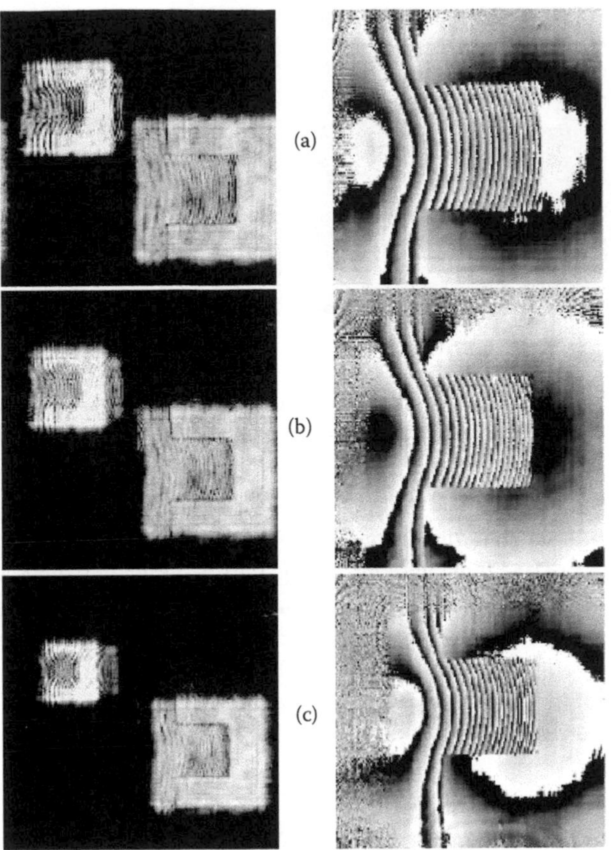

FIGURE 12.26 In-focus amplitude and phase contrast for the cantilever beam from three holograms [(a) 1, (b) 196, (c) 314] of the recorded sequence obtained by applying the focus-tracking procedure. (From Ref. 62.)

ten points, a hologram was digitized. Axial displacements were due to the overall thermal expansion of the metallic plate and the translation stage.

The signal of the inset in Figure 12.25 was analyzed by applying an FFT. For each added point recorded, the wrapped phase was detected and the unwrapped phase was calculated. The displacement was calculated from Equation 12.79, and Figure 12.25 shows the calculated displacement Δp as a function of the sampled point. The numerical reconstruction distance for each hologram was continuously updated. As shown in Figure 12.25, the last digital hologram reconstruction distance differs from the first one by ~40 mm. Figure 12.26a to Figure 12.26c show the amplitude and phase-contrast reconstructions for the cantilever beam from three different holograms of the recorded sequence of 314 holograms: n.1, n.196, and n.314, corresponding to three different temperatures. The reconstructions were performed automatically by applying the focus-tracking procedure. Figure 12.26a shows the reconstruction of the hologram n.1 at d = 100 mm; Figure 12.26b shows that of the hologram n.196 at d = 117.3 mm, and Figure 12.26c shows that of the hologram n.314 at d = 140.8 mm. In the phase-map image, the wrapped phase observed on the cantilever indicates it had an intrinsic out-of-plane deformation. As expected, the reconstructions in Figure 12.26 are all in focus. In Figure 12.26b and Figure 12.26c, the size of the reconstructed object is smaller because of the larger reconstruction distance. Furthermore, all 314 holograms in the sequence were reconstructed well in focus, which demonstrates the soundness of the technique. Figure 12.27a shows the amplitude of the last recorded hologram n.314 from another sequence at d = 138 mm, which resulted from the focus-tracking procedure applied to a different MEMS structure with a

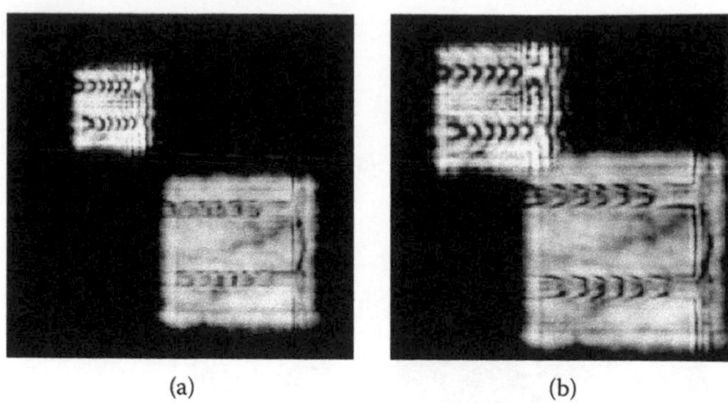

(a) (b)

FIGURE 12.27 In-focus amplitude reconstruction of a bridge: (a) at d = 138 mm, by the focus-tracking procedure; (b) at d = 100 mm without taking into account the focus change. (From Ref. 62.)

bridge shape and with smaller dimensions. In Figure 12.27b, the reconstruction of the same hologram n.314 at d = −100 mm, without taking into account the focus change, is shown for comparison. The effect of the focus tracking in the amplitude reconstruction is apparent. Figure 12.28a to Figure 12.28d show the phase of the lower MEMS of Figure 12.27, wrapped and unwrapped, respectively. Figure 12.29 shows that the consequence of defocus also influences the reconstructed phase map, thus affecting the quantitative information. It is clearly visible that the edges of the bridge are blurred when the focus-tracking method is not applied.

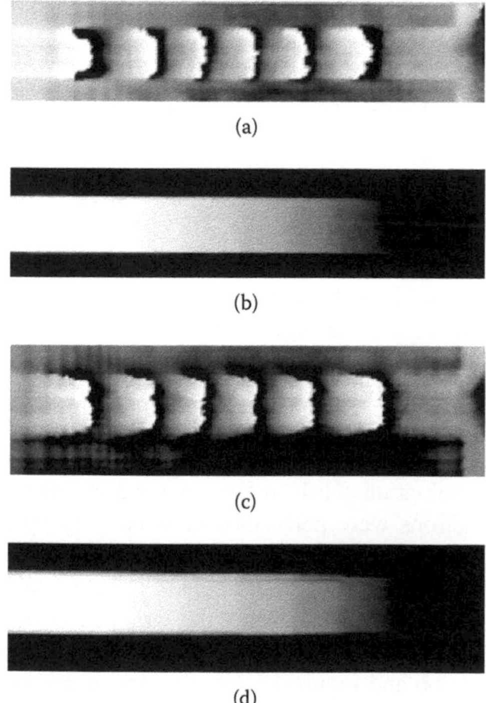

(a)

(b)

(c)

(d)

FIGURE 12.28 Wrapped and unwrapped phase-contrast reconstruction, respectively, of the lower MEMS bridge from Figure 12.27: (a) applying the focus-tracking procedure, (b) applying the focus-tracking procedure, (c) without focus tracking, (d) without focus tracking. (From Ref. 62.)

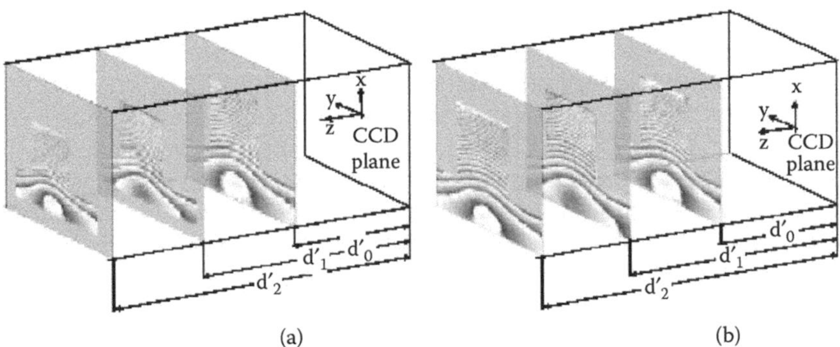

FIGURE 12.29 Wrapped image phases of a cantilever MEMS from three holograms of a sequence of 314 holograms reconstructed at different distances: (a) without application of padding operation, (b) with application of padding operation. (From Ref. 62.)

The method can still be applied as a quasi-real-time procedure. In principle, if three phase shifts are measured at three different points in the digital hologram, it would be also possible to track tilts, allowing focus tracking in tilted image planes.

12.4.5 Controlling Size Independently of Distance and Wavelength

As discussed earlier, with FTM reconstruction, the reconstruction pixel (RP) increases with the reconstruction distance so that the size of the image, in terms of number of pixels, is reduced for longer distances. In the convolution method, the RP does not change but remains equal to the pixel size of the recording array. The convolution method is more appropriate for reconstruction at small distances, whereas the FTM is useful for longer distances according to the paraxial approximation necessary to apply it [10,11].

As described in the previous paragraph, when DHM is applied to inspect silicon MEMS when subjected to thermal load, unwanted longitudinal motion of the structure can occur and, consequently, the focus can be lost. Although well-focused images can be obtained for each hologram, it is not possible to compare two of them directly as they have different sizes, owing to the different widths of the RP. For this reason, direct subtraction of unwrapped phase maps from two holograms at two different distances cannot be obtained easily. Consequently, quantitative information on the deformations caused by the thermal load cannot be obtained.

Similar difficulties arise in multiwavelength-DH (MWDH) used for color display and for applications in metrology. In MWDH, for each wavelength, the size of the RP increases with the reconstruction wavelength, for a fixed reconstruction distance. Therefore, digital holograms recorded with different wavelengths produce images with different sizes when numerically reconstructed by FTM. In fact, MWDH requires simultaneous reconstruction of images recorded with different wavelengths (colors), and the resulting reconstructed images must be perfectly superimposed to get a correct color display [58]. This is prevented by the differing image sizes, and this also prevents the phase comparison required for holographic interferometry [59]. Otherwise, as proposed in the literature, it is necessary to use a resizing operation of the reconstructed images at the end of the reconstruction process or a scaling operation on the hologram.

A method for controlling the image size of the reconstructed images generated by the FTM so that two images of the same object recorded at two different distances can be directly compared has been discovered [60]. Moreover, the same method can be applied in MWDH. The method is intrinsically embedded in the holographic reconstruction process without the need for image scaling at the end of the process.

The size can be controlled through the synthetic enlargement of the aperture of the CCD array, or in other words, by increasing the number of the pixels of the recorded digital holograms. Reconstruction is obtained through the pixel size ($\Delta\xi$, $\Delta\eta$) of the CCD array, which is different from that (Δx,Δy) in the image plane and related by $\Delta x = d\lambda / N\Delta\xi$ and $\Delta y = d\lambda / N\Delta\eta$, where $N \times N$ is the number of pixels in the CCD array and d is the reconstruction distance, namely, the distance measured backward from the hologram plane ($\xi - \eta$) to the image plane (x,y). It is clear that by FTM, the reconstructed image is enlarged or contracted according to the reconstruction distance d and that the size of the RP depends on the lateral number of the pixels N. Image size can be controlled by changing the RP, using a larger number of pixels in the reconstruction process. In fact, N can be enlarged by padding the matrix of the hologram with zeros in both the horizontal and vertical directions such that

$$N_2 = N_1 \left(d_2 / d_1 \right) \tag{12.81}$$

getting $\Delta x_1 = \Delta x_2 = (d_1\lambda)/(N_1\Delta\xi) = (d_2\lambda)/(N_2\Delta\xi)$, where N_1 is the number of pixels of the hologram recorded at the nearer distance, d_1. In a similar way in MWDH, if one hologram has been recorded with wavelength λ_1 and a second with λ_2, where $\lambda_2 > \lambda_1$, at the same distance, then the number of pixels of that hologram may be changed such that

$$N_2 = N_1 \left(\lambda_2 / \lambda_1 \right) \tag{12.82}$$

in order to obtain the same width for the RP:

$$\Delta x_1 = \Delta x_2 = (d\lambda_1)/(N_1\Delta\xi) = (d\lambda_2)/(N_2\Delta\xi)$$

Demonstration of the concept has been reported in [60], and the same approach has been found useful in recovering the required resolution for correct determination of the shape of highly curved MEMS [61]. Here we will discuss only the first case, namely, the control of the image size independent of distance adopted in the reconstruction process. By using the same 314-hologram sequence, digital holograms of a silicon MEMS cantilever structure were recorded while thermal load was applied, as described previously, in which the sample experienced an unwanted longitudinal displacement during the test. In fact, it was shown that through a real-time focus-tracking procedure, in-focus amplitude and phase images of the cantilever were obtained; however, the width of the RP Δx_1 was dependent on the distance. For that reason, a direct subtraction was not possible between the phase images of the sequence to evaluate deformations between two phase maps at different temperatures. Figure 12.29a shows three reconstructed phase maps of the MEMS at distances of $d_0' = 100$ mm, $d_1' = 120$ mm, and $d_2' = 140$ mm, respectively.

It is clear that the size of the cantilever, in terms of pixels, is reduced as expected, because the three holograms were recorded and reconstructed at different distances. The same reconstructions after the padding procedure with Equation 12.81 was applied to the holograms are shown in Figure 12.29b. The holograms reconstructed at distance d_0' had 512 × 512 pixels; the holograms at d_1' and d_2' were padded with zeros up to 614 × 614 pixels and 718 × 718 pixels, respectively, and the reconstructed phase images maintained the same size independently of the distance.

Direct phase subtraction of two reconstructed phase maps recorded at two distances is possible by adoption of this method, as demonstrated in Figure 12.30a to Figure 12.30d, where are shown (a) the unwrapped phase map of the hologram recorded at distance $d_1 = 115.5$ mm, (b) the unwrapped phase image of the MEMS at $d_2 = 139.5$ mm without zero padding, (c) the unwrapped phase of the MEMS at $d_2 = 139.5$ mm obtained by the padding operation, (d) the difference between the

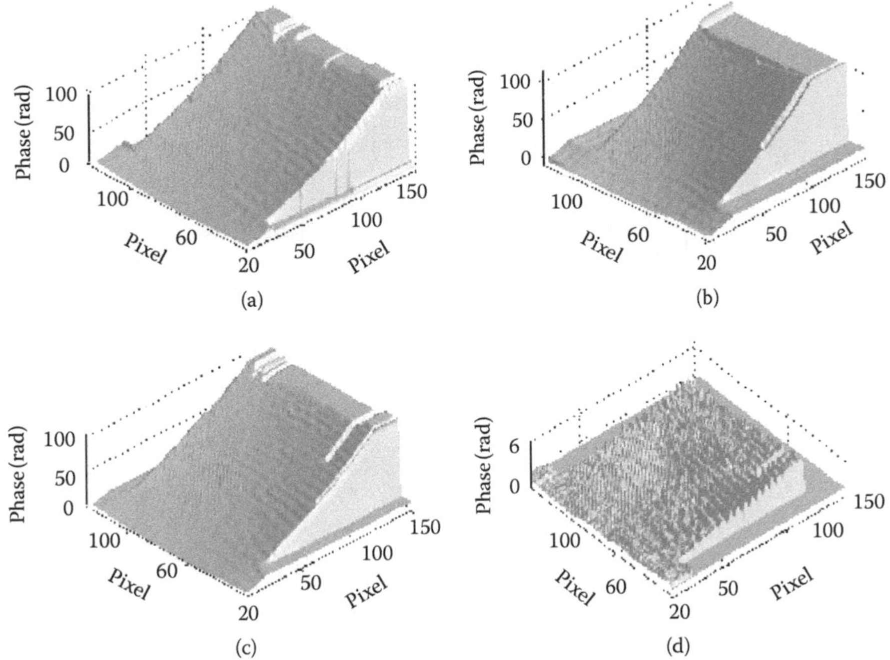

FIGURE 12.30 Unwrapped phase of MEMS at distance: (a) $d_1 = 115.5$mm, (b) $d_2 = 139.5$ mm, (c) $d_2 = 139.5$ mm with padding operation, (d) phase map subtraction between (a) and (c). (From Ref. 60.)

unwrapped phase maps with equal size, indicating the small deformation caused by thermal load. Thus, phase subtraction between the phase maps at different distances can be performed.

It is important to note that all the digital holograms were recorded by an off-axis setup and that the reconstructions were performed using a reference beam in the numerical reconstruction to get the virtual image centered in the reconstructed image plane by adopting the method discussed in a previous paper [62]. In all figures, the same number of pixels were extracted from reconstructed images.

12.5 THE APPLICATION OF DH TO THE INVESTIGATION OF MICROCOMPONENTS

Here, we describe the methods for the investigation of objects having surfaces that are optically rough and flat. Optically rough means that the statistical fluctuations in the height of the surface profile are higher than half of the wavelength of the applied light. Such surfaces are also called *technical surfaces*. Consequently, optically flat surfaces show height fluctuations smaller than half the wavelength. To this category belong, for instance, components such as microlenses made of glass or plastics and components with polished silicon surfaces.

12.5.1 Experimental Prerequisites for the Investigation of Microcomponents

During the past 30 years, the domain of optical metrology was the investigation of macroscopic or large-scale components. A simple, direct transfer of the experience acquired in this field to the inspection of microcomponents is difficult [8,54]. Before the description of the inspection methods, three problem classes are addressed briefly: *loading*, *preparation*, and *observation*.

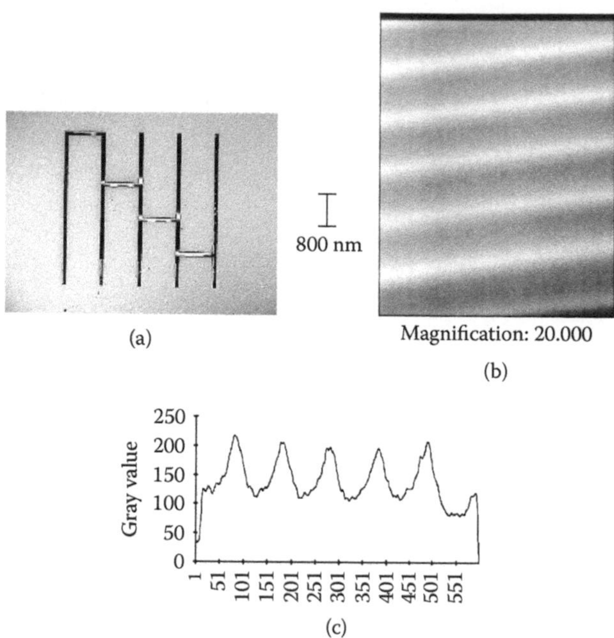

FIGURE 12.31 Silicon microbeams etched in silicon wafers: (a) silicon microbeams (length 6 mm, width 1 mm, and thickness 0.2 mm) made by SENTECH GmbH Berlin, (b) etched sinusoidal surface profile (1200 lines/mm, depth 200 nm), (c) plot of the surface profile.

12.5.1.1 The Loading of Microcomponents

It is old wisdom that the quality of the question contributes significantly to the quality of the answer. For instance, in systems engineering the information about a system delivered in response to an input signal depends to a great extent on the kind of input. The same applies to holographic nondestructive testing. Therefore, the design of the loading is an important prerequisite for testing any component and, in particular, for testing microcomponents. Unfortunately, the availability of loading equipment for microscale investigations remains very restricted. This concerns both the load and its measurement. For the design of loading equipment, three rules should be taken into account:

1. The load should be nondestructive.
2. The application of the load to the specimen should be very exact.
3. The amount of the load should be verified precisely.

Silicon microbeams, as shown in Figure 12.31a, are appropriate test structures for the design of inspection procedures in the microscale. Using these test structures, the suitability and efficiency of measuring techniques can be controlled [55]. The developed technology for the preparation of such test structures can be completed with an adapted surface-structuring process that makes it the possible to generate a very fine sinusoidal grid on the surface for the direct application of grating interferometry without additional preparation of the specimen (see Chapter 7.1 and Figure 12.31b and Figure 12.31c). For the investigation of such microbeams with respect to material parameters, two loading devices are constructed: one for the measurement of in-plane displacements (Figure 12.32a) and the other for the measurement of out-of-plane displacements (Figure 12.32c). Both devices ensure precise loading and its verification [55].

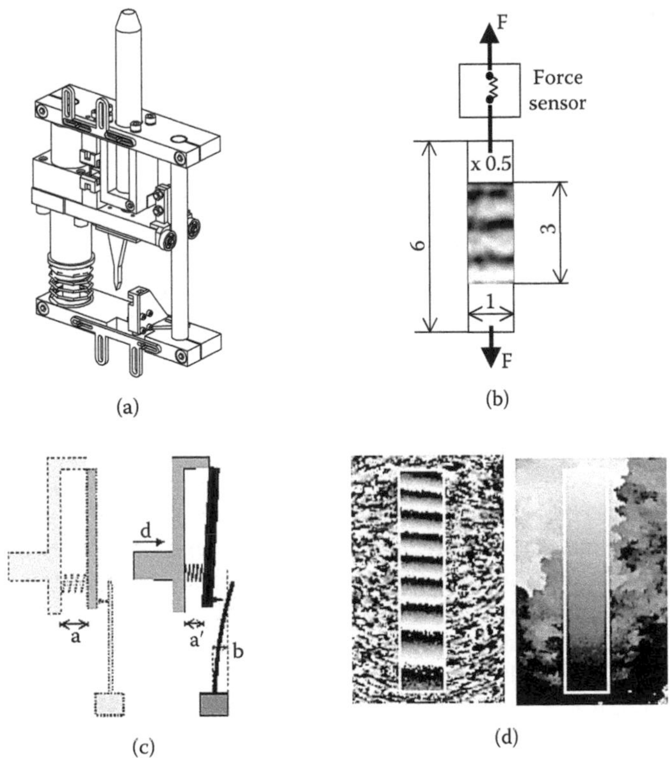

FIGURE 12.32 Loading equipment for in-plane and out-of-plane displacement measurement of silicon micro-beams: (a) loading machine (IMIF Warzaw) for in-plane displacement measurement; (b) in-plane displacement measurement by grating interferometry, the interferogram represents a displacement for F = 0.2 N; (c) loading device for out-of-plane displacement measurement; (d) phase distribution for out-of-plane displacement measurement made with digital holography.

With the first loading device, the beam was tensile-loaded in the range of F = 0.0 N to 0.8 N in steps of 0.1 N. The force was monitored by a specially designed strain-gauge-based sensor having a resolution of ±0.05 N. One example derived with grating interferometry is shown in Figure 12.32b. In the second loading device (Figure 12.32c), the out-of-plane loading is performed with a small tip pushed against the top of the silicon beam, which is fixed on the other side. Using a spring with a well-known spring constant and a known displacement of the holding mechanism, a defined force can be introduced into that beam. Knowing the force and the measured displacement of the surface of the beam, one can calculate several material properties (see Subsection 12.5.2.2.2). Figure 12.32d shows phase plots made with DH, using this loading principle. The applied load of 0.0938×10^{-6} N causes a displacement of 3.5 μm at the top of the beam.

Another important test for microcomponents and microstructures is the thermal test. Indeed, it is important for engineers to understand the behavior of such samples when they are subjected to thermal load, essentially for two main reasons. First, there is the need to investigate thermal expansion and deformation under thermal load to know if such effects can cause anomalies in the normal functionality of active structures. Second, to study possible mechanical failures induced by a high number of thermal cycles (thermal load fatigue). Investigation of microstructures under thermal load is not easy, and some problems could arise during the experimental tests.

The modalities and experimental difficulties connected with such kind of tests will be described and discussed. Furthermore, it will be shown that accurate interferometric measurement, even in quasi-real-time regimes, can be performed to obtain full-field quantitative information on deformation of microstructures under thermal test.

12.5.1.2 The Handling and Preparation of Microcomponents

Because of their smallness and fragility of the handling, manipulation and preparation of microcomponents within the interferometer setup have to be considered carefully. The handling and preparation of microcomponents for optical metrology are quite different in comparison to macrocomponents. Thus, for example, it is not recommended to coat their surfaces with white paint — a procedure that is quite usual in holographic interferometry to achieve a better reflectivity — or to shape the surface of the component after their production, as is done in grating interferometry. If the grating is applied on thin microbeams, serious problems may occur with the handling of the specimen during the process of photoresist spinning and during its exposure. The grating can become distorted and cause faulty and noisy fringe patterns. Therefore, a special technology was implemented to receive integrated microelement zero-thickness specimen gratings. Such gratings with a sinusoidal profile are generated directly on the surface of the silicon microbeams during their production [54,55]. The sinusoidal microgrid with a spatial frequency of 1200 mm^{-1} is imaged on the photoresist of the entire silicon wafer. Special etching procedures ensure a 1:1 reproduction of the sinusoidal intensity distribution in the depth of the material (depth $\approx$ 200 nm) (see Figure 12.31b and Figure 12.31c).

In contrast to most speckle-based techniques, grating interferometry can easily be adapted to long-term investigations, offering high sensitivity at the same time. On the other hand, DH is used to determine the deformation of a rough surface by measuring the complex amplitude of the scattered electric field incident on the camera target. Based on this hologram, the electric field in the object plane can be reconstructed using the Huygens–Fresnel approach. Thus, no lens or imaging device is needed. Additionally, in contrast to grating interferometry, DH provides information about the object's deformation in both directions, parallel and normal to the surface. The combination of these measurement techniques leads to a new complementary approach, offering the features of both methods at the same time [63]. It enables the operator to investigate the deformation of an object's surface under long-term conditions and at interferometric scale. Because no imaging device is needed, it allows the sensor design to be very compact. Finally, the reconstruction of the electric field amplitude is not limited to a single plane but can be adapted to the problem. In a polymeric substrate such as PMMA that is often used in microtechnology as housing material, the surface grating can be applied using the phase mask method [64].

Further problems have to be considered regarding the fixing of the sample into the clamping devices of the loading system. Inappropriate fixing can cause unwanted deformations, which distort the measurement result significantly. One example is the measurement of the Poisson's ratio of the microbeam material [65]. The measurement principle is described in Subsection 12.5.2.2.1. Pure bending is applied to the microbeams. The relevant contour lines of the surface displacements are hyperbolas that can be observed with DH. However, the application of the measurement principle requires a defined fixing of the beam and its symmetric loading. In the other case, the contour lines are distorted and cannot be evaluated correctly. Unfortunately, with the investigation of microbeams, strict compliance with these conditions is very difficult. Usually, the fringe pattern shows some slight distortions, for instance, as seen in Figure 12.32a.

In this case, DH can also help. Because all object states can be handled separately, the tilt can be compensated numerically. Such a correction of the distorted pattern, Figure 12.33a, is shown in Figure 12.33b. Using this pattern, the Poisson ratio can be calculated correctly.

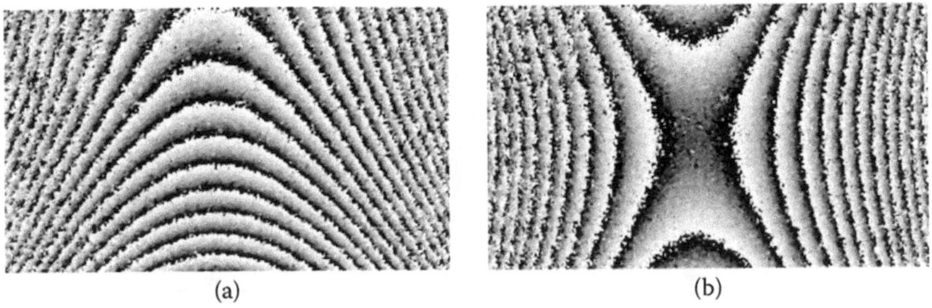

FIGURE 12.33 Numerical correction of distorted fringe patterns by digital holography: (a) distorted fringe pattern, (b) numerically corrected fringe pattern.

12.5.1.3 The Observation and Evaluation of Microcomponents

Optical methods with fringe patterns as primary output depend on the resolution of the fringes as the information carrier. Noisy fringes complicate considerably the reconstruction of the phase as the primary measuring quantity. Especially in coherent optical methods, a special kind of noise called *speckle* is always present. When any object that is not an ideal specular reflector is illuminated with a coherent beam, microscopic fluctuations in the surface randomly *dephase* the scattered and reflected light. When this light is imaged onto a sensor surface, interference between dephased components produces a random fluctuation in amplitude. Consequently, the resulting fringes are very noisy and in the most dramatic case, the pattern has a 100% modulation function and an average grain size that is determined by the wavelength of the light and the optical system (see Chapter 13). Although the size of the object has no direct influence on the speckle size, the signal-to-noise ratio is sometimes worse for small objects.

12.5.2 Investigation of Objects with Technical Surfaces

In general, the shape and the deformation are quantified with the help of different measuring devices. Consequently, the results are represented in different coordinate systems with different pitches between the data points and have to be matched together. A lot of calibration effort, manual interactions, and numerical matching errors arise from that. As shown in Section 12.3, DH offers an elegant approach to registering shape and deformation of arbitrary objects in a single setup, either with holographic contouring or with holographic deformation analysis. Because neither the camera nor the object is replaced, the pixelwise correspondence of the shape and the deformation data in the reconstruction procedure is insured. In combination with a physical model of the mechanical behavior of the loaded object — provided that the applied load and the boundary conditions are well known — material properties of microcomponents such the Young's modulus, the Poisson ratio, and the thermal expansion coefficient can be determined. The model must contain one or more of the material constants as parameters. A numerical fit to the measured data according to the physical model delivers the required parameters with an accuracy that is determined by the numerical reliability of the model. The whole evaluation process can be displayed as shown in Figure 12.34 [39].

12.5.2.1 Combined Shape and Displacement Measurement of Small Objects using DH

To demonstrate the combined shape and displacement measurement by DH, a small electrical 2.2-kΩ resistor is used as a test object (Figure 12.35a) [39]. Its shape is registered using the described multiwavelength technique. The differences of the wavelengths are $\Delta\lambda = 0.03$, 0.5, and 3.0 nm (at a working wavelength of $\lambda = 591.01$ nm). This results in synthetic wavelengths of 1.16 cm, 0.7 mm,

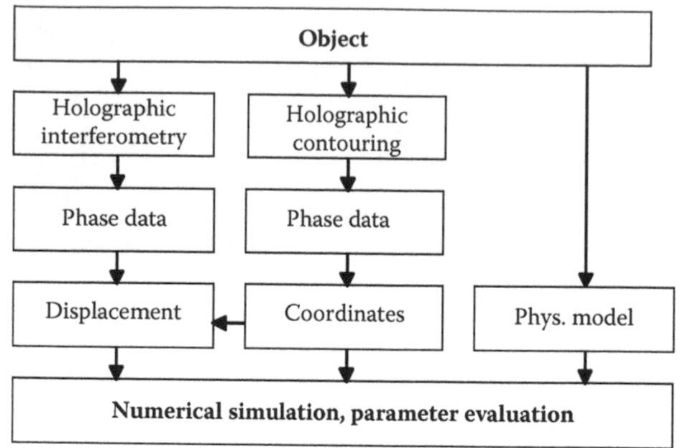

FIGURE 12.34 Evaluation scheme of material properties based on a combined shape and deformation measurement.

and 117 μm, (Figure 12.35b to Figure 12.35d). In Figure 12.35b to Figure 12.35d, one fringe represents approximately half of the synthetic wavelength. The described unwrapping algorithm was applied and, finally, the phase map was transformed into a height map (Figure 12.35e). The result is plotted in 3-D for better visualization (Figure 12.35f).

The measured data were compared with measurements preformed by a high-precision laser triangulation sensor. A match of the shape data sets shows that the noise level of the interferometric method is about 1:20 of the smallest synthetic wavelength Λ^k. In this example, the uncertainty in the measurement data is about ± 6 μm.

For the inspection of the thermal behavior of the resistor, an electrical voltage is applied that causes heating and deformation of the surface. Because of the applied voltage, the resistor was heated from 28.84 to 37.44°C. The corresponding wrapped and unwrapped interferograms show the phase difference of the digitally reconstructed phases before and after the thermal loading (Figure 12.36a). The appearing mod 2π fringes represent the deformation in the direction of one of the sensitivity vectors. Each fringe represents a deformation of about $\lambda/2$ or 295 nm. For the measurement of the 3-D displacement field **d**(u, v, w), four illumination directions are applied. The resulting four phase maps can be unwrapped and are, together with the shape data, the basis for the solution of the corresponding equation system (see Equation 12.45 with n = 4). The 3-D plot in Figure 12.36c shows the deformation field as small arrows on the surface of the object. The deformation is magnified by a factor 10^6 relative to the shape.

Another example deals with the inspection of the 3-D shape of a micro gas-flow sensor [66] (Figure 12.37a). The active part of this object has a size of 0.7×0.7 mm^2 (Figure 12.37b). The shape of the active part — the central paddle — is of special interest for the production of such a microsensor. Because of the simplicity of the geometry, a simple two-wavelength-contouring procedure was implemented. The used wavelengths are $\lambda_1 = 575$ nm and $\lambda_2 = 577$ nm. This results in a synthetic wavelength of $\Lambda = 166$ μm and a height difference Δz of two consecutive contour lines of $\Delta z = 83$ μm. Because the achieved phase resolution was approximately $2\pi/20$, the corresponding height resolution was about 4 μm. Using a 1000×1000 pixel camera with a target size of 7×7 mm^2, a lateral resolution of about 10 μm could be achieved.

Figure 12.38 shows the intermediate results of DH, starting from the two reconstructed phase distributions up to the plot of the continuous interference phase. In Figure 12.39, it can be seen that the surface of the sensor is slightly disturbed. This unwanted effect must be avoided by production control and construction optimization. Because the working principle of the sensor is

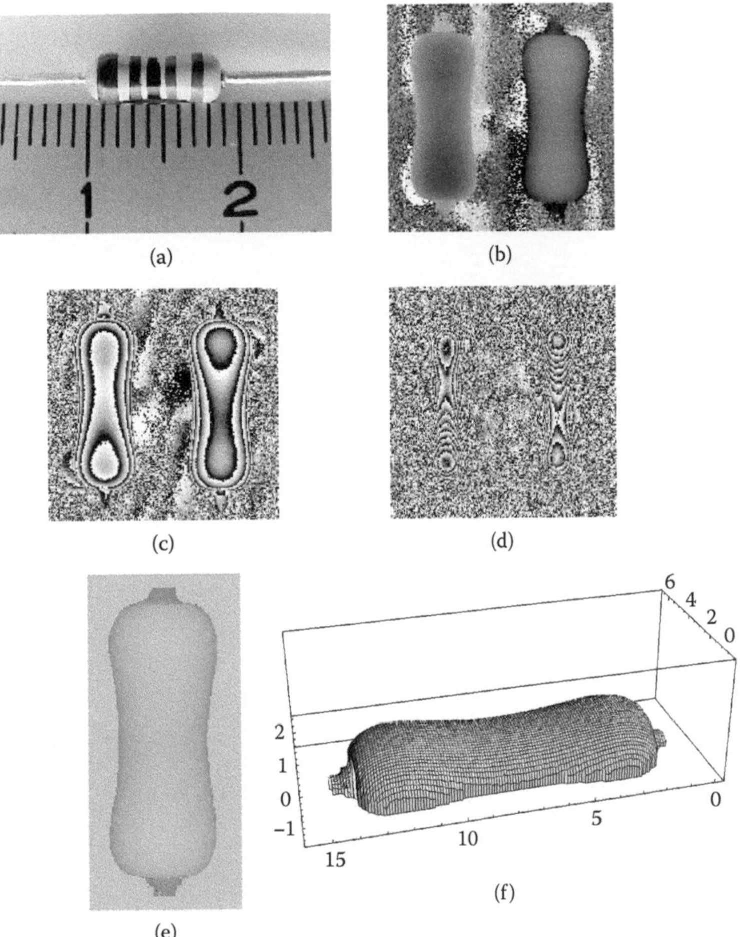

FIGURE 12.35 Shape measurement of a resistor using digital holography and multiwavelength contouring. (a) Photograph of the resistor (ruler scale is in cm), (b) wavelength contouring interferogram width $\Delta\lambda = 0.03$ nm, (c) wavelength contouring interferogram width $\Delta\lambda = 0.5$ nm, (d) wavelength contouring interferogram width $\Delta\lambda = 3.0$ nm, (e) unwrapped interference phase, (f) 3-D plot of the shape of the object. (From Seebacher, S. et al., The determination of material parameters of microcomponents using digital holography, *Opt. Laser. Eng*. 36, 103, 2001.)

based on a gas-dependent deformation within the gas stream, further experiments are directed to a combined shape and displacement measurement.

12.5.2.2 The Determination of Material Parameters of Microcomponents using DH

The implemented system described in the following subsection is a first step in the direction of reliable determination of material parameters with optical metrology [39]. It is based on a consistent interferometrical shape and deformation analysis in combination with adapted loading equipment. The precise measurement of coordinates, displacements, and forces allows the determination of several material parameters such as the Poisson-ratio, Young's modulus, and the thermal expansion coefficient. For the evaluation of the system performance, an appropriate class of test samples

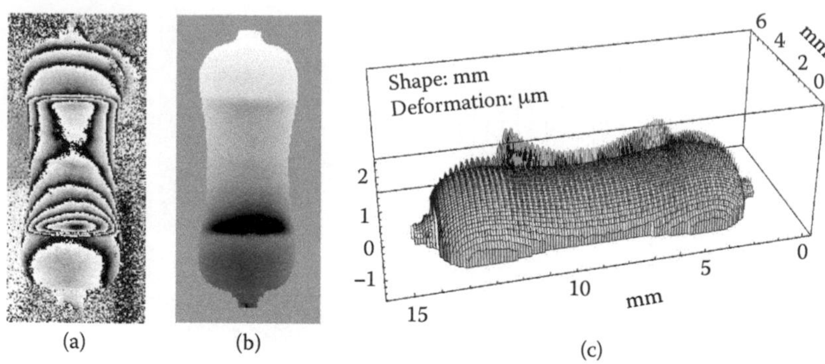

FIGURE 12.36 Displacement field of a heated resistor measured with digital holographic interferometry. (a) Mod 2π interference phase, (b) unwrapped interference phase, (c) combined shape and deformation plot of the heated resistor. (From Seebacher, S. et al., The determination of material parameters of microcomponents using digital holography, *Opt. Laser. Eng*. 36, 103, 2001.)

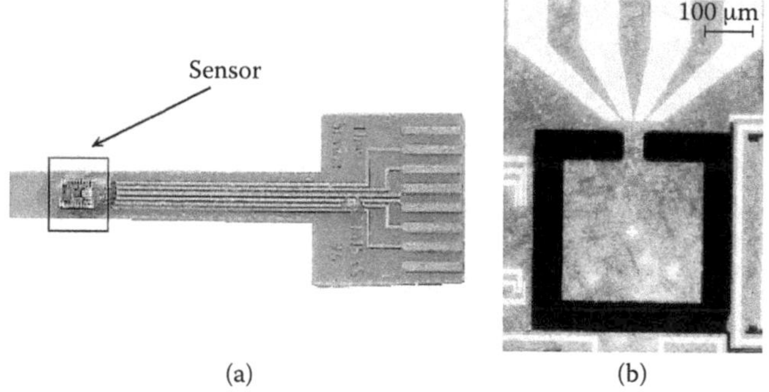

FIGURE 12.37 Investigated gas-flow sensor: (a) gas-flow sensor (courtesy of IMSAS Bremen), (b) investigated part (1 mm × 1 mm).

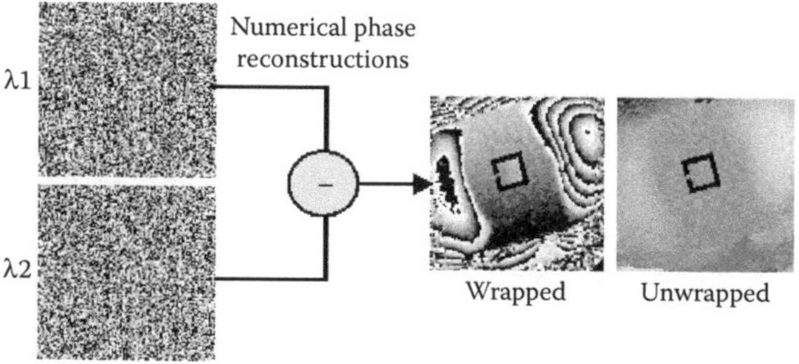

FIGURE 12.38 Two-wavelength contouring of a gas-flow microsensor using digital holography.

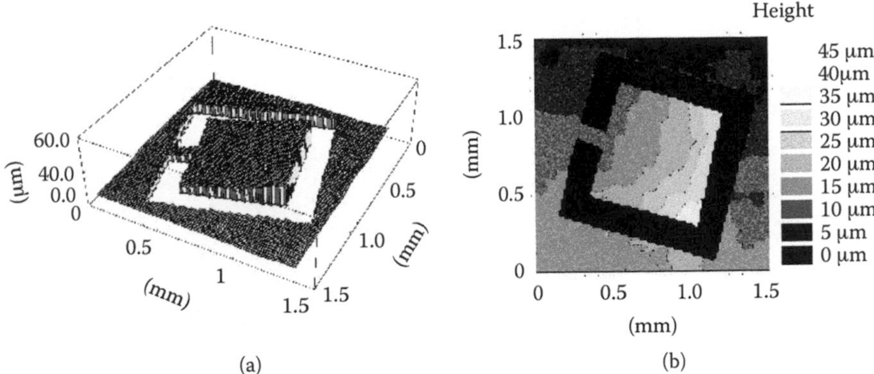

FIGURE 12.39 Reconstructed shape of the gas-flow sensor: (a) pseudo-3-D plot of the height profile, (b) contour line of the height plot.

was created consisting of silicon microbeams, membranes, and tongues [55] (Figure 12.40). The schematic setup of the optical part is shown in Figure 12.41a. The interferometer consists of an optimized arrangement with four illumination directions and one observation direction to measure the 3-D displacements and coordinates precisely. It includes a CCD camera, a laser, a beam splitter cube to guide the reference beam to the CCD target, and some optics for beam shaping. Optionally, a fiber coupler can be included to switch between several illumination directions for the derivation of vector displacements. Such a setup can be designed in a very compact way as shown in Figure 12.41b.

12.5.2.2.1 The Determination of the Poisson Ratio of Microbeams

The physical models used for the determination of the material's properties can be of any desired complexity. Simple models are easy to handle but imply errors in the evaluation of the physical quantities. Thus, material properties should always be specified together with the applied model. For the demonstration of the complete measurement procedure, we prefer rather simple models that assume linear elastic behavior on the part of the specimen.

The determination of the Poisson ratio by holographic interferometry is a well-known example of the efficiency of coherent optical methods [65]. DH provides a relatively simple way to determine this quantity. Moments of force are applied to a rectangular sample at opposite sides. Figure 12.42a shows a typical loading machine designed especially for small objects. The small dimensions of the samples demand a precise adjustment of all components: the "chop," which pushes against the

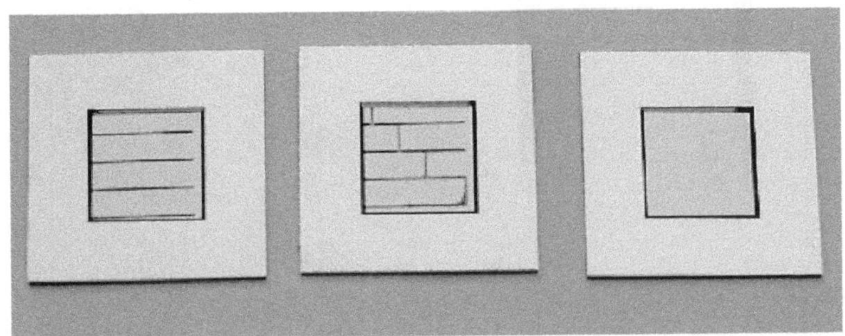

FIGURE 12.40 Test samples made from a 100-mm diameter silicon wafer. The size of the structure is 9 mm × 9 mm, the thickness of the beams vary from 10 μm to 40 μm.

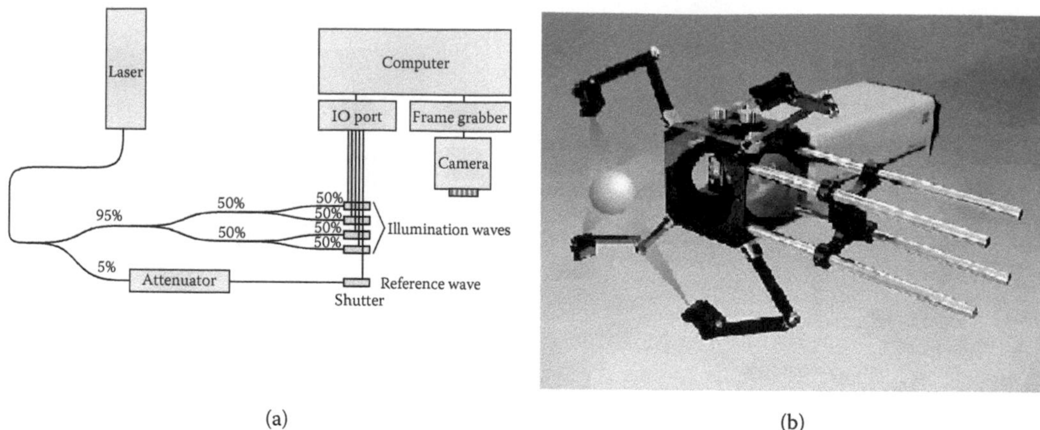

FIGURE 12.41 The holographic interferometer. (a) Schematic setup of the interferometer, (b) compact digital holographic interferometer with four illumination directions. (From Seebacher, S. et al., The determination of material parameters of microcomponents using digital holography, *Opt. Laser. Eng*. 36, 103, 2001.)

object from above, the support, and the sample, which has the shape of a small rectangular beam. In this way, a homogeneous deformation is achieved. Unwanted torsions of small magnitude are corrected numerically, which is easy with the use of the mod 2π phase maps from DH [54]. The resulting deformation is recorded and evaluated. The deflection causes hyperbolic fringe structures (Figure 12.42b). Conventionally, the Poisson ratio is derived numerically from the angle between

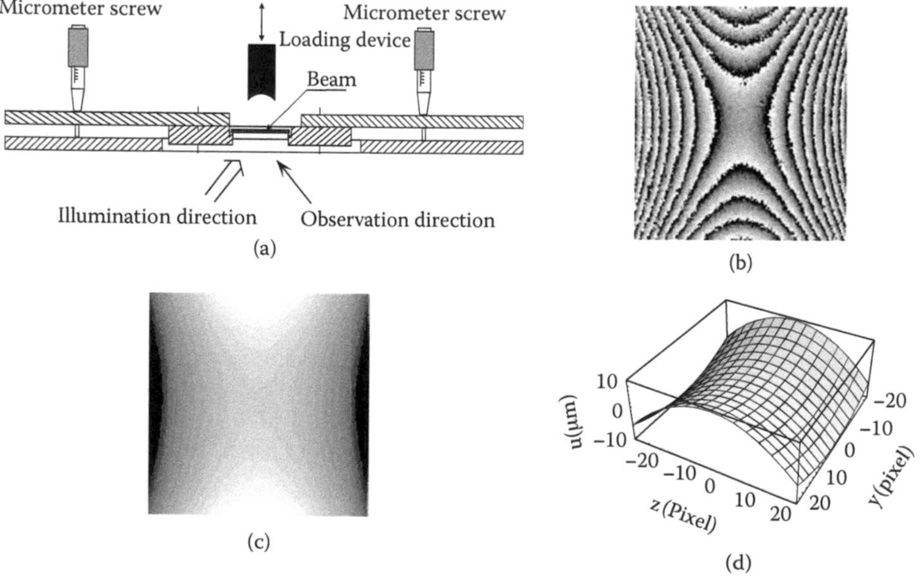

FIGURE 12.42 The measurement of the Poisson ratio by digital holography: (a) schematic experimental setup, (b) reconstructed mod 2π phase map, (c) unwrapped phase field, (d) approximated function according to Equation 12.83.

TABLE 12.1
Measured Poisson Ratio of Different Materials [39]

Material	Width (mm)	Thickness (mm)	Length (mm)	Poisson Ratio: Measured [Literature Value]
Spring steel	1.2	0.2	12	0.288 [0.29–0.31]
Spring steel	2.0	0.1	12	0.301 [0.29–0.31]
Structural steel	1.0	0.5	10	0.338 [0.29–0.31]
Structural steel	1.5	0.5	10	0.345 [0.29–0.31]
Titanium	2.0	0.8	10	0.359 [0.361]
Titan	1.5	0.8	10	0.381 [0.361]

the asymptotic lines of the fringes of equal phase [65]. The deformation can be formulated as a first-order approximation by the following equation:

$$u(y,z) = -\frac{1}{2R}[y^2 + \nu(a^2 - z^2)] \tag{12.83}$$

Here the u means the deformation in the x direction at position (y,z) on the surface of the object. ν stands for the Poisson ratio. R is the radius of curvature and a is a parameter that is not important for the following evaluation process. R results from a slight deflection in the initial state. This helps to ensure a proper mechanical contact between the support and the sample. Equation 12.83 shows that the upper and lower surfaces of the sample are deformed to parabolic curves where the inside is bent in a convex manner and the outside is curved in a concave manner. Because this analytical model contains the Poisson ratio as a parameter, it is possible to use the measured deformation for its evaluation. This is performed numerically by approximating the model to the data (Figure 12.42c) with a least-squares fit (Figure 12.42d).

The values obtained by this method show good reproducibility and accuracy in comparison to conventional optical techniques for small samples. Table 12.1 contains some of the results for beams made of spring steel, structural steel, and titanium. The values correlate with the values given by the manufacturers within the tolerances of the material batches.

12.5.2.2.2 The Determination of Young's Modulus of Microbeams

Young's modulus can be determined in a similar way as the Poisson ratio if the physical model contains this quantity as a parameter. We use small silicon beams as shown in Figure 12.40 that are clamped at one edge and mechanically loaded with a defined force at the opposite edge. The 3-D surface displacements **d**(u,v,w) (Figure 12.43c) can be measured with the interferometer by evaluating at least three interferograms (Figure 12.43b) made with different illumination directions [67]. A model of the bending of a beam having Young's modulus E as a free parameter is the basis for a numerical fit of the experimental values:

$$u(y) = \frac{F \cdot l^3}{6EI_y}\left(2 - \frac{3y}{l} + \frac{y^3}{l^3}\right) \tag{12.84}$$

u is the displacement in the x direction, and y is a position on the beam of the length l. I_y is the axial moment of inertia in the (x,z)-plane that can be estimated with the help of a shape measurement. F is the force applied to the tip of the beam. The applied forces are relatively small, so a special loading mechanism was developed (Figure 12.43a). The spring constant k is assumed to

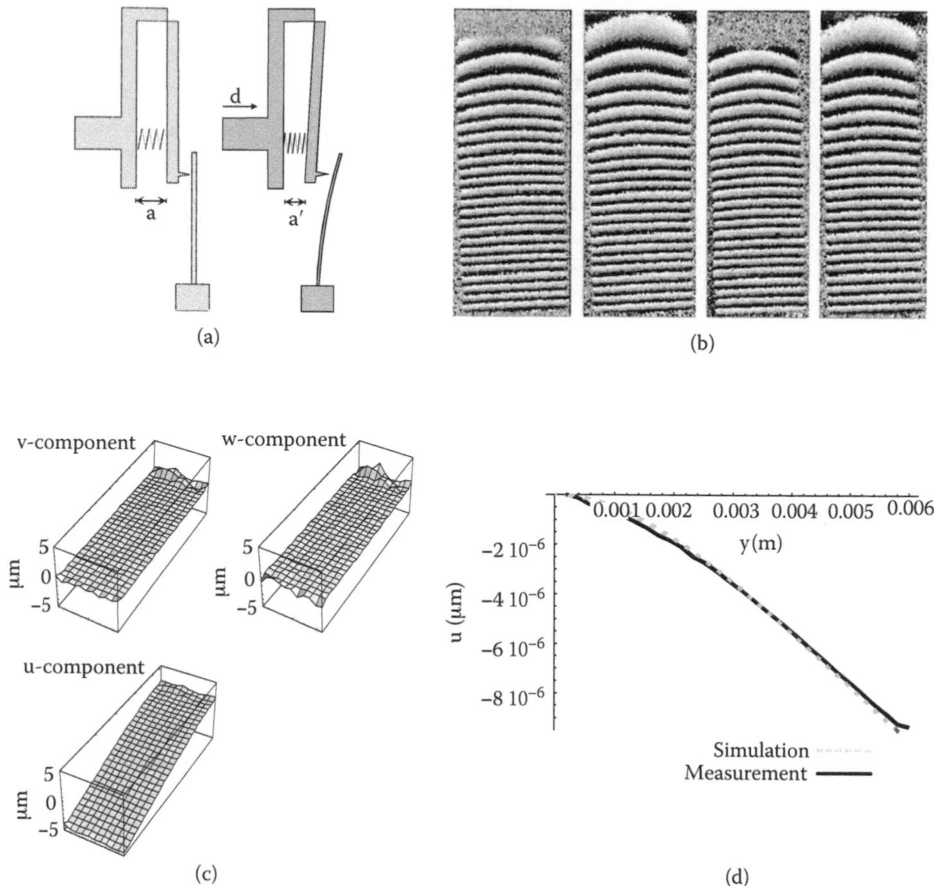

FIGURE 12.43 Determination of Young's modulus by digital holography: (a) working principle of the loading mechanism for the small samples, (b) four interferograms recorded from four different illumination directions, (c) deformation calculated in Cartesian coordinates (scale of the plots in μm), (d) profile of the deformation in the x direction.

be known precisely, as well as the displacement $\Delta a = a - a'$. With this information the force can be evaluated with the equation

$$F = k\,\Delta a \tag{12.85}$$

Several experiments with thin beams made of silicon (dimensions: length 3 mm and width 1 mm) delivered an average value of E = 162 MPa. The literature value (in the considered crystal direction) is about 166 MPa. These values can vary widely according to the material's history, treatment, and degree of impurity.

12.5.2.2.3 The Determination of the Thermal Expansion Coefficient of Microbeams

For the interferometric investigation of the thermal behavior of various specimens, it must be ensured that thermal turbulences and nonuniform temperature distributions are avoided. Therefore, a vacuum chamber that can be supplied with adapted loading devices was constructed (Figure 12.44a). The used thermal loading device is capable of maintaining a constant temperature to within an accuracy of 0.02°C in a range of about 20°C up to 180°C (Figure 12.44b). The holographic interferometer was mounted outside at the observation window of the chamber (Figure 12.44c).

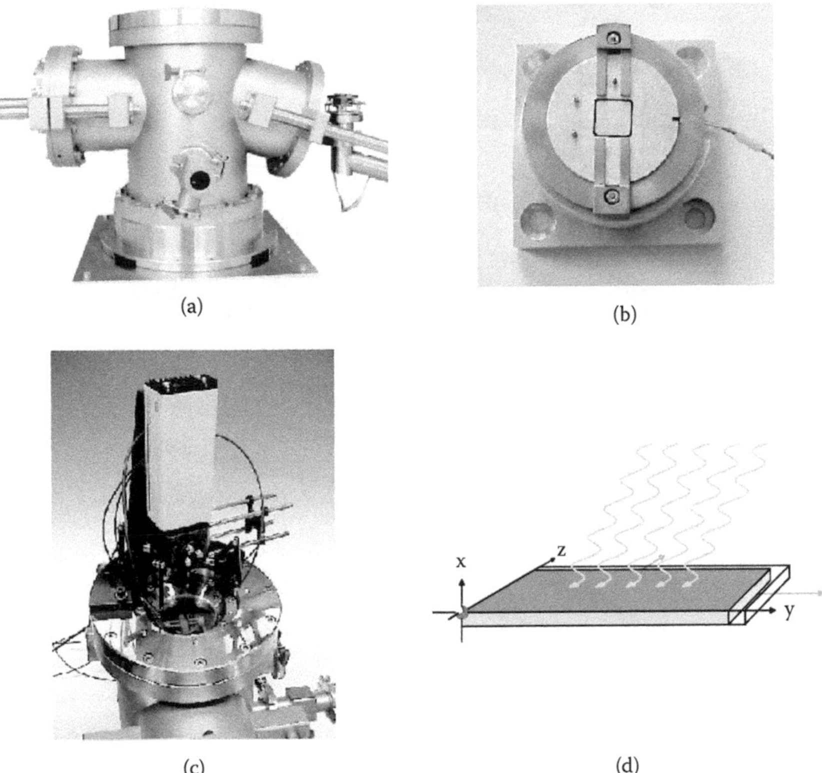

FIGURE 12.44 Inspection system for the investigation of microcomponents with respect to material parameters (the devices shown in Figure 12.44a and Figure 12.44b are made by CWM Chemnitz, Germany): (a) vacuum chamber with the supply channel, (b) equipment for thermal loading, (c) interferometer mounted on the inspection window, (d) coordinate system used for the calculation of the thermal expansion coefficient.

As a test object, a monocrystal silicon beam (Figure 12.44d) with 4.5 Ωcm phosphor coating was used. The interferograms are recorded at different temperature differences. The complete evaluation process can be summarized as follows:

- The geometry of the setup was measured to get the geometry matrix for the evaluation of the three displacement components.
- Four holograms are recorded in the initial state of the object.
- The object is loaded thermally and recorded holographically from four different illumination directions
- The displacement vector components (u,v,w) are calculated based on the evaluation of the four interferograms.
- Rigid-body motions are separated from internal deformations of the object itself by subtracting the mean movement from the displacement values.
- The absolute length change ΔL is determined as well as the total length of the beam L_0, which can be performed by using the imaging properties of DH. The thermal expansion coefficient in the y and z directions can be simply calculated using Equation 12.86 with ΔT as the temperature change.

$$\alpha = \frac{\Delta L}{L_0 \cdot \Delta T} \tag{12.86}$$

The extension in the x direction is too small to be detectable with the method applied.

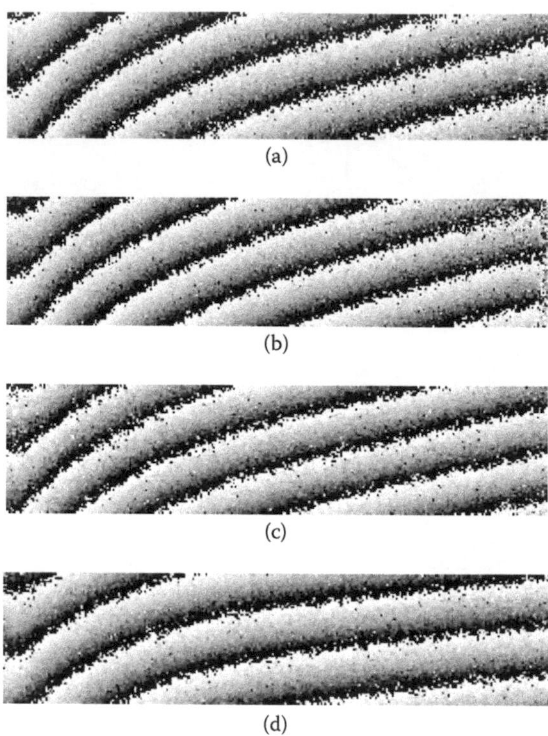

FIGURE 12.45 Four interferograms due the deformation of the object by thermal load using four different illumination directions: (a) illumination direction 1, (b) illumination direction 2, (c) illumination direction 3, (d) illumination direction 4.

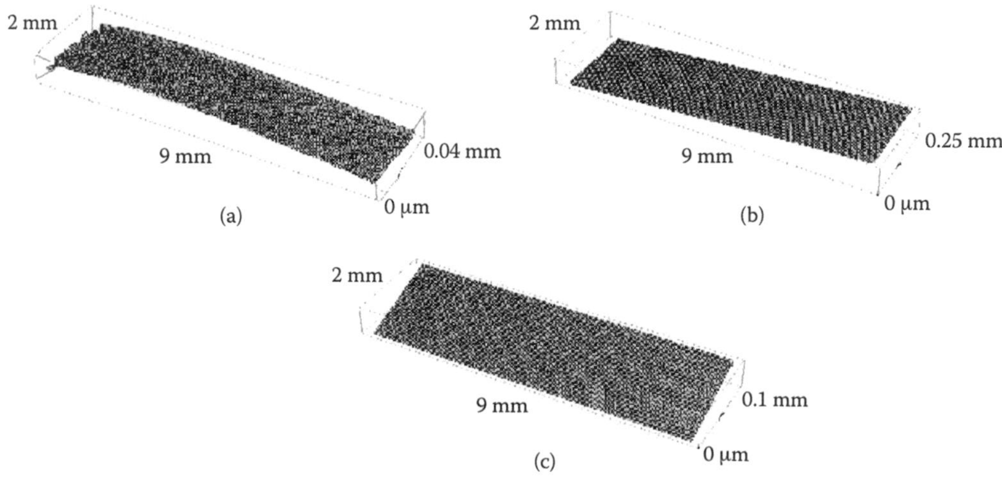

FIGURE 12.46 3-D-displacement components (u,v,w): (a) u-component, (b) v-component, (c) w-component.

As an example, the thermal expansion coefficient α of a small 2 mm × 9 mm × 100 μm monocrystal silicon beam is measured. Figure 12.45 shows the four resulting interferograms. The applied temperature difference ΔT was 30°C. After elimination of the rigid-body motion, the three deformation components are evaluated as shown in Figure 12.46. Considering the dimensions of the beam we obtain a value for α of about 2.92×10^{-6} 1/K. The literature values vary widely because of different measurement methods, conditions, and material batches: α varies from 2.4×10^{-6} to 6.7×10^{-6} 1/K.

12.5.3 Investigation of Microcomponents with Optical Surfaces

12.5.3.1 Testing Silicon MEMS Structures

The tested microstructures are made using bimorph structures with a silicon nitride (Si_3N_4) 140-nm-thick layer over a 1500-nm-thick polysilicon layer. The polysilicon layer can be doped or undoped, and it is deposited by low-pressure chemical vapor deposition (LPCVD) at 680°C, followed by a thermal annealing for 30 min at 1050°C. In some cases, an Si_3N_4 layer is deposited by the oven at 800°C. The lower sacrificial layer was made of a p-vapox (fully oxidized porous silicon) 1150-nm-thick film, deposited by atmospheric pressure chemical vapor deposition (APCVD), and then a thermal process at 1050°C was applied in order to remove hydrogen from the layer. The p-vapox was removed by a wet etching process. The differing physiochemical behavior of the various layers under the thermal processing causes the occurrence of residual stresses, inducing out-of-plane deformations of the MEMS. In particular, these stresses are due to the interaction between the polysilicon and Si_3N_4 that implies a deformation toward the external side. Here, we will show some examples of application of DH to measurement of profiles of silicon MEMS structures. In fact, the DH method can be applied to characterize MEMS with several different geometries and shapes, such as cantilever beams, bridges, and membranes. Dimensions of the inspected microstructures range from 1 to 50 μm.

12.5.3.1.1 Profile Reconstruction

Results obtained by applying DH to the samples described in the preceding text are reported. In order to correct the curvature phase introduced by the MO, a kind of double-exposure digital holographic interferometry has been employed [50,52]. The first exposure is recorded of the object under investigation, whereas the second one is of a flat reference surface in the proximity of the object. This is a typical case encountered, for example, when MEMS structures are inspected, because the micromachined parts have been realized on a flat silicon substrate. In that case, the area around the micromachined structure offers a very good plane surface that can be used as a reference. The experimental results give information on the global deformation experienced by the silicon structures because of the residual stress induced by the fabrication process.

Figure 12.47a shows a picture, obtained by the scanning electron microscope (SEM), of a group of cantilever beams that experienced a complex final deformation. In Figure 12.47b, the wrapped phase map image obtained from the holographic reconstruction of a single digital hologram is shown. Figure 12.47c shows the full-field deformation of the MEMS. In Figure 12.47d the deformation (z-axis) along a beam B has been reported to show details of the complex shape assumed by the cantilever beam. In the plot of Figure 12.47d, the anchor point of the beam demonstrating that some unwanted etching occurred there is identified.

From the profile along the x axis (Figure 12.47d), it can be noted that in the central region the cantilever, because of its weight, falls on the substrate. In Figure 12.47b, it is can be noted that the structures C and D are so highly deformed that the wrapped phase is undersampled after half their length. Furthermore, on the tip of the cantilevers, longitudinal fringes appear, probably because of the fact that light reflected back from the surface goes out of the useful aperture of the MO.

In Figure 12.48 the profile of the cantilever A of Figure 12.47b is shown. This full-field deformation can be segmented along the x and y axes to find the deformation profile along the two axes. In Figure 12.48a, the profile along the x axis is reported, and we note that the sacrificial layer etching spreads underneath the anchor area of the cantilevers; thus, this region is also subjected to

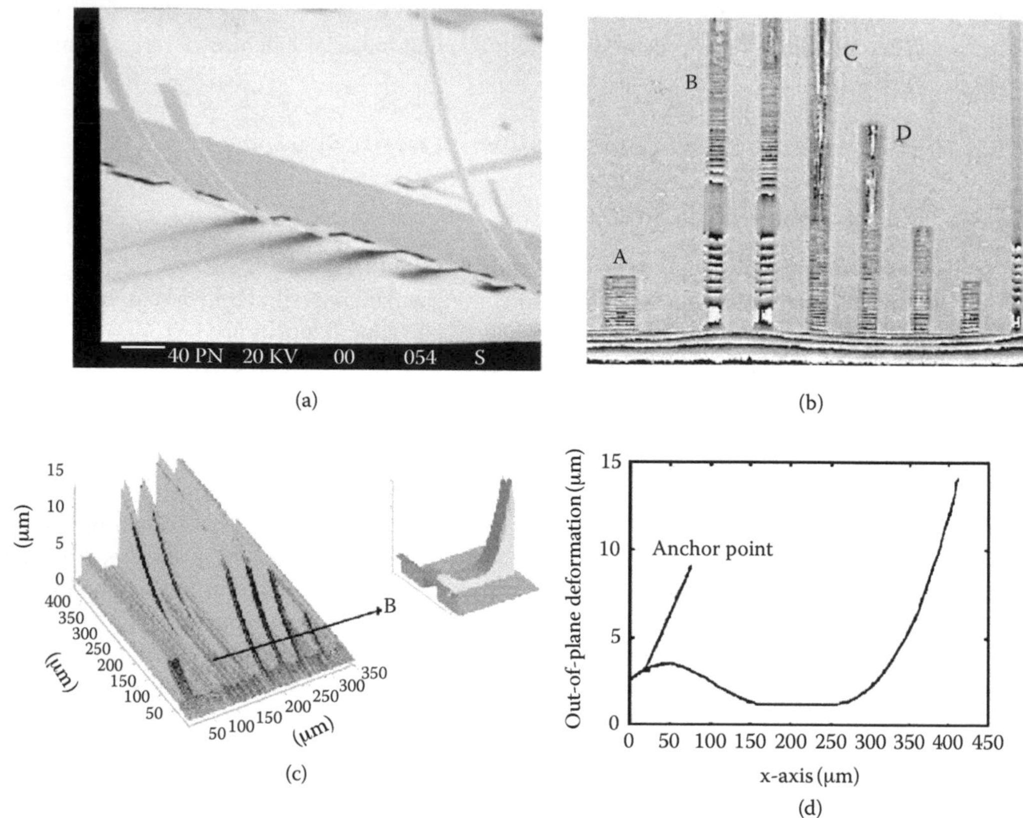

FIGURE 12.47 (a) SEM picture of a group of cantilever beams, (b) phase-contrast image of the cantilevers numerically reconstructed at distance $d = 100$ mm, (c) full-field deformation of the cantilever with detail of the deformation experienced by a single cantilever, (d) profile along the single cantilever showing the anchor point and the complex shape due to the residual stress. (From Ref. 52.)

an out-of-plane bend. The profile along the y axis (Figure 12.48b) shows a small warping along this axis. In Figure 12.49a, an SEM picture of a group of bridges is shown. The wrapped phase map is shown in Figure 12.49b; whereas in Figure 12.49c, the full-field image of the deformation of these regions is shown. Finally, in Figure 12.50, the result of a rectangular membrane inspection is reported. In this case, only the corner regions have been considered, and the fringes present in the wrapped phase indicate again out-of-plane displacements.

12.5.3.1.2 Residual Stress Calculation

From deformation profiles obtained by the use of DHM and applying analytical and numerical models [50–53], the residual stress inside the structure can be estimated [68].

From each profile, it is possible to understand both the constant average stress, which is uniform along the cantilever length, and the gradient stress, which is variable along the cantilever section. These two quantities allow evaluation of the overall residual stress field. In fact, the transverse cantilever deflections of length L in many cases can be expressed as:

$$y \cong \frac{1}{2R}x^2 + \left(\theta_0 + \theta_1\right)x \quad \text{for } x \in \left[0, L\right] \tag{12.87}$$

where $(\theta_0 + \theta_1)$ is the cantilever initial slope, and R is its curvature radius. The total angular rotation of the structure is the superimposition angular contribution θ_0 due to mean stress

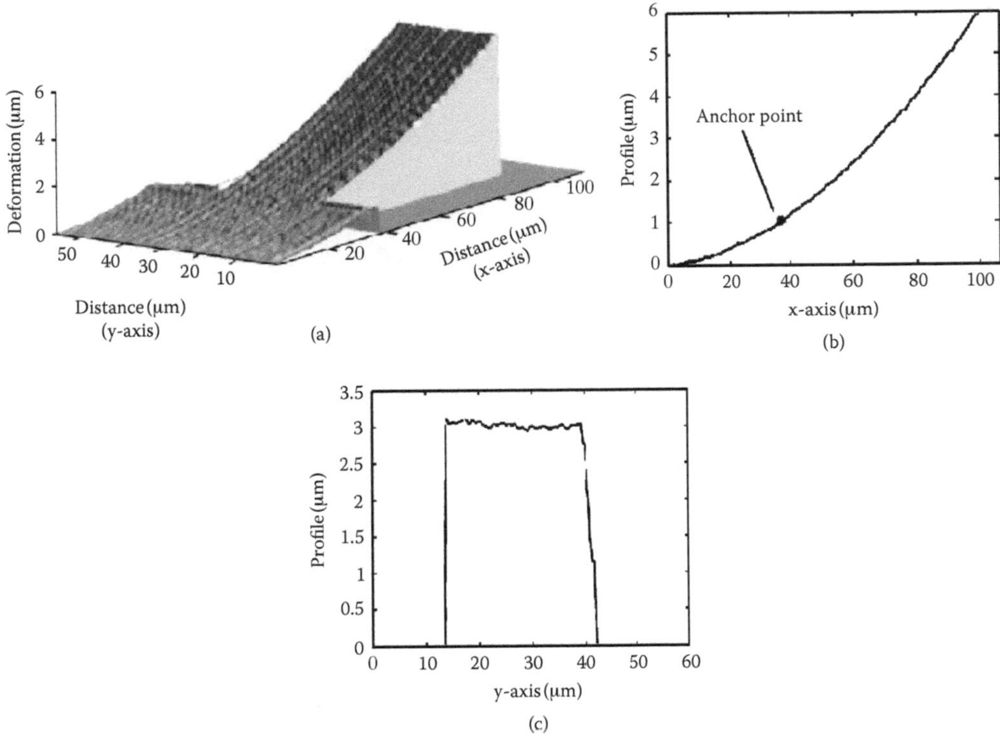

FIGURE 12.48 Full-field deformation (c) obtained from the unwrapped phase of the beam A in Figure 12.47b: (a) profile of the cantilever shown in Figure 12.47b along the x axis and (b) along the y axis. (From Ref. 52.)

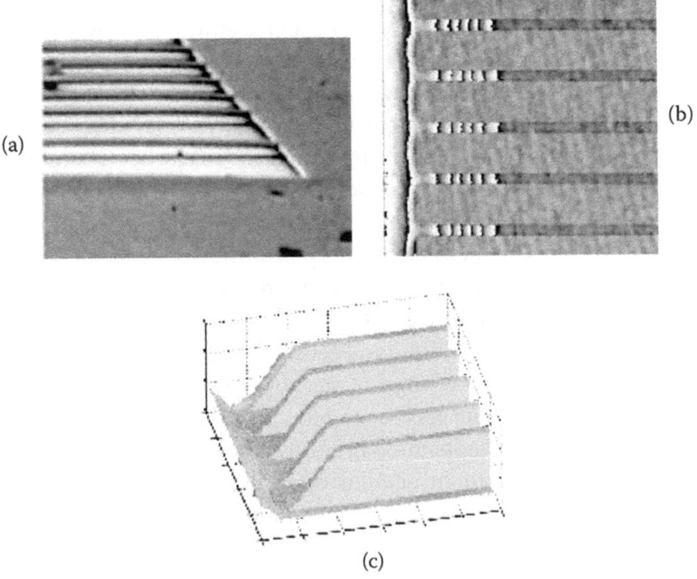

FIGURE 12.49 (a) An SEM image of a group of bridges, (b) the wrapped phase map image limited to anchor regions of the bridges, (c) the full-field deformation image of the bridges. (From Ref. 52.)

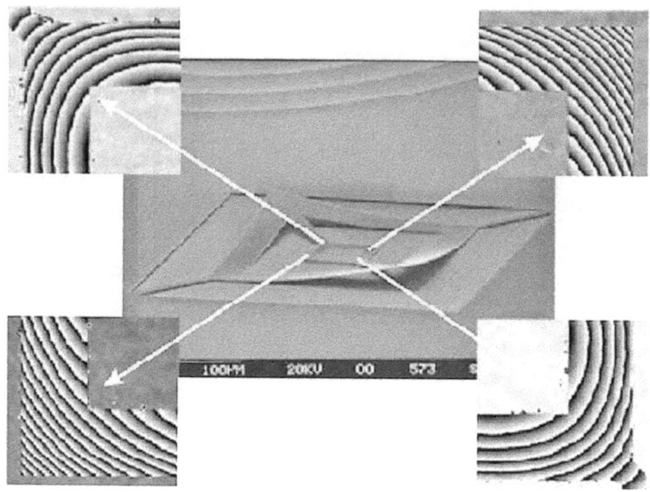

FIGURE 12.50 An SEM image of a membrane and the wrapped phase map of the corner regions. (From Ref. 52.)

and θ_1 due to stress gradient. From experimental fit [68], it is possible to model a linear dependence of θ_0 on h:

$$\theta_0 \approx \frac{\sigma_0}{E}\left(1.33+0.45\nu\right)\left(-0.014h+1.022\right) \tag{12.88}$$

and a quadratic dependence of θ_1:

$$\theta_1 \approx \frac{\sigma_1}{E}\left(0.0086h^2-0.047h+0.81\right) \tag{12.89}$$

Thus, with Poisson ratio ν, elastic modulus E, and cantilever thickness h being known, and calculating the peak value $\sigma_1 = Eh/2R$ of the gradient stress by measuring the structure curvature, from Equation 12.89 the stress gradient angular deflection θ_1 can be obtained. Finally, the residual stress component σ_0 can be determined using Equation 12.87.

An example of this methodology can be carried out utilizing the data related to the cantilever described earlier. Actually, the procedure is strictly valid only for monolayer structure, but it can be observed that the lower polysilicon layer is more than 10 times thicker than the Si_3N_4 upper layer. This situation allows us to suppose that bimorph cantilever mechanical properties are mainly related to the thick polysilicon lower layer, even if the resulting deformation and stress are determined by the upper thin compressive Si_3N_4 layer. Thus, following the procedure described earlier, the longitudinal deformation profile reported in Figure 12.51b can be fitted to the quadratic curve $y = 0.0005 \cdot x^2 + 0.01 \cdot x$. From knowledge of the total cantilever thickness h = 1640 nm, the polysilicon Poisson ratio $\nu = 0.279$, and elastic modulus E = 182 GPa [68], it is possible to calculate the constant and gradient stress components. The resulting stress components are $\sigma_0 = 1174$ MPa and $\sigma_1 = 149.24$ MPa, and the resultant angular rotation components at the cantilever root are $\theta_0 = 9.38$ mrad and $\theta_1 = 0.62$ mrad. Thus, the first approximation to the total residual stress field specifies a variation from a value of $\sigma_{total} = 1024$ MPa at the free surface y = h/2, to $\sigma_{total} = 1323$ MPa at the interface between the film and substrate.

By means of this procedure, a characterization of the fabrication processes, in terms of induced residual stresses, can be achieved. Thus, an accurate determination of the profile, obtained by means of DH, is essential to understanding the deformation due to the residual stress.

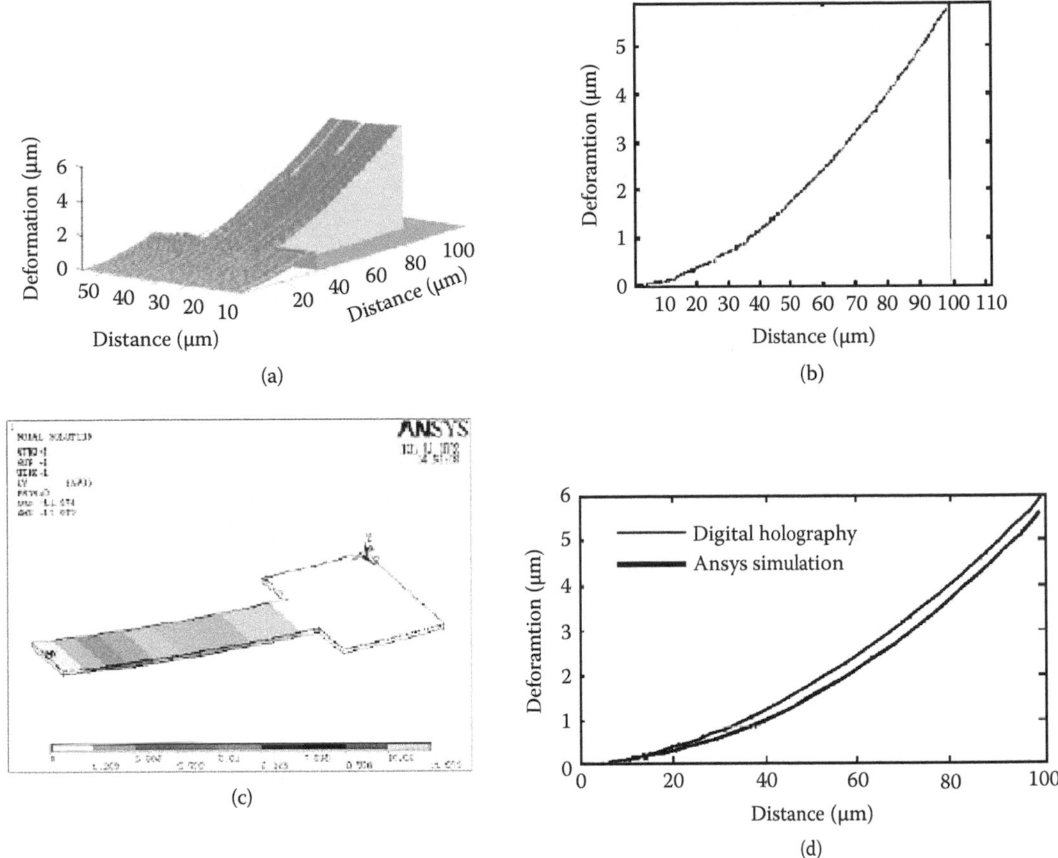

FIGURE 12.51 (a) Full-field deformation of a cantilever obtained by DH methods, (b) longitudinal cantilever profile, (c) deformation simulations of the cantilever by ANSYS software, (d) comparison between the profile obtained with the DH and simulated profile. (From Ref. 58.)

Nevertheless, if the profile of the deformation is too warped, the profile determination could be complicated. In fact, for a warped profile, the curvature of the cantilever could reflect the object beams outside the NA of the MO, or the fringes exceed the Nyquist limit imposed by the limited spatial resolution of the CCD array. These kinds of problems can be observed clearly in cantilevers C and D of Figure 12.47b. In order to avoid this problem, the chip can be slightly tilted. Residual stress could be so high that a significant deformation could cause a severe warping in the small, simple cantilever along the y axis. An extreme case is presented in Figure 12.52, in which the two cantilevers assume a "spoon" shape. It is not clear why one cantilever deforms in a symmetric way, whereas the other also experiences a rotation along one side.

12.5.3.2 Thermal Load Testing of MEMS

In the preceding paragraph, DHM was used to reconstruct static profiles. However, the DH technique can also be applied efficiently to evaluate structure profile variations. In other words, by means of a continuous recording of the hologram of the structure, it is possible to measure the variation of the profile due to some external effect, such as a temperature variation, a pressure change, static electrical field variation, and so on.

In the following sections, results of investigation of two typical silicon MEMS structures will be shown and described. The first is a passive silicon cantilever MEMS studied by application of

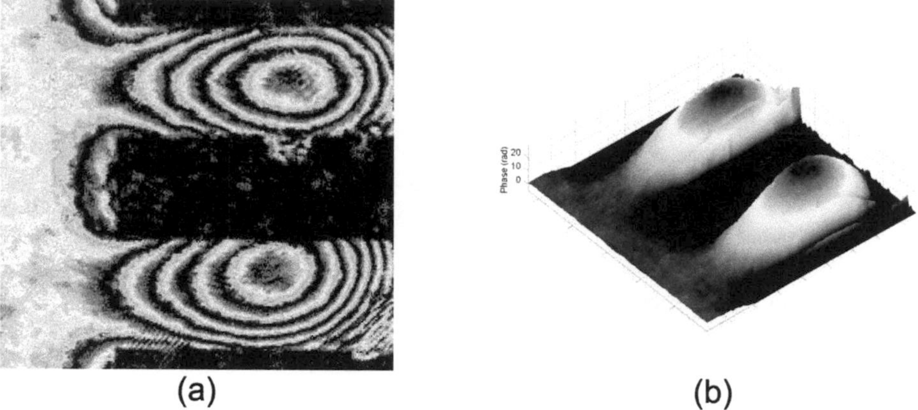

FIGURE 12.52 (See color insert following page 110.) (a) Wrapped phase map of a severely deformed cantilever silicon MEMS, (b) unwrapped phase map showing the actual 2D structure profile.

thermal load by an external source. The second is an active MEMS made of an Si membrane on which is a resistive layer through the which the thermal load is applied.

12.5.3.2.1 Cantilever Beams Subject to Thermal Load

This type of characterization has been performed on the bimorph cantilevers described in the preceding section. These structures were subjected to a thermal load, increasing their temperature from 23°C up to 130°C. In Figure 12.53a and Figure 12.53b the wrapped phase maps relative to a cantilever at temperatures of 23°C and 130°C, respectively, are reported. Actually, there is no difference between the two phases; therefore, the analyzed structures seem immune to a temperature variation of 107°C.

The same cantilever was subjected to potassium hydroxide (KOH) etching in order to remove most of the polysilicon layer under the Si_3N_4 layer. In Figure 12.54, the typical T-shaped cantilever obtained after KOH etching is reported. This kind of cantilever was subjected to the same

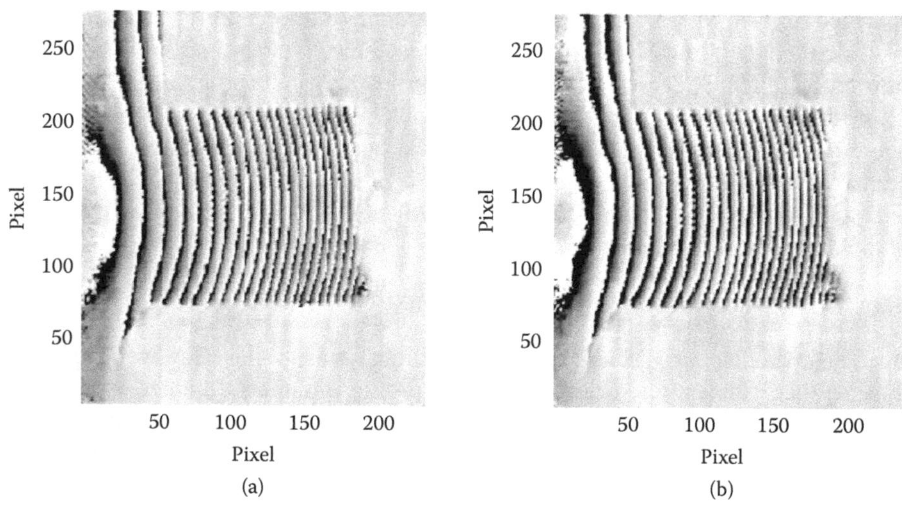

FIGURE 12.53 A phase map obtained at (a) 23°C, (b) 130°C. (From Ref. 52.)

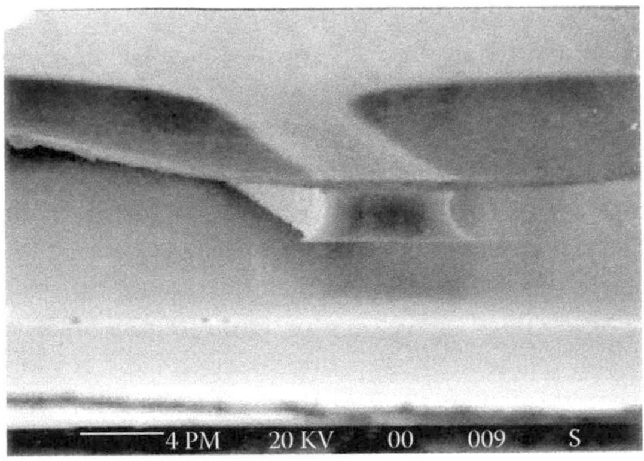

FIGURE 12.54 An SEM picture of the typical cantilever shape after KOH etching. (From Ref. 52.)

temperature variation of 107°C, and the obtained phase maps are reported in Figure 12.55a and in Figure 12.55b, where it can be noted that the wrapped phase map has about two fringes more than the phase map relative to a temperature of 23°C (Figure 12.55a); in other words, the cantilever at 130°C is more warped than that at 23°C.

This behavior can be strengthened through the observation of the full-field phase map difference at 130°C and at 23°C (Figure 12.55c). Finally, profiles of the cantilever for three different temperatures (23°C, 56°C, and 130°C), at the mid-section of the beam, are reported in Figure 12.55d.

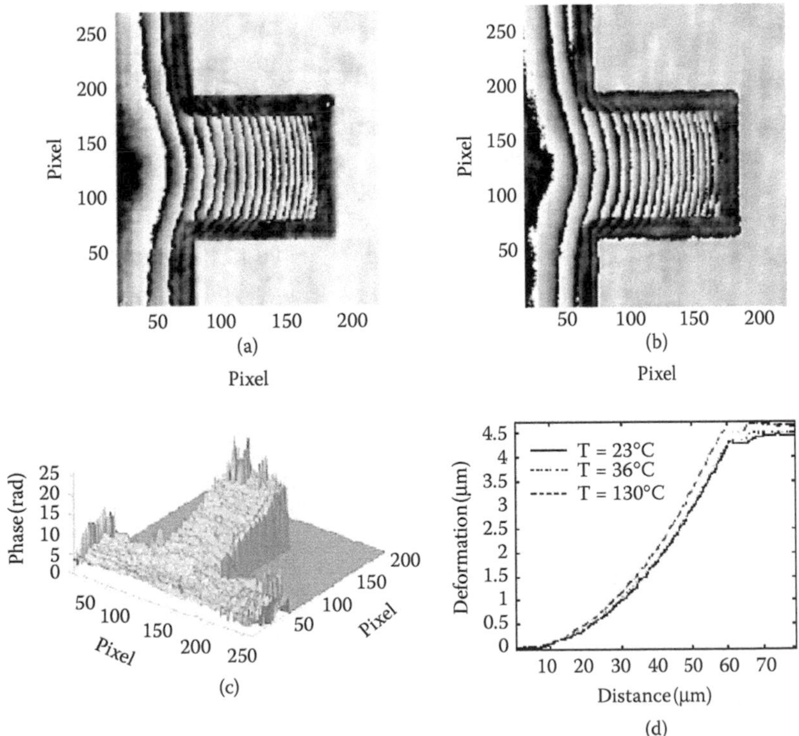

FIGURE 12.55 (a) A phase map for a T-shaped cantilever at 23°C, (b) a phase map for a T-shaped cantilever at 130°C, (c) the full field of the difference between the phase map at 130°C and that at 23°C, (d) cantilever profiles, at mid-section, for three different temperatures: 23, 56, and 130°C. (From Ref. 52.)

12.5.3.2.2 Testing Complex Si Micromembranes

Recent trends in gas sensors involve use of miniaturized devices characterized by very low power consumption and by fabrication processes compatible with IC silicon technology, which offer many potential advantages compared to standard sensor fabrication processes. Following these trends, many efforts have been made to develop gas sensors on Si, and several different microstructures have been conceived. They rely on a thin sensing film deposited on a suspended membrane having low thermal conductivity and realized using the Si bulk micromachining technique. The gas sensor fabrication needs different materials, with different properties, and different technological processes, which involve high-temperature treatments. Consequently, the structure is affected by the presence of residual stresses, appearing in the form of undesired bending of the membrane. Moreover, when the temperature of the heating resistor increases, a further warpage of the structure is induced. High thermal stress gradients and induced deformations can lead to mechanical failure of the structure.

Hence, for fabricating reliable gas sensors, it is of primary importance to investigate and to understand the behavior of the structure under actual working conditions. To this end, for this kind of structure two main nondestructive methods are of interest: thermographic and deformations analysis. Thermographic analysis can give information on the presence of hot-spots in the heating element that can induce a very high mechanical stress gradient. Deformation analysis is also of fundamental importance for giving information about the quantitative deformation caused by the thermal load. Here, we present how DHM can be applied to perform continuous monitoring of deformations occurring in such complex microstructures. Using DH, it is possible to determine, with high accuracy, the out-of-plane displacement map due to the residual stress in quasi-real-time and how this deformation is affected by thermal loads. In particular, a map of the profile of the structure has been evaluated both in static and quasi-static conditions. In the latter case, the profile changes due to the biasing of the heater resistor have been measured. The components of the test device structure are (see Figure 12.56a from top to bottom): the sensing layer, the passivation layer, the heater resistor, and the thin-film membrane on silicon substrate. In Figure 12.56b, a picture of the upper side of the structure as seen through an optical microscope is shown. The membrane dimension is 1.5×1.5 mm^2, whereas the device active area is about 0.8×0.8 mm^2. The heating resistor was fabricated by selective etching of a Pt layer sputtered onto the Si_3N_4 membrane. A layer of TiN was used to improve the Pt adhesion to the substrate. Finally, a passivation layer is necessary to electrically insulate the heater from the sensing element.

The heating element has a spiral shape. The spiral layout has a variable pitch, to account for the distribution of heat losses inside the active area [69]. An average gradient of about 0.05°C/μm

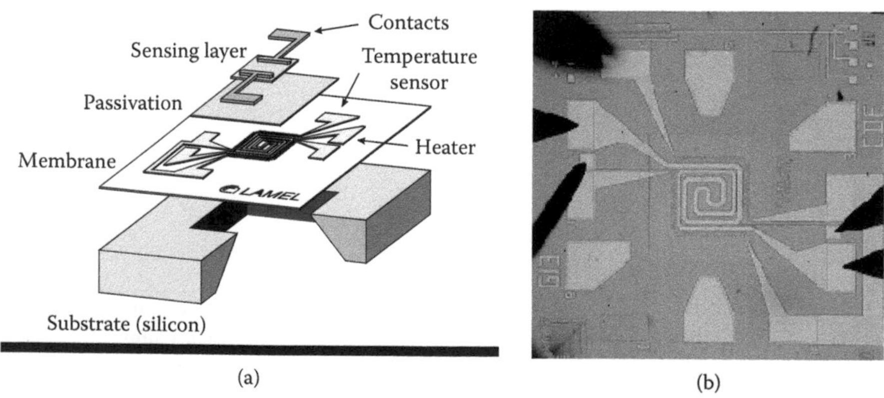

FIGURE 12.56 Sensor device structure: (a) layout, (b) picture as seen through an optical microscope.

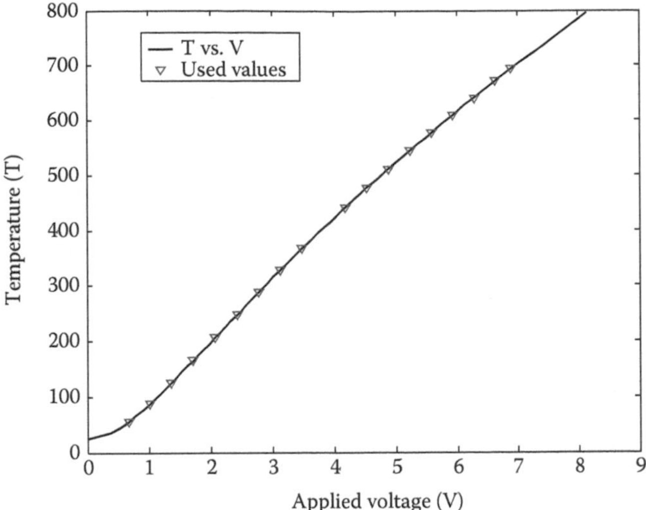

FIGURE 12.57 Temperature vs. applied voltage of the sensor.

has been obtained in the active area in the case of a peak temperature of 460°C. Out of the edge of the active area, the temperature decreases very rapidly, reaching the ambient temperature.

The microstructure was tested by DHM in off-axis configuration. The membrane was mounted on the optical bench and the temperature was changed by applying different voltages to the heating resistor (see Figure 12.57). First, a digital hologram was recorded at room temperature without application of voltage. Thereafter, while the voltage and, consequently, the temperature of the structure was raised, a sequence of holograms was recorded. In Figure 12.58a and Figure 12.58b, respectively, typical amplitude and phase reconstruction maps of the membrane are shown. The phase map is obtained by the difference between the phase at room temperature and at 400°C.

The phase map of each hologram was subtracted from the phase map of the first reference hologram to obtain the out-of-plane displacement occurring between each hologram with respect

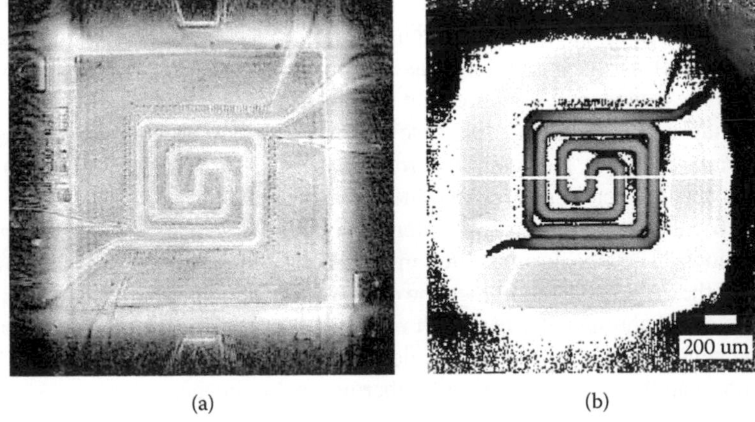

FIGURE 12.58 (a) Amplitude reconstruction, (b) phase map reconstruction at 400°C with respect to room temperature. (From Ref. 69.)

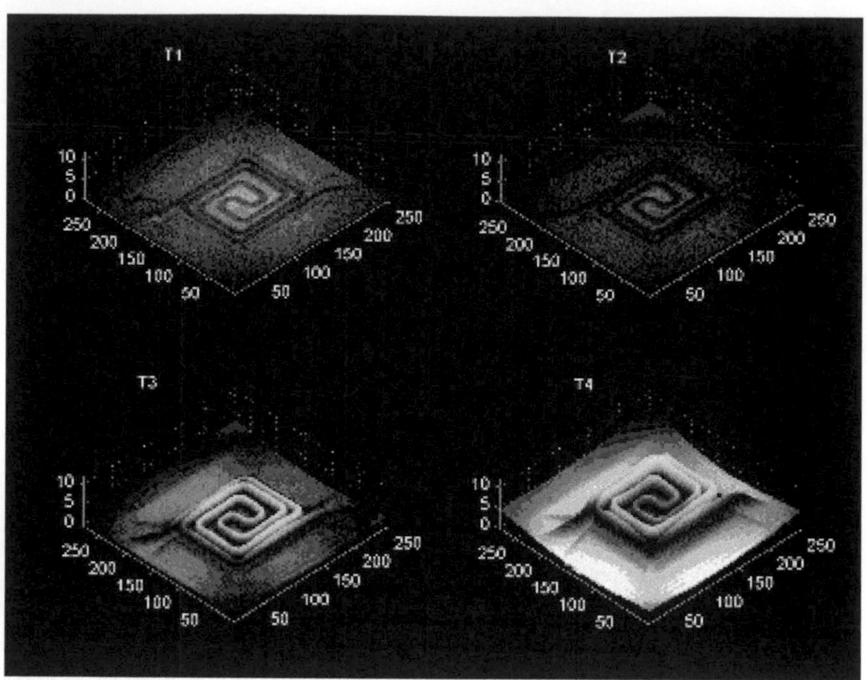

FIGURE 12.59 (See color insert.) Out-of-plane deformation of the membrane at different temperatures: T1 = 100°C, T2 = 200°C, T3 = 400°C, T4 = 700°C (Z-axis is in radians).

to the first one. From the phase differences, the deformation of the membrane can be readily extracted as a function of the temperature, because the relationship between the voltage applied and the temperature is known independently. In Figure 12.59, the phase differences between the hologram recorded at room temperature (V = 0.0 V) and four higher temperatures, respectively, at T1 = 100°C, T2 = 200°C, T3 = 400°C, and T4 = 700°C, corresponding to four holograms or the recorded sequence are shown. Deformations and distortions of the membrane are clearly visible when the temperature is increased. The thermal load was applied on such microstructures, some unpredictable longitudinal displacement of support could occur. In such cases, to obtain amplitude- and phase-contrast images reconstructed well in focus, the distance employed in the numerical reconstruction process has to be updated for each hologram. For this reason, even for the case described in this subsection, during the recording of the sequence, the phase-shift signal in a small flat area (4×4 pixels) of the holograms was detected. In Figure 12.60, the detected signal as recorded during the experimental measurement is shown.

From the signal, it is clear that in this case a very small overall longitudinal displacement of the sample occurred. However, the signal recorded in that case is very useful for compensating for the phase shift of the different maps for the different holograms.

In Figure 12.61, the out-of plane profile along a line at the center of the spiral for the different temperatures is shown. The signals are shifted in phase and demonstrate a very large deformation in correspondence to the metallic spiral. In fact, the deformation is mainly because of the higher thermal expansion coefficient of the metallic part with respect to the silicon membrane. A buckling of the membrane at the center is also very clear. The deformation map can be very useful when compared to the temperature map that can be obtained by thermographic images.

12.5.3.3 Microlens Testing

The results of DHM [43] technology to measure precisely the shape of microoptical components such as refractive and diffractive lenses, and also some examples of diffracting devices realized

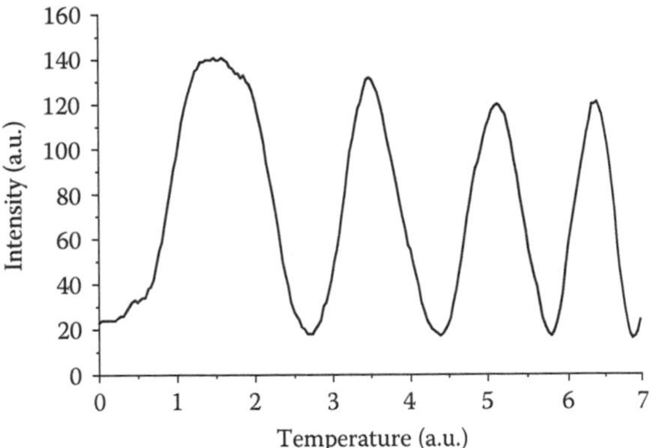

FIGURE 12.60 Phase-shift signal detected at the flat area close to the membrane, revealing a small longitudinal displacement.

with integrated optics technology, are reported in this subsection. The measurements were realized with an instrument developed previously in the microvision and microdiagnosis group1 at the Institute of Applied Optics of the Swiss Federal Institute of Technology in Lausanne (Switzerland), in collaboration with a start-up company Lyncée Tec SA2. Results reported in this subsection were obtained through the cooperation of the Disco program, where the performances of DHM applied to the measurement and quality control of microoptical components were evaluated.

As DHM is an imaging technique with high resolution and real-time observation capabilities, it appears as an imaging modality that offers broad and appealing perspectives in microscopy. The method makes available quantitative data that are derived from the digitally reconstructed complex wavefronts. Simultaneous amplitude and absolute phase are computed from a single hologram, acquired digitally with a video camera, and used to determine precisely the optical path length of the beam propagating in the microoptical device or reflected by it. Highly resolved images of the refractive indices and/or of the shape of the object can be derived from these data. The method requires the adjustment of several reconstruction parameters, which can be performed easily with a computer-aided method developed in our group. Using a high NA, submicron transverse resolution has been achieved. Accuracies of approximately half a degree have been estimated for phase

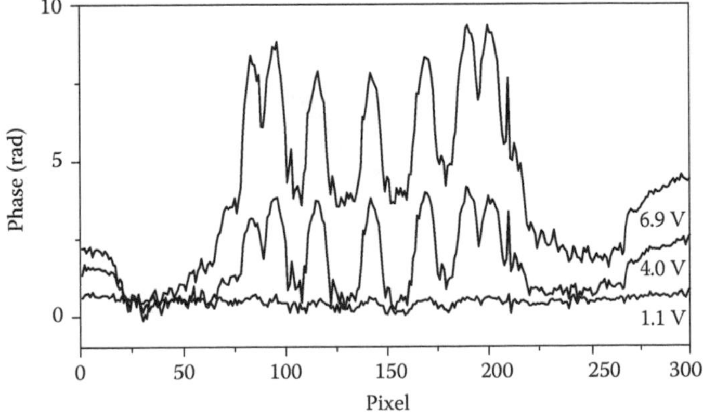

FIGURE 12.61 Example of profiles in radians obtained along the white line in Figure 12.58b for three different applied voltages.

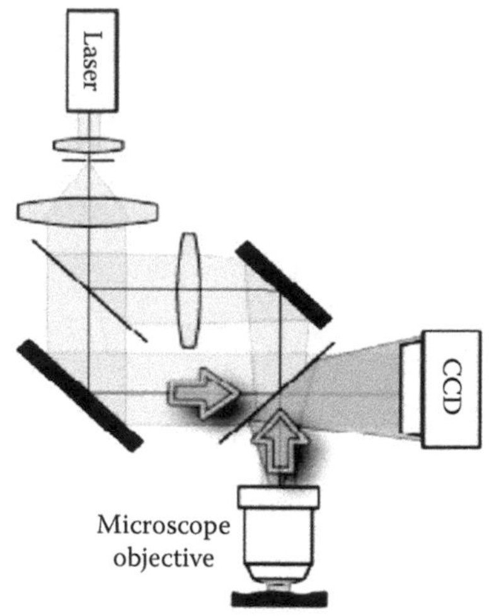

FIGURE 12.62 DHM in a reflection.

measurements. In reflection geometry, this corresponds to a vertical resolution less than 1 nm at a wavelength of 630 nm. In the transmission geometry, the resolution is limited to around 10 nm for an object with a refractive index of around 1.5 in air.

DHM was employed for the recording of a digital hologram of the specimen by means of a standard CCD camera inserted at the exit of a "Michelson-related" (see Figure 12.62) or Mach–Zehnder type (see Figure 12.63) interferometer. The geometries of these systems have been chosen for imaging the specimen in transmission and in reflection, respectively. Several other configurations are possible, according to the targeted applications. The reference beam is controllable both in intensity and polarization in order to optimize contrast and signal.

For the investigation of a microscopic specimen, an MO is introduced in the interferometer to adapt the sampling capacity of the camera to the information content of the hologram.

The reconstruction method is based on the Huygens–Fresnel theory of diffraction described earlier, which can be also considered as valid for this application. A magnifying lens is used to

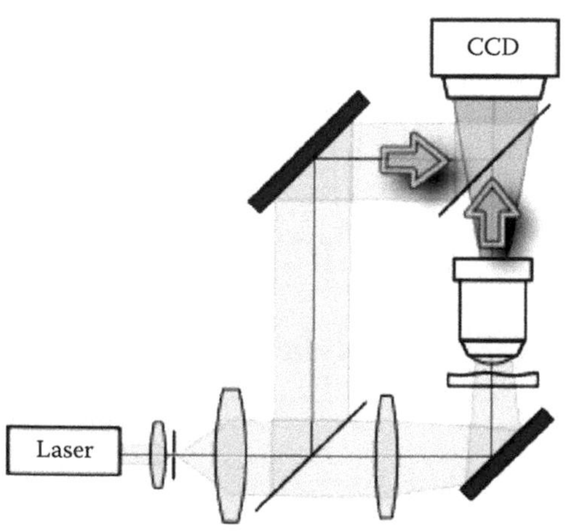

FIGURE 12.63 DHM in a transmission.

form the hologram on the CCD chip. The propagation of the complex optical wavefront can be simply computed as the Fresnel transform of the wave field.

To assess the precision of DHM imaging technology for the measurement of the profile of microoptical components, a comparison is made with the results of a scanning probe instrument. The instrument is an Alpha-step 200 from Tencor Instrument and yield measurements of profiles with an accuracy of 5 nm (vertical resolution) with a 12-μm tip diameter. Data were collected on a diameter of the lens. For DHM measurements, ten holograms providing ten profiles were registered and reconstructed to evaluate the temporal stability of the system. This gave a mean standard deviation on one pixel of 2.5° = 8.9 nm. This value can be considered our axial resolution for this sample. The comparison between the mean DHM profile and the alpha step profile has been carried out in detail. The matching of the two measurements has demonstrated that the DHM measurement provides profiles of the components with high accuracy (errors below 1%). Quantitative, high-accuracy data can be obtained on the entire lens from a single hologram, making DHM a unique, fast, and precise tool for refractive lens analysis.

12.5.3.3.1 Refractive Optics

A refractive lens of diameter ≈200 μm and thickness 18 μm has been measured. It is observed through a transmission DHM with a $\lambda = 658$ nm laser source and a 20X MO (NA = 0.5, f = 9.18 mm). Such a lens can be considered to be representative of a large category of refractive lenses manufactured in the microoptical industry. Figure 12.64 shows typical phase images obtained with DHM, without (a) and with (b) 2-D phase unwrapping. Without phase unwrapping, the phase jumps due to optical path lengths greater than the wavelength can be observed. The slopes of the sides of the lens are weak enough so that the transverse resolution is sufficient to numerically unwrap the phase image without errors. The corresponding 3-D pictures of Figure 12.65 illustrate the imaging capability of the DHM.

Comparative measurements between DHM and the scanning probe (Alpha-step 200 from Tencor Instrument) are shown in Figure 12.66. An excellent match between the two measurements is obtained, within 1%. DHM quantitative data can be derived from a single hologram and with high accuracy over a field of view covering the entire lens. Determination of the radius of curvature of the microlens is accurate to within less than 1%.

Comparison with standard models of microlens' shapes can be achieved numerically. In particular, a small deviation from an ideal spherical surface of the lens can be detected by fitting the experimental phase data to the mathematical expression. The residues are computed and presented in Figure 12.67 for five different refractive lenses, in which some defects appear regularly, in particular, at the center of the lens and on the border. To document the presence of such defects in more detail, in Figure 12.68, the residue measured on a profile taken on a horizontal diameter of the lens shown on Figure 12.67a is shown. Other defects appear more localized and are not repeated

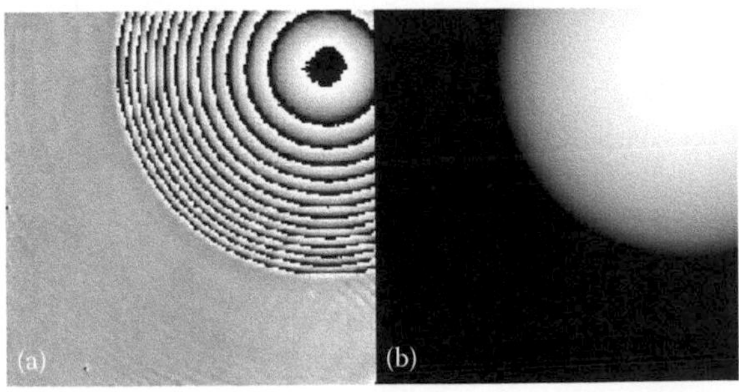

FIGURE 12.64 Refractive lens: (a) phase image, (b) unwrapped phase image.

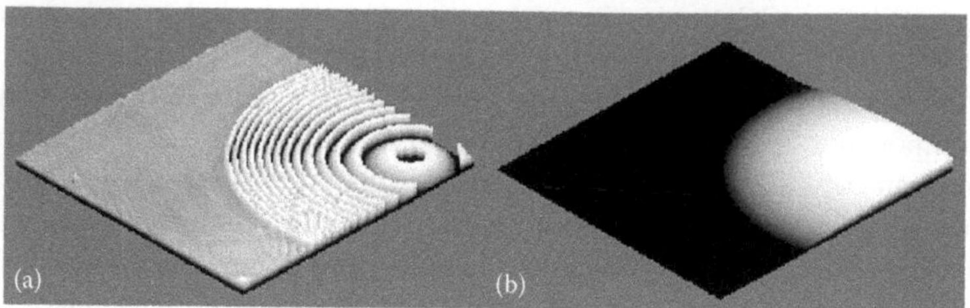

FIGURE 12.65 Refractive lens: (a) perspective phase image, (b) perspective unwrapped phase image.

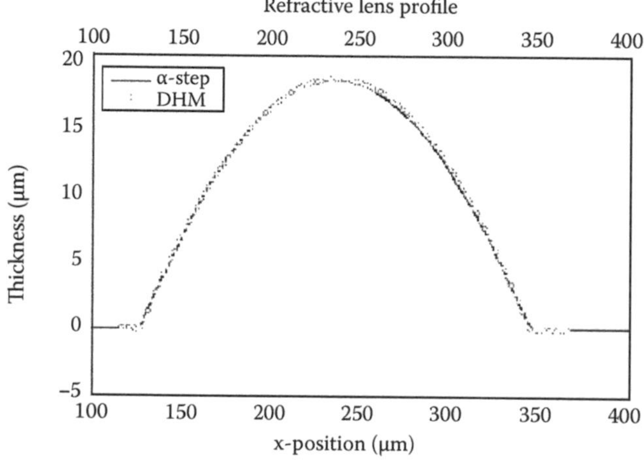

FIGURE 12.66 Comparison between DHM and an Alpha-step.

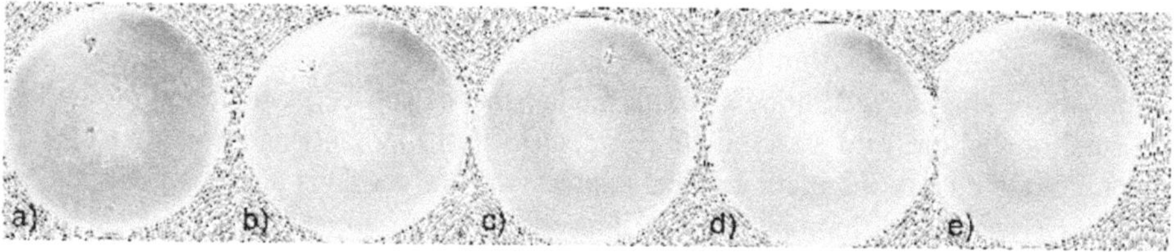

FIGURE 12.67 Five different lenses: a mathematical model for the spherical shape has been used to compare real and ideal profiles. After adjustment, the residue is obtained as the difference between the real and ideal profile and is presented in figure (a) to (e). These residues are called "defect of sphericity."

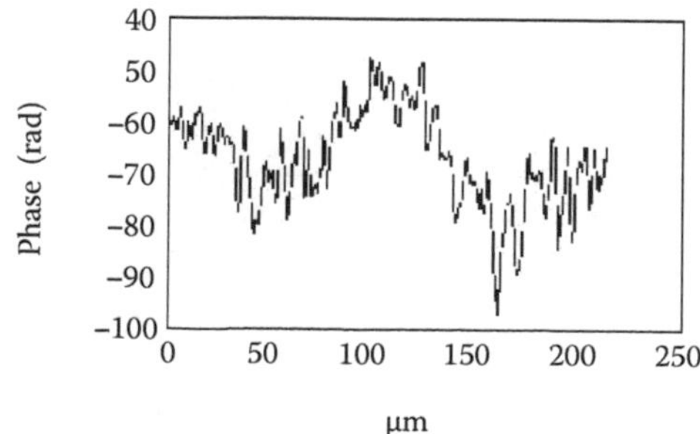

FIGURE 12.68 Typical residue: difference between real and ideal spherical profile of a lens.

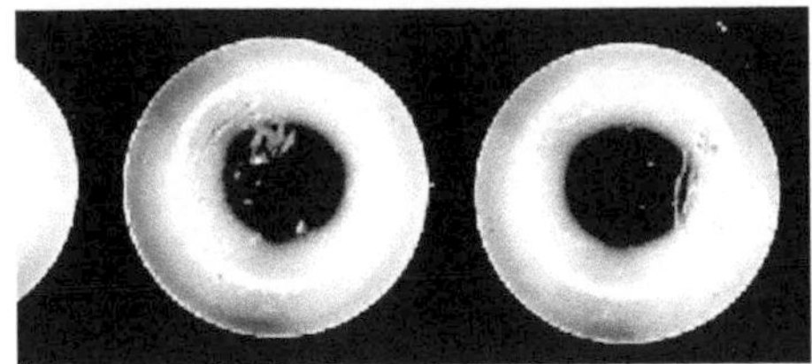

FIGURE 12.69 Refractive lens observed with dark field microscope 10X and showing the presence of localized defects.

on each investigated lens. Such defects can also be observed on a dark field image taken with a 10X MO (Figure 12.69).

12.5.3.3.2 Silicon Refractive Lens

Measurements carried out on silicon microlenses with a reflection DHM setup are reported. The diameter of the lens is around 250 μm, and it has a thickness of 4.2 μm. A (λ = 658 nm) laser source and a 20X MO were used. The image shown in Figure 12.70 presents the typical phase images obtained with DHM, in which it is possible to observe the phase jumps due to optical path lengths greater than one wavelength. In this case, the transverse resolution is sufficient to unwrap the phase image without loss of accuracy.

A comparison between DHM and scanning probe measurements (12-μm tip, 5-nm vertical resolution) were made on the profile of the lens. The comparison between the unwrapped DHM profile and the alpha step profile is shown on Figure 12.71. A high accuracy match (error less than 1%) of the two measurements was obtained. The great advantage of a DHM instrument is that quantitative and accurate topology can be derived from a single hologram over the entire lens, rendering DHM fast and precise for refractive lens analysis.

12.5.3.3.3 Fresnel Lens

A microoptical system composed of Fresnel lenses of diameter around ≈ 250 μm and thickness of 1 μm was imaged through a transmission DHM with a λ = 658 nm laser source and a 10X MO. The two first images, shown in Figure 12.72, present the typical phase images obtained with the DHM, in gray level (a) and in perspective (b).

Comparison between the DHM profile and that obtained with the scanning probe (alpha step, 12- μm tip,) is presented in Figure 12.73. The profiles show larger differences (a few percent) than

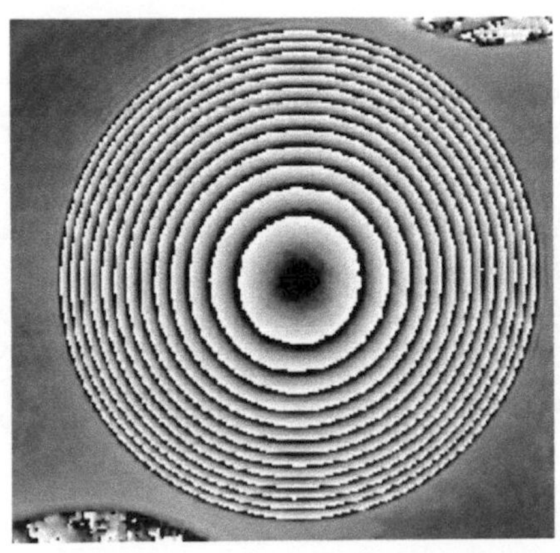

FIGURE 12.70 Si-refractive lens phase image taken with the reflection DHM, MO 20X.

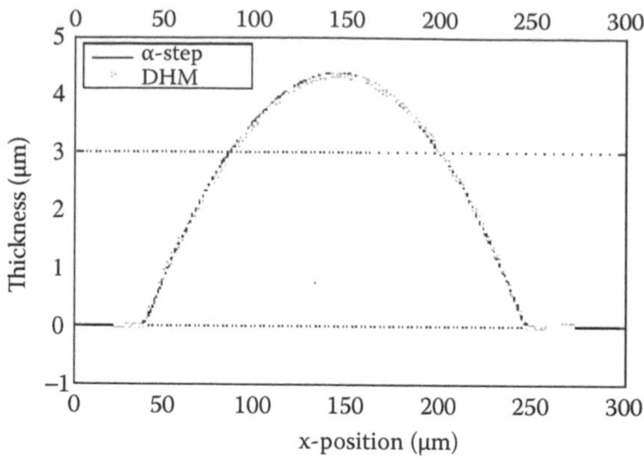

FIGURE 12.71 Comparison between the profiles measured of a Si refractive lens with DHM and with a scanning probe apparatus. Phase unwrapped from DHM data.

those obtained with refractive lenses (less than 1%). The observed variations of the DHM profile are stable and repetitive in time and cannot be attributed to some random fluctuations or noise. They are more likely because of the parasitic contribution of diffraction originating from rapid variations of the profile slope, near the steep edges of the Fresnel lens. Some corrections should be made to the calculation of the profile near these steep edges according to the Rayleigh–Sommerfeld formula. A much better estimate of the real profile can be achieved by taking into account a more exact expression of diffraction. Nevertheless, it appears from the comparison that some smoothing of the profile is observed with the scanning probe, which is due to the natural width of the probe tip: 12 μm. A major advantage of DHM is, therefore, its capacity to restore, after correction for the profile slope, the precise profile of the ridges in their dips and tips. Such precision can hardly be achieved with a stylus having a tip size larger than 10 μm.

12.5.3.3.4 Diffractive Optics

Transparent diffractive systems were imaged with a transmission digital holographic microscope with a $\lambda = 658$ nm laser source, a 20X and a 63X MO, and with a reflection digital holographic microscope having a $\lambda = 658$ nm laser source and a 20X MO.

Figure 12.74 and Figure 12.75 show the phase images of the diffractive system, with four different phase levels appearing, corresponding to four different thicknesses of the sample. The repetitive pattern is formed of crosses and squares diffracting light.

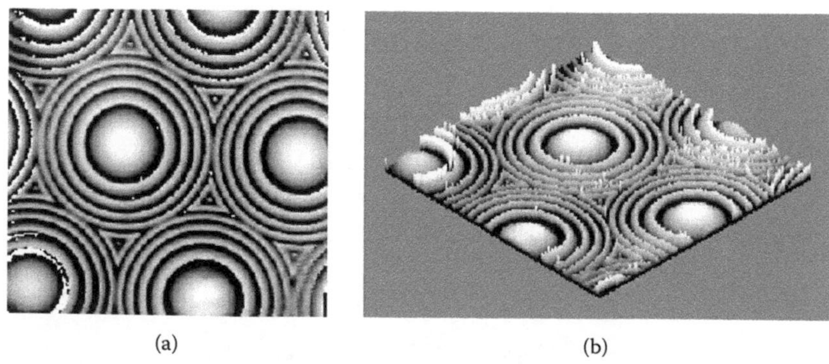

FIGURE 12.72 (a) Fresnel lens phase image, (b) Fresnel lens perspective phase image.

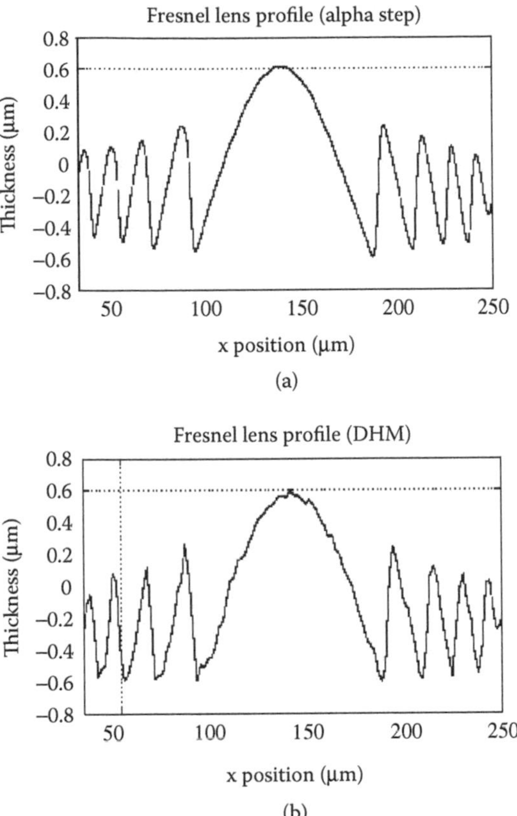

FIGURE 12.73 Comparison between DHM (lower) and an alpha step (upper) thickness measurement (12 μm tip, 5 nm vertical resolution) on the profile of a Fresnel lens.

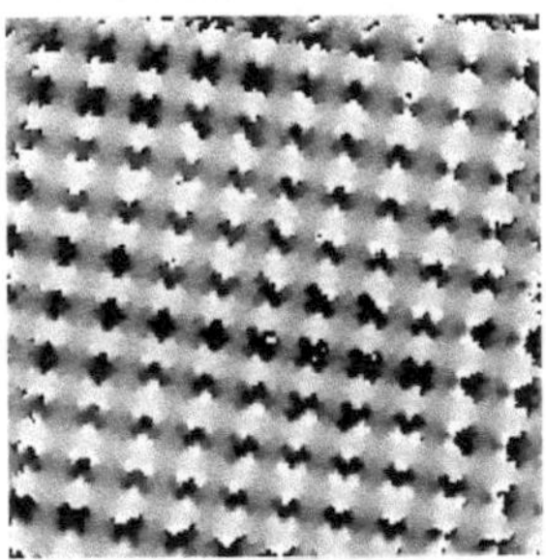

FIGURE 12.74 Phase image of the diffractive transmission 20X, FOV 240 μm.

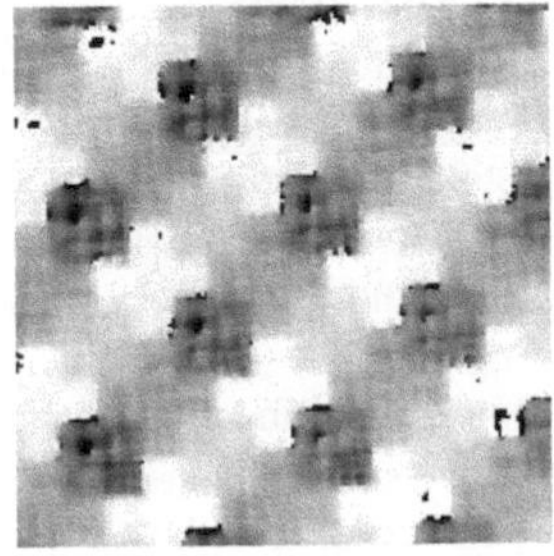

FIGURE 12.75 Phase image of the diffractive device observed in transmission 63X, FOV 70 μm.

Experimental data show excellent agreement with the profiles obtained with an instrument based on a mechanical probe scanning principle. The images shown here illustrate the great potential of DHM in the precise determination of the geometries and properties of microoptical components.

Some limitations on DHM technology are to be expected when highly diffracting elements are investigated: steep profiles generate wavefront components that cannot be taken into account accurately in wavefront reconstruction, at least in the present state of the art. These contributions introduce some small deviations from the real profile, which appears as a systematic error. More accurate modeling of the diffraction will be considered to improve profile measurement in strongly diffracting structures. On the other hand, the measurement of the profile of steep structures cannot be easily achieved with stylus-probe measuring equipment: the size of the stylus prevents the operator from reaching the steepest points and the vertical parts of the structures. Some improvement in DHM technology would therefore yield a unique measuring tool for investigations and quality control in microoptics.

12.6 CONCLUSION

In a few years, digital holography has become a valuable diagnostic tool for inspection and characterization of microcomponents and microsystems. A significant research effort has been put in by many scientists both in academic as well as in industrial groups. One of the key advantages of DH is the intrinsic flexibility offered by the numerical reconstruction process. In fact, the method is flexible in different ways: focus adjustment, aberration removal, and direct management of the phase map. All these features make DH complementary to the well-established interferometers in microscope configuration.

Although DH seems to be mature for a variety of applications in the industrial field, we are sure that DH will be the object of more and surprising developments in the near future. DH will also benefit from advancements in related fields such as improvements of resolution of solid-state arrays (CCD and CMOS), development of faster and more efficient algorithms for image reconstruction, and development of new display technologies with higher resolution.

Such developments will give more opportunities for DH to be used for even newer and more complicated situations in optical metrology, especially in the micro–nanoscale domains.

REFERENCES

1. Gabor, D., Microscopy by reconstructed wave-fronts, *Proc. R. Soc. A,* 197, 454, 1949.
2. Goodman, J.W. and Lawrence, R.W., Digital image formation from electronically detected holograms, *Appl. Phys. Lett.*, 11, 77, 1967.
3. Demetrakopoulos, T.H. and Mitra, R., Digital and optical reconstruction of images from suboptical patterns, *Appl. Opt.*, 13, 665, 1974.
4. Yaroslavsky, L.P. and Merzlyakov, N.S., *Methods of Digital Holography*, Consultants Bureau, New York, 1980.
5. Schnars, U., Direct phase determination in holographic interferometry using digitally recorded holograms, *J. Opt. Soc. Am. A*, 11, 2011, 1994.
6. Osten, W., Active metrology by digital holography, *Proc. SPIE*, 4933, 96, 2003.
7. Seebacher, S., Osten, W., and Jüptner, W., 3-D Deformation analysis of micro-components using digital holography, *Proc. SPIE*, 3098, 382, 1997.
8. Osten, W., Jüptner, W., and Seebacher, S., The qualification of large scale approved measurement techniques for the testing of micro-components, *Proc.* 18*th Symp. on Experimental Mechanics of Solids*, Jachranka, 1998, p. 43.
9. Butters, J.N. and Leendertz, J.A., Holographic and video techniques applied to engineering measurement, *J. Meas. Control*, 4, 349, 1971.
10. Schnars, U. and Jüptner, W., Direct recording of holograms by a CCD target and numerical reconstruction, *Appl. Opt.*, 33, 197, 1994.

11. Grilli, S. et al., Whole optical wavefields reconstruction by digital holography, *Opt. Exp.*, 9, 294, 2001.
12. Kreis, Th. and Jüptner, W., Principles of digital holography, Jüptner, W. and Osten, W., Eds., *Proc. Fringe'97, 3rd International Workshop on Automatic Processing of Fringe Patterns*, Akademie Verlag, Berlin, 1997, p. 353.
13. Hariharan, P., *Optical Holography*, Cambridge University Press, Cambridge, 1984.
14. Kreis, Th. and Jüptner, W., The suppression of the dc-term in digital holography, *Opt. Eng.*, 36, 2357, 1997.
15. Leith, E.N. and Upatnieks, J., New techniques in wavefront reconstruction, *J. Opt. Soc. Am.*, 51, 1469, 1961.
16. Wagner, C. et al., Digital recording and numerical reconstruction of lensless Fourier holograms in optical metrology, *Appl. Opt.*, 38, 4812, 1999.
17. Goodman, J.W., *Introduction to Fourier Optics*, McGraw-Hill, New York, 1996.
18. Takeda, M. et al., Single transform Fourier-Hartley fringe analysis for holographic interferometry, in *Simulation and Experiment in Laser Metrology*, Füzessy, Z., Jüptner, W., and Osten, W., Eds., Akademie Verlag, Berlin, 1996, p. 67.
19. Yamaguchi, I. and Zhang, T., Phase-shifting digital holography, *Opt. Lett.*, 22, 1268, 1979.
20. Zhang, Y. et al., Image reconstruction for in-line holography with the Yang-Gu algorithm, *Appl. Opt.*, 42, 6452, 2003.
21. Lai, S. et al., A deformation and 3D-shape measurement system based on phase-shifting digital holography, *Proc. SPIE*, 4537, 273, 2002.
22. Seebacher, S., Application of Digital Holography for 3D-Shape and Deformation Measurement of Micro Components, Ph.D. thesis, University of Bremen, 2001.
23. Vest, C.M., *Holographic Interferometry*, John Wiley & Sons, New York, 1979.
24. Kreis, T., *Handbook of Holographic Interferometry — Optical and Digital Methods*, Wiley VCH, Weinheim, 2005.
25. Robinson, D.W., Phase unwrapping methods, in *Interferogram Analysis,* Robinson, D.W. and Reid, G.T., Eds., IOP Publishing, Bristol and Philadelphia, 1993, p. 194.
26. Sollid, J.E., Holographic interferometry applied to measurements of small static displacements of diffusely reflecting surfaces, *Appl. Opt.*, 8, 1587, 1969.
27. Osten, W., Some considerations on the statistical error analysis in holographic interferometry with application to an optimized interferometer, *Opt. Acta*, 32, 827, 1987.
28. Osten, W. and Kolenovic, E., An Optimized interferometer for 3D digital holographic endoscopy, *Proc. SPIE,* 5144, 150, 2003.
29. Maack, Th., Notni, G., and Schreiber, W., Three-coordinate measurement of an object surface with a combined two-wavelength and two-source phase-shifting speckle interferometer, *Opt. Commun.*, 115, 576, 1995.
30. Wagner, C., Osten, W., and Seebacher, S., Direct shape measurement by digital wavefront reconstruction and multi-wavelength contouring, *Opt. Eng.*, 39, 79, 2000.
31. Huntley, J.M., Automated analysis of speckle interferograms, in *Digital Speckle Pattern Interferometry and Related Techniques*, Rastogi, P.-K., Ed., John Wiley & Sons, Chichester, U.K., 2001, 59.
32. Skudayski, U. and Jüptner, W., Synthetic wavelength interferometry for the extension of the dynamic range, *Proc. SPIE*, 1508, 68, 1991.
33. Huntley, J.M. and Saldner, H., Temporal phase-unwrapping algorithm for automated interferogram analysis, *Appl. Opt.*, 32, 3047, 1993.
34. Takeda, M. and Yamamoto, H., Fourier-transform speckle profilometry: three-dimensional shape measurement of diffuse objects with large height steps and/or spatially isolated surfaces, *Appl. Opt.*, 33, 7829, 1994.
35. Zou, Y., Pedrini, G., and Tiziani, H., Surface contouring in a video frame by changing the wavelength of a diode laser, *Opt. Eng.*, 35, 1074, 1996.
36. Osten, W., Nadeborn, W., and Andrä, P., General hierarchical approach in absolute phase measurement, *Proc. SPIE*, 2860, 2, 1996.
37. Nadeborn, W., Osten, W., and Andrä, P., A robust procedure for absolute phase measurement, *Opt. Laser. Eng.*, 24, 245, 1996.
38. Seebacher, S. et al., Combined 3D-shape and deformation analysis of small objects using coherent optical techniques on the basis of digital holography, *Proc. SPIE*, 4101B, 520, 2000.

39. Seebacher, S. et al., The determination of material parameters of microcomponents using digital holography, *Opt. Laser. Eng.* 36, 103, 2001.
40. Kolenovic, E. et al., Miniaturized digital holography sensor for distal three-dimensional endoscopy, *Appl. Opt.*, 42, 5167, 2003.
41. Osten, W., Baumbach, T., and Jüptner, W., Comparative digital holography, *Opt. Lett.*, 27, 1764, 2002.
42. Whitehouse, D.J., Review article: surface metrology, *Meas. Sci. Technol.* 8, 955, 1997.
43. Cuche, E. et al., Digital holography for quantitative phase contrast imaging, *Opt. Lett.*, 24, 291, 1999.
44. Ferraro, P. et al., Compensation of the inherent wave front curvature in digital holographic coherent microscopy for quantitative phase contrast imaging, *Appl. Opt.*, 42, 1938, 2003.
45. Pedrini, G. et al., Aberration compensation in digital holographic reconstruction of microscopic objects, *J. Mod. Opt.*, 48, 1035, 2001.
46. Stadelmaier, A. and Massig, A.H., Compensation of lens aberrations in digital holography, *Opt. Lett.*, 25, 1630, 2000.
47. De Nicola, S. et al., Wave front reconstruction of Fresnel off-axis holograms with compensation of aberrations by means of phase-shifting digital holography, *Opt. Laser. Eng.*, 37, 331, 2002.
48. De Nicola, S. et al., Correct-image reconstruction in the presence of severe anamorphism by means of digital holography, *Opt. Lett.*, 26, 974, 2001.
49. Lei, X. et al., Studies of digital microscopic holography with applications to microstructure testing, *Appl. Opt.*, 40, 5046, 2001.
50. Ferraro, P. et al., Digital holography for characterization and testing of MEMS structures, *Proc. IEEE/LEOS Int. Conf. on Optical MEMS*, New York, 2002, 125.
51. Lei, X. et al., Hybrid holographic microscope for interferometric measurement of microstructures, *Opt. Eng.*, 40, 2533, 2001.
52. Coppola, G. et al., A digital holographic microscope for complete characterization of microelectromechanical systems, *Meas. Sci. Technol.*, 15, 529, 2004.
53. Ferraro, P. et al., Recent advancements in digital holographic microscopy and its applications, in *Optical Metrology in Production Engineering*, Osten, W. and Takeda, M., Eds., *Proc. SPIE,* 5457, 481, 2004.
54. Osten, W. et al., The qualification of optical measurement techniques for the investigation of material parameters of microcomponents, *Proc. SPIE*, 3825, 152, 1999.
55. Jüptner, W. et al., Combined measurement of silicon microbeams by grating interferometry and digital holography, *Proc. SPIE*, 3407, 348, 1998.
56. Cuche, E. et al., Simultaneous amplitude-contrast and quantitative phase-contrast microscopy by numerical reconstruction of Fresnel off-axis holograms, *Appl. Opt.*, 38, 6994, 1999.
57. Gu, J. and Chen, F., Fast Fourier transform, iteration, and least-squaresfit demodulation image processing for analysis of single-carrier fringe pattern, *J. Opt. Soc. Am. A*, 12, 2159, 1995.
58. Yamaguchi, I., Matsumura, T. and Kato, J., Phase-shifting colour digital holography, *Opt. Lett.*, 27, 1108, 2002.
59. Demoli, N., Vukicevic, D., and Torzynski, M., Dynamic digital holographic interferometry with three wavelengths, *Opt. Express*, 11, 767, 2003.
60. Ferraro, P. et al., Controlling image size as a function of distance and wavelength in Fresnel transform reconstruction of digital holograms, *Opt. Lett.*, 29, 854, 2004.
61. Ferraro, P. et al., Recovering image resolution in reconstructing digital off-axis holograms by Fresnel-transform method, *Appl. Phys. Lett.*, 85, 2709, 2004.
62. Ferraro, P., Controlling images parameters in the reconstruction process of digital holograms, *IEEE J. Selected Top. Quantum Electron.*, 10, 829, 2004.
63. Falldorf, C. et al., Digital grating interferometry: a complementary approach, *Proc. SPIE*, 5457, 225, 2004.
64. Othonos, A. and Kalli, K., *Fiber Bragg Gratings*, Artech House, Norwood, MA, 1999.
65. Yamaguchi, I. and Saito, H., Application of holographic interferometry to the measurement of poisson's ratio, *Jpn. J. Appl. Phys.*, 8, 768, 1969.
66. Osten, W. et al., Absolute shape control of microcomponents using digital holography and multi-wavelengths-contouring, *Proc. SPIE*, 4275, 71, 2001.
67. Seebacher, S. et al., Combined measurement of shape and deformation of small objects using digital holographic contouring and holographic interferometry, Rastogi, P., Ed., *Proc. Int. Conf. on Trends in Optical Nondestructive Testing*, Lugano, 2000, 55.

68. Coppola, G., Digital holography microscope as tool for microelectromechanical systems characterization and design, *J. Microlithography, Microfabrication, Microsyst.*, 4, 013012, 2005.
69. Coppola, G., Digital holographic microscope for thermal characterization of silicon microhotplates for gas sensor, in *Optical Micro- and Nanometrology in Manufacturing Technology*, Gorecki, C. and Asundi, A.K., Eds., *Proc. SPIE*, 5458, 228, 2004.

13 Speckle Metrology for Microsystem Inspection

Roland Höfling and Petra Aswendt

CONTENTS

13.1 Introduction 427
13.2 Basics 428
 13.2.1 Properties of Speckles in Imaging Systems 428
 13.2.2 Extracting Information from Speckle Patterns 431
  13.2.2.1 Classification of Speckle Signals for Metrology 432
  13.2.2.2 Evaluation Techniques for Signal Extraction 434
  13.2.2.3 Optical Arrangements for Speckle Interferometry 442
13.3 Applications 443
 13.3.1 Quality Assurance on Wafer Level 443
 13.3.2 Characterization of Operational Behavior 446
  13.3.2.1 Visible Light Interferometry for a Wide Scale 446
  13.3.2.2 DUV-Interferometry for Improved Resolution 452
13.4 Conclusion 456
References 457

13.1 INTRODUCTION

The methods described in this chapter rely on a common phenomenon: the granularity observed whenever a rough surface is illuminated with coherent light. These speckles cover the image of an object as a noisy pattern of high contrast.

Recognized early by physicists, the speckle effect attracted the attention of the community of optical engineering scientists by the application of lasers in the 1960s. It became obvious that the speckle pattern not only represents image-degrading noise but also carries information about the scattering object itself. Based upon their detailed mathematical description [1], the study of the measuring capabilities of speckles started in many laboratories, and most measuring principles originate from that time. Dainty [2] and Erf [3] edited the contributions of pioneering authors in 1975 and 1978, respectively. One of the promising options in speckle methods is the use of electronic imaging, and this path was followed from the very beginning [4,5]. The progress in computer technology, allowing the presence of computing facilities in the optical laboratory, is another development in the history of speckle metrology, enabling the storing and processing of speckle patterns digitally [6–8]. In the following decades, advanced CCD technology, solid-state laser sources, and miniaturized computing paved the way for speckle metrology to move out of the laboratory and onto the factory floor. Meanwhile, a number of measuring systems are commercially available and have proved valuable for a variety of applications. Apart from a few exceptions [9], speckle metrology was mainly applied to objects with sizes ranging from a few centimeters to some meters [10]. During the past decade, novel measuring needs caused by the birth of microelectromechanical systems (MEMS) have emerged.

The following sections will describe how speckle metrology can be modified and adapted in order to serve investigations in the microscopic world, i.e., for structural dimensions from a few millimeters to several micrometers. Section 13.2 introduces the basic concepts of speckle metrology and gives a brief theoretical background of the methodology. Three major fields of application are treated in Section 13.3: nondestructive testing, analysis of MEMS operation, and the determination of mechanical parameters of microsystems.

13.2 BASICS

13.2.1 Properties of Speckles in Imaging Systems

The small size of the object under investigation in MEMS technology requires optical magnification to match the size of a CCD chip. Therefore, microscopic imaging is needed, and correspondingly, the following treatment of speckles is restricted to systems as shown in Figure 13.1 (subjective speckles).

In order to understand the process of speckle generation, the following assumptions are made:

1. A monochromatic light source is used.
2. The surface roughness equals or exceeds the wavelength of light.
3. The surface reflection does not change the polarization.
4. The object is composed of small individual scatterers not resolvable by the imaging system used.

Assuming a plane wave illuminating the surface, each scatterer will introduce a different phase shift to the light collected by the imaging lens. The limited lens aperture (diameter D) causes diffraction so that a single scatterer is imaged as a small diffraction pattern. The size σ of such an Airy disk is of the order

$$\sigma \cong \frac{2\lambda z}{D} \tag{13.1}$$

where λ is the wavelength of light and z is the distance between lens aperture and image plane. The crux is that the diffraction disks of neighboring scatterers overlap and allow for interference of light. Correspondingly, the intensity I in a given image point is the result of the coherent superposition of all contributing Airy disks (Figure 13.2, left). That means, their complex light amplitudes A_i

$$A_i = |A_i| e^{i\varphi_i} \tag{13.2}$$

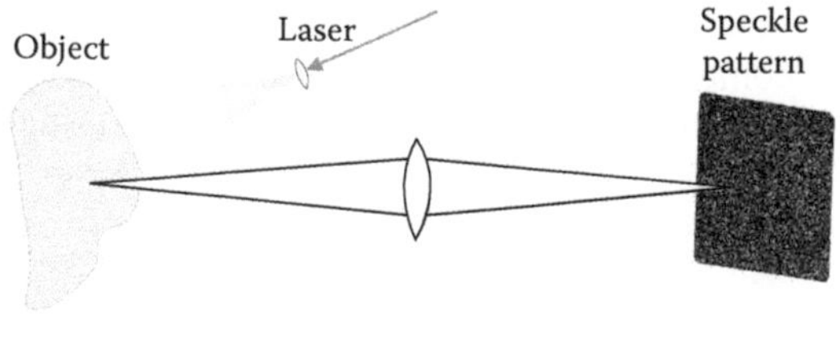

FIGURE 13.1 Generation of speckles in the image plane.

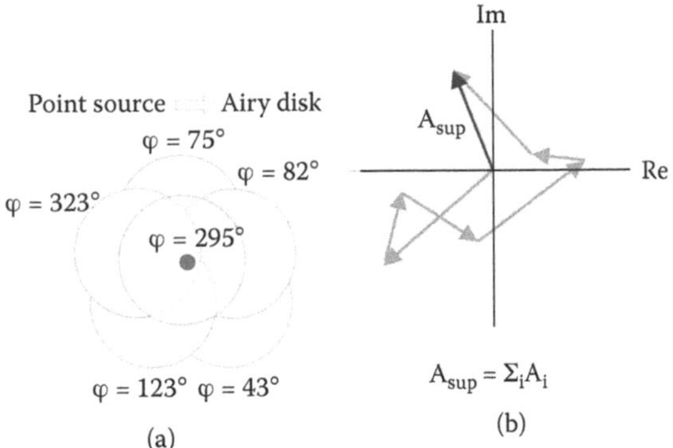

FIGURE 13.2 Speckle generation by interference of differently phased light from six scatterers (six phasors).

have to be added (Figure 13.2, right), taking into account their respective phase angles φ_i:

$$I = \left|A_{\text{sup}}\right|^2 = \left|\sum_i \left|A_i\right| e^{i\varphi_i}\right|^2 \tag{13.3}$$

For objects with uniform reflectivity, we have

$$\left|A_i\right|^2 = I_0 \tag{13.4}$$

and Equation 13.3 becomes

$$I = I_0 \cdot \left|\sum_i e^{i\varphi_i}\right|^2 \tag{13.5}$$

where I_0 is the object brightness under incoherent illumination.

Obviously, the intensity is mainly controlled by the phase mixture and can vary between 0 and I_0. The second term of Equation 13.5 represents multiplicative noise with maximum contrast: the speckle pattern. Figure 13.3 illustrates the effect for a microscopic component under white light and laser light, respectively.

FIGURE 13.3 Microstructure under white light (left) and laser illumination (right).

What happens when moving to a neighboring image point at a distance of one Airy disk diameter? Other scatterers will overlap there so that the phase mix is likely to be completely different at this position and the intensity can change by 100%. Consequently, the mean speckle size σ, i.e., the area of constant intensity, is of the size of the Airy disk (Equation 13.1). The full mathematical description of this phenomenon is given in Reference 1.

It was shown there that the statistical properties of speckles significantly differ from that of ordinary "noise" as it is present in detectors. The most prominent effect of speckle patterns is the outstanding contrast. Complete darkness is generated by destructive interference of light. Also of importance for optical metrology is the fact that the speckle modulation is multiplicative so that the high contrast is maintained also on the bright features of an object.

Another phenomenon can be understood from the complex amplitude vector sum shown in Figure 13.2. If the total amplitude A_{sup} is very low, only small changes in the overlap can cause high variations in speckle phase. In other words, sudden phase jumps can occur where the observed object points appears dark in the image. This effect has been investigated both theoretically [11–13] and experimentally [14,15]. An example is shown in Figure 13.14.

Aiming at investigations on a microscopic scale, it has to be considered whether high magnifications M influence the speckle size. A convenient formula was given by Ennos [11], using the aperture ratio F of the lens (f/number):

$$\sigma \cong (1+M)\lambda \cdot F \tag{13.6}$$

To find the resolution limit, it is useful to introduce the object-related speckle size Σ, i.e., the virtual projection of the speckle onto the object:

$$\Sigma \cong (1+M)\lambda F / M \tag{13.7}$$

By way of illustration, Equation 13.7 has been plotted for a system with F = 5.6 and a wavelength of 488 nm (Figure 13.4). The curve shows that the object-related speckle size decreases with increasing magnification. In other words, the speckles are not magnified with the object!

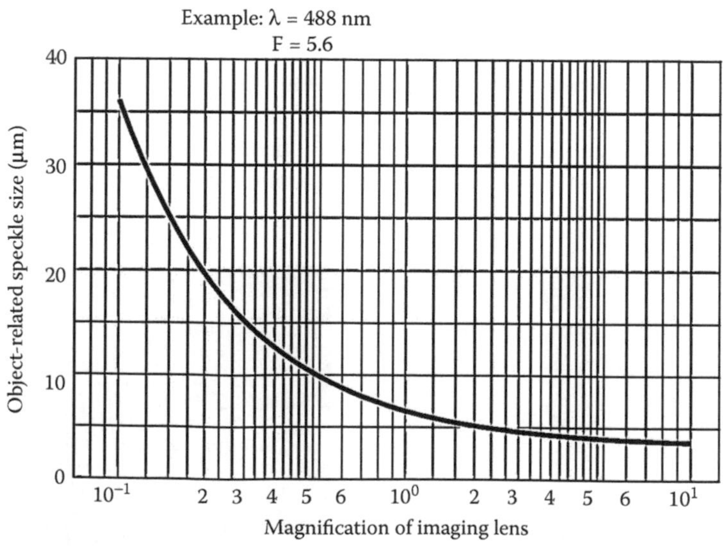

FIGURE 13.4 Object-related speckle size dependent upon magnification.

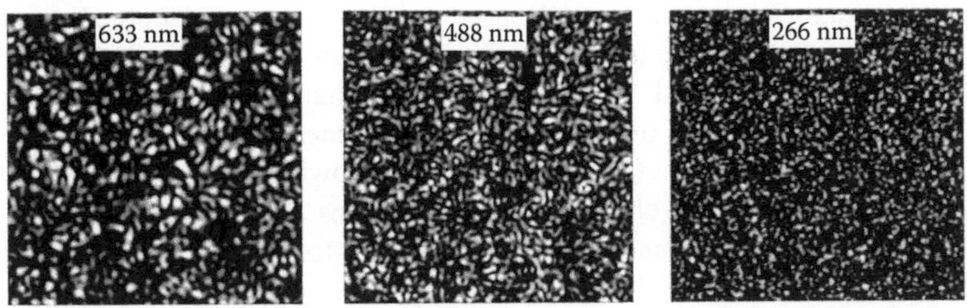

FIGURE 13.5 Speckle patterns for a microscopic field of view at three wavelengths.

The second parameter controlling the speckle size is the wavelength λ. It is quite clear that a shorter wavelength yields smaller diffraction disks and, therefore, smaller speckles in the image. These theoretical considerations have been proved experimentally, in particular for the microscopic range. Typical results of three different wavelengths are given in Figure 13.5.

An area of 75 μm is zoomed in order to make the speckles visible in the photographs. The mean speckle sizes observed are 2.0, 1.6, and 0.85 μm, respectively. Accordingly, microscopic features can be resolved also in the presence of speckle noise, and a short wavelength is promising for highest resolution.

Up to now, fully developed speckle fields have been considered for the case of a mean surface roughness greater than $\lambda/2$ so that the phase fluctuations caused by the scatterers cover the whole range from 0 to 2π. This condition is fulfilled for many engineering surfaces but is not evident for micromachining, and for this reason, the smoothness limitation has been studied. According to Asakura [17], the speckle contrast reduction starts at $R_q = \lambda/4$ and approaches 0 at about $\lambda/20$. Experiments at $\lambda = 266$ nm confirm these results. A silicon surface has been scanned by an atomic force microscope yielding a mean roughness of 20 nm. The same surface was subjected to a deep UV laser illumination and imaged to a CCD camera. The histogram of the image plane speckles is shown in Figure 13.6. Compared to the fully developed pattern obtained from an $R_q = 200$ nm surface, the contrast is reduced but still sufficient for measurements.

13.2.2 Extracting Information from Speckle Patterns

The speckle pattern is a unique fingerprint of the object surface under certain illumination and imaging conditions. The backtracing of information about the object and its mechanical behavior,

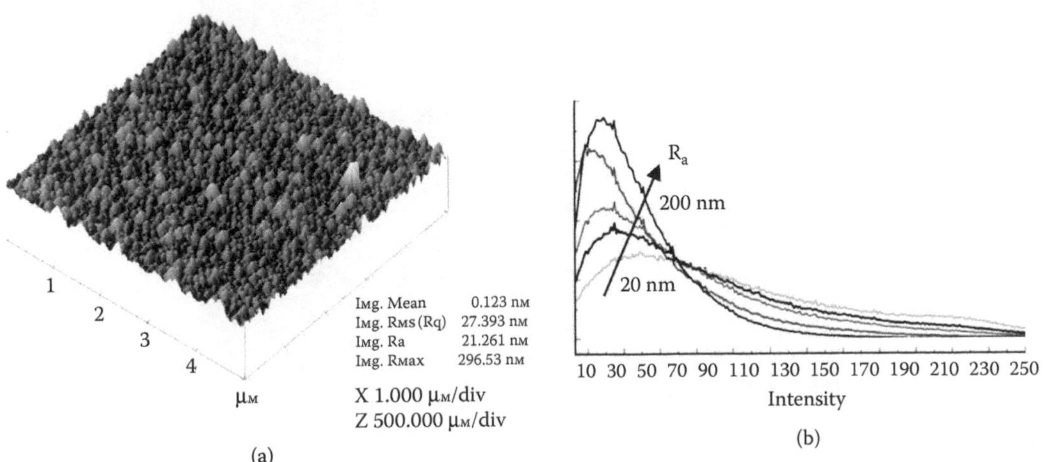

FIGURE 13.6 Microroughness $R_q = 20$ nm: AFM result (a), intensity histograms up to $R_q = 200$ nm (b).

however, are not straightforward. Frequently used methodologies and optical arrangements are given in the following text. All the techniques described aim at extracting mechanical quantities such as displacement, strain, and vibration amplitude that occur on opaque, solid objects. Approaches for the measurement of shape, surface roughness, or microstructural changes are published elsewhere [18] and are not considered in the following.

First, the primarily observable effects in speckle patterns are briefly classified to prepare the ground for the more detailed treatment of their exploitation for metrology purposes.

13.2.2.1 Classification of Speckle Signals for Metrology

There are three major quantities that have proved useful in gaining knowledge about the mechanical behavior of coherently imaged objects:

- Local speckle contrast
- Speckle shift in the image plane
- Speckle phase in the image plane

13.2.2.1.1 Speckle Contrast

The speckle pattern in the image plane is a 2-D cut through a 3-D speckle intensity structure. The 3-D speckle has a cigar-like shape with the axis along the line of sight and a length corresponding to the focal depth of the imaging system [19]. Displacing the object along the optical axis causes a so-called speckle boiling effect if the movement is large enough, i.e., speckles appear and disappear (decorrelate). The pattern changes its structure and for a vibrating surface, the time-averaged recording of the corresponding speckles is blurred; the speckle contrast is lost in the moving regions. Figure 13.7 shows an example: a disk vibrating at resonance. The nodal lines of the vibration mode can be identified as a region where the speckle contrast is maintained. It is the only effect that can be instantly observed by the eye without any additional recording and processing.

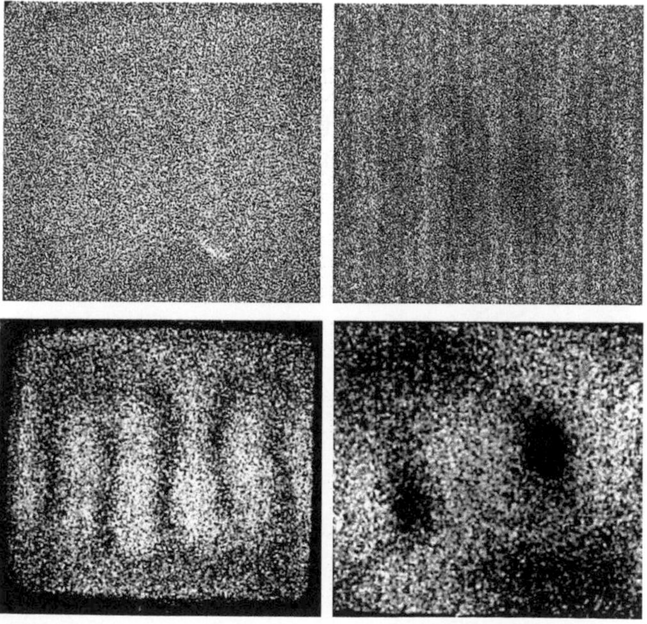

FIGURE 13.7 Directly observable disk vibration pattern by one-beam coherent illumination. Top: camera image (left), high-pass filtered results (right), at f = 2060 Hz; bottom: contrast enhancement for f = 2060 Hz (left) and 875 Hz (right).

This experiment had been described already in the very early days of speckle metrology [20]; it works for rather large amplitudes (a couple of microns). Comprehensive work followed, improving the sensitivity, the image quality, and the quantitative readout. Subsection 13.2.2.2 will explain sophisticated methods to enhance the results considerably.

13.2.2.1.2 Speckle Shift

What will happen to the speckles in the image plane if the object is moved only a very small amount, so that the pattern does not decorrelate? The intensity pattern is a result of a certain phase mix produced by the given surface structure in the vicinity of the imaged object point. Therefore, the speckles appear to be stuck to the object surface (in fact, they exist in the image only!). If the surface is displaced by $\mathbf{d} = (d_x, d_y, d_z)^T$, the resulting speckles will also shift $(D_x, D_y)^T$ according to the imaging magnification M. In particular, any in-plane object motion, i.e., perpendicular to the optical axis, is translated into a corresponding image displacement:

$$D_x = \frac{d_x}{M} + f(d_z)$$
$$D_y = \frac{d_y}{M} + f(d_z) \qquad (13.8)$$

The effect of out-of-plane displacements d_z, i.e., movements along the optical axis, is related to the change of the effective magnification M and is one order of magnitude smaller; it needs attention for error analyses in particular.

13.2.2.1.3 Speckle Phase

As shown in Figure 13.2, the speckle intensity in a given image point is the result of a coherent sum of complex amplitudes according to the phase mix of the overlapping diffraction patterns of the scatterers. If a displacement causes an optical path variation $\Delta\varphi$ between the object and the image point, the contribution of all scatterers will change by the same amount. Figure 13.8 (left) illustrates the effect: the resultant amplitude A is just rotated in the complex plane, and the intensity $I = A^*A$ is not modified. If, however, a reference wave is added coherently, the phase shift $\Delta\varphi$ causes a modulation of the speckle intensity by a factor

$$(1 + v \cdot \cos(\Delta\varphi))$$

where the visibility v controls the contrast ratio of the signal.

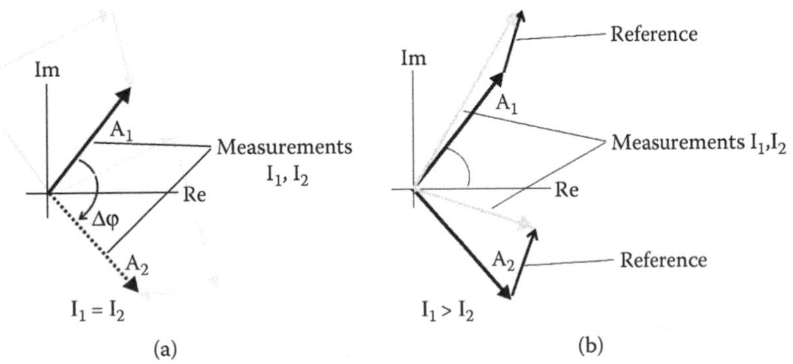

FIGURE 13.8 Speckle phase composition (three phasors) before and after a phase change (a), same situation with additional reference wave (b).

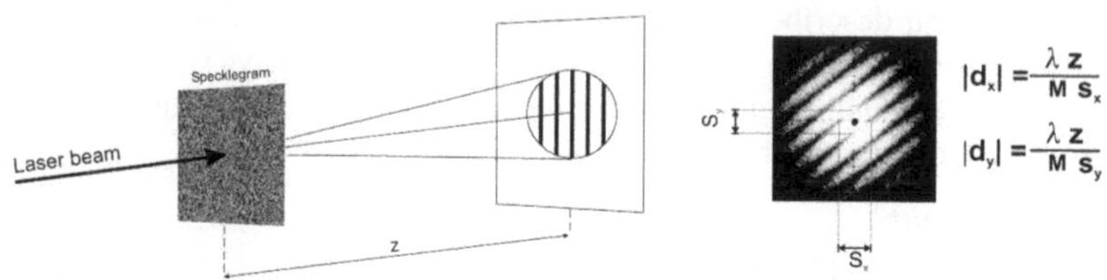

FIGURE 13.9 Displacement measurement from a double-exposure specklegram: point-by-point probing (left), Young's fringes (right).

13.2.2.2 Evaluation Techniques for Signal Extraction

There are a number of well-proven and sophisticated technologies for the evaluation of the information hidden in the speckle patterns. The topic of speckle contrast measurement is not further considered here; it concerns rather large displacements that are not of relevance for MEMS systems.

13.2.2.2.1 Speckle Photography (SPP)

Two major techniques were developed and applied for the reading of speckle shifts occurring during a mechanical experiment. The highly resolved recording of speckle patterns on holographic film or plates was very popular in the past owing to the simplicity and accuracy of evaluation. A double recording of the object before and after static load yields a "specklegram," and the displacement vector can be made visible by directing a thin laser beam at the photoplate. The diffraction pattern observed on a screen at some distance z (Young's fringes) reveals both displacement components for the addressed object point. Measurements can be taken by a simple ruling (Figure 13.9).

Today, the availability and performance of image processing computer technology has replaced this method almost completely. The speckle patterns I_1 and I_2 are recorded by a CCD or CMOS camera before and after the load is applied [21]. The calculation of the correlation function for a subimage (tile) with a certain mutual image shift between I_1 and I_2 and the detection of the correlation maximum yields the speckle displacement $(D_x, D_y)^T$ for the corresponding subimage. Improvements can be achieved by optimizing a complete plane transformation including rotation, shear, and strain.

13.2.2.2.2 Speckle Interferometry (SPI)

Introducing an additional coherent wave, progressing in the direction of **h,** makes the speckle pattern, observed from direction **k**, interferometrically sensitive (see Figure 13.8). Any small object displacement will cause the speckle to decorrelate periodically according to the wavelength of laser light.

The phase difference $\Delta\varphi$ due to an object displacement **d** can be expressed by the sensitivity vector **S**[22]:

$$\Delta\varphi = \frac{2\pi}{\lambda}\mathbf{S}\bullet\mathbf{d} = \frac{2\pi}{\lambda}(\mathbf{h}-\mathbf{k})\bullet\mathbf{d} \tag{13.9}$$

First, the sensitivity is driven by the wavelength of light λ so that a blue or UV laser will be advantageous for the small displacements expected in MEMS analyses. Second, the sensitivity direction (Figure 13.10) is given by the difference between the unit vectors of the illumination and observation direction (**h** − **k**) and this is an important issue for the understanding of SPI interferometers discussed in the following text. It is worth mentioning that the displacement vector is not a directly measurable value; only the inner product **S•d** is related to the phase difference, which has to be obtained from the experimental patterns.

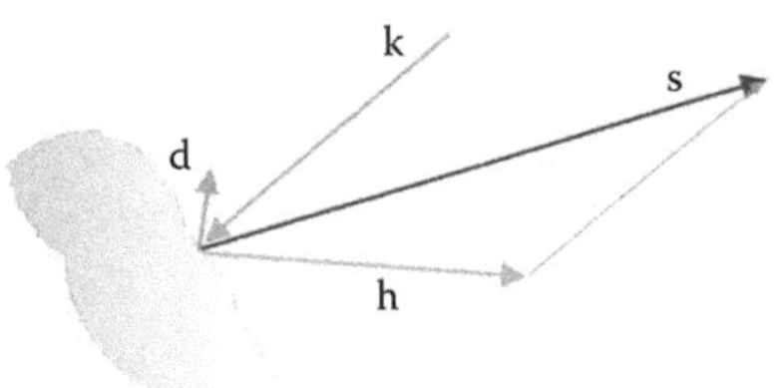

d...Vector of object displacement
h...Unit vector of observation direction
k...Unit vector of illumination direction

FIGURE 13.10 Composition of the sensitivity vector in speckle interferometry.

There are two basic approaches for the evaluation of phase differences $\Delta\varphi$ from interferometric speckle patterns:

1. Difference of phases: calculate the phase independently for the speckle field before and after load.
2. Phase of difference: calculate a fringe pattern by correlating intensities, and evaluate the phase change from the fringe pattern.

Both methods have their advantages and limitations, and both are currently used in laboratories and commercial systems.

In the phase of difference method, a secondary interference pattern I_{sec} (correlogram) is calculated according to

$$I_{sec} \propto \left\langle (I_1 - I_2)^2 \right\rangle = 2\left\langle I_0 \right\rangle^2 \cdot \left(1 - \rho_{12}\right) \tag{13.10}$$

where the correlation ρ_{12} is directly related to the phase change $\Delta\varphi$ by

$$\rho_{12} = \frac{\left\langle \left(\left\langle I_1 - \left\langle I_1 \right\rangle \right\rangle\right) \cdot \left(\left\langle I_2 - \left\langle I_2 \right\rangle \right\rangle\right) \right\rangle}{\left\langle I_1 \right\rangle \cdot \left\langle I_2 \right\rangle} = \frac{\left|\left\langle A_1^* A_2 \right\rangle\right|^2}{\left\langle I_1 \right\rangle \cdot \left\langle I_2 \right\rangle} \tag{13.11}$$

with

$$\left\langle A_1^* A_2 \right\rangle = e^{i\Delta\varphi} \tag{13.12}$$

and finally yields:

$$I_{sec} \propto \left\langle (I_1 - I_2)^2 \right\rangle = \left\langle I_0 \right\rangle^2 \cdot \left(1 - \mathrm{v} \cdot \cos\left(\Delta\varphi\right)\right) \tag{13.13}$$

The term "phase of difference" becomes obvious from this latter formula. The application of Equation 13.13 needs only a few processor operations for $(I_1 - I_2)^2$, and current PC equipment can do these at video rate even for megapixel camera inputs. This real-time capability is an important practical advantage; the temporal development of the interference pattern can be directly observed on the system monitor during the experiment, and a qualitative impression of the mechanical behavior of the specimen or component is instantly available.

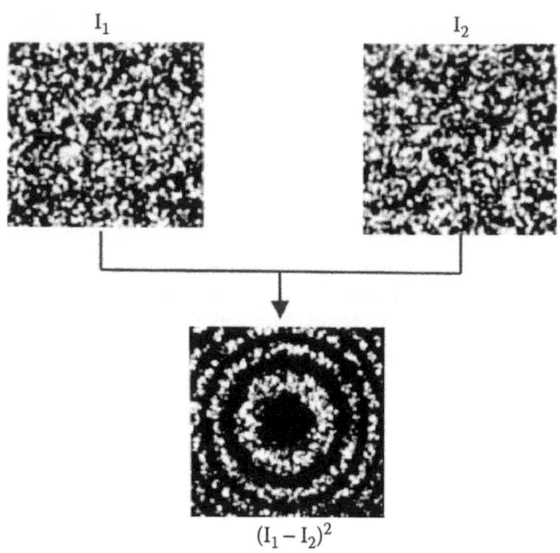

FIGURE 13.11 Generation of secondary interference fringes I_{sec} in the "phase of difference" approach.

The brackets ⟨...⟩ denote an ensemble mean that is the average of all possible speckle patterns that can occur. In practice, only one speckle pattern is usually available, and Equation 13.13 has to be extended by a "noise" term, the speckle intensity itself. The noise is clearly visible in Figure 13.11, which demonstrates the process of generating SPI fringes. The coarse fringes indicate that speckle suppression is the most demanding task for the phase of difference method.

The further quantitative evaluation of the SPI fringes takes advantage of the numerous techniques available for interferometry in general [23]. The quantity of interest in SPI measurements is the phase difference $\Delta\varphi$ that is wrapped into the fringe-generating $(1 + v \cos(\Delta\varphi))$ function.

A rough estimate of the phase values is immediately given by manual fringe counting; this is not very precise but is fast and reliable. Automatic fringe tracking can be used after a careful fringe preprocessing (noise reduction) in order to identify the maxima and minima of the $\cos(\Delta\varphi)$ function. The most powerful solution is the phase-shifting approach; it is briefly described now, using a graphic explanation rather than an extended mathematical proof.

It is a general task in interferometry to determine the phase of a complex light amplitude $\mathbf{A} = |A|e^{i\varphi}$. It is a challenge because the frequency of light is very high and, therefore, all available light detectors are intensity based; they provide the absolute value of the amplitude $|A|$ but not the phase φ. Figure 13.12 illustrates the situation in the complex plane: only the length of $\mathbf{A}$ (or $\mathbf{A} + \mathbf{A}_{Ref}$)

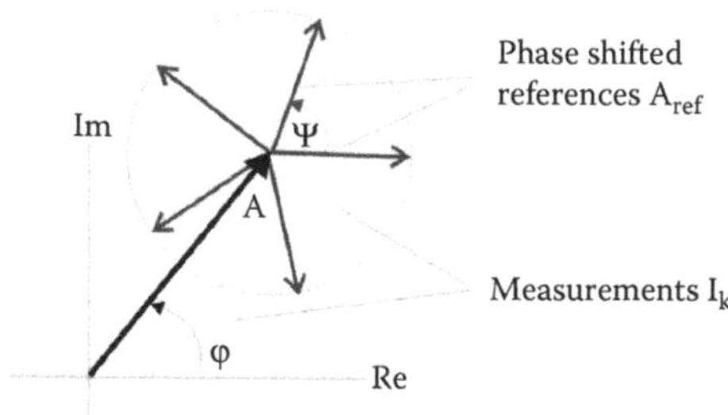

FIGURE 13.12 Principle of phase-shifting interferometry, k = 5 phase steps.

can be directly measured. Required, however, is the angle of vector **A** or the location of its ending point, respectively. For that reason, a series of intensity recordings $I^k = |\mathbf{A} + \mathbf{A}_{Ref}(\psi^k)|^2$ is taken with the reference wave $\mathbf{A}_{Ref}$ modified at intervals. Phase steps ψ^k are applied to the reference in order to gain the desired information.

The schematic shows that the resulting complex amplitude vectors $\mathbf{A} + \mathbf{A}_{Ref}(\psi^k)$ will move in a circle by shifting the reference phase; the center of this circle is the desired ending point of vector **A**. That means: measuring the corresponding intensities I^k is nothing but sampling circle line positions. At least three points are required to define a circle, and so the desired phase of **A** can be calculated if three or more phase-shifted intensities have been recorded. Many different algorithms have been published in the past using more than k = 3 phase steps in order to suppress experimental errors caused by nonlinearities and disturbances. An overview can be found in References 24–26. The mathematics becomes quite simple for the case of k = 4 steps of $\pi/2$ each:

$$\begin{aligned}
I^1 &= I_0 \cdot \left(1 + \mathrm{v} \cdot \cos\left(\varphi + 0 \cdot \frac{\pi}{2}\right)\right) = I_0 \cdot (1 + \mathrm{v} \cdot \cos(\varphi)) \\
I^2 &= I_0 \cdot \left(1 + \mathrm{v} \cdot \cos\left(\varphi + 1 \cdot \frac{\pi}{2}\right)\right) = I_0 \cdot (1 - \mathrm{v} \cdot \sin(\varphi)) \\
I^3 &= I_0 \cdot \left(1 + \mathrm{v} \cdot \cos\left(\varphi + 2 \cdot \frac{\pi}{2}\right)\right) = I_0 \cdot (1 - \mathrm{v} \cdot \cos(\varphi)) \\
I^4 &= I_0 \cdot \left(1 + \mathrm{v} \cdot \cos\left(\varphi + 3 \cdot \frac{\pi}{2}\right)\right) = I_0 \cdot (1 + \mathrm{v} \cdot \sin(\varphi))
\end{aligned} \tag{13.14}$$

and it is easy to verify that the phase value can be obtained for each individual image point (i,j) by

$$\varphi(i,j) = \arctan \frac{I^4(i,j) - I^2(i,j)}{I^1(i,j) - I^3(i,j)} + 2 \cdot N(i,j) \cdot \pi \tag{13.15}$$

Owing to the nature of the arctan function (and consideration of the sign of nominator and denominator), all phase values are mapped into the $(-\pi, +\pi)$ interval. An additional integer function, the fringe count $N(i,j)$, is needed to "unwrap" the phase field. Comprehensive work on the unwrapping methodology in interferometry is available [27] and can be used in SPI. Recently, the temporal follow-up of the mechanical experiment [28] was proposed and proved to be also a very powerful approach.

Phase-shift technology has been successfully applied to the phase of difference fringes in SPI. Figure 13.13 shows an example for a microsystem in which the displacement is small and the phase varies by less than one period over the field of view. Any other fringe evaluation method would fail in this case.

How does the methodology work in practice? For the sake of simplicity, the following discussion will be restricted to k = 4, although k > 4 solutions are also very promising in some applications.

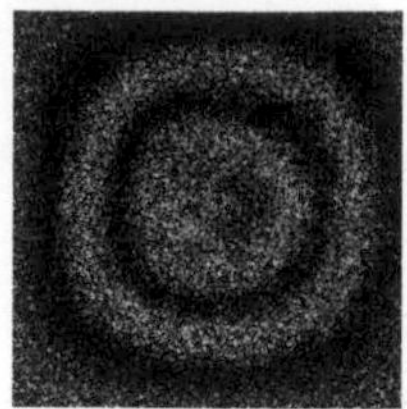 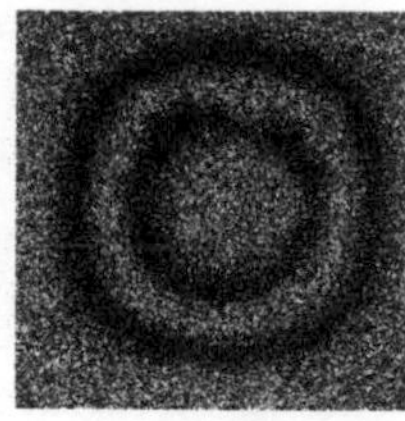 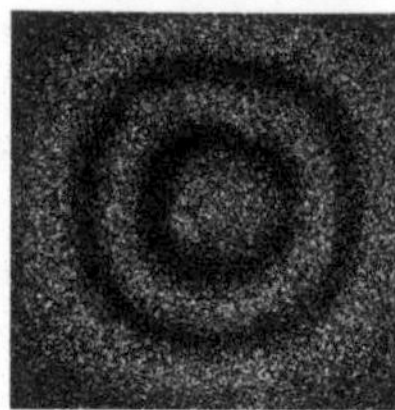 

FIGURE 13.13 Phase-shifting SPI for a microcomponent, k = 4.

It has to be noted that the secondary fringes I_{sec} according to Equation 13.13 can be constructed from a variety of recording schemes:

- 1 + 4 Method: One speckle pattern is taken before loading, and four phase-shifted patterns are taken after loading.
- 4 +1 Method: Four phase-shifted patterns are taken before loading, and one speckle pattern is taken after loading.
- 2 + 2 Method: Two phase-shifted patterns are taken before loading, and two phase-shifted speckle patterns are taken after loading.

In any case, four image differences I^1_{sec}, ..., I^4_{sec} are calculated that represent a $\psi^k = k \cdot \pi/2$ phase-shift series of secondary fringes (see Table 13.1). The choice of method is driven by practical needs and experimental conditions.

Added in the last row of Table 13.1 is the 4 + 4 method, in which a full set of k = 4 phase-shifted speckle patterns is recorded before and after load application, respectively. This will take more time during the experiment but provides additional information and can be used to compensate for phase-shifting errors.

On the other hand, this 4 + 4 recording scheme opens the doors to the second evaluation method in speckle interferometry: the difference of phase approach. In this case, the wrapped phase distribution within the speckle pattern is calculated before and after loading, separately, and the phase difference in each image point (i, j) is given as

$$\Delta\varphi(i,j) = \arctan\frac{I_1^4(i,j) - I_1^2(i,j)}{I_1^1(i,j) - I_1^3(i,j)} - \arctan\frac{I_2^4(i,j) - I_2^2(i,j)}{I_2^1(i,j) - I_2^3(i,j)} + 2 \cdot N(i,j) \cdot \pi \tag{13.16}$$

TABLE 13.1
Recording Schemes for k = 4 Phase-Shifting SPI

Method	Phase-Shifted Records (ψ) Before Loading	Phase-Shifted Records (ψ) After Loading	Calculation of I^1_{sec} ($\psi = 0$)	Calculation of I^2_{sec} ($\psi = \pi/2$)	Calculation of I^3_{sec} ($\psi = \pi$)	Calculation of I^4_{sec} ($\psi = 3\pi/2$)
1 + 4	$I^1_1(0)$	$I^1_2(0)$, $I^2_2(\pi/2)$, $I^3_2(\pi)$, $I^4_2(3\pi/2)$	$(I^1_1 - I^1_2)^2$	$(I^1_1 - I^2_2)^2$	$(I^1_1 - I^3_2)^2$	$(I^1_1 - I^4_2)^2$
4 + 1	$I^1_1(0)$, $I^2_1(\pi/2)$, $I^3_1(\pi)$, $I^4_1(3\pi/2)$	$I^1_2(0)$	$(I^1_1 - I^1_2)^2$	$(I^2_1 - I^1_2)^2$	$(I^3_1 - I^1_2)^2$	$(I^4_1 - I^1_2)^2$
2 + 2	$I^1_1(0)$, $I^2_1(\pi/2)$	$I^1_2(0)$, $I^2_2(\pi)$	$(I^1_1 - I^1_2)^2$	$(I^2_1 - I^2_2)^2$	$(I^1_1 - I^2_2)^2$	$(I^2_1 - I^1_2)^2$
4 + 4	$I^1_1(0)$, $I^2_1(\pi/2)$, $I^3_1(\pi)$, $I^4_1(3\pi/2)$	$I^1_2(0)$, $I^2_2(\pi/2)$, $I^3_2(\pi)$, $I^4_2(3\pi/2)$	$(I^1_1 - I^1_2)^2 +$ $(I^2_1 - I^2_2)^2 +$ $(I^3_1 - I^3_2)^2 +$ $(I^4_1 - I^4_2)^2$	$(I^1_1 - I^2_2)^2 +$ $(I^2_1 - I^3_2)^2 +$ $(I^3_1 - I^4_2)^2 +$ $(I^4_1 - I^1_2)^2$	$(I^1_1 - I^3_2)^2 +$ $(I^2_1 - I^4_2)^2 +$ $(I^3_1 - I^1_2)^2 +$ $(I^4_1 - I^2_2)^2$	$(I^1_1 - I^4_2)^2 +$ $(I^2_1 - I^1_2)^2 +$ $(I^3_1 - I^2_2)^2 +$ $(I^4_1 - I^3_2)^2$

If the phase change $\Delta\varphi$ is very small, the fringe count N can be set to zero, and the need for unwrapping is avoided by incremental recording of speckle patterns [29]. Otherwise, a phase-unwrapping procedure has to be applied to $\Delta\varphi$.

Figure 13.14 shows an example of a wrapped phase distribution in a speckle field. It is a typical phenomenon of speckle patterns that there are regions where the intensity is permanently low. Adding a phase-shifted reference will not produce any usable phase information, and there is simply no measurement possible at these points without completely changing the whole speckle pattern. For computerized evaluation, these dark points have to be detected automatically and removed from the result; they are indicated as black areas in the phase map of Figure 13.14.

13.2.2.2.3 Time-Averaged Recordings

Static loading conditions have been discussed so far, i.e., experiments in which the object is not in motion during the CCD camera frame recording, typically a 20–40 msec period of time. MEMS are made to move, so the dynamic loading case has to be taken into account for analyzing their operational behavior. Vibrating components are common in many microsystems; the corresponding natural frequencies are rather high, ranging from a couple of kilohertz up to several megahertz, due to the small structural dimensions. For this reason, freezing the motion by choppers or laser pulses is difficult, if not impossible, in this case. Even the stroboscopic operation, i.e., temporal modulation of the illumination or reference wave, is quite limited at frequencies above 1 MHz. So, time-averaged recordings are preferred, and they are described in the following in more detail. The typical vibration amplitude in MEMS is on the order of the wavelength of light or below, and the use of a reference wave is mandatory; it generates interferometrically sensitive speckle fields suited to read out the nanometer motions. Three modes of operation will be discussed that differ with respect to their experimental complexity and signal quality, respectively:

- Single-frame SPI
- Subtraction SPI
- Moving phase reversal reference

13.2.2.2.3.1 Single-Frame SPI

The main effect in single-frame SPI is the steady change of the speckle pattern during a period of vibration. The camera will integrate all the temporary patterns, and the resulting image is blurred in regions where the object is vibrating; high speckle contrast will be found along the nodal lines only. This happens any time the camera takes a new frame, and there is no other information used

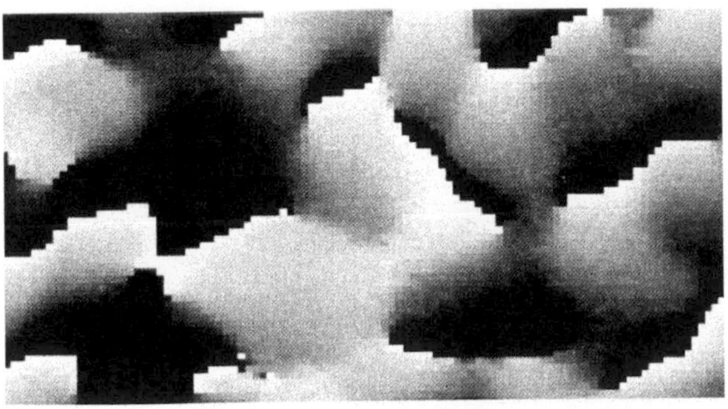

FIGURE 13.14 Phase distribution within a speckle pattern.

but the current CCD image. This is a big advantage because the stability requirements for the whole setup are relaxed and concern just the exposure time of the camera. From this point of view, the technique is very robust. On the other hand, the fringe quality is limited if not poor, even if there are image processing filters applied to enhance the difference between blurred and speckled image areas.

13.2.2.2.3.2 Subtraction SPI

Vibration-mode pictures of much better contrast are obtained if the speckle pattern of the object at rest is taken into account as the initial state. The formula for the static load case (Equation 13.13) can be used with the modification that the loaded state speckles do not represent a single snapshot but a time average over the phases obtained from the sinusoidal vibrations with the angular frequency ω.

$$I_{\sec} \propto \left\langle (I_1 - I_2)^2 \right\rangle = \left\langle I_0 \right\rangle^2 \cdot \int \left(1 - \mathrm{v} \cdot \cos\left(\Delta\varphi_{vib} \sin\left(\omega t\right)\right)\right) dt \tag{13.17}$$

The maximum phase change $\Delta\varphi_{vib}$ corresponds to the vibration amplitude vector in the same way as the static displacement vector yields the static phase change $\Delta\varphi_{stat}$. It is obvious that in MEMS the vibration period is much shorter than the camera framing time so that the limit T→ ∞ can be taken in summing up infinite states of the phase change:

$$\frac{1}{T} \cdot \int_0^T e^{i\Delta\varphi_{vib} \cdot \sin(\omega t)} \cdot dt = J_0(\Delta\varphi_{vib}) \tag{13.18}$$

where J_0 is the zero-order Bessel function of the first kind. The secondary fringes are modulated now by the Bessel function instead of the cosine:

$$I_{\sec} \propto \left\langle (I_1 - I_{vib})^2 \right\rangle = \left\langle I_0 \right\rangle^2 \cdot \left(1 + v \cdot J_0(\varphi_{vib})\right) \tag{13.19}$$

The most important difference between the cosine and the Bessel function is the reduced fringe visibility for higher phase values (see Figure 13.15).

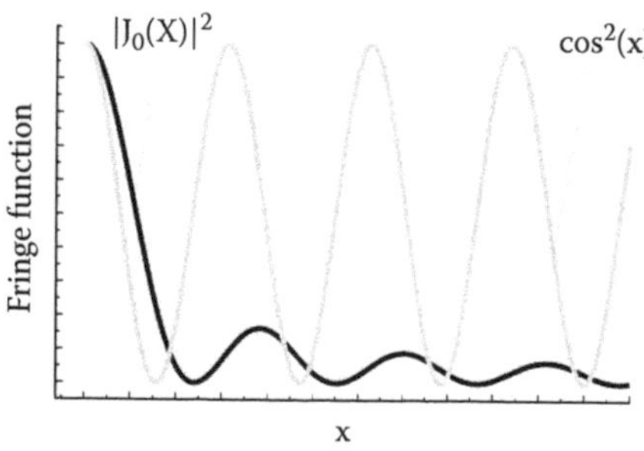

FIGURE 13.15 Fringe-generating functions for the static load (light gray), for time-averaged vibrations (black).

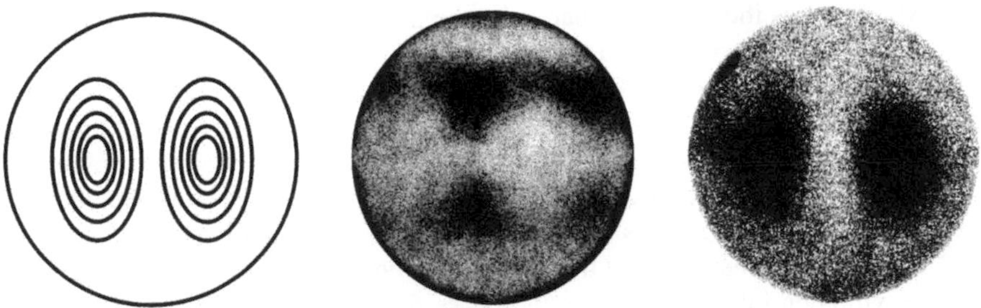

FIGURE 13.16 Membrane vibration plus static tilt: schematic of the mode shape (left), result of subtraction SPI (middle), result of moving phase reversal reference SPI (right).

Under real-world conditions, disturbances will occur: slow, quasi-static phase changes $\Delta\varphi_{stat}$ between the reference state (object at rest) and the respective object state under vibrational load. This yields a modified relation in which both phase terms appear simultaneously:

$$I_{sec} \propto \left\langle (I_1 - I_{vib})^2 \right\rangle = \left\langle I_0 \right\rangle^2 \cdot \left(1 + v \cdot J_0(\varphi_{vib}) \cdot \cos\left(\Delta\varphi_{stat}\right)\right) \tag{13.20}$$

The overlap of the two contributions $\Delta\varphi_{vib}$ and $\Delta\varphi_{stat}$ is likely to make the measurement difficult to interpret; an example is given in Figure 13.16. A micromembrane is excited at resonance, and an additional tilt occurs owing to movements in the fixture. The expected vibration mode is shown schematically on the left. It is hard to recognize in the subtraction SPI image in the middle, in which both phase contributions $\Delta\varphi_{vib}$ and $\Delta\varphi_{stat}$ are mixed.

13.2.2.2.3.3 Moving Phase Reversal Reference

The misleading effects described in the preceding text can be excluded by applying the moving phase reversal reference [30,31]. Figure 13.17 illustrates the principle of this technique. The main idea is to use the phase modulation capability of the time-averaged speckle patterns as the measuring signal. A phase reversal occurs if the reference wave is phase-shifted by π. The corresponding

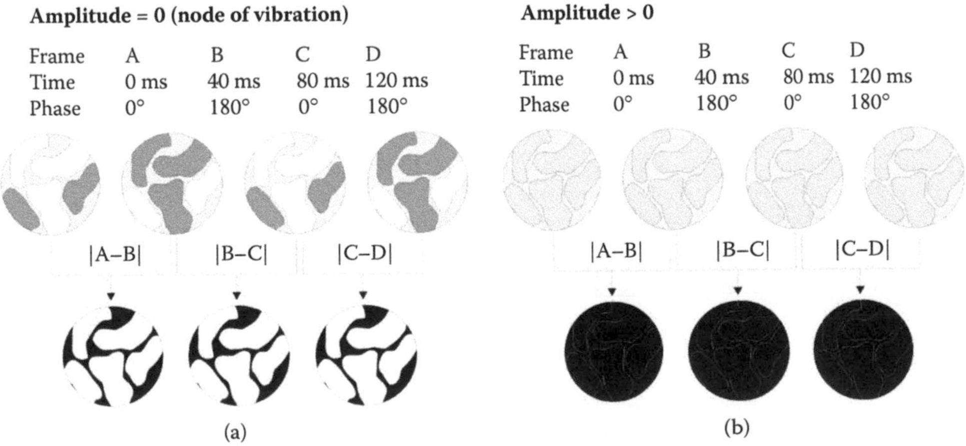

FIGURE 13.17 Principle of moving phase reversal reference SPI: intensity output inside vibration node (a), outside vibration mode (b).

modulation, which means the intensity change by the phase reversal, is well developed in regions where the speckles are preserved, i.e., in the areas of the vibration nodes. On the other hand, on vibrating parts of the object, the speckle intensity varies with high frequency and is averaged out during a video frame, so that the phase reversal has no effect. In order to make the vibration mode visible, a phase reversal is applied between adjacent video frames. Taking moving differences $|A - B|^2$, $|B - C|^2$, $|C - D|^2$, etc. produces a high intensity in areas of good modulation, i.e., on the nodes. In contrast, the vibrating zones that do not show any modulation appear dark. The example in Figure 13.17 (right) demonstrates that the moving reference is no longer sensitive to disturbances; it separates the vibration information from the phase disturbances and gives a clear image. This processing technique combines the fringe quality of the subtraction mode with the robustness of the single-frame technique.

It is important to note that the phase-shift operation is applied between two consecutive camera frames, so that the required phase modulation frequency is low; it equals the camera frame rate (<100 Hz), whereas the MEMS vibration can be in the megahertz range.

Further improvements in fringe quality can be achieved if two more patterns are recorded at rest and under vibration, respectively, that have an additional relative 90° phase shift. The full mathematical description of this approach has been published as *electrooptic holography* (*EOH*) [32,33].

13.2.2.3 Optical Arrangements for Speckle Interferometry

There are a number of interferometers available for speckle pattern interferometry [34], and they are used in various applications. It is a significant advantage of SPI that displacement components in the surface plane (in-plane) and parallel to the surface normal (out-of-plane) can be separated by the optical arrangement chosen. The sensitivity vector **S** has been introduced in Equation 13.9 as a mathematical description of the interferometers. It is given by the difference between the vectors of illumination **h** and observation **k**, respectively, and defines the displacement direction measured:

$$\mathbf{S} = \mathbf{h} - \mathbf{k} \tag{13.21}$$

Two arrangements are described in more detail; they proved to be well suited for measurements in MEMS. The optical arrangement shown in Figure 13.18 senses the out-of-plane displacements or vibrations of a microstructure. It is very similar to a Twyman–Green interferometer, but with the difference that the reference mirror is replaced by an optically rough scattering surface. In this way, the reference wave is also a speckle field, and the interferometer does not need any time-consuming object alignment. One fringe period corresponds to an object displacement of $\lambda/4$.

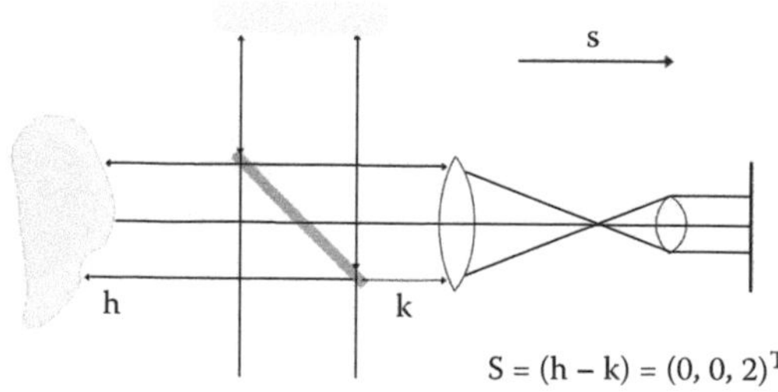

FIGURE 13.18 Speckle interferometer for out-of-plane measurements.

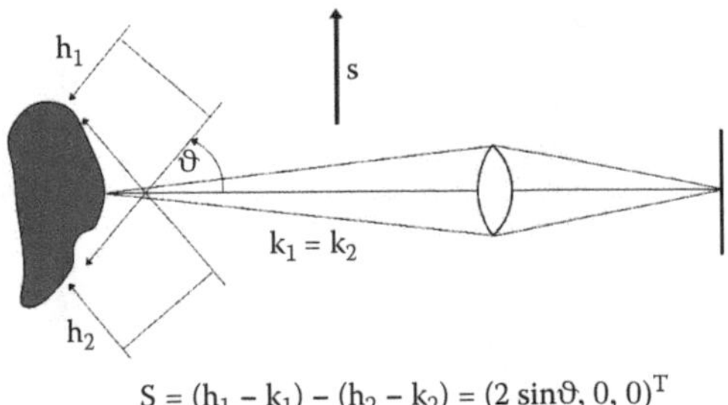

FIGURE 13.19 Speckle interferometer for in-plane measurements.

The ability to sense in-plane displacements with interferometric sensitivity is remarkable in the speckle method. This can be achieved by dual illumination as shown in Figure 13.19. Two coherent beams illuminate the object symmetrically at a given angle. Each illumination generates an independent coherent speckle pattern; they interfere in the image plane. The total sensitivity results from the difference

$$\mathbf{S} = (\mathbf{h}_1 - \mathbf{k}_1) - (\mathbf{h}_2 - \mathbf{k}_2) \tag{13.22}$$

where the observation direction $\mathbf{k}$ is the same for both parts and gives

$$\mathbf{S} = (\mathbf{h}_1 - \mathbf{h}_2) \tag{13.23}$$

In other words, the speckle field generated by one illumination direction serves as the reference for the other. In symmetric arrangements, the difference vector $(\mathbf{h}_1 - \mathbf{h}_2)$ lies perpendicular to the optical axis (in-plane), and the observation direction is eliminated. Parallel illuminating beams provide constant sensitivity for the whole field of view without the need for telecentric imaging. The displacement sensitivity depends upon the angle of incidence only.

In both interferometers, the waves can be manipulated independently, so that phase-shifting capabilities may be easily incorporated by a mirror mounted on a piezo transducer. The detailed implementation of these interferometers in certain applications is described in Section 13.3.

13.3 APPLICATIONS

13.3.1 Quality Assurance on Wafer Level

The specific problem of MEMS testing very close to the manufacturing process is the fact that there is no electrical contact and no mechanical interface established yet at this stage. However, MEMS quality becomes obvious by the resonant properties of a single element on the wafer. Speckle interferometry guarantees high efficiency and reliability for this task owing to its ability to yield resonant frequencies as well as information about the vibration modes. The main aspects for quality assurance are related to the ability to perform the following:

- Measure on wafer level directly
- Check a number of single elements in parallel
- Visualize vibration modes that correspond to a resonance

Figure 13.20 illustrates a laboratory system specifically designed for testing 4-in. wafers. The main part is a fiber-optic speckle interferometer equipped with an infrared laser diode and a phase-shifting unit [35]. Two modes of operation can be chosen. The interferometer is sensitive to in-plane movements if the wafer is illuminated symmetrically from two points (Figure 13.20, right upper). In the other configuration, one beam serves as a reference wave so that out-of-plane deformations

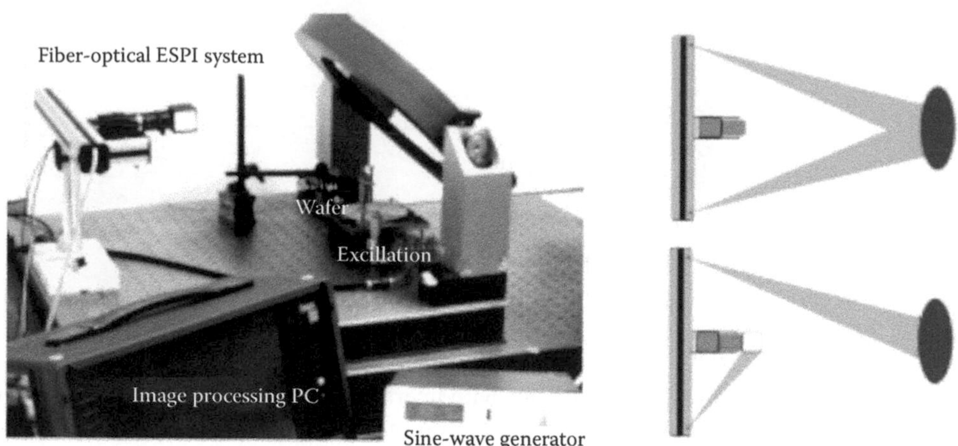

FIGURE 13.20 Laboratory setup for wafer testing (left), in-plane and out-of-plane configurations (right).

are detected (Figure 13.20, right lower). A specific wafer mounting has been designed using a circumferential vacuum clamping to guarantee defined boundary conditions and to suppress resonant modes of the whole disk. Excitation of the single elements is realized by a computer-controlled piezo transducer or loudspeaker via solid-state or acoustic waves, respectively.

In the case of wafer testing, the phase reversal technique (see Subsection 13.2.2.2.3.3) is applied, and the SPI signal is displayed on the system monitor at video rates, so that resonant vibrations can be found by frequency scanning in the numerically predefined range. Typical fringe patterns for wafer testing are shown in Figure 13.21. The advantage of the full-field-of-view technique is obvious: resonant vibrations of the disk as a whole (Figure 13.21, left) can be reliably distinguished from the resonances of the single elements. Even different modes of the elements can be clearly seen (zoomed parts). Amplitudes of 200 nm already produce maximum contrast, so that the energy of excitation may be kept very small, and the microparts are not subjected to dangerous loads.

Systematic tests were carried out using the setup described in order to detect the resonance of each single micropart. Wafers with different thickness, variations in the fabrication process, and local defects have been investigated. For different types of MEMS, a respective set of control parameters is chosen at first based upon numerical analysis; for example, 370–570 Hz and excitation voltage of 15 mV. Then, the frequency is scanned automatically in 1-Hz steps, and the SPI pattern is displayed for every frequency. All microparts can be observed simultaneously, and the resonances are identified by appropriate software automatically. One frequency scan takes less than 1 min and

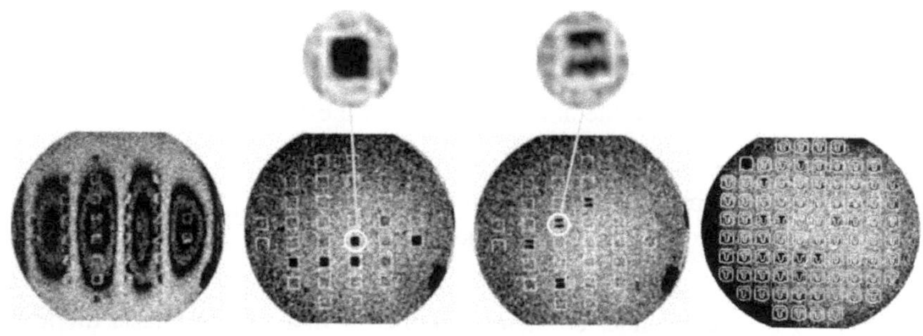

FIGURE 13.21 Time-averaged SPI patterns of MEMS wafers: whole disk vibration (left), first and second out-of-plane mode (middle), and in-plane mode (right).

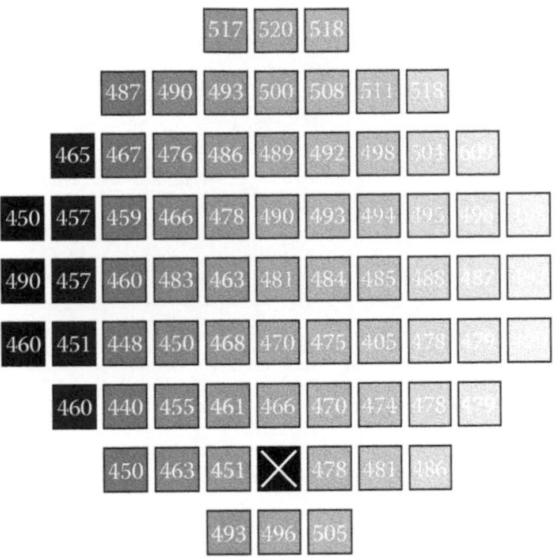

FIGURE 13.22 First mode resonances across a wafer.

is sufficient for testing the whole wafer. The result of such a test run is shown in Figure 13.22. A resonant frequency is assigned to each micropart. In the example, one of the elements did not vibrate at all in the frequency range of interest.

It is clearly visible that the frequencies are systematically distributed across the wafer. Two effects can be observed:

1. The frequency increases from lower left to upper right owing to the wedge shape of the wafer caused by wafer cutting.
2. Frequencies of outer elements are higher because of inhomogeneous etching.

From a mechanical point of view, spring thickness is important for a quality assessment. Therefore, a polygon fit is used to establish a relation between the resonant frequency f and the spring or membrane thickness D. Figure 13.23 shows an example in which the nominal thickness is 34 μm with a permissible variation of ± 1 μm.

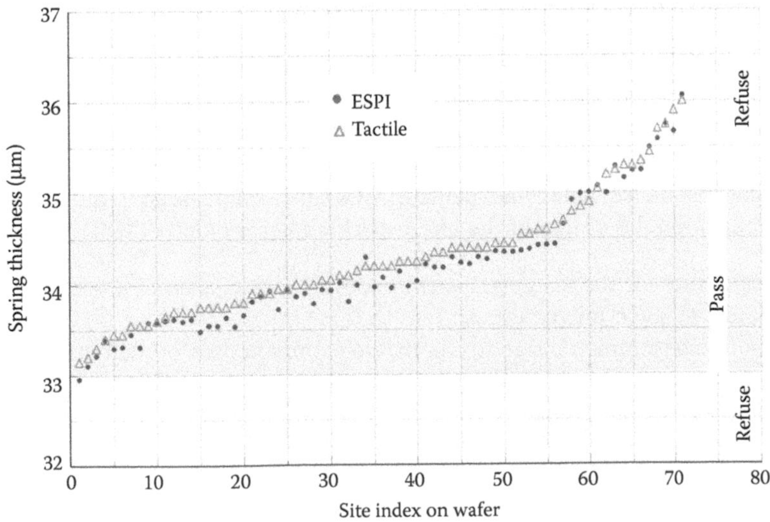

FIGURE 13.23 Spring thickness for wafer in Figure 13.22, ESPI result compared with destructive tactile measurements.

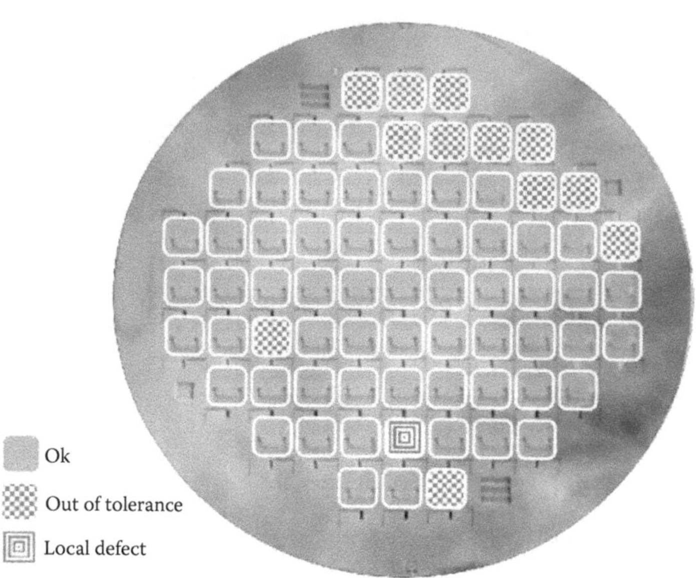

FIGURE 13.24 Final result of a SPI test run: quality assessment for a MEMS wafer.

In addition, tactile measurements have been performed for this wafer in order to verify the approach. The plot illustrates the very good agreement between the SPI result and the corresponding tactile measured value for every single spring or mass structure on the wafer. The standard deviation for the difference is as small as $\sigma = 0.1\ \mu$m.

The final quality assessment is that a number of microparts on the wafer have to be sorted out because of variations in feature dimension, and others are refused because of local defects. All other micromechanical parts are qualified to operate as expected. In the final testing sheet (Figure 13.24), the MEMS on the wafer are marked by appropriate colors of quality.

13.3.2 Characterization of Operational Behavior

13.3.2.1 Visible Light Interferometry for a Wide Scale

Deformation, vibration, and strain are used to characterize the operational behavior of microstructures, just as it is common for macroscopic components. Most current MEMS are manufactured with an overall size up to several millimeters, but their smallest parts, for example, springs, have nominal dimensions of some micrometers. Therefore, one goal of the development was the design of a system capable of changing the field of view on a large scale. Altogether, four aspects significantly influenced the design of the microspeckle interferometer (MSI):

1. High flexibility for the imaging system to cover a wide scale of object size from a few millimeters to some micrometers
2. Maximum interferometric stability in the noncommon path of the optical setup
3. Capability to use different types of laser sources
4. Miniaturized, mobile design

A Twyman–Green type interferometer was chosen as a platform to fulfill these requirements. Figure 13.25 shows the optical layout.

The incoming laser beam is adapted to the corresponding field of view. In this way, various types of laser can be used; three different sources have been tested: a He–Ne @ 633 nm, an Ar–Ion @488 nm, and a Nd–YAG laser @532 nm. After passing the beam adapter, the light illuminates

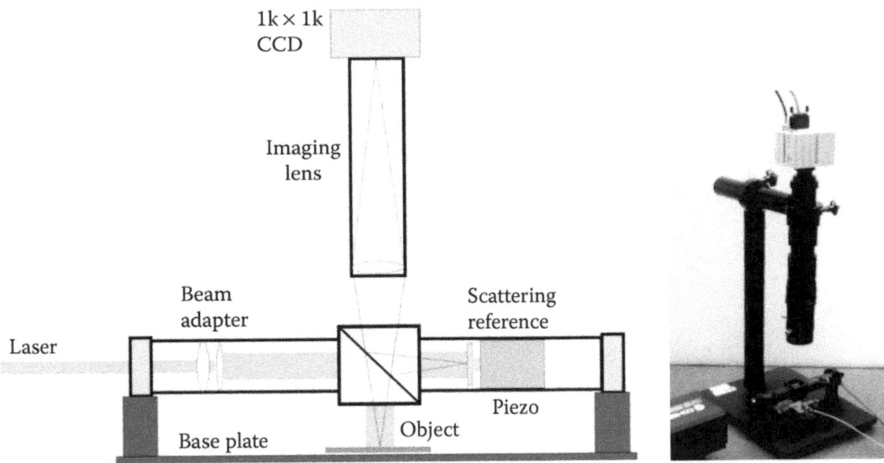

FIGURE 13.25 VIS-MSI layout (left), photograph (right).

both the object and the reference via a beam splitter cube. The scattering surface is mounted on a piezo translator (PZT) to provide phase-shifting capability. All these elements are fixed in a tube system that is in rigid connection to a baseplate. The light scattered from the object and reference, respectively, leaves the beam splitter unit on a common path. Various objective lenses can be chosen in order to image the object onto the CCD target. The Infinity K2 long-distance microscope has an interchangeable fixed-focus front objective lens providing high magnification from M = 2 to M = 20. Usual CCD zoom lenses are sufficient at low magnifications. High-resolution imaging is realized by an Adimec MX12 camera recording 1024 × 1024 pixels at video rate.

The described MSI has been extensively tested for different fields of view and two types of load: dynamic and static. A bulk micromachined gyroscope was selected as the object under investigation. This sensor has a multilayer design and is produced in two steps: wet anisotropic etching and wafer bonding. The basic design of the gyroscope is a tuning fork driven at resonance. From a mechanical point of view, three loading conditions are of interest:

1. Resonant vibration
2. Static bending
3. Static torsion

Figure 13.26 shows an SEM view of one layer of the resonator structure.

In all cases, the microstructure has to be studied at different scales. Magnification of M = 0.6 allows for measurements on the resonator as a whole, whereas details can be observed at M = 2.

FIGURE 13.26 SEM photographs of one gyrolayer.

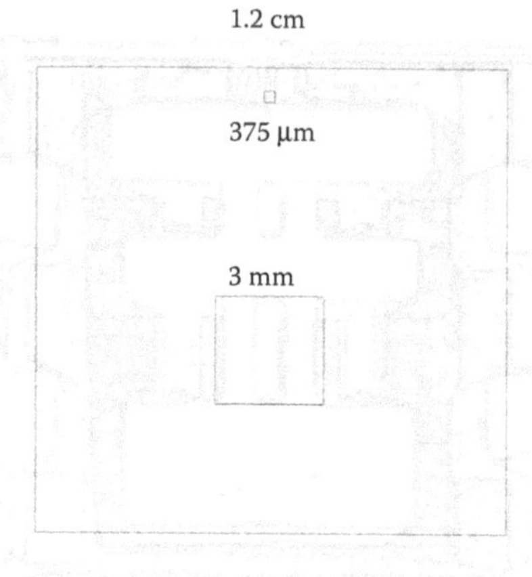

FIGURE 13.27 Different fields of view for the VIS-MSI.

The highest magnification of M = 20 has to be used to resolve deformation and vibration of the smallest features. Figure 13.27 illustrates the changeable size of the field of view on the sensor.

Dynamic loading was the first application case in the series of verification tests of the MSI. The resonant structure is attached to a thin piezoelectric plate. A Wavetek model 23 generates harmonic oscillations serving as an input to a Brüel & Kjæer type 2713 voltage amplifier that drives the PZT. Typical natural frequencies are in the range of 1 to 500 kHz. The pictures in Figure 13.28 make it obvious that the sensor has a variety of modes, for example, a pronounced torsional vibration at 5.1 kHz imaged at M = 0.6. Smaller beams are in resonance at higher frequencies, but they are difficult to see at low magnification. Therefore, the same region was analyzed with better resolution using the long-distance microscope at M = 2.

One quantitative result of this closer look is given in Figure 13.29; the amplitude of displacement is on the order of nanometers. Both bending springs have slightly different resonances in the bending as well as in the torsional modes. This indicates the influence of thickness variations, residual stresses, or other imperfections due to the etching process.

The active micromembranes described in Subsection 13.3.2.2 are other examples. The moving phase reversal reference method yields stable and high-contrast interference patterns that are visible, for example, in Figure 13.30. This signal processing scheme is based upon phase-switching operations that are synchronized with the camera recording rate; it does not, however, depend upon the vibration of the object and therefore is unlimited in frequency. The series of fringe patterns gives an impression of the performance of the VIS-MSI. Starting with the first basic mode at 62 kHz, resonant vibrations can be identified up to high modes; the maximum frequencies approach 4 MHz in this example.

  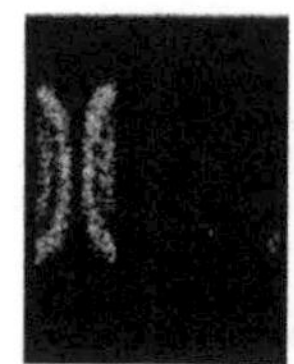 

FIGURE 13.28 Selected vibration modes of the gyrolayer: 5.1 kHz (left), 148.5 kHz (middle), 334/339 kHz (right).

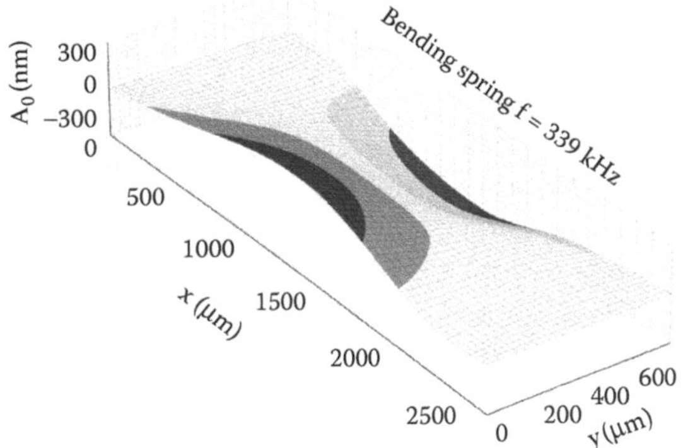

FIGURE 13.29 Torsional vibration-mode shape of one spring.

The second important load case for the MEMS structures is static deflection. For the gyrolayer, this was achieved by rigidly affixing it to a specific counter electrode assembly positioned in the MSI. Four electrodes are arranged in such a way that electrostatic forces can be applied at different positions of the gyrolayer. Bending and torsional forces can be introduced by individual control of the corresponding electrodes. A high-voltage supply allows for the precise control of the acting force. The MSI scaling potential is demonstrated in the fringe patterns in Figure 13.31, in which the magnification increases from 0.6 to 20. The smallest structure in the series is a part of the torsional spring with a width of about 90 μm. Only a fraction of a fringe appears across this object, so that the displacement field is visualized instead of the fringe pattern. The corresponding quantitative results obtained by automated analysis are shown in the following text. The overview image at M = 0.6 is well suited to identifying regions of maximum deformation. What cannot be seen directly from the fringes is the S-shaped bending line of the spring visible as a result of quantitative evaluation. The bending springs can be studied in more detail at higher magnification (M = 2). Another area of interest is the torsional spring at the top of the sensor. Fringe analysis yields a

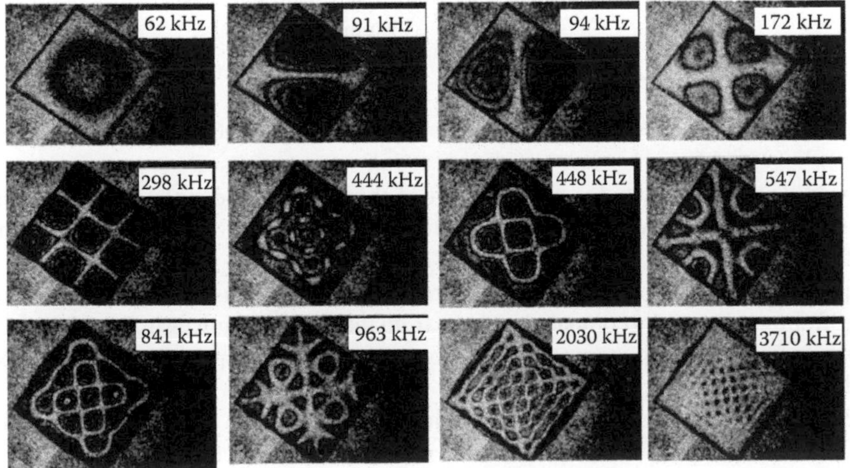

FIGURE 13.30 Series of resonances of an active micromembrane 1 × 1mm in size.

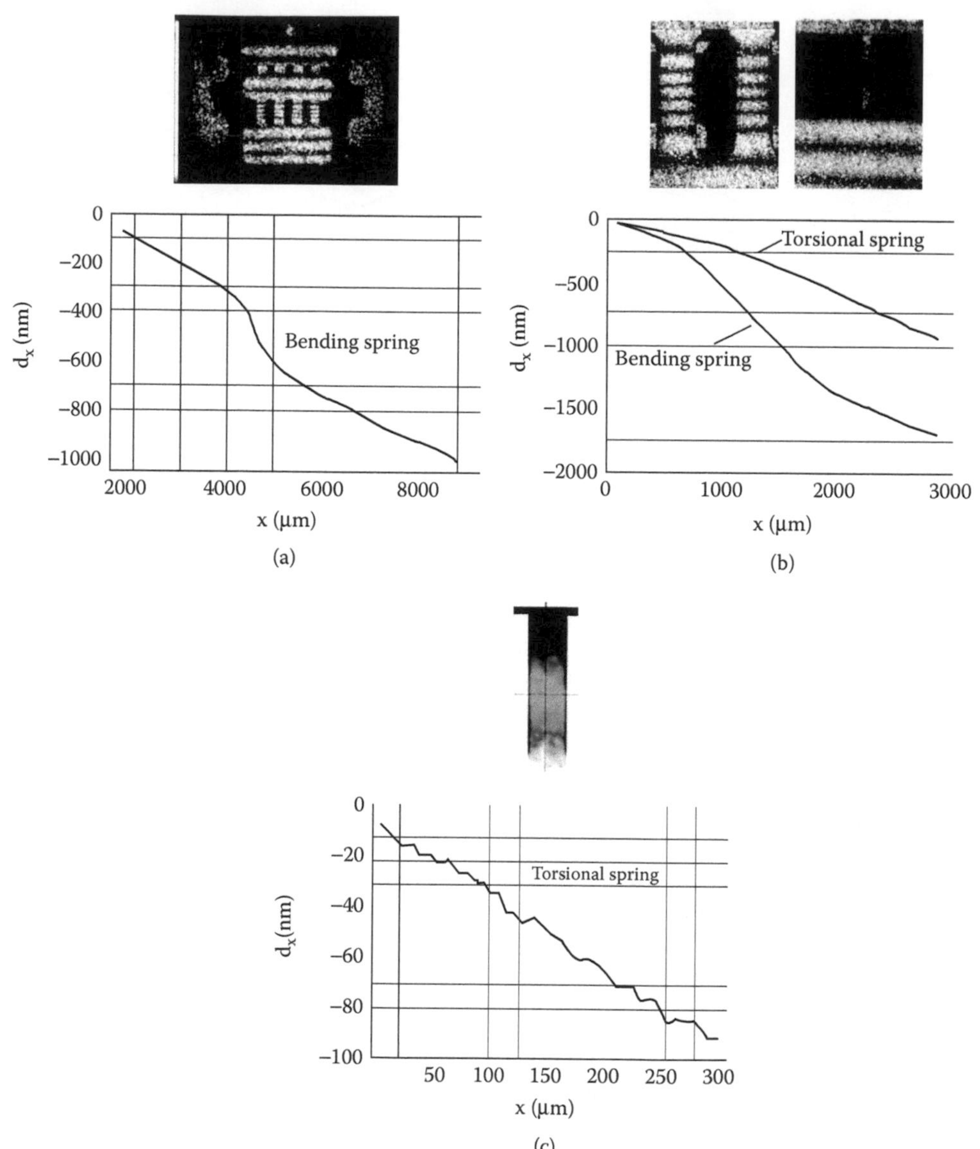

FIGURE 13.31 Fringe patterns and deformation due to bending load: magnification M = 0.6(a), M = 2(b), M = 20(c).

uniform bending along the whole spring. A part of the 90-μm-wide structure can be resolved by the highest magnification (M = 20). The quantitative analysis reveals the ability of the MSI to measure displacement fields on small structural details.

An equivalent series of experiments was carried out with the sensor subjected to torsional load. Figure 13.32 shows the results compared to those from the pure bending tests. The different behavior clearly appears in the fringe patterns; and even for the 90-μm-wide spring, a torsion is recognizable from the high-resolution displacement field.

The 3-D graph of the small spring together with a part of the mass is very illustrative (Figure 13.33) of its operational behavior and demonstrates that microparts can be studied in detail by means of microscopic speckle techniques.

Finally, a rigid body tilt is used to estimate the rms error at such nanodisplacements. Figure 13.34 illustrates that the measurement data scatter with a standard deviation of 2 nm,

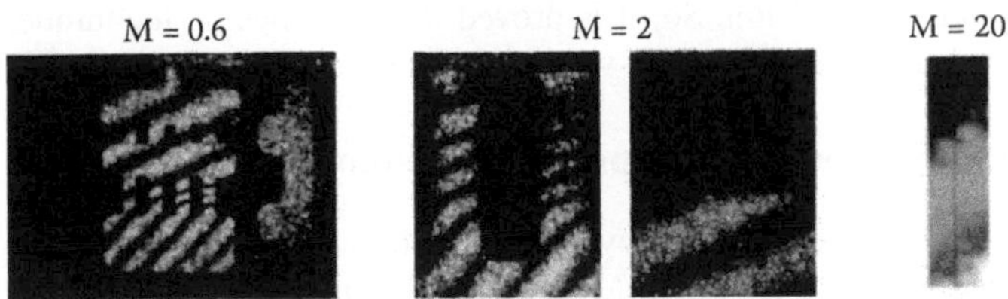

FIGURE 13.32 Fringe patterns due to torsional load.

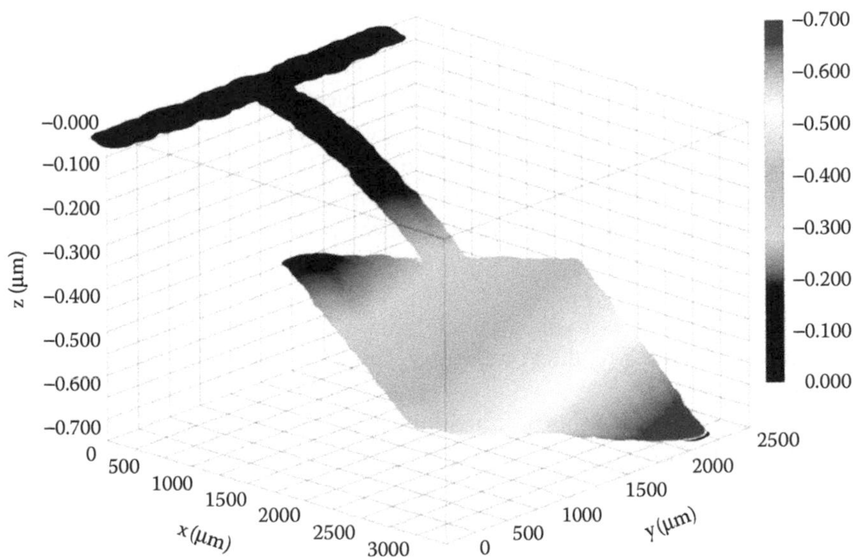

FIGURE 13.33 Microtorsion and bending of the 90-μm width microspring.

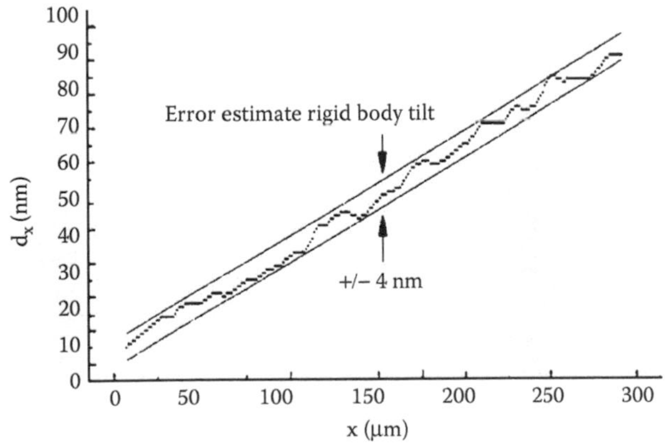

FIGURE 13.34 Estimation of random error of deformation.

resulting in an rms error of ±4 nm. So, it is proved that the speckle technique is very well suited for measuring deformation and vibration with high accuracy.

13.3.2.2 DUV-Interferometry for Improved Resolution

Although the results described in the previous subsection are very promising, the wavelength of laser light has to be decreased to well below the visible range if:

- The object size scales down to several micrometers.
- The surface is highly glossy.
- The motion of the microstructures is in the nanometer range.

The potential of SPI can be fully exhausted when a deep-UV laser source is used. However, because of the short wavelength, all optical and optoelectronic elements had to be carefully selected and tested. For example, commercial high-quality optics, specifically coated, can still cause a considerable loss of intensity. Therefore, an interferometer has been designed that uses only as many optical elements as are absolutely necessary and that can be easily reconfigured to provide both in-plane and out-of-plane sensitivity. The in-plane operation of the setup is achieved using the well-known dual-illumination setup (Figure 13.35a). For the study of out-of-plane motions, a special beam layout has been developed in order to meet the following two objectives:

1. Glossy MEMS surfaces (with a certain part of scattered light) shall be feasible to investigate and, therefore, direct reflections from the object into the camera have to be kept out.
2. Transmissive optical elements in front of the microscopic lens shall not be used to avoid image aberrations.

The solution is sketched in Figure 13.35b. The basic principle of a speckled reference wave is maintained. A UV mirror in front of the microscope lens covers half of the aperture and collects the scattered reference wave while the object beam enters directly.

The detector is a modified PULNIX TM1040 CCD camera that provides the desired resolution of about 1k × 1k pixel. The window has been removed for improved UV performance. The digital LVDS signal output of the camera is fed to the Matrox MeteorDig PCI interface board. A specific suprasil2-quartz dedicated optics has been developed, providing a very low

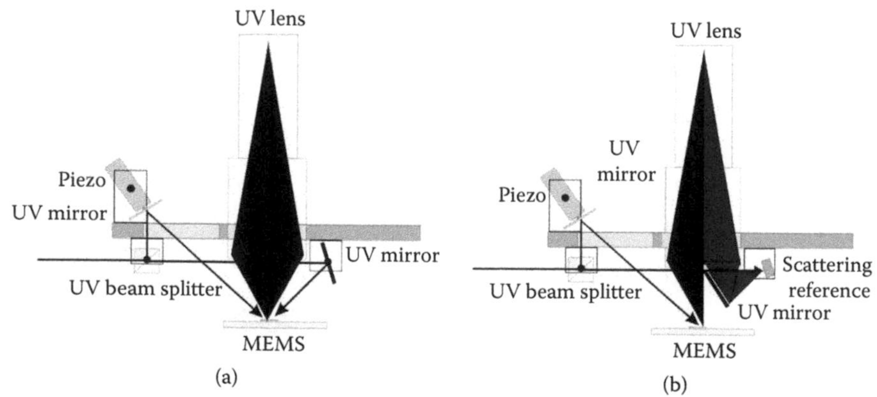

FIGURE 13.35 DUV-MSI layout: in-plane sensitivity (a), out-of-plane sensitivity (b).

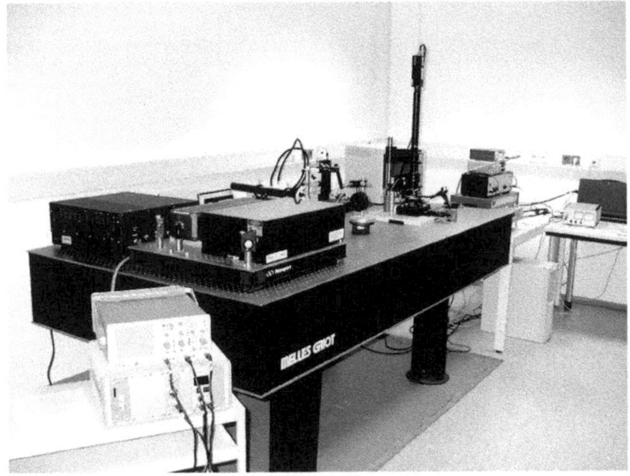

FIGURE 13.36 DUV laboratory.

absorption at the desired 266 nm. The primary magnification is M = 20, and the resolution of 1000 line pairs per millimeter is close to the diffraction limit. Under these conditions, the object-related speckle size obtained is in the range of 0.5 μm. Figure 13.36 shows the DUV-MSI in the laboratory.

Active silicon micromembranes developed within the European-funded OCMMM project [36,37] have been investigated in this study. All of them were micromachined with high precision using an etch stop to control membrane thickness at exactly 5 μm. The piezoelectric actuator for the membrane is integrated into the microsystem as shown in the schematic in Figure 13.37.

The piezoactuator is placed on top of the membrane with the PZT layer sandwiched between two electrodes. The lower electrode is connected to a bonding pad, and the upper one is floating. If a voltage is applied, the electric field in the PZT causes membrane deflection. A set of membranes varying in size and shape, put in both single and ensemble arrangements, has been investigated. Figure 13.38 gives examples of square- and octagon-shaped membranes before and after bonding.

Applying a DC voltage to the piezo via the contacting pads causes a static deflection of the membrane. The value of maximum deflection w depends upon the electric load U, and the characteristic function w(U) has to be determined for each micromembrane. A clear and well-defined

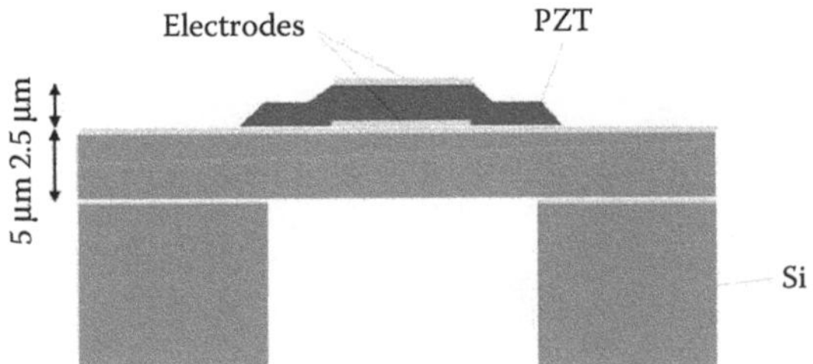

FIGURE 13.37 Schematic of an active micromembrane.

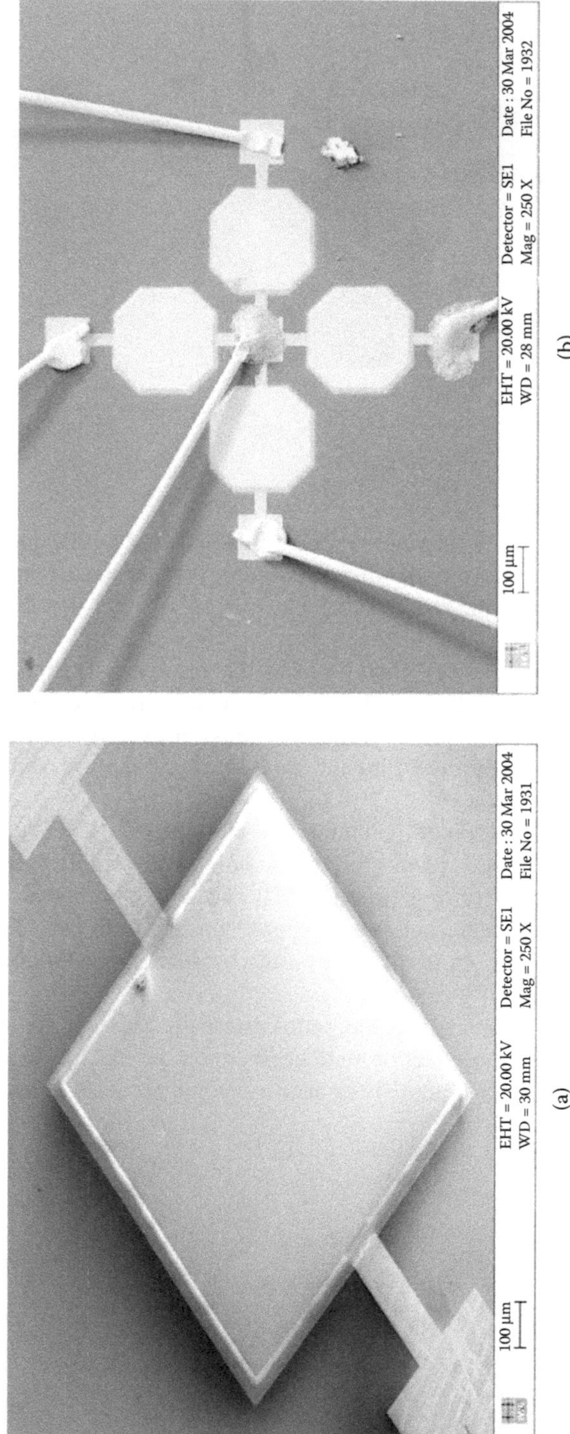

FIGURE 13.38 Scanning eletrode microscope (SEM) photographs of active micromembranes: square-shaped (a), octagon-shaped (b).

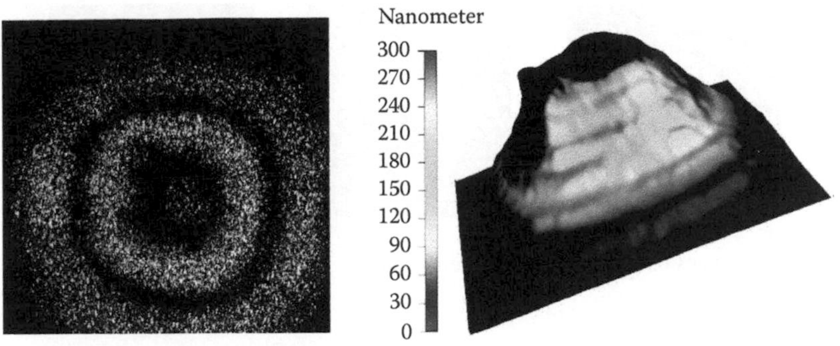

FIGURE 13.39 (See color insert following page 110.) Static deflection of a 250-μm piezomembrane at U = 30 V: fringe pattern (left), displacement field (right).

fringe pattern has been obtained, and the quantitative evaluation by phase shifting yields a full field of displacement, revealing small local imperfections of the structure (Figure 13.39). It is worth mentioning that the maximum deflection does not exceed 300 nm.

Scaling down the dimensions of the membrane results in even smaller displacements. The results in Figure 13.40 were highly reproducible. The deformation is nonhomogeneous in this case, and it becomes obvious that the strip conductor also undergoes a nanometer deformation owing to the electric field applied. The displacement profile along the intersection reveals that the optical systems have the ability to measure displacement fields with values below 10 nm.

In a second step, the piezomembranes were subjected to sinusoidal signals (high-frequency AC voltages) in order to investigate their dynamic operational behavior. Low excitation voltages of about 10 V peak to peak proved sufficient to generate clearly detectable vibration-mode shapes at natural frequencies. The smallest membranes studied were 100 μm in size and octagonal in shape. One example is shown in Figure 13.41. The basic vibration mode is found at rather high frequency, and the first fundamental mode occurs at 879 kHz (left). Increasing the frequency yields a vibration pattern with a circular nodal line in the center (middle), and the highest resonance is at 6.5 MHz, where the signal is too weak to recognize the mode shape.

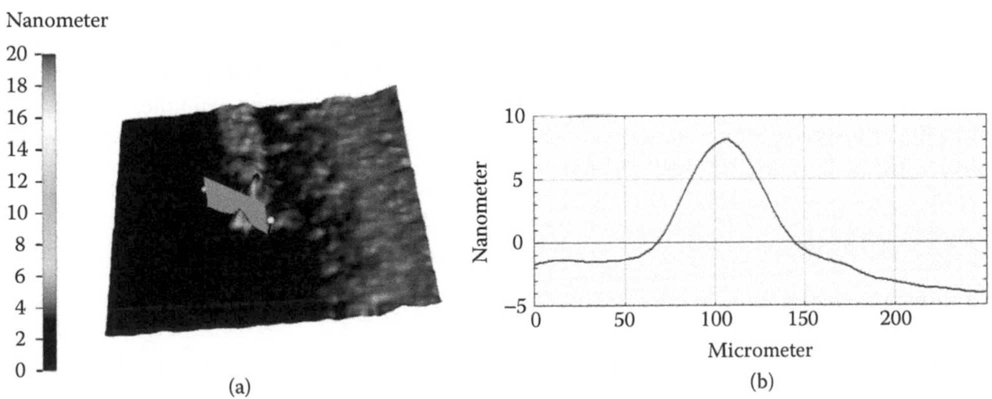

FIGURE 13.40 Static deflection of a 120-μm piezomembrane at U = 30 V: displacement field (a), profile across cutting line (b).

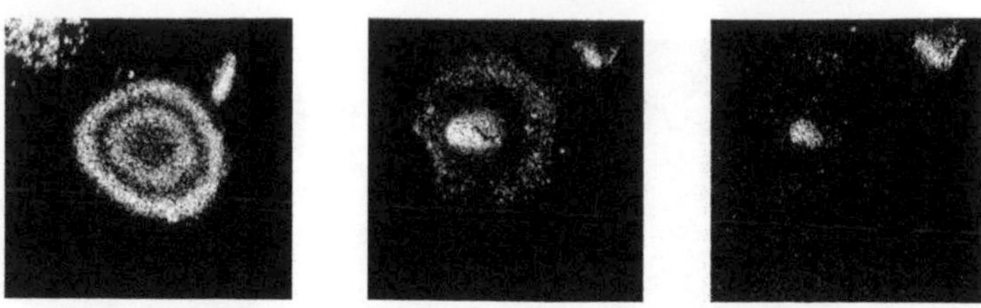

FIGURE 13.41 Resonances of a 100-μm micromembrane.

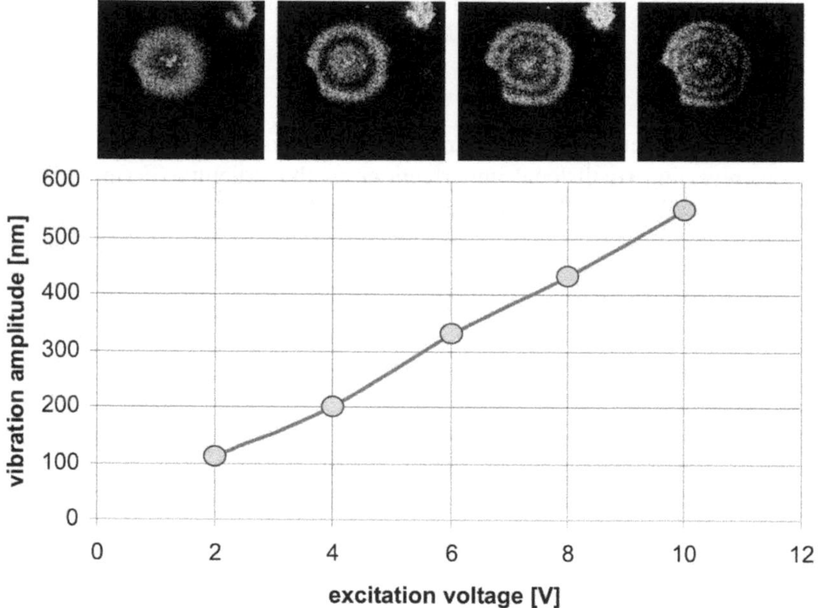

FIGURE 13.42 Calibration of a 100-μm active micromembrane at 879 kHz.

For the final application of the active micromembranes, it is necessary to predict the vibration amplitude for a certain voltage applied, i.e., the MEMS have to be calibrated. Such calibration can be carried out by the DUV-MSI very efficiently and Figure 13.42 gives an example. The membrane shows a linear response for a peak-to-peak voltage between 2 and 10 V. The corresponding vibration amplitude varies between 100 and 550 nm.

13.4 CONCLUSION

The following conclusions are important:

- The object-related speckle size does not scale with the magnification of the imaging system. Therefore, microscopic views can be produced with an object-related speckle size of less than 1μm.
- Micro-speckle-interferometers can be built, providing either high flexibility or a maximum of measuring sensitivity.

- The use of advanced deep ultraviolet laser sources has significant advantage for the study of microsystems. Speckle interferometry becomes feasible also on high-gloss surfaces, and both lateral resolution as well as measuring sensitivity are improved.
- Micro-speckle-interferometry is a valuable tool for the analysis of the operational behavior of Micro-Electro-Mechanical Systems. It provides clear advantages if the MEMS surface topography is complex including steep regions, if the vibration frequencies exceed the MHz range, or, if time-consuming object alignment has to be avoided.

REFERENCES

1. Goodman, J.W., Statistical properties of laser speckle patterns, in Dainty, J.C., Ed., *Laser Speckle and Related Phenomena*, Berlin: Springer-Verlag, 1975, pp. 9–75.
2. Dainty, J.C., *Laser Speckle and Related Phenomena*, Berlin: Springer-Verlag, 1975.
3. Erf, R.K., *Speckle Metrology*, New York: Academic Press, 1978.
4. Butters, J.N., Leendertz, J.A., Holographic and video techniques applied to engineering measurements, *Meas. Cont.*, 4: 349–354, 1971.
5. Macovski, A., Ramsey, S.D., Schaefer, L.F., Time lapse interferometry and contouring using television systems, *Appl. Opt.*, 10: 2722–2727, 1971.
6. Nakadate, S., Yatagai, T., Saito, H., Electronic speckle pattern interferometry using digital image processing techniques, *Appl. Opt.*, 19: 1879–1883, 1980.
7. Creath, K., Digital speckle pattern interferometry (DSPI) using a 100 × 100 imaging array, *Proc. SPIE*, 501: 292–298, 1984.
8. Höfling, R., *Digitale Specklemusterinterferometrie mit einem parallelen Bildprozessor*, Düsseldorf: VDI Verlag, 1988, Reihe 8 Nr. 145.
9. Løkberg, O.J., Høgmoen, K., Holje, O.M., Vibration measurement on the human ear drum *in vivo*, *Appl. Opt.*, 18: 763–765, 1979.
10. Stetson, K.A., Pryputniewicz, R.J., Eds., *Proc. SEM Hologram Interferometry and Speckle Metrology*, Baltimore, 1990.
11. Nye, J.F., Berry M.V., Dislocations in wave trains, *Phys. Rev. Soc. London A*, 336: 165–190, 1974.
12. Kolenovic, E., Osten, W., Jüptner, W. Non-linear speckle phase changes in the image plane caused by out-of-plane displacement, *Opt. Comm.*, 171/4-6: 333–344, 1999.
13. Freund, J. Optical vortices in Gaussian random wavefields: statistical probability densities, *J.O.S.A.*, A 11, 1644–1652.
14. Sotomaru, T., Tsuchiya, H., Miyamoto, Y., Takeda, M., Evolution of optical vortices and phase singularities in scattered random fields: Experimental observation of their 3D-structure, in Osten, W., Jüptner, W., Eds., *Proc. HoloMet 2001,* International Balatonfüred Workshop, 75–81, Bremen 2001 BIAS.
15. Huntley, J.M., Noise immune phase unwrapping algorithm, *Appl. Opt.,* 28: 3268–3270, 1989.
16. Ennos, A.E., Speckle interferometry, in Dainty, J.C., Ed., *Laser Speckle and Related Phenomena*, Berlin: Springer-Verlag, 1975, pp. 203–253.
17. Asakura, T., Surface roughness measurement, in Erf, R.K., Ed., *Speckle Metrology*, New York: Academic Press, 1978, pp. 11–49.
18. SPIE Milestone Series MS 132, *Electronic Speckle Pattern Interferometry*, Bellingham, WA, 1996.
19. Gastinger, K., Lokberg, O.J., Winter, S., Eds., *Proc. SPIE,* Vol. 4933, Speckle Metrology 2003, Trondheim.
20. Tiziani, H.J., Vibration analysis and deformation measurement, in Erf, R.K., Ed., *Speckle Metrology*, New York: Academic Press, 1978, pp. 73–110.
21. Sjödahl, M., Digital speckle photography, in Rastogi, P., Ed., *Digital Speckle Pattern Interferometry and Related Techniques*, Chichester: Wiley, 2001, pp. 289–336.
22. Sollid, J.E., Holographc interferometry applied to measurements of small static displacements of diffusely reflecting surfaces. *Appl. Optics,* 8: 1587–1595, 1969.
23. Yamaguchi, I., Fringe formations in deformation and vibration measurements using laser light, in Wolf, E., Ed., *Progress in Optics*, XXII, Amsterdam: Elsevier, 1985, pp. 271–340.

24. Osten, W., *Digitale Verarbeitung und Auswertung von Interferenzbildern*, Berlin: Akademie Verlag, 1991.
25. Kreis, Th., Osten, W., Automatische Rekonstruktion von Phasenverteilungen aus Interferogrammen, *tm-Techn. Messen.*, 58: 235–246, 1991.
26. Huntley. J.M., Automated analysis of speckle interferograms, in Rastogi, P.K., Ed., *Digital Speckle Pattern Interferometry and Related Techniques.* New York: Wiley and Sons, 2001, pp. 59–139.
27. Ghiglia, D.C., Pritt, M.D., *Two-Dimensional Phase Unwrapping: Theory, Algorithms, and Software,* New York: Wiley, 1998.
28. Huntley, J.M., Saldner, H., Temporal phase-unwrapping algorithm for automated interferogram analysis, *Appl. Opt.*, 32: 3047–3052, 1993.
29. Coggrave, C.R., Huntley, J.M., Real-time visualization of deformation fields using speckle interferometry and temporal phase-unwrapping, *Opt. Laser. Eng.*, 41: 601–620, 2004.
30. Aswendt, P., Schmidt, C.D., Zielke, D., Schubert, S.T., Inspection system for MEMS characterization on wafer level using ESPI, in *Proc. SPIE*, Vol. 4400, 2001, pp. 43–50.
31. Chen, F., Griffen, C.T., Allen, T.E., Digital speckle interferometry: some developments and applications for vibration measurement in the automotive industry, *Opt. Eng.*, 37: 1390–1397, 1998.
32. Stetson, K.A., Prohinsky, W.R., Electrooptic holography and its application to hologram interferometry, *Appl. Opt.*, 24: 3631–3637, 1985.
33. Pryputniewicz, R.J., *Holographic Numerical Analysis*, Worcester: Worcester Polytechnic Intstitute, 1992.
34. Jones, R., Wykes, C., Holographic and speckle interferometry, Cambridge: Cambridge University Press, 1983.
35. Höfling, R., Aswendt, P., Fiber Optic DSPI Strain Gauge System for Engineering Applications, VDI Berichte Nr. 118: 51–56, 1994.
36. http://www.yole.fr/ocmmm
37. Gorecki, C., Sabac, A., Bey, P., Gut, K., Jacobelli, A., Dean, T., An integrated opto-mechanical sensor for in situ characterisation of MEMS: the implementing of a reas out architecture, Bellingham, WA: SPIE Vol. 5145, 2003, pp. 189–195.

14 Spectroscopic Techniques for MEMS Inspection

Ingrid De Wolf

CONTENTS

14.1 Introduction 459
14.2 Raman Spectroscopy (RS) 460
    14.2.1 Principle 460
    14.2.2 Instrumentation 461
    14.2.3 Application to Microsystems 463
        14.2.3.1 Stress 463
        14.2.3.2 Coatings 471
14.3 Spectroscopic Ellipsometry (SE) 472
    14.3.1 Principle 472
    14.3.2 Applications to MEMS 472
14.4 Dual-Beam Spectroscopy (DBS) 474
    14.4.1 Principle 474
    14.4.2 Applications to MEMS 474
14.5 X-Ray Photoelectron Spectroscopy (XPS) 476
    14.5.1 Principle 476
    14.5.2 Applications to MEMS 476
14.6 High-Resolution Electron Energy Loss Spectroscopy (HREELS) 476
    14.6.1 Principle 476
    14.6.2 Applications to MEMS 477
14.7 Auger Electron Spectroscopy (AES) 478
    14.7.1 Principle 478
    14.7.2 Applications to MEMS 478
14.8 Brillouin Scattering (BS) 479
    14.8.1 Principle 479
    14.8.2 Applications to MEMS 479
14.9 Conclusions 479
References 480

14.1 INTRODUCTION

Spectroscopic techniques such as x-ray photoelectron spectroscopy (XPS), spectroscopic ellipsometry (SE), dual-beam spectroscopy (DBS), high-resolution electron energy loss spectroscopy (HREEL), Auger electron spectroscopy (AES), Raman spectroscopy (RS), and Brillouin scattering (BS) have found several applications in the microelectronics world. Most of these techniques are used for materials analysis, some for film thickness measurements, others for mechanical analysis.

This chapter describes applications of these spectroscopic techniques for the study of microelectromechanical systems (MEMS). For each technique the principle and measurement setup is briefly described and examples of applications to microsystems are discussed.

The main focus is on Raman spectroscopy because this technique can be used to measure local mechanical stress. During the fabrication of MEMS, residual stress or stress gradients are often induced in the thin films. When these films are released to become movable parts of a device such as a bridge, cantilever, or membrane, this stress often causes the device to fail owing to buckling, curling, or even fracture. RS is one of the few techniques that can be used to study this stress.

Although not commonly applied to MEMS, BS also gives information on mechanical properties; it can be used to determine the elastic constants of thin films. XPS, HREEL, and AES are mainly used to determine film composition and chemical bonds on MEMS surfaces. They are typically applied to investigate self-assembled monolayers (SAM) or thin surface films on MEMS. SE and DBS are used to study the thickness of such films and their optical parameters.

14.2 RAMAN SPECTROSCOPY (RS)

14.2.1 PRINCIPLE

Cardona and co-workers used RS to study phonons (lattice vibrations) in a large range of semiconductors. In particular, Anastassakis [1,2] focused on the effect of external perturbations (mechanical stress, temperature, and electrical fields) on the Raman peaks. Especially, the sensitivity of the Raman signal to mechanical stress caught the interest of microelectronics researchers, and from the 1990s on, the technique has been used to study mechanical stress induced by different processing steps in semiconductor devices and films [3].

RS is a nondestructive optical analysis technique. It measures the frequency of molecular or lattice vibrations in gases, fluids, or solids through the interaction of laser light, used as excitation source, with these vibration modes.

Vibrations in a crystal lattice (phonons) are classically described in terms of collective motions in the form of waves (lattice vibrations) with a certain frequency ω_j. When monochromatic light, as from a laser source, is incident on the crystal, the associated electric field **E** (frequency ω_i) will induce an electric moment $\mathbf{P} = \varepsilon_o \chi \mathbf{E}$, where χ is the susceptibility tensor. If the atoms are vibrating, χ may change as a function of these vibrations. As a result, the induced moment will reradiate light, ω_s, with frequency components $\omega_l + \omega_j$ and $\omega_i - \omega_j$. These are called first-order anti-Stokes and Stokes Raman scattering, respectively. Raman scattering can only be observed if the susceptibility tensor changes because of the vibrations. This is described by the so-called Raman tensors, R_j. These second-rank tensors are obtained from group theoretical considerations.

The Raman scattering efficiency of a given crystal depends on the Raman tensors and on the polarization vector of the incident ($\mathbf{e}_i$) and scattered ($\mathbf{e}_s$) light, and is given by

$$\mathrm{I} = \mathrm{C}\sum_j |\mathbf{e}_i.\mathbf{R}_j.\mathbf{e}_s|^2 \tag{14.1}$$

where C is a constant and $\mathbf{R}_j$ is the Raman tensor of the phonon j. For example, in the crystal coordinate system x = [100], y = [010], and z = [001], the Raman tensors of diamond-type (O_h point group) or zinc-blende-type (T_d point group) semiconductors are given by

$$\mathbf{R}_x = \begin{pmatrix} 0 & 0 & 0 \\ 0 & 0 & d \\ 0 & d & 0 \end{pmatrix} \quad \mathbf{R}_y = \begin{pmatrix} 0 & 0 & d \\ 0 & 0 & 0 \\ d & 0 & 0 \end{pmatrix} \quad \mathbf{R}_z = \begin{pmatrix} 0 & d & 0 \\ d & 0 & 0 \\ 0 & 0 & 0 \end{pmatrix} \tag{14.2}$$

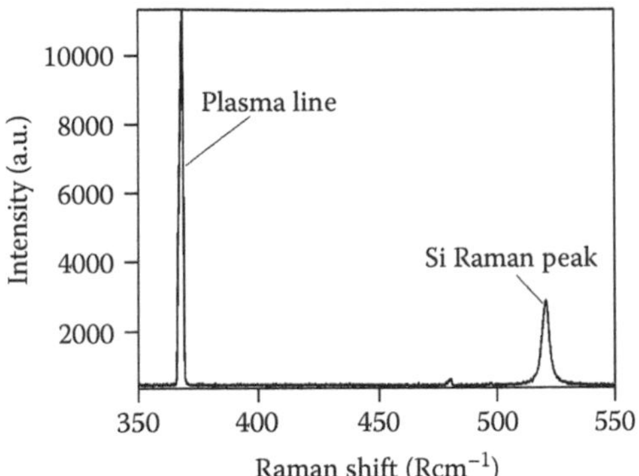

FIGURE 14.1 Raman spectrum of crystalline silicon.

As most semiconductors are opaque for the visible light used for Raman scattering, Raman experiments are mostly performed in a backscattering configuration, in which the incident light is perpendicular to the surface (z-direction), and the backscattered light is collected and analyzed along the same direction (–z). For backscattering from a (001) surface of, for example, crystalline Si, only the longitudinal optical (LO) phonon (described by R_z) can be observed (Equation 14.1 and Equation 14.2), and this only when $\mathbf{e}_i$ is perpendicular to $\mathbf{e}_s$. The two other tensors describe the transversal optical phonons (TO). Figure 14.1 shows the LO-Raman peak (Stokes) of crystalline silicon.

The most straightforward application of RS is identification of the sample. The frequency and number of the Raman peaks is a unique fingerprint for the material. There exist several catalogues listing the Raman spectra from fluids, semiconductors, polymers, etc. Raman signals from crystalline solids exhibit very sharp peaks. Irregular crystals, such as amorphous ones, have broad asymmetrical peaks. So, the Raman spectrum gives direct information on the crystallinity of the sample. However, the most interesting property of Raman spectra for the analysis of MEMS is the dependence of the frequency of the peaks on mechanical strain. By monitoring this frequency at different positions on the sample, a "strain map" can be obtained. Raman instruments dedicated to stress measurements are able to measure frequency changes as small as 0.02 cm^{-1}. For silicon, this corresponds to a stress sensitivity of about 10 MPa. The relation between strain or stress and the Raman frequency is rather complex [1–4], and depends on the so-called phonon deformation potentials (PDPs) of the investigated material. All strain tensor components influence the frequency, resulting in rather complex relations between the Raman peak frequencies and the strain tensor elements. Compressive stress results in an increase of the Raman frequency, whereas tensile stress results in a decrease. But these relations are mostly too complex to offer quantitative information on the stress or strain in a certain sample. In some circumstances, however, the relation becomes simply linear. This is, for example, the case for uniaxial stress (σ) or biaxial stress ($\sigma_{xx} + \sigma_{yy}$) in silicon:

$$\sigma\ (\text{MPa}) = -435\Delta\omega\ (\text{cm}^{-1}) \quad \text{or} \quad \sigma_{xx} + \sigma_{yy}\ (\text{MPa}) = -435\Delta\omega\ (\text{cm}^{-1}) \tag{14.3}$$

where $\Delta\omega$ is the shift of the Raman frequency from its stress-free value. This relation is often used to translate a measured Raman frequency shift into a stress value.

14.2.2 Instrumentation

Figure 14.2 shows the setup of a classical RS instrument. Monochromatic light from a laser is focused on the sample. Depending on the application, the wavelength of the laser light can range from UV to IR. The light passes through two filters, F1 and F2. F1 is a plasma filter that is used to block the plasma lines (see Figure 14.1) from the laser light, if required. In some cases, plasma

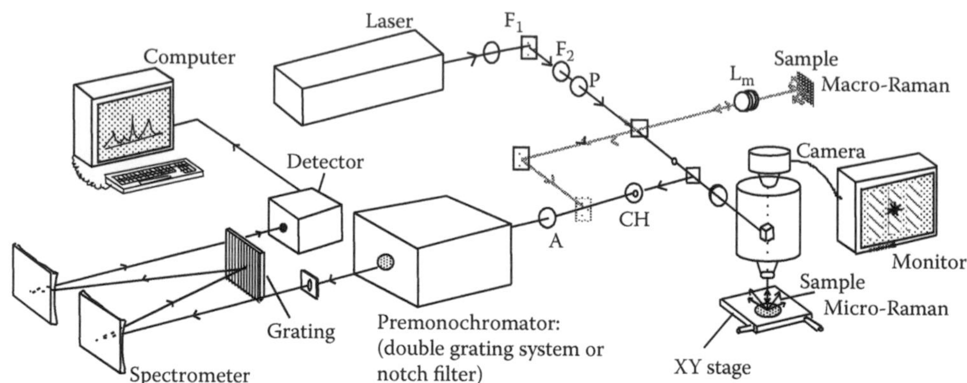

FIGURE 14.2 Conventional Raman spectroscopy instrument with macro and micro options.

lines are allowed and used as calibration peaks. F2 is a neutral density filter that can be used to reduce the laser power. The polarizer P is used to control the polarization of the incident light.

In the instrument depicted in Figure 14.2, two measurement modes are possible: macro-RS and micro-RS (μRS). If the light is focused through a conventional lens (L_m) on the sample, one speaks of macro-RS. The probed area on the sample depends on the lens L_m and is typically larger than 10 μm. If on the other hand the light is focused on the sample through an optical microscope, one speaks of μRS. The latter results in spot sizes down to about 1 μm, which makes it very useful for the study of MEMS. The sample is mounted on a computer-controlled X-Y stage, which allows it to be scanned in small steps (typical 0.1 μm) in any direction. The backscattered light is collected, directed into the premonochromator and spectrometer, and focused on a detector.

The main purpose of the premonochromator is to reject the laser light that has a much larger intensity than the Raman signals. In modern systems, this premonochromator is replaced by a small notch filter, enabling an important reduction in the size and complexity of the instrument.

μRS systems are mostly equipped with a confocal hole (CH). This allows performing confocal measurements by controlling the slice of the sample contributing to the detected Raman signal, enabling depth measurements in transparent samples.

An analysis technique is of interest to MEMS when it can provide information on material properties in samples with micrometer or even submicrometer dimensions. The spatial resolution of a Raman instrument is mainly defined by the size of the focused laser spot on the sample. It is often defined as the diameter of the ring at which the first zero of the Bessel function occurs (Rayleigh criterion) [43]. However, a definition of the spot size for RS applications that fits the experimental observations better is given by the diameter of a diffraction-limited spot at which its intensity has decreased to $1/e^2$ of its value in the middle of the spot [16]. It is given by

$$\varnothing = \frac{0.88\lambda}{NA} \tag{14.4}$$

where λ is the wavelength of the light and $NA = n \sin(\theta)$ is the numerical aperture of the objective. The angle θ is the angle the outer rays make with the optical axis, and n is the refractive index of the material surrounding the object. The spatial resolution can be improved by using a shorter wavelength and/or a higher NA. Ordinary objectives (with the sample in air) are limited by $n = 1$ (refractive index of air) to a maximum $NA \approx 1$ ($\sin\theta \leq 1$). With a standard 100X lens, $NA = 0.95$, a spot diameter of about 0.9 μm is obtained for 458-nm laser light. Increasing the breaking index n can be done by using, for example, an oil immersion objective [5]. With a 100X oil objective, $NA = 1.4$, spot sizes as small as 0.30 μm can be obtained.

14.2.3 Application to Microsystems

The vibration frequencies measured by RS are sensitive to internal and external perturbations. As a result, the technique can be used to study the composition, phase, crystallinity, crystal orientation and, in some cases, doping of materials [6]. Because of its good spatial resolution, RS has also found many applications in the study of microsystems [7]. These include the measurement of stress and stress gradients, film thickness and film composition, as are discussed in the following text.

14.2.3.1 Stress

Residual stresses and stress gradients that are present in freestanding films in MEMS can have a detrimental effect on the performance of these systems. They may result in buckling of cantilevers and beams, curling of membranes, fracture, and even (at high temperatures) creep in metal films. It is rather difficult to control or predict these stresses; they highly depend on the used film materials and on the deposition methods and parameters. Several publications discuss the use of RS to study these stresses in semiconductor films [8]. In the following text, we discuss some case studies in which RS is applied to analyze MEMS.

14.2.3.1.1 Stress in Membranes

Figure 14.3 shows an example in which μRS was used to measure the stress in the square Si membrane of a pressure sensor [9]. This sensor was processed by etching cavities in a silicon wafer and bonding the wafer anodic to a glass substrate (Figure 14.3a). This bonding introduced a negative pressure inside the chamber. Using laser-based profilometry measurements it was found that the membrane was deflected inward by about 5 μm (Figure 14.3b). A μRS system equipped with an autofocus module was used to scan the surface and to measure the Si Raman peak at different points on the square membrane.

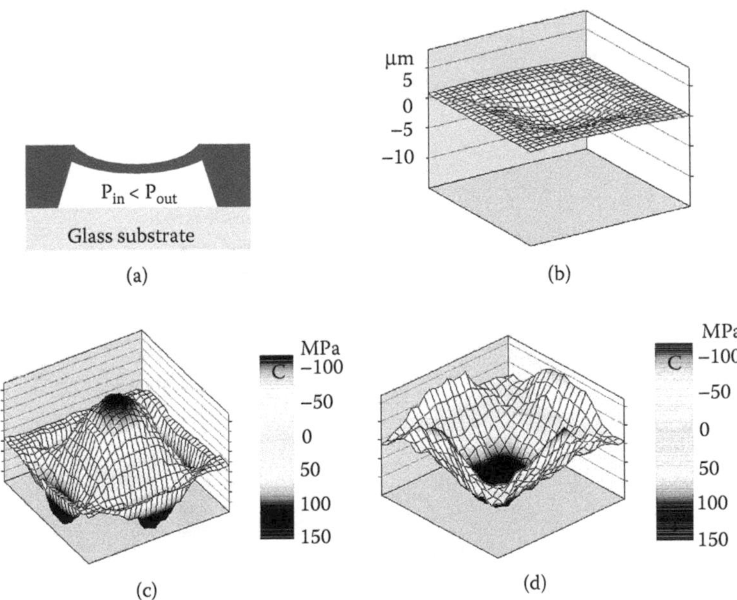

FIGURE 14.3 (a) Cross section of an Si membrane of a pressure sensor bonded to a glass substrate. The membrane is under pressure; (b) deflection of the membrane measured using laser interference, (c) biaxial mechanical stress in the top surface of the membrane measured using micro-Raman spectroscopy, (d) biaxial mechanical stress in the bottom surface of the membrane measured using micro-Raman spectroscopy. (From van Spengen, W.M., De Wolf, I., and Knechtel, R., Experimental two-dimensional mechanical stress characterization in a silicon pressure sensor membrane using micro-Raman spectroscopy, *Proc. Int. Symp. on Photonics and Applications (SPIE), Micromachining and Microfabrication*, p. 104, 2000.)

The Raman peaks were fitted with a Lorentzian function to determine their frequency. A plasma line was also fitted, using a Gaussian function for calibration. And a stress-free Si wafer was measured to obtain the stress-free Raman frequency. Figure 14.3c shows the mechanical stress in the top surface of the membrane, calculated from the change of the Raman frequency ($\Delta\omega$) from its stress-free value and assuming biaxial stress (Equation 14.3). It is clear that this Raman experiment did provide a very good image of the stress in the silicon membrane. In this picture, a positive value indicates compressive stress, such as at the center of the membrane. A negative value, as can be seen near the sides of the membrane, indicates tensile stress. The RS technique also allows probing of the silicon through a glass substrate. This makes it possible to study, for example, the glass–silicon interface, or the stress at the bottom of the membrane. The result of such an experiment for the same pressure sensor is shown in Figure 14.3d. It shows that stress at the bottom surface of the membrane is opposite to that measured at the top surface, as expected.

Similar measurements as those discussed in the preceding text were reported by Zhao et al. [10] for membranes of silicon-based micropumps. They show the stress map obtained in a quarter diaphragm region under 40 psi, as shown in Figure 14.4. They also observed compressive stress

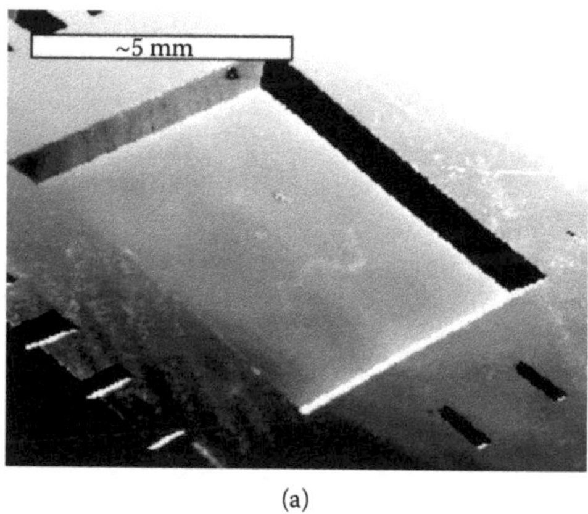

(a)

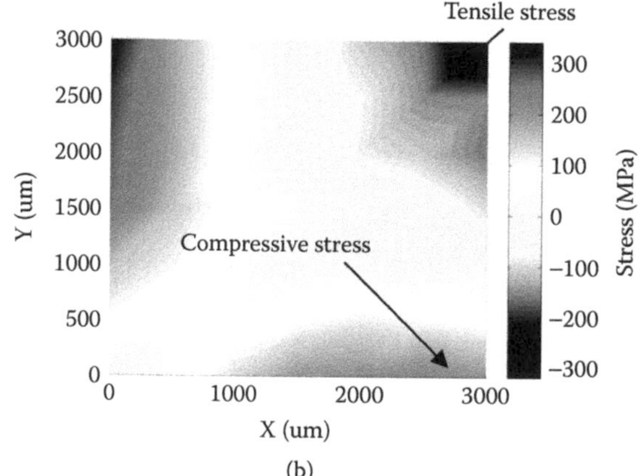

(b)

FIGURE 14.4 (a) 6 mm × 6 mm Si diaphragm of a micropump. (b) Stress distribution in a quarter part of the diaphragm interpreted from the measured Raman shift. (From Zhao, Y., Ludlow, D.J., Li, B., and Zhang, X., Evaluation of design and performance of a cryogenic MEMS micropump realized with Si-based diaphragm using combination of zygo/wyko interferometer and Raman spectroscopy, *Micro- and Nanosystems, Materials Research Society Symposium Proceedings*, Vol. 782, A5.52.1–A5.52-6, 2003. With permission from the authors and the Materials Research Society.)

near the center of the edges of the membrane and tensile stress around the center point of the membrane. They compared these data with finite-element modeling (FEM) results, and a very good correlation was observed. RS data can indeed be very useful for verifying and optimizing numerical and finite element models (FEMs) of stress [4]. Accurate modeling of mechanical stresses and deformations is crucial for the successful design, fabrication, and operation of micromachined devices. However, FEMs are susceptible to errors from several sources, including uncertainties in the used values of materials' properties, improper identification of boundary conditions, and imprecise metrology. For this reason, validation of the models using an experimental technique is mandatory. RS is one of the few techniques that can be used for this purpose. Other techniques that can be used are x-ray diffraction and moiré [19] or speckle [41] analysis (see also other chapters in this volume).

Several other authors did report on the use of RS to measure stress in membranes. For example, Starman et al. [11] used RS to study stresses in polysilicon MEMS micromirrors fabricated using the MUMP's process and also compared the results with FEM results. They concluded that RS can measure both residual and externally induced stresses in a MEMS device. Cho et al. [12] applied RS to study stress induced by electroplated permalloy films on Si membranes and used the results to optimize their processing.

14.2.3.1.2 Stress in Beams

Also, the stresses in cantilever and clamped-clamped beams were investigated by RS. For example, Srikar et al. [13] report on RS measurements of bending stresses in deep-reactive-ion-etched (DRIE) single-crystal Si beams of length 2950 μm, width 480 μm, thickness 150 μm, and fillet radius 65 μm. The structures are shown in Figure 14.5a. In addition to a thorough RS investigation of stresses in these structures, they also compared the measurement results with FEM. Errors in the FE model can easily occur in DRIE structures because of etching nonuniformities associated with this processing technique. The authors performed line scans measuring Raman spectra across the thickness of the flexure at different distances, x, from the support. Some results are shown in Figure 14.5b. Their FE models match the Raman measurements very well if the PDPs of Anastassakis [2] are

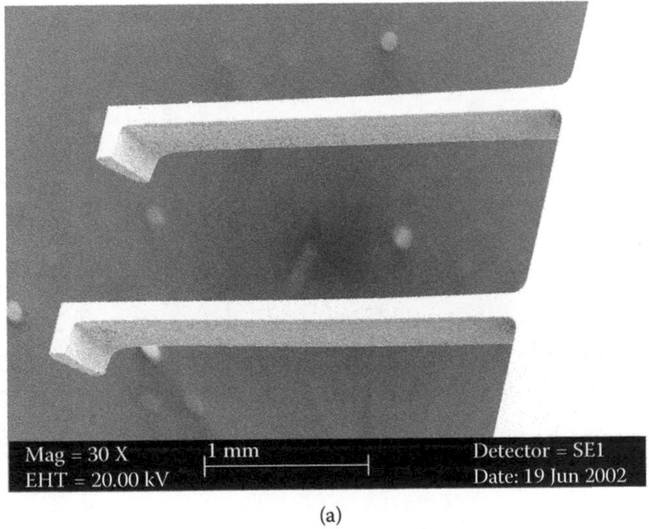

(a)

FIGURE 14.5a Scanning electron micrograph of deep-reactive-ion-etched single-crystal silicon. (From Srikar, V.T., Swan, A.K., Ünlü, M.S., Goldberg, B.B., and Spearing, S.M., Micro-Raman measurement of bending stresses in micromachined silicon flexures. *IEEE J. Microelectromech. Syst.*, Vol. 12(6), 779–787, 2003. With permission from authors.)

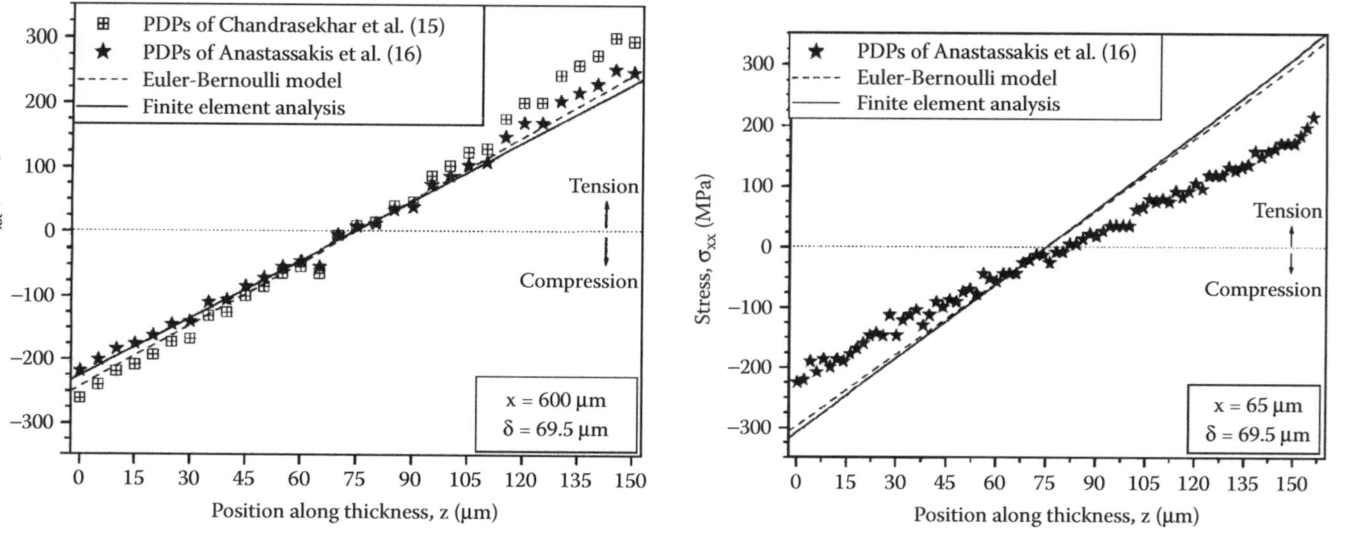

FIGURE 14.5b Stress profile across the thickness of the beam at (left) x = 600 μm. The FEM is in very good agreement with the predictions of the beam model. At (right) x = 65 μm, closer to the beam near the root, the FEM results deviates from the Raman data. (From Srikar, V.T., Swan, A.K., Ünlü, M.S., Goldberg, B.B., and Spearing, S.M., Micro-Raman measurement of bending stresses in micromachined silicon flexures. *IEEE J. Microelectromech. Syst.*, Vol. 12(6), 779–787, 2003. With permission from authors.)

used for the calculations (Figure 14.5b, left). The authors also demonstrated that the interpretation of the Raman signals is more complicated at places with complex stresses, such as, for example, at the root of the beam (Figure 14.5b, right).

Also, poly-Si clamped-clamped beams were studied using RS. Starman et al. [14] studied the stress in poly-Si microbridges before release and its dependence on phosphorous implantation and anneal steps. They found that these steps can alter and reduce the stresses and that this can be monitored by RS. In this way, RS can directly contribute to an improved yield, reliability, and functionality of the MEMS.

14.2.3.1.3 Pitfalls

There is one important drawback of RS when measuring stress in MEMS. It is well known that the laser light, focused on the sample to a spot size of about 1 μm, results in some local heating of the sample. The Raman frequency is also dependent on the temperature, and local heating will result in a downshift of the signal. For example, the relation between temperature variation T and Raman peak position of silicon is given by [6]:

$$\Delta\omega = -0.0242\ \Delta T \tag{14.5}$$

As long as the thermal conduction and thickness of the sample are large enough, as was, for example, the case for the study of the Si beams of Figure 14.5a [13], this heating is very small and its effect on the Raman signal can be neglected. However, for thin or small samples, as is often the case in MEMS, the heat cannot easily flow away in the sample and the local heating may dominate the effect of mechanical stress on the Raman signal. An example of this was presented by van Spengen et al. [15] for poly-Si cantilever beams (Fig. 14.6). A Raman scan was performed along the length of narrow polysilicon beams. A large downshift of the Raman frequency was found, as is shown in Figure 14.6. If this downshift were due to stress, tensile stress values up to 400 MPa should be present in the beams. Also, this maximal stress value was the same for the three longest beams. It was clear from this experiment that the measured Raman shift was not stress related, but caused by local heating of the beams by the focused laser light. Close to the bondpads at the left and right side of the beams, the heat can get away and the measured Raman shift is small. Closer to the center of the beam, the heat cannot get away easily and the heating effect is maximal.

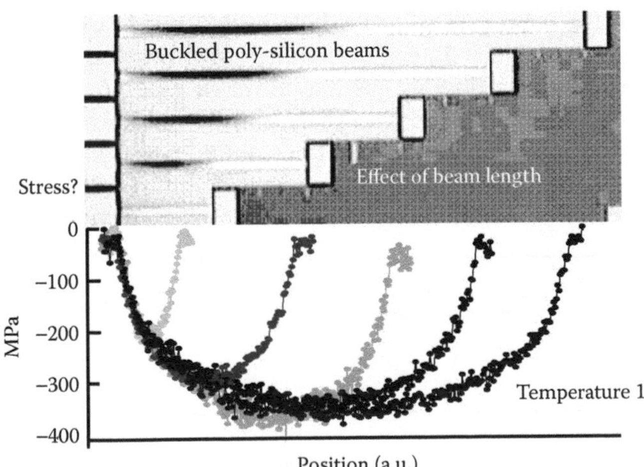

FIGURE 14.6 Stress (tensile) in clamped-clamped poly-Si beams calculated from the shift of the Raman peak frequency from the stress-free value. In this case, the stress values are not correct: the shifts are caused by local heating induced in the beams by the focused laser beam, not by stress. (From van Spengen, M., Reliability of MEMS, Ph.D. dissertation, ISBN 90-5682-504-6, Katolieke Universiteit Leuven, 2004.)

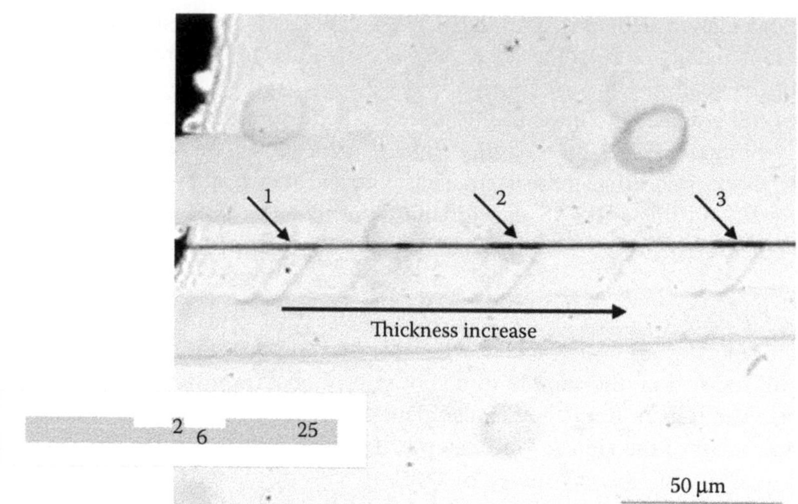

FIGURE 14.7 TEM sample in which the Raman measurements are performed. The 2-μm active regions that are measured by Raman are indicated (regions 1, 2, and 3). Inset: dimensions and topology of the three regions. (From EC project STREAM IST-1999-10341.)

From this experiment it is clear that care has to be taken when measuring stress in MEMS using μRS. One should ensure that the shifts measured are not related to local heating effects.

A question that can be asked is: for which sample thickness might local heating influence the Raman shift? This, of course, depends not only on the thickness, but also on the width of the investigated structures and on the materials' properties. For crystalline Si membranes, we deduced that problems start for membranes thinner than 3 μm. This number was determined from an experiment on a sample that was thinned for transmission electron microscopy (TEM) investigation (Figure 14.7). The sample consisted of two Si pieces glued together and locally thinned until a hole is formed in the center. The hole is at the left in this picture. Next to the hole, the sample is very thin (see fringes). A Raman scan was performed along a line passing close to 2-μm-wide silicon lines isolated by 6-μm-wide trenches filled with SiO_2 (see inset). They can be distinguished at three positions on the sample: close to the hole in the sample where the sample is thinnest (position 1), farther away (position 2), and still farther away (position 3).

The line scan was done as close as possible to the surface of the Si piece containing the trenches, so that stress induced by the trenches in the silicon could be measured. The measurement started at the side of the hole. The power of the laser on the sample was about 1.5 mW (concentrated in the focused laser beam, wavelength 458 nm). Figure 14.8 shows the intensity

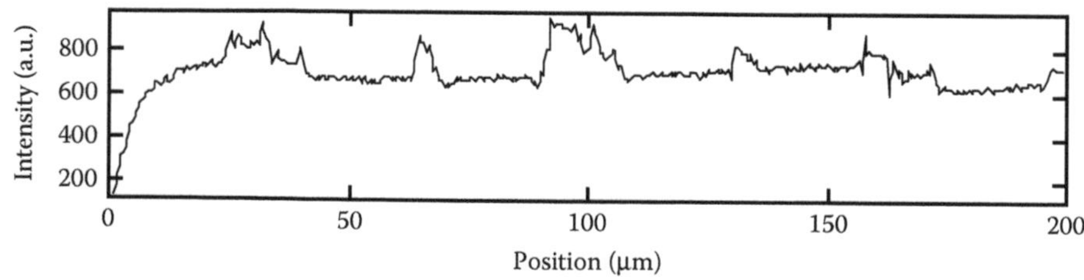

FIGURE 14.8 Raman intensity as a function of distance from the hole in the TEM sample. (From EC project STREAM IST-1999-10341.)

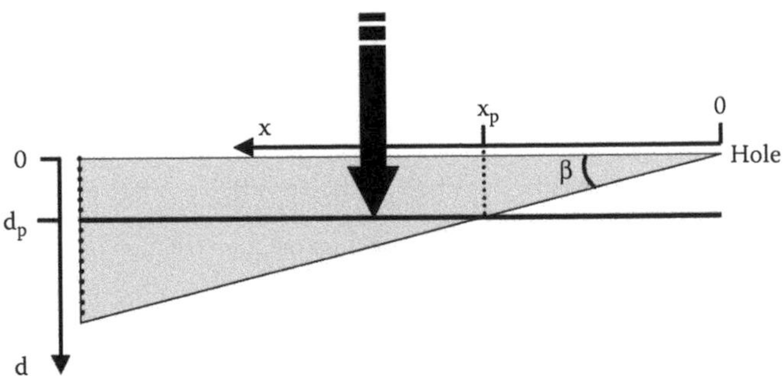

FIGURE 14.9 Relation between depth d_p of Raman signal, distance x from the hole in the sample, and angle β of the sample. (From EC project STREAM IST-1999-10341.)

of the silicon Raman peak measured during this line scan. It shows some local variations near the structures, but its baseline level stays, far from the edge of the hole, constant at about 700 a.u. Near the hole (position 0), the intensity is very small and increases with distance from the hole up to a distance of about 22 μm.

This can be explained by an increase in the thickness of the sample until it is larger than the penetration depth of the laser light, as shown in Figure 14.9. As long as the sample thickness is smaller than the penetration depth d_p of the laser light in the sample, the Raman intensity will increase with distance x. As soon as $x > x_p$, i.e., when the thickness of the sample is larger than d_p, the intensity will remain constant. From Figure 14.8 it follows that x_p is about 22 μm.

So, from the change in the intensity with distance x from the hole, it is possible to calculate the thickness "d" of the sample. The total Raman scattered light intensity integrated from the surface of the sample to the bottom of the sample can simply be calculated as follows [42]:

$$I_S = I_0 D \int_0^d e^{-2\alpha d} dx = \frac{I_0 D}{2\alpha}(1 - e^{-2\alpha d}) \tag{14.6}$$

where I_0, D, and α are the incident light intensity, the Raman scattering cross section, and the photoabsorption coefficient of silicon, respectively. The factor "2" arises because the light has to go in and come out of the material to be detected by the Raman system. For the 457.9-nm wavelength of Si, $\alpha = 3.666\ 10^6\ \text{m}^{-1}$ [42]. If we assume that the shape of the sample is a sharp triangle with angle β (see Figure 14.9), then the relation between the thickness "d" of the sample and the distance "x" from the edge is given by d = tgβ x and Equation 14.6 becomes

$$I_s(x) = \frac{I_0 D}{2\alpha}(1 - e^{-2\alpha . tg(\beta) x}) \tag{14.7}$$

Figure 14.10 shows the fit of an exponential relation to the part of the Raman intensity curve where the intensity changes (hence, for $x < x_p$) close to the hole. As can be seen in this figure, a very good fit is obtained, giving a value for tg(β) = 0.0270 ± 0.0006 or β = 1.54° ± 0.03°. From this fit, we know the relation between sample thickness and position (Equation 14.7).

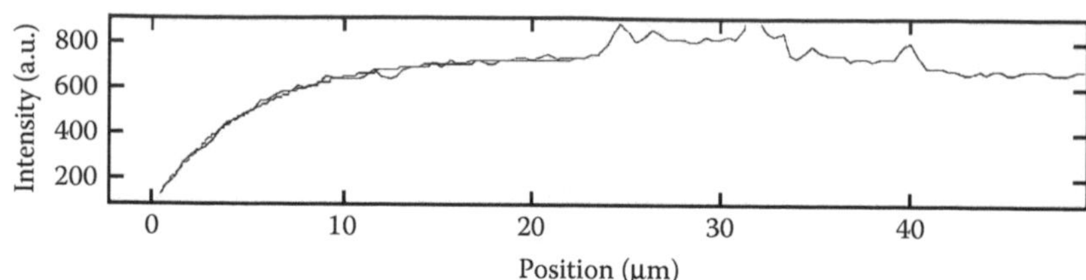

FIGURE 14.10 Change of the Raman intensity as a distance from the hole in the TEM sample and fit of Equation 14.1 to the Raman intensity as a function of the position. (From EC project STREAM IST-1999-10341.)

Figure 14.11 shows the variation of the Raman peak position (Raman frequency, ω) with distance from the hole in the TEM sample. On top of the figure, a schematic drawing indicates the position of the SiO_2 trenches. The straight line in the figure indicates the thickness variation of the sample, obtained from the preceding fit. We clearly see two kinds of variation of ω:

1. A local variation in ω near the trench structures. This is due to the local stress variation near the structures.
2. An overall decrease in ω toward the hole in the sample, starting at about 75 to 100 μm away from the hole. This is not due to a stress variation, but to a change in the local temperature of the sample. The position where this decrease starts corresponds to a sample thinckness of 2 to 3 μm.

So, the decrease in the Raman frequency close to the hole in the TEM sample is predominantly due to local heating of the thin sample, i.e., due to a temperature increase. This phenomenon occurs in crystalline Si for samples thinner than about 3 μm.

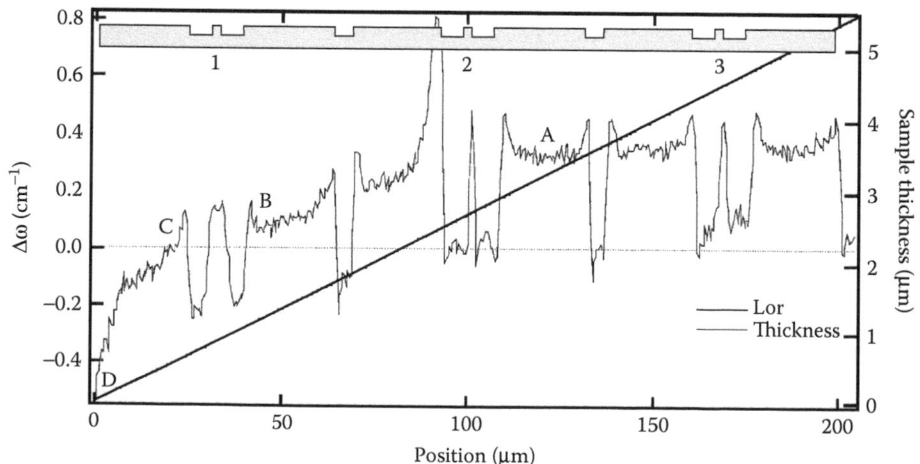

FIGURE 14.11 Raman frequency shift as a function of distance from the hole in the TEM sample. Measurement along the top surface on the cross section of the sample. Straight line: thickness variation of the sample as calculated from the variation of the Raman intensity and Equation 14.6 and Equation 14.7. (From EC project STREAM IST-1999-10341.)

In Figure 14.11 three points, A, B, and C, are indicated. We assume that the stress in these points is the same. So, the change in $\Delta\omega$ between these points is due to a change in local heating because of sample thinning. Assume that at point A ($\Delta\omega = 0.328$ cm^{-1}) there is no heating effect yet (sample thick enough to dissipate heat). We take that point as reference. Close to the hole (point D), $\Delta\omega = -0.54$ cm^{-1}, indicating a temperature increase of 36°C (Equation 14.5).

One can expect that this effect becomes worse in materials with a larger absorption coefficient, such as polycrystalline Si, which is often used in MEMS. In addition, these calculations were done for membranes; for beams, the effect will be even larger.

In conclusion, one should be very careful with the interpretation of the data when using μRS to extract values of stress in MEMS. Local heating caused by the probing laser beam can falsify the measurement results.

14.2.3.2 Coatings

Stiction and friction are typically of concern for MEMS with moving and touching parts, such as microgears and switches. These problems can often be prevented by the application of a coating. μRS can in some cases be used to study the composition of such a coating [17]. However, a coating is in general only effective if it is uniformly applied, i.e., also in high-aspect-ratio areas such as between gear teeth. Not many techniques allow probing in such small, deeper areas, but RS is an exception to this, owing to its small probing spot size. And it can also be applied to study the thickness uniformity. For example, Ager et al. [18] used the intensity of the Raman signal of an antifriction carbon (DLC) coating to study its thickness variations on Ni-alloy gears (Figure 14.12). They showed that the coating was uniform on the top gear surface, but somewhat thinner at a certain depth inside the gear teeth. This only works if the penetration depth of the light used for the Raman signal excitation is larger than the thickness of the investigated layer. In that case, the intensity of the Raman signal is directly related to the thickness of the coating.

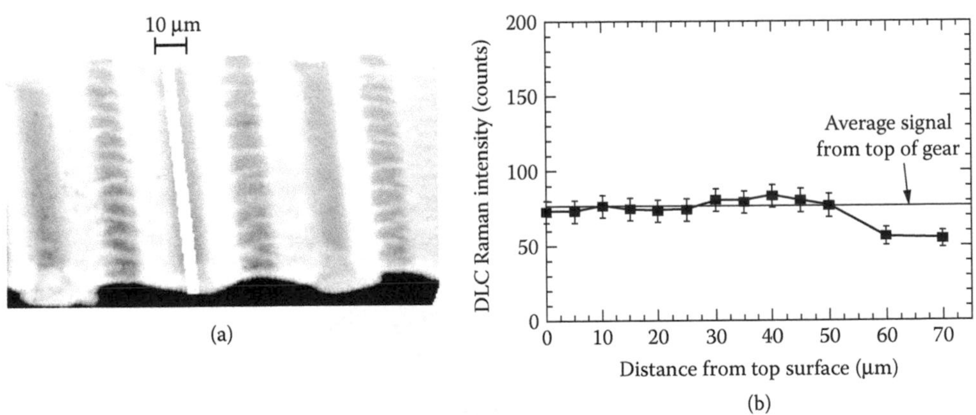

FIGURE 14.12 (a) SEM detail of area between gear teeth. The width of the Raman probe (5 μm) and the probed region are indicated by the white line. (b) DLC Raman intensity between gear teeth as a function of the distance from the top surface of the microgear. The intensity observed from scans on the top of the gear is indicated by the solid straight line. (From Ager, J.W., III, Monteiro, O.R., Brown, I.G., Follstaedt, D.M., Knapp, J.A., Dugger, M.T., and Christenson, T.R., Performance of ultra hard carbon wear coatings on microgears fabricated by LIGA, Vol. 546, *Proc. Materials Research Society*, 1998. With permission from the authors and the Materials Research Society.)

14.3 SPECTROSCOPIC ELLIPSOMETRY (SE)

14.3.1 Principle

SE is an optical technique that provides information on optical parameters of thin films. It is used to determine the index of refraction, *n*, and the extinction coefficient, *k*, of the film. If these are known, the film thickness or variations in this thickness can be extracted from the SE data. Also, information on the interface roughness can be obtained. The thickness sensitivity goes from submonolayers to millimeters. SE is often used to study the formation and changes in properties of thin films on thick substrates, for example, SiO_2 on silicon.

Figure 14.13 shows the principle of SE. Polarized light with wavelengths ranging from ultraviolet to near-infrared (depending on the used detector and on the spectral region of interest) is reflected from the sample surface, and the change in the polarization is measured. After reflection on a surface, a linearly polarized light beam becomes, in general, elliptically polarized. The reflected light has phase changes that are different for electric field components polarized parallel (p) and perpendicular (s) to the plane of incidence. The ellipsometer measures this state of polarization. From this measurement, the so-called ellipsometry parameters Δ and ψ are obtained as a function of the wavelength of the light. This experiment can be done in any ambient transparent to the used wavelengths, i.e., vacuum, gases, and liquids. An advantage of this is that SE can be applied to monitor *in situ* the deposition of films, their changes upon annealing or other processing steps, etc.

The result of an SE measurement of a nitrided SiO_2 film on a silicon substrate is shown in Figure 14.14 (Courtesy of Hugo Bender, IMEC). The ellipsometry parameters are measured as a function of the wavelength of the incident light. In this case, the measurement is performed for three different incident angles of the light (72°, 75°, and 78°). A computer model was fitted to the obtained results in order to determine the thickness of the film (full lines in Figure 14.14). It was estimated to be 2.80 ± 0.04 nm if the measurement was only performed at 75°. By performing the measurement at three angles, the sensitivity was increased to 2.86 ± 0.02 nm. This can also be done for multilayer structures. The technique requires samples with parallel interfaces and smooth surfaces.

14.3.2 Applications to MEMS

SE is applied in several studies to analyze self-assembled monolayers (SAM) on MEMS. These are layers of organic molecules that are used to modify the chemical and physical properties of insulator, semiconductor, or metal surfaces. In microsystems, such layers are, for example, used

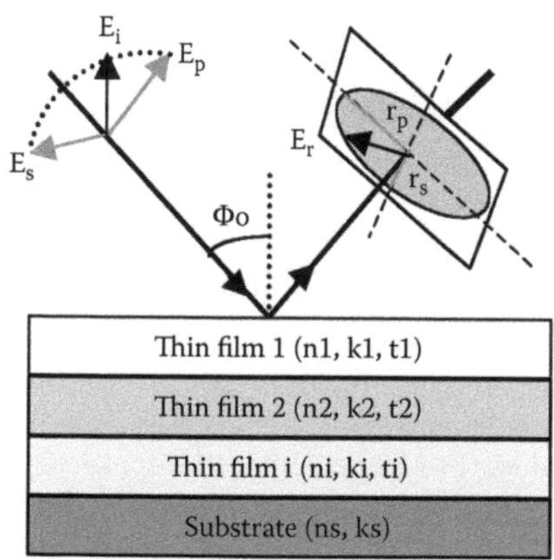

FIGURE 14.13 Principle of spectroscopic ellipsometry.

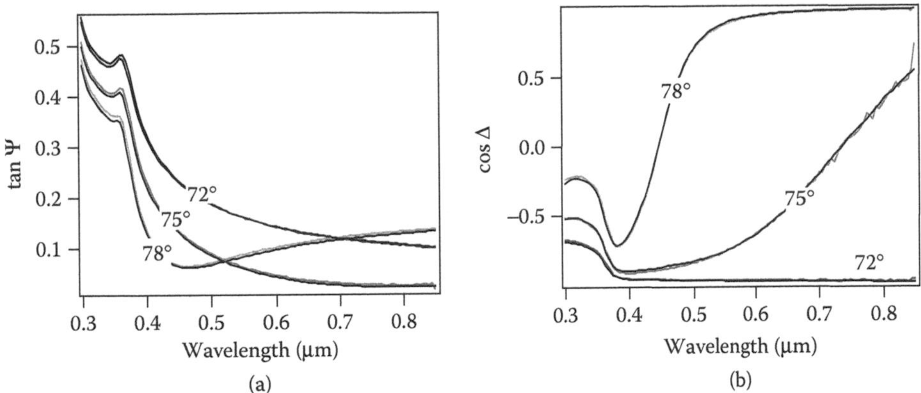

FIGURE 14.14 Plot of the spectroscopy ellipsometry parameters Δ and ψ for a nitrided SiO_2 film on silicon substrate. (Courtesy of Hugo Bender, IMEC.)

for microtribological lubrication [20,21], biosensors [22–24], corrosion barriers on metals [24–26], and anticapillary stiction coatings offering safe structural release [21] and good lifetime.

Han et al. [27] applied SE to study the thickness of an alkanethiol SAM film on Ge (111) surfaces. They assumed that the optical parameters are similar to those of polyethylene films, and they used a two-layer model. Figure 14.15 shows the SE data and the very good fit of the computer model to these data. From these measurements the thickness of the SAM film was estimated to be 25 Å.

A very interesting application of SE was discussed by Schutte and Martin at ISTFA2004 [28]. The reliability and performance of many MEMS is directly influenced by their surface characteristics. The packaging process can affect this surface, either by contamination due to outgassing of the package materials or due to the influence of the high-temperature steps that may occur during the packaging process. Even surface films of only a few angstroms can alter the performance and wear life in inertial MEMS. The authors installed microspot optics on a

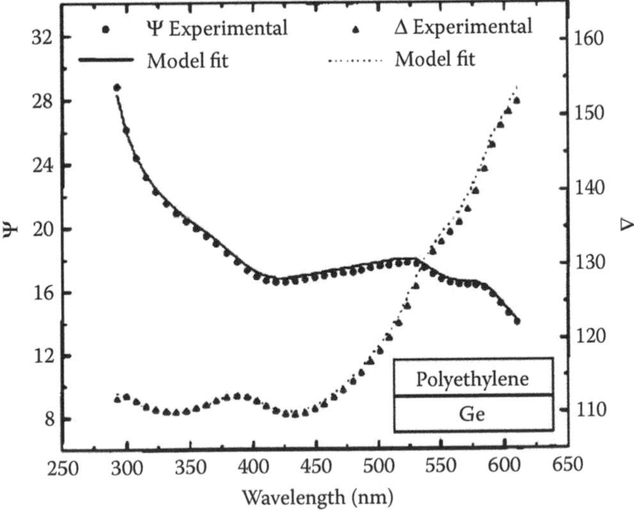

FIGURE 14.15 Ellipsometric angles ψ and Δ of the octadecanethiol SAM film on Ge(111). (From Han, S.M., Ashurst, W.R., Carraro, C., and Maboudian, R., Formation of alkanethiol monolayer on Ge(111), *J. Am. Chem. Soc.*, 123(10): 2422–2425, 2001. With permission from the authors and the American Chemical Society.)

manual ellipsometer and developed a sample preparation and test protocol to measure surface films after packaging. Experiments were conducted to characterize the effects of outgassing from die-attached materials and the influence of antistiction coatings within the package environment. They showed that such a micro-SE is a very valuable tool for MEMS devices and process characterization after packaging.

14.4 DUAL-BEAM SPECTROSCOPY (DBS)

14.4.1 Principle

A technique that is closely related to SE, called DBS, was successfully applied by AMD to control the curvature of poly-Si MEMS structures [29]. DBS uses nonpolarized polychromatic light (wavelengths, for example, between 220 and 840 nm) to illuminate the sample (Figure 14.16).

The intensity of the reflected light is measured as a function of the wavelength of the incident light and normalized with respect to a standard, such as a bare silicon surface. As in SE, each film of the sample has a certain refractive index, n, and extinction coefficient, k. Some of the light incident on the sample will be reflected from the surface, and another part will pass through the film and reflect/adsorb on the next interface. As in SE, if n and k for each layer are known, the thickness of each layer can be extracted using the appropriate software and models.

14.4.2 Applications to MEMS

Curvature of functional parts of MEMS can be detrimental to the device performance. Several techniques exist to measure this curvature, such as optical interferometry, slope measurements, and dynamic focus. However, they cannot distinguish between changes in thickness of the moving part, or changes in the gap between the part and the substrate. Sledziewski et al. [29] demonstrated that DBS is a very attractive technique to measure the curvature of released polysilicon MEMS structures. They show that DBS can accurately measure both the void space between the polysilicon beam and the substrate, and the thickness of the beams. They applied the technique to an integrated microelectromechanical accelerometer (*i*MEMS®) manufactured by Analog Devices Inc. (ADI) that uses polycrystalline silicon as the sense element [29]. It has cantilever sense beams that are attached to a suspended proof mass. In order to have optimal performance, no bending (curvature) is allowed in the polysilicon film. An extreme case (not from the standard production) of curvature is illustrated in Figure 14.17 [29].

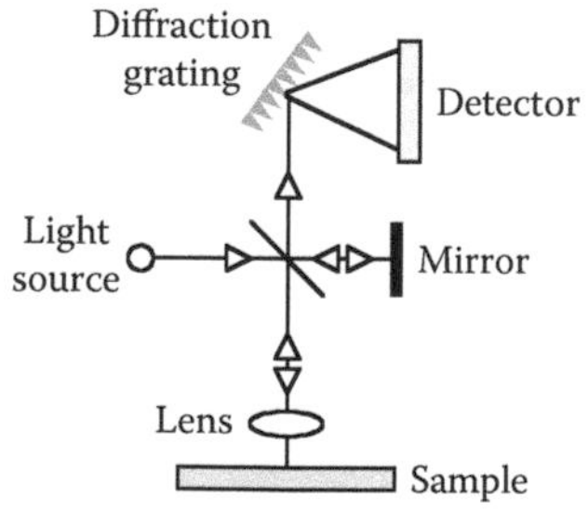

FIGURE 14.16 Principle of the dual-beam spectroscopy technique.

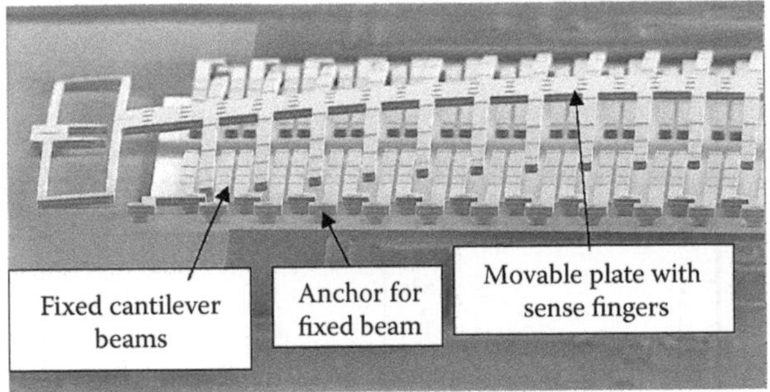

FIGURE 14.17 Severe curvature of polysilicon movable proof mass. (From Sledziewski, J.M., Nunan, K., Robinson, T., and Bar-on, I., Curvature measurement of polysilicon MEMS structures using spectroscopic reflectometry, SEM, Society for Experimental Mechanics, *Annu. Conf. Expo. on Exp. Appl. Mech.*, 332–335, 2001. With permission from the Society for Experimental Mechanics.)

DBS was applied along the length of a movable 2.0-μm-thick polysilicon proof mass. The spacing between it and the single-crystal silicon substrate, i.e., the void space, was approximately 1.6 μm. Figure 14.18 shows the result of the measurements. It is clear from this figure that the polysilicon film thickness is uniform over the measurement area, but the void thickness does vary, indicating curvature of the beam.

The results obtained by DBS were compared with results obtained using an interferometer technique. The authors showed that the latter was less precise, less reproducible, and less reliable.

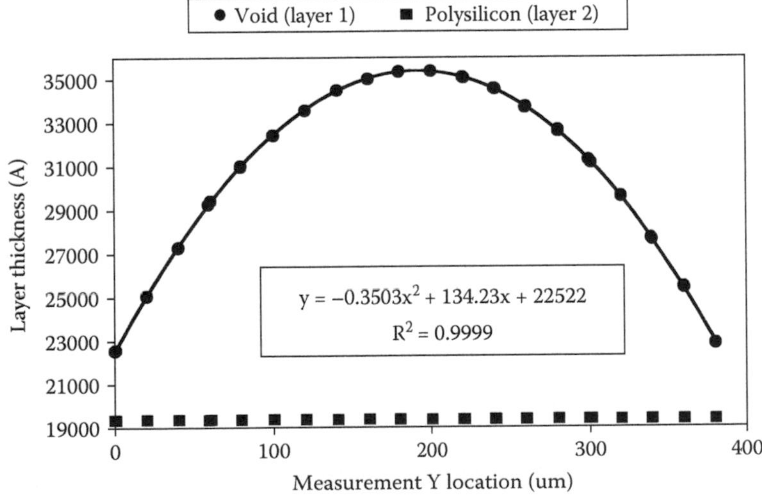

FIGURE 14.18 Polysilicon and void thickness measurement using DBS illustrating beam curvature. (From Sledziewski, J.M., Nunan, K., Robinson, T., and Bar-on, I., Curvature measurement of polysilicon MEMS structures using spectroscopic reflectometry, SEM, Society for Experimental Mechanics, *Annu. Conf. Expo. on Exp. Appl. Mech.*, 332–335, 2001. With permission from the Society for Experimental Mechanics.)

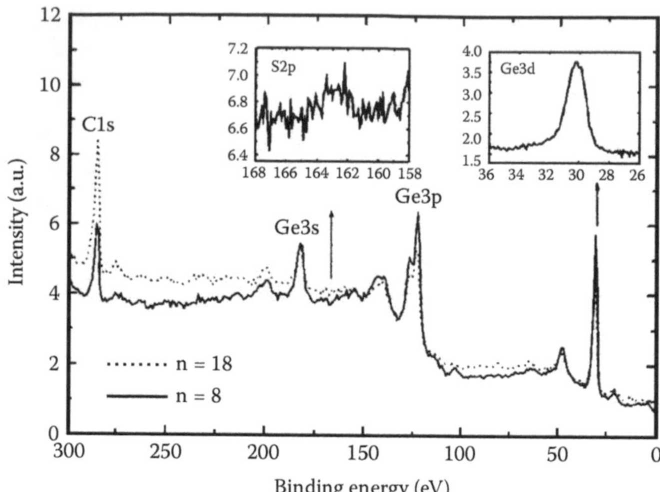

FIGURE 14.19 X-ray photoelectron spectra of octane- and octadecanethiol on a Ge(111) surface. The insets show the S(2p) and Ge(3d) close-ups for octadecanethiol. (From Han, S.M., Ashurst, W.R., Carraro, C., and Maboudian, R., Formation of alkanethiol monolayer on Ge(111), *J. Am. Chem. Soc.*, 123(10): 2422–2425, 2001. With permission from the authors and the American Chemical Society.)

14.5 X-RAY PHOTOELECTRON SPECTROSCOPY (XPS)

14.5.1 Principle

XPS is mainly used to obtain information on the chemical bonds on surfaces and to determine the thickness of thin films. In the XPS technique, the sample is radiated in vacuum with x-rays that cause emission of photoelectrons at the surface (depth 1 to 5 nm). These emitted photoelectrons have a characteristic energy for each chemical element and its bonding state, allowing identification of the element. XPS can distinguish chemical arrangements such as silicon-to-silicon bonds from silicon-to-oxygen bonds. These differ in binding energy, giving rise to a shift in the energy of the photoelectrons. XPS can do imaging and, combined with sputtering, depth profiling up to 1-μm depth. The size of the analysis area can be as small as 26 μm, which makes the technique applicable to MEMS.

14.5.2 Applications to MEMS

XPS is especially valuable for analyzing functional groups in different materials (insulators, semiconductors, metals), including polymers and other organic materials. Han et al. [27] used, in addition to the previously mentioned SE technique, XPS also to analyze the SAM alkanethiol monolayer on Ge(111) (Figure 14.19). XPS showed the presence of C and Ge and the presence of S atoms at the monolayer/Ge interface. Very little O was observed at the surface and attributed to contamination. The authors demonstrated with this study that hydrophobic alkanethiol monolayers readily form on HF-treated Ge(111) surfaces.

14.6 HIGH-RESOLUTION ELECTRON ENERGY LOSS SPECTROSCOPY (HREELS)

14.6.1 Principle

HREELS is high-resolution EELS, i.e., electron energy loss spectroscopy [30]. In EELS, a sample is subjected to a monoenergetic beam of electrons. This can be done in a transmission electron microscope (TEM) or scanning TEM (STEM). The electrons lose energy during their path through the thin sample owing to various interactions. These losses are measured at the detector on the other side of the sample. They can reveal the composition of the sample. EELS can also be performed on the surface of the sample. In that case the electron beam is reflected, mostly without loss of energy (elastically scattered electrons) but also for a small part with loss of energy owing to interactions with plasmons or other excitations, giving a number of peaks at lower energy. In HREELS, energy losses are studied at a high resolution (about 30 meV). This provides data on the vibrations of molecules on surfaces. HREELS is often used for the analysis of molecular films and polymers. The technique is particularly sensitive to the outermost groups of the film with a mean depth of analysis of some angstroms.

14.6.2 Applications to MEMS

As can be expected from the common applications of HREEL, it is also very useful in microsystems for the study of surface layers such as self-assembled monolayers (SAM). An example is the study of Kluth et al. [31]. They investigated the behavior of alkylsiloxane SAM on oxidized Si(100) at high temperatures. Figure 14.20 shows HREEL spectra of the C–H stretch and the overtone of the Si–O–Si asymmetric stretch for an OTS-coated oxidized Si(100) surface. The spectra remain unchanged up to 740K, indicating that the monolayer remains stable up to that temperature. Upon annealing to 780K, the intensities of the C–H modes decrease slightly relative to the Si–O modes, indicating a reduction in the monolayer coverage. Annealing to 815K results in a further reduction of the C–H intensities and the appearance of two peaks in the C–H stretch at 2920 and 2980 cm^{-1}, along with the appearance of a single peak at 1400 cm^{-1}. These peaks are consistent with methyl groups directly bonded to silicon atoms, suggesting that the Si–C

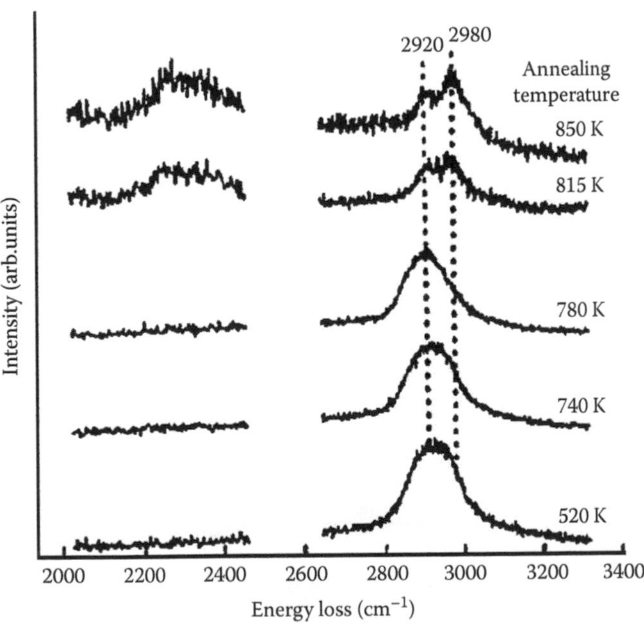

FIGURE 14.20 HREEL spectra of the C–H stretch for the OTS-coated oxidized Si(100) surface at different temperatures. (From Kluth, G.J., Sander, M., Sung, M.M., and Maboudian, R., Study of the desorption mechanism of alkylsiloxane self-assembled monolayers through isotopic labeling and high resolution electron energy-loss spectroscopy, *J. Vac. Sci. Technol. A*, 16(3): 932–936, 1998. With permission from the authors.)

bond remains intact even while the chains have begun to desorb; thus, desorption must occur through C–C bond cleavage. Such studies offer important information on the reliability of such SAM films.

The same group also used HREEL to investigate the interaction of hydrogen/deuterium atoms [32] and of sulfur dimmers [33] with SAMs. The technique was also applied to study the formation of SAMs on Ge(111) [27].

14.7 AUGER ELECTRON SPECTROSCOPY (AES)

14.7.1 Principle

During the interaction of an electron beam (energy 3 to 20 keV) with a conducting sample, Auger electrons are generated. This is a three-step process (Figure 14.21). The incident electrons cause emission of core electrons (K) from the atoms of the sample. This leaves the atom with a hole that is filled by electrons from a higher level (L1 to K). The energy thus released can be converted into an x-ray (EDX) or transferred to another electron that is emitted. This electron is called an Auger electron. The energy of the Auger electron is characteristic of the atom that emitted it and can thus be used to identify the composition of the sample. As such, the technique is used for the analysis of the surface of samples; it has a submonolayer sensitivity. It cannot detect hydrogen or helium, but it is sensitive to all other elements, being most sensitive to the low-atomic-number elements. By combining it with sputtering, a depth analysis can be performed. The electron beam can be focused to sizes as low as 10 nm, giving the technique a high spatial resolution.

14.7.2 Applications to MEMS

Because the technique is sensitive to thin surface layers, it is as such used to control thin layers on MEMS or contamination of an MEMS surface from, for example, other contacting surfaces or packages. Stoldt et al. [34] used AES, for example, to confirm the chemical composition of an SiC coating on MEMS.

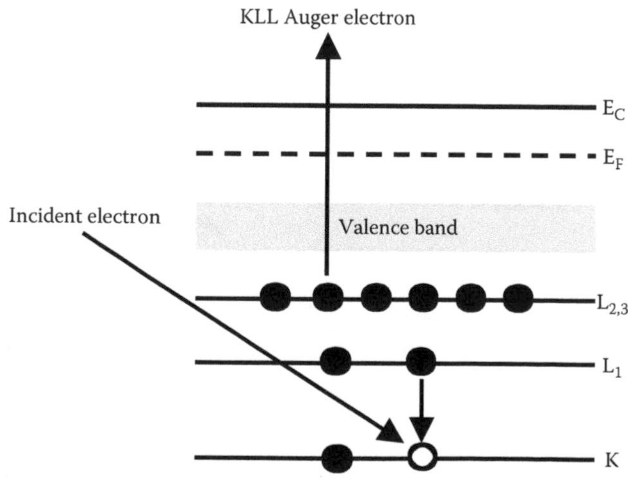

FIGURE 14.21 Origin of the Auger electron.

14.8 BRILLOUIN SCATTERING (BS)

14.8.1 Principle

Surface Brillouin scattering* (SBS) [35–37] is a nondestructive technique similar to RS: a beam of monochromatic light is used as a probe to reveal phonons that are naturally present in the medium under investigation. SBS measures acoustic phonons (surface acoustic waves), whereas RS mainly detects optical phonons. It is sensitive to residual stress, annealing effects, and mass density. From SBS measurements the complete set of elastic constants of the investigated material can be determined. The probed area of the sample is of the order of 10×10 μm, which makes the technique useful for MEMS also. It can characterize films as thin as a few tens of nanometers.

BS makes use of a high-resolution spectrometer, mostly a Fabry–Perot interferometer. This system consists of two very flat mirrors mounted precisely parallel to each other with a variable spacing; a special setup is the so-called Sandercock interferometer [38].

The determination of the elastic constants of thin films by SBS is mostly done on layers of thickness less than the acoustic wavelength (0.3 to 0.4 μm), supported by substrates with acoustic phase velocities higher than that of the films (slow film/fast substrate). Under these conditions, a number of discrete acoustic modes, namely the Rayleigh and the Sezawa modes [39], are revealed and the corresponding phase velocity can be measured. The measurements are usually performed on films of different thickness and with different angles of incidence.

14.8.2 Applications to MEMS

The application of BS is well known for uniform thin films, but not for MEMS. However, the technique certainly is expected to make important contributions to this domain. It can be used for the determination of the elastic constants, i.e., to obtain information on mechanical characteristics of films, which is of great importance for MEMS. During the past 15 years, the group at the GHOST laboratory, University of Perugia [40], extensively exploited SBS to characterize semiconductor and/or dielectric films of different materials such as C_{60}, AlN, GaSe, InSe, InGaAs, SiO_2, SnO_2, a-Ge:H, a-C, and even multilayered metallic structures (Ag/Ni, Nb/Fe, Ta/Al, FeNi/Cu, and FeNi/Nb). They applied the technique for example to about 1-μm-thick ZnO films, deposited by magnetron sputtering on Si(100) substrates. They demonstrated that Brillouin spectra from a single transparent film, about 1 μm thick, can provide the amount of information necessary for the unambiguous determination of the whole set of independent elastic constants.

14.9 CONCLUSIONS

It is clear that most of the spectroscopic techniques that are commonly applied to the study of semiconductor and metal surfaces can also be used to study MEMS. A large variety of material properties can be investigated. This chapter could not address all spectroscopic techniques; its main focus is on RS because one of the most important functionality and reliability issues in MEMS is the mechanical stress in the free structures. RS offers some unique possibilities in this field because of its small probing spot (μm spatial resolution) and small, controllable penetration depth and has as such already been applied to various MEMS applications. It is expected that RS will find many more applications for the study of mechanical stresses in MEMS. Although this was not demonstrated yet, it should be possible to study, *in situ*, the stress near the hinges of moving beams, for example, Si resonators, and to correlate the results with FEM and with expected failure mechanisms such as fatigue.

Another important issue in MEMS is surface contamination, especially when dealing with contacting surfaces. Contamination can, for example, affect the contact resistance in MEMS

* Most of the information on Brillouin scattering in this chapter is contributed by G. Carlotti from the GHOST laboratory, University of Perugia, Italy.

switches or the sensitivity to wear in rotating structures. It can be studied using AES, XPS, HREELS, and, in some cases, RS also.

Stiction, wear, and friction problems in MEMS are often reduced by coating the structures with a protective layer such as a carbon coating or a SAM. The composition of these coatings can be studied by the preceding techniques, but the thickness also has to be controlled. If the film is Raman active, this can sometimes (depending on the thickness of the film) be studied using RS. SE can also be applied for this purpose and a special micro-SE setup with a smaller probing size to study MEMS was already demonstrated. Another variation of SE, DBS, even allows studying both the thickness of free MEMS structures and the gap below them.

Spectroscopic techniques can be applied to study MEMS; they are, in general, easy, nondestructive, and offer information on structural (thickness) and material properties (both chemical and mechanical). But it is clear that the MEMS also drives the development of these techniques toward higher sensitivity, a better spatial resolution, and a larger application area.

REFERENCES

1. Anastassakis, E., Pinczuk, A., Burstein, E., Pollak, F.H., and Cardona, M., Effect of uniaxial stress on the Raman spectrum of silicon, *Solid State Commun.*, 8: 133–138, 1970.
2. Anastassakis, E., Canterero, A., and Cardona, M., Piezo-Raman measurements and anharmonic parameters in silicon and diamond, *Phys. Rev. B*, 41: 7529–7535, 1990.
3. De Wolf, I., Stress measurements in Si microelectronics devices using Raman spectroscopy, *J. Raman Spectrsc.*, 30: 877–883, 1999.
4. De Wolf, I., Maes, H.E., and Jones, S.K., Stress measurements in silicon devices through Raman spectroscopy: bridging the gap between theory and experiment, *J. Appl. Phys.*, 79: 7148–7156, 1996; De Wolf, I., Anastassakis, E., Addendum, *J. Appl. Phys.*, 85(10): 7484–7485, 1999.
5. De Wolf, I., Chen, J., Rasras, M., van Spengen, W.M., and Simons, V., High-resolution stress and temperature measurements in semiconductor devices using micro-Raman spectroscopy, *Proc. Int. Symp. on Photonics and Applications (SPIE)*, 3897: 239, 1999.
6. De Wolf, I., Semiconductors, in Pelletier, M.J., Ed., *Analytical Applications of Raman Spectroscopy*, Blackwell Science, 1999, pp. 435–472.
7. De Wolf, I., Chen, J., van Spengen, W.M., The investigation of microsystems using Raman spectroscopy, *Opt. Laser. Eng.*, 36: 213–223, 2001.
8. De Wolf, I., Topical review: micro-Raman spectroscopy to study local mechanical stress in silicon integrated circuits, *Semicond. Sci. Technol.*, 11: 19, 1996.
9. van Spengen, W.M., De Wolf, I., and Knechtel, R., Experimental two-dimensional mechanical stress characterization in a silicon pressure sensor membrane using micro-Raman spectroscopy, *Proc. Int. Symp. on Photonics and Applications (SPIE), Micromachining and Microfabrication*, 2000, p. 104.
10. Zhao, Y., Ludlow, D.J., Li, B., and Zhang, X., Evaluation of design and performance of a cryogenic MEMS micropump realized with Si-based diaphragm using combination of zygo/wyko interferometer and Raman spectroscopy, *Micro- and Nanosystems, Materials Research Society Symposium Proceedings*, 782: A5.52.1–A5.52-6, 2003.
11. Starman, L.A., Jr., Busbee, J., Reber, J., Lott, J., Cowan, W., and Vandelli, N., Stress measurement in MEMS devices, *Modeling and Simulation of Microsystems*, ISBN 0-9708275-0-4, 2001, pp. 398–399.
12. Cho, H.J., Oh, K.W., Ahn, C.H., Boolchand, P., and Nam, T.-C., Stress analysis of silicon membranes with electroplated permalloy films using Raman scattering, *IEEE Trans. Magn.*, 37(4): 2749–2751, 2001.
13. Srikar, V.T., Swan, A.K., Ünlü, M.S., Goldberg, B.B., and Spearing, S.M., Micro-Raman measurement of bending stresses in micromachined silicon flexures, *IEEE J. Microelectromech. Syst.*, 12(6): 779–787, 2003.
14. Starman, L.A., Jr, Ochoa, E.M., Lott, J.A., Amer, M.S., Cowan, W.D., and Bushbee, J.D., Residual stress characterization in MEMS microbridges using micro-Raman spectroscopy, *Modeling and Simulation of Microsystems* 2002, ISBN 0-9708275-7-1, 2002, pp. 314–317.
15. van Spengen, M., Reliability of MEMS, Ph.D. dissertation, ISBN 90-5682-504-6, Katolieke Universiteit Leuven, 2004.

16. EC project STREAM IST-1999-10341.
17. Maeda, Y., Yamamoto, H., and Kitano, H., *J. Phys. Chem.*, 99(13): 4837–4841, 1995.
18. Ager, J.W., III, Monteiro, O.R., Brown, I.G., Follstaedt, D.M., Knapp, J.A., Dugger, M.T., and Christenson, T.R., Performance of ultra hard carbon wear coatings on microgears fabricated by LIGA, Vol. 546, *Proc. Materials Research Society*, 1998.
19. Voloshin, A., *Thermal Stress and Strain in Microelectronics Packaging*, John H. Lau, Ed., Van Nostrand Reinhold, New York, 1993, pp. 272–304, chap. 8.
20. Bhushan, B., Kulkarni, A.V., Koinkar, V.N., Boehm, M., Odoni, L., Martelet, C., and Belin, M., *Langmuir*, 11: 3189–3198, 1995.
21. Maboudian, R., *Surf. Sci. Rep.*, 30: 207–270, 1998.
22. Rehak, M., Snejdarkova, M., and Otto, M., *Biosensor. Bioelectron.*, 9: 337–341, 1994.
23. Rickert, J., Weiss, T., Kraas, W., Jung, G., and Gopel, W., *Biosensor. Bioelectron.*, 11: 591–598, 1996.
24. Haneda, R., Nishihara, H., and Aramaki, H., *J. Electrochem. Soc.*, 144: 1215–1221, 1997.
25. Laibinis, P.E. and Whitesides, G.M., *J. Am. Chem. Soc.*, 114: 9022–9028, 1992.
26. Zamborini, F.P. and Crooks, R.M., *Langmuir*, 14: 3279–3286, 1998.
27. Han, S.M., Ashurst, W.R., Carraro, C., and Maboudian, R., Formation of alkanethiol monolayer on Ge(111), *J. Am. Chem. Soc.*, 123(10): 2422–2425, 2001.
28. Schutte, E.K. and Martin, J., Detecting the 10 Angstroms that can change MEMS performance, *Conf. Proc. from the 30th Int. Symp. for Testing and Failure Analysis (ISTFA)*, 2004, 216–220.
29. Sledziewski, J.M., Nunan, K., Robinson, T., and Bar-on, I., Curvature measurement of polysilicon MEMS structures using spectroscopic reflectometry, SEM, Society for Experimental Mechanics, *Annu. Conf. Expo. on Exp. Appl. Mech.*, 2001, 332–335.
30. Ibach, H., *Electron Energy Loss Spectrometers: The Technology of High Performance*, in Hawkes, P.W., Ed., Springer Series in Optical Sciences, Vol. 63, New York, 1991.
31. Kluth, G.J., Sander, M., Sung, M.M., and Maboudian, R., Study of the desorption mechanism of alkylsiloxane self-assembled monolayers through isotopic labeling and high resolution electron energy-loss spectroscopy, *J. Vac. Sci. Technol A*, 16(3): 932–936, 1998.
32. Kluth, G.J., Sander, M., Sung, M.M., and Maboudian, R., Interaction of H(D) atoms with octadecylsiloxane self-assembled monolayers on the Si(100) surface, *Langmuir*, 13(24): 6491–6496, 1997.
33. Kluth, G.J., Carraro, C., and Maboudian, R., Direct observation of sulfur dimers in alkanethiol self-assembled monolayers on Au(111), *Phys. Rev. B*, 59(16) R10: 449–452, 1999.
34. Stoldt, C.R., Carraro, C., Ashurst, W.R., Fritz, M.C., Gao, D., and Maboudian, R., Novel low-temperature CVD process for silicon carbide MEMS, *Transducers '01 Eurosensors XV, The 11th International Conference on Solid-State Sensors and Actuators*, Munich, Germany, June 10–14, 2001.
35. Comins, J.D., Surface Brillouin scattering, in Levy, M., Bass, H., Stern, R., and Keppens, V., Eds., *Handbook of Elastic Properties of Solids, Liquids, and Gases, Vol. I: Dynamic Methods for Measuring the Elastic Properties of Solids*, New York: Academic Press, 2001, pp. 349–378.
36. Grimsditch, M., Brillouin scattering, in Levy, M., Bass, H., Stern, R., and Keppens, V., Eds., *Handbook of Elastic Properties of Solids, Liquids, and Gases, Vol. I: Dynamic Methods for Measuring the Elastic Properties of Solids*, New York: Academic Press, 2001, pp. 331–347.
37. Mutti, P., Bottani, C.E., Ghislotti, G., Beghi, M., Briggs, G.A.D., and Sandercock, J.R., Surface Brillouin scattering — extending surface wave measurements to 20 GHz, in Briggs, A., Ed., *Advances in Acoustic Microscopy*, New York: Plenum Press, 1995, pp. 249–300.
38. Sandercock, J.R., in Cardona, M. and Güntherodt, G., Eds., *Light Scattering in Solids III*, Berlin: Springer-Verlag, 1982, p. 173.
39. Farnell, G.W. and Adler, E.L., in Mason, W.P. and Thurston, R.N., Eds., *Physical Acoustics*, Vol. 9. New York: Academic Press, 1972, pp. 35–127.
40. Carlotti, G., Link to the full list of publications and to the activity of the group: http: // ghost. fisica. unipg.it/.
41. Chiang, F.P., Wang, Q., and Lehman, F., Nontraditional methods of sensing stress, strain and damage in materials and structures, ASTM STP 1318, American Society for Testing and Materials, 1997. Speckle metrology, in *Metals Handook*, 9th ed., Vol. 17, Nondestructive Evaluation and Quality Control.
42. De Wolf, I., Jimenez, J., Landesman, J.-P., Frigeri, C., Braun, P., Da Silva, E., and Calvet, E., Raman and Luminescence Spectroscopy for Microelectronics, European Commission, 1998 — V, 98, EUR 18595.
43. Born, M.A. and Wolf, E., *Principles of Optics*, 4th ed., Oxford: Pergamon Press, 1970.

Index

A

Abbe's theory, 191
Aberrations
digital holographic microscopy, compensation of, 378, 379
digital holographic microscopy, removal of, 378–385
circular fringe carrier removal with phase-shifting method, 381–382
comparison of methods, 384–385
double exposure technique, 382–384
successive approximations method, 384
phase, correction of, 376
Absolute phase measurement, digital holography, 373
Accelerometers
MEMS devices, 326
OEHM testing, 334–338
Accuracy
continuous spatial phase distribution, 333
image correlation techniques, 66
laser Doppler vibrometry, 274–275, 278–279
OEHM testing, 333, 334
Twyman-Green microinterferometer, 296
Acoustooptical modulator, laser Doppler vibrometers, 253
Active membranes, out-of-plane deformations, 314–316, 317, 318
ActiveMIL, 50
ActiveX controls, 49
Activ Vision Tools, 50
Actuators
dynamic evaluation of active MOEMS with stroboscopic technique, 314–321
scratch drive, out-of-plane microdisplacement measurements, 308–314
Additive effect, colors, 6
Adimec MX12 camera, 447
AFM, *see* Atomic force microscopy (AFM)
Airy disk, 428, 430
Alias effect, laser Doppler vibrometry, 259, 260
Amplifiers, CCD sensors, 17
Amplitude maps, interference
fringe pattern demodulation, 218
vibration measurements, stroboscopic interference microscopy, 236
Amplitude measurement uncertainty, laser Doppler vibrometry, 278
Analog cameras, 20
Analog decoding techniques, laser Doppler vibrometry, 255–256
Analog demodulation, laser Doppler vibrometry, 278
Analog interfaces, frame grabbers, 21, 22
Analog modulation, OEHM, 348
Analog-to-digital converters, laser Doppler vibrometry, 260
Analysis methods, image processing, 31–36
Angle resolution, digital holography, 352
Angle resolved scattering (ARS), 103, 105; *see also* Light scattering, inspection techniques
light scattering measurement systems, 108, 111
micromirrors for DUV application, 114
standardization, 109–110
total integrated scattering, 106–107
Animation, laser Doppler vibrometry ODS, 286
ANSYS software, 312, 314
Aperture
digital holography
controlling size independently of distance and wavelength, 390
numerical reconstruction by lensless Fourier approach, 361
interferometer, 294
laser Doppler vibrometry, 269
detector, 250, 251
and measurement accuracy, 278, 279
numerical, 251, 269, 278, 279
speckle metrology, 428, 430
submicron transverse resolution, 415
Applied MEMS Inc., 282–283
Arctangent phase method, 256–259, 437
Area, connectivity analysis, 35
Argon ion laser, 446–447
Aspect ratio, OEHM, 348
ASTM standard E1392, 109–110
Atmospheric pressure chemical vapor deposition (APCVD), 405
Atomic density, silicon oxinitride films, 297–300
Atomic force microscopy (AFM), 55, 122–143
applications, case study, 134–139
atomic force profilometer, 139–141
complementary techniques, 142
components and operating principles, 122–125
controller, 124
detection, input signal, set point, and error signal, 124–125
probe, 123
scanner, 123, 124
Z feedback loop, 125
image correlation techniques
deformation measurement, 57
displacement and strain fields, 69
finite element (FE) modeling, 68

instrumentation, 70, 71, 72
microcrack evaluation, 89, 90–92, 93
requirements, 72, 73
three-D deformation analysis, 92–95
imaging modes, primary, 125–127
imaging modes, secondary, 127–134
conductive AFM (CAFM), 130, 131
electric force microscopy (EFM) and surface potential imaging, 131, 132
force modulation imaging, 132, 133
lateral force microscopy (LFM), 127–128
magnetic force microscopy (MFM), 129–130, 131
phase imaging, 129
scanning capacitance microscopy (SCM), 132–133, 134
scanning spreading resistance microscopy (SSRM), 133
tunneling AFM (TUNA), 133
light scattering techniques combined with, 115, 116
MEMS devices, 325–326
Moiré methods using high-resolution microscopy, 175–176, 180, 182
nonimaging modes, 134
surface roughness measurements, 37
topography scans, 56
Atomic force profilometer, 139–141
Atomic force spectroscopy, 134
Auger electron spectroscopy (AES), 478
Automated AFMs, 134
Automatic fringe pattern analyzer (AFPA), grating (Moiré) interferometry, 207–208
AutoProbe M5 AFM system, 71
Axis of inertia, principal, 41

B

Backscattering, 104–105
laser Doppler vibrometry errors, 278
light scattering measurement systems, 107
standardization, 110
Ball grid array (BGA) package, 179, 180
Band-pass filters, 7, 8, 219
Beam (light source)
Gaussian, Guoy phase delay, 279
speckle metrology, DUV interferometry, 452
Beam ratio, optoelectronic holography microscopy, 331
Beams/cantilevers/microbeams, 210, 211
cantilever beam acceleration sensor, laser Doppler vibrometry, 283–289
in-plane analysis, 287–289
out-of-plane analysis, 284–286
digital holography, 392, 399, 405, 406
controlling size independently of distance and wavelength, 390
interferometry, 365
microscopy, 382, 389
padding operation, 389
Poisson ratio determination, 399–401
residual stress determination, 408, 409
thermal expansion coefficient, 402–405
thermal load testing, 409–411
Young's modulus determination, 401–402
polysilicon microcantilever, holography of, 382
Beam splitter
digital holographic microscopy, 376
digital holography, 354, 399
laser Doppler vibrometry
integrated techniques, 279
nonpolarizing, 252–253
polarizing, 251–253
optoelectronic holography microscope, 329, 330
speckle metrology, 452
Twyman-Green microinterferometer, 294, 295
Bertrand lens, 191–192, 194
Bessel function
optoelectronic holography microscopy, 332
speckle metrology, 440
time-averaged interference microscopy, 239
Bessel function-based calibration, laser Doppler vibrometry, 277–278
Best fit method
DIC-FEA, 68
geometric feature measurement, 44–45
BIAS Fringe Processor, 50
Bicell PSPD, 127
Binarization
grayscale, 28
histogram, 26–27
Binary pixel counting, 33
Bipolar transistors, 19
Blooming effect, 19
Blur, laser Doppler vibrometry, 281
Bode plot, laser Doppler vibrometry, 287, 288
Boiling effect, speckle, 432
Bore holes
position recognition, 40
presence verification, 45–46
Bounding box
connectivity analysis, 35
finding test object or region of interest, 39
Bragg cell, 253, 258
Bright-field illumination, 10–12
Brightness, presence verification, 45
Brillouin scattering (BS) spectroscopy, 479
Broadband illumination
laser Doppler vibrometry, 261
two-beam homodyne interference microscopy, 223, 224
Buckled membranes
critical stress of buckling, 306
out-of-plane analysis, 293
silicon oxinitride films, 300
Built-in stress sensors, 56
Bulge tests, interference microscopy applications, 232
Bulk micromachining, 327
Bump deformation, heating, 77–81

C

Calibration
digital holography, phase measurement, 375
laser Doppler vibrometry, 274–279

accuracy of, 278
Bessel function-based, 277–278
general aspects, 274–275
influences on measurement accuracy, 278–279
mechanical comparison calibration, 275
with synthetic Doppler signals, 276–277
light scattering measurement systems, 108
optoelectronic holography microscopy, 330, 331, 333
Camera Link(TM), 21, 51
Cameras, *see also* CCD detectors/sensors
computer vision components
color, 19–20
types and interfaces, 20–21
optoelectronic holography microscope, 329, 330
preprocessing, 24
speckle metrology, 447
Cantilevers, *see* Beams/cantilevers/microbeams
Carré algorithm, 194
Carrier frequency method, 50, 51
Carrier Moiré method, 177
Cartesian coordinates, phase data transformation, 375
Cavity waveguide, grating (Moiré) interferometry, 206–207
CCD detectors/sensors
computer vision components, 16–18
confocal microscope, 150–151
digital holography, 353, 354, 355, 360, 399, 417
controlling size independently of distancc and wavelength, 390
microscopy, 376, 377, 378
numerical reconstruction by lensless Fourier approach, 361–362
grating (Moiré) interferometry, 207
grid diffraction method, 189, 191, 192, 193, 195, 197
laser Doppler vibrometry
combined techniques, 279
integrated techniques, 280
out-of-plane analysis, 284–285
OEHM, 348
optoelectronic holography microscope, 329, 330
speckle metrology, 430, 434
DUV interferometry, 452
time-averaged recordings, 439
spectral response plot, 8
stroboscopic interferometry, contrast modulation, 316, 317
Twyman-Green microinterferometer, 296
CCIR/BAS/FBAS, 20
Center of gravity, connectivity analysis, 35
Character-coded devices, defect and fault detection, 48
Characteristic curve, CMOS photodiodes, 19
Chebyshev method, best fit, 44, 45
Chemical force microscopy, 128
Chirp, laser Doppler vibrometry, 261, 282–283
Chromatic confocal sensor principle, 149, 150, 152
Chromatic error, 7
Circular fringe carrier removal, digital holographic microscopy, 381–382
Circular polarization, laser Doppler vibrometry, 270
Clocking, 17, 262
Closing, morphological operators, 30, 31
CMOS detectors/sensors
computer vision components, 18–19
digital holographic microscopy, 377
digital holography, 353, 355
OEHM, 348
optoelectronic holography microscope, 329, 330
speckle metrology, 434
Coake(R), 50
Coatings/coated surfaces
light scattering measurement systems, 105
presence verification, 45
Raman spectroscopy applications, 471
Coaxial bright-field illumination, 11–12
Coaxial connectors, camera interfaces, 20
Coblentz spheres, 106, 107, 108, 109
Coefficient of thermal expansion (CTE), 56
image correlation techniques, 64–65, 66
accuracy of, 88
correlation technique, 87
finite element (FE) modeling, validation of, 81
foil thickness and, 84, 86
three-D deformation analysis, 94
MEMS silicon materials mechanical properties, 328
OEHM testing, 343
Coffin Manson approach, 77, 82
Cognex Vision Library (CVL), 49
Coherence function, two-beam interferometry with broadband illumination, 223
Coherence probe microscopy (CPM), 219; *see also* Interference microscopy techniques
Coherence radar, 219; *see also* Interference microscopy techniques
Coherent light sources, OEHM, 348
Collimator
grating (Moiré) interferometry, 207, 208
illumination, 9
Twyman-Green microinterferometer, 295
Color
behavior of, 5–8
digital holography, multiwavelength (MWDH), 389
spectral reflectivity mapping, 231
Color-coded components, defect and fault detection, 48
Colored light, 6–7
Color scale coded displacement plots, 60
Color sensors/color imaging
computer vision components, 19–20
light scattering measurement systems, 110, 111
presence verification, 46
Comb drive, 308
Comb function, 363
Common Vision Blox, 50
Compactness, connectivity analysis, 35
Complex geometries, digital holography, 373, 375
Composition, silicon oxinitride films, 297–300
Compressive stress, silicon oxinitride films, 300, 301, 302
Computer aided design (CAD), 326
Computer-aided manufacturing (CAM), 326
Computer processing/reconstruction, *see also* Image processing and computer vision
digital holographic microscopy, 375–376
speckle metrology, 452–453

Computer Vision library, 49
Concave lens, 9
Conductive atomic force microscopy (CAFM), 130, 131
Confocal microscopy
  AFM, complementary methods, 142
  laser Doppler vibrometry, integrated techniques, 281, 282
  principles, 145–156
    measurement with, 151–153
    MEMS measurement applications, 153–155
    microscopes, 150–151
    point sensors, 145–149, 150
Conic equation, general, 42
Connectivity analysis, 35
  finding test object or region of interest, 38, 39
  position recognition, 41
Contact interactions, scratch drive actuators, 309
Contact mode AFM, 125, 134
Continuity, fringe, 51
Continuous spatial phase distribution, 333
Continuous-wave mode, stroboscopic interferometry, 316
Contour/contouring
  connectivity analysis, 35
  contour-based pattern matching, 27
  contour extraction, position recognition, 40–41
  digital holographic interferometry, 370
  geometric feature measurement, 43
  holographic, 366, 396, 397
    multiwavelength, 373, 375
  image correlation techniques, 75
Contour roughness, dilation filters and, 30
Contrast, 33
  finding test object or region of interest, 37–38
  fringe, 51
  position recognition, 40, 42
  presence verification, 45
  speckle metrology, 430
Contrast modulation, stroboscopic interferometry, 316, 317
Contrast ratio, 33, 40
Controller, atomic force microscopy (AFM), 124
Conversion parameter, laser Doppler vibrometry, 249, 250
Convex hull
  connectivity analysis, 35
  position recognition, 41, 42
Convolution (spatial filtering)
  digital holography, 359, 360, 363
  image processing, 27–31
Coordinates
  connectivity analysis, 35
  position recognition, 40, 41
Correction, digital holography, 379–381, 394, 395
Correlation analysis, *see also* Cross-correlation algorithms on digitized micrographs
  DIC-FEA, 68, 69, 70
  finite element, 68, 69
  image correlation techniques, 74, 75–76
  laser Doppler vibrometry, combined techniques, 279
  OEHM testing, wafer level, 341, 342
  pattern matching, 36
Correlation microscopy, 219; *see also* Interference microscopy techniques
Crack loading analysis, 58, 59, 61
Crack opening, mode I, 61–62
Crack opening displacement (COD) field, 68
Creep, 64, 77
Critical stress of buckling, silicon oxinitride films, 306
Cross-correlation algorithms on digitized micrographs
  analysis scheme, 74
  deformation analysis, 57–60
Cross-spectrum, laser Doppler vibrometry, 261
CTE, *see* Coefficient of thermal expansion (CTE)
Curvature of wavefront, digital holographic microscopy, 379–381

D

Dark-field illumination, 10
Data acquisition, laser Doppler vibrometry
  synchronization of, 289
  techniques, 259–262
Data interfaces, DIC-FEA, 68
Data point correction, 75
Data smoothing, 75
Decoding, laser Doppler vibrometry
  analog techniques, 255–256
  digital, 258
  in-phase and quadrature (I&Q) format, 257
Deep reactive ion etching (DRIE), 315
Deep ultraviolet (DUV) light sources
  light scattering measurement systems, 103, 108
  micromirrors for application, 114
  speckle metrology, 452
Defect detection, image correlation techniques, 80–81
Defect of sphericity, 418
Deflection shapes, laser Doppler vibrometry, 263
Deflectometry, 314
Deformation analysis, image correlation techniques based on AFM micrographs, 92–95
Deformation field, coefficient of thermal expansion, 65
Deformation measurement, 56
  digital holography
    advantages of, 375
    focus tracking in DHM, 385–389
    profile reconstruction, 405–406, 407
  image correlation techniques, 57–70
    capabilities and limits, 66–67
    cross-correlation algorithms on digitized micrographs, 57–60
    derived properties determination, 63–65
    extraction of displacement and strain fields, 60–63
    finite element (FE) simulation, combining with, 68–70
    instrumentation, 71
Deformations, mechanical properties defined by, 56
Delta-sigma converters, laser Doppler vibrometry, 260
Demodulation/unwrapping
  digital holographic interferometry, 373, 374, 375
  laser Doppler vibrometry, 256–259
    analog, 255–256
    data acquisition, 259
    drop-out, 251
    and phase noise, 251

Density, MEMS silicon materials mechanical properties, 328
Dephasing, digital holography, 395
Depth of field, OEHM, 348
Depth-scanning fringe projection (DSFP), 145, 156–159, 160
  experimental realization, 158–159, 160
  intensity model, 156–157
Derived properties determination, image correlation techniques, 63–65, 76
Destructive interference, speckle metrology, 430
Detection, *see also specific techniques*
  atomic force microscopy (AFM), 124–125
  digital holographic microscopy, 377
  grid diffraction method, 189
  laser Doppler effect, 249–251
    interferometry, 250
    shot noise in detection of light, 249
    wavefront aberrations and laser speckle, 250–251
  laser Doppler vibrometry
    interferometry, 250
    noise sources, 271
    shot noise in detection of light, 249
    wavefront aberrations and laser speckle, 250–251
DHM, *see* Digital holography, microscopy (DHM)
DIC, *see* Image correlation techniques
Dielectric stack, light scattering measurement systems, 105
Differential Doppler technique, 268
Diffraction, digital holography, 356
Diffraction method, 191
  grid line pattern analysis with FT method, 188–190
  microscopic grid methods, 183
Diffractive optics, digital holographic microscopy technology for testing, 420, 421, 422
Diffusely scattered illumination
  disadvantage of, 9
  ground glass screen, 8
Digital cameras, 20
Digital decoding, laser Doppler vibrometry, 258
Digital demodulation, laser Doppler vibrometry, 256–259, 278
Digital holography, 352–422
  applications, experimental techniques, 391–395, 396
    handling and preparation, 394, 395
    loading, 392–394
    observation and evaluation, 395, 396
  applications, objects with optical surfaces, 405–422
    microlens testing, 414–422
    testing silicon MEMS structures, 405–409
    thermal load testing, 409–414
  applications, objects with technical surfaces, 395–405
    combined shape and displacement measurement of small objects, 395–397, 398, 399
    material parameters determination, 397, 399–405
  interferometry (DHI), 364–375
    advantages, 375
    displacement measurement, 366–367
    phase measurement, direct and absolute, 372–375
    principles, 364–366
    shape measurement, 368–372
  microscopy (DHM), 375–391
    aberrations, compensation of, 378, 379
    aberrations, removal of, 378–385
    controlling size independently of distance and wavelength, 389–391
    focus tracking, 385–389
    optical setup, 376–377
  theory and basic principle, 353–363
    convolution approach, 360
    digital recording and reconstruction of wavefronts, 353–358
    discretization, influences of, 363–364
    Fresnel approximation, 358–360
    lensless Fourier approach, 361–362
    phase-shifting approach, 362–363
    reconstruction principles, 358–363
Digital image correlation (DIC) techniques, *see* Image correlation techniques
Digital interfaces, frame grabbers, 21
Digital light projection (DLP) projectors, 2
Digital (micro)mirror devices (DMDs), 2; *see also* Mirrors/micromirror devices
Digital modulation, OEHM, 348
Digital signal processor (DSP), laser Doppler vibrometry, 256
Digital speckle pattern interferometry (DSPI), 353
Digitized micrographs, deformation measurement, 57–60
Dilation filter, 30
Dimensions, connectivity analysis, 35
Direct phase modulation, interference microscope, 221
Direct phase subtraction, digital holography, 390
Discretization, digital holography, 363–364
Dispersion, 5
Dispersion effects, two-beam interferometry with broadband illumination, 223
Displacement decoder
  laser Doppler vibrometry, 255–256
  noise spectra, 272–273
Displacement maps, from AFM scans, 56
Displacement measurement
  digital holography, 352, 395–397, 398, 399
  digital holography interferometry, 366–367
Displacement measurement for kinematic and mechanical description, 56
Displacement noise, 272, 273
Displacement and strain fields
  digital holographic microscopy focus tracking, 385–389
  image correlation techniques, 60–63
    finite element (FE) modeling, 69, 70
  Moiré patterns, 168
Distance measurements, position recognition, 40
Doppler vibrometry, *see* Laser Doppler vibrometry
Double-exposure specklegram, 434
Double-exposure technique
  digital holographic interferometry, 364, 405
  digital holographic microscopy, aberration removal, 382–384
Double-sided telecentric lenses, 14–15
Downscaling, image correlation techniques, 66
Drift, scanning-type imaging, 72–73
Drop-out, laser Doppler vibrometry, 251
Dual-beam spectroscopy (DBS), 474–476

Dual illumination, speckle metrology, 452
Dual mode MEMS mirror, laser Doppler vibrometry, 282–283
Dynamic evaluation of active MOEMS with stroboscopic technique, 314–321
  active membranes, 314–316, 317, 318
  torsional micromirrors, 317, 319, 320, 321
Dynamic measurements
  interference microscopy techniques, 233–240
    applications of dynamic IM in MEMS field, 240
    interferometric signal in dynamic case, 233–235
    vibration measurements by stroboscopic IM, 235–238
    vibration measurements by time-averaged IM, 238–239
  laser Doppler vibrometry, *see* Laser Doppler vibrometry
  OEHM, *see* Optoelectronic holography/holographic microscopy (OEHM)
  out-of-plane deformations, 314–321
Dynamic range, laser Doppler vibrometry, 269
Dynamic speckle correlation interferograms, 332

E

Edge detection, 33–35
  finding test object or region of interest, 38
  geometric feature measurement, 43
  position recognition, 40
  presence verification, 46
Edge effects, interference microscopy, 226, 227
Edge filters, 8
Edge smoothing, presence verification, 47
Effective thickness, silicon oxinitride films, 301
Efficiency, heterodyning, 253
EIA-644, 21
Eight-bit depth digitizing, 60
Elastic constant evaluation, 240
Elastic properties, 56, 328
Elastic recoil detection analysis (ERDA), 297, 298, 299
Electrical field depletions, 19
Electrical hum, laser Doppler vibrometers, 253
Electric force microscopy (EFM), 129, 131–132
Electrodeposition, fabrication process issues, 231
Electromagnetic spectrum, 5
Electromagnetic wave equation, 247
Electromechanical methods, interference microscopy combined with, 232
Electron beam lithography, grid and grating fabrication, 165–166
Electron beam (E-beam) Moiré system, 174–175, 179, 180, 182
Electronic signal substitution, laser Doppler vibrometry calibration, 276–277
Electronics noise, image preprocessing, 24
Electroplated structures, MEMS fabrication process, 327
Electrostatic field interactions, actuation source, 308
Elementary wave, digital holography, 356
Ellipsometry, spectroscopic techniques, 472–474
Ellpisoids, digital holographic interferometry, 371
Erosion filters, 30
Error-free interferometer, laser Doppler vibrometry, 278
Errors
  atomic force microscopy, 73
  laser Doppler vibrometry, 278
  speckle metrology, 433
Error signal, atomic force microscopy (AFM), 124–125
Error squares, minimizing, 34–35
Etching
  fabrication process issues, 231
  presence verification, 45
Exposure, laser Doppler vibrometry integrated techniques, 280, 281
Extended-focus image, confocal microscope, 153
External photo-effect, 15

F

Fabrication processes
  interference microscopy, 231
  MEMS (microelectromechanical systems), 327, 328
Fast Fourier transform (FFT) analysis
  digital holographic microscopy, 378, 387
  laser Doppler vibrometry, 247, 260–261, 284, 289
  vibration measurements, stroboscopic interference microscopy, 236–238
Fatigue models, 56, 77, 82
Fault detection, 4, 47–48
Feedback loop, atomic force microscopy, 125
Fermat, principle of, 269
Fiber-based vibrometer, 246
Fiber-optic micro-Moiré interferometer, 169–170, 171
Field integration, 18, 19
Field-of-scan length, image correlation techniques, 66
Field-of-view parameters, image correlation techniques, 66
Field programmable gate arrays, 21
FIFO memory, 21
Films, thin films, membranes, micromembranes
  digital holography, complex silicon micromembranes, 412–414
  interference microscopy applications, 232
    dynamic, 240
    film thickness mapping, 228–230
    monochromatic or white light scanning interferometry, 229–230, 232
  laser Doppler vibrometry ranges and limits, 268
  light scattering measurement systems, 105, 110–111
  out-of-plane analysis, 293
  speckle metrology, 440, 448, 449
    DUV interferometry, 453, 454, 455, 456
Filter cores, 27–28
Filters
  behavior of, 5–8
  digital holographic microscopy, 377
  laser Doppler vibrometry, 259, 260
  presence verification, 47
  spatial filtering (convolution), image preprocessing and processing methods, 28–31, 32
    gradient, 31, 32
    high-pass and low-pass, 28–29
    median, 29–30
    morphological, 30–31

Finite difference modeling (FDM), OEHM modeling and testing, 343, 344, 345
Finite element modeling (FEM), 56
analytical stress function calculation, 294
image correlation techniques, 76
deformation analysis, 68–70
three-D deformation analysis, 93
validation of, 81–82, 83, 84
laser Doppler vibrometry, out-of-plane analysis, 284–285
OEHM modeling and testing, surface mount technology, 343, 344, 345
OEHM testing, microaccelerometers, 335, 336
scratch drive actuators, 312
silicon oxinitride films, 301
Finite elements method (FEM), 201
Finite Fresnel transform, digital holography, 359
Finite impulse response (FIR) filters, laser Doppler vibrometry, 259
FireWire, 21
Five-frame temporal phase-shift algorithm, 317
Five-frame temporal phase stepping method, 316, 318
Fizeau objective, 220
Flip chips, 77–80
Focal length, digital holographic microscope lens, 379
Focused ion beam (FIB) systems
grid and grating fabrication, 166, 167
image correlation techniques
instrumentation, 70, 71, 72
residual stress determination, 95, 97
Moiré methods using high-resolution microscopy, 177, 179, 181, 182
Focusing, laser beam, 264, 265
Focus tracking, digital holography microscopy, 385–389
Foils, coefficient of thermal expansion, 64
Force modulation imaging, 132, 133
Force spectroscopy, AFM, 134
Forward scattering, 104–105
Four-bucket integration technique, 235–236
Fourier approach, lensless, 361–362
Fourier filtering, TEM Moiré method, 178
Fourier holography, lensless Fourier approach, 361–362
Fourier plane imaging, 192
Fourier series, surface roughness measurements, 104
Fourier transform
digital holography, 359, 360, 362
fringe processing, 50, 51
grid line pattern analysis with, 184–188
phase shifting methods versus, 188
Fourier transform spectrometer, spectrally resolved profilometer, 221
Foveon X3 color sensor, 20
Fracture mechanics, 56, 64, 66, 68
Frame grabbers
computer vision components, 21–22
and electronic noise, 24
grating (Moiré) interferometry, 207–208
Twyman-Green microinterferometer, 296
Frame integration, 19
Frame-transfer CCD, 17, 18, 19
Franhofer diffraction, 191
Franhofer Institute IOF, 106–107, 109
Frequency, accelerometer fundamental natural frequency, 338, 340
Frequency distribution, fringe, 51
Frequency modulated signal, laser Doppler vibrometry, 272
Frequency response, *see also* specific methods
laser Doppler vibrometry, 278
video microscopy, 281
Frequency response function (FRF), laser Doppler vibrometry, 284, 285
Frequency spectrum, laser Doppler vibrometry, 289
Frequency voltage converter, 246–247
Fresnel approximation, digital holography, 358–360
Fresnel diffraction, 362
Fresnel hologram, 353, 355, 358
Fresnel lens, 419–420, 421
Fresnel transform, 359, 360
Fringe coherence envelope function, 223
Fringe continuity, 51
Fringe count, speckle metrology, 437
Fringe-counting problem, DHI, 364
Fringe frequency, 226
Fringe-locus function, 348
Fringe modulation function, 231
Fringe order, 225
Fringe pattern demodulation techniques, 218
Fringe peak scanning microscopy, 219; *see also* Interference microscopy techniques
Fringe projection, 376
Fringes, *see also* Grid and Moiré methods
depth-scanning fringe projection (DSFP), 145, 156–159, 160
digital holographic interferometry, 364
digital holographic microscopy, 376, 378
laser Doppler vibrometry, 267, 268
Moiré patterns, 168
pattern processing in optical metrology, 50–51
Fringe spacing
digital holography, recording and reconstruction of wavefronts, 353
interference microscopy, measurements on heterogeneous surfaces, 227
laser Doppler vibrometry, 278–279
Fringe tracking, 51
Full-field vibrometry, 262–263
Full-frame CCD, 17, 18, 19
Full width at half maximum (FWHM)
confocal response curve, 147
laser Doppler vibrometry, 270
Fundamental natural frequency, accelerometers, 338, 340

G

Gas flow sensor, 397, 398, 399
Gas sensor, 93, 94, 412–414
Gate arrays, field programmable, 21
Gaussian beam
digital holographic microscopy, 376
laser Doppler vibrometry, 263–264, 269, 279
Gauss method, best fit, 44, 45

Geometric conditions, digital holographic interferometry, 365, 366
Geometric feature measurement, 3, 42–45, 352
Geometric features, OEHM modeling and testing of surface mount technology, 343–347
Geometric matching, 68
Geometric model matching, 36
Geometric pattern matching, 38
Geometry data, oscillation data combined with, 279
Glossy surfaces, speckle metrology, 452
Goniometers, 106, 107, 108, 109
Goniophotometers, 106
Gradient filters, 31, 32
Graphical representations, displacement and strain fields, 60
Graphics, software packages, 61
Grating diffraction, 110–111
Grating equation, roughness, 104
Grating (Moiré) interferometry, 201–214; *see also* Grid and Moiré methods
- measurement system, 207–209
- principle of, 202–204
- specimen grating technology, 209–210
- waveguide, 204–207
  - concept of waveguide grating interferometer head, 204–207
  - modified WGI for 3D components of displacement vector measurements, 207
- waveguide grating interferometry applications, 210–214
  - electronic packaging, 212–214
  - material constants determination, 210, 211
  - polycrystalline materials analysis, 210, 212
  - semiconductor microlaser matrix testing, 210, 212, 213

Gratings, atomic force microscopy, 175
Grayscale
- illumination, 6–7
- mean grayscale intensity, 32
- stretching or mapping, 25–26

Grid diffraction method, 188–190
Grid line pattern analysis, 183, 184
- with FT method, 184–188
  - accuracy of, 187–188
  - sensitivity, 186–187
  - spatial resolution, 185–186
- with phase-shifting method, 188

Grid and Moiré methods, 56, 163–197; *see also* Grating (Moiré) interferometry
- fabrication, grids/gratings, 164–167
  - electron beam lithography, 165–166
  - focused ion beam (FIB) milling, 166, 167
  - moving point source holographic interferometer, 165
  - photoresist, 164–165
- micro-Moiré interferometer, 167–174
  - application, microelectronic packaging, 171–172, 173
  - fiber-optic, 169–170, 171
  - principle, 167–169
- microscopic grid methods, 182–197
  - applications, 191–197
  - grid diffraction method, 188–190
  - grid line pattern analysis with FT method, 184–188
  - grid line pattern analysis with phase-shifting method, 188
- Moiré methods using high-resolution microscopy, 174–182
  - AFM, 175–176
  - applications, 179–182
  - electron beam, 174–175
  - focused ion beam (FIB), 177
  - SEM, 176–177
  - TEM, 178–179

Grids, *see also* Grid and Moiré methods
- deformation, 60
- displacement and strain fields, 61
- image correlation techniques, 75

Ground glass screen, 8
Guoy phase delay, 279
Gyroscope testing, speckle metrology, 447–452

H

HALCON, 49
Half-frame transfer CCD, 18
Halogen lamp, 8
Hardness
- MEMS silicon materials mechanical properties, 328, 329
- nanoindentation technique, 294
- silicon oxinitride films, 300, 304, 305, 306

Hariharan algorithm, 194
Harmonic electrostatic actuators, 308
Helium-Neon laser, 251, 446–447
Heterodyne detection, laser Doppler vibrometry, 251–252
Heterodyne high-frequency displacement decoder, 255–256
Heterodyne interferometers, laser Doppler vibrometry, 274
Heterodyne shift, laser Doppler vibrometry, 268
Heterodyne signal, laser Doppler vibrometry, 246–247
Heterodyning, spatial, 50
Heterogeneous surface measurements, interference microscopy, 226–228
Heurisko, 49
Hidden-Markov model, 48
Hierarchical demodulation/unwrapping, digital holography, 373, 374
High-aspect ratio packages, OEHM, 348
High-frequency vibrations, Doppler modulation, 254
High-pass filters, 28–29, 377
High-reflective multilayer system, light scattering measurement systems, 112–114
High-resolution electron energy loss spectroscopy (HREELS), 476–478
High-resolution scanning electron microscopy, 56
High-resolution stitching, OEHM wafer level testing, 342–343
High-speed events, OEHM, 348
Hill-climbing search algorithm, 342
Histogram analysis, 33
Histogram equalization, 26, 27
Histograms, 24–27
Holographic contouring, 366, 396, 397

Holographic interferometer, 165; *see also* Digital holography; Optoelectronic holography/holographic microscopy (OEHM)
Homodyne detection, laser Doppler vibrometry, 252–253
Homodyne interferometry
  interference microscopy, *see* Interference microscopy techniques
  laser Doppler vibrometry, 278
Homogenous strain
  coefficient of thermal expansion, 65
  image correlation techniques, 66
Hull points, 41
Huygens-Fresnel principle, 356, 357, 416
Huygens principle, 361
Hyperboloid systems, digital holography, 368, 375

I

IEEE1394, 21, 51
Illumination, 4–5; *see also* Light sources
  AFM, complementary methods, 142
  computer vision components, 8–12
  digital holography
    directions, 375
    wave, 354
  finding test object or region of interest, 37, 39
  grating (Moiré) interferometry, 203, 204, 207
  laser Doppler vibrometry
    integrated techniques, 279
    ranges and limits, 268
    shot noise in detection of light, 249
  presence verification, 45
Image acquisition, computer vision process, 22
Image analysis, computer vision process, 23–24
Image capture, image correlation techniques, 87–88
Image correlation techniques, 55–99
  applications, 76–99
    defect detection, 80–81
    finite element (FE) modeling, validation of, 81–82, 83, 84
    material properties measurement, 82–88
    microcrack evaluation, 88–92, 93
    residual stresses in microcomponents, determination of, 95–99
    strain analysis on microcomponents, 76–80
    three-D deformation analysis based on AFM micrographs, 92–95
  deformation measurement, 57–70
    capabilities and limits, 66–67
    cross-correlation algorithms on digitized micrographs, 57–60
    derived properties determination, 63–65
    extraction of displacement and strain fields, 60–63
    finite element (FE) simulation, combining with, 68–70
  equipment/instrumentation, 70–76
    high-resolution scanning microscope requirements, 72–73
    measurement system components, 70–72
    software, 73–76
Image points, 13
Image processing and computer vision, 2–52
  analysis and processing, 22–48
  analysis methods, 31–36
  classification of tasks, 2
  components, 4–22
    behavior of light, colors, and filters, 5–8
    illumination, 8–12
    lens systems, 12–15
    sensors, 15–22; *see also* Sensors, computer vision components
  computer vision process, 22–24
  fringe pattern processing in optical metrology, 50–51
  measurement and testing tasks, 36–49
    defect and fault detection, 47–48
    finding test object or region of interest, 36–39
    geometric feature measurement, 42–45
    position recognition, 40–42
    presence verification, 45–47
  optoelectronic holography microscope, 329, 330
  preprocessing and processing methods, 24–31, 32
    histograms, 24–27
    point transformations, 27–28
    spatial filtering, 28–31, 32
  software, 49–50
  spectral operations, 32–36
Image resolution, image correlation techniques, 66, 67
Image size, digital holography, 389
Impinging, declined vibrometer laser beams, 247
Incident light arrangement
  grating (Moiré) interferometry specimen grating spatial frequency, 203, 204
  illumination, 8
Incremental stress intensity factors, 92
Inertia, principal axis of, 41
Inertial sensors, OEHM testing, 334
Infinity K2 microscope, 447
Infrared light
  LED, 7
  light scattering measurement systems, 103, 106, 108
Inhomogeneities, and local strain variations, 63
Injection molding, presence verification, 45
In-phase and quadrature (I&Q) signal, laser Doppler vibrometry, 252, 256, 257, 258, 259
In-plane configuration
  digital holographic interferometry, 370
  speckle metrology, 444
In-plane displacements and measurements
  digital holography, loading of microcomponents, 392, 393
  grating (Moiré) interferometry, *see* Grating (Moiré) interferometry
  laser Doppler vibrometry, 267, 289
    beam acceleration sensor, 287–289
    ranges and limits, 268–269
  speckle metrology, 442, 444, 452
Input auto-spectrum, laser Doppler vibrometry, 261
Input signal, atomic force microscopy (AFM), 124–125
Inspection
  light scattering, *see* Light scattering, inspection techniques
  OEHM wafer level testing, 341–342

Integration techniques, four-bucket, 235–236
Intensity distribution, Twyman-Green microinterferometer system, 295
Intensity map, interferometric profilometers, 225
Intensity model, DSFP principles, 156–157
Intensity scale linearity, fringe, 51
Intensity signal, confocal microscope, 148, 151
Interfaces
cameras, 20–21
frame grabbers, 21, 22
Interference
digital holography, 355
laser Doppler vibrometry detector, 250
speckle metrology, 429, 430
Interference fringes, *see* Fringes
Interference microscopy techniques, 218–240
applications of interferometric profilometers in MEMS field, 231–233
dynamic measurements, 233–240
applications of dynamic IM in MEMS field, 240
interferometric signal in dynamic case, 233–235
vibration measurements by stroboscopic IM, 235–238
vibration measurements by time-averaged IM, 238–239
instrumentation, 218–222
interferometers, 219–220
light sources, 219
microscopes with direct phase modulation, 221
microscopes with OPD modulation, 220
microscopes with wavelength modulation, 221
principles of operation, 218–219
spectrally resolved microscopes, 221–222
performance and issues of, 226–230
edge effects, 226, 227
film thickness mapping, 228–230
measurements on heterogeneous surfaces, 226–228
spectral reflectivity mapping, 231
static measurements by, 224–226
surface profiling by low-coherence interferometry, 225–226
surface profiling by monochromatic IM, 224–225
two-beam homodyne, modeling, 222–224
with broadband illumination, 223, 224
with monochromatic illumination, 222–223
two-beam interference microscopy, 223–224
Interferometers
grid and grating fabrication, 165
interference microscopy instrumentation, 219–220
laser Doppler vibrometry, 250
out-of-plane deformations, platform architecture and, 294–296
Interferometric homodyne profilometry/vibrometry, 219; *see also* Interference microscopy techniques
Interferometric signal in dynamic case, 233–235
Interferometry, 56
AFM, complementary methods, 142
digital holographic, 364–375
advantages, 375
displacement measurement, 366–367
phase measurement, direct and absolute, 372–375
principles, 364–366
shape measurement, 368–372
grating (Moiré), 201–214; *see also* Grating (Moiré) interferometry
image correlation techniques versus, 66
laser Doppler effect detection, 249–251
interferometry, 250
shot noise in detection of light, 249
wavefront aberrations and laser speckle, 250–251
laser Doppler vibrometry, combined techniques, 279
laser Moiré, 182
light scattering techniques combined with, 115
Interlaced mode, sensors, 18–19
Interline transfer (IT) CCD sensors, 18–19
Internal photo-effect, 15
Internal residual stresses, interference microscopy, 231
Inverse square law, 5
Inverse trigonometric functions, digital holographic interferometry, 373
IR cutting filter, 8
ISO 13696, 110
ISO standards, 109–110
Isotropy, 2-D, 104
IT (interline transfer) CCD sensors, 18–19

J

Johnson noise, 271

K

Kinematic description, 56
Knoop hardness, 328, 329
Kohler illumination, 279

L

Lambertian standard, 108
Laser diodes
optoelectronic holography microscope, 329, 330
Twyman-Green microinterferometer, 294–295
Laser Doppler vibrometry, 246–290
combination with other techniques, 279–281, 282
dynamic interference microscopy with, 240
examples, cantilever beam acceleration sensor, 283–289
in-plane analysis, 287–289
out-of-plane analysis, 284–286
examples, dual mode MEMS mirror, 282–283
full-field vibrometry, 262–263
laser Doppler effect, 247–249
laser Doppler effect, interferometric detection, 249–251
interferometry, 250
shot noise in detection of light, 249
wavefront aberrations and laser speckle, 250–251
measurement accuracy and calibration of vibrometers, 274–279
calibration based on Bessel functions, 277–278
calibration using synthetic Doppler signals, 276–277
general aspects, 274–275

influences on measurement accuracy, 278–279
mechanical comparison calibration, 275
measuring on microscopic structures, 263–270
optical arrangements, 263–266
ranges and limits, 268–270
three-D techniques, 266–268
resolution, noise-limited, 270–274
techniques, 251–262
data acquisition, 259–262
homodyne and heterodyne detection, 252–253
optical arrangements, 251–252
techniques, signal processing, 254–259
analog decoding techniques, 255–256
digital demodulation by arctangent phase method, 256–259
fundamental relationships of Doppler modulation, 254–255
Laser illumination, 8
Laser light sources, 8
atomic force microscopy, 124–125
digital holographic microscopy, 376, 377
digital holography, 353, 354
Doppler vibrometry, *see* Laser Doppler vibrometry
interference microscopy, 219
light scattering measurement systems, 106, 109
speckle metrology, 429, 430, 434, 446–447, 452
Laser Moiré interferometry, 182
Laser scanning microscopy, light scattering techniques combined with, 115
Laser speckle, laser Doppler vibrometry, 250–251
Lateral force microscopy (LFM), 127–128
Leakage effect, laser Doppler vibrometry, 261
Least-squares method
DIC-FEA, 68
digital holographic interferometry, 367
Leitz Orthoplan II Pol polarizing microscope, 191–192
Lensless Fourier approach, digital holography, 361–362
Lens systems and optics components, *see also specific techniques*
Bertrand, 191–192, 194
computer vision, 9, 12–15
confocal microscope, 151
digital holographic microscope, 377, 379
digital holographic microscope, technology for testing microlens, 414–422
diffractive optics, 420, 421, 422
Fresnel lens, 419–420, 421
refractive optics, 417–419
silicon refractive lens, 419
digital holographic microscopy, 376, 378
digital holography, 354, 355, 416
distortions
aberration removal, digital holographic microscopy, *see* Aberrations
preprocessing, 24
grating (Moiré) interferometry, 207
interference microscopy, 376
interferometric, 220, 221
laser Doppler vibrometry, 263–264, 278–279
OEHM, 329–330
speckle metrology, 428, 430, 447
Libraries, image processing and computer vision software, 49
LIGA (Lithographie Galvanik and Abformung), 266
Light, behavior of, 5–8
Light dome, 11
Light-emitting diode (LED) light sources
colored light, 7
illumination, 8–9
interference microscopy, 219
laser Doppler vibrometry, integrated techniques, 279–281
Light-induced noise, laser Doppler vibrometry, 249, 271
Lighting, *see* Illumination
Light panel, 8–9
Light pulse delay shifting (LPDS), 236
Light scattering
ground glass screen, 8
inspection techniques, 103–117
applications, 110–114
combined with profilometric techniques, 115, 116, 117
instrumentation, 106–109
standardization, 109–110
theoretical background, 104–105
laser Doppler vibrometry, 250–251
optoelectronic holography microscopy, 330
speckle metrology, 428, 429, 430
Light sources, 8; *see also* Illumination
interference microscopy instrumentation, 219
laser, *see* Laser light sources
optoelectronic holography microscope, 329, 330, 348
speckle metrology, 428, 434
Twyman-Green microinterferometer, 294–295
Linear elastic fracture mechanics (LEFM), microcrack evaluation, 90
Linearity, intensity scale, 51
Linearly polarized light, laser Doppler vibrometry, 270
Line sensors, CCD, 16
Line-shift registers, 17
Linnik microscope/interferometer, 220, 221, 278, 376
Load displacement curve, silicon oxinitride films, 300
Loading
digital holography, 375, 392–394
image correlation techniques, 67
speckle metrology applications, 447, 448
Loading modules, image correlation techniques, 70
Load state images, 56
Local pattern size, image correlation techniques, 75
Location of region of interest, *see* Region of interest location
Location of test objects, 3, 36–39
Low-coherence interferometry, 219; *see also* Interference microscopy techniques
Low-pass filters, 28–29, 259
LVDS data channels, 21

M

Machine vision measurements, laser Doppler vibrometry, integrated techniques, 280
Mach-Zehnder interferometer, 220, 221, 416

digital holographic microscopy, 376
laser Doppler vibrometry, 251, 252
Magnetic force microscopy (MFM), 129–130, 131
Magnification
grid diffraction method, 195
OEHM, 348
Maps
contour line, image correlation techniques, 75
displacement, from AFM scans, 56
Masks, spatial, 27, 28, 75
Material inhomogeneities, and local strain variations, 63
Material/mechanical properties
digital holography, 395, 397, 399–405
Poisson's ratio, 399–401
thermal expansion coefficient, 402–405
Young's modulus, 401–402
image correlation techniques, 56
measurement of, 82–88
microcrack evaluation, 88–92
waveguide grating interferometry applications, 210, 211
Matrix notation, digital holographic interferometry, 367–368
Matrox Imaging Library (MIL), 49
Matrox MeteorDig PCI interface, 452
Maximum distance sets, position recognition, 42
Mean grayscale intensity, 32
Mean square noise current, laser Doppler vibrometry, 271
Measurement time, laser Doppler vibrometry, 289
Measurement uncertainty, laser Doppler vibrometry, 278
Mechanical comparison calibration, laser Doppler vibrometry, 275
Mechanical profilometry, light scattering techniques combined with, 115
Mechanical resonances, 279
Median filters, 29–30
Membranes, *see* Films, thin films, membranes, micromembranes
Memory
frame grabbers, 21, 22
image transfer with CCD sensors, 17
M(O)EMS [micro(opto)electromechanical systems], *see also specific techniques*
applications, 2, 3
confocal microscopy, 153–155
geometric feature measurement, 42–43
image correlation techniques, derived properties determination, 63–65
interference microscopy applications, 232
dynamic IM, 240
interferometric profilometers, 231–233
in motion, *see* Laser Doppler vibrometry
presence verification, 45
two-dimensional, illumination for imaging, 10
Meshes, image correlation techniques, 75
Method of minimizing error squares, 34–35
Michelson microscope/interferometer, 156, 278, 416
digital holographic microscopy, 376
laser Doppler vibrometry, 251–252
white light scanning interferometry, 229
Microaccelerometers, OEHM testing, 334–338
Microbeams, *see* Beams/cantilevers/microbeams
Microcracks
image correlation techniques, 93
silicon oxinitride films, 306
MicroDAC (microdeformation analysis by means of correlation algorithms), 57
Micrographs
image correlation techniques
deformation measurement, 57–60
three-D deformation analysis, 92–95
Micrograting components, light scattering measurement systems, 111
Microinterferences, digital holography, 353
Microlens systems, 15, 414–422
Micromachining, 327, 328
Micromembranes, *see* Films, thin films, membranes, micromembranes
Microroughness
power spectral density (PSD), 104
speckle metrology, 430
Microscope/microscopy
digital holography, 375–391
aberrations, compensation of, 378, 379
aberrations, removal of, 378–385
controlling size independently of distance and wavelength, 389–391
focus tracking, 385–389
optical setup, 376–377
image correlation techniques, 70
interferometry, 219; *see also* Interference microscopy techniques
speckle metrology, 447
Microscope scanning vibrometer (MSV), 264, 265, 266, 269, 279, 282–283
Microscopic depth-scanning fringe projection (DSFP), 156–159
experimental realization, 158–159, 160
intensity model, 156–157
Microscopic structures, laser Doppler vibrometry, 263–270
optical arrangements, 263–266
ranges and limits, 268–270
three-D techniques, 266–268
Microspeckle interferometer, 446–452
Microstructure, light scattering measurement systems, 110
Microtensile module, FIB crossbeam, 71
Minimizing error squares, 34–35
Min/max coordinates, connectivity analysis, 35
Mireau microscope/interferometric objective, 220, 222, 278, 378
Mirrors/micromirror devices
AFM imaging, 134–139
laser Doppler vibrometry, 282–283
light scattering measurement systems, 106, 114
out-of-plane deformations, 321
out-of-plane deformations, torsional micromirrors, 317, 319, 320
Twyman-Green microinterferometer, 294
Mirror tilts, piezo-bimorph-driven, 264, 265, 266
Mixing of colors, 5
Modal analysis (time-averaged mode), OEHM, 331–333
Mode I crack opening, 61–62

Modulation
Doppler vibrometry, 254–255
interference microscope
direct phase, 221
OPD, 220
wavelength, 221
Modulation index, Doppler modulation, 255
MOEMS, *see* M(O)EMS [micro(opto)electromechanical systems]
Moiré techniques, 56, 66; *see also* Grating (Moiré) interferometry; Grid and Moiré methods
Monochromatic light
grid diffraction method, 188
interference microscopy
applications, 231
measurements on heterogeneous surfaces, 227
light sources, 7
two-beam homodyne interference microscopy, 222–223
Twyman-Green microinterferometer, 294–295
Monochrome cameras, interaction of colored light with colored objects, 6–7
MOSFET transistors, 19
Moving phase reversal reference speckle interferometry, 441–442
Moving point source holographic interferometer, 165
Multiple point measurements, image correlation techniques, 67
Multiplexers, frame grabbers, 21
Multiwavelength contouring, digital holography with, 373
Multiwavelength digital holography (MWDH), 373, 375, 389, 395
MUMPS polysilicon technology, spectral reflectivity mapping, 231
mvIMPACT, 50

N

Nabla operator, 369, 372
NanoDAC interfaces, 68, 69, 70
Nanofocus AG, 155
Nanoindentation technique, 294, 297, 300, 301, 304, 305
Nano Indenter II, 304
Nano-Moiré, TEM, 182
Narrow band-pass filters, 7
National metrology institutes, 274
Natural frequency, accelerometers, 338, 340
Nd-YAG laser, 446–447
Near-IR
LED light sources, 7
light scattering measurement systems, 108
Negative lens, digital holography, 355
NeuroCheck, 50
Neuronal nets, defect and fault detection, 48
Nipkow disk, 150, 151, 155
NIST traceable gauge testing, 333–334
Noise
decoders, 272–273
grid diffraction method, 195–196
image preprocessing, 24
laser Doppler vibrometry, 251, 253, 271–274
phase, digital holography, 373
presence verification, 47
speckle metrology, 429, 436
speckles versus, 430
Noise-limited resolution, laser Doppler vibrometry, 274
Noncontact mode AFM, 126, 127
Nonhomogeneous specimen, image correlation techniques, 64
Nonhomogeneous strain, 65, 66
Noninterlaced mode, sensors, 17, 18
Nonlinearity, Doppler vibrometry, 246
Normal equation system, digital holographic interferometry, 367
Normalized histograms, 24, 25
Normalized spectral reflectance curve, 230
NTSC/EIA RS170, 20
Numerical aperture
digital holography, numerical reconstruction by lensless Fourier approach, 361
laser Doppler vibrometry, 251, 269, 278–279
submicron transverse resolution, 415
Twyman-Green microinterferometer, 294
Numerical oscillator (NCO), laser Doppler vibrometry, 258
Nyquist sampling theorem, 281

O

Object areas, image correlation techniques, 75
Objectives/lens systems, *see* Lens systems and optics components
Object points, 13
Object-sided telecentric lens, 13–15
Object wave, digital holography, 353, 355
OEHM, *see* Optoelectronic holography/holographic microscopy (OEHM)
Off-axis configuration, digital holographic microscopy, 377
One-axis scanner, x-y scanner elements, 264
One-dimensional problem, laser Doppler vibrometry, 247
Opening, morphological operators, 30, 31
Open Source image processing library, 49
Operational deflection shapes (ODS), 247, 263, 284, 285, 286
Optical character recognition, 48
Optical components, instrumentation, *see* Lens systems and optics components; *specific methods*
Optical diffraction strain sensor, 189, 190
Optical loading, interference microscopy combined with, 232
Optically induced errors, laser Doppler vibrometry, 278
Optical microscopes, image correlation techniques, 70
Optical path difference (OPD) modulation, interferometer, 218, 220, 221
Optical profilometry, 145–159, 160
AFM, complementary methods, 142
confocal microscopy, principles of, 145–156
measurement with, 151–153
MEMS measurement applications, 153–155
microscopes, 150–151
point sensors, 145–149, 150

laser Doppler vibrometry with, 279
microscopic depth-scanning fringe projection (DSFP), principles of, 156–159, 160
experimental realization, 158–159, 160
intensity model, 156–157
Optical surfaces, digital holography, 405–422
advantages of, 375
microlens testing, 414–422
testing silicon MEMS structures, 405–409
thermal load testing, 409–414
Optimization algorithm, laser Doppler vibrometry, 281
Optoelectronic holography/holographic microscopy (OEHM), 325–348
applications, 333–343
study and characterization of accelerometers, 334–338
testing NIST traceable gauges, 333–334
applications, measurement and simulation of surface mount technology components, 343–347
computational modeling, 343–345
experimental results, 345–346, 347
applications, testing at wafer level, 338–343
high-resolution stitching, 342–343
inspection procedure, 341–342
fabrication process overview, 327–329
instrumentation and technique, 329–333
microscope, 329, 330
static mode, 330–331
time-averaged mode, 331–333
silicon mechanical properties, 328–329
Optomechanical characteristics of membranes by pointwise deflection method, 296–308
composition and atomic density of silicon oxinitride thin films, 297–300
experimental results, 302–308
mechanical properties of, 300–302
Orientation recognition, 40, 41, 42
Oscillators, laser Doppler vibrometry, 246, 271
Out-of-plane components, DHI, 366
Out-of-plane coordinates, DHI, 372
Out-of-plane deformations/displacements, 293–321
digital holography, loading of microcomponents, 392, 393
dynamic evaluation of active MOEMS with stroboscopic technique, 314–321
active membranes, 314–316, 317, 318
torsional micromirrors, 317, 319, 320, 321
grating (Moiré) interferometry, 203, 207
interference microscopy, 231
interferometric platform architecture, 294–296
laser Doppler vibrometry, beam acceleration sensor, 284–286
optomechanical characteristics of membranes by pointwise deflection method, 296–308
composition and atomic density of silicon oxinitride thin films, 297–300
experimental results, 302–308
mechanical properties of, 300–302
scratch drive actuator microdisplacements, 308–314
experimental results, 310–314
operation, 309–310
speckle interferometry, 442, 443
speckle metrology, 433, 444
DUV interferometry, 452
stroboscopic interference microscopy, 237, 238
two-beam interference microscopy, 224
Oversampling, laser Doppler vibrometry, 260

P

Packaging
micro-Moiré interferometer, 171–172, 173
waveguide grating interferometry applications, 212–214
Padding operation, digital holography, 388, 389, 390, 391
Parabolic phase correction factor, 379
Parabolic phase factor, 378, 379
Parabolic subpixel algorithm, 60
Parallel clocking, 17
Parallel light rays, 9
Parallel transmission, image information, 21
Particles
high-reflective multilayer system, 113
light scattering measurement systems, 112
and power spectral density, 115
Pattern analysis, grid line, 183
with Fourier transform, 184–188
with phase-shifting method, 188
Pattern matching, 35–36
contour-based, 27
finding test object or region of interest, 38, 39
position recognition, 42
Pattern size, image correlation techniques, 75
Periodic chirp, laser Doppler vibrometry, 261
Periodic function, digital holography, 353
Perturbation models, first-order, 104
Phase aberrations, correction of, 376
Phase angle
Doppler modulation, 254
laser Doppler vibrometry, 271
Phase contrast image, DHM, 376
focus tracking, 388
reconstruction of, 378
Phase delay, laser Doppler vibrometry, 279, 281
Phase detection microscopy (PDM), 73
Phase deviation
Doppler modulation, 254
laser Doppler vibrometry, 272
Phase difference
digital holographic interferometry, 370, 372
speckle interferometry, 435
Phase discontinuities, 51
Phase distribution
optoelectronic holography microscopy, 330
Twyman-Green microinterferometer, 296
Phase factor, digital holographic microscopy, 378, 379
Phase history, laser Doppler vibrometry, 278
Phase imaging
AFM imaging modes, secondary, 129
digital holography, 417, 418
Fresnel lens, 419, 420
Phase lag, AFM phase imaging, 129

Phase-locked loop (PLL) circuit, laser Doppler vibrometry, 256, 269, 270
Phase maps, interference
fringe pattern demodulation, 218
silicon oxinitride films, 307
Phase measurement, digital holography interferometry, 372–375
Phase modulation
digital holographic microscopy, unwrapped, 381, 382
interference microscope, 221
laser Doppler vibrometry, 273
Phase modulation method
fringe processing, 51
speckle metrology, 441
Phase noise, laser Doppler vibrometry, 251, 271
Phase of difference method, speckle interferometry, 435, 436
Phase sampling method, fringe processing, 50, 51
Phase shift
digital holographic microscopy, 377
grid diffraction method, 193–194
laser Doppler vibrometers, homodyne, 253
speckle interferometry, 443
speckle metrology, 433, 441–442
two-beam interferometry, with broadband illumination, 223
Phase shifter, white light interferometry, 221
Phase shift error, grid diffraction method, 195–196
Phase-shifting method
applications, 232
digital holographic microscopy, circular fringe carrier removal, 381–382
digital holography, 362–363
fringe processing, 50, 51
grating (Moiré) interferometry, 208
grid line pattern analysis, 188
speckle metrology, 436, 437–439
stroboscopic, 236–238
Phase stepping
fringe processing, 51
optoelectronic holography microscopy, 330, 331
Phase subtraction, digital holography, 390
Phase wrapping/unwrapping, *see* Unwrapped phase/unwrapping; Wrapped phase
Photocurrent shot noise, laser Doppler vibrometry, 249, 271
Photodiodes, 15
CMOS sensors, 19–20
laser Doppler vibrometry, 246
Photo-effect, 15–16, 19
Photomultiplier tubes, light scattering measurement systems, 109
Photoresist method, grid and grating fabrication, 164–165
Piezo-actuated micromechanical devices, interference microscopy applications, 232
Piezo-based two-axes scanners, 264
Piezo-bimorph-driven mirror tilts, 264, 265, 266
Piezoelectric phase stripper, optoelectronic holography microscope, 329
Piezoelectric scanner, atomic force microscopy, 124
Piezoelectric transducer (PZT)
digital holographic microscopy, 377
dynamic evaluation of active MOEMS with stroboscopic technique, 314–321
interference microscopy applications, 232
optoelectronic holography microscopy, 333
speckle interferometry, 443
speckle metrology, 447, 448, 453
Twyman-Green microinterferometer, 296
Piezomembranes, speckle metrology, 453, 454, 455
Pinhole
confocal microscopy, 148, 149
laser Doppler vibrometry, 269
Pixel counting, binary, 33, 46
Pixels, 17
Planarity, ADM measurements, 137, 138, 139
Planar motion analyzer, 280
Plasma-enhanced chemical vapor deposition (PECVD), 293–294, 297, 304, 308
Plasmons, surface, 104
Plastic behavior, 56
Plastics, image correlation techniques, 85, 86
Plates, coefficient of thermal expansion, 64
Point sensors, confocal microscopy, 145–149, 150
Point transformations, image preprocessing and processing methods, 27–28
Pointwise deflection methods, optomechanical characteristics of membranes by, 296–308
composition and atomic density of silicon oxinitride thin films, 297–300
experimental results, 302–308
mechanical properties of, 300–302
with nanoindentation, 301
Poisson process, laser Doppler vibrometry, 249
Poisson ratio, 56, 64
digital holography, 394, 395, 397
microbeam, 399–401
image correlation techniques, 66, 87–88, 90
silicon oxinitride films, 300
thin film materials, 296–297
Polarization
digital holographic microscopy, 376
grating (Moiré) interferometry, 207
laser Doppler vibrometry, 270
light scattering measurement systems, 106
scattering properties, 105
Polarizing beam splitter
interferometer with direct phase modulation, 221
laser Doppler vibrometry, 251–252
Polarizing microscope, 191
Polished surfaces, light scattering measurement systems, 110–113
Polycrystalline materials analysis, waveguide grating interferometry applications, 210, 212
Polytec microscope scanning vibrometer, 266, 281
Position, laser Doppler vibrometry measurement, 289
Position recognition, 3, 40–42
Position sensing detectors (PSDs), grid diffraction method, 189
Power-on thermal deformations, surface mount technology, 345, 346
Power ratio, laser Doppler vibrometry, 271–272
Power spectral density (PSD), light scattering measurement systems, 116

particle contaminations and, 115
roughness, 104
Power spectrum, laser Doppler vibrometry, 273
Precision, OEHM testing, 333, 334
Presence verification, 3–4, 45–47
Pressure measurement, silicon oxinitride films, 308
Principal axis of inertia, 41
Prism, 5
Probe, atomic force microscopy (AFM), 123
Profilometry
AFM, complementary methods, 142
atomic force profilometer, 139–141
grid, 188
holographic reconstruction, silicon MEMS structures, 405–406, 407
interference microscopy
low-coherence, 225–226
monochromatic, 224–225
laser Doppler vibrometry with, 279
light scattering inspection techniques combined with, 115, 116, 117
Progressive scan CCD sensors, 19
Projection, homogenous light source, 9
Projection Moiré pattern, 183
Projectors, DLP (digital light projection), 2
PSPDs, 127, 128
PULNIX TM1040 CCD camera, 452
Pulsed light source, interference microscopy, 219

Q

Quadcell PSPD, 127, 128
Quad flat pack (QFP) package, 168, 171–172, 173, 180
Quadratic masks, 27, 28
Quadrature signal pair, laser Doppler vibrometry, 278
Quantization noise, laser Doppler vibrometry, 271
Quarter wave plate, laser Doppler vibrometry, 252–253
Quasi-monochromatic light source, interference microscopy, 219
Quasi-static actuator, 293, 308–314
Quasi-static evaluation of out-of-plane deformations, *see* Out-of-plane deformations/displacements

R

Radiofrequency devices
laser Doppler vibrometry, 258, 259, 278
MEMS devices, 326
Radiofrequency-plasma-assisted reaction, chemical vapor deposition, 297–298
Raman spectroscopy, 56, 460–471
applications
coatings, 471
stress, 463–471
instrumentation, 461–462
principle, 460
Random phase relation, laser Doppler vibrometry detector, 251
Raster scanning, atomic force microscopy, 123, 124, 125
Rayleigh scatter, light scattering measurement systems, 108
Rayleigh-Sommerfeld diffraction formula, 357
Real-time technique, digital holographic interferometry, 364
Reconstruction principles, digital holography, 358–363
Reconstruction wave, digital holography, 355, 356
Recording and reconstruction of wavefronts, digital holography, 353–358
Reference beam wavefront, OEHM, 330
Reference grating
AFM, 175
FIB, 177
microscopic grid methods, 182
References, Doppler vibrometry measurements, 274–275
Reference wave, digital holography, 353, 354, 355, 357
Reflective surfaces
illuminating, 11
light scattering measurement systems, 112–114, 115, 116
presence verification, 46
speckle metrology, 428, 429
Refraction slab rotation, grid diffraction method, 195
Refractive index, 5
silicon oxinitride films, 297, 298, 306, 307, 308
Young's modulus as function of, 304, 305, 306
Refractive optics, digital holographic microscopy technology for testing, 417–419
Region of interest location, 3
image correlation techniques, 74
image processing and computer vision, 47–48
laser Doppler vibrometry, 287
measurement and testing tasks, 36–39
OEHM testing, 333, 337, 340, 341, 342
presence verification, 46
Registers, line-shift, 17
Regression methods, edge detection, 34
Reliability, OEHM testing, 333
Reproducibility, image correlation techniques, 67
Residual stresses
digital holographic microscopy, 406, 408–409
image correlation techniques, 66, 95–99
silicon oxinitride films, 304, 306
thin film materials, 296–297
Resolution
grid diffraction method, 195
image correlation techniques, 66–67
laser Doppler vibrometry, 247, 274
frequency spectrum, 289
noise-limited, 270–279
OEHM, 348
Resonance frequencies, cantilever beam acceleration sensor, 282–283
Resonances, mechanical, 279
Resonance triplet, 284, 285, 286
RGB image
color sensors and camera, 20
spectral reflectivity mapping, 231
Root mean square average roughness, 296
Root mean square shot noise current, 249
Rotation, position recognition, 42
Roughness, 105
atomic force microscopy, 122
connectivity analysis, 35

interference microscopy applications, 232
laser Doppler vibrometry, 247, 289
ranges and limits, 268
wavefront aberrations, 250–251
light scattering measurement systems, 110–111, 115, 116
speckle metrology, 428, 430
Twyman-Green microinterferometer, 296
RS422, 21
Rutherford backscattering spectroscopy (RBS), 297, 298, 299

S

Sampling rate, laser Doppler vibrometry, 262
Sampling theorem
digital holography, 353
laser Doppler vibrometry, 260
Nyquist, 281
Saturation effects, fringe, 51
Saw-tooth image, 51
Scaling
actuator downsizing, 308
digital holography, controlling image size, 389
image correlation techniques, 75
laser Doppler vibrometry, 278
position recognition, 42
Scanner, atomic force microscopy (AFM), 123, 124
Scanning capacitance microscopy (SCM), 132–133, 134
Scanning electron microscopy (SEM), 56
bridges, 406, 407
cantilever beams, 405, 406
deformation measurement, 57
digital micromirror device (DMD), 136
electron beam lithography, 166
image correlation techniques, 67
instrumentation, 70, 72
microcrack evaluation, 89
residual stress determination, 97
Moiré methods using high-resolution microscopy, 176–177, 179, 180, 181, 182
Scanning force microscope (SFM), 56, 89
Scanning Kelvin probe force microscopy (electric force microscopy), 131, 132
Scanning laser Doppler vibrometer (SLDV), 262–263, 282–283
Scanning Moiré method, 164
Scanning near-field optical microscopes (SNOM), 57
Scanning probe microscopes (SPMs), 122; *see also* Atomic force microscopy (AFM)
Scanning spreading resistance microscopy (SSRM), 133
Scanning white light interferometry (SWLI), 219; *see also* Interference microscopy techniques
Scattering point, laser Doppler vibrometry, 268
Scorpion Vision Software, 50
Scratch drive actuator, 294
Scratch drive actuator microdisplacements, 308–314
experimental results, 310–314
operation, 309–310
Search algorithm
correlation approach, 58, 59
OEHM testing, wafer level, 341, 342
Search matrix, 59
Self-calibrating algorithms, grid profilometry, 188
Semiconductor microlaser matrix testing, waveguide grating interferometry applications, 210, 212, 213
Semi-Linnik objectives, 220, 221
Sensitivity
digital holographic interferometry, 370
microaccelerometers, OEHM testing, 335
OEHM testing, 334
Sensitivity vector
digital holographic interferometry, 366, 369–370, 371–372
speckle metrology, 434, 435
Sensors, *see also* CCD detectors/sensors
computer vision components, 15–22
camera types and interfaces, 20–21
CCD, 16–18
CMOS, 18–19
color sensors and camera, 19–20
frame grabbers, 21–22
digital holography, complex silicon micromembranes, 412–414
Serial clocking, 17
Serial image processing, 75
Serial transmission, image information, 21
Setpoint, atomic force microscopy, 124–125
Shadows
digital holographic interferometry, 370
presence verification, 46
Shape characteristics
laser Doppler vibrometry, 289
spatial operations, 33
Shape measurement, digital holography
advantages of, 375
combined with displacement measurement, 395–397, 398, 399
interferometry, 368–372
Shear strain, image correlation techniques
finite element (FE) modeling, validation of, 82, 83
thermal cycling, 78, 80
Shot noise in detection of light, laser Doppler vibrometry, 249, 271
Sidebands, laser Doppler vibrometry, 273
Signal processing
interferometer with wavelength modulation, 221
laser Doppler vibrometry, 254–259
analog decoding techniques, 255–256
digital demodulation by arctangent phase method, 256–259
fundamental relationships of doppler modulation, 254–255
noise sources, 271
Signal substitution, laser Doppler vibrometry calibration, 276–277
Signal-to-noise ratio
fringe, 51
laser Doppler vibrometry, 271–272
Silicon die and die-paddle interface, 169, 172, 173
Silicon mechanical properties, 328–329
Silicon microbeam testing, 210, 211
Silicon micromembranes, 412–414; *see also* Films, thin films, membranes, micromembranes

Silicon nitride materials, 405–409
Silicon oxinitride films, 293; *see also* Out-of-plane deformations/displacements
  composition and atomic density of, 297–300
  nanoindentation technique, 294
Silicon refractive lens, digital holographic microscopy technology for testing, 419
Single-chip color sensors, 19, 20
Single-frame speckle interferometry, 439–440
Single-point measurements, 56, 247
Single-shot resolution, laser Doppler vibrometry, 274
Single-wavelength interferometry, 225
Sinusoidal waves, surface roughness measurements, 104
Skeleton method, 50
Smart cameras, 20
Smoothing, data, 75
Smoothing effect, 30, 47
Sobel filter, 34
Software
  correlation, 61
  finite element (FE) analysis, 70
  image correlation techniques, 64, 68, 70, 73–76
Solder fatigue, 77
Space-bandwidth product (SPB), digital holography, 352, 360, 362
Spatial carrier phase-shifting (SCPS) method, 208
Spatial filter cores, 29
Spatial filtering
  digital holography, 359, 360, 363
  image preprocessing and processing methods, 28–31, 32
Spatial frequency
  digital holography, recording and reconstruction of wavefronts, 353
  grating (Moiré) interferometry specimen grating, 203
Spatial frequency distribution, fringe, 51
Spatial heterodyning, 50
Spatial masks, 28
Spatial operations, 33
Spatial phase distribution, continuous, OEHM testing, 333
Spatial phase modulation, fringe processing, 51
Spatial phase shifting (SPS) method, 188
Spatial resolution
  digital holography, recording and reconstruction of wavefronts, 353
  OEHM, 348
Specimen grating, AFM, 175
Speckle
  digital holography, 395
  laser, 250–251
Speckle boiling effect, 432
Speckle metrology, 376, 427–457
  applications, characterization of operational behavior, 446–456
    DUV interferometry for highest resolution, 452–456
    visible light interferometry for wide scale, 446–452
  applications, quality assurance on wafer level, 443–446
  classification of speckle signals
    contrast, 432–433
    phase in image plane (speckle phase), 433
    shift in image plane (speckle shift), 433
  dynamic speckle correlation interferograms, 332
  evaluation techniques for signal extraction
    speckle interferometry (SPI), 434–439
    speckle photography (SPP), 434
    time-averaged recordings, 439–441
  extraction of information from patterns, 431–443
    classification of speckle signals, 432–433
    evaluation techniques for signal extraction, 434–442
    optical arrangements, 442–443
  imaging system properties, 428–431
Speckle pattern interferometry, 353
Spectrally resolved interference microscopes, 221–222
Spectral noise density, laser Doppler vibrometry, 273
Spectral operations, image preprocessing and processing methods, 32–36
Spectral reflectivity mapping, white light scanning interferometry, 228, 230, 231
Spectral response plots, 8
Spectroscopic techniques, 459–480
  atomic force microscopy, 134
  Auger electron, 478
  Brillouin scattering, 479
  dual-beam, 474–476
  ellipsometry, 472–474
  high-resolution electron energy loss (HREELS), 476–478
  Raman spectroscopy, 460–471
    applications, coatings, 471
    applications, stress, 463–471
    instrumentation, 461–462
    principle, 460
  X-ray photoelectron (XPS), 476
Spectrum analyzer, laser Doppler vibrometry, 247
Spherical profile of lens, real versus ideal, 418
Sphericity, defect of, 418
Spot diameter, laser Doppler vibrometry, 269–270
Spotlight source, 9, 11
Standards
  image acquisition hardware and interfaces, 51
    analog cameras, 20–21
    image transmission modes, 21
  light scattering measurement systems, 108, 109–110
Static evaluation of out-of-plane deformations, *see* Out-of-plane deformations/displacements
Static measurements, by interference microscopy techniques, 224–226
  surface profiling by low-coherence interferometry, 225–226
  surface profiling by monochromatic IM, 224–225
Static mode, optoelectronic holography/holographic microscopy (OEHM), 330–331
Statistical properties of speckles, 430
Statistics, strain data points, 64
Stefan-Boltzmann constant, 345
Stitching, OEHM, 341, 348
Strain measurements, 56
  grating (Moiré) interferometry, *see* Grating (Moiré) interferometry
  image correlation techniques, 65
    data point statistics, 64
    derived properties determination, 63–65

microcomponent analysis, 76–80
microcrack evaluation, 92, 93
strain fields, 60–63, 75
strain versus temperature curve, 86, 87
Moiré patterns, 168
Strain sensor, optical diffraction, 188–190
Stress
Raman spectroscopy applications, 463–471
silicon oxinitride films, 308
Stress function
interferometry with nanoindentation data, 294
silicon oxinitride films, 301–302
Stress intensity factors, 89, 92, 93
Stress measurements
AFM-Moiré measurements, 56
interference microscopy applications, 232
Stress relaxation, 64
Stress sensors, 56
Stroboscopic illumination, laser Doppler vibrometry, 279–281
Stroboscopic interference microscopy
laser Doppler vibrometry, 286, 289
light sources, 219
vibration measurements, 235–238
Stroboscopic interferometry, out-of-plane deformations, 314–321
active membranes, 314–316, 317, 318
torsional micromirrors, 317, 319, 320, 321
Stylus profiler, AFM, 140–141
Subpixel algorithm, parabolic, 60
Substrate deformation, interference microscopy applications, 232
Subtraction operation, digital holographic microscopy, 378
Subtraction speckle interferometry, 440–441
Subtractive effect, color, 6
Successive approximations method, 384
Suprasil2 quartz dedicated optics, 452–453
Surface micromachining, 327, 328
Surface mount technology components, OEHM, 343–347
computational modeling, 343–345
experimental results, 345–346, 347
Surface plasmons, 104
Surface potential imaging, AFM imaging modes, 131–132
Surface properties, *see also* Light scattering, inspection techniques
atomic force microscopy, *see* Atomic force microscopy (AFM)
interference microscopy, heterogeneous surface measurements, 226–228
interference microscopy profiling
low-coherence interferometry, 225–226
monochromatic illumination, 224–225
Synchronization, laser Doppler vibrometry, 280, 289
Synthetic Doppler signals, laser Doppler vibrometer calibration, 276–277

T

Tandem confocal microscope, 150
TappingMode AFM, 126–127, 132
Technical surfaces
defined, 391
digital holography, 395–405
combined shape and displacement measurement of small objects, 395–397, 398, 399
material parameters determination, 397, 399–405
Telecentric lenses, 12, 13–15
Telecentric observation and illumination, digital holography, 369
Television technology, interlaced mode and, 19
Temperature, *see also* Thermal load
image correlation techniques, 66
OEHM modeling and testing, surface mount technology, 343–347
silicon oxinitride films, 308
Temperature coefficient, scaling factor, 278
Temperature regulation, gas sensor, 93
Template matching, 35–36
Temporal modulation techniques, fringe processing, 51
Temporal phase shifting (TPS) method, 188, 316, 318
Temporal phase stepping, fringe processing, 51
Temporal unwrapping, 373
Tensile stress/strain
finite element (FE) modeling, validation of, 82, 83
image correlation techniques, instrumentation, 72
silicon oxinitride films, 301–302
Test objects location, 3, 36–39
Texture analysis, defect and fault detection, 47–48
Thermal conduction, MEMS silicon materials mechanical properties, 328
Thermal cycling, image correlation techniques, 77
Thermal expansion, coefficient of, 56
digital holography, 395, 397, 398, 402–405
image correlation techniques, 56, 64–65
interference microscopy applications, 232
Thermal load
digital holography, 352, 393, 396, 409–414
cantilever beam, 410–411
complex silicon micromembranes, 412–414
microscope focus tracking, 385–389
image correlation techniques
analysis scheme, 74
defect detection, 80–81
microcrack evaluation, 93
temperature versus strain curve, 86, 87
three-D deformation analysis, 93–94, 95
Thermal noise, laser Doppler vibrometry, 271
Thermomechanical analysis (TMA)
image correlation techniques, defect detection, 80–81
image correlation techniques versus, 87
interference microscopy combined with, 232
surface mount technology, OEHM modeling and testing, 343–347
thin foils and, 83
Thickness, sample
image correlation techniques, 83–84, 86
interference microscopy, 228–230
light scattering measurement systems, 105
silicon oxinitride films, 301
speckle metrology, 445
Thin lens, imaging behavior, 13

Thin materials, *see also* Films, thin films, membranes, micromembranes
foils, classical thermomechanical analysis and, 83
image correlation techniques, 63, 64
Three-chip color sensors, 19, 20
Three-D techniques
deformation analysis based on AFM micrographs, 92–95
digital holography, 396, 397, 417
advantages of, 375
interferometry, 372
microscopy (DHM), 375, 376
laser Doppler vibrometry, 266–268, 279
OEHM, 347
scratch drive actuators, 311
Three-phase clocking, 17
Time-averaged interference microscopy, vibration measurements, 238–239
Time-averaged mode/technique
digital holographic interferometry, 364
optoelectronic holography/holographic microscopy (OEHM), 331–333
speckle interferometry, 441–442
speckle metrology, 444
moving phase reversal reference, 441–442
single-frame SPI, 439–440
subtraction SPI, 440–441
stroboscopic interferometry, 316, 317
Time history, laser Doppler vibrometry, 278
Time series accumulation, 25
Time window functions, laser Doppler vibrometry, 261
Toolboxes, programming, 49–50
TOPCON SEM SX-40A, 165–166
Topography, atomic force microscopy, 125
Torsional micromirrors, out-of-plane deformations, 317, 319, 320, 321
Torsional vibrations-measuring-vibrometer, 246
Torsion resonance mode (TRMode) AFM, 127
Total integrated scattering, 106–107, 110
Total internal reflection, waveguide grating microinterferometer head, 206
Total scattering, 103, 105; *see also* Light scattering, inspection techniques
high-reflective multilayer system, 112–114
light scattering measurement systems, 108
Tracking, fringe, 51
Trajectory analysis, laser Doppler vibrometry, 287, 288
Transistors, CMOS sensors, 19–20
Transmission electron microscopy (TEM), 178–179, 182
Transmissive elements, speckle metrology, 452
Transmissive light arrangement
instrumentation, 8, 9
position recognition, 40
Trigonometric functions, digital holographic interferometry, 373
Triplet, resonance, 284, 285, 286
Tungsten halogen white light source, 219
Tunneling AFM (TUNA), 133
Two-axes scanners, 264
Two-beam homodyne interference microscopy, 222–224
with broadband illumination, 223, 224
with monochromatic illumination, 222–223
Two-beam spectroscopy, 474–476
Two-dimensional detector, spectrally resolved profilometer, 222
Two-dimensional discrete field of correlation coefficients, 59
Two-dimensional phase wrapping, digital holography, 417
Two-dimensional strain measurement, optical diffraction strain sensor, 190
Twyman-Green (micro)interferometers, 207, 208, 294, 442, 446–447
digital holographic microscopy, 376
out-of-plane analysis, 293; *see also* Out-of-plane deformations/displacements
two-beam, 222

U

U-field Moiré fringe patterns, 169, 175–176, 180
Ultralow expansion substrate, QFP package, 171
Ultrasonic displacement decoder, 255–256
Ultrasonic phase-locked loop (PLL), laser Doppler vibrometry, 269
Ultraviolet cutting filter, 8
Ultraviolet region
deep ultraviolet (DUV) light sources
light scattering measurement systems, 103, 108
micromirrors for application, 114
speckle metrology, 452
laser, speckle metrology, 430, 434
LED light sources, 7
light scattering measurement systems, 103, 106, 108
photoresist grating fabrication, 164
Uncertainty, laser Doppler vibrometry measurement, 278
Uniaxial tensile testing, 64
Unwrapped phase map, 197
laser Doppler vibrometry, 258
silicon oxinitride films, 307
Unwrapped phase/unwrapping, 51
digital holography 388, 417, 418
advantages of, 375
hierarchical, 374
interferometry, 364, 373
microscopy, 381, 382
speckle metrology, 437
USB 2.0, 21, 51

V

Vacuum ultraviolet (VUV), light scattering measurement systems, 103, 108, 109
Van der Graff accelerator, 298
Vector plots
displacement and strain fields, 60
image correlation techniques, 75
Velocity decoders, noise spectra, 272
Velocity noise, 272, 273
Verification of presence, 3–4, 45–47
Vertical scanning interferometry, 219; *see also* Interference microscopy techniques
Vertical strain fields, image correlation techniques, 83, 84

V-field fringe patterns, 171, 175–176, 180
Vibration isolation, image correlation techniques, 66
Vibration measurements/vibrometry
interference microscopy
stroboscopic IM, 235–238
time-averaged IM, 238–239
vibration map, 218
laser Doppler, *see* Laser Doppler vibrometry
speckle metrology, 432
Video microscope, laser Doppler vibrometry, integrated techniques, 279–280
Video processing, OEHM, 329, 330
Video rate, stroboscopic interference microscopy, 238
Video signal, analog cameras, 20–21
Vignetting, 15
Virtual displacement field, 197
Viscoelastic behavior, 56
Viscoplastic behavior, 56
Visible light interferometry, speckle metrology, 446–452
Visible ultraviolet, light scattering measurement systems, 108
VisionPro, 50
VisualBasic, 49
Voltage controlled oscillator (VCO), 256
Volume scattering, 105
VULSTAR system, 109

W

Wafer level testing
interference microscopy, wafer-level encapsulated devices, 232
optoelectronic holography/holographic microscopy (OEHM), 338–343
high-resolution stitching, 342–343
inspection procedure, 341–342
speckle metrology, 443–446
Wavefront, OEHM reference beam, 330
Wavefront aberrations, laser Doppler vibrometry, 250–251
Wavefront curvature, digital holographic microscopy, 379–381
Wavefront function, Twyman-Green microinterferometer, 295–296
Waveguide grating interferometry (WGI)
applications, 210–214
electronic packaging, 212–214
material constants determination, 210, 211
polycrystalline materials analysis, 210, 212
semiconductor microlaser matrix testing, 210, 212, 213
concept of waveguide grating interferometer head, 204–207
grating interferometry, principle of, 202–204
measurement system, 207–209
modified WGI for 3D components of displacement vector measurements, 207
specimen grating technology, 209–210
Wavelength (λ), 5
digital holography, multiwavelength (MWDH), 389
grating (Moiré) interferometry specimen grating spatial frequency, 203, 204
interference microscopy
light sources, 219
measurements on heterogeneous surfaces, 227
light scattering measurement systems, 103, 106, 107–108, 115, 117
OEHM testing, 333
speckle metrology, 430, 434
Wavelength modulation, interference microscope, 221
White light
behavior of light, colors, and filters, 5, 6
speckle metrology, 429
White light interferometry
direct phase modulation, 221
light scattering techniques combined with, 115
White light scanning interferometry (WLSI), 219, 225; *see also* Interference microscopy techniques
applications, 231
film thickness mapping, 228–230
measurements on heterogeneous surfaces, 228
Whole-field mapping, grating (Moiré) interferometry, 201
Width analysis, ADM measurements, 139
Wit image processing software, 50
Working distance, interferometer, 294
Wrapped phase, 51, 196
digital holography, 417
interferometry, 364
microscopy, focus tracking, 388
profile reconstruction, 406, 407
silicon oxinitride films, 307
speckle metrology, 439

X

X-ray photoelectron spectroscopy (XPS), 476
X-Y scanner, 264, 265

Y

Yield strength, MEMS silicon materials mechanical properties, 328
Young's fringes, specklegram, 434
Young's modulus, 56, 64
digital holography, 395, 397, 401–402
interference microscopy applications, 232
nanoindentation technique, 294
silicon oxinitride films, 300, 301, 304, 305, 306, 308
thin film materials, 296–297

Z

Zero thickness specimen gratings, 209–210
Z feedback loop, AFM, 125, 127, 129

Milton Keynes UK
Ingram Content Group UK Ltd.
UKHW052025071024
449327UK00027B/2434